Advanced Composites Manufacturing and Plastics Processing

Advanced Composites Manufacturing and Plastics Processing

Topic Editors

Patricia Krawczak
Ludwig Cardon

MDPI

Basel • Beijing • Wuhan • Barcelona • Belgrade • Novi Sad • Cluj • Manchester

Topic Editors
Patricia Krawczak
Centre for Materials and Processes
IMT Nord Europe, Institut Mines-Télécom
University of Lille
Douai
France

Ludwig Cardon
Centre for Polymer and Material Technologies
Department of Materials, Textiles and Chemical Engineering
Ghent University
Zwijnaarde
Belgium

Editorial Office
MDPI AG
Grosspeteranlage 5
4052 Basel, Switzerland

This is a reprint of the Topic, published open access by the journals *Fibers* (ISSN 2079-6439), *Journal of Composites Science* (ISSN 2504-477X), *Journal of Manufacturing and Materials Processing* (ISSN 2504-4494), *Materials* (ISSN 1996-1944) and *Polymers* (ISSN 2073-4360), freely accessible at: https://www.mdpi.com/topics/N0VZ7O2N31.

For citation purposes, cite each article independently as indicated on the article page online and as indicated below:

Lastname, A.A.; Lastname, B.B. Article Title. *Journal Name* **Year**, *Volume Number*, Page Range.

ISBN 978-3-7258-4237-7 (Hbk)
ISBN 978-3-7258-4238-4 (PDF)
https://doi.org/10.3390/books978-3-7258-4238-4

Cover image courtesy of IMT Nord Europe

Contents

About the Editors

Patricia Krawczak

Patricia Krawczak graduated with an engineering degree (1989) from Mines Douai, France, then with a MSc (1990) and a PhD in organic and macromolecular chemistry (1993), and finally a diploma of habilitation in physics (1999) from the University of Lille, France. She is currently a Full Professor at IMT Nord Europe, a top-level graduate school of engineering and a research center of the Institut Mines-Télécom (IMT)—the number one public group of engineering and management graduate school in France. After having managed IMT Nord Europe's research and education department "Polymers and Composites Technology and Mechanical Engineering" for almost 20 years (2000–2019), she has been in charge of the strategic program "Technology Platforms" of the Institut Mines-Télécom (IMT) since 2019. She has gained deep experience in the field of plastics and composites engineering through numerous collaborations with industry and academia, participated as a principal investigator in numerous national and European industry-oriented projects, and supervised a number of master's and PhD students. She has co-authored over 400 journal or conference publications and book chapters and is ranked among the World's top 2% of scientists (according to Stanford University/Elsevier data updates). Her research interests cover advanced processing/manufacturing technologies and the physics and mechanics of polymers and polymer composites (including bio-based materials). Additionally, she is a member of the Executive Management Board of the French Society of Plastics Engineers (SFIP) and a member of the Scientific and Strategic Advisory Board of the French Technical Center for Plastics and Composites (IPC). She is also a Knight of two National Orders: the Academic Palms (decoration for valuable services to the universities in education and teaching) and the Legion of Honor (highest French order of merit, decoration for public service and professional activity with eminent merits).

Ludwig Cardon

Ludwig Cardon obtained a master's degree in mechanical engineering. To improve his professional skills, until 1990 he was quality manager at MG Fasteners, with afterwards assistant at University College Ghent, Belgium. To strengthen his academic career, he obtained a PhD in Engineering at BCU, UK, in 2006. In 2007 he became an assistant professor and associate professor in 2011 at University College Ghent. He was an invited lecturer on polymer engineering at UFCS Brazil, a visiting researcher on polymer engineering and additive manufacturing at University of Minho (Portugal), the University of Maribor (Slovenia), and the Norwegian University of Science and Technology (Norway). Since 2011 he has been Associate Professor at Ghent University, Full Professor as from 2022, and head of the Centre for Polymer and Material Technologies CPMT. He is a coauthor of >120 peer-reviewed research articles, 5 book chapters, 4 books, >180 peer-reviewed full-length conference articles, and 3 patents. His research focuses on the investigation of advanced and sustainable (3D) processing of (bio)polymers/composites, including characterization, performed up to industrial scale, using both industrial and unique in-house developed equipment in combination with experimental validation. His research is related to various interconnected fields, including (bio)polymer and mechanical engineering, materials science, and processing technologies, both for conventional and sustainable thermoplastic polymers.

Preface

This reprint focuses on the latest scientific development of advanced composites manufacturing and plastics processing. It contains research articles addressing new process developments, modeling/simulation, monitoring/control, and performance or application issues, with either experimental or numerical approaches.

The contributions are clustered in three fields:

(i) Advanced composites manufacturing (coupling/hybridization of processes; automated fiber, yarn or tape placement and additive manufacturing; long fiber thermoplastics compounding; fibers and fabrics production; manufacturing of sustainable bio-based composites; laminated glass/polymer composites).

(ii) Machining, milling and drilling of polymers, composites and multi-materials laminates.

(iii) Polymer processing and compounding (functionalization; plastics recycling).

Patricia Krawczak and Ludwig Cardon
Topic Editors

materials

MDPI

Editorial

Advanced Composites Manufacturing and Plastics Processing

Patricia Krawczak [1,*] and Ludwig Cardon [2,*]

1 Centre for Materials and Processes, IMT Nord Europe, Institut Mines-Télécom, University of Lille, 941 Rue Charles Bourseul, 59508 Douai, France
2 Centre for Polymer and Material Technologies, Department of Materials, Textiles and Chemical Engineering, Ghent University, Technologiepark 130, 9052 Zwijnaarde, Belgium
* Correspondence: patricia.krawczak@imt-nord-europe.fr (P.K.); ludwig.cardon@ugent.be (L.C.)

1. Introduction

Environmental and energy concerns and digitalization are currently profoundly reshaping the plastics and composites industry. Manufacturing processes and systems evolve accordingly in order to cost-effectively produce high-performance, high-quality, lightweight, and multifunctional parts with a reduced carbon footprint. All composites manufacturing and polymer processing technologies are concerned with this trend: liquid composite molding (e.g., resin transfer molding and resin infusion/vacuum infusion), automated lay-up (e.g., automated fiber placement and automated tape laying), filament winding, prepreg technology, pultrusion, autoclave, compression molding, film stacking, additive manufacturing/3D printing, injection molding, over-molding/back-molding, extrusion, blow molding, thermoforming, rotational molding, foaming, coating, preforming of textile reinforcement, joining/welding, and mold technologies (i.e., mold making and design).

This topic gathers 25 original research articles and one technical note on the latest advances in composites manufacturing and plastics processing. These contributions address new process developments, modeling/simulation, monitoring/control, and performance or application issues, with either experimental or numerical approaches. All types of polymers (thermoplastics, thermosets, and elastomers) and fibers/fillers (glass, carbon, ceramic, mineral, and vegetal) are concerned, whether they come from recycled, bio-based, or fossil feedstocks.

2. Advanced Composites Manufacturing

2.1. Coupling/Hybridization of Processes

Hirsch et al. [1] have optimized the manufacturing of hybrid fiber-reinforced polyamide composite structures by coupling fused granular fabrication (FGF)—a large format additive manufacturing (LFAM) technology that focuses on cost-effective granulate-based manufacturing by eliminating the need for semi-finished filaments—and automated tape laying (ATL). A significant improvement in the flexural stiffness and strength of the manufactured FGF structures was observed by hybridization with 60% glass fiber-reinforced polyamide 6 unidirectional tapes.

Atzler et al. [2] have investigated the overprinting of continuous fiber-reinforced laminates, aiming at producing high-performance, functional structures by means of a hybrid process combining fused granular fabrication (FGF) with automated fiber placement (AFP), where an additively manufactured structure is bonded in situ onto a thermoplastic laminate. The authors have proposed a workflow for the compensation of substrate defects when overprinting in extrusion-based processes. The novel process flow uses a 3D scan of a laminate to adjust the geometry of the additively manufactured structure in order to achieve a constant layer height in the 3D print and, thus, constant mechanical properties.

Kizak et al. [3] have combined additive manufacturing and injection molding and focused on the interfacial fusion bonding of hybrid additively manufactured components made of polycarbonate (neat, with glass or carbon fiber reinforcement) under torsional loading. The effect of various surface treatments (sandpaper, sandblasting, plasma) on injection-molded parts and the influence of different build chamber temperatures during additive manufacturing on torsional strength were specifically examined. The findings emphasize the critical role of surface treatment for the injection-molded components before additive manufacturing. Also, the torsional strength could be increased by up to 87% by actively heating the build chamber.

Picard et al. [4] have explored the manufacturing of cost-effective and customized composites by strategically placing carbon fiber-reinforced thermoplastics in multi-material designs. A model for the simultaneous processing of non-reinforced and reinforced thermoplastic layers, also known as insert overmolding, was developed with the aim of identifying essential parameters to minimize insert flow and ensure desired fiber orientation and positional integrity. A process window for tailored reinforcements in overmolding processes can be easily defined that way.

Ebert et al. [5] have addressed the challenging issue of the functionalization of highly filled continuous glass fiber-reinforced thermoplastic pultruded profiles by welding. For this purpose, the classic hot-tool welding process was adopted, and an infrared emitter with line focus was integrated for heating the polyamide 6 thermoplastic profiles. Despite a very high fiber content of 70 vol.%, a strong welded joint between the unreinforced tension rod and the pultruded plate could be achieved.

2.2. Automated Fiber, Yarn, or Tape Placement and Additive Manufacturing

Legenstein and Fauster [6] have studied the effect of flashlamp heating system parameters in order to maximize the wedge peel strength of thermoplastic carbon fiber tape in the automated tape placement (ATP) process. Flashlamp heating systems are rather new in this field of application and offer high energy density with low safety requirements and moderate costs compared to laser-assisted automated tape placement systems.

Shi et al. [7] have proposed a peel test method to characterize the decay law of prepreg tape tack, a key factor affecting the automatic tape laying process, at different temperatures. A new tack test device was designed, and the decay rate of prepreg tack at different temperatures was tested. Based on the experimental results, a prepreg tack decay model was also proposed. Another achievement of the study is a new statistical unit for prepreg tack, which can establish the relationship between the tack of prepreg and its remaining storage time and reduce prepreg management costs.

Knoch et al. [8] have implemented a novel approach in modeling the system limits of a braked, high-speed direct yarn-laying process with in situ impregnation of carbon fiber heavy tows as textile reinforcements for concrete parts. In particular, the yarn spool overrun after the robot has come to a standstill was investigated and modeled using physical equations, taking into account the travel speed, acceleration of the robot, and braking force of the spool brake. Furthermore, models for robot braking time, carbon spool diameter, and spool mass were developed.

Consul et al. [9] have explored large format additive manufacturing (LFAM) with carbon fiber-reinforced polyaryletherketones (PAEK), particularly a slow crystallizing grade, and investigated how extrusion parameters (including line width, layer height, layer time, and extrusion temperature) affect the mechanical properties of the printed parts. Thermal history during printing was monitored using thermocouples and infrared cameras.

The study suggests aggregated metrics, enthalpy deposition rate, and shear rate under the nozzle that should be maximized to enhance mechanical performance. Interestingly, the common practice of setting fixed layer times does not ensure repeatable part quality.

Baddour et al. [10] have designed prepregnation and fused filament fabrication parameters for recycled PP- and PA-based continuous carbon fiber (cCF) composites. Plasma treatment of the cCF was also explored, as was the annealing of the produced parts to enhance the flexural properties. Eventually, overall guidelines were formulated for the successful production of cCF-based composites.

Antolin-Urbaneja et al. [11] have addressed the screw-extrusion of carbon fiber-reinforced polyamide in pellet form in order to design and 3D-print aeronautical mold preforms, based on an extensive experimental characterization campaign and a multi-physics computational guidance. The designed mold, around 90% lighter than the original invar design, was numerically proven to fulfill thermal and mechanical requirements with high performance.

2.3. LFT Compounding

Aiming at improving the mechanical properties of injection-molded long fiber thermoplastic pellets produced by a thermoplastic pultrusion process, Tipboonsri and Memon [12] have added polypropylene-graft-maleic anhydride (PP-g-MAH) as a coupling agent during the injection molding process. A 4 wt.% concentration of PP-g-MAH was found to be optimal to reach the highest tensile, flexural, and impact strengths.

2.4. Fibers and Fabrics Production

Semitekolos et al. [13] have developed a pilot-scale fiber sizing line, including its initial design and installation, operational phases, and optimization of key process parameters. Furthermore, a range of sizing solutions was investigated and formulated, exploring carbon-based nanomaterial types with different surface functionalization and concentrations to evaluate their impact on the surface morphology and mechanical properties of carbon fibers. The incorporation of nanomaterials, specifically N2-plasma-functionalized carbon nanotubes and few-layer graphene, has demonstrated notable improvements (90% increase) in interfacial shear properties.

Ivars et al. [14] have investigated the impact of carding and blending recycled carbon fibers (rCF) with crimped thermoplastic polypropylene (PP) fibers on the mechanical properties of rCF using a Weibull statistical approach. Overall, carding rCF with PP fibers helps in the mechanical property uniformity of the resulting carded webs without compromising tensile performance. The work also highlights the potential of the carding process with or without thermoplastic fibers to efficiently realign and give continuity to discontinuous recycled carbon fibers.

2.5. Manufacturing of Sustainable Bio-Based Composites

Pisupati et al. [15] have elucidated the influence of the cooling rate, and thus of the crystallinity, on the bending and impact properties of compression-molded non-woven flax/polylactic acid (PLA) biocomposites with different flax fiber contents. An improvement of 25% and 100% in flexural modulus was observed for the composites with 40 wt.% and 50 wt.% of flax fiber, respectively, when subjected to a lower cooling rate. This makes such flax/PLA composites promising for semi-structural applications, providing a sustainable alternative with enhanced lightweight performance.

Li et al. [16] have compared the physical properties of coconut fiber-reinforced polypropylene composite materials, produced through a hybrid injection molding process and a layered hot-pressing process. The layered hot-pressed composite material exhibited optimal comprehensive mechanical performance and could be applied to airbag covers in the field of automotive safety, as confirmed by the modeling and finite-element simulation carried out.

2.6. Laminated Glass/Polymer Composites

Jeong et al. [17] have proposed an equivalent model of polyvinyl butyral (PVB) laminated glass to simulate the head injury criterion (HIC) when a pedestrian collides with a tram rail vehicle. Whereas the traditional PLC model replicates the multi-layer structure of PVB-laminated glass, including tempered glass and PVB film, the new simplified equivalent model considers tempered glass and PVB film as a single shell element layer, incorporating the multi-layer characteristics via integration points. As analytical equivalency with the PLC model is maintained, the equivalent model produces results comparable to the PLC model while significantly reducing analysis time from 8 h to 1 h under the same hardware and software conditions.

3. Machining, Milling, and Drilling of Polymers, Composites, and Multi-Material Laminates

Nargis et al. [18] have optimized the drilling quality of unidirectional carbon fiber-reinforced polymer composites, as well as hybrid Al_2O_3 alumina and hybrid SiC silicon carbide nanocomposites, through an experimental exploration using step, core, and twist drills. Response surface methodology (RSM) and statistical tools were used to analyze the surface roughness of the hole, and two machine learning models—artificial neural network (ANN) and random forest (RF)—were used for predictive analysis. Both models have shown robust predictive capabilities, with RF demonstrating superior performance over ANN and RSM.

Bolar et al. [19] have analyzed the performance of helical milling and drilling operations while machining carbon fiber-reinforced aluminum laminates. Indeed, being a difficult-to-cut material, fiber/metal laminates often pose challenges during conventional drilling and require judicious selection of machining parameters to ensure defect-free laminates. Helical milling appears to be a promising technique for producing good-quality holes, having a lower surface roughness than that achieved by conventional drilling.

Wu et al. [20] have investigated the machinability and surface properties of cryogenically cooled poly(methyl methacrylate) machined via single-point diamond turning (SPDT). Cryogenic machining was shown to be useful for improving the form accuracy and reducing the surface roughness of PMMA.

4. Polymer Processing and Compounding

4.1. Functionalization

Andrade-Guel et al. [21] have elaborated by melt-blowing functional technical nonwoven fabrics from polymer nanocomposites made of polyamide 6 and amine-modified nanoclay having adsorbent properties of toxins and dyes and also showing an antibacterial behavior.

Alkarri et al. [22] have paid attention to the antimicrobial, thermal, mechanical, and gas barrier properties of linear low-density polyethylene extrusion blow-molded bottles. New methods of incorporating antimicrobial particles into the bottles were developed. The antimicrobial particles were thermally embossed on the external surface of the bottle through two particle deposition approaches (spray and powder) over the mold cavity. Both deposition approaches have led to a significant enhancement in antimicrobial activity, as well as barrier properties, while maintaining thermal and mechanical performance.

Prihandana et al. [23] have investigated the influence of activated carbon (AC) particle size on the properties and performance of polysulfone composite membranes for protein separation. The polymeric composite membranes were fabricated via the phase-inversion method, employing water as the coagulant and using variable loadings of AC particle sizes. The impact of the AC powder particle sizes on membrane morphology, water contact angle, porosity, average pore size, molecular weight cutoff, pure water flux, and protein rejection was examined. On this basis, recommendations could be provided for the selection of AC particle sizes for protein separation in conjunction with polysulfone ultrafiltration membranes.

Al Mais et al. [24] have highlighted the importance of diverse forms, e.g., morphology and particle size, and concentrations of graphitic carbon nitride (g-C_3N_4) as strengthening elements in thermosetting epoxy resin-based composites. Optimal mechanical properties were achieved with a 0.5 wt.% bulk g-C_3N_4 filler, enhancing tensile strength by 14%. An increased toughness, no significant impact on the degradation temperature, a 17% increase in glass transition temperature, and an improvement in thermal breakdown up to 600 °C were also noticed.

4.2. Plastics Recycling

Gall et al. [25] have compared end-of-life vehicle (ELV) and packaging-based recyclates as components in polypropylene (PP)-based compounds designed for automotive applications. Two ELV recyclate grades based on bumper recycling were analyzed in comparison to a packaging-based post-consumer recyclate (PCR) and compounded with virgin PP base material and mineral reinforcement. While the targeted mechanical properties could nearly be reached, further development work is still necessary to improve the flowability of the PCR compounds.

Through a twin-screw extrusion process, Kharmoudi et al. [26] have formulated blends based on post-consumer recycled high-density and high-molecular-weight polyethylene used for manufacturing rainwater drainage pipes and investigated the effect of graphene addition on mechanical and thermal behavior. Adding low amounts of graphene to recycled polyethylene blends could be an interesting alternative to ensure good mechanical properties and a resistance to crack propagation comparable to virgin polyethylene compounds.

Author Contributions: Conceptualization, P.K.; writing—original draft preparation, P.K.; writing—review and editing, P.K. and L.C. All authors have read and agreed to the published version of the manuscript.

Funding: This research received no external funding.

Conflicts of Interest: The authors declare no conflicts of interest.

References

1. Hirsch, P.; Scholz, S.; Borowitza, B.; Vyhnal, M.; Schlimper, R.; Zscheyge, M.; Kotera, O.; Stipkova, M.; Scholz, S. Processing and Analysis of Hybrid Fiber-Reinforced Polyamide Composite Structures Made by Fused Granular Fabrication and Automated Tape Laying. *J. Manuf. Mater. Process.* **2024**, *8*, 25. [CrossRef]
2. Atzler, F.; Hümbert, S.; Voggenreiter, H. A Workflow for the Compensation of Substrate Defects When Overprinting in Extrusion-Based Processes. *J. Manuf. Mater. Process.* **2024**, *8*, 147. [CrossRef]
3. Kizak, M.; von Bartschikowski, A.; Trauth, A.; Heigl, C.; Drechsler, K. Investigation of the Interfacial Fusion Bonding on Hybrid Additively Manufactured Components under Torsional Load. *Polymers* **2024**, *16*, 2719. [CrossRef]
4. Picard, P.K.W.; Osswald, T.A.; Zaremba, S.; Drechsler, K. Modeling of a Process Window for Tailored Reinforcements in Overmolding Processes. *J. Compos. Sci.* **2024**, *8*, 65. [CrossRef]
5. Ebert, C.; Dürr, M.N.; Bonten, C. Functionalization of Continuous Fiber-Reinforced Thermoplastic Pultrusion Profiles by Welding. *J. Compos. Sci.* **2025**, *9*, 6. [CrossRef]
6. Legenstein, A.; Fauster, E. Effect of Flashlamp Heating System Parameters on the Wedge Peel Strength of Thermoplastic Carbon Fiber Tape in the Automated Tape Placement Process. *J. Manuf. Mater. Process.* **2024**, *8*, 91. [CrossRef]

7. Shi, J.; Wang, W.; Wang, Y.; Qi, J.; Xiao, J. A Peel Test Method to Characterize the Decay Law of Prepreg Tape Tack at Different Temperatures. *Materials* **2024**, *17*, 2449. [CrossRef]
8. Knoch, E.; Rittner, S.; Holschemacher, K. Analysis and Modeling of the System Boundaries of a High-Speed Direct-Yarn-Placement System for In Situ Impregnation of Carbon Fibre Heavy Tows as Textile Reinforcements for Concrete Parts. *Fibers* **2024**, *12*, 47. [CrossRef]
9. Consul, P.; Feuchtgruber, M.; Bauer, B.; Drechsler, K. Influence of Extrusion Parameters on the Mechanical Properties of Slow Crystallizing Carbon Fiber-Reinforced PAEK in Large Format Additive Manufacturing. *Polymers* **2024**, *16*, 2364. [CrossRef]
10. Baddour, M.; Garcia-Campà, R.; Reyes, P.; D'hooge, D.R.; Cardon, L.; Edeleva, M. Designing Prepregnation and Fused Filament Fabrication Parameters for Recycled PP- and PA-Based Continuous Carbon Fiber Composites. *Materials* **2024**, *17*, 1788. [CrossRef]
11. Antolin-Urbaneja, J.C.; Vallejo Artola, H.; Bellvert Rios, E.; Gayoso Lopez, J.; Hernández Vicente, J.I.; Luengo Pizarro, A.I. Experimental Characterization of Screw-Extruded Carbon Fibre-Reinforced Polyamide: Design for Aeronautical Mould Preforms with Multiphysics Computational Guidance. *J. Manuf. Mater. Process.* **2024**, *8*, 34. [CrossRef]
12. Tipboonsri, P.; Memon, A. The Impact of PP-g-MAH on Mechanical Properties of Injection Molding of Long Glass Fiber/Polypropylene Pellets from Thermoplastic Pultrusion Process. *J. Manuf. Mater. Process.* **2024**, *8*, 53. [CrossRef]
13. Semitekolos, D.; Papadopoulos, I.; Anagnou, S.; Dashtbozorg, B.; Li, X.; Dong, H.; Charitidis, C.A. Nanomaterial-Enhanced Sizings: Design and Optimisation of a Pilot-Scale Fibre Sizing Line. *Fibers* **2024**, *12*, 16. [CrossRef]
14. Ivars, J.; Labanieh, A.R.; Soulat, D. Influence of the Thermoplastic Fiber Ratio on the Mechanical Properties of Recycled Carbon Fibers During the Carding Process. *Materials* **2025**, *18*, 302. [CrossRef] [PubMed]
15. Pisupati, A.; Curto, M.; Laurent, T.; Cosson, B.; Park, C.H.; Dhakal, H.N. Influence of Cooling Rate on the Flexural and Impact Properties of Compression Molded Non-Woven Flax/PLA Biocomposites. *Polymers* **2025**, *17*, 493. [CrossRef] [PubMed]
16. Li, J.; Zhou, Y.; Chen, J.; Hu, H.; Sun, M. A Comparative Study of Airbag Covers for Automotive Safety Using Coconut Shell Fiber/PP Composite Materials. *J. Compos. Sci.* **2024**, *8*, 328. [CrossRef]
17. Jeong, Y.; Jeon, Y.; Lee, W.; Yoon, J. Development of an Equivalent Analysis Model of PVB Laminated Glass for TRAM Crash Safety Analysis. *Polymers* **2025**, *17*, 25. [CrossRef]
18. Nargis, T.; Shahabaz, S.M.; Acharya, S.; Shetty, N.; Malghan, R.L.; Shetty, S.D. A Comprehensive Study on the Optimization of Drilling Performance in Hybrid Nano-Composites and Neat CFRP Composites Using Statistical and Machine Learning Approaches. *J. Manuf. Mater. Process.* **2024**, *8*, 67. [CrossRef]
19. Bolar, G.; Dinesh, A.A.; Polishetty, A.; Shetty, R.; Hiremath, A.; Neelakantha, V.L. Performance Analysis of Helical Milling and Drilling Operations While Machining Carbon Fiber-Reinforced Aluminum Laminates. *J. Manuf. Mater. Process.* **2024**, *8*, 113. [CrossRef]
20. Wu, X.; Kang, Q.; Jiang, X.; Fang, X. Machinability and Surface Properties of Cryogenic Poly(methyl methacrylate) Machined via Single-Point Diamond Turning. *Materials* **2024**, *17*, 866. [CrossRef]
21. Andrade-Guel, M.; Cabello-Alvarado, C.J.; Ávila Orta, C.A.; Cadenas-Pliego, G.; Cruz-Ortiz, B. Functional Technical Textile-Based Polymer Nanocomposites with Adsorbent Properties of Toxins and Dyes also Have Antibacterial Behavior. *Materials* **2024**, *17*, 3007. [CrossRef] [PubMed]
22. Alkarri, S.; Naveed, M.; Alali, F.; Vachon, J.; Walworth, A.; Vanderberg, A. Anti-Microbial, Thermal, Mechanical, and Gas Barrier Properties of Linear Low-Density Polyethylene Extrusion Blow-Molded Bottles. *Polymers* **2024**, *16*, 1914. [CrossRef] [PubMed]
23. Prihandana, G.S.; Muthi'ah, A.D.; Sriani, T.; Mahardika, M. The Influence of Activated Carbon Particle Size on the Properties and Performance of Polysulfone Composite Membrane for Protein Separation. *J. Compos. Sci.* **2024**, *8*, 483. [CrossRef]
24. Al Mais, D.; Mustapha, S.; Baghdadi, Y.N.; Bouhadir, K.; Tehrani-Bagha, A.R. Various Morphologies of Graphitic Carbon Nitride (g-C_3N_4) and Their Effect on the Thermomechanical Properties of Thermoset Epoxy Resin Composites. *Polymers* **2024**, *16*, 1935. [CrossRef]
25. Gall, M.; Mileva, D.; Stockreiter, W.; Salles, C.; Gahleitner, M. Comparing End-of-Life Vehicle (ELV) and Packaging-Based Recyclates as Components in Polypropylene-Based Compounds for Automotive Applications. *Polymers* **2024**, *16*, 1927. [CrossRef]
26. Kharmoudi, H.; Lamtai, A.; Elkoun, S.; Robert, M.; Diez, C. Effect of Graphene on the Mechanical Properties of Recycled High-Density and High-Molecular-Weight Polyethylene Blends. *Materials* **2024**, *17*, 4733. [CrossRef]

Journal of
Manufacturing and Materials Processing

MDPI

Article

Processing and Analysis of Hybrid Fiber-Reinforced Polyamide Composite Structures Made by Fused Granular Fabrication and Automated Tape Laying

Patrick Hirsch [1,*], Simon Scholz [1], Benjamin Borowitza [1], Moritz Vyhnal [1], Ralf Schlimper [1], Matthias Zscheyge [1], Ondrej Kotera [2], Michaela Stipkova [2] and Sebastian Scholz [2]

1 Fraunhofer Institute for Microstructure of Materials and Systems IMWS, Walter-Hülse-Straße 1, 06120 Halle (Saale), Germany; simon.scholz@imws.fraunhofer.de (S.S.); benjamin.borowitza@imws.fraunhofer.de (B.B.); moritz.vyhnal@imws.fraunhofer.de (M.V.); ralf.schlimper@imws.fraunhofer.de (R.S.); matthias.zscheyge@imws.fraunhofer.de (M.Z.)
2 Fraunhofer Institute for Machine Tools and Forming Technology IWU, Reichenhainer Straße 88, 09126 Chemnitz, Germany; ondrej.kotera@iwu.fraunhofer.de (O.K.); michaela.stipkova@iwu.fraunhofer.de (M.S.); sebastian.scholz@iwu.fraunhofer.de (S.S.)
* Correspondence: patrick.hirsch@imws.fraunhofer.de

Abstract: Fused granular fabrication (FGF) is a large format additive manufacturing (LFAM) technology and focuses on cost-effective granulate-based manufacturing by eliminating the need for semifinished filaments. This allows a faster production time and a broader range of usable materials for tailored composites. In this study, the mechanical and morphological properties of FGF test structures made of polyamid 6 reinforced with 40% of short carbon fibers were investigated. For this purpose, FGF test structures with three different parameter settings were produced. The FGF printed structures show generally significant anisotropic mechanical characteristics, caused by the layer-by-layer building process. To enhance the mechanical properties and reduce the anisotropic behavior of FGF structures, continuous unidirectional fiber-reinforced tapes (UD tapes), employing automated tape laying (ATL), were subsequently applied. Thus, a significant improvement in the flexural stiffness and strength of the manufactured FGF structures was observed by hybridization with 60% glass fiber-reinforced polyamide 6 UD tapes. Since the effectiveness of UD-tape reinforcement depends mainly on the quality of the bond between the UD tape and the FGF structure, the surface quality of the FGF structure, the interface morphology, and the tape-laying process parameters were investigated.

Keywords: additive manufacturing; fused granular fabrication; automated tape laying; composite; polyamide

1. Introduction

In recent decades, additive manufacturing (AM) based on the targeted deposition of a polymer melt has focused on the production of small-format components with very high surface quality and optimal geometric brilliance [1]. The fused filament fabrication (FFF) printing systems used are typically based on filament processing with typical filament diameters of 1.75–2.85 mm and allow extrusion rates of up to 6000 mm/min or build rates of 100–480 g/h, depending on the printed object and the type of filament used [2]. However, these filament-based printers usually have small printing space of max. approximately 0.5 m × 0.5 m × 0.5 m and printing rates of 20 cm^3/h [3]. As a result, the requirement for larger component volumes, an associated larger construction space, and an increase in printing speed was derived [4,5]. This led to a new term for this type of printing technology, large format additive manufacturing (LFAM), which, however, has not yet been able to establish itself consistently in the industry. The aim of LFAM is to significantly reduce costs to produce larger components more quickly [6].

Fused granular fabrication (FGF) is a special technology within the framework of LFAM, which focuses on granulate or pellet-based manufacturing processes [7]. In comparison to FFF, this procedure can reduce costs by eliminating the need for semifinished filaments, since, for example, the costs for a filament compared to granules are in a ratio of around 10:1 [8]. Moreover, the use of adapted extruders with significantly larger discharge nozzles is possible. This means that production times can be reduced by a factor of up to 200 [9]. Further advantages also lie in the use of a significantly broader range of materials, since all industrially usable thermoplastic polymers are available in the form of granules. These can also be adapted to the respective requirements using fibers, particles, and additives and thus compounded into precisely fitting composite materials with adjustable material properties [10]. This also includes recycled materials from other processing methods, e.g., composite manufacturing, as recently also reported for FFF processes [11,12].

FGF printing processes can be described as young in relation to additive manufacturing processes and have several specific features in the quality of the components produced and in relation to the rheological behavior of the granules in the extrusion process. Transferring the experience from classic extrusion technology is only possible to a limited extent, as systems with movable print heads have limitations in terms of weight and screw length, which play a rather subordinate role in stationary machines. For this reason, several studies have been carried out for a variety of materials (unreinforced/reinforced) to investigate the influence of processing parameters on rheological behavior and glass transition temperature [13–16]. The degradation behavior also plays a role that needs to be considered. However, due to the reduced screw lengths, shorter residence times of the material are to be expected and thus also reduced damage/degradation of the polymers. Regarding the mechanical properties of the printed components, as with FFF-manufactured structures, there is a clear anisotropy with respect to the layers in the Z-axis [17]. This is further increased when using fiber-reinforced material [18]. In addition to conventional synthetic reinforcing fibers, natural fibers are also being used to increase the mechanical properties of the resulting parts [19]. Furthermore, the quality of the surface and the geometry of components manufactured using FGF is significantly worse than components manufactured using FFF due to the nozzle size and the material quantities deposited. This requires postprocessing steps such as simple milling or more complex processes such as grinding, filling, polishing, and painting [20]. Another solution described in the literature is printing with two differently fine printing nozzles [21].

A promising strategy for improving the mechanical properties of components from the FGF process is the subsequent or simultaneously application of continuously fiber-reinforced filaments or tapes [22,23]. In comparison to continuously fiber-reinforced filaments, higher fiber volume proportions up to 40–50 vol.-% and the associated significantly higher strength and stiffness can be achieved by using continuously fiber-reinforced tapes. The term tape laying generally refers to the automated, direction- and position-variable placement of unidirectional fiber-reinforced thermoplastic tapes (UD tapes) on flat or curved substrates [24]. Thus, local reinforcements or fiber placement according to the load path have already been implemented for many different applications [25]. In particular, 3D tape laying enables targeted reinforcement of the components in accordance with the load path [26]. Typically, the UD tapes are additively processed in layers to form thin-walled, shell-shaped composite structures. This requires a so-called laying head, with which the UD tapes are heated, laid down, and trimmed. The geometry of the later component is dictated by the substrate, onto which the previously melted UD tapes are applied under light contact pressure during the deposition process.

The current state of the art is represented here by the technologies of automated tape laying (ATL) and automated fiber placement (AFP), which were developed in the aerospace industry and are currently finding their way into a wide variety of industries [27]. In the ATL process, UD tapes are unwound from a spool by a storage unit, heated to processing temperature, then placed on a tool contour with a defined contact pressure and finally cut off at the end of the laying path. After a layer of tape has been deposited, the storage unit is realigned, and the next layer of tape can be deposited. Typical laying speeds are approximately 1 m/s [28]. In this way, the laminate is built up step by step according to a given individual layer and laminate definition. Depending on the feeding of the UD tapes, unassembled from a reel or preassembled from a cassette, a distinction is made between single-stage and two-stage or dual systems, as a combination of both process options. This technology is currently mainly used to produce load-resistant lightweight structures with adapted fiber orientation in the aviation industry [29].

A major disadvantage of the current ATL systems is, on the one hand, their size and the uneconomical way of working compared to the processing of conventional semifinished textile products. In addition, tape laying in curved paths is not easily possible with ATL technology. For this reason, the AFP technology was developed, which enables, for example, the shaping of radii or the fiber-friendly design of hole edges and openings [30]. The AFP depositing unit divides the tape into several individual strands of around 10 mm width before depositing it, which means that the fiber path can also be set in a defined manner over certain radii and deposited individually [31]. However, the laying speed is reduced compared to the ATL method.

To partially overcome the previous mentioned disadvantages of ATL, a very compact innovative ATL system, F3-Compositor, with reduced weight of the moving machine parts using a six-axis robot for 3D tape laying, has been recently developed and applied in this work. The innovative approach of the F3-Compositor combines the high-speed tape deposition at a constant laying speed of up to 2 m/s with high reproducibility and material efficiency [32]. The tape-laying head with an integrated heat source based on a hydrogen–oxygen gas mixture enables the heat energy to be targeted in just the right amount and place required for the tape consolidation, thus avoiding unwanted large heat-affected areas. This technology represents an optimal complement to the FGF technology and can therefore be used for the production of hybrid 3D printed components with local fiber reinforcement and significantly improved mechanical properties.

2. Materials and Methods

2.1. Fiber-Reinforced Polyamide Composite Materials

The fiber-reinforced polyamide material used for the FGF process was a short carbon fiber-reinforced polyamide 6 compound (PA6/CF40, AKROMID® B3 ICF 40 black 5020) produced by AKRO-PLASTIC GmbH (Niederzissen, Germany). This material contains 40% recycled content and was used in the supplied granular form. A continuous glass fiber-reinforced polyamide 6 (PA6/GF60-UD, Celstran® CFR-TP PA6-GF60) supplied from TICONA GmbH (Sulzbach, Germany) was used for the tape-laying process in the form of tapes with a width of 3 mm and a thickness of 0.3 mm. Both materials were dried at 80 °C for 4 h prior to processing. The main properties of the used materials derived from the manufacturer's data sheets are summarized in Table 1.

Table 1. Properties of the used fiber-reinforced polyamide composite materials.

Composite Material	PA6/CF40	PA6/GF60-UD
Grade	AKROMID® B3 ICF 40 black	Celstran® CFR-TP PA6-GF60
Supplier	AKRO-PLASTIC GmbH	TICONA GmbH
Reinforcement	carbon fiber	glass fiber
Fiber Weight Content (%)	40	60
Density (g/cm^3)	1.31	1.69
Tensile Modulus (GPa)	32 (dry) 13.6 (conditioned)	29.7 (dry) 27.5 (conditioned)
Tensile Strength (MPa)	220 (dry) 135 (conditioned)	679 (dry) 642 (conditioned)
Melting Temperature (°C)	220	220

2.2. Fused Granular Fabrication

The used FGF technology, shown in Figure 1, consists of an extrusion unit with a mass output of up to 5 kg/h, a nozzle set with diameters of 1, 1.5, 3, and 6 mm, and integrated sensors (thermal camera μ-Epsilon TIM640). The extrusion unit was adapted to an industrial six-axis robot (Stäubli, RX160) with Siemens Sinumerik NC control for three-dimensional processing. A melt temperature of 280 °C was set for the printing process. A heatable aluminum plate with an area 1 × 1 m^2 was used as a print bed, and the temperature of the heating system was set to 120 °C. The resulting substrate temperatures were measured in the range 70–75 °C.

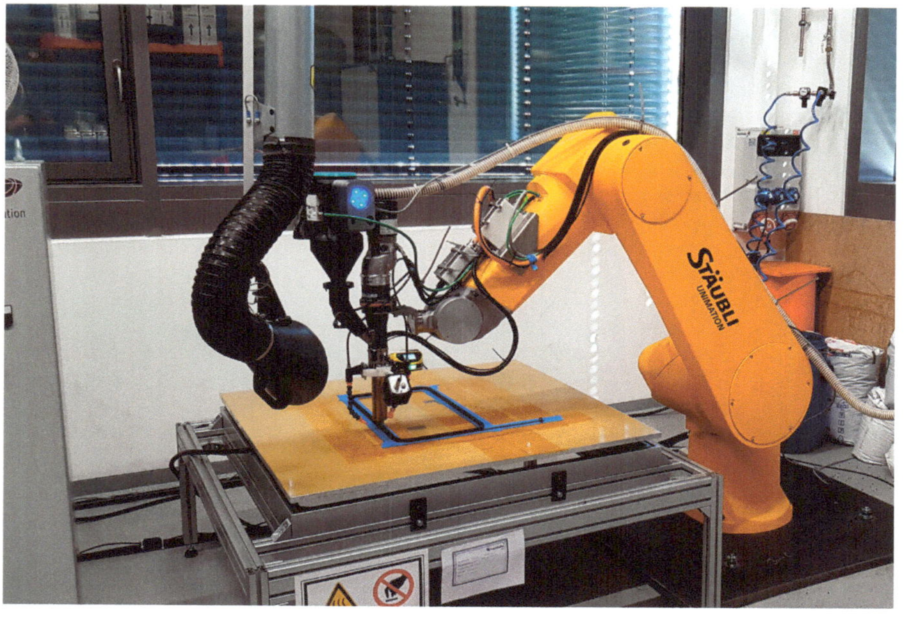

Figure 1. Fused granular fabrication unit used to produce fiber-reinforced polyamide test structures.

With this setup, specific test structures in the shape of a rib structure with 6 parallel walls and 1 perpendicular wall were printed with a one-way printing path direction, as shown in Figure 2. After printing, the structure walls were marked with letters from A to G according to the printing order and cut into 7 single plates (150 × 305 mm) that were used as blanks for manufacturing the test samples. The test structures were printed with three different parameter setups; the resulting wall thicknesses are given in Table 2.

Table 2. Processing parameters and resulting wall thicknesses of the fused granular fabrication process.

Plate Type	Printing Speed v (mm/s)	Layer Height h (mm)	Nozzle Diameter d (mm)	Extrusion Width w (mm)	Ratio w/d (-)	Ratio h/d (-)
FGF-1	400	0.50	1.5	2.7	1.80	0.33
FGF-2	200	0.75	1.5	4.3	2.87	0.5
FGF-3	125	1.00	3.0	5.2	1.73	0.33

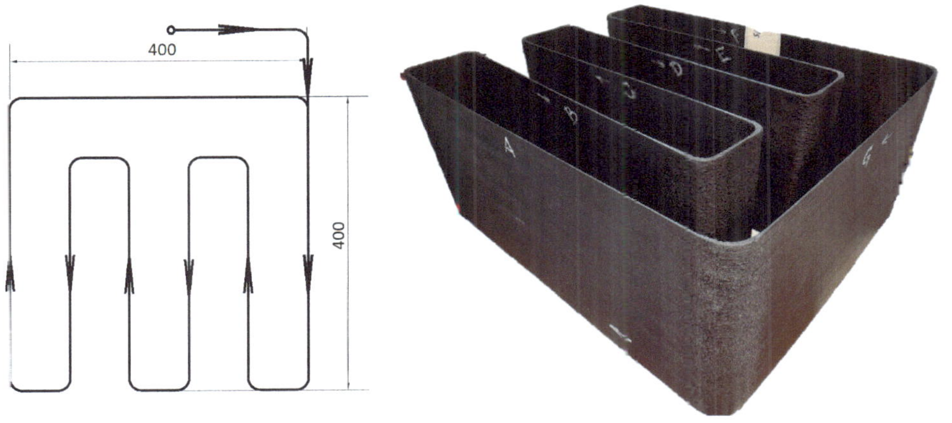

Figure 2. Schematic illustration of the fused granular fabrication printing path direction (**left**, top view) and the resulting "ribbed ridge" test structure (**right**).

2.3. Preparation of Fused Granular Fabrication Test Specimen

The test samples for the mechanical analysis were manufactured by water jet cutting from the processed FGF plates in horizontal and vertical directions (see Figure 3). Apart from the difference in the extrusion width of the three plate types and therefore the thickness of the test specimen, the dimensions used for tensile and bending test samples stayed the same and are shown in Figure 4. The width x of the bending test specimen was selected based on the plate thickness following DIN EN ISO 178:2019 [33]. The resulting widths for FGF-1, 2, and 3 were 25, 10, and 15 mm, respectively. The 80 mm sample length and the 64 mm length of the support span were constant, as shown in Figure 4. Accelerated conditioning according to DIN EN ISO 1110 was performed prior to testing [34]. At 70 °C and 62% relative humidity, the weight gain was controlled regularly to validate the conditioning duration based on sample thickness.

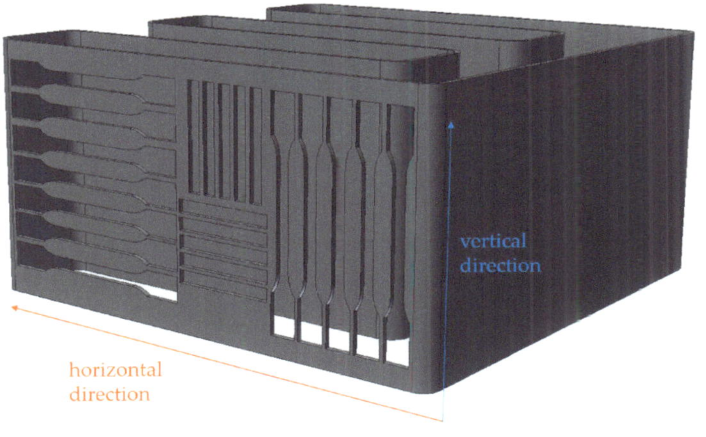

Figure 3. Demonstration of the test sample preparation from the fused granular fabrication structures in horizontal (printing) and vertical direction.

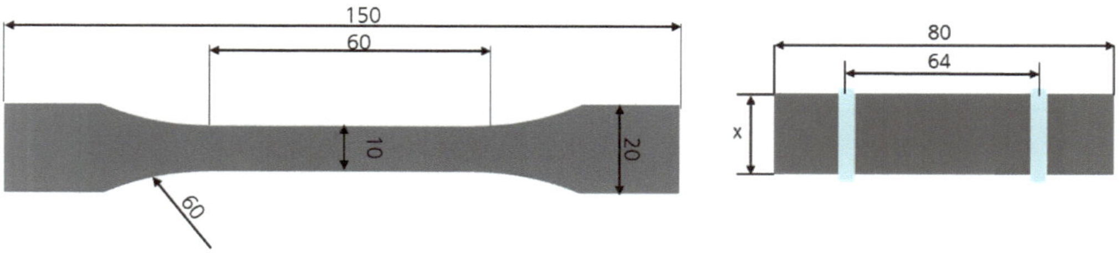

Figure 4. Dimensions of the fused granular fabrication tensile (**left**) and bending test specimen (**right**).

2.4. Automated Tape Laying

Processing of the PA6/GF60-UD materials was performed on the different FGF plates with an automated tape-laying unit (F3-Compositor, ASH Automation Steeg und Hoffmeyer GmbH) with a maximum layup speed of 250 mm/s, as shown in Figure 5. In this technology, the UD tape is fed with a constant velocity and locally heated above its melting temperature just before deposition. This is achieved by an open-flame heating mechanism, consisting of a hydrogen and oxygen gas mixture. The gas mixture can also be diluted with normal air. At the deposition point, the UD tape is compacted and consolidated in situ by a roll with a predetermined force. Additionally, heat is extracted by water cooling of the roll. This leads to solidification and the tape is cut at the end of a track. The processing parameters used are given in Table 3. For each FGF plate, two configurations with the UD-tape direction parallel or perpendicular to the printing direction were processed (see Figure 6 for illustration).

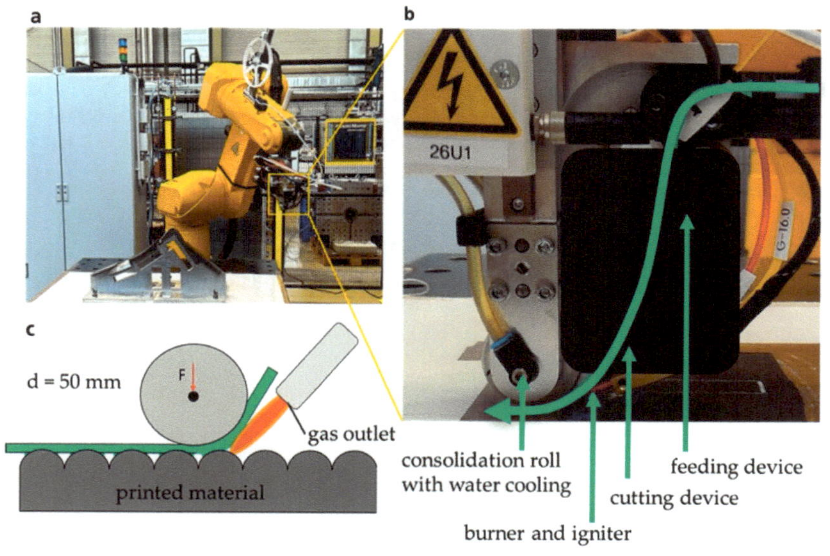

Figure 5. Automated tape-laying unit used to produce fiber-reinforced polyamide test structures (**a**), close-up of the tape-laying head (**b**), and schematic illustration of the tape-laying process (**c**).

Table 3. Processing parameters of the automated tape-laying process.

Laying Speed v (mm/s)	Force on Tape F (N)	Gas Flow Rate V (Nl/min)	Gas Composition $2\,H_2 + O_2$ (%)	Roll Temperature T (°C)
250	60	2.0/2.5	100	20

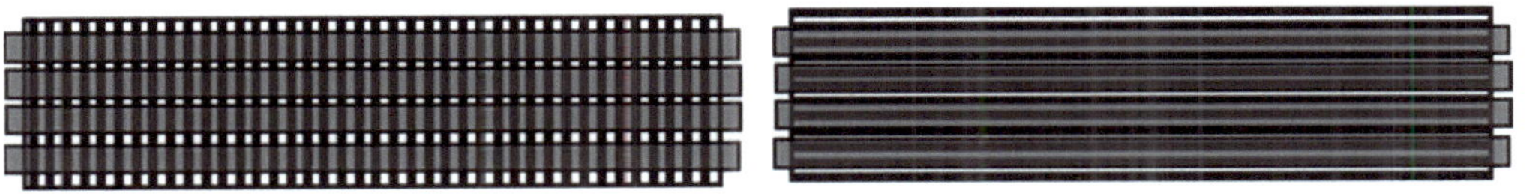

Figure 6. Perpendicular (**left**) and parallel (**right**) configuration of the UD tapes on the fused granular fabrication printing direction.

2.5. Preparation of Automated Tape-Laying Test Specimen

The bending test samples for the mechanical analysis of the automated tape-laying process were also manufactured by water jet cutting parallel to the tape direction from the processed hybrid test plates in accordance with DIN EN ISO 178:2019 (see Figure 4 (right) for illustration) [33]. According to the varied thickness due to the tape laying, the width x of the samples was modified to 26.6, 13.8, and 17 mm.

2.6. Morphological Analysis

For the detailed examination of print quality, fiber alignment and the bonding between the UD tapes and the printed FGF material, test specimens 15 × 15 mm in size were embedded in a clear and low-viscosity epoxy resin with a curing time of 12 h (EpoFix) and polished with a diamond polishing solution to a 0.25 μm finish. Micrographs were taken with an Olympus BX51 optical microscope. Figure 7 shows the schematic illustration of the sectional view where the samples were cut and polished, and the resulting micrograph of these samples. Additionally, a 3D-laser scanning microscope (Keyence VK-X1050) was used to measure surface roughness, which is determined by the geometry of the FGF layer lines. For this, an area of 70 × 10 mm was measured for each FGF plate type. For the density measurement by ethanol immersion, 15 × 15 mm samples for each FGF plate type were prepared using a band saw.

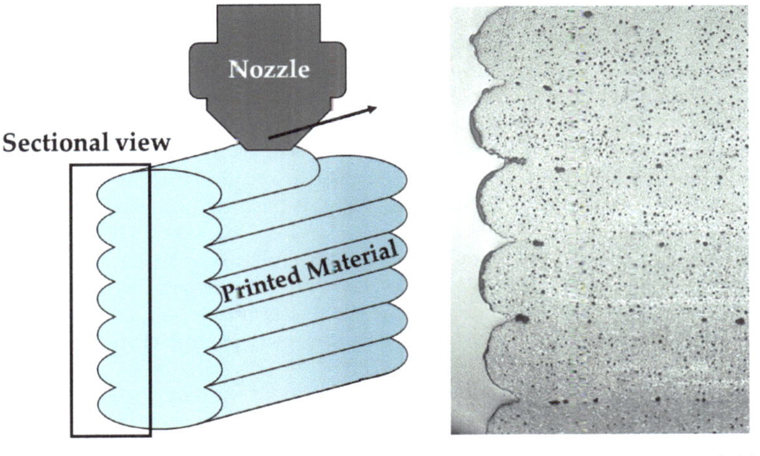

Figure 7. Schematic illustration of the sectional view (**left**) and resulting micrograph (**right**) of the fused granular fabrication test samples

2.7. Mechanical Analysis

Tensile testing according to DIN EN ISO 527 was performed for FGF samples without UD tape, as shown in Figure 8 [35]. In this, a constant test speed of 1 mm/min was used for testing the tensile properties. Three-point bending tests were conducted on the FGF and hybrid test specimens with UD tapes according to DIN EN ISO 178:2019 (see Figure 8) [33]. A Z050 Zwick/Roell testing machine was used for the tensile and three-point bending tests with a load cell of 20 kN. A constant test speed of 2 mm/min was used to determine the flexural modulus, while the rest of the test was conducted at 10 mm/min. The 80 mm sample length and the 64 mm length of the support span always stayed the same. All hybrid samples were tested with the UD tapes on the underside of the samples loaded in tension mode. In the case of the hybrid samples and for the samples without any UD-tape reinforcement, a homogeneous cross-section was assumed for calculating bending stress.

Figure 8. Experimental setup of the three-point bending test (**left**) and the tensile test (**right**) for the test specimen from fused granular fabrication and automated tape laying.

3. Results and Discussion

3.1. Composite Structures by Fused Granular Fabrication

3.1.1. Morphological Analysis

Figure 9 shows a comparison of the micrograph cross-sections of the printed FGF structures. The gray semicircles represent the outer surface edge of the printed structures and the vertical thickness between the layers is referred to as layer height. The geometry of the printed layers is the result of the ratios of the extrusion width and layer height to printing nozzle diameter (Table 2). The bigger ratios of the layer width and height to the nozzle diameter lead to greater layer irregularity and to reduction in the ideal semicircularity in the cross-section of the layers. FGF-2 is the most irregular one and has much enclosed air between the layers. Therefore, FGF-2 is expected to have a reduced bonding quality of the layers. In contrast, FGF-1, with similar layer height, width, and nozzle diameter ratios as FGF-3, has almost no visible air gaps between layers, which is probably related to the higher temperatures of the previous printed layer due to the higher printing speed. FGF-2 and FGF-3 both have a greater density of air pockets than FGF-1. They can be seen as small black circles in Figure 9.

Table 4 shows the measurements of the density and the surface roughness for the three plate types. As expected, FGF-1 has the smallest roughness while R_a for FGF-2 and FGF-3 are equal. The high roughness of FGF-2 is because of the extrusion width to printing nozzle diameter ratio. FGF-1 has the highest density close to the value in the datasheet. The larger number of the air pockets in FGF-2 and FGF-3 are the reason for their decreased density. However, the observed surface roughness values R_a of 63 μm for FGF-1 and 95 μm for FGF-2 and FGF-3 samples are much higher than values reported in the literature for fused deposition modeling test samples with R_a < 50 μm [36]. Significant reduction in surface roughness can be achieved by postprocessing, e.g., by laser polishing [37].

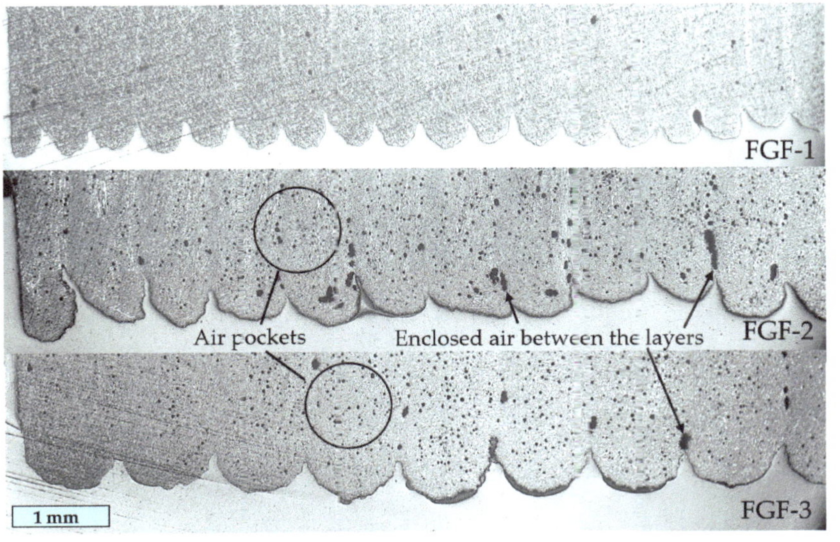

Figure 9. Micrographs showing the surface of the fused granular fabrication test samples.

Table 4. Surface roughness and density of the fused granular fabrication test samples.

Sample	Surface Roughness R_a (µm)	Surface Roughness R_z (µm)	Density ρ (g/cm^3)
FGF-1	63	533	1.301
FGF-2	95	1475	1.273
FGF-3	95	703	1.277

3.1.2. Mechanical Analysis

Depending on the different layer orientation, significant different fracture behavior was observed in the manufactured FGF samples (Figure 10). In tensile and flexural specimens with vertical layer orientation, interlayer fracture always occurred. Samples with horizontal layer orientation broke and created sharp edges. Figure 11 shows the mean stress/strain curves for the FGF test samples obtained during tensile testing. Samples of FGF-2 with horizontal layer configuration have a noticeable difference in modulus of elasticity, which is addressed to their different fiber orientation in the layers. The direction of the fibers directly influences the modulus of elasticity [38].

Because of the high ratio of extrusion width to nozzle diameter, many carbon fibers are not aligned with the printing direction. This can be seen in Figure 12. The fibers should barely be visible when directed out of plane. The areas with a high density of misaligned fibers appear white; in the case of FGF-2, the white areas are much more numerous.

The analysis of the tensile test (Figures 13–15) shows that the tensile behavior is strongly dependent not only on test direction (horizontal vs. vertical) but also on the used parameters in manufacturing of the fused granular fabrication test samples, especially for the tensile strength and the elastic modulus of the FGF test samples. The mechanical properties of FGF-3 are similar to prior analysis of the same material using the same fabrication process [39]. The highest value of elastic modulus and tensile strength in the horizontal direction was achieved by sample FGF-1, with 17,440 MPa and 126 MPa, respectively. In this case, it is evident that the mechanical properties of samples in horizontal direction can even exceed the values specified in the data sheet for an injection molded specimen [40]. However, significantly lower tensile strength and elastic modulus values were recorded in the vertical direction, indicating strong anisotropic behavior of the FGF parts. Nevertheless, there were no significant differences in the values between all three types of FGF tested.

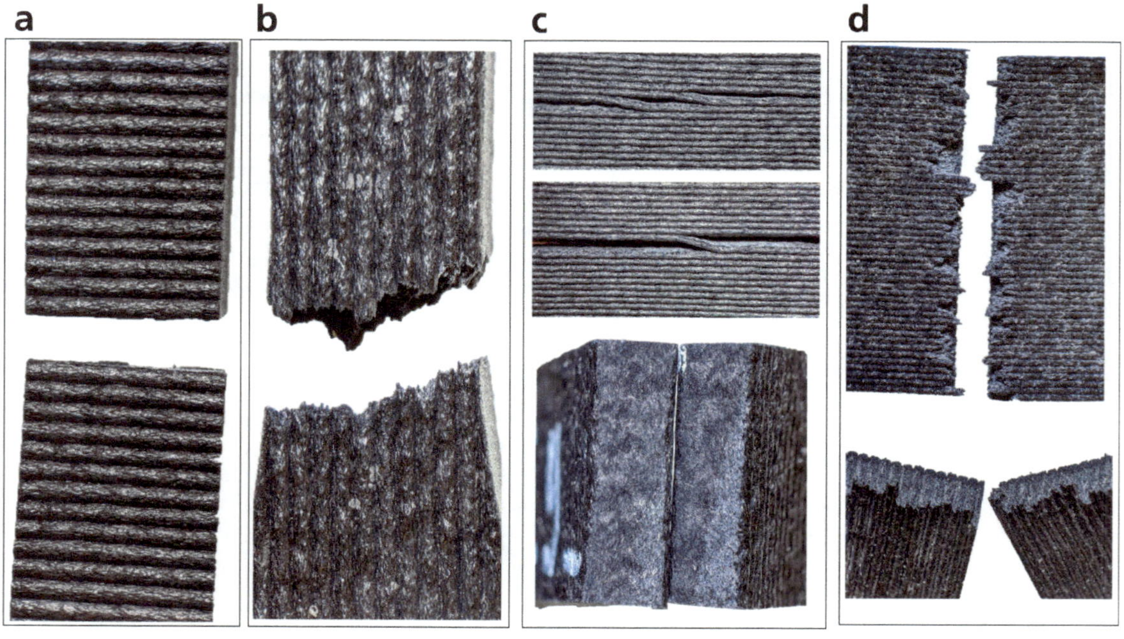

Figure 10. Fracture behavior of the fused granular fabrication tensile samples in vertical (**a**) and horizontal (**b**) direction and bending samples in vertical (**c**) and horizontal (**d**) direction.

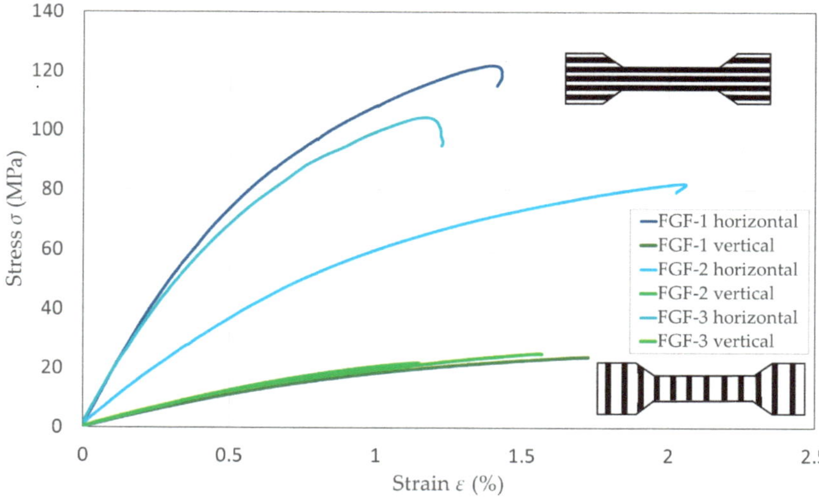

Figure 11. Stress/strain curves for the fused granular fabrication tensile test samples with horizontal and vertical layer configuration.

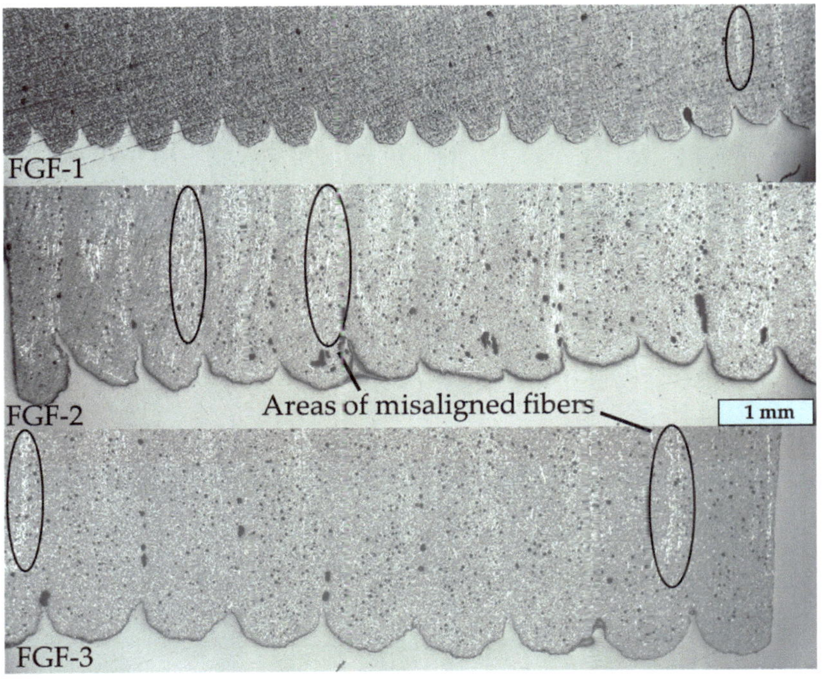

Figure 12. Micrographs of the fused granular fabrication test samples showing areas with a high density of carbon fibers that are not aligned with the print direction. Misaligned fibers appear as thin white streaks.

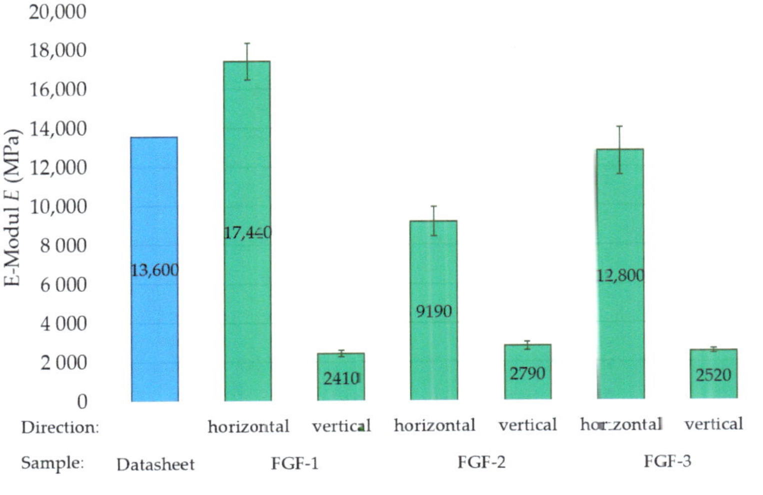

Figure 13. Elastic modulus of the fused granular fabrication test samples (green columns) in comparison to the datasheet value for injection-molded test samples (blue column).

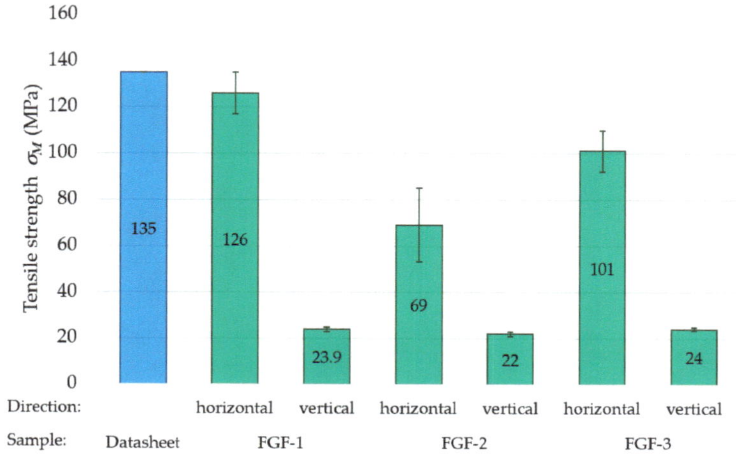

Figure 14. Tensile strength of the fused granular fabrication test samples (green columns) in comparison to the datasheet value for injection-molded test samples (blue column).

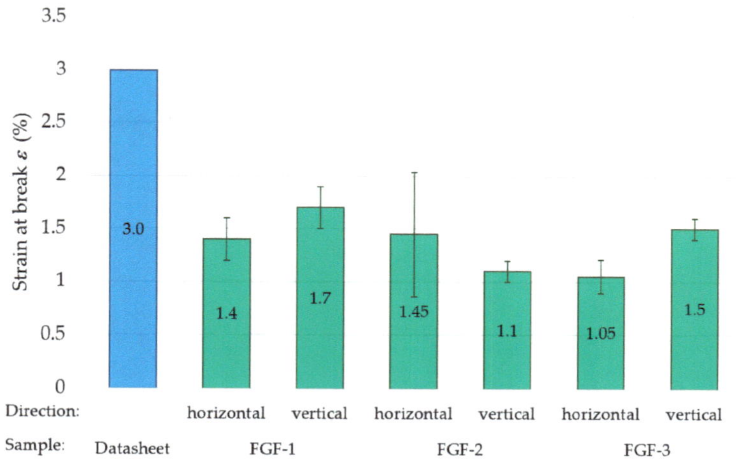

Figure 15. Strain at break of the fused granular fabrication test samples (green columns) in comparison to the datasheet value for injection-molded test samples (blue column).

The FGF specimens in the horizontal and vertical directions achieved relatively similar values of elongation at break. The FGF-2 samples tended to have a lower elongation at break compared to other tested samples in the vertical direction. The reason could be the irregularity of the layers and the resulting air gaps in between them, which results in strong local differences in the interlayer strength within the same specimen. This explanation is also consistent with the exceptionally high variability in tensile strength and strain of FGF-2. In comparison to the horizontal direction, slightly higher values of elongation at break were measured by FGF-1 and FGF-3 samples in the vertical direction. These results may be in contrast with several studies, where the samples indicated much higher elongation in the layer direction than in the perpendicular direction [39,40]. In contrast, according to another studies, also investigating the tensile properties of the PA6/CF additively manufactured specimens, elongation at break values in the vertical direction were similar or even higher than in the horizontal direction [41,42]. This may suggest a more complex issue in the effects on elongation at break. A possible explanation is high diffusion of the PA6/CF40 matrix molecules between the layers leading to an increased interlaminar strength. Additionally, the significant anisotropy in the stiffness and strength can lead to higher stress concentrations and crack propagation in the samples of the horizontal direction, e.g., at fibers or air pockets, thus resulting in a lower strain at break. However, to the authors' best knowledge, there are no recent studies addressing this issue in more detail for short-fiber-reinforced FGF samples with respect to different print build-up orientation.

3.2. *Hybrid Composite Structures from Fused Granular Fabrication and Automated Tape Laying*

3.2.1. Morphological Analysis

Micrographs of the interface of the hybrid test samples manufactured by fused granular fabrication and additional automated tape laying are shown in Figure 16. As can be seen, there are significant differences in the interface quality between the samples, depending on the surface roughness of the FGF plates. If the shape of the FGF layer lines is unaltered and the matrix material of the UD tapes fails to fill the gaps, the UD tape will not attain full contact with the entire FGF surface. This is because of the limited flexibility of the UD tapes and the fixed diameter of the consolidation roll. As also can be seen in Figure 16, with the height variation in the different FGF layer lines some are not in contact with the UD tape at all. A possible reason could be that the gas volume used during the ATL process was too low, or the layup speed was too high to completely melt the FGF surface. In consequence, the matrix material of the tape is pushed to the side and sometimes completely fills the gaps between the layers, as can be seen for FGF/ATL-1. However, both good and bad connections between the UD tapes and the FGF material can be found in every sample and the gaps in between are always present. Because of the structural irregularity of FGF-2, the resulting gaps vary the most in size.

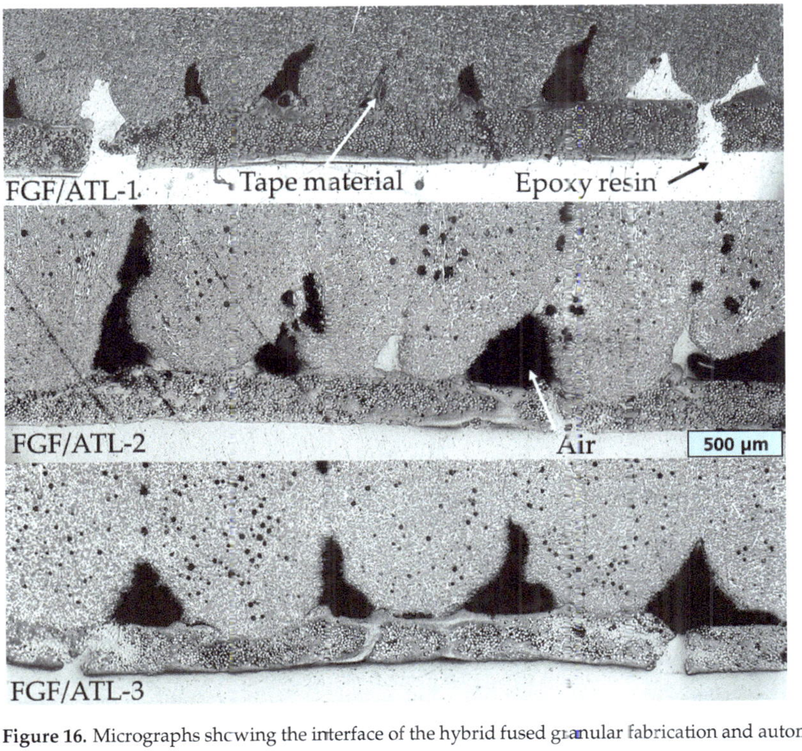

Figure 16. Micrographs showing the interface of the hybrid fused granular fabrication and automated tape-laying test samples.

3.2.2. Analysis of Process Parameters

The contact area between the UD tape and the printed material increases as the UD tape and the FGF surface melt to a greater depth. Thus, two different gas volume flow rates were tested for the ATL process to analyze the effect on the resulting interface to the FGF structures. An increase in the gas volume flow rate results in higher heat generation and more effective melting of the UD tape and FGF surface, as can be seen in Figure 17 for the test series FGF/ATL-3 with 2.0 and 2.5 Nl/m gas volume flow rates. The FGF material in both micrographs is on the top half and the white dots are carbon fibers directed out of plane. In the bottom half, the UD tape with the glass fibers, which are also directed out of plane, can be seen. The FGF layer lines seem unaltered and push the matrix material of the tapes to the side.

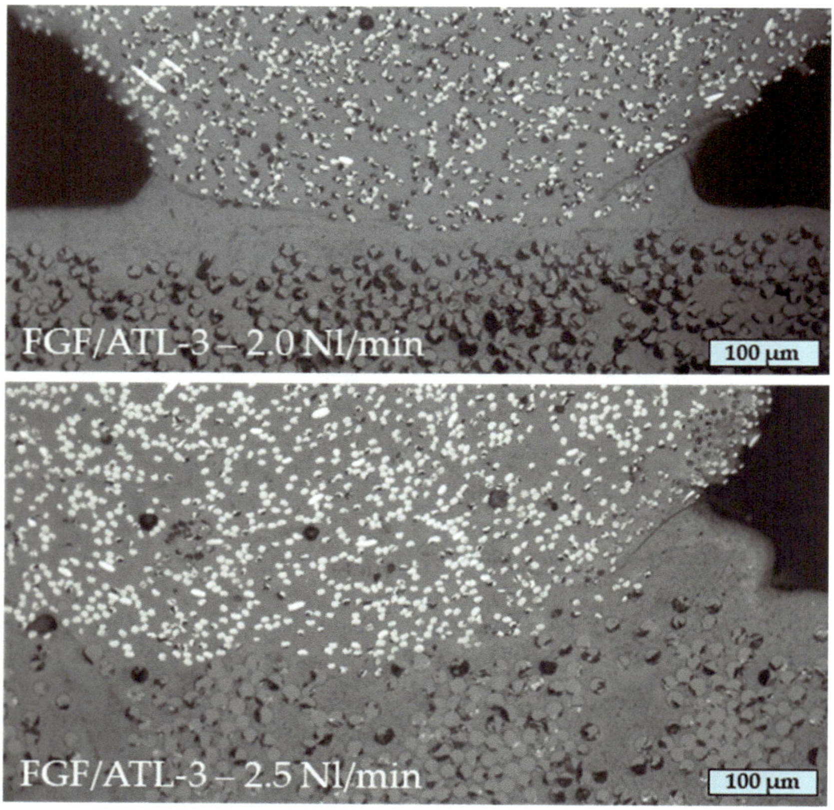

Figure 17. Micrographs showing the interface of hybrid fused granular fabrication and automated tape-laying test samples with different gas volume flow rates during the tape-laying process.

The interface has a significant effect on the resulting mechanical properties in these hybrid fiber-reinforced structures. When the amount of heat absorbed by the UD-tape surface is too low, the diffusion of the matrix polymers in the FGF material is hindered and results in adhesion that is too weak to strengthen the material significantly. This can be seen in the stress/strain diagrams in Figure 18 that show the bending tests for the two analyzed gas volume flow rates of FGF/ATL-3. The drops in stress of the samples manufactured with 2.0 Nl/m gas volume flow rate are exactly where the UD tapes became detached from the FGF material without breaking. The increase of 0.5 Nl/m in the gas volume flow rate resulted in doubling the maximum stress. During hybrid additive manufacturing with ATL and gas-assisted heating, it is therefore essential to find an optimum gas volume flow rate at a given process speed for each UD tape and FGF material combination. The amount of thermal energy that is introduced into the interface in the tape-laying process has been shown to be the most critical parameter on the mechanical properties and tape adhesion [43,44]. When the adhesion between the UD tape and the printed material is sufficiently robust, our observations suggest that the UD tape will experience failure through tearing rather than detachment from the substrate.

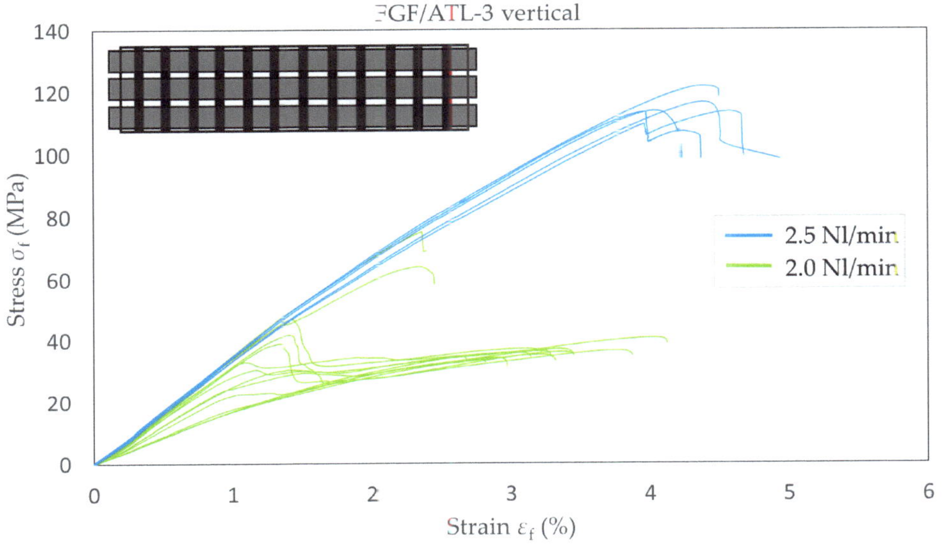

Figure 18. Stress/strain curves from bending tests of hybrid fused granular fabrication (FGF-3) and automated tape-laying test samples with different gas volume flow rates during the tape-laying process.

3.2.3. Mechanical Analysis

Figures 19 and 20 show the stress/strain curves from the bending tests of hybrid FGF and ATL test samples in both layer configurations, with and without UD-tape reinforcement of FGF-1 and FGF-2. The UD tapes were located on the bottom side of the samples during the test and therefore experienced a tension load. For both plate types and in both testing directions, a significant increase in mechanical properties was found.

The resulting mechanical properties from the bending test of FGF/ATL-1 and FGF/ATL-2 are shown in the Figures 21–23. There is a significant increase in flexural modulus, flexural strength, and strain at break when the UD tapes are added in the hybrid samples. For FGF/ATL-1, the flexural strength increased by a factor of five in the vertical testing direction and is close to the horizontal value without tape. The low flexural modulus of FGF/ATL-2 can be contributed to the fiber misalignment in FGF-2, as discussed previously. For both FGF/ATL-1 and FGF/ATL-2, the increase in mechanical properties by the UD tapes is higher for the vertical testing direction. FGF-1, possessing the smallest layer height and therefore the lowest surface roughness, is expected to have the largest contact area with the tape. FGF/ATL-1 also has the highest UD tape to FGF plate thickness ratio. Because the UD tape has a much higher tensile strength than the printed FGF material, an increase in flexural strength is expected with the increase in tape to plate thickness ratio when the tape is on the bottom side and experiences a tension load during the bending test. However, if the tape is on the top side of the specimen during the bending test or is perpendicular to the testing direction, then it does not increase the mechanical properties in that significant range.

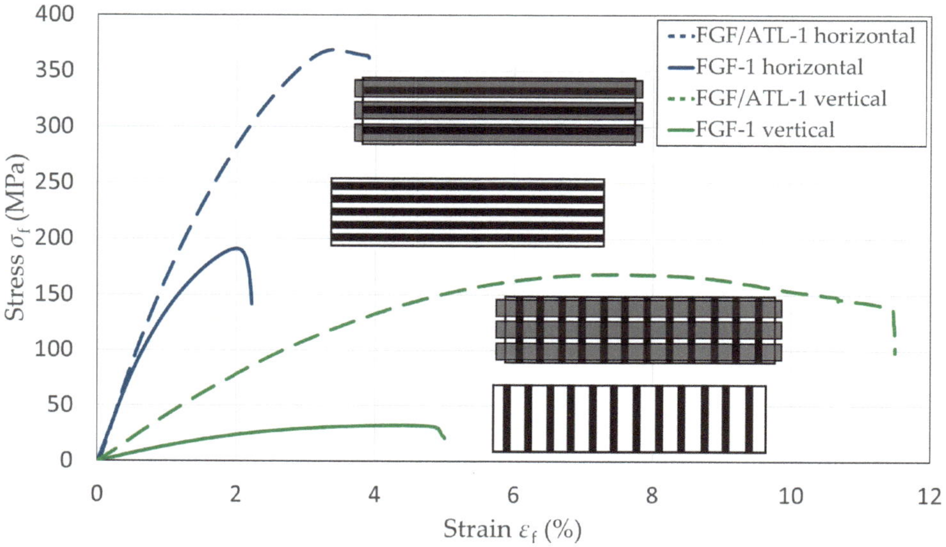

Figure 19. Stress/strain curves from bending tests of hybrid fused granular fabrication (FGF-1) and automated tape-laying test samples in horizontal and vertical layer configuration with and without UD-tape reinforcement.

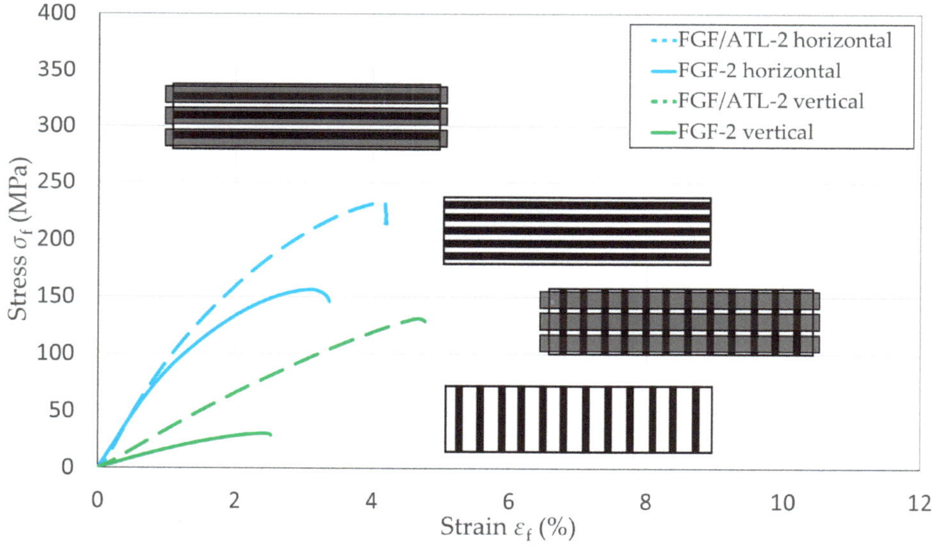

Figure 20. Stress/strain curves from bending tests of hybrid fused granular fabrication (FGF-2) and automated tape-laying test samples in horizontal and vertical layer configuration with and without UD-tape reinforcement.

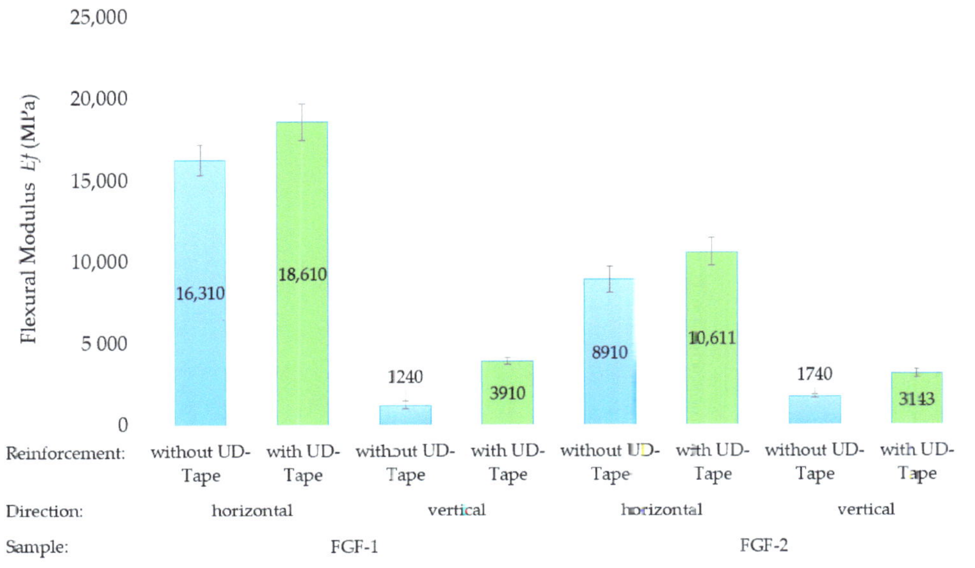

Figure 21. Comparison of flexural modulus of hybrid fused granular fabrication and automated tape-laying test samples (green columns) with fused granular fabrication test samples without UD tapes (blue columns).

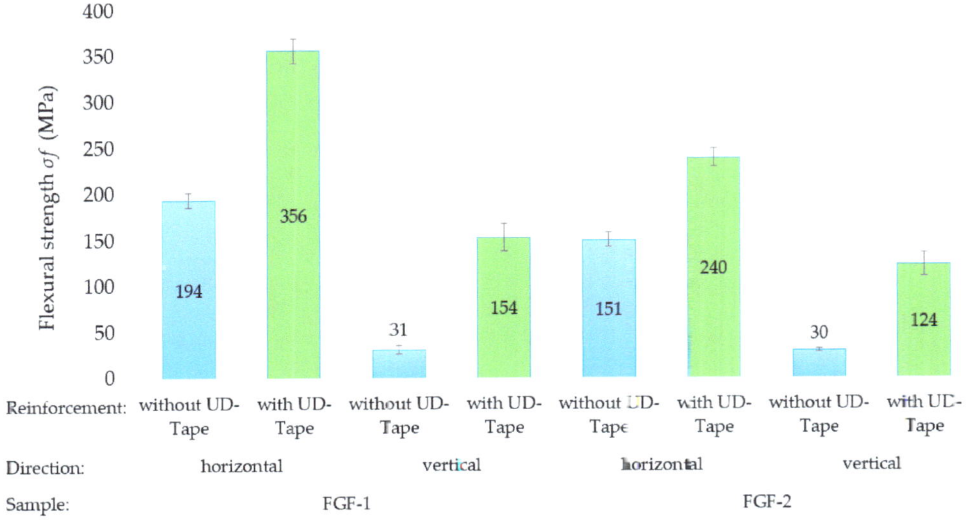

Figure 22. Comparison of flexural strength of hybrid fused granular fabrication and automated tape-laying test samples (green columns) with fused granular fabrication test samples without UD tapes (blue columns).

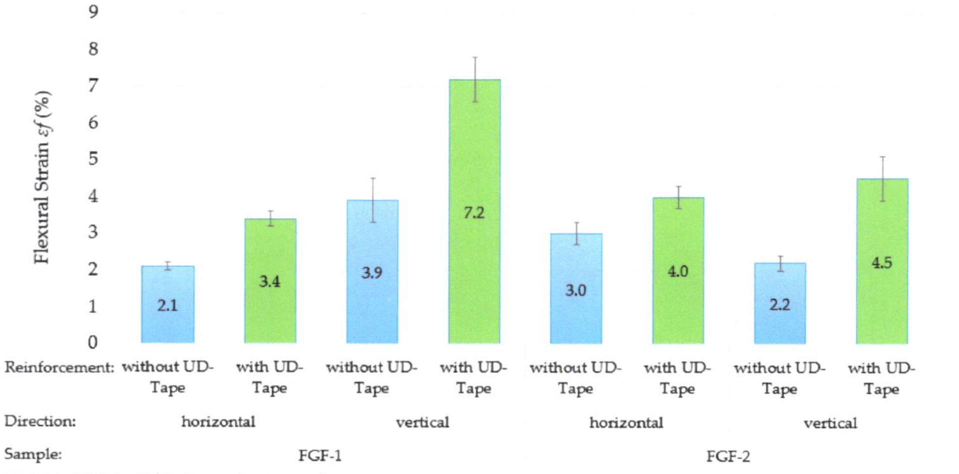

Figure 23. Comparison of strain at break for hybrid fused granular fabrication and automated tape-laying test samples (green columns) with fused granular fabrication test samples without UD tapes (blue columns).

4. Conclusions

Three different plates of carbon fiber-reinforced polyamide 6 were fabricated by fused granular fabrication (FGF) and characterized by morphological and mechanical analysis. The following conclusions were found:

- The process parameters have a strong influence on the surface roughness and mechanical properties. Especially the extrusion width, the layer height, and nozzle diameter ratios were found to be the essential factors affecting the regularity of printed geometry, surface quality, and misalignment of the reinforcing fibers in the FGF layers. This fact has a significant impact on the mechanical properties, particularly in the horizontal test direction, as found in the tensile tests of the FGF test samples.
- By respecting optimal printing process parameters, comparable values for the elastic modulus and tensile strength of the manufactured FGF materials to the datasheet values from injection-molded samples could be achieved.
- Additional automated tape laying (ATL) can significantly increase the mechanical properties of FGF structures, in some cases by several times. This was demonstrated by tape-laying glass-fiber-reinforced polyamide 6 UD tapes on the FGF plates.
- For ATL using gas as the heat source, it is essential to find the optimum gas volume flow rate at a given speed and UD tape and FGF material combination. An amount of heat absorbed by UD tape that is too low does not ensure sufficient diffusion of the matrix polymers into the FGF structure. As a result, it does not lead to effective reinforcement of the FGF material as the adhesion to the UD tapes is too weak.
- Lower surface roughness of the FGF materials leads to fewer gaps in the UD tape and therefore better stress transfer and higher mechanical values of resulting flexural modulus and strength, as demonstrated by the bending tests. Further improvement and analysis of the adhesion between the FGF materials and UD tapes are needed to determine optimal process parameters for both processes. Additionally, for multilayer components the tape–tape interface needs to be investigated.

Author Contributions: Conceptualization, P.H.; data curation, B.B., O.K. and M.S.; formal analysis, S.S. (Simon Scholz), B.B. and M.S.; funding acquisition, M.Z., R.S. and S.S. (Sebastian Scholz); investigation, B.B. and M.V.; methodology, P.H.; project administration, P.H. and S.S. (Sebastian Scholz); supervision, P.H.; validation, M.V.; writing—original draft, P.H., S.S. (Simon Scholz) and

O.K.; writing—review and editing, P.H., R.S., M.S. and O.K. All authors have read and agreed to the published version of the manuscript.

Funding: This research received no external funding.

Data Availability Statement: Data is contained within the article.

Conflicts of Interest: The authors declare no conflicts of interest.

References

1. Kristiawan, R.B.; Imaduddin, F.; Ariawan, D.; Ubaidillah, U.; Arifin, Z. A review on the fused deposition modeling (FDM) 3D printing: Filament processing, materials, and printing parameters. *Open Eng.* **2021**, *11*, 639–649. [CrossRef]
2. Love, L.J.; Duty, C.E.; Post, B.K.; Lind, R.F.; Lloyd, P.D.; Kunc, V.; Peter, W.H.; Blue, C.A. Breaking Barriers in Polymer Additive Manufacturing. In Proceedings of the SAMPE Conference, Baltimore, MD, USA, 19–20 May 2015.
3. Doshi, M.; Mahale, A.; Singh, S.K.; Deshmukh, S. Printing parameters and materials affecting mechanical properties of FDM-3D printed Parts: Perspective and prospects. *Mater. Today Proc.* **2022**, *50*, 2269–2275. [CrossRef]
4. Ahuja, B.; Karg, M.; Schmidt, M. Additive manufacturing in production: Challenges and opportunities. In Proceedings of the SPIE LASE: Laser 3D Manufacturing II, San Francisco, CA, USA, 7–12 February 2015.
5. Huang, Y.; Leu, M.C.; Mazumder, J.; Donmez, A. Additive Manufacturing: Current State, Future Potential, Gaps and Needs, and Recommendations. *J. Manuf. Sci. Eng.* **2015**, *137*, 014001. [CrossRef]
6. Fleury, A.; Gregory, M.; Bennett, D. The future of manufacturing. *J. Manuf. Technol. Manag.* **2007**, *18*, 323–325. [CrossRef]
7. Pignatelli, F.; Percoco, G. An application- and market-oriented review on large format additive manufacturing, focusing on polymer pellet-based 3D printing. *Prog. Addit. Manuf.* **2022**, *7*, 1363–1377. [CrossRef]
8. Post, B.K.; Lind, R.F.; Lloyd, P.D.; Kunc, V.; Linhal, J.M.; Love, L.J. The economics of big area additive manufacturing. In Proceedings of the 27th Annual International Solid Freeform Fabrication Symposium—An Additive Manufacturing Conference, Austin, TX, USA, 8–10 August 2016.
9. Duty, C.E.; Kunc, V.; Compton, B.; Post, B.; Erdman, D.; Smith, R.; Lind, R.; Lloyd, P.; Love, L. Structure and mechanical behavior of Big Area Additive Manufacturing (BAAM) materials. *Rapid Prototyp. J.* **2017**, *23*, 181–189. [CrossRef]
10. Silva, R.; Sereno, P.; Mateus, A.; Mitchell, G.R.; Carreira, P.; Santos, C.; Vitorino, J.; Domingues, J. Adaptive Platforms and Flexible Deposition System for Big Area Additive Manufacturing (BAAM). *Appl. Mech. Mater.* **2019**, *890*, 3–20. [CrossRef]
11. Sam-Daliri, O.; Ghabezi, P.; Flanagan, T.; Finnegan, W.; Mitchell, S.; Harrison, N. Recovery of Particle Reinforced Composite 3D Printing Filament from Recycled Industrial Polypropylene and Glass Fibre Waste. In Proceedings of the 8th World Congress on Mechanical, Chemical, and Material Engineering MCM'22, Prague, Czech Republic, 31 July–2 August 2022.
12. Sam-Daliri, O.; Ghabezi, P.; Steinbach, J.; Flanagan, T.; Finnegan, W.; Mitchell, S.; Harrison, N. Experimental study on mechanical properties of material extrusion additive manufactured parts from recycled glass fibre-reinforced polypropylene composite. *Compos. Sci. Technol.* **2023**, *241*, 110125. [CrossRef]
13. Mohamed, O.A.; Masood, S.H.; Bhowmik, J.L. Optimization of fused deposition modeling process parameters: A review of current research and future prospects. *Adv. Manuf.* **2015**, *3*, 42–53. [CrossRef]
14. Ajinjeru, C.; Kishore, V.; Lindahl, J.; Sudbury, Z.; Hassen, A.A.; Post, B.; Love, L.; Kunc, V.; Duty, C. The influence of dynamic rheological properties on carbon fiber-reinforced polyetherimide for large-scale extrusion-based additive manufacturing. *Int. J. Adv. Manuf. Technol.* **2018**, *99*, 411–418. [CrossRef]
15. Ajinjeru, C.; Kishore, V.; Chen, X.; Hershey, C.; Lindhal, J.; Kunc, V.; Hassen, A.A.; Duty, C. Rheological survey of carbon fiber-reinforced high-temperature thermoplastics for big area additive manufacturing tooling applications. *J. Thermoplast. Compos. Mater.* **2021**, *34*, 1443–1461. [CrossRef]
16. Billah, K.M.M.; Lorenzana, F.A.R.; Martinez, N.L.; Wicker, R.B.; Espalin, D. Thermomechanical characterization of short carbon fiber and short glass fiber-reinforced ABS used in large format additive manufacturing. *Addit. Manuf.* **2020**, *35*, 101299. [CrossRef]
17. Ngo, T.D.; Kashani, A.; Imbalzano, G.; Nguyen, K.T.Q.; Hui, D. Additive manufacturing (3D printing): A review of materials, methods, applications and challenges. *Compos. B Eng.* **2018**, *143*, 172–196. [CrossRef]
18. Zohdi, N.; Yang, R. Material Anisotropy in Additively Manufactured Polymers and Polymer Composites: A Review. *Polymers* **2021**, *13*, 3368. [CrossRef]
19. Haque, A.N.M.A.; Naebe, M. Tensile Properties of Natural Fibre-Reinforced FDM Filaments: A Short Review. *Sustainability* **2023**, *15*, 16580. [CrossRef]
20. Liu, X.; Chi, B.; Jiao, Z.; Tan, J.; Liu, F.; Yang, W. A large-scale double-stage-screw 3D printer for fused deposition of plastic pellets. *J. Appl. Polym. Sci.* **2017**, *134*, 45147. [CrossRef]
21. Chesser, P.; Post, B.; Roschli, A.; Carnal, C.; Lind, R.; Borish, M.; Love, L. Extrusion control for high quality printing on Big Area Additive Manufacturing (BAAM) systems. *Addit. Manuf.* **2019**, *28*, 445–455. [CrossRef]
22. Wickramasinghe, S.; Do, T.; Tran, P. FDM-Based 3D Printing of Polymer and Associated Composite: A Review on Mechanical Properties, Defects and Treatments. *Polymers* **2020**, *12*, 1529. [CrossRef]
23. Raspall, F.; Velu, R.; Vaheed, N.M. Fabrication of complex 3D composites by fusing automated fiber placement (AFP) and additive manufacturing (AM) technologies. *Adv. Manuf. Polym. Compos. Sci.* **2019**, *5*, 6–16. [CrossRef]

24. Yassin, K.; Hojjati, M. Processing of thermoplastic matrix composites through automated fiber placement and tape laying methods: A review. *J. Thermoplast. Compos.* **2018**, *31*, 1676–1725. [CrossRef]
25. Voelkl, H.; Kießkalt, A.; Wartzack, S. Design for Composites: Derivation of Manufacturable Geometries for Unidirectional Tape Laying. *Proc. Des. Soc. Int. Conf. Eng. Des.* **2019**, *1*, 2687–2696. [CrossRef]
26. Kropka, M.; Muehlbacher, M.; Neumeyer, T.; Altstaedt, V. From UD-tape to Final Part—A Comprehensive Approach towards Thermoplastic Composites. *Procedia CIRP* **2017**, *66*, 96–100. [CrossRef]
27. Brasington, A.; Sacco, C.; Halbritter, J.; Wehbe, R.; Harik, R. Automated fiber placement: A review of history, current technologies, and future paths forward. *Compos. Part C* **2021**, *6*, 100182. [CrossRef]
28. Link, T.; Rosenberg, P.; Henning, F. Prediction of Gaps in Automated Tape Laying and Their Influence on Porosity in Consolidated Laminates. *J. Compos. Sci.* **2022**, *6*, 207. [CrossRef]
29. Jayasekara, D.; Lai, N.Y.G.; Wong, K.-H.; Pawar, K.; Zhu, Y. Level of automation (LOA) in aerospace composite manufacturing: Present status and future directions towards industry 4.0. *J. Manuf. Syst.* **2022**, *62*, 44–61. [CrossRef]
30. Zhang, L.; Wang, X.; Pei, J.; Zhou, Y. Review of automated fibre placement and its prospects for advanced composites. *J. Mater. Sci.* **2020**, *55*, 7121–7155. [CrossRef]
31. Dhinakaran, V.; Surendar, K.V.; Hasunfur Riyaz, M.S.; Ravichandran, M. Review on study of thermosetting and thermoplastic materials in the automated fiber placement process. *Mater. Today Proc.* **2020**, *27*, 812–815. [CrossRef]
32. F3-Compositor Enables 3D Deposition and Simultaneous Joining of Thermoplastic UD Tape Semi-Finished Products. Available online: https://www.imws.fraunhofer.de/en/presse/pressemitteilungen/f3--compositor-enables-3d-deposition-and-simultaneous-joining-of.html (accessed on 9 November 2023).
33. *DIN EN ISO 178:2019-08*; Plastics—Determination of Flexural Properties (ISO 178:2019). German version EN ISO 178:2019; ISO: Geneva, Switzerland, 2019.
34. *DIN EN ISO 1110:2019-09*; Plastics—Polyamides—Accelerated Conditioning of Test Specimens (ISO 1110:2019). German version EN ISO 1110:2019; ISO: Geneva, Switzerland, 2019.
35. *DIN EN ISO 527-1:2019-12*; Lastics—Determination of Tensile Properties—Part 1: General Principles (ISO 527-1:2019). German version EN ISO 527-1:2019; ISO: Geneva, Switzerland, 2019.
36. Ahn, D.; Kweon, J.-H.; Kwon, S.; Song, J.; Lee, S. Representation of surface roughness in fused deposition modelling. *J. Mater. Process. Technol.* **2009**, *209*, 5593–5600. [CrossRef]
37. Maier, R.; Bucaciuc, S.-G.; Mandoc, A.C. Reducing Surface Roughness of 3D Printed Short-Carbon Fiber Reinforced Composites. *Materials* **2022**, *15*, 7398. [CrossRef]
38. Consul, P.; Beuerlein, K.; Luzha, G.; Drechsler, K. Effect of Extrusion Parameters on Short Fiber Alignment in Fused Filament Fabrication. *Polymers* **2021**, *13*, 2443. [CrossRef]
39. Tagscherer, N.; Bär, A.; Zaremba, S.; Drechsler, K. Mechanical Analysis of Parameter Variations in Large-Scale Extrusion Additive Manufacturing of Thermoplastic Composites. *J. Manuf. Process.* **2022**, *6*, 36. [CrossRef]
40. Barera, G.; Dul, S.; Pegoretti, A. Screw Extrusion Additive Manufacturing of Carbon Fiber Reinforced PA6 Tools. *J. Mater. Eng. Perform.* **2023**, *32*, 9579–9597. [CrossRef]
41. König, M.; Diekmann, J.; Lahres, M.; Middendorf, P. Experimental investigation of process-structure effects on interfacial bonding strength of a short carbon fiber/polyamide composite fabricated by fused filament fabrication. *Prog. Addit. Manuf.* **2022**, *7*, 593–607. [CrossRef]
42. Kepenekci, M.; Gharehpapagh, B.; Yaman, U.; Özerinc, S. Mechanical performance of carbon fiber-reinforced polymer cellular structures manufactured via fused filament fabrication. *Polym. Compos.* **2023**, *44*, 4654–4668. [CrossRef]
43. Atzler, F.; Raps, L.; Freund, J.; Tröger, S.; Hümbert, S. Bonding of Low-Melting Polyaryletherketone onto Polyamide 6: A Concept for Molds for Automated Fiber Placement. *J. Compos. Sci.* **2023**, *7*, 371. [CrossRef]
44. Janssen, H.; Peters, T.; Brecher, C. Efficient Production of Tailored Structural Thermoplastic Composite Parts by Combining Tape Placement and 3d Printing. *Procedia CIRP* **2017**, *66*, 91–95. [CrossRef]

Journal of
Manufacturing and Materials Processing

MDPI

Article

A Workflow for the Compensation of Substrate Defects When Overprinting in Extrusion-Based Processes

Fynn Atzler, Simon Hümbert * and Heinz Voggenreiter

German Aerospace Center (DLR), Institute for Structures and Design (BT), 70569 Stuttgart, Germany
* Correspondence: simon.huembert@dlr.de; Tel.: +49-711-6862-8044

Abstract: Fused granular fabrication (FGF) is used in industrial applications to manufacture complex parts in a short time frame and with reduced costs. Recently, the overprinting of continuous fibre-reinforced laminates has been discussed to produce high-performance, functional structures. A hybrid process combining FGF with Automated Fibre Placement (AFP) was developed to implement this approach, where an additively manufactured structure is bonded in situ onto a thermoplastic laminate. However, this combination places great demands on process control, especially in the first printing layer. When 3D printing onto a laminate, the height of the first printed layer is decisive to the shear strength of the bonding. Manufacturing-induced surface defects of a laminate, like thermal warpage, gaps, and tape overlaps, can result in deviations from the ideal geometry and thus impair the bonding strength when left uncompensated. This study, therefore, proposes a novel process flow that uses a 3D scan of a laminate to adjust the geometry of the additively manufactured structure to achieve a constant layer height in the 3D print and, thus, constant mechanical properties. For the above-listed surface defects, only thermal warpage was found to have a significant effect on the bonding strength.

Keywords: digital twin; fused granular fabrication; thermoplastic laminates; in situ bonding; PEEK

1. Introduction

Additive Manufacturing (AM) has become significantly more widespread in industrial applications in recent years. AM of high-temperature thermoplastics like polyether ether ketone (PEEK) is used to produce complex and lightweight parts more quickly and at a lower cost than traditional manufacturing technologies, while having mechanical properties comparable to light metals like aluminium [1,2]. AM using a robotic arm with six degrees of freedom (DOF) is especially suited for industrial applications and processes since it can be applied to a variety of complex geometries [3]. This is particularly attractive in the context of the repairing of defects.

The combination of 3D printing and in situ Automated Fibre Placement (AFP) is one method to achieve high-performance parts with the utilisation of the design freedom of AM. Laminates manufactured by AFP can withstand significantly more force than 3D-printed structures and can therefore be used as skins of sandwich structures. Extrusion-based 3D printing, like fused granular fabrication (FGF), can be used to manufacture a core structure of a sandwich. By using FGF to manufacture the core structure, these structures can be designed with integrated electronics or radiation shielding for aerospace applications and bonded onto the laminate in situ. This enables a compact design, e.g., satellites [4,5].

This overprinting of thermoplastic laminates has already been demonstrated [6,7]. A prior study showed that the height of the first printed layer significantly influences the shear strength of the bonding with a laminate [6]. A thin layer height results in improved shear strength in comparison with a thicker layer height.

Defects in the thermoplastic laminate can lead to geometric deviations from the ideal geometry and thus influence the bonding strength. Several defects in laminates manufactured by AFP are known and are schematically shown in Figure 1. When not placed precisely enough, tapes can overlap or form gaps between them. Tapes on a double-curved surface have to form those defects at some point. Gaps and overlaps can also occur on the edge of a part when the ends of the tapes do not meet up perfectly. Using steering to lay tapes on a curved path can result in wrinkles. Those defects result in a deviation of the laminate surface from the ideal surface by the thickness of the tape. Depending on the lay-up of the laminate, a waviness and thermal warpage add to the deformation of a laminate [8–10].

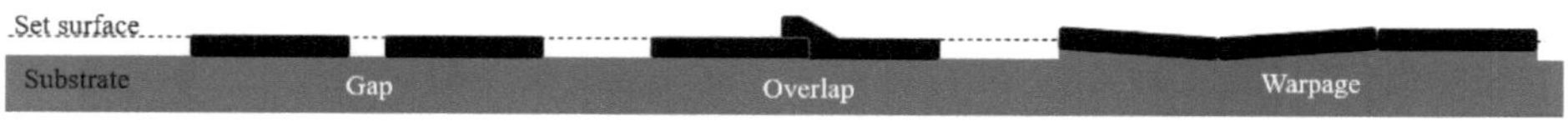

Figure 1. Schematic depiction of AFP defects.

The multiple parameters of the AFP process make it difficult to give exact dimensions and shape characteristics of defects due to the high degree of variability in the process. This fact is also reflected in the results of review articles highlighting different effects of the same defect category [11]. However, multiple studies have been published, giving information on certain defects' exact shape and effect. Rectangular gap defects with a width of 0.76 mm–2.50 mm reduced the compressive strength of laminates by up to 27% [12]. Triangular gap defects with an opening angle of 12° and a base width of 0.25 inches decreased tensile strength by up to 22% [13]. A combination of multiple gap and overlap defects in one laminate can result in decreased mechanical properties or no effect of those defects [11,14].

Since the mechanical properties of the bond between the 3D-printed structure and the laminate is significantly influenced by the layer height of the first 3D-printed layer, and the above-listed defects result in a deviation of the real geometry from the ideal geometry of the laminate, it is crucial to account for those defects in the FGF process.

The repair of defects with robotic AM has been widely researched in medical engineering. Due to the restricted field of view in medical applications, tactile scanning methods or manual reverse engineering are often used to digitalise defects [15–17]. Process chains using optical scanning methods were only demonstrated in ex vivo applications [18,19]. Since optical scanning allows for the scanning of large areas with high scanning resolution, it is more suitable for non-medical industrial applications. For industrial applications, manual measuring and reverse engineering of gap defects in laminates have already been demonstrated for high-performance composites [20]. However, inherent to this approach are high processing times and a need for extensive manual work. Using in-process images of parts and in-process monitoring of parameters enables the detection and correction of defects in the 3D print using a machine-learning approach [21].

This study aimed to develop a workflow that considers the relevant defects in the laminate for overprinting and compensates for them through adapted toolpath planning. The effects of certain defects of AFP laminates on the bonding with AM structures are unknown. Hence, this study was divided into two parts: Firstly, the effect of gaps and overlaps on the shear strength of the bonding was investigated. Using these results, a suitable resolution of a digital model could be defined as only including defects that needed to be accounted for in the FGF process. By excluding other defects, the file size of the digital twin could be minimal. Afterwards, a process flow was developed and tested to adjust the geometry of the 3D-printed structure based on the defects of the thermoplastic laminate. A constant height of the first layer needed to be achieved to realise optimal mechanical properties of the bonding.

2. Materials and Methods

2.1. Manufacturing Facilities

For the in situ bonding of PEEK and Carbon-Fibre-Reinforced Polymer (CFRP) laminates with a matrix of a polyaryl ether ketone, a DXR system developed by Hans Weber Maschinenfabrik GmbH (Kronach, Germany) was used, and is shown in Figure 2a. It consists of a single screw extruder mounted on a robotic arm with six DOF. The extruder is divided into three heating zones, which can be heated to a temperature of up to 450 °C. The nozzle diameter can be set to 1 mm, 2 mm, or 3 mm. The print bed can be heated to a temperature of up to 300 °C.

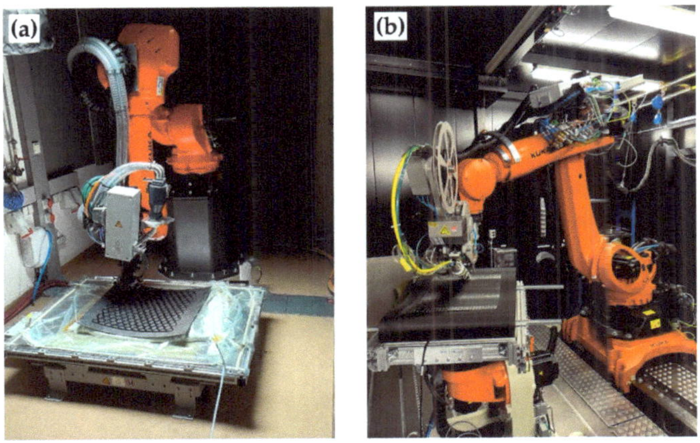

Figure 2. Robotic FGF system DXR (**a**) and AFP facility (**b**).

The AFP facility (AFPT GmbH, Dörth, Germany) of the DLR in Stuttgart was used for the fabrication of the thermoplastic laminates (Figure 2b). The setup consists of a six-DOF robotic arm with a Multi-Tape Laying Head (MTLH) as its end effector. The MTLH is manufactured by AFPT GmbH and can lay up to 3 tapes with a width of 0.5 in (12.7 mm) simultaneously.

2.2. Materials

The PEEK granulate TECACOMP PEEK 150 CF30 was used as a feedstock material for the FGF process. The material, fabricated by Ensinger Plastics (Nufringen, Germany), consists of carbon-fibre-reinforced PEEK with a fibre mass fraction of 30%. The properties of the material are listed in Table 1. The process parameters for the FGF process are stated in Table 2.

Table 1. Material properties of carbon fibre-reinforced PEEK [22].

Property	TECACOMP PEEK 150 CF30
Density/g/cm^3	1.38
Young's modulus/MPa	17,500
Ultimate strength/MPa	190
Nozzle temperature/°C	400–440
Bed temperature/°C	130–150

Table 2. Used parameters of the 3D printing process for the manufacturing of specimens.

Parameter	Value
Nozzle temperature/°C	420
Temperature of heating zone1/°C	420
Temperature of heating zone2/°C	420
Temperature of print bed/°C	280
Ambient temperature/°C	30
Printing speed/mm/s	20
Layer height/mm	0.6
Nozzle diameter/mm	3.0

The CFRP laminates were made from the unidirectional thermoplastic prepreg tape Toray Cetex TC1225, produced by Toray Advanced Composites (Morgan Hill, CA, USA). The prepreg has a PAEK as the matrix material. The PAEK has a lower bulk melting temperature than PEEK and is called low-melting PAEK (LM-PAEK). The melt temperature and glass transition temperature of the LM-PAEK are 305 °C and 147 °C, respectively. The prepreg tapes have a fibre mass fraction of 66% [23].

2.3. Methodology

In order to develop an effective workflow for the overprinting of laminates, the influence of the various defects in the laminate must be known, and a manageable digital process chain for web planning must be developed. This study, therefore, consisted of two main steps. Firstly, the influence of possible sources of defects on the joining of the 3D print to the laminate was analysed. The focus here was on defects from the AFP process. These were divided into local defects, in particular, gaps and overlaps, and global defects, in particular, warpage and waviness. The error potential of digital data acquisition and handling had to also be taken into account.

The second step followed a workflow that included measuring both the real geometry and the compensating defects. The results of the previous series of tests determined, in particular, which defects had to be taken into account and the necessary level of detail of the 3D measurement. Finally, the workflow was verified using a simple test print. The schematic structure of the study is shown in Figure 3.

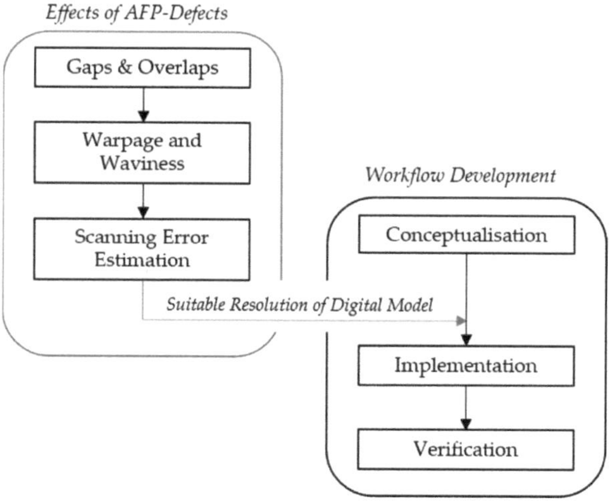

Figure 3. Schematic structure of the study.

2.3.1. Determining the Effect of Defects

To determine the effect of certain surface defects of a thermoplastic laminate on the bonding with additively applied PEEK, single-lap shear specimens based on ASTM D3846-08 [24] were used. Five specimens were produced for each investigated defect. Specimens were manufactured by printing a cuboid per specimen onto the defect of a laminate (Figure 4b). Afterwards, the specimens were separated and notched using a disk saw (Figure 4c).

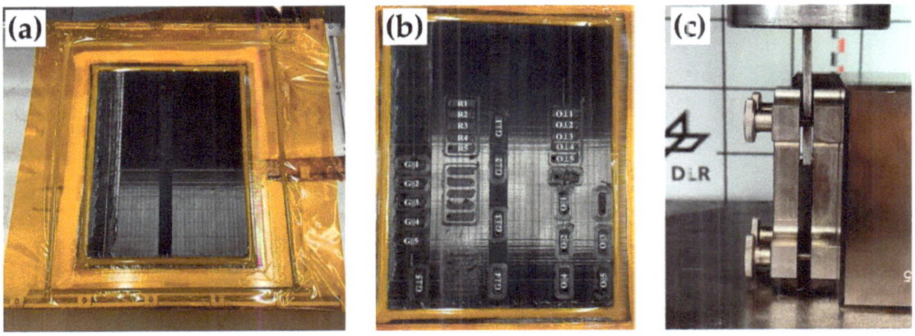

Figure 4. Print and test setup for single-lap shear specimens: (**a**) laminate before printing, mounted on the print bed; (**b**) printed specimens on laminate; (**c**) compression test for shear specimen.

As the substrate for the printing of specimens, a laminate made from 0.5 in wide unidirectional carbon fibre prepreg tapes with LM-PAEK as the matrix material was used. The laminate had a $[0°/90°]_n$ layup and was manufactured using the AFP facility of the DLR in Stuttgart.

Using literature [8,9] and analysis of the used laminates, three surface defects were identified as the most common. Gaps between tapes and overlapping tapes were two of those. Thermal stress in a laminate can result in warpage and, depending on the layup, waviness. Gaps and overlaps can result in the laminate surface deviating from the ideal plane by up to 0.4 mm. The waviness of a laminate can result in a deviation of up to 0.6 mm.

For this study a laminate was produced which had gaps and overlaps in various places. The top layer intentionally missed some gaps to allow the placement of specimens in different orientations onto defects, as shown in Figure 4a. Five specimens were printed in an area with no identifiable defects, as a reference, and named in the following R-specimens. Gaps and overlaps were placed parallel and orthogonal to the axis of the applied force in specimens. For each variation, five specimens were produced and tested. Specimens with gaps and overlaps parallel to the direction of load are named G || and O ||, respectively (see Figure 5b). Specimens with gaps and overlaps orthogonal to the direction of load are named G ⊥ and O ⊥ (see Figure 5a).

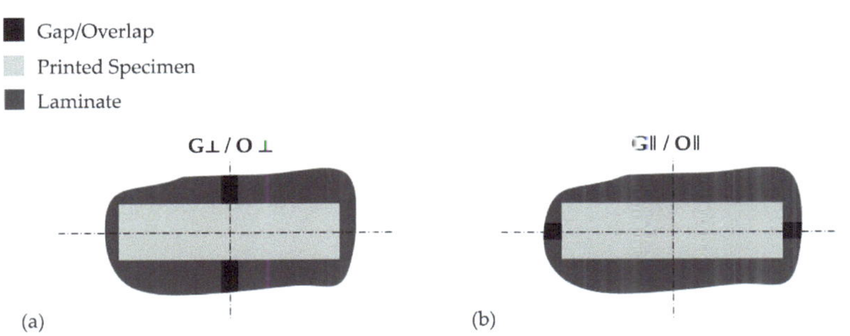

Figure 5. Schematic representation of specimen configuration for laminate defects. (**a**) Defect perpendicular to the test direction. (**b**) Defect parallel to the test direction.

Global warpage or waviness of a laminate results in a global inconsistent layer height. This can impair the bonding strength. To investigate the effect of a global warpage or a large systematic defect, like waviness, of LM-PAEK laminates on the bonding with PEEK in extrusion-based processes, single-lap shear specimens based on ASTM D3846-08 [24] were produced. The toolpath was generated for the printing of specimens with a set layer height of 0.6 mm. Since the warpage of a substrate will result in over- or underextrusion, the layer height was manually changed to 0.4 mm and 0.8 mm. For both variations, five specimens were produced and tested. The specimens were notched using a disk saw.

If local or global defects need to be accounted for in the bonding process, the deviation of the real geometry from the ideal geometry progresses into the AM part. When successively changing the form of layers from the real geometry of the substrate to the ideal geometry of the AM part, changes in layer height in the individual layers themselves are the result. This leads to an over- and/or underextrusion, which can influence the interlaminar shear strength of the AM part. This effect was, therefore, investigated using single-lap shear specimens based on ASTM D3846-08 [24]. Specimens were printed with a nozzle size of 3 mm and a layer height of 0.6 mm, which is the usual used setting. Without changing the amount of extrusion, specimens were produced with a layer height of 0.4 mm and 0.8 mm, resulting in an over- and underextrusion. For all variations, five specimens were produced and tested. The specimens were notched using a disk saw.

Table 3 summarises the defects considered, their effects on the printing process, and the corresponding tests carried out.

Table 3. AFP defects and the corresponding tests performed.

Laminate Defect	Effect on Overprinting	Tested Property
Gaps	Local underextrusion in the first layer	Bond strength parallel to the defect Bond strength perpendicular to the defect
Overlaps	Local overextrusion in the first layer	Bond strength parallel to the defect Bond strength perpendicular to the defect
Warpage/Waviness	Over-/underextruson in the first layer Over-/underextruson within 3D print	Bond strength at ±0.2 mm layer height error Inter-laminar strength of 3D print at ±0.2 mm layer height error

Finally, data handling during path planning can also lead to deviations between the real laminate geometry and the robot's path planning during 3D printing. In order to capture the real geometry of the laminate, including the defects, the geometry was scanned in 3D. This measurement resulted in a mesh-based representation of the laminate surface. The more precise this measurement is, the larger the amount of data. However, very fine meshes are difficult to process in toolpath planning. It, therefore, makes sense to reduce the resolution of the mesh and, therefore, the amount of data. This reduction in the digital model can lead to a deviation between the real geometry and the digital model. The test results for over- and underextrusion were used to assess which deviations and, therefore, which data reduction can be tolerated. For this investigation, a scan of a laminate with significant waviness was used. The number of vertices was reduced to 90%, 80%, 70%, 60%, 50%, 40%, 30%, 20%, 10%, 5%, 2%, and 1% of that in the original scan. The data sets were then exported as .stl-files and compared to the original scan data. The maximal deviations for each file were noted.

2.3.2. Workflow Implementation

The basic structure of the here-developed process chain is shown in Figure 6. Firstly, the geometry of a CFRP laminate has to be digitalised. An optical scanner, GOM ATOS 5, was used for this purpose. Afterwards, the scanning data are reduced to the necessary degree of detail using the mechanical test and used to modify the ideal geometry of the AM structure to account for the deviations between the ideal and the real geometry of the substrate. Using this modified geometry, a toolpath can be created. To achieve a constant layer height in the AM process, the toolpath needs to be non-planar, utilising three degrees of freedom of a 3D printer.

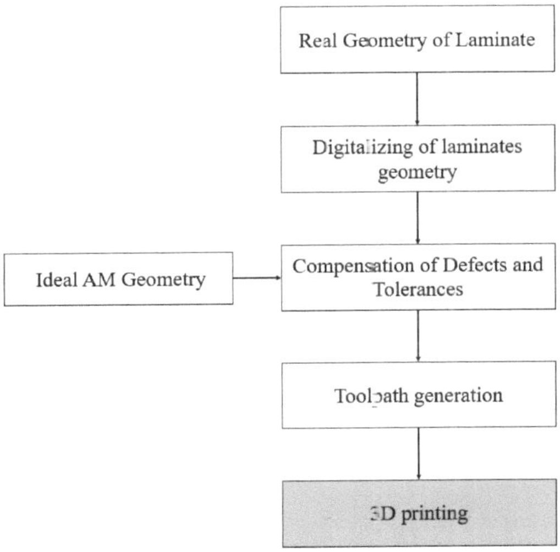

Figure 6. Basic structure of the developed process chain.

A demonstrator was produced to validate the workflow. Two lines with a width of 3 mm and a length of 500 mm were printed onto a laminate with significant warpage and surface waviness. One line was applied with an ideal planar geometry, while the other's geometry was changed using the described workflow. The height profiles of the lines were measured to calculate the layer thickness over the length of the lines. The measurements were then compared using a paired *t*-test.

2.4. Software

For the editing of 3D scanning data and .stl-files, the software GOM Inspect Suite 2020 (Carl Zeiss IQS Deutschland GmbH, Oberkochen, Germany) and Blender Version 3.4.1 by the Blender Foundation (Amsterdam, Netherlands) was used. The toolpath was generated using a modified version of the software Slic3r (Version 1.2.9) [25].

2.5. Testing and Statistics

The single-lap shear tests were performed using a RetroLine 1475 testing machine built by Zwick Roell GmbH & Co. KG. (Ulm, Germany). The testing machine is equipped with a 5 kN load cell. The testing speed was 1 mm/s. The crosshead travel and the measured force were recorded throughout the test.

For the evaluation of the process flow, the surfaces of specimens were scanned using a Keyence VR5000 profilometer (Keyence Deutschland GmbH Leinfelden-Echterdingen, Germany).

An unpaired two-sided *t*-test was used to compare the two groups and evaluate the effect of defects. A paired *t*-test was used to later evaluate the effect of the process chain on the continuity of the layer height. p values ≤ 0.05 were considered statistically significant. The Bonferroni–Holm correction was used to adjust for alpha errors.

3. Results

3.1. Effects of Defects

Gaps had an average width of 0.43 mm, and overlaps had an average width of 0.03 mm with a standard deviation of 0.15 mm and 0.06 mm. R-specimens achieved a mean shear strength of 18.87 MPa with a standard deviation of 4.78 MPa.

The adjusted p values for the comparison of the reference values with those of O ||, O ⊥, G ||, and G ⊥-specimens were 0.62, 0.31, 0.11, and 0.62 and thus indicated no statistically significant difference in shear strength between the reference specimens and the specimens with defects. The arithmetic mean values of the shear strength of the O ||, O ⊥, G ||, and G ⊥-specimens were 21.17 MPa, 23.38 MPa, 26.08 MPa, and 21.34 MPa and are shown with the corresponding standard deviation in Figure 7.

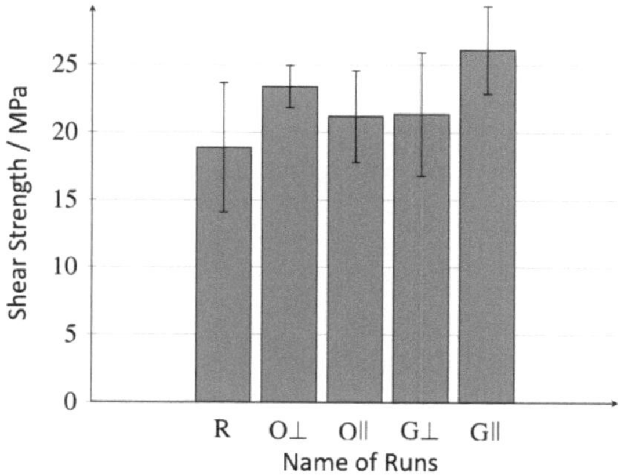

Figure 7. Shear strength of the bonding of PEEK on certain surface defects.

An overextrusion and underextrusion resulting from a deviation of the actual layer height from the set layer height significantly influenced the shear strength of the bonding. The results of the mechanical tests are shown in Figure 8 and indicate an effect of -18.28 MPa mm^{-1}.

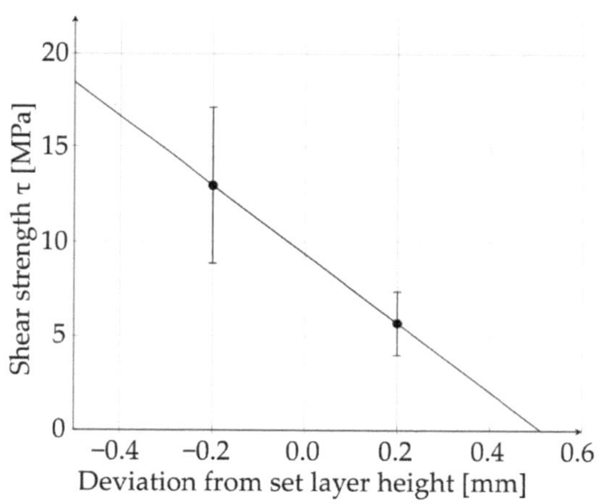

Figure 8. Effect of under-/overextrusion on the shear strength of the bonding of PEEK on PEEK/CF-laminates.

The effect of over- and underextrusion is shown in Figure 9 and was determined to be -54.45 MPa mm^{-1}.

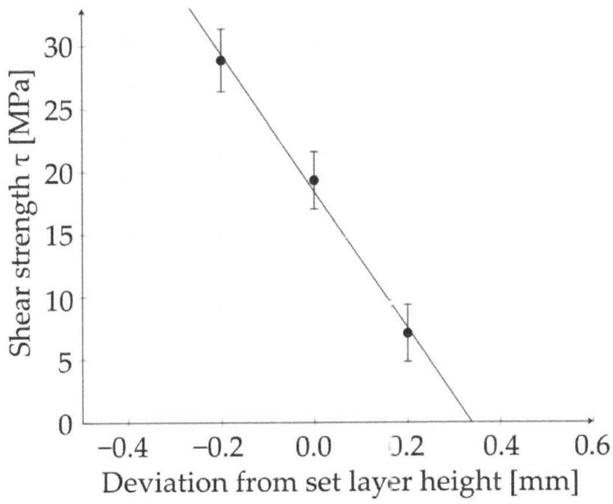

Figure 9. Effect of over- and underextrusion on the interlaminar shear strength in 3D-printed structures out of CF-PEEK.

3.2. Process Flow to Account for Surface Defects

3.2.1. Conceptualizing the Process Chain

To compensate for surface defects of laminates while in situ bonding with PEEK in the FGF process, the following process flow (Figure 10) was used:

Firstly, a laminate was held on the print bed using vacuum bagging and was heated to 280 °C. A GOM ATOS 5 was then used to scan the laminate's surface, creating a point cloud. The scanning setup is shown in Figure 11. To include reference points in the scan, the tip of the extruder was positioned in three different positions near the laminate's surface, and is also scanned, which can be seen in Figure 12b. Using the robotic arm software, the coordinates of these reference points could be noted and later used to transform the point cloud coordinates using the GOM Inspect Suit software. Besides the laminate's surface, the scan also included the surrounding print bed and extruder parts, as seen in Figure 12a. By deleting these, only the laminate's surface was left, and the file size was reduced. The cleansed scan is shown in Figure 12c. To further reduce the amount of data, the GOM Inspect Suit software can be used to set a specific number of points used to describe the scanned surface.

Points were not distributed evenly. The point cloud was denser in areas of high detail. Gaps and overlaps are local defects, and thus, more points are necessary to describe those areas. Warpage and waviness of laminates affect the geometry of the whole laminate but still allow a continuous surface in contrast to gaps and overlaps.

Considering that gaps and overlaps do not have a significant influence on the bonding of PEEK on the laminate, these defects did not have to be accounted for in the described process flow and, thus, did not need to be included in the digital twin. By reducing the number of points used to describe the laminate surface, small details like gaps and overlaps are removed while preserving more global defects like warpage and waviness of the laminate.

To evaluate the effect of the reduction in vertices, scan data of the laminate shown in Figure 12c was used to compare .stl-files of the original scan with .stl-files with reduced vertices. Down to a reduction of 40% of the original number of vertices, no deviation was measurable, as shown in Figure 13d. When reduced to 10% of the original data, an .stl-file showed local deviations of up to ±0.01 mm (Figure 13c). Those deviations increased to ±0.02 mm when reducing the data to 5% (Figure 13b). At 1% of the original number of vertices, the deviations extended over larger areas and amounted to up to ±0.05 mm (Figure 13a).

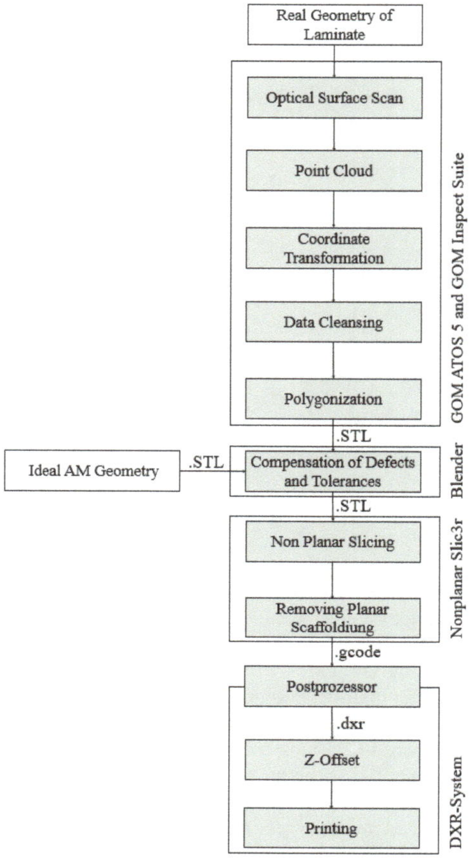

Figure 10. Process flow used in this study.

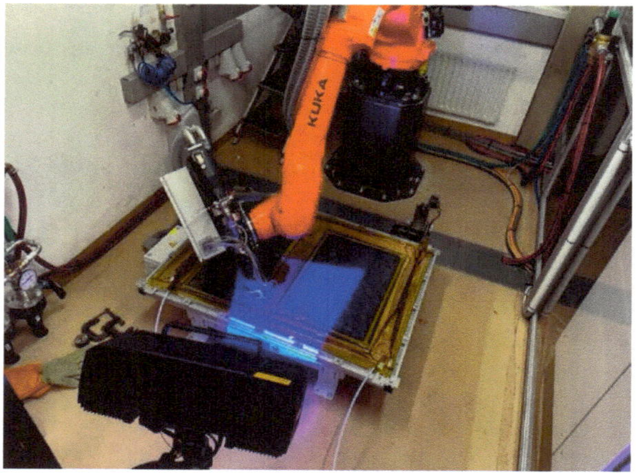

Figure 11. Scanning of a laminate on the print bed.

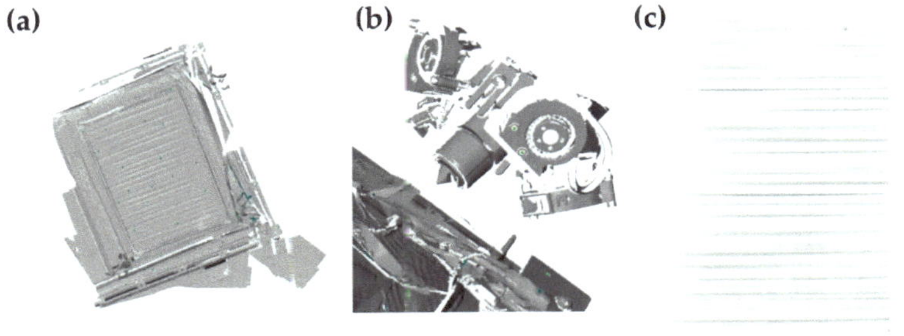

Figure 12. Scan of the laminate on the print bed: (**a**) unedited, (**b**) extruder tip included in the scan, and (**c**) cleaned scan data.

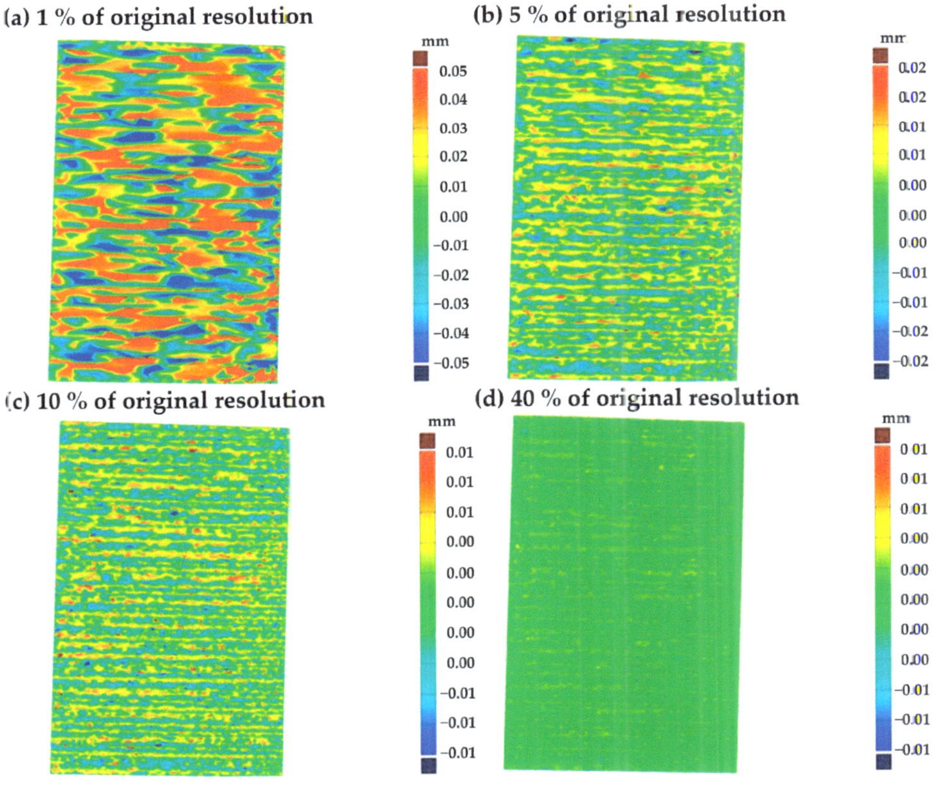

Figure 13. Geometric deviations of .stl-files from an original geometry when reduced to (**a**) 1%, (**b**) 5%, (**c**) 10%, and (**d**) 40% of the original amount of data.

Using these results and the effect of under-/overextrusion (see Section 3), an expected reduction in shear strength when reducing the number of vertices of an .stl-file could be estimated. The corresponding diagram is shown in Figure 14. Since the effect of over- and underextrusion was greater on the interlaminar shear strength of the AM part, this effect was used in the model shown in Figure 14.

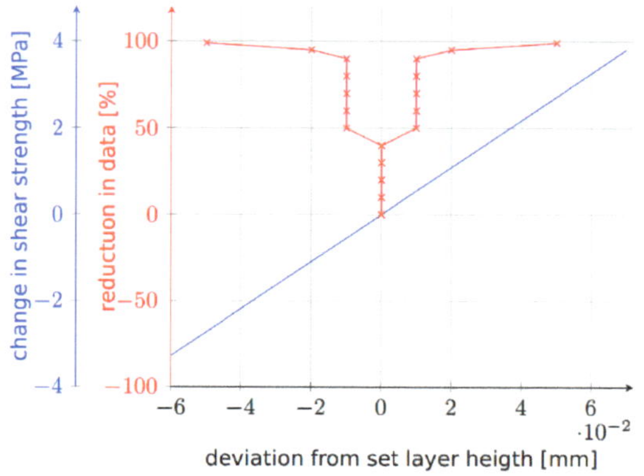

Figure 14. Geometric deviations of a scan from the real geometry dependent on the relative reduction in vertices in an .stl-file and the subsequent change in shear strength of the bonding of PEEK on planar CF/PEEK laminates.

Dependent on the required mechanical properties of a part, Figure 14 can be used to lower the resolution of an .stl-file to ease the further use of the data without risking the impairment of the bonding strength in an unsuitable manner.

The digital twin of the laminate can then be exported as an .stl-file and imported in the software blender together with the geometry of the structure to be 3D printed, the so-called core geometry. When designing the core geometry, attention should be paid to placing the geometry on the xy-plane and letting it extend in z-direction. Using the 'Boolean' operator, the digital twin of the laminate can be used to remove everything from the core geometry above the laminate surface. The modified core geometry can then be exported as an .stl-file.

To generate the toolpath for the 3D printer, a modified version of the software Slic3r is used. This slicer divides the part into two parts: one printed using a traditional planar toolpath and the other printed using a nonplanar toolpath. The planar printed segment functions as a scaffolding for the nonplanar top layers. By deleting the planar scaffolding, a purely nonplanar toolpath can be achieved easily and exported as a .gcode-file.

The robotic FGF system of the DLR requires a postprocessor to convert the .gcode-file into a .dxr-file, which is specific to the system. Lastly, the toolpath needs to be shifted in the z-direction since the top layer of the modified core geometry is at the same height as the laminate surface. The distance the toolpath needs to be shifted depends on the number of layers and the layer height.

3.2.2. Testing the Process Flow

A demonstrator was manufactured to evaluate the benefit of the developed process flow. It consisted of two 3 mm wide lines, which were printed onto a laminate. The laminate had a unidirectional layup, resulting in a substantial waviness of the surface, as shown in Figure 12c. To compare the profile of the laminate with that of the 3D-printed lines, the demonstrator was scanned using the Keyence VR-5000 profilometer. Figure 15a shows the 3D surface scan of the demonstrator. The profile of the laminate was measured between the two printed lines. Based on the scan shown in Figure 12c, it was assumed that the profile of the laminate between the lines was similar to that under the lines. In Figure 15b, the profile of the uncorrected line is shown. It can be seen that in places where the laminate is elevated, the layer height of the 3D print is thinned. The result of the corrected toolpath is shown in Figure 15c. The layer height is more constant compared with the uncorrected line.

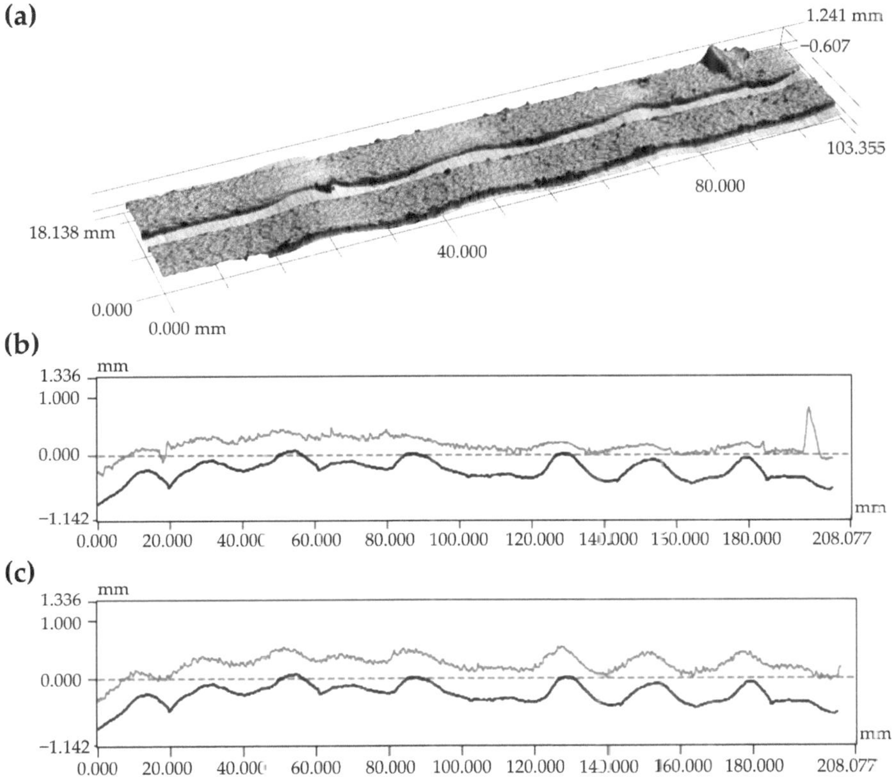

Figure 15. Data of a scan of the demonstrator with the profilometer VR-5000: (**a**) surface scan of the demonstrator, (**b**) surface profiles along a line on the laminate (black) and the printed structure (grey) without usage of the process flow, and (**c**) surface profiles along a line on the laminate (black) and the printed structure (grey) with the usage of the process flow.

4. Discussion

Preliminary studies [4] have shown that the thickness of the first 3D-printed layer is decisive for the shear strength of the bonding on the laminate. The influence of layer thickness on the mechanical properties of 3D-printed PEEK had already been researched and found to be significant [26]. Ref. [27] was able to show, both experimentally and simulatively, that initial contact is an essential component of bond formation in 3D printing. These results are in good agreement with the experiments on over- and underextrusion shown in this study. Gaps and overlaps of tapes led to a local deviation from the target layer height. Therefore, the bonding has to be impaired in these areas. A form closure which improves the relative shear strength could be possible but can only be proven by further testing with a high sample size. The results of the single-lap shear tests had, in relation to the mean value, a high standard deviation of 3.12 MPa on average. With a significance value of 0.05 and a statistical power of 0.8, a sample size of 156 specimens would be required to detect a difference of 1 MPa between two groups with statistical significance. It is still important to note that even with no significant influence of gaps and overlaps of tapes on the strength of the bonding with PEEK, a layer height which is smaller than the tape thickness would lead to a collision of the extruder with the part. Further, it could be possible that the investigated effects of defects increase with smaller nozzle diameters. Further research is needed to characterise the effect of defects with their spectrum of different shapes and characteristics. This includes, in particular, investigating the effects of defects on curved structures and more complex dynamic profiles of path planning. The investigations must also be extended to other material systems in order to be able to use the process on a large scale.

The principle suitability of the developed process chain was demonstrated in this study. By using open-source software for the editing of .stl-files and for generating the toolpath, this process flow is adaptable for a variety of FFF/FGF-machines, and costs are reduced. It was possible to adjust for the waviness and general warpage of a laminate with the developed process flow, as shown in Figure 9. This can be used to realise layer heights smaller than the height of a defect without the risk of a collision between the print head and substrate.

Reducing the detail of a scan reduces the file size, easing further data processing. However, reducing the detail of a scan increases the deviation from the ideal geometry, which can negatively influence the bonding of PEEK on the laminate. The software GOM Inspect Suite can be used to compare the original scan with the edited data, which allows one to estimate the change in bonding strength.

The used slicing software is only capable of generating a toolpath with three DOF, thus only being usable to account for small defects on a planar laminate. Using other toolpath-generating software, like AI Sync, all six degrees of freedom of the robotic arm can be utilised to orient the nozzle always orthogonal to the substrate. Furthermore, more advanced software can be used to adjust the toolpath successively to an ideal geometry. This is to prevent the surface defects of the laminate from progressing in other regions of a part. However, it is noteworthy that the used software, excluding those specific to hardware, was open-source and, thus, enables the applicability of the proposed workflow to a variety of AM processes. The part of the workflow used to compensate surface defects and generate a toolpath uses .stl-files as input, and .gcode-files are used as output. Thus, this process chain is adaptable for using any surface measuring system that can export the collected data as an .stl-file and with any robotic manufacturing system that accepts .gcode-files as an input.

Current systems for the compensation of substrate defects in 3D printing are either integrated into specific products, and thus expensive and hard to modify, or require extensive manual data collection and processing [20,28]. Future research should focus on automating the presented workflow and integrating multiple systems into one software solution. An algorithm for deleting defects, which do not need to be compensated for, out of the data set, without lowering the resolution of the data set in general, might be necessary when overprinting complex curved surfaces.

5. Conclusions

In this thesis, a novel process chain for the overprinting of AFP laminates is proposed, which takes into account, and compensates for, surface defects in the FGF process. For this purpose, the different defects that occur in AFP laminates were first analysed with regard to their effect on overprinting in the FGF process. It was shown that deviations from the set layer height, which result in an underextrusion of the first layer, lead to a weakened bond between FGF-printed structures and laminates. This effect mainly occurs with global defects such as warpage or waviness. However, the results of this work show that gaps and overlaps of tapes in AFP laminates have no significant influence on the bond strength despite the unavoidable local deviation from the target layer height. Overall, there are three main factors that must be taken into account for toolpath planning:

- Global underextrusion due to excessive distance between the nozzle and the laminate in the order of the layer height leads to poor bonding;
- Local defects such as gaps and overlaps have no significant influence on bonding;
- Recording the real geometry and taking it into account in the path planning is essential for a stable process.

A global deviation from the ideal geometry could be successfully compensated with the help of 3D scanning of the real geometry of an open-source software for path planning. The workflow was visualised on a wavy surface, which led to a significantly more constant layer height. By keeping the layer thickness and the extrusion ratios constant on flawed surfaces, the presented workflow is suitable for implementing a stable process on large structures.

The demonstrated process chain is adaptable for a variety of AM methods where a constant distance between the print head and the substrate is required to control a part's properties. For non-planar parts or more complex defects, software utilising all six DOF of a robotic arm must be used, which will be the subject of further study.

Author Contributions: Conceptualization, F.A., S.H. and H.V.; methodology, F.A. and S.H.; validation, F.A.; formal analysis, F.A.; investigation, F.A.; resources, F.A., S.H. and H.V.; data curation, F.A.; writing—original draft preparation, F.A.; writing—review and editing, S.H.; visualization, F.A.; supervision, S.H. and H.V.; project administration, F.A. and S.H.; funding acquisition, H.V. All authors have read and agreed to the published version of the manuscript.

Funding: This research was funded by the Ministry of Economic Affairs, Labour and Tourism Baden-Württemberg reference numbers 3-4332.62-DLR/59 (IRAS III) and WM3-4332-157/68 (IRAS IV).

Data Availability Statement: The data presented in this study are available on request from the corresponding author.

Conflicts of Interest: The authors declare no conflicts of interest.

References

1. Shekar, R.I.; Kotresh, T.M.; Rao, P.M.D.; Kumar, K. Properties of high modulus PEEK yarns for aerospace applications. *J. Appl. Polym. Sci.* **2009**, *112*, 2497–2510. [CrossRef]
2. Zalaznik, M.; Kalin, M.; Novak, S. Influence of the processing temperature on the tribological and mechanical properties of poly-ether-ether-ketone (PEEK) polymer. *Tribol. Int.* **2016**, *94*, 92–97. [CrossRef]
3. Wang, H.Y.; Li, C.Y.; Ding, Y.M.; Li, J.; Chen, S.C. Experimental study on repairing of damaged cast iron cylinder heads by 3D printing arc welding system. In Proceedings of the 7th Global Conference on Materials Science and Engineering, Wuhan, China, 19–20 April 2019. [CrossRef]
4. Hümbert, S.; Meth, J.; Echsel, M.; Lengowski, M.; Stäbler, T. Additve manufacturing of radiation shielding for small satellites. In Proceedings of the 72nd International Astronautical Congress (IAC), Dubai, United Arab Emirates, 25–29 October 2021.
5. Matkovic, N.; Kupzik, D.; Steidle-Sailer, C.; Friedmann, M.; Fleischer, J. Novel Robot-Based Process Chain for Flexible Production of Thermoplastic Components with CFRP Tape Reinforced Structures. *Proc. CIRP* **2022**, *106*, 21–26. [CrossRef]
6. Hümbert, S.; Schmidt, I.; Atzler, F.; Lengowski, M. Mechanical characterization of in-situ bonding between PEEK filaments and laminates in the FFF process. In Proceedings of the ECCM20, Lausanne, Switzerland, 26–30 June 2022.
7. Caprais, I.; Joyot, P.; Duc, E.; Deseur, S. Bonding between high-performance polymer processed by Fused Filament Fabrication and PEEK/carbon fiber laminate. In Proceedings of the ESAFORM2021, Online, 14–16 April 2021. [CrossRef]
8. Harik, R.; Saidy, C.; Wiliams, S.J.; Gurdal, Z.; Grimsley, B. *Automated Fiber Placement Defect Identity Cards: Cause, Anticipation, Existence, Significance, and Progression*; University of South Carolina: Columbia, SC, USA, 2018.
9. Heinecke, F.; Willberg, C. Manufacturing-Induced Imperfecions in Composite parts Manufactures via Automated Fiber Placement. *J. Compos. Sci.* **2019**, *3*, 56. [CrossRef]
10. Zenker, T.; Bruckner, F.; Drechsler, K. Effects of defects on laminate quality and mechanical performance in thermoplastic Automated Fiber Placement-based process chains. *Advanced manufacturing: Polym. Compos. Sci.* **2019**, *5*, 184–205. [CrossRef]
11. Fereidouni, M.; Hoa, S.V. In-situ consolidation of thermoplastic composites by automated fiber placement: Characterization of defects. *J. Thermoplast. Compos. Mater.* **2024**. [CrossRef]
12. Sawicki, A.; Minguett, P. The effect of intraply overlaps and gaps upon the compression strength of composite laminates. In Proceedings of the 39th AIAA/ASME/ASCE/AHS/ASC Structures, Structural Dynamics, and Materials Conference and Exhibit, Long Beach, CA, USA, 20–23 April 1998.
13. Falcó, O.; Lopes, C.S.; Mayugo, J.A.; Gascons, N.; Renart, J. Effect of tow-drop gaps on the damage resistance and tolerance of Variable-stiffness panels. *Compos. Struct.* **2014**, *116*, 94–103. [CrossRef]
14. Raps., L.; Schiel, I.; Chadwick, A.R. Effect of gap defects on in-situ AFP-manufactured structures. In Proceedings of the 20th European Conference on Composite Materials, Lausanne, Switzerland, 26–30 June 2022
15. Lipskas, J.; Deep, K.; Yao, W. Robotic-Assisted 3D Bio-printing for Repairing Bone and Cartilage Defects through a Minimal Invasive Approach. *Sci. Rep.* **2019**, *9*, 3746. [CrossRef] [PubMed]
16. Ma, K.; Zhao, T.; Yang, L.; Wang, P.; Jin, J.; Teng, H.; Xia, D.; Zhu, L.; Jiang, Q.; Wang, X. Application of robotuicassited in situ 3D printing in cartilage regeneration with HAMA hydrogel: An in vivo study. *J. Adv. Res.* **2020**, *23*, 123–132. [CrossRef] [PubMed]
17. Li, L.; Shi, J.; Ma, K.; Jin, J.; Wang, P.; Liang, H.; Cao, Y.; Wang, X.; Jiang, Q. Robotic in situ 3D bio-printing technology for repairing large segmental bone defects. *J. Adv. Res.* **2021**, *30*, 75–84. [CrossRef] [PubMed]

18. Li, L.; Yu, F.; Shi, J.; Shen, S.; Tang, H.; Yang, J.; Wang, X.; Jiang, Q. In situ repair of bone and cartilange defects using 3D scanning and 3D printing. *Sci. Rep.* **2017**, *7*, 9416. [CrossRef] [PubMed]
19. Zhang, Y.; Qiao, J.; Thang, G.; Tian, H.; Li, L. Artifical Intelligence-Assited Repair System for structural and Electrical Restoration Using 3D Printing. *Adv. Intell. Syst.* **2022**, *4*, 2200162. [CrossRef]
20. Raps, L.; Atzler, F.; Chadwick, A.R.; Voggenreiter, H. In-situ automated fiber placement gap defect filled by fused granular fabrication. *Manuf. Lett.* **2024**, *40*, 125–128. [CrossRef]
21. Jin, L.; Zhai, X.; Wang, K.; Zhang, K.; Wu, D.; Nazir, A.; Jiang, J.; Liao, W.-H. Big data, machine learning, and digital twin assisted additive manufacturing: A review. *Mater. Des.* **2024**, *244*, 113086. [CrossRef]
22. Ensinger GmbH. TECACOMP PEEK 150 CF30 Black 1015086—Compounds. Available online: https://www.ensingerplastics.com/en/compounds/cf-peek-compound-tecacomp-peek-150-cf30-black (accessed on 14 December 2022).
23. Toray Advanced Composites Toray Cetex® TC1225—Product Data Sheet. Available online: https://www.toraytac.com/product-explorer/products/gXuK/Toray-Cetex-TC1225 (accessed on 10 March 2024).
24. *ASTM D3846-08(2015)*; Standard Test Method for In-Plane Shear Strength of Reinforced Plastics. ASTM International: West Conshohocken, PA, 2015.
25. Ahlers, D.; Zhang, J.; Hendrich, N. 3D Printing of Nonplanar Layers for Smooth Surface Generation. Master's Thesis, University of Hamburg, Hamburg, Germany, 2018.
26. Wu, W.; Geng, P.; Li, G.; Zhao, D.; Zhang, H.; Zhang, J. Influence of layer thickness and raster angle on the mechanical properties of 3d-printed PEEK and a comparative study between PEEK and ABS. *Materials* **2015**, *8*, 5271. [CrossRef] [PubMed]
27. Coogan, T.J.; Kazmer, D.O. Prediction of Interlayer Strength in Material Extrusion Additive Manufacturing. *Addit. Manuf.* **2020**, *35*, 101368. [CrossRef]
28. Yi, N.; Chen, Y.; Shen, J.; Davies, R.; Ghita, O. Correlation between interfacial bond strength and degree of healing in overprinting PAEK on CF/PAEK composites. *Compos. Part A* **2024**, *183*, 108217. [CrossRef]

polymers

MDPI

Article

Investigation of the Interfacial Fusion Bonding on Hybrid Additively Manufactured Components under Torsional Load

Melike Kizak [1,*], Anna von Bartschikowski [1], Anna Trauth [2], Christian Heigl [3] and Klaus Drechsler [1]

1 Chair of Carbon Composites, Technical University of Munich, Boltzmannstr. 15, 85748 Garching, Germany; anna.bartschikowski@tum.de (A.v.B.); klaus.drechsler@tum.de (K.D.)
2 Institute of Materials Resource Management, University of Augsburg, Am Technologiezentrum 8, 86159 Augsburg, Germany; anna.trauth@mrm.uni-augsburg.de
3 Toray Automotive Center Europe, Am Gfild 6, 85375 Neufahrn bei Freising, Germany; christian.heigl.t8@mail.toray
* Correspondence: melike.kizak@tum.de

Abstract: Hybrid manufacturing processes integrate multiple manufacturing techniques to leverage their respective advantages and mitigate their limitations. This study combines additive manufacturing and injection molding, aiming to efficiently produce components with extensive design flexibility and functional integration. The research explores the interfacial fusion bonding of hybrid additively manufactured components under torsional loading. Specifically, it examines the impact of various surface treatments on injection molded parts and the influence of different build chamber temperatures during additive manufacturing on torsional strength. Polycarbonate components, neat, with glass or carbon fiber-reinforcement, are produced and assessed for dimensional accuracy, torsional strength, and fracture behavior. The findings emphasize the critical role of surface treatment for the injection molded components before additive manufacturing. Additionally, the study identifies the influence of chamber temperatures on both dimensional accuracy and torsional strength. Among all investigated materials, plasma-treated neat samples exhibited the best torsional strength. The torsional strength was increased by up to 87% by actively heating the build chamber to 186 °C for neat polycarbonate. These insights aim to advance the quality and performance of hybrid additively manufactured components, broadening their application potential across diverse fields.

Keywords: composites; additive manufacturing; fused filament fabrication; hybrid materials; injection molding; interface bonding; mechanical characterization; torsion

1. Introduction

Hybrid manufacturing utilizes multiple manufacturing techniques to produce a component, strategically leveraging the advantages of each method. Combining additive manufacturing (AM) and injection molding (IM) opens new possibilities for producing complex parts and has gained increasing importance. This combination allows for the productivity and precision of IM to be paired with the design freedom of AM. In recent years, hybrid manufacturing processes that combine AM and IM have garnered interest across various industries [1,2].

In integrating AM structures as inserts in IM processes, complex or lightweight structures are created using AM and placed into IM tools before injecting liquid material. These inserts can enhance the strength and stiffness of the final IM component without significantly increasing its overall weight [3]. Additionally, hybrid manufacturing methods can be employed to repair existing parts, where defective or worn-out sections are repaired or supplemented with metal or plastic, enabling the reuse of components with economic and environmental benefits [4–7].

Another application area is printing on IM parts with structurally optimized or functionally integrated features. AM techniques are used to add additional structures to already molded parts, referred to as substrates. These AM structures can include reinforcements [8], cooling channels, or sensors integrated into the component [9,10], enabling the production of complex parts that are both lighter and more functional [11].

Regardless of the application and production process, the bonding between the IM part and the AM structure is crucial for the quality of the manufacturing process [12,13]. To understand how the printed material adheres to the substrate, the interactions at the material surfaces must be considered, primarily involving adhesion and diffusion processes. A high adhesion strength between the substrate surface and the first printed layer correlates with good wettability [14]. A well-wettable surface achieves a larger contact area between the plastic melt and the solid surface, promoting diffusion processes at the interface. During the printing process, the surfaces of the substrate and the molten filament each possess a specific energy due to the free bonds on the surfaces. The stronger these free bonds, the higher the surface energy of the materials and the greater the tendency for adhesion, resulting in better bonding. Thus, higher surface energy leads to better adhesion between the AM and the IM part. Good adhesion can only be achieved if molecular segments at the respective surfaces of the substrate and AM structure interact, requiring a similarity in the strength and nature of the molecular forces at both surfaces [15,16].

The energy of plastics is structurally low [15], but wettability can be improved by increasing surface energy, as seen in plasma treatments. Plasma, an ionized gas consisting mainly of positively charged ions and free electrons, can split chemical bonds on the surface and create new functional groups, altering the surface's chemical properties [17,18].

Penter et al. [12] demonstrated the effectiveness of plasma using modified tensile tests, where a plasma-treated IM plate was subsequently printed using the fused filament fabrication process. A smaller contact angle indicated increased surface energy, confirmed by higher mechanical properties in the tensile test.

The second phenomenon relevant to bonding between the IM part and the AM structure is diffusion [19]. Plastics already possess a certain permeability for liquids and gases due to the free volume between macromolecules. When plastics are heated, the mobility of the molecules increases, allowing adjacent molecules of the plastic melt to diffuse across the interface. The exact diffusion rate depends on various factors, such as temperature and time. Better interfacial bonding between the IM part and the AM structure can be achieved by increasing the build chamber temperature, thus enhancing molecular mobility and slowing the cooling of the extruded melt, giving the molecules more time to diffuse [20–22].

Various mechanical tests can be employed to characterize the adhesion between two layers. According to Grellmann and Seidler [23], these tests can be categorized based on the type of stress applied: tensile, compressive, bending, torsional, and shear stress. This work focuses exclusively on torsional stress.

In the study by Gong et al. [24], a hybrid manufacturing process combining AM and IM was investigated to improve the mechanical properties of acrylonitrile butadiene styrene (ABS) samples. Additively manufactured ABS inserts were placed in a T-shaped injection mold before the casting process. The study found that hybrid samples, especially those with a T-shape and a medium filling density of 50%, exhibited improved strength and potential cost savings in producing customized products.

Moritzer et al. [8] aimed to enhance the strength and stiffness of thin-walled plastic components. These components often have poor mechanical properties, so the thin-walled areas were reinforced with specially adapted structures using fused filament fabrication. The material used was the high-performance thermoplastic polyetherimide. Tensile, compressive, bending, and torsion tests were conducted to determine the static strength properties. The resulting hybrid structure exhibited higher strength or stiffness depending on the reinforcement structure's shape compared to components without reinforcement structures. While following existing DIN standards for tensile, compressive, and bending tests, they developed a unique approach for torsion testing, fixing the test specimen in a special device and loading it in a screw testing machine until failure.

Weaver et al. [25] characterized the interface of a hybrid component by the first part being wrought and then building the second part using AM. Although a powder-bed-based method was used instead of filament printing, interlayer adhesion remained crucial. The test specimens had threaded ends to apply torque during torsion testing, revealing that the additive part generally had higher strength and lower ductility than the rolled specimens. No failure occurred at or near the interface between the rolled substrate and the AM material, highlighting the potential of AM for adding features or repairing existing structures.

The study by Guo et al. [26] explores the use of ultrasonic additive manufacturing to create high-strength joints between carbon-fiber-reinforced polymer and aluminum alloy for vehicle structures. The process enabled the embedding of carbon fibers into an aluminum matrix, resulting in improved mechanical performance. The hybrid structures were tested in four-point bending, dynamic axial crush, and quasi-static torsion tests. The results demonstrated that UAM-based joints exhibit 13% higher peak torque and fail by a gradual shearing of the interface and thus allow structural integrity compared to conventional riveted joints.

2. Research Question

Despite the extensive research on hybrid AM, there remains a significant gap in the fundamental understanding of the adhesion mechanisms between AM structures and IM parts. The previous studies assume good adhesion between the components but have not isolated the adhesion scenario. The systematic investigation of interface bonding between AM and IM parts is crucial. This research aims to address this gap by conducting a detailed and systematic investigation of the interface bonding between AM and IM parts. Unlike previous studies, which have largely focused on the overall performance of hybrid components, this work isolates the adhesion process to uncover the underlying mechanisms that govern the strength and reliability of the bond. Therefore, this research provides fundamental insights into the adhesion mechanisms, which are crucial for enhancing the reliability and performance of hybrid manufactured components. These insights not only contribute to the existing body of knowledge but also enable the development of innovative solutions for a wide range of applications, thereby advancing the field of hybrid manufacturing.

This study investigates the influence of various surface treatments on the IM substrate and the effect of different build chamber temperatures during AM. The primary focus is on how these factors impact the strength and adhesion quality between the IM and AM layers by means of torsion tests and fracture analysis. The specific research questions addressed in this study are:

- How do different surface treatments of the IM substrate affect the torsional strength of the hybrid component?
- How do varying chamber temperatures during hybrid manufacturing affect the torsional strength of the hybrid component?
- How do these factors influence the fracture patterns observed in torsion tests?

The study involves a series of torsion tests to evaluate the adhesion quality under different conditions, followed by a detailed analysis of the resulting fracture patterns to understand the failure mechanisms at the interface.

3. Materials and Methods

3.1. Materials

This study used three material combinations: polycarbonate (PC) printed segment onto PC IM substrate, glass-fiber-reinforced PC (G-PC) printed segment onto G-PC IM substrate, and carbon-fiber-reinforced PC (C-PC) printed segment onto C-PC IM substrate. The PC IM substrate plates were produced from XANTAR 18 UR-PC granulate provided by Mitsubishi Engineering-Plastics Corporation (Minato City, Japan), while the printed segments were created using PolyLite PC filament from Polymaker (Changshu, China), with a diameter of 1.75 mm. DAHLTRAM C-250GF granulate from Airtech Europe Sarl (Differdingen, Luxemburg), which contains 20% glass fiber, was utilized for the G-PC substrate plates. Ultrafuse PC GF30 filament from BASF (Ludwigshafen, Germany), featuring a 1.75 mm diameter, was used for the printing. The C-PC substrate plates were made using DAHLTRAM C-250CF granulate from Airtech, which has a 20% carbon fiber content, and the AM structures were produced with CarbonX Fiber ezPC filament, also 1.75 mm in diameter, from 3DXTech (Grand Rapids, MI, USA).

3.2. Differential Scanning Calorimetry (DSC)

Prior to sample fabrication, the base materials were analyzed using DSC. A DSC Q200 device and an RCS90 cooling unit from TA Instruments (New Castle, DE, USA), were employed to determine the materials' glass transition temperature (T_g). This temperature is crucial for sample production as the experimental series investigated manufacturing at various build chamber temperatures in relation to T_g. Two filament samples and two granulate samples were analyzed for each material type (PC, G-PC, and C-PC). Each sample underwent two heating cycles from 30 °C to 200 °C at a rate of 10 K/min. Only the second cycle was used for evaluation, as it eliminated effects such as residual stresses or moisture in the material. T_g was determined according to DIN EN ISO 11357-2 [27].

3.3. Sample Preparation

Before manufacturing the samples, it was necessary to dry both the granulate for the IM substrate parts and the filaments for the printed segment. A Memmert UF 110 Plus universal oven was used for this purpose, and it was also employed to dry the three granulates before IM and the plates directly before printing. As recommended by the manufacturers, the drying process was carried out at 120 °C for at least four hours.

The substrate plates were manufactured using an ENGEL (Schwertberg, Austria), tie-bar-less injection molding machine VC 330/90 tech. The injection unit was equipped with an all-purpose 35 mm screw, heated by four independent zones, with a temperature range of 315 °C to 330 °C from the feeder to the nozzle. The nozzle itself was maintained at a constant temperature of 315 °C. Following the injection of molten material into the heated tool at a temperature of 105 °C, the injection unit applied back pressure for a period of 25 s until the injection gate was frozen. The back pressure was 700 bar. Subsequently, the injection unit plasticized a second shot, which was then cooled for 8 s. During this time, the injection unit detached from the tool, enabling the tool to open and the press side to eject the plate. Thereafter, the sprue was removed using a band saw to prepare the plates for the subsequent AM process.

To investigate the influence of different surface treatments on the substrate plates' interfacial bonding, the substrate plates underwent sandblasting, manual sanding, and plasma treatment, with untreated plates serving as a reference.

For sandblasting, a Sandmaster AG (Zofingen, Switzerland) machine was used with glass beads sized between 90 μm and 150 μm as the abrasive medium, applied at a pressure of 3 bar. Manual sanding was performed using 180-grit sandpaper, involving circular motions with even pressure until a visually homogeneous roughness was achieved. These processes aimed to increase surface roughness by removing material from the plates' surfaces. After the respective surface treatment, each sample was cleaned with compressed air and isopropanol to remove any abrasive residues and ensure precise measurement of the achieved surface roughness. The final method, plasma treatment, employed the piezobrush PZ3 plasma pen from relyon plasma GmbH. The standard module, suitable for non-conductive substrates like plastics, treated the PC and G-PC plates, while the nearfield module, designed for conductive materials, was used for the C-PC plates. The plasma treatment was manually conducted by holding the plasma pen at full intensity over the area where an AM sample would later be printed for 30 s per interface. In this case, the substrate plates were cleaned with isopropanol before the plasma treatment.

Moreover, four different chamber temperatures were examined: no active tempering (T_0), T_g (T_1), T_g + 20 °C (T_2), T_g + 40 °C (T_3).

The test specimens were fabricated using the filament printer GEWO Performer 260 (Woerth/Hoerlkofen, Germany). The slicing software Simplify3D (version 4.1.2) was used to prepare the print files. The specific printing parameters are listed in Table 1. The geometry of the specimens, with a height of 11 mm and base diameter of 10 mm, is shown in Figure 1.

Table 1. Printing parameters.

Parameter	Value
Nozzle diameter	0.4 mm
Layer width	0.4 mm
Layer height	0.2 mm
Heatbed temperature	-
Nozzle temperature PC	260 °C
Nozzle temperature G-PC	300 °C
Nozzle temperature C-PC	270 °C

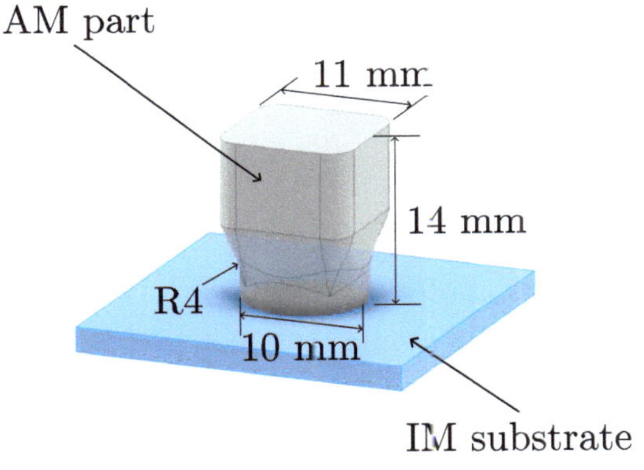

Figure 1. Geometry of the hybrid sample.

3.4. Roughness Measurement and Fracture Behavior

To determine the average surface roughness (R_a) of both treated and untreated injection-molded plates, as well as for the subsequent analysis of fracture patterns and their maximal roughness value R_z, a Keyence VR-5000 profilometer (Osaka, Japan) was used. Multiple line roughness measurements were taken in both vertical and horizontal directions along 22 lines on three randomly selected sample plates.

3.5. Dimensional Accuracy

To assess the accuracy of the printing process, all printed specimens were measured once using a digital caliper with an accuracy of ±0.03 mm. The measurements included the height of the specimens and the diameter of the first printed layers.

3.6. Torsion Test

The torsion tests were conducted using the ElectroPuls E10000 Linear-Torsion machine from Instron GmbH (Darmstadt, Germany), equipped with a load cell capable of measuring up to 25 Nm. The tests were performed using the manufacturer's WaveMatrix 2 materials testing software. A pre-load tensile force of 1 N was applied. The AM part of the hybrid specimen was rotated at a rate of 1 °/s until it reached an angle of 360° or until the test was manually stopped upon failure. Five samples per configuration were tested. Figure 2 shows the test setup, including the adapter.

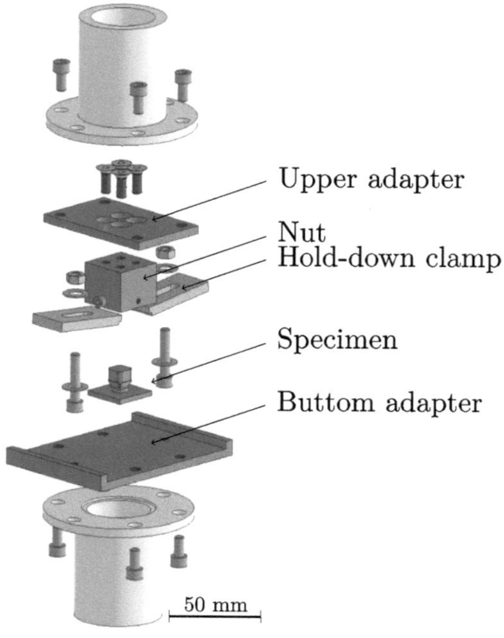

Figure 2. Torsion test set-up.

4. Results

4.1. DSC

Table 2 presents the arithmetic mean of T_g for each material (PC, G-PC, and C-PC) and material form (substrate and filament). A comparison of the average T_g between granulate and filament of the same material reveals a maximum difference of 35 °C for PC, while the difference for G-PC is 7 °C and 3 °C for C-PC.

Table 2. Glass transition temperature T_g of PC, G-PC, and C-PC substrate and filament, respectively, as measured with DSC.

	Substrate T_g [°C]	Filament T_g [°C]
PC	146	111
G-PC	149	142
C-PC	149	146

4.2. Roughness Measurement

The results of the surface roughness analysis, categorized by material and surface treatment, are displayed in Figure 3. The arithmetic mean of the R_a values and the standard deviation are shown. Untreated PC plates had the lowest roughness at 38.6 µm. Manual sanding increased the roughness of PC substrates to 81.4 µm, and sandblasting further increased it to 296.4 µm. An opposite trend was observed for fiber-reinforced plates: untreated plates had the highest roughness. Surface treatments reduced roughness, with G-PC substrates showing a reduction of 10.6% after manual sanding and 10.9% after sandblasting. For C-PC plates, sanding resulted in a decrease of 1.7%, and sandblasting led to a decrease of 0.3%.

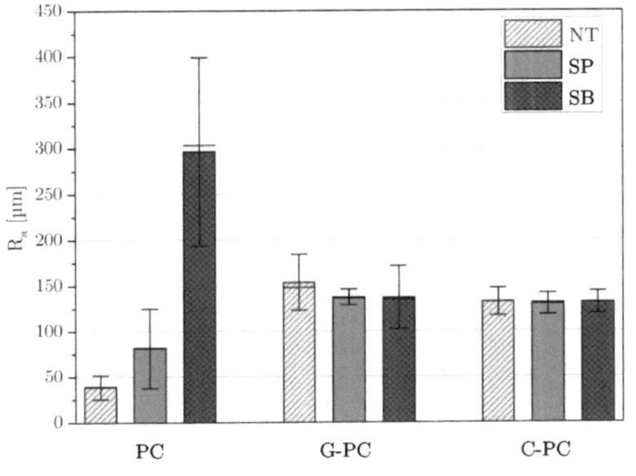

Figure 3. Average roughness value R_a of PC, G-PC, and C-PC for surface treatments no treatment (NT), sandpaper (SP), and sandblasting (SB).

4.3. Dimensional Accuracy

The measurement results of the height and base diameter of the printed segment are graphically presented in Figure 4. The red dotted horizontal line indicates the target dimension of 14 mm for the height and 10 mm for the base diameter. For all materials, it was observed that height decreased and base diameter increased with rising build chamber temperature. Samples of the same material printed at the same build chamber temperature showed consistently low variations, regardless of the surface treatment.

For PC, the highest geometric accuracy in height was observed at T_g. Deviations from the target value were −2.6% for PC, −1.0% for G-PC, and −2.7% for C-PC. An unheated build chamber provided the second-best accuracy for PC and C-PC, while for G-PC, a build chamber temperature 20 °C above T_g was optimal. The lowest heights for all materials were recorded at 40 °C above T_g, with deviations of −8.7% for PC, −4.8% for G-PC, and −4.9% for C-PC. Overall, the fiber-reinforced samples showed smaller deviations compared to the neat PC samples.

Regarding the diameter of the first printed layers, the highest accuracy was achieved in an unheated build chamber, with accuracy decreasing as the temperature increased. The smallest deviation for PC was +0.9%, for G-PC +1.3%, and for C-PC +2.1%. PC samples consistently exhibited the highest deviation at higher temperatures, while C-PC samples had the smallest deviation among the materials tested.

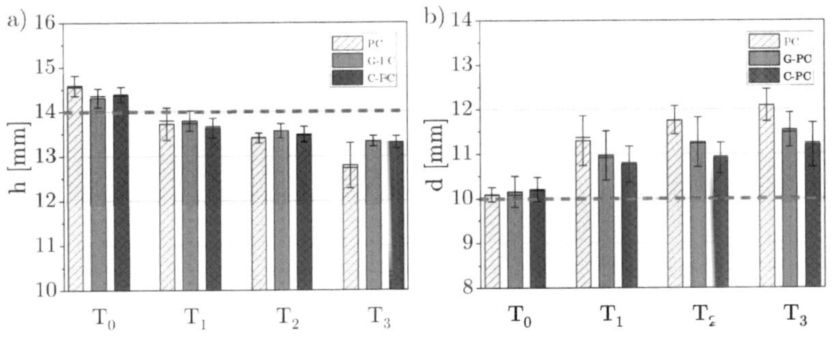

Figure 4. (**a**) Height h and (**b**) base diameter d measurements of PC, G-PC, and C-PC for chamber temperatures T_0–T_3 with the target dimensions of 14 mm for the height and 10 mm for the base diameter marked as dashed red lines.

4.4. Torsion Test

The results of the torsion tests are presented in Tables 3–5. The lowest torsional stress and failure angle in PC samples were observed for samples printed in an unheated chamber. The failure angle of the samples for all treatments increased until chamber temperature T_2. Untreated and plasma-treated samples printed at T_2 showed the highest torsional stress before decreasing for samples printed at T_3. The samples with sandblasted and sanded PC plates showed an increase in strength for samples printed up to T_1, followed by a decrease for samples printed up to T_3.

Table 3. Average ($\bar{x}$) and standard deviation (σ) of torsional stress (τ) and failure angle (α) of PC samples printed at chamber temperatures T_0–T_3 and surface treatments no treatment (NT), plasma (PL), sandpaper (SP), and sandblasting (SB).

		T_0		T_1		T_2		T_3	
		τ [N/mm^2]	α [°]	τ [N/mm^2]	α [°]	τ [N/mm^2]	α [°]	τ [N/mm^2]	α [°]
NT	$\bar{x}$	4.95	7.61	36.70	32.83	39.05	40.37	25.30	25.37
	σ	1.68	2.73	4.05	4.30	1.73	4.39	5.23	13.52
PL	$\bar{x}$	12.84	16.90	37.16	26.87	37.26	45.69	33.99	39.61
	σ	2.91	3.52	5.12	6.34	3.53	4.50	2.28	7.44
SP	$\bar{x}$	1.96	2.92	37.41	31.83	34.56	37.60	28.01	25.14
	σ	0.78	1.49	1.47	4.00	1.32	5.13	5.28	4.89
SB	$\bar{x}$	3.08	5.26	32.47	28.09	29.13	39.90	24.38	25.14
	σ	2.57	2.02	4.01	6.09	4.57	4.59	6.86	8.63

Untreated, plasma-, and sandpaper-treated G-PC samples showed maximal torsional stress for samples printed at T_3. The samples with the sandblasted G-PC substrate exhibit the maximal strength printed with chamber temperature T_2. Higher chamber temperatures reduced the results' scatter. The fracture angles for G-PC samples also increased with chamber temperature, from an average of 4.30° for samples printed in an unheated chamber to 31.35° for samples printed at the highest temperature.

Table 4. Average ($\bar{x}$) and standard deviation (σ) of torsional stress (τ) and failure angle (α) of G-PC samples printed at chamber temperatures T_0–T_3 and surface treatments no treatment (NT), plasma (PL), sandpaper (SP), and sandblasting (SB).

		T_0		T_1		T_2		T_3	
		τ [N/mm^2]	α [°]	τ [N/mm^2]	α [°]	τ [N/mm^2]	α [°]	τ [N/mm^2]	α [°]
NT	$\bar{x}$	17.46	6.50	36.53	18.12	35.45	20.72	38.51	29.75
	σ	10.71	4.59	4.80	4.72	2.23	4.49	1.61	5.54
PL	$\bar{x}$	18.64	8.39	36.08	16.00	34.34	13.76	38.89	29.93
	σ	5.59	2.97	1.88	2.59	1.50	2.73	2.45	3.64
SP	$\bar{x}$	8.37	4.30	35.99	14.01	35.63	18.70	37.27	31.35
	σ	2.34	1.72	2.18	1.31	2.49	3.26	1.59	2.78
SB	$\bar{x}$	14.07	5.59	31.90	14.68	36.93	18.32	33.72	24.98
	σ	8.83	4.89	9.90	5.91	2.25	1.85	1.64	3.33

The C-PC samples' lowest strength and failure angle were also observed at T_0. The maximal failure angle for all treatments was at chamber temperature T_3. The samples with untreated and plasma-treated substrates showed the highest torsional stress at T_3, while sandpaper-treated and sandblasted samples peaked at T_2.

Table 5. Average ($\bar{x}$) and standard deviation (σ) of torsional stress (τ) and failure angle (α) of C-PC samples printed at chamber temperatures T_0–T_3 and surface treatments no treatment (NT), plasma (PL), sandpaper (SP), and sandblasting (SB)

		T_0		T_1		T_2		T_3	
		τ [N/mm^2]	α [°]	τ [N/mm^2]	α [°]	τ [N/mm^2]	α [°]	τ [N/mm^2]	α [°]
NT	$\bar{x}$	10.72	10.33	30.27	15.43	30.83	17.43	31.35	22.53
	σ	3.05	3.52	1.83	2.14	3.51	5.48	3.60	3.03
PL	$\bar{x}$	13,67	6.10	31.51	16.23	30.22	17.18	32.11	22.62
	σ	5.11	6.10	1.69	1.77	0.92	2.40	1.27	2.96
SP	$\bar{x}$	7.42	7.34	23.56	16.90	31.03	15.96	24.93	17.26
	σ	4.52	5.38	0.84	1.09	0.60	2.66	2.23	6.01
SB	$\bar{x}$	9.63	2.54	28.51	14.39	30.82	18.97	29.85	22.78
	σ	3.91	0.39	2.64	2.64	2.60	3.58	2.08	3.11

4.5. Fracture Behavior

Figure 5 shows representative fracture surfaces of plasma-treated substrates arranged from left to right in order of increasing build chamber temperature. The fracture surface roughness for all materials increased with rising build chamber temperature. In PC samples, a spiral shape was visible on the fracture surfaces. The maximal roughness value R_z of the respective circular fracture surface are shown in Table 6. It is evident that the roughness consistently increases with temperature, being highest in PC and lowest in C-PC. While the roughness from T_0 to T_1 increases approximately two to four times, the difference among the three higher temperatures is comparatively small.

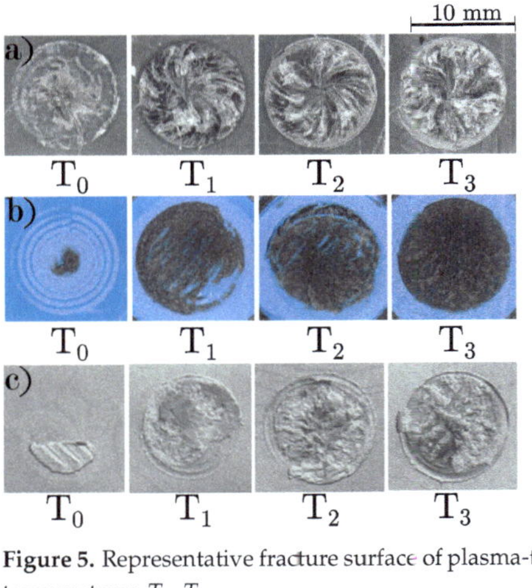

Figure 5. Representative fracture surface of plasma-treated (**a**) PC, (**b**) G-PC, and (**c**) C-PC for chamber temperatures T_0–T_3.

Table 6. Maximal roughness value R_z of the representative fracture surface of plasma-treated PC, G-PC, and G-PC for chamber temperatures T_0–T_3 in µm

	PC	G-PC	C-PC
T_0	611 µm	285 µm	22 µm
T_1	1952 µm	608 µm	44 µm
T_2	2096 µm	634 µm	50 µm
T_3	2108 µm	642 µm	92 µm

5. Discussion

To determine T_g, both granulate and filament underwent DSC analyses. The PC filament displayed a T_g that was 34.2 °C lower than that of the granulate, attributed to additives that improve printability. Consequently, the build chamber temperature T_3 was already 74.2 °C above the filament's T_g, leading to poor dimensional accuracy of PC samples.

The surface roughness analysis showed that sandblasting created a rougher surface on PC than manual sanding, likely due to the fine 180-grit sandpaper used. The G- and C-PC substrates initially have a higher roughness as a result of their fiber reinforcement. The roughness of glass-fiber-reinforced plates decreased with sanding and sandblasting. This might be because the fibers form a harder surface than the PC matrix material, making them less susceptible to abrasion. Carbon fibers, being harder than glass fibers, showed less reduction in roughness from surface treatments.

The dimensional accuracy analysis revealed that the height of all three materials decreased and the base diameter increased with rising temperatures, regardless of surface treatment. The smaller deviations in fiber-reinforced materials can be attributed to the fiber-reinforcement and the lower T_g of the PC filament. For PC, the build chamber temperature T_1 was already 35 °C above the filament's T_g, whereas the differences were significantly smaller for G-PC and C-PC.

Torsion tests on PC samples indicated that the interfacial strength increased until a chamber temperature of T_2. Beyond this critical temperature, both the torsional stress and fracture angle decreased. This suggests that both the thermal behavior of the injection-molded granulate and the filament are crucial for interfacial strength. The high build chamber temperature for the filament caused significant geometric deviations and reduced mechanical properties, likely due to filament degradation. For PC, the sanded samples in a heated chamber consistently performed better than sandblasted ones. Untreated and plasma-treated samples generally showed the highest mechanical properties. Sanding and sandblasting introduced microstructures and irregularities, leading to local stress concentrations and lower mechanical properties. The higher roughness of sandblasted samples compared to sanded ones resulted in lower fracture moments. Plasma treatment tended to increase strength, aligning with findings from Penter et al. [12], by activating the surface and improving wettability, leading to better adhesion.

G-PC samples, similar to PC, showed higher strength in untreated and plasma-treated specimens, with the highest strength achieved at T_3. However, there were minimal differences between the fracture moments, likely due to similar R_a values.

Comparing the results across the three materials, PC samples achieved the highest strength of 39.05 N/mm^2 with untreated samples printed with chamber temperature T_2, followed by G-PC samples with a maximum of 38.89 N/mm^2 for plasma-treated samples printed with T_3. Neat materials achieved lower maximal torsional stress: 32.11 N/mm^2 for (plasma-treated and printed with chamber temperature T_3) and 9.0 Nm for C-PC (plasma-treated and printed with chamber temperature T_3). The carbon fibers appeared to negatively impact torsional strength. The previous studies by Tekinalp et al. [28] and Liao et al. [29] showed that fibers increase strength in tensile and bending tests when aligned with the printing and loading directions. In this study, the printing direction and, thus, the fiber orientation differed. The filament paths were laid in concentric circles around the hybrid specimen's longitudinal axis, aligning fibers parallel to the interface and along concentric circles around the longitudinal and torsional axes. This likely resulted in lower fracture moments due to asymmetric stress distributions, as described by Du et al. [30]. Multiple studies have shown that crack propagation in fiber-reinforced plastics under shear and torsional loads mainly occurs through delamination at the fiber–matrix interfaces, explaining the lower torsional strength of carbon-fiber-reinforced composites compared to PC parts.

Furthermore, glass and carbon fibers have higher stiffness and strength than the PC matrix, leading to a more brittle failure behavior of the samples. This behavior was evident in the optical analysis of fracture surfaces. Ductile fractures in PC samples showed significant deformation and crack propagation. In contrast, brittle fractures in fiber-reinforced samples were flatter and lacked visible deformation features. Images of PC samples indicated that the fracture surfaces became more uniform with increasing temperature. However, the highest temperature sample showed irregularities, likely due to filament embrittlement from degradation. The maximal roughness value of the fracture surface increased with temperature across all materials, likely due to more intense diffusion at higher temperatures.

6. Conclusions

This study investigated the effects of different surface treatments and build chamber temperatures on the torsional strength of hybrid samples produced through a combination of injection molding and additive manufacturing. The following are the conclusions:

- Among the surface treatments, untreated and plasma-treated samples exhibited the best torsional strength, while abrasive methods like sanding and sandblasting reduced strength. This reduction in strength may be attributed to the formation of stress concentrations from these abrasive treatments, which hinder the diffusion process and consequently lower the torsional strength.
- PC's ideal build chamber temperature was identified at 166 °C, yielding a maximum torsional strength of 12.3 Nm. The fiber-reinforced samples achieved lower maximum torques of 11.7 Nm for G-PC and 9.0 Nm for C-PC at T_3. While PC samples displayed ductile behavior with a smooth, spiral fracture pattern at the optimal temperature, fiber-reinforced samples failed abruptly and brittlely without noticeable deformation features.
- Increasing build chamber temperatures led to a decrease in sample height and an increase in base diameter across all materials, affecting the overall dimensional accuracy.
- The fracture surfaces became rougher with increasing temperature for all materials, with PC samples showing a distinctive spiral pattern.

Author Contributions: M.K.: conceptualization, project administration, visualization, writing—original draft, A.v.B.: investigation, writing—review and editing, A.T.: scientific discussion, writing—review and editing, C.H.: injection molding, writing—review and editing, K.D.: funding acquisition, resources, supervision. All authors have read and agreed to the published version of the manuscript.

Funding: This research is supported by the German Federal Ministry for Economic Affairs and Climate Action (BMWK) on the basis of a decision by the German Bundestag.

Institutional Review Board Statement: Not applicable.

Data Availability Statement: Additional data, such as profilometer scans and raw data, are available from the corresponding author upon reasonable request.

Acknowledgments: The authors thank the Automotive Center Europe (AMCEU) of Toray Industries Europe GmbH for producing the IM substrate. Further, the authors thank Stefan Schmitt und Kay Weidenmann from the research group Hybrid Composite Materials at University of Augsburg for providing access to the torsion testing machine.

Conflicts of Interest: Author Christian Heigl was employed by the company Toray Automotive Center Europe. The remaining authors declare that the research was conducted in the absence of any commercial or financial relationships that could be construed as a potential conflict of interest.

References

1. Gaub, H. Customization of mass-produced parts by combining injection molding and additive manufacturing with Industry 4.0 technologies. *Reinf. Plast.* **2016**, *60*, 401–404. [CrossRef]
2. Boros, R.; Rajamani, P.K.; Kovács, J.G. Combination of 3D printing and injection molding: Overmolding and overprinting. *EXPRESS Polym. Lett.* **2019**, *13*, 889–897. [CrossRef]
3. O'Quinn, R.; Pradeep, S.A.; Pilla, S.; Farahani, S. Additive insert molding (AIM)—A practical paradigm for mass customization of multi-material/functional parts. In Proceedings of the International Manufacturing Science and Engineering Conference, New Brunswick, NJ, USA, 12 June 2023; pp. 1–19 [CrossRef]
4. Leino, M.; Pekkarinen, J.; Soukka, R. The Role of Laser Additive Manufacturing Methods of Metals in Repair, Refurbishment and Remanufacturing—Enabling Circular Economy. *Phys. Procedia* **2016**, *83*, 752–760. [CrossRef]

5. Bi, G.; Gasser, A. Restoration of Nickel-Base Turbine Blade Knife-Edges with Controlled Laser Aided Additive Manufacturing. *Phys. Procedia* **2011**, *12*, 402–409. [CrossRef]
6. Walachowicz, F.; Bernsdorf, I.; Papenfuss, U.; Zeller, C.; Graichen, A.; Navrotsky, V.; Rajvanshi, N.; Kiener, C. Comparative Energy, Resource and Recycling Lifecycle Analysis of the Industrial Repair Process of Gas Turbine Burners Using Conventional Machining and Additive Manufacturing. *J. Ind. Ecol.* **2017**, *21*, S203–S215. [CrossRef]
7. Rahito, D.A.W.; Azman, A.H. Additive Manufacturing for Repair and Restoration in Remanufacturing An Overview from Object Design and Systems Perspectives. *Processes* **2019**, *7*, 802. [CrossRef]
8. Moritzer, E.; Hirsch, A.; Bürenhaus, F. Development and modeling of design and process guidelines for FDM structures for the partial reinforcement of hybrid structures. *AIP Conf. Proc.* **2019**, *2065*. [CrossRef]
9. Shinde, M.S.; Ashtankar, K.M. Additive manufacturing–assisted conformal cooling channels in mold manufacturing processes. *Adv. Mech. Eng.* **2017**, *9*, 1–14. [CrossRef]
10. Chen, D.; Han, Z.; Zhang, J.; Xue, L.; Liu, S. Additive Manufacturing Provides Infinite Possibilities for Self-Sensing Technology. *Adv. Sci.* **2024**, *11*, 2400816. [CrossRef]
11. Lachmayer, R.; Lippert, R.B.; Kaierle, S. *Additive Serienfertigung*; Springer: Berlin/Heidelberg, Germany, 2018. [CrossRef]
12. Penter, L.; Maier, J.; Kauschinger, B.; Lebelt, T.; Modler, N.; Ihlenfeldt, S. *3D Printing Technology for Low Cost Manufacturing of Hybrid Prototypes from Multi Material Composites*; Springer: Berlin/Heidelberg, Germany, 2020. [CrossRef]
13. Rajamani, P.K.; Ageyeva, T.; Kovács, J.G. Personalized mass production by hybridization of additive manufacturing and injection molding. *Polymers* **2021**, *13*, 309. [CrossRef]
14. Wool, R.P.; Yuan, B.L.; McGarel, O.J. Welding of Polymer Interfaces. *Polym. Eng. Sci.* **1989**, *29*, 1340–1367. [CrossRef]
15. Doobe, M. *Kunststoffe Erfolgreich Kleben*; Springer: Wiesbaden, Germany, 2018. [CrossRef]
16. Hornbogen, E. *Werkstoffe*; Springer: Berlin/Heidelberg, Germany, 2019. [CrossRef]
17. Vesel, A.; Mozetic, M. New developments in surface functionalization of polymers using controlled plasma treatments. *IOP Publ.* **2017**, *50*, 293001. [CrossRef]
18. López-García, J.; Primc, G.; Junkar, I.; Lehocký, M.; Mozetič, M. On the Hydrophilicity and Water Resistance Effect of Styrene-Acrylonitrile Copolymer Treated by CF4 and O2 Plasmas. *Plasma Process. Polym.* **2015**, *12*, 1075–1084. [CrossRef]
19. Prager, S.; Tirrell, M. The healing process at polymer-polymer interfaces. *J. Chem. Phys.* **1981**, *75*, 5194–5198. [CrossRef]
20. Awaja, F. Autohesion of polymers. *Polymer* **2016**, *97*, 387–407. [CrossRef]
21. Butler, C.A.; Mccullough, R.L. An Analysis of Mechanisms Governing Fusion Bonding of Thermoplastic Composites. *J. Thermoplast. Compos. Mater.* **1998**, *11*, 338–363. [CrossRef]
22. De Gennes, P.G. Reptation of a Polymer Chain in the Presence of Fixed Obstacles. *J. Chem. Phys.* **1971**, *55*, 572–579. [CrossRef]
23. Grellmann, W.; Seidler, S. *Kunststoffprüfung*; Carl Hanser Verlag GmbH & Co. KG: Munich, Germany, 2011. [CrossRef]
24. Gong, K.; Liu, H.; Huang, C.; Cao, Z.; Fuenmayor, E.; Major, I. Hybrid Manufacturing of Acrylonitrile Butadiene Styrene (ABS) via the Combination of Material Extrusion Additive Manufacturing and Injection Molding. *Polymers* **2022**, *14*, 5093. [CrossRef]
25. JWeaver, M.; Linn, J.R.; Miles, M.P. Interface Joint Strength between SS316L Wrought Substrate and Powder Bed Fusion Built Parts. *Materials* **2021**, *14*, 3041. [CrossRef]
26. Guo, H.; Gingerich, M.B.; Headings, L.M.; Hahnlen, R.; Dapino, M.J. Experimental investigation of CFRP-AA structures joined by ultrasonic additive manufacturing (UAM) and resistance spot welding (RSW). *Compos. Part B Eng.* **2023**, *260*, 110768. [CrossRef]
27. *DIN EN ISO 11357-2*; Kunststoffe—Dynamische Differenzkalorimetrie (DSC)—Teil 2: Bestimmung der Glasübergangstemperatur und der Glasübergangsstufenhöhe. DIN Deutsches Institut für Normung e.V.: Berlin, Germany, 2020.
28. Tekinalp, H.L.; Kunc, V.; Velez-Garcia, G.M.; Duty, C.E.; Love, L.J.; Naskar, A.K.; Blue, C.A.; Ozcan, S. Highly oriented carbon fiber–polymer composites via additive manufacturing. *Compos. Sci. Technol.* **2014**, *105*, 144–150. [CrossRef]
29. Liao, G.; Li, Z.; Cheng, Y.; Xu, D.; Zhu, D.; Jiang, S.; Guo, J.; Chen, X.; Xu, G.; Zhu, Y. Properties of oriented carbon fiber/polyamide 12 composite parts fabricated by fused deposition modeling. *Mater. Des.* **2018**, *139*, 283–292. [CrossRef]
30. Du, F.; Alghamdi, S.; Riabbans, B.; Tan, T. An experimental study on the fracture of a unidirectional carbon fiber-reinforced composite under quasistatic torsion. *Compos. Part B* **2019**, *172*, 547–554. [CrossRef]

Journal of
Composites Science

MDPI

Article

Modeling of a Process Window for Tailored Reinforcements in Overmolding Processes

Philipp K. W. Picard [1], Tim A. Osswald [2], Swen Zaremba [1,*] and Klaus Drechsler [1]

1 TUM Department of Mechanical Engineering, Chair of Carbon Composites, Technical University of Munich, 85748 Garching, Germany; info.lcc@ed.tum.de (K.D.)
2 Polymer Engineering Center (PEC), University of Wisconsin-Madison, 1513 University Ave, Madison, WI 53706, USA; tosswald@wisc.edu
* Correspondence: zaremba@tum.de

Abstract: This study explores cost-effective and customized composite applications by strategically placing carbon fiber-reinforced thermoplastics in multi-material designs. The focus is on developing a model for the simultaneous processing of non-reinforced and reinforced thermoplastic layers, with the aim of identifying essential parameters to minimize insert flow and ensure desired fiber orientation and positional integrity. The analysis involves an analytical solution for two layered power-law fluids in a squeeze flow setup, aiming to model the combined flow behavior of Newtonian and pseudo-plastic fluids, highlighting the impact of the non-Newtonian nature. The behavior reveals a non-linear trend in the radial flow ratio towards the logarithmic consistency index ratio compared to a linear trend for Newtonian fluids. While a plateau regime of consistency index ratios presents challenges in flow reduction for both layers, exceeding this ratio, depending on the height ratio of the layers, enables a viable overmolding process. Therefore, attention is required when selectively placing tailored composites with long-fiber-reinforced thermoplastics or unidirectional reinforcements to avoid operating in the plateau region, which can be managed through appropriate cavity or tool designs.

Keywords: squeeze flow; overmolding; power-law fluid; fluid array; pseudo-plastic; process window; tailored reinforcements

1. Introduction

Carbon fiber-reinforced composites offer exceptional strength-to-weight ratios, making them highly desirable for lightweight applications in the automotive [1] and aerospace [2] industries. However, the high cost of carbon fibers limits their economic feasibility, particularly in the automotive sector [3]. To overcome this, load-adapted reinforcements are proposed [4], selectively placing composites in load-bearing areas and tailoring fiber direction and length accordingly [5]. Thermoplastics, with their weldability, enable these multi-material designs by combining reinforced and non-reinforced inserts [6]. The advancement of part design optimizations requires varying overmolding layer thicknesses and employing load-adapted reinforcements such as unidirectional and long fiber reinforcements. Modeling the flow of such materials as power-law fluids facilitates understanding of part design's impact on the transverse flow of composite inserts during processing. The aim is to explore analytically the interplay between part and insert design, processability, and design limitations, offering insights for cost-effective and customized composite applications.

2. A Model for the Transverse Flow of Composite Inserts

The shear flow behavior of filled and unfilled polymers can be modeled as a power-law fluid. A power-law fluid exhibits shear rate-dependent behavior, which characterizes its non-Newtonian nature and plasticity. In the case of a power-law coefficient of 1, the fluid is considered Newtonian, where the shear stress is directly proportional to the shear rate [7]. Conversely, for a pseudo-plastic fluid with a power-law coefficient of zero, the shear stress remains independent of variations in the shear rate [8]. In this context, the shear stress maintains a constant magnitude. Below the critical shear stress, flow ceases, while above it, flow is initiated, marking the transition from a solid-like to a fluid-like state [9]. This critical shear stress is commonly known as the yield stress [10]. Polymers typically exhibit power-law coefficients ranging between 0 and 1, rendering them pseudo-plastic fluids. Notably, an augmentation in filler content amplifies the material's plastic behavior [11].

When implementing local load-adapted reinforcements, the integration of non-reinforced and reinforced thermoplastic polymer layers occurs via a unified molding process. This process necessitates the establishment of a pressure gradient towards the flow front to facilitate the welding of layers [12–14] and mold filling, typically employing the non-reinforced material. These layers are typically arranged in close proximity to each other. It is noteworthy that the pressure gradient can induce transverse flow not only in the non-reinforced layer but also in the reinforced layer [15,16]. Additionally, the presence of the non-reinforced layer generates shear stress on the reinforced layer as an insert [17]. The transverse flow of the reinforced layer, influenced by these processing parameters, might disrupt the desired fiber orientation and positional integrity. Consequently, the derivation of an analytical or semi-analytical model based on constitutive equations for a representative configuration becomes indispensable. Thus far, only analytical models for non-interacting power-law layers [18] have been published. The model in this paper facilitates the identification of relevant process, part design, and material parameters crucial for minimizing insert flow.

3. Constitutive Equations for Squeeze Flow of Two Layered Power-Law Fluids

For the development of an analytical model, the constitutive equations are derived from a squeeze flow experimental arrangement (Figure 1), wherein two circular-shaped specimens with equal radii are positioned between two rigid plates at the center. The variation in inter-plate distance corresponds to the applied closing force. In this context, the closing speed is maintained at a constant value, and the resultant closing force is calculated as the output parameter.

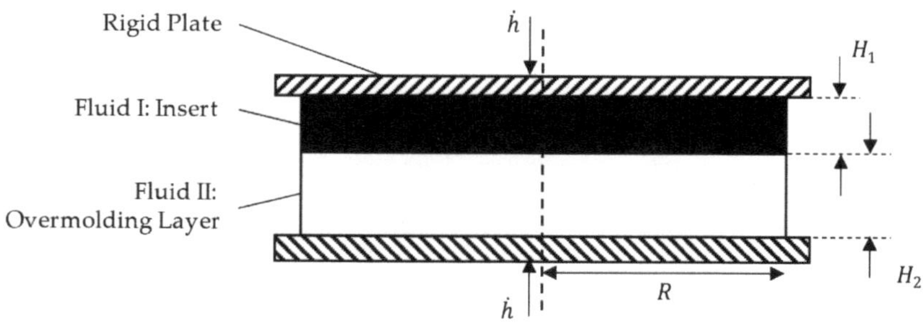

Figure 1. A schematic representation of a squeeze flow setup consisting of two circular-shaped adjacent fluid layers with different flow properties and heights between two moving rigid plates.

The volume of the specimen remains constant throughout the process. Consequently, the radii of the layers progressively increase as the height decreases. The continuity equation, along with the equations of motion in the radial (r) and axial (z) directions, retain their form as in a single-layer squeeze flow setup. Moreover, the boundary conditions at the wall and the plane of symmetry remain unchanged, as schematically shown in Figure 2.

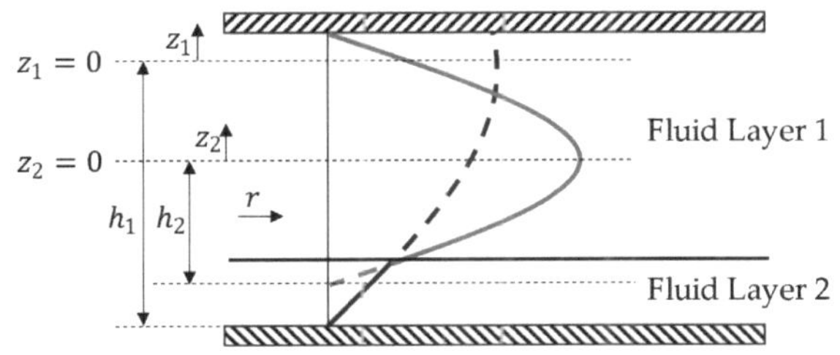

Figure 2. A schematic representation of a radial velocity profile for two adjacent fluid layers with their representative layer heights and coordinate systems.

A coordinate system is introduced for each layer, positioned at the flow field's line of symmetry. Consequently, the actual layer height is substituted with a representative layer height, influenced by the interaction with the second layer. The closing rate at the boundaries of the representative layers undergoes the same transformation. This results in two sets of constitutive equations, which are identical but separately defined for each layer. Their solution is presented in Appendix A, which employs a power-law fluid material model.

The determination of representative heights and their time derivatives for each layer necessitates the inclusion of four additional boundary conditions. These boundary conditions are specified at the interface between the two fluid layers as well as at each position along the radial direction (r).

$$\frac{\partial p_1}{\partial r} = \frac{\partial p_2}{\partial r} \tag{1}$$

$$\tau_1 = -\tau_2 \; at \; z_1 = h_1 - H_1 \text{ and } z_2 = h_2 - H_2 \tag{2}$$

$$f_1 = f_2 \; at \; z_1 = h_1 - H_1 \text{ and } z_2 = h_2 - H_2 \tag{3}$$

$$\dot{H}_1 + \dot{H}_2 = 2\dot{h} \tag{4}$$

The first boundary condition (Equation (1)) asserts that the pressure remains solely a function of r, and a force balance between layer 1 and layer 2 must be maintained. Consequently, the pressure and pressure gradient in both layers must be equal at every r position. The second (Equation (2)) and third (Equation (3)) boundary conditions account for shear force and velocity balance between the first and second layers. These conditions ensure that the shear forces and velocities at the contact of the two layers are consistent with each other. Lastly, the fourth boundary condition (Equation (4)) arises from a mass balance consideration, indicating that the sum of the closing rates of layer 1 and layer 2 at each position along the z-axis must equal the total closing rate.

These boundary conditions yield four equations, and the four unknowns are the representative height and their time derivatives for each layer. Therefore, the term "representative" signifies that these values correspond to a single-layer setup that yields the same pressure gradient, effectively representing the equal flow resistance in radial flow.

4. Solution and Model Validation

The utilization of the boundary conditions yields a solution for each position along the radial direction (r), which is discussed first with the resulting dimensionless numbers and their interpretation. The solution is then validated by the investigation of exemplary resulting flow fields and the overall flow rate of each layer for a variation of the consistency index ratios.

4.1. Resulting Solution and Dimensionless Numbers of the Model

The solution for each position along the radial direction (r) is denoted by Equation (5). The derivation of this equation is given in Appendix B. The factors X_1 and Z (Equations (6) and (7)) in the equation are functions of the ratio between the representative height and the actual height, as well as the ratio of the actual heights of layer 1 and layer 2 at the given radial position. These functions, raised to the power-law index difference between the layers, are interconnected with the overall process setup parameters to the power of the power-law difference, incorporating variables such as the closing speed, radius, and total height. Further parameters are the consistency index ratio and a factor N that results from the power-law coefficients as depicted in Equation (6).

$$\left(\frac{\dot{h}r}{2H^2}\right)^{n_1-n_2}\frac{m_1}{m_2}\left(\left(1+\frac{H_1}{H_2}\right)^2\left(\frac{H_2}{h_2}\right)^2\frac{2X_1}{Z}\right)^{n_1-n_2}N-1=0 \quad (5)$$

$$N=\left(\frac{n_2}{1+2n_2}\right)^{n_2}\left(\frac{1+2n_1}{n_1}\right)^{n_1} \quad (6)$$

$$X_1,\ Z=f\left(\frac{H_1}{H_2};\frac{H_2}{h_2}\right) \quad (7)$$

Equation (15) demonstrates that the obtained results rely on the radial position, especially when dealing with unequal power-law coefficients. This implies that the height change ratio between layers 1 and 2 varies at each radial position. Consequently, the boundary layer between the two layers deviates from a horizontal line and becomes a function of r over time.

The corresponding rate of change along r of the layer height reduction is not considered within the finite difference analysis in r. However, it is considered in a global mass balance calculation to determine the total change in the layer radii. In other words, height reduction is not treated as a function of r within a finite radial element. This simplification is employed in this model to facilitate the utilization of the analytical solution of the squeeze flow equation, leading to a semi-analytical model that allows the identification of dimensionless numbers and their interrelation.

The resulting dimensionless numbers are given by π_1, π_2, and π_3 in Equations (8)–(10).

$$\pi_1=\frac{H_1}{H_2} \quad (8)$$

$$\pi_2=\frac{m_1}{m_2} \quad (9)$$

$$\pi_3=\left(\frac{\dot{h}r}{H^2}\right)^{n_2-n_1} \quad (10)$$

The first dimensionless number, π_1, in Equation (8) is the ratio of the actual layer heights, representing a key design parameter for the part design. The second dimensionless number, π_2, in Equation (9) is the ratio of the consistency index ratios, representing a key design parameter for the material selection. The third dimensionless number, π_3, in Equation (10) incorporates global parameters, such as the overall dimensions of the setup and the closing speed as a processing parameter.

Furthermore, the resulting equations show that the power-law indices remain independent parameters that cannot be summarized as a dimensionless number, which means that the different combinations of power-law indices will lead to non-similar behaviors in the setup.

4.2. Model Validation—Flow Fields

To validate that the model fulfills the boundary conditions and corresponds to a setup as shown in Figure 2, the flow field in the radial direction at a fixed position in r from a representative setup is investigated. The flow field is a visual representation of the results and allows us to visually evaluate if the results are reasonable and valid combinations of representative single-layer flow fields. For a representative flow field, the consistency indices and the power-law indices for both layers are different. Layer one represents a pseudo-plastic layer with a high consistency index, while layer two represents a Newtonian fluid layer with a low consistency index. The resulting flow field is shown in Figure 3, left.

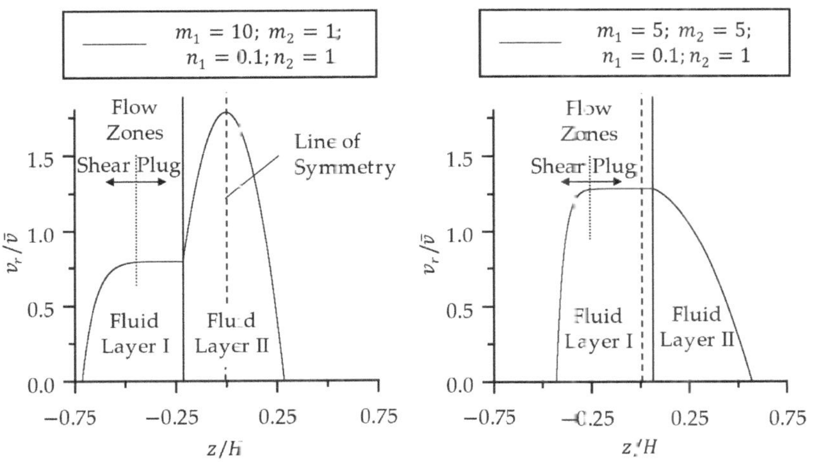

Figure 3. Dimensionless radial velocity by the dimensionless position in z for fluid layer 1 and fluid layer 2 with the line of symmetry for both flow fields for two different consistency index ratios.

The line of symmetry for both layers coincides, defining the zero position as per Equation (A12) in Appendix B. At the contact point of the layers, the velocity equals, as indicated by Equation (A3). The velocity gradients differ due to the distinct power-law indices and consistency indices. The pseudo-plastic fluid exhibits a typical shear zone at the wall and a flat "plug-flow" zone in the center, while the Newtonian layer shows the usual parabolic shape. Additionally, the Newtonian layer, with its lower consistency index, demonstrates a higher overall radial flow rate compared to the pseudo-plastic layer. Despite these differences, the flow fields resemble those of single layers, albeit with a shift in the representative height and flow rate, i.e., closing rate, determined by the pressure gradient balance between the layers as per Equation (A1). When the ratio of consistency indices is decreased (Figure 3, right), the line of symmetry shifts towards the insert layer. In such cases, the Newtonian layer approaches a linear "shear-like" velocity profile. The overall radial flow ratio of the two layers serves as an output for subsequent parameter investigations.

4.3. Effect of Newtonian and Non-Newtonian Layer Combinations

In Figure 4, the ratio of radii is shown for two power-law fluids in a squeeze flow setup, with a constant closing speed applied after a fixed period of time The square symbols represent the combination of two shear-thinning fluids, approximating pseudo-plastic behavior, while the triangular symbols depict the combination of two Newtonian fluids with a power-law exponent of 1.

In the case of two Newtonian fluids, the radii ratio exhibits a linear decrease with a logarithmic increase in the consistency indices. This reasonably implies that the fluid with a higher consistency index results in a reduced radial extension of the layer compared to the fluid with lower consistency. When both fluids have the same flow resistance, the flow field becomes symmetric, and both layers have identical radial extensions.

In contrast, when dealing with two pseudo-plastic fluids, the radii ratio deviates from a linear relationship and shows a non-linear trend towards the logarithmic consistency index ratio. This results in a plateau region where the dependence between the radii ratio and consistency index ratio strongly reduces for two pseudo-plastic layers compared to two Newtonian layers. Therefore, within a specific range of consistency index ratios, the flow of the pseudo-plastic layer cannot be reduced to the same extent as with two Newtonian fluids, presenting a potential challenge in processing. Only when a sufficient consistency index ratio is reached is the flow of the mixture minimized and becomes comparable to that of two Newtonian layers.

Overall, the diagrams highlight the influence of consistency indices on the radial distribution of power-law fluids in a squeeze flow setup, showcasing the differences between shear-thinning (pseudo-plastic) and Newtonian behavior. Their implications for processing are further analyzed by varying the set of parameters.

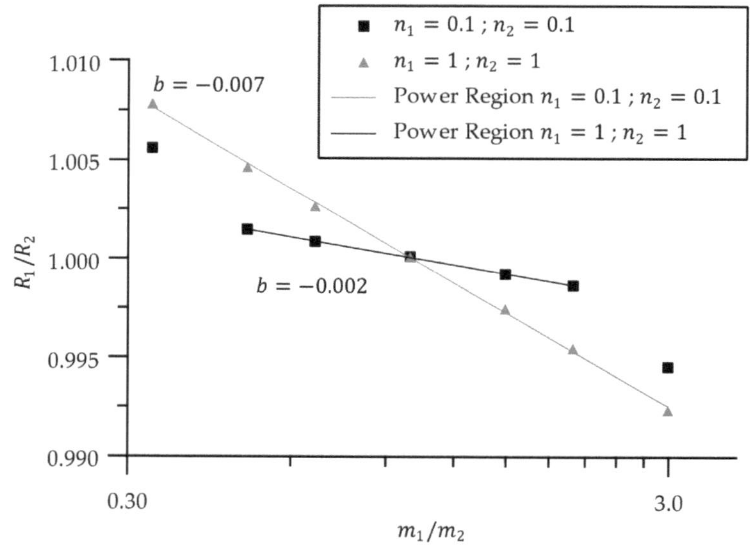

Figure 4. Ratio of the radii of two pseudo-plastic and two Newtonian fluid layers as a ratio of their consistency indices.

5. Implications for the Application of Tailored Reinforcement Inserts

In this chapter, the results are detailed for combinations that are most relevant for typical applications in overmolding and therefore lead to model-based design optimization guidelines for processing and part design.

5.1. Flow Behavior of Pseudo-Plastic Inserts

To investigate the behavior in a representative setup, a combination of a pseudo-plastic layer ($n_1 = 0.1$, $H_1 = 0.25$) representing a highly filled insert and a thicker Newtonian layer ($n_2 = 1$, $H_2 = 1.5$) representing an overmolding layer is examined. Figure 5 illustrates the resulting height reduction rate of each layer, plotted against the logarithm of the consistency ratio at the initial state (specifically at $r = 0.5R$). As the experiment maintains a constant volume, the height reduction rate corresponds to the radius ratio as well. The graph exhibits similar non-linear relations as observed for two pseudo-plastic layers (Figure 4).

Three characteristic regimes can be observed. In Regime 1, the insert has a lower consistency index compared to the overmolding layer, leading to a correspondingly higher height reduction rate. As the consistency index ratio increases, the height reduction rate of the insert decreases until it is surpassed by the height reduction rate of the overmolding layer. This marks the transition to Regime 2, where a plateau region follows. In this region, the height reduction rates of both layers remain within a close range for a wide range of consistency index ratios. Only when the consistency index ratio exceeds a ratio of approximately 3, which marks the beginning of Regime 3, does the height reduction ratio change again. In this zone, the height reduction rate of the overmolding layer increases while the height reduction rate of the insert decreases. At the final consistency index ratios, the height reduction becomes limited to the overmolding layer.

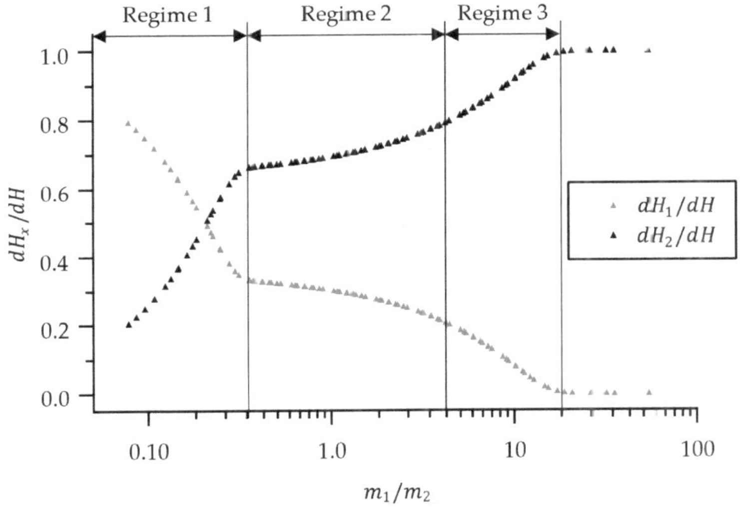

Figure 5. The height change of the two fluid layers scaled by their height change in a single-layer setup as a ratio of their consistency indices with the indication of three flow zones.

This analysis provides insights into the behavior of the two layers in the representative setup, highlighting the impact of consistency index ratios on the height reduction rates and the dominant flow of the overmolding layer in the later stages of the process.

To investigate the underlying flow principles, additional outputs are presented in Figure 6. The top diagram shows the ratio of the pressure gradient in a combined layer setup as a ratio of a single layer pressure gradient, where the boundary conditions match those of a single layer setup. Thus, this ratio indicates how much of the flow resistance is due to a pressure gradient. The bottom diagram illustrates the corresponding representative heights of the layers, as introduced in the constitutive equations. These representative heights serve as indicators of the flow being driven by drag, as they can only be increased by altering the radial velocity at the contact surface, thereby inducing a shear flow in addition to a pressure-driven squeeze flow.

In Regime 1, the ratios of h_1 and the pressure gradient indicate that the flow is similar to the single-layer flow of the insert. The velocity profile remains entirely within the layer height, and the resistance corresponds to the insert's single-layer flow resistance. The overmolding layer interacts with the shear layer of the insert and is solely dragged, as shown by the lower pressure gradient compared to a single-layer setup.

In Regime 2, the representative height of the overmolding layer decreases and the pressure gradient increases, indicating a shift from drag-driven to pressure-driven squeeze flow. The velocity profile of the insert corresponds to the right side of Figure 3. The overmolding layer is now interacting with the plug flow regime of the insert. The velocity profile shape of the overmolding layer remains relatively similar in this regime, with a change in the representative height. Only the curvatures towards the walls are changing slightly, creating the plateau region in Figure 5. When the velocity of the overmolding layer exceeds that of the inserts, as indicated by the crossing lines in Figure 6, the flow ratio changes more rapidly. This shift is accompanied by a transition from shear-dominated to pressure-dominated squeeze flow. Towards the end of Regime 3, the overmolding layer interacts with the second shear layer of the insert near the wall, which leads to a rapid reduction in the insert flow due to the high slopes in the insert velocity profile.

These findings shed light on the flow principles underlying the interaction between the layers, emphasizing the significance of different zones and their impact on flow resistance. By staying within Zone 3, the processing conditions can be optimized to achieve the desired flow behavior.

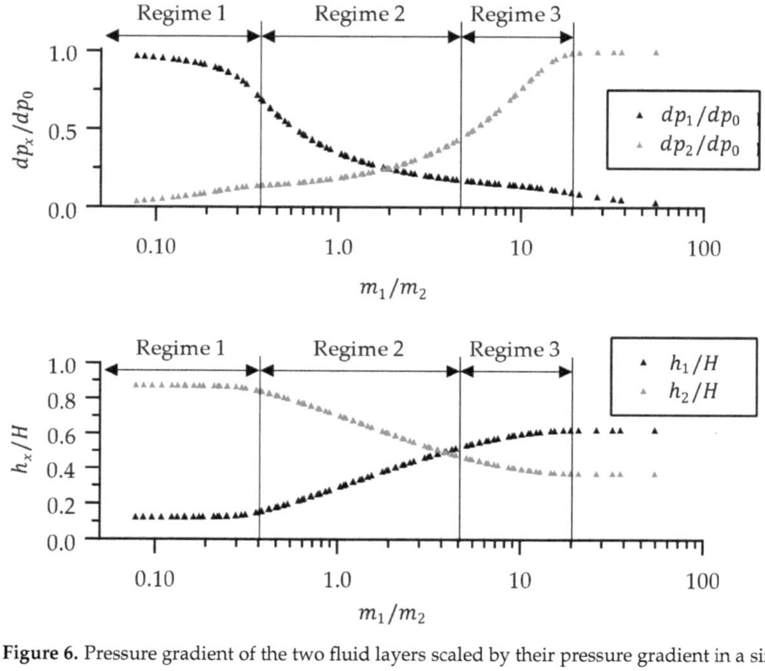

Figure 6. Pressure gradient of the two fluid layers scaled by their pressure gradient in a single layer setup (**top**) and their representative height scaled by the total height (**bottom**) as a ratio of their consistency indices with the indication of three flow zones.

5.2. Processing Window Resutling from Part and Material Designs

In composite processing, the design of the insert plays a crucial role in achieving the desired properties. Therefore, a processing window is defined for inserts with varying layer heights. Figure 7 displays the radii ratios after the initial timestep, representing the initial flow ratios of the layers while neglecting edge effects. Three setups are examined: two setups with a total height of 1, for which the insert and overmolding layers are interchanged, and one setup with a total height of 2 and a height ratio of 1.

The graph exhibits the same three zones as observed in the previous results for all setups. This similarity underscores the consistency of the findings based on the chosen dimensionless numbers. The boundaries of the zones are approximated by a linear relationship between the radius ratio and the consistency index ratio. This implies that the transition from one zone to another depends on both the height ratio and the consistency index ratio. Higher height ratios lead to shifts in the transition points towards higher radii ratios and consistency index ratios. Remarkably, the change in total height aligns with the previous results, indicating that the height ratio plays a more significant role in determining the flow ratio than the total height.

By setting the minimization of insert flow as a processing condition, a process window can be defined. It is evident that at a certain consistency index ratio, the flow of the insert ceases. This critical consistency index ratio depends on the height ratio and is higher for thin inserts combined with thick overmolding layers.

These findings emphasize the influence of height ratio and consistency index ratio on flow behavior and provide insights into defining process windows for composite processing. The results highlight the importance of optimizing the design parameters, particularly the height ratio for a given material, to control the flow ratios and achieve the desired flow characteristics in composite materials.

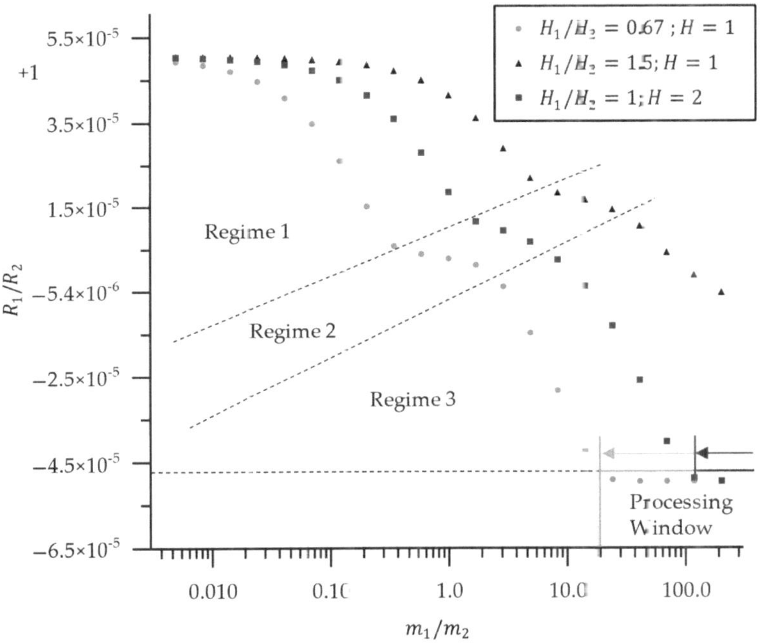

Figure 7. Ratio of the radii for the combination of a pseudo-plastic and a Newtonian fluid layer at different height ratios as a ratio of their consistency indices with the indication of three flow zones and a process window.

5.3. Model-Based Optimization of Process and Part Design

The model shows two extreme cases. The first is a combination of a thin insert with a high consistency index and a thick overmolding layer. The combination of a sufficient consistency index ratio and height ratio can lead to a complete cessation of flow in the insert. This corresponds to the typical application of an organo sheet based on woven continuous fibers, yielding a high consistency index and a thick overmolding layer, such as for a rib structure.

When the material design is fixed, such as in tailored materials, the height ratio becomes increasingly important, especially for lower consistency index ratios. This means that for well-known examples of inserts like woven or cross-ply configurations, the flow of the insert is not expected to occur as the critical consistency index ratio is not exceeded. However, for selectively placing composites in load-bearing areas and tailoring fiber direction and length accordingly, such as for inserts made of long fiber thermoplastics (LFT) or unidirectional (UD) reinforcements, special attention must be given to the process and part design to prevent insert flow.

In that case, the thickness can be reduced by changing the processing conditions. The thickness can be further reduced by using pre-heated inserts, where only the surface is molten in a pre-heating step. This is depicted in Figure 8. The overmolding layer is not affected by the pre-heating step, but the insert is. Only the molten layer that exceeds the melting temperature is considered part of the squeeze flow setup. Therefore, the layer height that counts for the model is reduced significantly, reducing the expected flow of the insert.

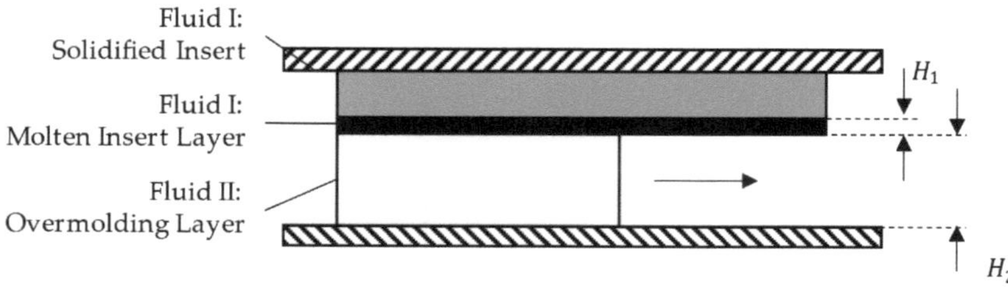

Figure 8. Schematic representation of an overmolding setup with an overmolding layer and an insert, where the insert consists of a molten and a solidified layer.

Furthermore, the flow of the insert can be reduced by lowering the pressure gradient in directions in which no flow is required and which are critical for the insert, such as perpendicular to the fiber direction for UD materials. This can be controlled by a suitable cavity or tool design that lowers the gap height in the desired directions. Figure 9 shows the assembly with a tool that partly covers the edges of the insert. The tool blocks the insert from flowing sideways. However, it remains to be demonstrated whether the reduction of the gap height directly leads to a decrease in flow resistance.

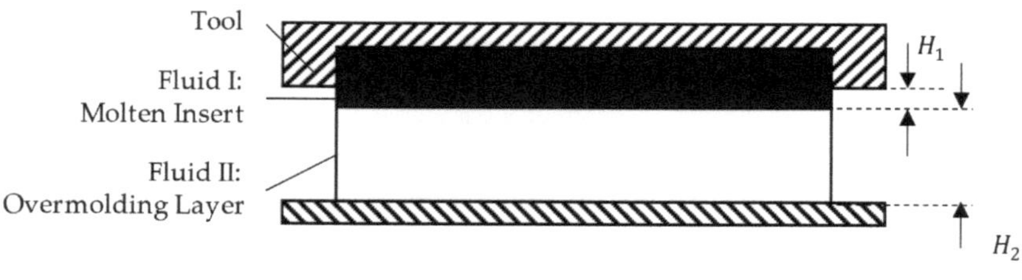

Figure 9. Schematic representation of an overmolding setup with an overmolding layer and an insert, where part of the insert height is covered by a tool.

5.4. Comparison to Applications and Experimental Results

Several studies have investigated the overmolding process with varying materials and part geometries [19]. The main focus of these studies is the mechanical properties resulting from the buildup of an interface between the overmolding material and the insert. However, the flow of the insert is not being observed or reported for short fiber reinforced, long fiber reinforced, UD, or multi-axial fiber reinforced inserts for different cavity designs. Therefore, these setups should represent processing windows that fall under the above model. Three reasons have been identified for this.

First, the ratio of fiber content between the overmolding material and the insert. The consistency index increases with filler content, as empirically studied for different matrix systems and fillers [20]. This, in principle, increases the consistency index ratio for combinations of unfilled and fiber-filled materials. The same can account for combinations of filled materials with different fiber content, such as LFT overmolding materials combined with UD inserts with a 50% higher filler content [21]. In both cases, the consistency index ratio is moved towards the processing window in Figure 7.

Second, a common shape for the overmolding material is ribs locally attached to a thin laminate, e.g., with height ratios above 10 [6,21,22]. This corresponds to the high height ratios in Figure 7 and the tool design in Figure 9. This shifts the process window to even higher consistency index ratios and enables the combination of a thermoforming process in which the laminate is completely heated through the thickness [22].

Third, for complete overmolding or lower height ratios, the insert is not thermoformed and only heated to the temperature of the mold [23–25] or at the surface [26]. This corresponds to a thin molten layer on the insert for welding it with the overmolding layer, as shown schematically in Figure 8.

Further experimental investigation is required to identify the limits of tailoring composite inserts in overmolding parts and material designs. This can be achieved by exploring a broader set of materials and geometry.

6. Conclusions and Outlook

The analytical solution for the squeeze flow of two layers of a power-law fluid shows the interplay between consistency index ratios, height ratios, material design, and part design, which plays a crucial role in ensuring successful processing and achieving the desired properties in the manufacturing and overmolding of thermoplastic composite parts. Composite inserts, modeled as pseudo-plastic materials, show extensive flow compared to the overmolding layer when a critical consistency index ratio is reached. Exceeding this critical ratio is particularly significant for low height ratios between the overmolding layer and the insert. As this combination of parameters is more typical for tailored inserts, the right processing becomes crucial for their implementation. The right processing can include the reduction of flowable layer height by local heating of the insert or an optimized cavity design for a reduced radial pressure gradient.

The model includes some simplifications to reach an analytical solution. Therefore, the results should be seen as a guideline for the most important parameters rather than for predicting the deformation of an insert. To account for the non-isothermal effects of injection overmolding, it is important to consider the effective layer thicknesses in non-steady-state conditions and temperature gradients. These effects arise from using inserts that are pre-heated and molds that are cooler than the solidification temperatures of the materials. In addition, the model can be linked to models for the formation of an interface between the overmolding material and the insert. For more predictive models, the material model can be extended to advanced pseudo-plastic material models and evaluated by finite element models and experiments. Furthermore, this approach can address anisotropic material behavior, particularly for unidirectional inserts, which play a crucial role in load-adapted reinforcements.

Author Contributions: Conceptualization, P.K.W.P., T.A.O. and S.Z.; methodology, P.K.W.P. and T.A.O.; validation, P.K.W.P.; formal analysis, P.K.W.P.; investigation, P.K.W.P.; resources, S.Z. and K.D.; data curation, P.K.W.P.; writing—original draft preparation, P.K.W.P.; writing—review and editing, P.K.W.P., T.A.O., S.Z. and K.D.; visualization, P.K.W.P.; supervision, T.A.O., S.Z. and K.D.; All authors have read and agreed to the published version of the manuscript.

Funding: This research received no external funding.

Data Availability Statement: Data are available on request from the authors.

Conflicts of Interest: The authors declare no conflict of interest.

Appendix A

In this appendix, the solutions for the two sets of constitutive equations are given. The two sets of constitutive equations, which are identical but separately defined for each layer as denoted by the index "i", are given in Equations (A1)–(A3).

$$\text{Scaling}: \tilde{z}_i = \frac{z_i}{h_i}; \tilde{v}_{zi} = \frac{v_{zi}}{\dot{h}_i}$$

$$\text{Continuity}: \frac{1}{r}\frac{\partial r v_{ri}}{\partial r} + \frac{\dot{h}_i}{h_i}\frac{\partial v_{zi}}{\partial \tilde{z}_i} = 0 \tag{A1}$$

$$\text{Motion } (r): 0 = -\frac{\partial p}{\partial r} + \frac{1}{h_i}\frac{\partial \tau_{z_i r}}{\partial \tilde{z}_i} \tag{A2}$$

$$\text{Motion } (z): 0 = -\frac{\partial p}{\partial \tilde{z}_i} \tag{A3}$$

The boundary conditions encompass a no-slip condition (Equation (A5)), denoted by a radial velocity of zero, at the wall. At the centerline, there exists no velocity gradient in the radial direction (Equation (A4)), resulting in a zero z-velocity component at the same location (Equation (A6)). Additionally, the z-velocity assumes a value equal to the closing speed at the wall (Equation (A7)).

$$\text{I}: \frac{\partial v_{ri}}{\partial \tilde{z}_i} = 0 \; at \; \tilde{z}_i = 0 \tag{A4}$$

$$\text{II}: v_{ri} = 0 \; at \; \tilde{z}_i = 1 \tag{A5}$$

$$\text{III}: \tilde{v}_{z_i} = 0 \; at \; \tilde{z}_i = 0 \tag{A6}$$

$$\text{IV}: \tilde{v}_{z_i}\left(\tilde{z}_i = -1\right) = 1 \tag{A7}$$

Consequently, a solution is obtained for the pressure gradient (Equation (A8)), velocity in the radial direction (Equation (A9)), and velocity in the z-direction (Equation (A10)). The representative heights and closing speeds of each layer serve as unknown variables. To determine the shear stress, a material model based on a power-law fluid is employed. Consequently, the resulting equations (Equations (A8)–(A10)) are analogous to the Stefan equation [27] obtained in a single-layer configuration, but they incorporate a representative layer height h_1 and a respective closing rate.

$$\frac{\partial p}{\partial r} = \left(-\dot{h}_i\right)^{n_i}\left(\frac{r}{2}\right)^{n_i}\left(\frac{1+2n_i}{n_i}\right)^{n_i}\left(\frac{1}{h_i}\right)^{1+2n_i} m_i \tag{A8}$$

$$v_{ri} = \frac{1}{2}\frac{1+2n_i}{n_i-1}\frac{\left(-\dot{h}\right)}{h} r\left(\left|\tilde{z}_i\right|^{\frac{1}{n_i}+1} - 1\right) \tag{A9}$$

$$\tilde{v}_{z_i} = \frac{1+2n_i}{n_i+1}\left(\frac{n_i}{1+2n_i}\left|\tilde{z}_i\right|^{\frac{1}{n_i}+1}\tilde{z}_i - \tilde{z}_i\right) \tag{A10}$$

Appendix B

In this appendix, the mathematical proof of Equation (5) is provided.

From Equations (1) and (A8):

$$\begin{aligned} -\dot{h}_1^{\,n_1}\left(\tfrac{r}{2}\right)^{n_1}\tfrac{m_1}{H_1^{1+2n_1}}\left(\tfrac{1+2n_1}{n_1}\right)^{n_1}\tfrac{H_1}{h_1}^{1+2n_1} \\ = -\dot{h}_2^{\,n_2}\left(\tfrac{r}{2}\right)^{n_2}\tfrac{m_2}{H_2^{1+2n_2}}\left(\tfrac{1+2n_2}{n_2}\right)^{n_2}\tfrac{H_2}{h_2}^{1+2n_2} \end{aligned} \tag{A11}$$

From Equations (2) and (A8):

$$\frac{H_1}{h_1} = Y\frac{H_2}{h_2} \tag{A12}$$

$$Y = -\frac{\frac{H_1}{H_2}}{1 - \frac{H_2}{h_2}\left(1 + \frac{H_1}{H_2}\right)} \tag{A13}$$

From Equations (4) and (A10):

$$\begin{aligned} &\frac{\dot{h}_1}{\dot{h}}\left(1 + \frac{1+2n_1}{1+n_1}\left(1 - \frac{H_1}{h_1}\right)\left(\frac{n_1}{1+2n_1}\left|1 - \frac{H_1}{h_1}\right|^{\frac{1}{n_1}+1} - 1\right)\right) \\ &\quad + \frac{\dot{h}_2}{\dot{h}}\left(1 + \frac{1+2n_2}{1+n_2}\left(1 - \frac{H_2}{h_2}\right)\left(\frac{n_2}{1+2n_2}\left|1 - \frac{H_2}{h_2}\right|^{\frac{1}{n_2}+1} - 1\right)\right) \\ &\quad = 2 \end{aligned} \tag{A14}$$

From Equations (3) and (A9):

$$\frac{\dot{h}_1}{\dot{h}}\frac{1+2n_1}{1+n_1}\frac{H_1}{h_1}\left(\left|1 - \frac{H_1}{h_1}\right|^{\frac{1}{n_1}+1} - 1\right) = \frac{\dot{h}_2}{\dot{h}}\frac{1+2n_2}{1+n_2}\frac{H_1}{H_2}\frac{H_2}{h_2}\left(\left|1 - \frac{H_2}{h_2}\right|^{\frac{1}{n_2}+1} - 1\right) \tag{A15}$$

From Equations (A15) and (A12):

$$\frac{\dot{h}_1}{\dot{h}} = -\frac{\dot{h}_2}{\dot{h}}X_1X_2 \tag{A16}$$

$$X_1 = \frac{\frac{1+2n_2}{1+n_2}\left(\left|1 - \frac{H_2}{h_2}\right|^{\frac{1}{n_2}+1} - 1\right)}{\frac{1+2n_1}{1+n_1}\left(\left|1 - \frac{H_2}{h_2}\right|^{\frac{1}{n_1}+1} - \left|1 - \frac{H_2}{h_2}\left(1 + \frac{H_1}{H_2}\right)\right|^{\frac{1}{n_1}+1}\right)} \tag{A17}$$

$$X_2 = \left|1 - \frac{H_2}{h_2}\left(1 + \frac{H_1}{H_2}\right)\right|^{\frac{1}{n_1}+1}\left(1 - \frac{H_2}{h_2}\left(1 + \frac{H_1}{H_2}\right)\right) \tag{A18}$$

$$\left(1 - \frac{H_2}{h_2}\left(1 + \frac{H_1}{H_2}\right)\right) < 0 \tag{A19}$$

From Equations (A19) and (A18):

$$X_2 = -\left|1 - \frac{H_2}{h_2}\left(1 + \frac{H_1}{H_2}\right)\right|^{\frac{1}{n_1}+2} \tag{A20}$$

From Equations (A14) and (A16):

$$\frac{\dot{h}_2}{\dot{h}} = \frac{2}{Z} \tag{A21}$$

$$\begin{aligned} Z = &\left(\left(1 + \frac{1+2n_2}{1+n_2}\left(1 - \frac{H_2}{h_2}\right)\left(\frac{n_2}{1+2n_2}\left|1 - \frac{H_2}{h_2}\right|^{\frac{1}{n_2}+1} - 1\right)\right)\right. \\ &\left. - X_1X_2\left(1 + \frac{1+2n_1}{1+n_1}\left(1 - Y\frac{H_2}{h_2}\right)\left(\frac{n_1}{1+2n_1}\left|1 - Y\frac{H_2}{h_2}\right|^{\frac{1}{n_1}+1} - 1\right)\right)\right) \end{aligned} \tag{A22}$$

From Equations (A11), (A21) and (A16):

$$\begin{aligned} -\left(-\tfrac{2}{Z}X_1X_2\right)^{n_1} &\left(\frac{\dot{h}r}{2}\right)^{n_1}\frac{m_1}{H_1{}^{1+2n_1}}\left(\frac{1+2n_1}{n_1}\right)^{n_1}\left(Y\frac{H_2}{h_2}\right)^{1+2n_1} \\ &= -\left(\tfrac{2}{Z}\right)^{n_2}\left(\frac{\dot{h}r}{2}\right)^{n_2}\frac{m_2}{H_2{}^{1+2n_2}}\left(\frac{1+2n_2}{n_2}\right)^{n_2}\frac{H_2}{h_2}^{1+2n_2} \end{aligned} \tag{A23}$$

From Equations (A13), (A18) and (A23):

$$\begin{aligned}\left(\frac{\dot{h}r}{2H^2}\left(\frac{H}{H_1}\right)^2\right)^{n_1} & \frac{m_1}{m_2}\left(\frac{1+2n_1}{n_1}\right)^{n_1}\left(\frac{2X_1}{Z}\right)^{n_1}\frac{{H_2}^{2n_2}}{{H_1}^{2n_1}}\left(\frac{H_1}{H_2}\right)^{2n}\frac{H_2}{h_2}^{2n_1} \\ &= \left(\frac{\dot{h}r}{2H^2}\left(\frac{H}{H_2}\right)^2\right)^{n_2}\left(\frac{1+2n_2}{n_2}\right)^{n_2}\left(\frac{2}{Z}\right)^{n_2}\frac{H_2}{h_2}^{2n_2}\end{aligned} \tag{A24}$$

$$H = H_1 + H_2 \tag{A25}$$

From Equations (A24) and (A25):

$$\left(\frac{\dot{h}r}{2H^2}\right)^{n_1-n_2}\frac{m_1}{m_2}\left(\left(1+\frac{H_1}{H_2}\right)^2\left(\frac{H_2}{h_2}\right)^2\frac{2X_1}{Z}\right)^{n_1-n_2}\left(\frac{n_2}{1+2n_2}\right)^{n_2}\left(\frac{1+2n_1}{n_1}\right)^{n_1} - 1 = 0 \tag{A26}$$

References

1. Ahmad, H.; Markina, A.A.; Porotnikov, M.V.; Ahmad, F. A review of carbon fiber materials in automotive industry. *IOP Conf. Ser. Mater. Sci. Eng.* **2020**, *971*, 032011. [CrossRef]
2. Mehl, K.; Schmeer, S.; Motsch-Eichmann, N.; Bauer, P.; Müller, I.; Hausmann, J. Structural Optimization of Locally Continuous Fiber-Reinforcements for Short Fiber-Reinforced Plastics. *J. Compos. Sci.* **2021**, *5*, 118. [CrossRef]
3. Lowe, J. *Woodhead Publishing Series in Textiles, Design and Manufacture of Textile Composites*; Long, A.C., Ed.; Woodhead Publishing: Cambridge, UK, 2005; Chapter 11; pp. 405–423.
4. Jespersen, S.T. Methodology for evaluating new high volume composite manufacturing technologies. Ph.D. Thesis, École Polytechnique Fédérale de Lausanne, Lausanne, Switzerland, 2008. [CrossRef]
5. Lässig, R.; Eisenhut, M.; Mathias, A.; Schulte, R.T.; Peters, F.; Kühmann, T.; Waldmann, T.; Begemann, W. *Serienproduktion von hochfesten Faserverbundbauteilen: Perspektiven für den Deutschen Maschinen- und Anlagenbau*; Roland Berger: Munich, Germany, 2012.
6. Akkerman, R.; Bouwman, M.; Wijskamp, S. Analysis of the Thermoplastic Composite Overmolding Process: Interface Strength. *Front. Mater.* **2020**, *7*, 27. [CrossRef]
7. Bird, R.B.; Armstrong, R.C.; Hassager, O. *Dynamics of Polymeric Liquids*, 2nd ed.; Wiley: New York, NY, USA, 2010.
8. Gibson, A.G.; Toll, S. Mechanics of the squeeze flow of planar fibre suspensions. *J. Non-Newton. Fluid Mech.* **1999**, *82*, 1–24. [CrossRef]
9. Nadai, A.; Hodge, P. Theory of Flow and Fracture of Solids, vol. II. *J. Appl. Mech.* **1963**, *30*, 640. [CrossRef]
10. Servais, C.; Luciani, A.; Manson, J.E. Squeeze flow of concentrated long fibre suspensions: Experiments and model. *J. Non-Newton. Fluid Mech.* **2002**, *104*, 165–184. [CrossRef]
11. Coussot, P. Yield stress fluid flows: A review of experimental data. *J. Non-Newton. Fluid Mech.* **2014**, *211*, 31–49. [CrossRef]
12. Yang, F.; Pitchumani, R. Healing of Thermoplastic Polymers at an Interface under Nonisothermal Conditions. *Macromolecules* **2002**, *35*, 3213–3224. [CrossRef]
13. Tierney, J.; Gillespie, J.W. Modeling of In Situ Strength Development for the Thermoplastic Composite Tow Placement Process. *J. Compos. Mater.* **2006**, *40*, 1487–1506. [CrossRef]
14. Lee, W.I.; Springer, G.S. A Model of the Manufacturing Process of Thermoplastic Matrix Composites. *J. Compos. Mater.* **1987**, *21*, 1017–1055. [CrossRef]
15. Barnes, J.A.; Cogswell, F.N. Transverse flow processes in continuous fibre-reinforced thermoplastic composites. *Composites* **1989**, *20*, 38–42. [CrossRef]
16. Jones, R.S.; Wheeler, A.B. Transverse Flow of Fibre-Reinforced Composites. In *Third European Rheology Conference and Golden Jubilee Meeting of the British Society of Rheology*; Springer: Berlin/Heidelberg, Germany, 1990; pp. 258–260.
17. Heinle, M.; Drummer, D. Measuring mechanical stresses on inserts during injection molding. *Adv. Mech. Eng.* **2015**, *7*, 1–6. [CrossRef]
18. Lee, S.J.; Peleg, M. Lubricated and nonlubricated squeezing flow of a double layered array of two power-law liquids. *Rheol. Acta* **1990**, *29*, 360–365. [CrossRef]
19. Aliyeva, N.; Sas, H.S.; Okan, B.S. Recent developments on the overmolding process for the fabrication of thermoset and thermoplastic composites by the integration of nano/micron-scale reinforcements. *Compos. Part A* **2021**, *149*, 106525. [CrossRef]
20. Polychronopoulos, N.D.; Vlachopoulos, J. Polymer Processing and Rheology. In *Polymers and Polymeric Composites: A Reference Series Book Series*; Springer: Berlin/Heidelberg, Germany, 2019. [CrossRef]
21. Giusti, R.; Lucchetta, G. Analysis of the welding strength in hybrid polypropylene composites as a function of the forming and overmolding parameters. *Polym. Eng. Sci.* **2017**, *58*, 592–600. [CrossRef]
22. Joppich, T.; Menrath, A.; Henning, F. Advanced Molds and Methods for the Fundamental Analysis of Process Induced Interface Bonding Properties of Hybrid, Thermoplastic Composites. *Procedia CIRP* **2017**, *66*, 137–142. [CrossRef]

23. Fu, L.; Ding, Y.; Weng, C.; Zhai, Z.; Jiang, B. Effect of working temperature on the interfacial behavior of overmolded hybrid fiber reinforced polypropylene composites. *Polymer Testing* **2020**, *91*, 106870. [CrossRef]
24. Andrzejewski, J. The Use of Recycled Polymers for the Preparation of Self-Reinforced Composites by the Overmolding Technique: Materials Performance Evaluation. *Sustainability* **2023**, *15*, 11318. [CrossRef]
25. Joo, S.-J.; Yu, M.-H.; Kim, W.S.; Lee, J.-W.; Kim, H.-S. Design and manufacture of automotive composite front bumper assemble component considering interfacial bond characteristics between over-molded chopped glass fiber polypropylene and continuous glass fiber polypropylene composite. *Compos. Struct.* **2020**, *236*, 111849. [CrossRef]
26. Matsumoto, K.; Ishikawa, T.; Tanaka, T. A novel joining method by using carbon nanotube-based thermoplastic film for injection over-molding process. *J. Reinf. Plast. Compos.* **2019**, *38*, 616–627. [CrossRef]
27. Stefan, J. Versuche über die scheinbare Adhäsion. *Ann. Physik.* **1875**, *230*, 316–318. [CrossRef]

Journal of
Composites Science

Article

Functionalization of Continuous Fiber-Reinforced Thermoplastic Pultrusion Profiles by Welding

Calvin Ebert *, Marcel Nick Dürr and Christian Bonten

Institut für Kunststofftechnik (IKT), University of Stuttgart, Böblingerstraße 70, 70199 Stuttgart, Germany; christian.bonten@ikt.uni-stuttgart.de (C.B.)
* Correspondence: calvin.ebert@ikt.uni-stuttgart.de

Abstract: Highly filled thermoplastic profiles, produced by in situ pultrusion, offer excellent mechanical properties, but further processing is necessary to expand the range of their applications. Due to the thermoplastic matrix, these materials are particularly well-suited for thermal welding processes. However, the high fiber content of up to 70 vol.-% presents a significant challenge in welding, an aspect that has not yet been thoroughly investigated in the existing literature. This study focuses on the further processing of the highly-filled profiles by adapting the classic hot tool welding process with the aim of investigating the underlying welding mechanism. An IR line-emitter is used to melt the PA6 matrix of the fiber-reinforced plastic component while the second adherend (unfilled PA6) is melted with a classic heating element. Afterward, the joints are tested for tensile and bending strength. The results of these mechanical tests demonstrate that a strong bond can be formed between the adherends. The joint strength reached values of up to 39 MPa, which corresponds to a welding factor of 0.81. Optical examination of the weld seam reveals a reason for the mechanical performance. At high joining pressures, a form-fit is created between the continuous fibers in the profile and the welded-on unfilled PA6.

Keywords: thermoplastic welding; in situ pultrusion; fiber-reinforced plastics

1. Introduction

Even though sales of fiber-reinforced plastics have not increased as much as in recent years [1] due to the numerous political and economic crises, these materials remain a key technology in the fight against global warming. Particularly remarkable are the advantages of thermoplastic fiber-reinforced plastics, which not only combine excellent mechanical properties but also offer recyclability and remelting possibilities. Long-fiber-reinforced or even continuous-fiber-reinforced plastics offer high tensile strength values in the fiber direction and can compete with metallic materials [2]. Thanks to these outstanding properties, these materials are already being used in a wide range of applications. For instance, fiber-reinforced plastics find extensive application in wind turbines and are also gaining increasing prominence in the automotive industry.

One way to produce continuous fiber-reinforced plastics is by means of the so-called melt-pultrusion process. In melt-pultrusion, the fiber bundles, usually glass or carbon fibers, are pulled through a shaping tool and impregnated with the highly viscous plastic matrix. In addition to optimized temperature control, the viscosity is also important for good adhesion between the matrix and fibers. The molten plastic hardens in the mold due to cooling and is pulled off and subsequently cut into the desired length. The result is a two-dimensional continuous fiber-reinforced profile [3]. Besides melt-pultrusion, there

 https://doi.org/10.3390/jcs9010006

also is a pultrusion process where the fibers are impregnated with low-molecular-weight monomers and the polymerization takes place within the shaping tool. While mainly thermosets (e.g., based on epoxy resins) are used in classic pultrusion, the production of pultruded profiles from thermoplastics has become the focus of interest in recent years. The reason for this is the possibility of remelting and mechanical recycling of the profiles. As the viscosity of a conventional thermoplastic melt is generally high, it poses difficulties during fiber impregnation. Thus, low-viscosity monomer systems that polymerize within the pultrusion process are beneficial for proper fiber-matrix adhesion. This form of reactive pultrusion is, therefore, known as thermoplastic in situ pultrusion [4]. The reaction of polyamide 6 (PA6) from ε-caprolactam is an important example of this process.

Even though the production of continuous fiber-reinforced profiles using the in situ process is already possible on a larger scale, the use of these profiles in the form of industrial components is not widely established. One reason for this is the low geometrical freedom because of the complex fiber guidance through the tool. Therefore, in addition to improving the pultrusion process, it is necessary to examine the further processing of the profiles more closely in order to enable the production of more complex components through intelligent joining processes. In addition to investigating solutions for connecting profiles, for example, to form larger surface elements, it can also be useful to investigate joining solutions for applying shape and functional elements such as brackets, snap hooks, or screw domes. When looking at the joining of the in situ manufactured profiles, another advantage compared to thermoset-based pultrusion processes becomes clear: due to the meltable thermoplastic matrix, thermal welding processes arise as a possible joining technique. In the following work, the joining of unfilled PA6 on continuous fiber-reinforced in situ pultruded profiles is investigated in detail by means of thermal welding.

1.1. Hot Tool Welding and Welding Mechanisms

Welding is one of the most important joining processes for thermoplastic components. In contrast to adhesive bonding or mechanical joining, the plastic must be transferred to the thermoplastic state by applying energy, which can be done in a variety of ways. As a result, numerous plastic welding processes have been developed and established [5–7].

One of the most common welding processes is hot tool welding. The process is not only widespread but also simple: The adherends are held to a certain pressure p_H against a PTFE-coated heating device, which is heated to a defined temperature T_H. This forms a melt film with a characteristic thickness, the so-called melting length L_0. After the heating time t_H has elapsed, the adherends are retracted from the hot plate and are brought together under a defined load p_J. Subsequently, the formed interface solidifies due to cooling. It is important that the joining pressure is not too high in order to prevent the plastic melt from being excessively squeezed out of the weld area. This would result in a weld seam that can only transfer significantly lower mechanical forces [8]. Many systems, therefore, use position control to better manage this issue. A defined displacement of the adherends, the so-called joining length s_F, is approached and kept constant. In some work, this is also used analogously in the heating phase to set a constant melting length L_0. In addition to many other applications, welding polyethylene (PE) pipes is one of the most common applications on an industrial scale [9].

In the past, many scientists have studied the factors that influence the joint quality in hot tool welding. Barber and Atkinson [10] used tensile tests to determine the best welding parameters for welding PE, polypropylene (PP), and polybutylene. For this purpose, various process parameters, such as the temperature of the heating plate, the joining pressure and the heating and joining times, were systematically varied. To evaluate their data, they used the so-called short-time welding factor f, which was first introduced by Menges and Zohren [11]:

$$f = \frac{\text{yield strength of welded material}}{\text{yield strength of basic material}} \quad (1)$$

Even if the tensile test often cannot adequately represent the loading that acts on the weld seam in real applications, these tests help to describe the quality of the welded joint and to quantitatively evaluate the influence of the process parameters. Bucknall et al. [12] carried out similar tests with PP and other materials, some of which were also filled with fibers. With their experiments, they were able to show that there is an optimum joining length for achieving high strengths. This optimum joining length depends on the heating time and the respective material. It ensures that there is no excessive orientation of the polymer chains orthogonal to the weld seam.

Of particular interest for the present work are the results of Stokes [13,14], who dealt with the hot tool welding of polyamide (PA). For PP and PA, the most relevant values from Stokes and other named researchers are summarized in Table 1. It is noticeable that there is hardly any literature on hot tool welding of unfilled PA and that the achievable welding factors are lower than that for PP, where welded samples can reach the same strength values as the unwelded sample (welding factor f = 1). In addition to these findings, it appears that the ratio ξ between the joining length s_F and the melting length L_0 ($\xi = s_F/L_0$) is an important parameter because it was found that the achievable mechanical properties strongly correlate with ξ [15,16]. Certain theories about the acting mechanisms during welding attempt to provide an explanation for this relationship. A thorough overview of the various hypotheses can be found in the Ph.D. theses of Frick [17] and Bonten [18]. The three most widely known hypotheses are briefly explained below.

One of the most important theories is the diffusion theory, which was first postulated by Voyutstkii and Vakula [19] in 1967. It simply states that the polymer chains can migrate across the interface under favorable conditions and that the chains of the adherends become entangled after a certain time. The basis of this theory is Fick's first law of diffusion [20]:

$$\frac{\partial m}{\partial t} = -D\frac{\partial c}{\partial x} \quad (2)$$

On the left-hand side of the equation, you can see the mass flow, and on the right-hand side, the diffusion coefficient D multiplied by the concentration gradient. Because D depends on the temperature and viscosity, the mass transfer processes are faster at higher temperatures and lower viscosities.

With their theory of viscoelastic contact, Anand et al. [21] have extended the diffusion theory to include the pressure dependency of the auto-adhesion processes. They indicated that the surfaces within the weld seam are deformed by the effect of the welding pressure and postulated that this deformation leads to the formation of additional secondary valence forces. The viscoelastic material properties of plastics are responsible for the fact that this theory is also time-dependent, so the final joint strength of the weld seam is only achieved after a relaxation process.

In his work, Potente [22] stated that, on the basis of these two theories, the maximum joint strengths would be reached only if the joining is performed on a time scale of a few minutes to enable a proper molecular entanglement across the interface. In contrast, the processes in industrial hot tool welding are much faster, but still, very high strengths can be achieved after a few seconds. From this, Potente [22] derived the minimum flow velocity hypothesis, which states that the pronounced flow of the melt in the joining phase not only leads to mixing due to thermally induced diffusion processes but that the molecules are also mechanically entangled with each other. All three theories can be used to justify the significance of ξ. Optimal thermal conditions at the interface favor diffusion and viscoelastic processes, and an optimal ξ ensures better mixing of the adherends.

In addition to these theories, it is important to mention that the crystallization behavior of semi-crystalline polymers is a parameter to be considered. In several publications, cocrystallization inside the weld seam is assigned an important role in the efficiency of welded joints [23–25]. Because welding is conducted above the crystallite melting temperature, the polymer chains exist in an amorphous state. During the joining phase, as the chains cool, mixed recrystallization between the adherends can occur, forming a region of cocrystallization that enhances force transmission within the weld seam [25].

1.2. Welding of Fiber-Reinforced Plastics

The strength of fiber-reinforced plastics is significantly influenced by the mechanical properties of the fibers and their orientation in the component. The fiber orientation in the weld seam is, therefore, also decisive for the welded joint and its strength in load direction. Table 1 lists welding factors for reinforced plastics from the literature. When analyzing the data from the literature, it becomes clear that the achievable strength is significantly lower than the strength for unreinforced plastics. The reason for this is the transverse flow in the joining phase, which is responsible for the fibers being aligned transversely to the seam. As a result, the reinforcing effect of the fibers is lost, and significantly lower weld seam factors can be achieved. It also shows that the optimum ξ for fiber-reinforced plastics is significantly lower than for unreinforced plastics. This means that s_F should be smaller in relation to the melting length L_0. In his work on the minimum flow velocity hypothesis, Potente [22] calculated the average flow velocity v_{m2} transverse to the weld seam:

$$v_{m2} = \frac{8}{\pi}\frac{p_J\ L_0^2}{\eta_m\ d} \tag{3}$$

d stands for the width of the weld seam and η_m is the average shear viscosity along the cross-section. In order to achieve the lowest possible velocity transverse to the welding direction, the joining pressure and, thus, the joining length must be limited with a fixed melt thickness. However, a material bond must still be created during welding, which also results in an optimum for ξ while welding fiber-reinforced plastics.

In addition to optimizing the process parameters, another way of increasing the weld seam strength is to improve the initial fiber orientation. This means to align as much of the fiber content as possible along the welding direction during the manufacturing process of the component, e.g., in the injection molding process. Fiebig and Schöppner [26] were able to achieve an improvement in weld seam strength of up to 45% for long-fiber-reinforced PP.

Table 1. Process parameters and welding factors for the hot tool welding of PP and PA found in the literature.

Material	T_H in °C	t_H in s	p_H in MPa	t_J in s	p_J in MPa	σ_t in MPa	f
			L_0 in mm		s_F in mm		
PP	240	25 (35)	0.02 MPa	180	0.1 MPa	n/a	0.99 [10]
PP	260	30	0.2 mm	n/a	0.8 mm	34	0.99 [12]
PP	240	25	n/a	90	1.0 mm	30.8	0.86 [27]
PA66	350	20	0.13 mm	10	0.66 mm	36.7	0.58 [13]

Table 1. *Cont.*

Material	T_H in °C	t_H in s	p_H in MPa	t_J in s	p_J in MPa	σ_t in MPa	f
			L_0 in mm		s_F in mm		
PPGF25	220	60	2.3 mm	n/a	0.5 mm	29	0.43 [15]
PAGF35	280	60	2.7 mm	n/a	1.0 mm	33	0.41 [15]
PA6GF15	270	25	n/a	15	1.0 mm	53.3	0.71 [28]
PA6GF33	380	30	n/a	15	n/a	n/a	0.48 [29]

1.3. Further Processing of Continuous Fiber-Reinforced Profiles

Even though thermoplastic pultrusion has some advantages over classic pultrusion on a thermoset basis, it is still rarely used in industry. Most applications are limited to structural elements such as aircraft floor coverings or bumpers in the automotive sector [30]. In addition, there are a few scientific studies on the use of pultruded materials in medical technology, for example as orthodontic wires [31] or as magnet resonance (MR) guidewire [32]. Another important application is the production of prepregs that are used as the basis for long fiber-reinforced thermoplastics (LFT).

What most applications have in common is that the profiles no longer need to be processed further for integration into an assembly. The main problem is that there is hardly any literature and preliminary work dealing with the behavior of thermoplastic pultruded materials and their further processing. The production of curved profiles is an exception [33]. However, the authors are not aware of any published work that deals with the welding of thermoplastic pultrusion profiles. One notable exception is a prior study carried out by the authors, which investigated functionalization using ultrasonic welding. This approach yielded maximum tensile strengths of up to 12 MPa [34]. This gap in the literature illustrates the necessity for further research in the field of joining thermoplastic pultrusion profiles.

Research on the joining of fiber-reinforced plastic structures has been extensive and multifaceted. However, the majority of studies concentrate on laminate systems manufactured by compression molding [35]. Similar to pultrusion, a significant challenge in the production of these components is the high melt viscosity of the polymer matrix. As a result, complete fiber impregnation can only be achieved up to a fiber volume fraction of 60% [36], which remains below that attainable with in situ pultrusion. Therefore, the results are only comparable to a limited extent. Moreover, these studies primarily focus on joining profiles or other geometries in the form of single lap shear specimens, with which the lap shear strength (LSS) can be determined [37,38]. The functionalization of continuous fiber-reinforced parts or profiles by welding on additional elements has not yet been investigated. Only a limited number of studies have conducted tensile tests. For instance, Goto et al. [39] joined carbon fiber-reinforced PA6 laminates with a fiber volume content of 50% using ultrasonic welding. The tensile strength of the joints was evaluated using the so-called cross-tensile test, which achieved a maximum strength of 7 MPa. Similar tests were carried out by Jandali et al. [40] for vibration welding, although the test specimen geometry was slightly different. For glass fiber-reinforced PA with a fiber volume content of 41.6%, a maximum tensile strength of 11.4 MPa was achieved.

1.4. Objective

In addition to the various theories on welded joints, the state of research also highlighted the special features of welding fiber-filled plastics. In contrast, the fiber content of the profiles used in this work is much higher (approx. 70 vol.-%) than the ones investigated in the literature. There, only the welding behavior of systems up to approx. 40–60 vol.-%

has been investigated in more detail. The weldability of materials with such high fiber contents has not yet been clarified in detail. If there is not enough plastic matrix, the diffusion and relaxation processes of the polymer chains can only take place to a limited extent and are simultaneously hindered by the immobile fibers. Secondly, the fiber orientation is always transverse to the welding direction. Due to the high length of the fibers and the high viscosity of the plastic matrix, they are also unable to change their orientation. It is, therefore, not possible to reinforce the weld seam by reorienting the fibers. The following investigation intends to provide information for the first time on whether welding is possible with such high fiber contents and whether and how the previous explanations of the welding mechanisms need to be expanded to this specific case. Therefore, the objective of this study is to clarify the underlying mechanisms governing the welding process of highly filled thermoplastic materials by systematically varying the process parameters. Additionally, the study aims to establish an initial correlation between the mechanical properties of the welded joint and the selected process parameters.

2. Materials and Methods

2.1. Used Materials and Sample Size

The pultruded profiles made of PA6 were provided by the Fraunhofer Institute for Chemical Technology (ICT), Pfinztal, Germany, and have a glass fiber content of 71 vol.%. The composite material, stored under standard conditions, has a tensile strength orthogonal to the fiber direction of $\sigma_{90°,PA6GF}$ = 32 MPa. Comparable profiles were able to achieve tensile strengths in the fiber direction of $\sigma_{0°,PA6GF}$ = 1182 MPa [41]. The cross-section of the profiles is rectangular with dimensions of 30 mm × 4 mm. The profiles were shortened to a length of 30 mm using a band saw in order to obtain a square sample for the welding tests. Subsequently, the second adherend, which is the unfilled PA6, was injection molded and then cut in two halves. Its geometry is based on the standard for tensile test specimens [42]. The material used for this unfilled PA6 part was of the type Ultramid B3S, manufactured by BASF, Ludwigshafen, Germany. Because the pultruded composite profile, as well as the unfilled adherend consisted of PA6, good chemical compatibility was ensured between both components. Moreover, the used PA6 has good flow properties, which also make it suitable for 3D printing applications [34]. However, it should be noted that the two PA6 types differ in molecular weight and, therefore, in flow behavior. In in situ pultrusion, the PA is produced by anionic polymerization. This brings the decisive advantage that polymerization is significantly faster than hydrolytic polymerization. As a result, the PA hardens faster and can be processed better in a continuous process such as pultrusion. Hydrolytic polymerization is often used for injection molding grades, including Ultramid B3S [43]. Another difference between the two types of polymerization is the temperature. As anionic polymerization takes place at significantly lower temperatures, the amount of low molecular weight components is less, compared to the hydrolytic polymerization [44]. The number average molar mass M_n of the PA6 matrix of the profiles is, therefore, approx. five times greater at 110,640 g/mol than for Ultramid B3S, which has an M_n of around 20,000 g/mol [45,46]. The different molar masses of the adherends will have an influence on the diffusion and flow processes during welding. However, this influence is not investigated in this work and should be subjected to future research.

2.2. Setup and Weld Parameters

In contrast to classic hot tool welding, in which both adherends are heated on the same heating plate, two different heating concepts were used in this work. Figure 1 shows the welding setup used for the experiments. At the top left corner, the setup is shown in the heating phase. The first adherend, the unfilled halved PA6 tension rod, is clamped in sledge 1 (S1). The pultrudate plate is clamped in sledge 2 (S2). A dedicated device was designed to clamp the pultrudate plate, facilitating both the welding tests and the mechanical testing of the joint. The plate is positioned within a square cavity and secured using a metal plate and four screws (see also in Figure 3, left). The sledges and the heating devices can be moved by pneumatic cylinders. The entire process is force-controlled. During the heating phase, the tension rod is pressed against the heating element with a specified force. This force is maintained constant by pneumatic cylinders and can be related to the initial cross-sectional area of the tension rod to determine the heating pressure p_H. S2 is fixed during the heating phase. The pultrudate plate is heated using an IR line focus. This IR-emitter from Opteon GmbH, Garbsen, Germany, has a maximum radiant power per area of I_{max} = 11.5 W/m^2, which is focused on a line at a correctly set focus distance between 50 and 60 mm. The advantage of IR heating is the uniform and focused heating of the pultrudate surface. As this surface is not perfectly flat due to the high production tolerances in pultrusion, uniform heat input with a heating plate is only possible to a limited extent. After the heating time has elapsed, the heaters retract and the adherends can be joined together (see bottom left in Figure 1).

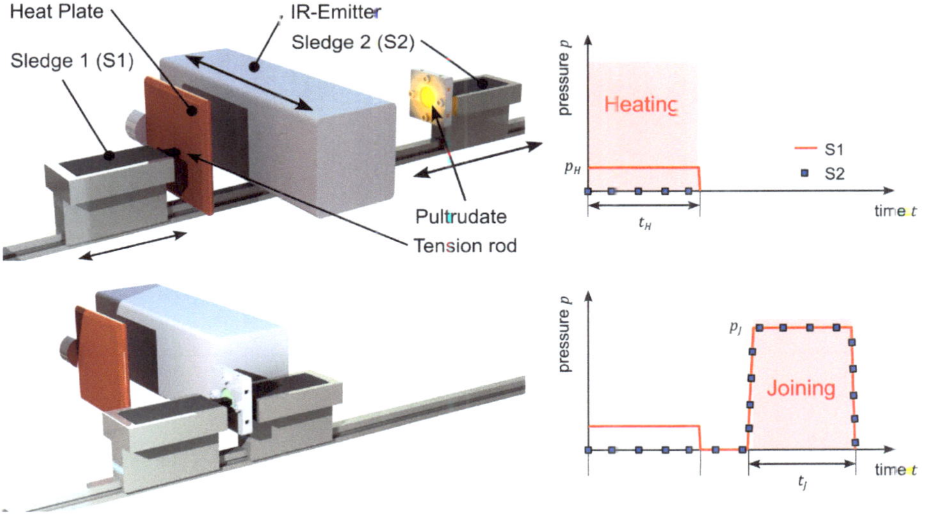

Figure 1. Welding setup in the heating phase (top left corner) and the joining phase (bottom left corner). The diagrams on the right show the corresponding pressure curves of the two pneumatic cylinders of the sledges.

To analyze the influence of the process parameters on the quality of the weld seam, the heating time t_H, the surface power of the IR-emitter P and the joining pressure p_J were varied. P is given in percent of the maximum power of the IR-emitter. In contrast to the heating times and plate temperatures from Table 1, the parameters here have been adapted to the application. The IR-emitter requires some time to bring the pultrudate plate into a molten state. This determines the minimum heating time as examined in preliminary tests. It was not possible to realize a welded joint below a surface power of the IR-emitter of 70%. If P is 75%, a heating time of at least 60 s is required to obtain a weld. These findings from preliminary tests determine the lower limit of the process parameters. A closer examination of the samples shown in Figure 2 allows the determination of the upper limit. The circular discoloration (blue line) on the pultruded plate is visible on both samples. This area, corresponding to the recess in the metal plate that fixes the pultrudate plate during welding, was exposed to the radiant heat from the IR-emitter during the heating phase. The sample on the right, exposed to higher radiant heat for a longer duration than the left one, exhibits pronounced brown discoloration on the surface, indicating decomposition and burn effects. Therefore, no longer heating times or higher radiant heat than shown here were investigated. The parameter set for the conducted tests is derived from these preliminary results and is presented in Table 2. Three joining pressures, two heating times and two different surface powers of the IR-emitter were tested in a full factorial test plan. The joining pressure of 1.5 MPa represents the maximum force achievable with the current setup.

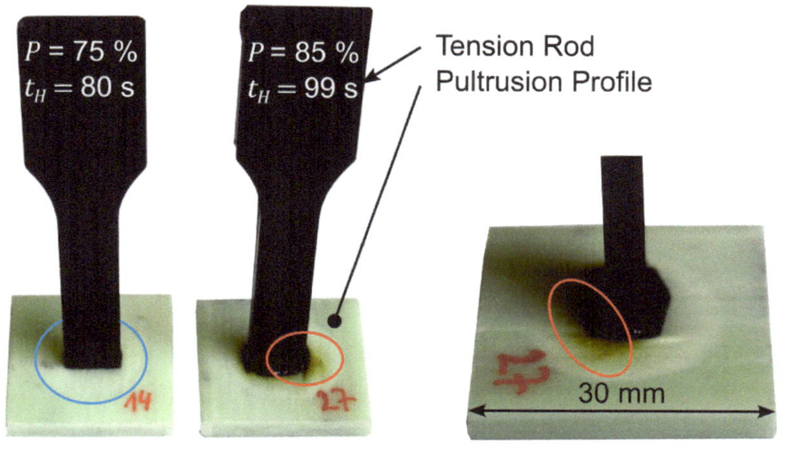

Figure 2. Photographs of welded samples. A detailed view of the right-hand sample is shown on the right.

Table 2. Process parameters for welding.

t_J	in s	30	
p_H	in MPa	0.2	
T_H	in °C	215	
No.	t_H in s	p_J in MPa	P in %
1	80	0.3	75
2	80	0.3	85
3	99	0.3	75
4	99	0.3	85
5	80	0.75	75
6	80	0.75	85
7	99	0.75	75
8	99	0.75	85
9	80	1.5	75
10	80	1.5	85
11	99	1.5	75
12	99	1.5	85

To minimize the influence of the extended heating time on the welding process, a lower heating temperature T_H was chosen for heating the tension rod, as compared to the values reported in literature. In differential scanning calorimetry (DSC) measurement, the melting range of the unfilled PA6 was determined (onset temperature 219.4 °C, end temperature 223.1 °C, start of melting 200 °C) and T_H was selected so that the temperature lies within this range. The decision was made to only vary the process parameters associated with heating the pultrudate plate. This approach was chosen because the influence of heating time and applied radiation on achieving a processable state for the profiles has not yet been systematically investigated. Nevertheless, in future research tests with different T_H and p_H are also necessary to determine the optimal set of process parameters.

2.3. Characterization of the Welded Samples

Mechanical tests were carried out to assess the quality of the welded joint. Most of the previous work in this area examined the strength in the tensile direction. As the results in this work will be compared with the literature values, tests were also carried out under this type of load with the present samples. In addition, the welded joints were tested under a combined bending and shear load. On the one hand, this type of load is closer to the application, i.e., impacts on functional elements during use are very likely. On the other hand, the comparison of both types of load and their fracture patterns should provide insight into the force transmission inside the weld seam. Both test devices are shown in Figure 3. The welded specimen is positioned and fixed within the custom-designed device. Subsequently, the device is pneumatically clamped into the tensile testing machine.

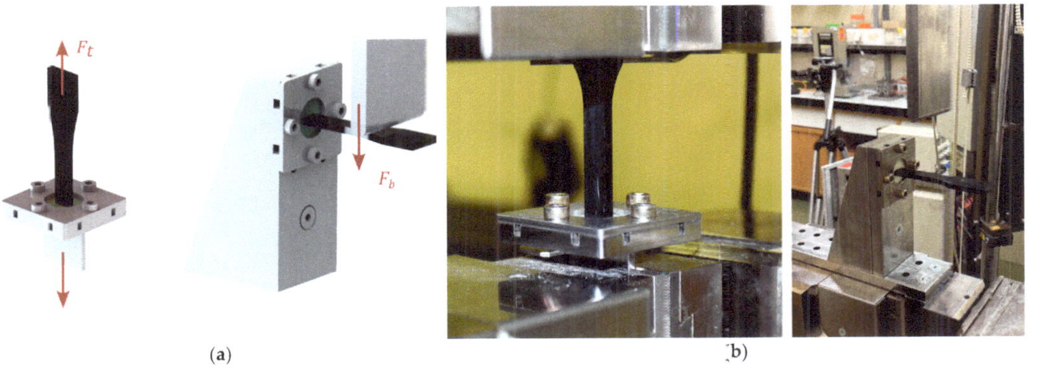

Figure 3. (**a**) Test setups for mechanical characterization: testing under tensile load (**left**) and bending load (**right**); (**b**) photographs of the setups in the mechanical laboratory at IKT.

For the evaluation of the mechanical tests, the tensile strength was calculated with the initial cross-sectional area of 10 mm × 4 mm. Six samples were welded per test series, three of which were subjected to tensile stress and three to bending stress. A lever arm of 30 mm was selected for the bending tests. All samples were stored in a standard climate (23 °C, 50% humidity) for at least 88 h before mechanical testing. Two welding factors are considered in the test evaluation:

$$f_{xu} = \frac{\sigma_{Weld}}{\sigma_{Ref}}, \quad f_{xf} = \frac{\sigma_{Weld}}{\sigma_{90^\circ,PA6GF}} \tag{4}$$

f_{xu} relates the weld seam strength to the strength of the unfilled partner, whereas f_{xf} relates to the strength of the pultrudate. x indicates the load direction with b for bending and t for tensile load. The mechanical properties were characterized using the UPM 1455 universal testing machine from Zwick Roell AG, Ulm, Germany. The used local cell has a maximum force of 20 kN. The test speeds for the tensile tests were kept constant at 5 mm/min and for the bending tests at 2 mm/min.

Optical analyses of the weld seam were carried out using the Leitz DMRXP microscope from Leitz Camera AG, Wetzlar, Germany.

3. Results and Discussion

Figure 4 shows the results of the tensile tests. In Figure 4a, the heating time is plotted on the x-axis and the calculated tensile stress on the y-axis for P = 75%. Similarly, Figure 4b for P = 85%. The lower limit of the error bars represents the lowest absolute value, while the upper limit indicates the highest value. The different joining pressures are highlighted in different shades of blue, and the tensile strength of the unfilled PA6 is shown in black as a reference. The results clearly show that high strengths can be achieved for almost all process parameters. At the shortest heating time and lowest radiation intensity, the tensile strength significantly decreases. One sample even broke before the mechanical test. Obviously, this low thermal energy, in combination with low joining pressure, was not sufficient to produce a good weld quality. The highest strength of 39 MPa is achieved with the highest investigated joining pressure of 1.5 MPa. It can also be seen that the joining pressure has a significant influence on the strength of the joint. The greater the joining pressure, the higher the strength. The best weld strengths were realized at the highest joining pressure of 1.5 MPa. This contradicts the current state of research, in which an optimum for the joining pressure for unreinforced thermoplastics is achieved at significantly lower values of around 0.1–0.5 MPa [6,10]. As mentioned in Section 1.2, this applies in particular to fiber-reinforced plastics. In contrast, there is no clear influence of the heating time and the radiation intensity on the strength of the joint. If the obtained results are compared with the reference sample and the welding factor f_{tu} is calculated, the highest value is 0.81. Values greater than 1 are reached for f_{tf}. The absolute values of the tensile strengths are roughly within the range of literature.

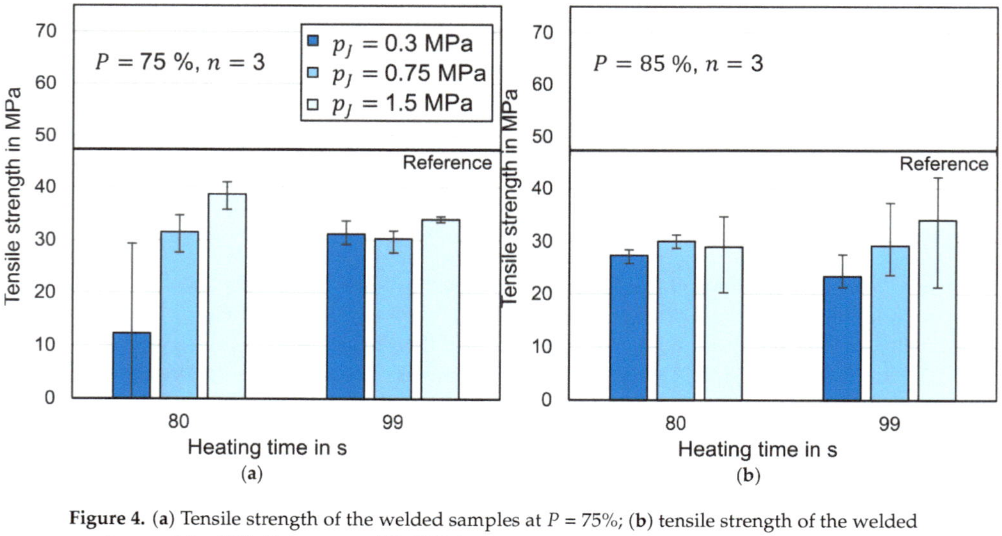

Figure 4. (**a**) Tensile strength of the welded samples at P = 75%; (**b**) tensile strength of the welded specimens at P = 85%. Reference unfilled PA6.

As announced in Section 2.3, not only the load in the tensile direction was considered. Figure 5 shows the results under bending load. The energy introduced by the IR-emitter into the pultrudate is plotted on the x-axis and is calculated as follows:

$$E = \frac{P \cdot I_{max} \cdot t_H}{100} \tag{5}$$

The bending strength is plotted on the y-axis. As for the tensile tests, reference values for the unwelded material were also determined, which are shown in black in the diagram.

For this purpose, a tension rod was clamped on one side, and a two-point bending test was carried out. The diagram shows a similar behavior for all three joining pressures. A maximum is reached at 782 J/m^2. Subsequently, the achievable bending strength drops for a higher applied energy. For the joining pressure of 1.5 MPa, there is a single outlier at 968 J/m^2. The results from the bending test are consistent with the results from the tensile test: The highest strengths were achieved at the highest joining pressures. It is also noticeable that even weld factors f_{bu} above 1 can be achieved. The maximum welding factor was 1.4. This means that the strength of the welded joint is higher than that of the base material. This observation is a strong indication that fibers must be present within the weld interface.

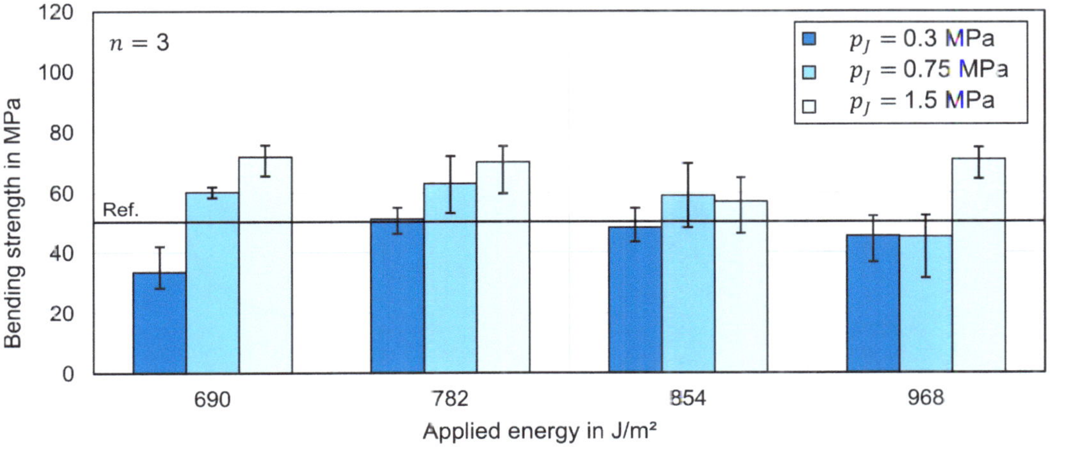

Figure 5. Bending strength of the welded samples plotted against the applied energy on the pultrudate plate.

Figure 6a shows the tensile stress plotted over the deformation for four exemplary samples. The reference sample is in black alongside a sample from series 1 (yellow), a sample from series 3 (red), and a sample from series 11 (blue). The path is plotted logarithmically. Typically, the stress–strain curve is used to determine the stiffness of a material in the linear elastic range. However, comparing the elongation in the weld seam with the elongation of the reference specimen is difficult because the initial lengths are not standardized across both specimens. This discrepancy makes it challenging to directly compare elongation. Therefore, it is more appropriate to compare the stress-deformation curves and their slopes, as they provide a more consistent and meaningful measure of the material's behavior under stress. The diagram shows that the reference specimen is significantly more ductile. It breaks after about 100 mm, whereas the welded samples already break at about 1 mm. The deformation at the yield stress of the reference is an order of magnitude higher than that of the weld seams. In contrast, as already shown in Figure 4, the yield strength itself only differs by approx. 20% to 40% for samples 3 and 11. In addition, it is noticeable that the curves of the welding samples lie exactly on top of each other despite being welded with different welding parameters and differ only in the deformation at the break. The gradient of the reference curve is lower. This indicates that the weld seam behaves more rigidly than the reference material.

To get a more accurate picture of the failure mode, the stress-deformation diagram for the bending test is shown in Figure 6b, analogous to Figure 6a. The reference is shown in black. In contrast to the tensile test, the reference test specimens (unfilled PA6) did not fail under the bending load. After falling below a relative force minimum, the test was aborted. In the welded joints, on the other hand, there was a component failure in the weld seam. Although the resistance to the bending load and, thus, the bending stress are increased, it is no longer possible to achieve such high bending strains. In comparison to the tensile tests, the curves of the welded samples are again similar and reach the stress maximum at smaller deformations than the reference specimen. This indicates a stiffening, which probably occurs due to the influence of the glass fibers. The drop in bending stress after failure is faster for the welded samples, which shows that the welded samples exhibit a fast and an "all at once" failure.

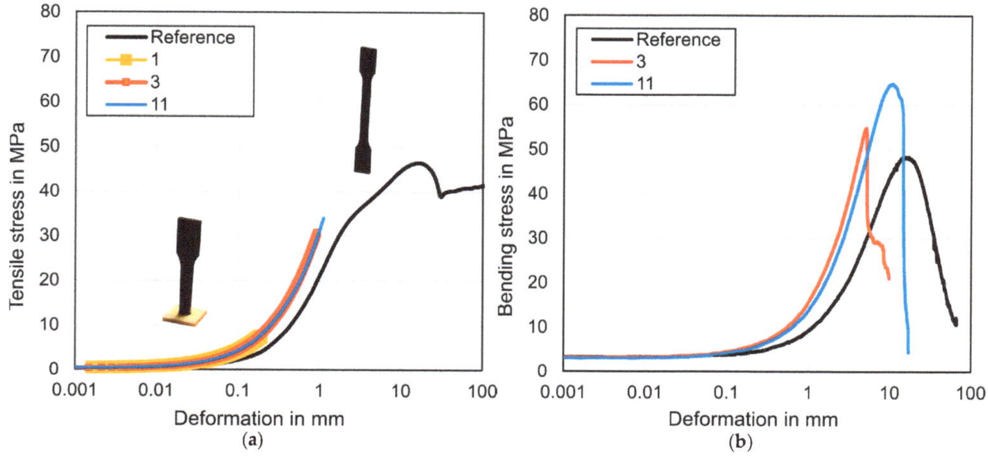

Figure 6. (**a**) Stress-deformation diagram under tensile load; (**b**) stress-deformation diagram under bending load. Reference unfilled PA6.

In order to understand how the welded joints work and how the forces can be transferred, it is necessary to take a look inside the weld seam. Figure 7 shows the corresponding microscopy images of a sample from series 9. This series was selected because it achieved the highest strength in the tensile test (see Figure 4a). In Figure 7a,b, the sample is shown before destruction. Figure 7c–e show the corresponding images after the tensile test and Figure 7f–h after the bending test. To create the images, the samples must first be embedded in resin. That is why three different samples that were all welded with the same parameters are shown. The samples are prepared in such a way that images are taken in the middle of the seam with the fibers pointing into the plane of the sheet.

Figure 7a shows the welded tension rod. The glass fibers can easily be identified as bright spots. It is noticeable that the tension rod has penetrated a few micrometers into the pultrudate. It can also be seen that a weld bead is formed in the edge area, and the pultrudate bulges upwards at this point. This bulging is better visible in the zoomed-in view in Figure 7b. The most outstanding detail is also visible there: The unfilled PA6 of the tension rod and the pultrudate mix during the welding process. This is clearly visible in the red ellipse, where the melt of the tension rod is present between individual glass fibers. This mixing is barely visible in the middle of the tension rod. During welding, the application of the joining pressure results in a cross-flow of the melt, which could be responsible for the good mixing in the boundary area.

Looking at the images taken after a mechanical failure in Figure 7c–h, it can be seen that the fracture zone runs along the area where the glass fibers have mixed with the polyamide of the tension rod. In the areas with hardly any mixing, the tension rod material is pulled away from the pultrudate, and an adhesion fracture occurs. The image in Figure 7d also clearly shows that although the fracture edge is at the edge of the mixing zone, the force transmission within the weld seam is so good that cracking occurs in the pultrudate. This can be seen from the black voids. The cohesive and adhesive fracture behavior is depicted in the image in Figure 7e. This image and the one below Figure 7h show the fracture surface taken with a standard digital camera. In Figure 7e, it can be seen that the material of the tension rod remains a black frame at the edge (cohesive failure). The bright surface of the pultrudate is visible in the center (adhesive failure).

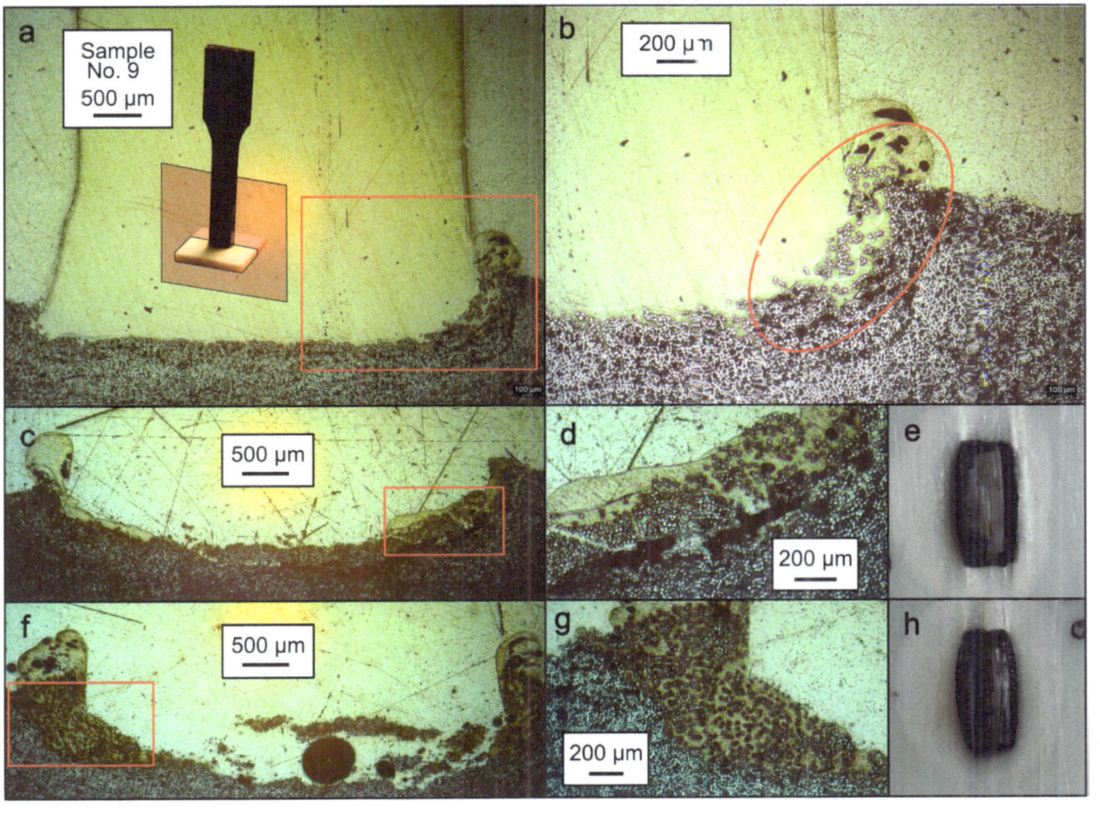

Figure 7. Microscopy images of the weld seam in sectional view. Test samples welded with the parameters of sample No. 9. (**a**,**b**) before applying mechanical stress, (**c**–**e**) after the tensile test, (**f**–**h**) after the bending test.

In contrast to the tensile load, the weld seam is loaded asymmetrically during bending. This asymmetry can also be seen in the images. The left-hand mixing zone (see Figure 7g), which is subjected to compressive stress during bending, is even more pronounced after the fracture than on the right-hand side, which was subjected to tensile stress. On the right, individual fiber strands were even pulled out. The black air bubble, which could not be avoided while embedding the sample in epoxy resin, directly lies under these detached fiber strands.

In order to investigate the influence of the process parameters on the penetration depth and the formation of the mixing zone, samples from series 1 to 10 were embedded and optically examined. Figure 8 shows the corresponding images of the three embedded samples. For the sample from series 1, which had the lowest energy input of 690 J/m^2

and a joining pressure of 0.3 MPa, it is evident that the tension rod did not penetrate the pultrudate plate. The large black air bubbles in the area of the weld seam indicate that the adherends were not aligned during welding. By increasing the joining pressure to 1.5 MPa, the test specimen penetrates into the pultrudate plate, and a mixing zone is formed (see Figure 7b). By increasing the applied radiation energy to 782 J/m^2, the penetration depth can significantly be increased. The penetration depth of the sample from series 9 is around 0.2 mm, while the penetration depth of the sample from series 10 increases to over 0.5 mm. To quantitatively analyze the relationship between penetration depth and process parameters, as well as the relationship between penetration depth and mechanical performance, all sample variations should be embedded. Multiple measurements should be performed to ensure statistical evidence.

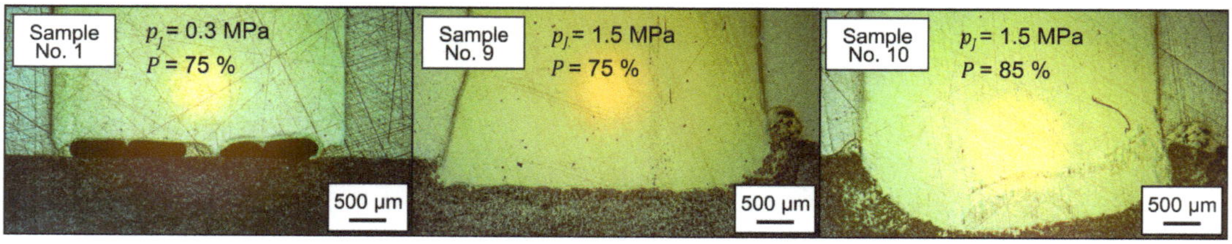

Figure 8. Microscopy images of the weld seam in sectional view. Test samples welded with the parameters of samples No. 1, 9, and 10. Heating time $t_H = 80$ s for all samples.

4. Conclusions

This study investigated the functionalization of continuous fiber-reinforced thermoplastic pultrusion profiles. For this purpose, the classic hot-tool welding process was adopted, and an IR-emitter with line focus was integrated for heating the thermoplastic profiles. The following are the specific conclusions:

- Despite a very high fiber content of 70 vol.-%, a strong welded joint between the unreinforced tension rod and the pultruded plate is possible. The maximum tensile strength achieved, 39 MPa (Figure 4), corresponds to a welding factor of 0.81, surpassing the welding factors for PA reported in the literature [15,28,29]. Moreover, significantly higher strengths were achieved compared to similar studies involving fiber-reinforced composite laminates [39,40].
- The results under tensile and bending loads have shown that the joining pressure is of decisive importance. The strength increases with increasing joining pressure. Heating time and applied radiant heat show no clear influence on the mechanical properties.
- The stress-deformation diagrams (Figure 6) show a steeper gradient for the welded samples compared to the reference specimens. This suggests a stiffening effect due to fiber interaction.
- Unlike classical welding theories for polymers, which focus on molecular interactions and chain entanglements, the present case involves a macroscopic mechanism. During welding, cross-flow causes mixing between the continuous fiber-reinforced pultrudate and unfilled PA6, creating a form fit between the fibers and the PA6 (Figure 7).
- The penetration depth depends on the process parameters. It increases with higher joining pressure and greater energy input.

These findings open up new questions for investigation. Accordingly, the authors would like to propose several suggestions for future research:

- The viscosity will have an influence on the formation of the mixing zone and the penetration depth. In addition, the crystallization behavior should be examined more closely in the future. Both parameters depend crucially on the temperature conditions during the process. In order to better understand the processes and force transmission, investigations with polyamides of different molecular weights are required here.

- A deeper analysis of the dependence between the penetration depth and the mechanical properties is needed. For this purpose, the penetration depth should be quantitatively determined as a function of the process parameters. This could help to define an optimal set of process parameters. The inclusion of T_H and p_H would enhance the experimental design.
- The welding of material systems, which consist of chemically diverging polymers, is usually a challenge since these systems often are incompatible on a molecular scale. Because the revealed welding mechanism acts on a macroscopic level, the welding of fiber-reinforced systems where both adherends consist of a different polymer might still result in acceptable weld quality.

Author Contributions: Conceptualization, C.E.; methodology, C.E.; formal analysis, C.E.; investigation, C.E. and M.N.D.; writing—original draft preparation, C.E.; writing—review and editing, M.N.D. and C.B.; visualization, C.E.; supervision, C.B.; project administration, C.B.; funding acquisition, C.B. All authors have read and agreed to the published version of the manuscript.

Funding: The results were generated as part of the "ModE-SoFa" project. This project (FKZ 01MV22016F) is funded by the German Federal Ministry for Economic Affairs and Climate Action (BMWK) on the basis of a decision by the German Bundestag.

Data Availability Statement: Data can be made available upon request to the authors.

Acknowledgments: The authors would like to thank the project partners from the Fraunhofer Institute for Chemical Technology (ICT), Pfinztal, Germany, for providing the pultrusion profiles.

Conflicts of Interest: The authors declare no conflicts of interest. The funders had no role in the design of the study, in the collection, analyses, or interpretation of data, in the writing of the manuscript, or in the decision to publish the results.

References

1. Witten, E.; Mathes, V. *The European Market for Fiber-Reinforced Plastics/Composites 2023: Market Developments, Trends, Challenges and Prospects*; AVK: Frankfurt, Germany, 2024
2. Bonten, C. *Plastics Technology: Introduction and Fundamentals*; Hanser Publishers: Munich, Germany, 2019; ISBN 9781569907689.
3. Meyer, R.W. *Handbook of Pultrusion Technology*; Springer: Boston, MA, USA, 1985.
4. Epple, S. In-Situ Pultrusion von Polymaid 6 zur lokalen, Lastpfadgerechten Verstärkung von Polyamid-Spritzgießbauteilen. Ph.D Thesis, University of Stuttgart, Stuttgart, Germany, 2018.
5. Rotheiser, J. *Joining of Plastics: Handbook for Designers and Engineers*, 3rd ed.; Hanser: Munich, Germany, 2009; ISBN 9783446445956.
6. Bonten, C.; Benatar, A.; Grewell, D.; Tuechert, C. *Welding*; Hanser: Munich, Germany, 2001; ISBN 3-446-21695-2.
7. Stokes, V.K. Joining methods for plastics and plastic composites: An overview. *Polym. Eng. Sci.* **1989**, *29*, 1310–1324. [CrossRef]
8. Grewell, D.; Benatar, A. Welding of Plastics: Fundamentals and New Developments. *Int. Polym. Process.* **2007**, *22*, 43–60. [CrossRef]
9. Troughton, M. *Handbook of Plastics Joining. A Practical Guide*; William Andrew: Norwich, NY, USA; TWI The Welding Institute: Cambridge, UK, 2008; ISBN 978-0-8155-1581-4.
10. Barber, P.; Atkinson, J.R. The use of tensile tests to determine the optimum conditions for butt fusion welding certain grades of polyethylene, polybutene-1 and polypropylene pipes. *J. Mater. Sci.* **1974**, *9*, 1456–1466. [CrossRef]
11. Menges, G.; Zohren, J. Untersuchung zum Heizelementschweißen von Polyolefinen. *Plastverarbeiter* **1967**, *18*, 165–171.
12. Bucknall, C.B.; Drinkwater, I.C.; Smith, G.E. Hot plate welding of plastics: Factors affecting weld strength. *Polym. Eng. Sci.* **1980**, *20*, 432–440. [CrossRef]
13. Stokes, V.K. The vibration and hot-tool welding of polyamides. *Polym. Eng. Sci.* **2001**, *41*, 1427–1439. [CrossRef]
14. Stokes, V.K. Comparison of vibration and hot-tool thermoplastic weld morphologies. *Polym. Eng. Sci.* **2003**, *43*, 1575–1602. [CrossRef]
15. Gehde, M.; Giese, M.; Ehrenstein, G.W. Welding of thermoplastics reinforced with random glass mat. *Polym. Eng. Sci.* **1997**, *37*, 702–714. [CrossRef]

16. Potente, H.; Kreiter, J.; Mohrmann, A. Heizelementstumpfschweißen mit konstanter Fügegeschwindigkeit. *Kunststoffe* **1988**, *78*, 84–87.
17. Frick, T. *Untersuchung der Prozessbestimmenden Strahl-Stoff-Wechselwirkungen Beim Laserstrahlschweißen von Kunststoffen*; Meisenbach: Bamberg, Germany, 2007; ISBN 978-3-87525-268-2.
18. Bonten, C. Beitrag zur Erklärung des Wirkmechanismus in Schweißverbindungen aus Teilkristallinen Thermoplasten. Ph.D. Thesis, University of Essen, Essen, Germany, 1998.
19. Voyutskii, S.S.; Vakula, V.L. The role of diffusion phenomena in polymer-to-polymer adhesion. *J. Appl. Polym. Sci.* **1963**, *7*, 475–491. [CrossRef]
20. Fick, A. Ueber Diffusion. *Ann. Der Phys.* **1855**, *170*, 59–86. [CrossRef]
21. Anand, J.N.; Kabam, H.J. Interfacial Contact and Bonding in Autohesion: I-Contact Theory. *J. Adhes.* **1969**, *1*, 16–23. [CrossRef]
22. Potente, H. Zur Theorie des Heizelement-Stumpfschweißens. *Kunststoffe* **1977**, *67*, 98–102.
23. Xue, Y.-Q.; Tervoort, T.A.; Rastogi, S.; Lemstra, J. Welding Behavior of Semicrystalline Polymers. 2. Effect of Cocrystallization on Autoadhesion. *Macromolecules* **2000**, *33*, 7084–7087. [CrossRef]
24. Frederix, C.; Beauchene, P.; Seguela, R.; Lefebvre, J.M. Kinetics of the non-isothermal fusion-welding of unlike ethylene copolymers over a wide crystallinity range. *Polymer* **2013**, *54*, 2755–2763. [CrossRef]
25. Bonten, C.; Schmachtenberg, E. A new hypothesis to describe the mechanisms acting in a welded joint of semicrystalline thermoplastics. *Polym. Eng. Sci* **2001**, *41*, 475–483. [CrossRef]
26. Fiebig, I.; Schoeppner, V. Influence of the Initial Fiber Orientation on the Weld Strength in Welding of Glass Fiber Reinforced Thermoplastics. *Int. J. Polym. Sci.* **2016**, *2016*, 7651345. [CrossRef]
27. Taşkıran, E.; Sayer, S.; Özes, Ç.; Yeni, Ç.; Ülker, A. Effect of process parameters and talc ratio on hot plate welding of polypropylene. *Mater. Und Werkst.* **2015**, *46*, 860–872. [CrossRef]
28. Yeni, Ç.; Ülker, A.; Sayer, S.; Özdemir, U.; Kocatüfek, U. Optimization of hot plate welding parameters of glass fibered reinforced Polyamide 6 (PA6 GF15) composite material by Taguchi method. *Usak Univ. J. Mater. Sci.* **2014**, *3*, 69. [CrossRef]
29. Kagan, V.A. Joining of nylon based plastic components—Benefits of vibration and hot plate welding technologies. *J. Inject. Molding Technol.* **2000**, *4*, 8.
30. Minchenkov, K.; Vedernikov, A.; Safonov, A.; Akhatov, I. Thermoplastic Pultrusion: A Review. *Polymers* **2021**, *13*, 180. [CrossRef] [PubMed]
31. Tanimoto, Y.; Inami, T.; Yamaguchi, M.; Nishiyama, N.; Kasai, K. Preparation, mechanical, and in vitro properties of glass fiber-reinforced polycarbonate composites for orthodontic application. *J. Biomed. Mater. Res. B Appl. Biomater.* **2015**, *103*, 743–750. [CrossRef] [PubMed]
32. Krueger, S.; Schmitz, S.; Weiss, S.; Wirtz, D.; Linssen, M.; Schade, H.; Kraemer, N.; Spuentrup, E.; Krombach, G.; Buecker, A. An MR guidewire based on micropultruded fiber-reinforced material. *Magn. Reson. Med.* **2008**, *60*, 1190–1196. [CrossRef] [PubMed]
33. Talabi, S.I.; Tobin, J.; Strom, B.; Brownstein, I.; Kunc, V.; Hassen, A.A. Recent and future developments in pultrusion technology with consideration for curved geometries: A review. *Compos. Part B Eng.* **2024**, *283*, 111678. [CrossRef]
34. Ebert, C.; Decker, H.; Tönnishoff, L.; Bonten, C.; Wilhelm, M.; Wruck, L. Modularer Batteriebaukasten für ein Feuerwehrfahrzeug. *Kunststoffe* **2024**, *7*, 82–85. [CrossRef]
35. Reis, J.P.; de Moura, M.; Samborski, S. Thermoplastic Composites and Their Promising Applications in Joining and Repair Composites Structures: A Review. *Materials* **2020**, *13*, 5832. [CrossRef]
36. Lengsfeld, H.; Lacalle, J.; Neumeyer, T.; Altstädt, V. *Faserverbundwerkstoffe: Prepregs und ihre Verarbeitung*; Hanser: München, Germany, 2020.
37. Turvey, G.J. Single-bolt tension joint tests on pultruded GRP plate—Effects of tension direction relative to pultrusion direction. *Compos. Struct.* **1998**, *42*, 341–351. [CrossRef]
38. Keller, T.; Vallée, T. Adhesively bonded lap joints from pultruded GFRP profiles. Part I: Stress–strain analysis and failure modes. *Compos. Part B Eng.* **2005**, *36*, 331–340. [CrossRef]
39. Goto, K.; Imai, K.; Arai, M.; Ishikawa, T. Shear and tensile joint strengths of carbon fiber-reinforced thermoplastics using ultrasonic welding. *Compos. Part A Appl. Sci. Manuf.* **2019**, *116*, 126–137. [CrossRef]
40. Jandali, G.; Mallick, P.K. Vibration Welding of Continuous-Fiber Thermoplastic Matrix Composites. *J. Thermoplast. Compos. Mater.* **2004**, *17*, 343–358. [CrossRef]
41. Behnisch, F.; Wilhelm, M.; Janssen, M. Kostengünstige, großserien- und recyclingfähige Bauweise für hochbelastete Bauteile: Oral Presentation. *Werkst. Auto* **2024**.
42. Deutsches Institut für Normung e. V. *Bestimmung der Zugeigenschaften: Teil 2: Prüfbedingungen für Form- und Extrusionsmassen*; Beuth: Berlin, Germany, 2012; (DIN EN ISO 527-2).
43. BASF. Ultramid (PA): Product Brochure. Available online: https://s3.amazonaws.com/entecpolymers.com/v3/uploads/BASF-Ultramid-PA-%E2%80%92-Product-Brochure.pdf (accessed on 23 August 2024).

44. Becker, G.W.; Binsack, R.; Bottenbruch, L.; Braun, D. *Kunststoff Handbuch: Technische Thermoplaste Polyamide*; Neuausg. Brussels, Belgium; Hanser: München, Germany, 1998; ISBN 9783446164864.
45. Oshinski, A.J.; Keskkula, H.; Paul, D.R. The role of matrix molecular weight in rubber toughened nylon 6 blends: 1. Morphology. *Polymer* **1996**, *37*, 4891–4907. [CrossRef]
46. Chen, J.; Radke, W.; Pasch, H. Analysis of polyamides by size exclusion chromatography and laser light scattering. *Macromol. Symp.* **2003**, *193*, 107–118. [CrossRef]

Journal of
Manufacturing and Materials Processing

MDPI

Article

Effect of Flashlamp Heating System Parameters on the Wedge Peel Strength of Thermoplastic Carbon Fiber Tape in the Automated Tape Placement Process

Alexander Legenstein and Ewald Fauster *

Processing of Composites and Design for Recycling, Montanuniversität Leoben, 8700 Leoben, Austria; alexander.legenstein@unileoben.ac.at
* Correspondence: ewald.fauster@unileoben.ac.at

Abstract: Laser-assisted automated tape placement systems are currently the state of the art regarding thermoplastic tape placement. Flashlamp heating systems are rather new in this field of application and offer high energy density with low safety requirements and moderate costs compared to laser-assisted automated tape placement systems. In this study, the effect of processing parameters on interlaminar bonding of carbon fiber-reinforced polyamide 6 tapes is investigated using a flashlamp heating system. The temperature during placement is monitored using an infrared camera, and the bonding strength is characterized by a wedge peel test. The bonding quality of the tapes placed between 210 °C and 330 °C at a lay-up speed of 50 mm/s is investigated. Thermogravimetric analysis, differential scanning calorimetry, and micrographs are used to investigate the material properties and effects of the processing conditions on the thermophysical properties and geometric properties of the tape. No significant changes in the thermophysical or geometric properties were found. Moisture within the tapes and staining of the quartz guides of the flashlamp system have significant influence on the bonding strength. The highest wedge peel strength of dried tapes was found at around 330 °C.

Keywords: automated tape placement; thermoplastic resin; bonding; polymer matrix composites; material characterization

1. Introduction

Automated Tape Placement (ATP) can be used to manufacture lightweight and high-performance composites structures in a single-step process using thermoplastic polymer matrix composite materials. ATP in this context is defined as the automated positional- and directional-variable placement of a single unidirectional reinforced preimpregnated polymer tape on flat or curved surfaces with defined laminate thickness [1]. This methodology does not require an autoclave cycle and is therefore faster, cleaner, and more energy efficient compared to thermoset ATP processes. This is particularly of interest in the aerospace industry, as the autoclave limits the dimensions of the part that can be manufactured [2,3]. Thermoplastic polymers not only help with realizing broader design spaces; they are also easier to repair and recycle compared to thermoset polymers due to their ability to be remelted. Thermoset polymers are brittle and are irreversibly crosslinked. Thermoplastic polymers are already fully polymerized in their resin state and do not exhibit crosslinking. The structural stability is achieved by enabling the polymer chains to move at high temperature and form entangled structures [4]. Most research on thermoplastics polymers for ATP focuses on high-performance matrix materials such as PEEK [4–20], PEKK [21–23], or PPS [24–27]. Engineering or commodity thermoplastics have received far less attention within the recent years. Some studies exist investigating PA6 [16,28–30], PA66 [31], PA12 [32,33], or PP [34–37]. Low-cost thermoplastic matrix systems are particularly of interest for cost-sensitive and high-volume industries such as the automotive sector [38], as they offer easy processability, high toughness, damage tolerance, and better recyclability compared to thermoset structures [39]. The most common methods of processing fiber-reinforced thermoplastic tapes with ATP are with laser [5,8,10,12,13] or hot gas torch heating systems [6,7,40–44]. Hot gas torch heating systems have a low energy density and long response time compared to laser systems. This limits the possible processing speed and restrains the design space. Laser systems come with higher equipment costs and need to be in an isolated room to guarantee safety [45]. Flashlamp systems can bridge the gap between hot gas torch and laser heating systems for thermoplastic ATP. These systems are based on pulsed light technology and utilize high-energy pulsed flashes for heating the tapes. A high voltage is used to ionize the xenon gas in the lamp to conduct electricity. The capacitors are discharged at regular intervals to generate flashes. These flashes deliver the heat to the target. Quartz guides are used as an optical medium to focus the flashes. Based on the geometry of the quartz, the energy can be distributed between the incoming tape and the substrate. These systems have lower acquisition and operation costs, as well as safety requirements, while offering equal power as laser systems [46,47]. Research on flashlamp systems is currently limited to reflectivity and emissivity analyses of composites heated by those systems [48], thus benchmarking those systems for CF/PPS [49] or facilitating the development of heat transfer models for dry fiber and thermoplastic materials [46,50]. Pulsed operation of the flashlamp system can cause high levels of material temperature (above the thermal degradation threshold of the materials) [46] This effect could influence the bonding strength of the laminate. Especially at complex geometries, local overheating can damage the matrix [51]. The bonding of thermoplastic materials is affected by the crystallinity level, the void content, or thermal degradation. All of those effects are dependent on the thermal history during processing. The process window for ATP needs to be defined precisely in order to guarantee high bonding quality and avoid altering the composite due to high temperature [52]. In ATP, small surface areas are typically heated with heat rates of up to 500 °C/s but with short exposure time (<0.2 s) depending on the placement speed [16]. High lay-up speeds need higher power inputs to sufficiently melt the matrix and bond the tapes. Furthermore, sufficient consolidation force is needed to ensure enough time for the molecular chains to reptate between the layers for optimal bonding [42]. Bonding occurs above the glass transition temperature (T_g) for amorphous thermoplastics and above the melting temperature (T_m) for crystalline polymers. However, the bonding of semicrystalline thermoplastics was also shown to occur below the melting point [53].

The work at hand aims to experimentally investigate the temperature distribution of unidirectional tapes during ATP with a flashlamp heating system. The temperature of the incoming tape was measured close to the nip point using an IR camera. Different processing conditions (pulse width and frequency of the heating system) were used to manufacture samples and evaluate the resulting temperature close to the nip point. The samples were made from unidirectional CF/PA6 tapes due to the low cost of this material compared to high-performance tapes (e.g., CF/PEEK) and the lack of ATP-based studies regarding CF/PA6, particularly with flashlamp heating. All samples consisted of four tape layers resulting in a $[0]_4$ laminate, which in turn was used to determine the wedge peel strength. The moisture within the tapes was measured. The impact of different drying conditions on the wedge peel strength was tested. Thermogravimetric Analysis (TGA) was used to determine the degradation point of the investigated polymers, and Differential Scanning Calorimetry (DSC) was used to study the influence of processing conditions on the thermophysical properties of the polymers. The fiber–matrix distribution and the bonding interface were investigated with optical microscopy.

2. Materials and Methods

2.1. Materials

The material used for this study is a fully impregnated thermoplastic CF/PA6 tape from SGL Carbon SE (Sigapreg® TP C U157-0/NF-T340/46%) with a width of 25.4 mm and a thickness of 0.2 mm. The fiber volume content is 42%, and the mass density is 1.42 g/cm^3. The fiber type is Sigrafil® C T50-4.4/255-T140. The matrix has a melting temperature T_m of 220 °C and a T_g of 58 °C.

2.2. ATP System and Placement Trials

The ATP system used in this study is shown in Figure 1d and consists of a placement head with feeding unit, a silicone compaction roller with 70 Shore A hardness, 30 mm width, and a diameter of 50 mm. The setup is able to place 25.4 mm wide tapes on a heated aluminum tool measuring 700 mm × 350 mm. The tool can be heated with an accuracy of ±1 °C.

The consolidation force was measured using a multicomponent sensor with an accuracy of 4 N and a resolution of 0.2 N. A hygrometer is used to measure room temperature and humidity. It was placed right next to the ATP rig and the material storage. The hygrometer has an accuracy of ±0.5 °C and a resolution of 0.1% for the temperature measurement, whereas the humidity can be measured with an accuracy of ±3% and a resolution of 0.1%. All placement trials were performed with a *Heraeus humm3* flashlamp system as heating source (see Figure 1c). The quartz guide has an angle of 23° to the substrate and a width of 30 mm. Two different chamfered quartz guides were used in this study. Figure 1a shows the geometry of a chamfered quartz guide with one radiating surface, and Figure 1b shows the geometry with three radiating surfaces. The configuration of the quartz guide in the processing environment is shown in Figure 1c,d. All tapes were placed directly on the aluminum tool. The tool was heated to 50 °C to ensure equal placement conditions for all trials. The lay-up speed was set to 50 mm/s, and the consolidation force was set to 500 N for all tests.

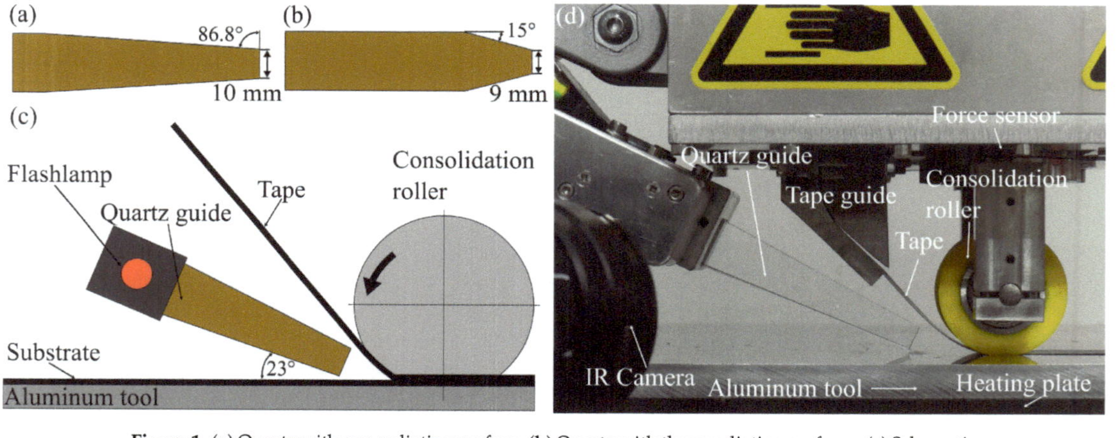

Figure 1. (**a**) Quartz with one radiating surface. (**b**) Quartz with three radiating surfaces. (**c**) Schematic of the flashlamp system placement within the ATP system. (**d**) ATP setup used in this study.

The samples for the peel test were manufactured at different flashlamp parameters as listed in Table 1. Three peel test samples were manufactured for each parameter configuration. For the first experimental set, the pulse width was kept constant, and for the other set, the frequency was kept constant. For both sets, the power was increased from around 2150 W to the maximum power setting of the flashlamp system (4400 W). To compare the temperature between the experimental sets, uniform power levels were chosen (ranging from 1 to 5, with 5 being the highest power setting, as shown in Table 1). The power levels differ with a maximum value of 26 W at the highest power setting.

Table 1. Experimental plan.

Exp. Set	Frequency [Hz]	Pulse Width [μs]	Power [W]	Power Level [-]
1	40	2200	2153	1
	50	2200	2691	2
	60	2200	3229	3
	70	2200	3767	4
	80	2200	4306	5
2	30	2950	2159	1
	30	3700	2676	2
	30	4550	3236	3
	30	5400	3779	4
	30	6200	4280	5

2.3. Temperature Measurement

In order to determine the associated temperature of the incoming tape for the parameter combinations shown in Table 1, a series of tapes was placed onto the aluminum tool to form a $[0]_4$ laminate. The temperatures were measured on the incoming third layer due to the tapes being pulled apart between the second and third layer for the wedge peel test. The temperature of the incoming tape was measured using an InfraTec Image IR 8300 IR camera. The IR camera has a detector format of 640 × 512 pixels with a temperature resolution of 0.025 K at 30 °C. The camera uses a lens with 25 mm focal length. The ni-point temperature cannot be measured directly with the IR camera. Combinations of surface emittance and specular reflections of the opposing surfaces near the nip point distort the temperature measurement [54]. Therefore, the temperature was measured closely before the nip point. It needs to be noted that the temperature could differ from the actual processing temperature, because it is not measured directly at the nip point. The emissivity was set to 0.91, as the values typically range from 0.8 to 0.95 depending on the material and the geometrical configuration of the setup [30].

2.4. Wedge Peel Strength Measurement System

The bonding quality of composite laminates are typically characterized by the crack propagation resistance with Double Cantilever Beam (DCB) tests for mode I and End Load Split (ELS) for mode II crack initiation [55]. Another common test method for evaluating of the bonding strength is the Short Beam Strength (SBS) test [4,21,53,56,57]. High-performance thermoplastics can be tested with SBS and show valid failure modes [4]. However, it was shown by Schaefer [58] that SBS tests on CF/PA6 result in wrong values, because the samples did not experience interlaminar crack formation and failed due to plastic shear. Stokes-Griffin et al. [16] also listed poor fiber–matrix bonding as a possible contributing factor for the failing test. Wedge peel tests have been widely used [13,16,18,59] to compare the interlaminar bonding of different composites and are an alternative to SBS or DCB tests for the evaluation of the bonding strength, as they show good correlation with DCB tests [60]. Although it is not yet standardized for this kind of test, wedge peel testing is conceptually comparable to standardized tests for adhesive bonding, e.g., ASTM D3762. The wedge peel test is able to continuously test the bonding strength over the whole sample length and is therefore suitable to identify any differences in bonding strength occurring from process conditions. Due to the noncontinuous heating behavior of the flashlamp system, different bonding levels over the length of the tape can occur. The peel test setup used for this study is shown in Figure 2. It consists of a wedge with a thickness of 10 mm and a wedge angle of 40° The tip of the wedge is rounded and has a radius of 0.5 mm. All surfaces of the wedge have a mean roughness depth (difference between peak and valley height) of Rz = 0.63 μm. The tapes are pulled over the wedge with the interface in between the second and third ply. The peel speed was set to 1 mm/s. The peel force is measured by two load cells (located underneath the peel wedge) with a capacity of 300 N and a resolution of 0.009 N at an acquisition rate of 10 Hz. The sample width was measured with a caliber at 10 points (same positions for each tape on the third layer) and averaged to calculate the arithmetic mean of the tape width. The digital caliber has a measurement accuracy of 0.03 mm.

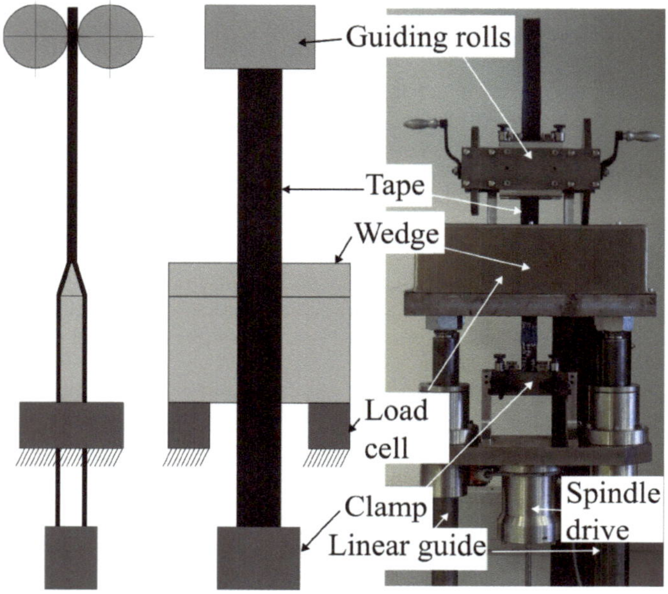

Figure 2. Side and front view of the wedge peel strength measuring rig used in this study.

In order to investigate the influence of humidity, the moisture of the tapes was measured before placement and before the wedge peel test. The moisture is measured using a Kern DAB 100-3 moisture analyzer. The moisture analyzer measures mass reduction with an accuracy of 0.003 g and a reproduceability of 0.15%.

2.5. Thermogravimetric Analysis

Thermogravimetric analysis was used to determine the degradation onset of the PA6 used for this study. The measurements were performed with a Mettler Toledo TGA/DSC 3+ STARe system. Samples were taken from random places and different positions along the length of multiple tapes in order to avoid systematic influences on the results. The weight of the samples ranged from 7 to 12 mg. The samples were heated from 25 °C up to 650 °C at a heating rate of 10 °C/min under a nitrogen atmosphere with a nitrogen flow of 50 mL/min.

2.6. Differential Scanning Calorimetry

In order to measure the melting point and crystallization temperature of the material used for this study, DSC measurements were conducted. These measurements were used to link potential changes in peel strength to a change in molecular structure (e.g., melting and crystallization temperature). The DSC tests for this study were run on a Mettler Toledo DSC 1 device. The samples were heated from 30 °C to 280 °C (above the infinite polymer melting temperature of 270 °C [16]) with a heating rate of 10 °C/min and held at 280 °C for 5 min to erase the process history and remaining crystal seeds. After holding the samples at 280 °C, they were cooled to 30 °C at a cooling rate of 10 °C/min and then heated up to 280 °C at 10 °C/min again to evaluate any changes in the molecular structure due to degradation. All tests were performed under nitrogen atmosphere.

3. Results and Discussion

3.1. Temperature Evaluation

Figure 3a shows the third layer of a tape placed at 50 mm/s onto two other tapes forming a $[0]_4$ substrate. The evaluation area for the temperature measurement is shown with a yellow ellipse. All data were collected with an acquisition frequency of 250 Hz. An uneven heat distribution across the tape width can be seen in the Figure, as well as an uneven heat distribution along the tape length. Figure 3b,c show thermographs with a chamfered quartz guide, as shown in Figure 1b. It can be observed that the matrix residue of the tape was collecting on the chamfered sides of the quartz guide due to the tape touching the quartz guide during the end phase of the placement. This led to different heating characteristics in the subsequent tapes. The chamfered quartz guides led to a high variability in temperature over the tape length with increasing amounts of residue buildup. Figure 3c shows the quartz guide heating up to over 400 °C, which led to a heating of the tape right after exiting the guiding system and a lower temperature in front of the nip point compared to Figure 3b. The geometries with chamfered sides were found to be more sensitive to residue buildup on the quartz guides. The geometry in Figure 1a was therefore been chosen for all further experiments.

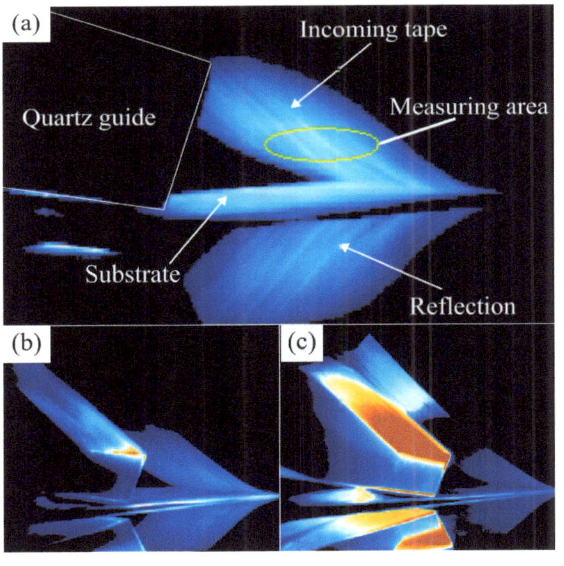

Figure 3. (**a**) Evaluation area for the temperature measurement. (**b**) Chamfered quartz guide after placing 3 tapes with 70 Hz and a pulse width of 1700 µs. (**c**) Chamfered quartz guide after placing 11 tapes with heating of the quartz guide due to residue collection on the surface.

Figure 4 shows the temperature evolution for the different power settings. Each entry in the plot was derived from temperature measurements on three different tapes. The temperature was measured at uniform locations along the tape length in sections where samples for subsequent wedge peel tests were extracted. It can be seen that the constant frequency with varying pulse width samples experienced higher temperature with higher power compared to the constant pulse width samples, except for 2691 W (50 Hz). This could be explained by the frequency shift, which leads to a higher energy, as the energy of a photon is directly proportional to the wavelength. Furthermore, the higher frequency leads to a slight change in the spectrum emitted by the quartz guide. This different spectrum could lead to a change in heat transfer within the tape.

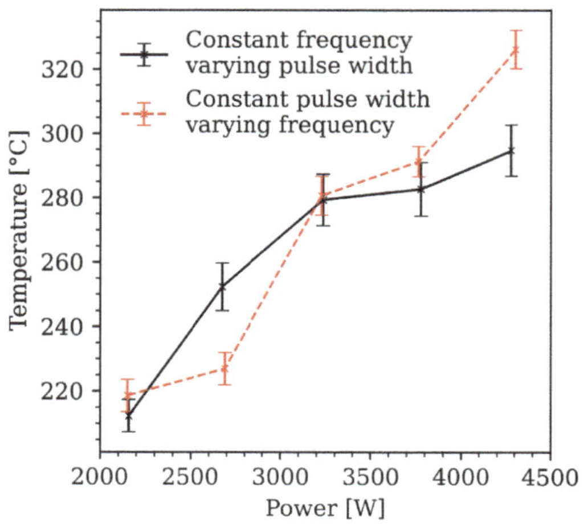

Figure 4. Process temperature for all power settings used in this study for constant frequency with varying pulse width (solid curve) and constant pulse width and varying frequency (dashed curve).

Figure 5 shows the temperature evolution over the tape length section relevant for the subsequent wedge peel tests for the tapes placed at a constant pulse width with varying frequency for one tape each. The spindle position 0 mm was at the beginning of the bonding between the tapes shortly after the start of the placement when the temperature reached a temperature level high enough to bond the tapes together. It can be seen that the temperatures for the experiments at 40 Hz and 50 Hz were closer to each other compared to the other values. The first peak in the temperature can be explained by the startup of the placement with the tape being closer to the quartz guide and therefore having a higher temperature. The temperature evolution for the experiments at 60 to 80 Hz showed higher variability within the peel test length. The 80 Hz sample showed the highest variability and increasing temperature over the peel test length, which could lead to different bonding within the tape over the length of the placement.

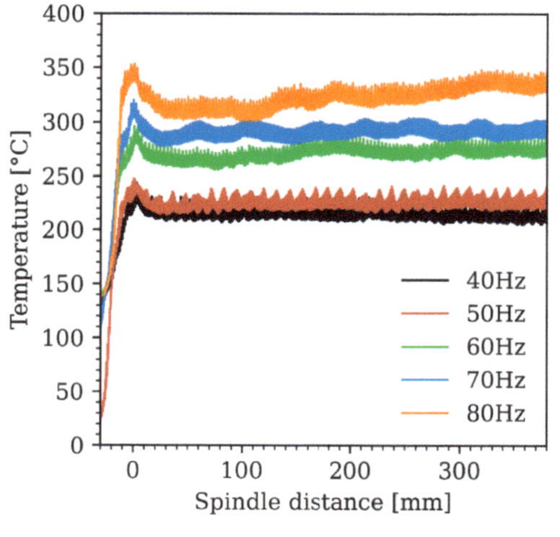

Figure 5. Temperature evolution over the wedge peel strength measuring distance for the tapes placed at constant pulse width with varying frequency.

3.2. Peel Strength

Figure 6 shows the results of the wedge peel strength tests and the tape width used for normalizing the peel strength data. The tape width after consolidation was measured at uniform locations for each tape with a caliber at 10 positions. The average of these measurements was used to calculate the wedge peel strength in the gray section seen in Figure 7. The temperature evaluation was also done with the temperature values in the gray section. The tape width after consolidation measured for the third layer was smaller compared to the first two layers. Those two layers experienced more heating and consolidation cycles and were therefore wider due to no squeeze flow restriction on the edges (e.g., adjacent tapes). The evaluation section differed for each peel test and was determined by a steady state section within the peel strength by using a moving average on the data. No other data cleaning was necessary, as no outliers could be found in the steady state section for the evaluation.

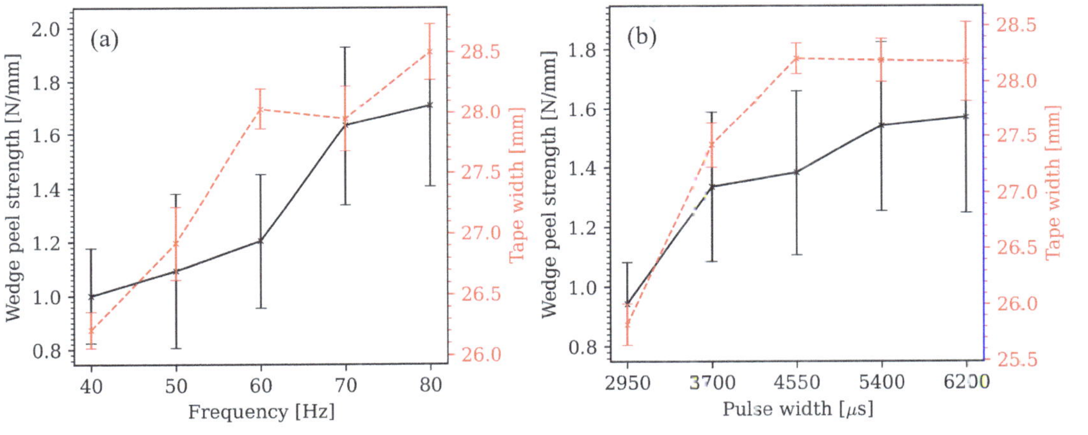

Figure 6. (**a**) Effects of the frequency on the wedge peels strength and the tape width. (**b**) Effects of the pulse width on the wedge peels strength and the tape width.

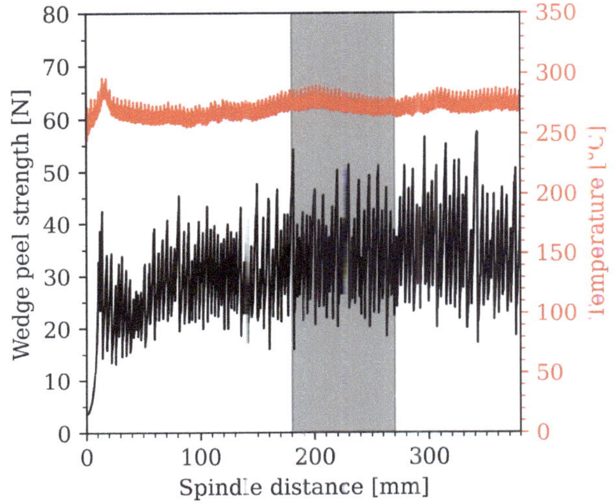

Figure 7. Peel strength and temperature over the peel test length for the 60 Hz sample.

All tapes were placed directly after they were taken out of the oven. The moisture was measured before the placement on 53 tapes. The tapes were dried in the moisture analyzer for 7 min at 120 °C. The average moisture was 1.61 m% with a standard deviation of 0.23 m%. This resulted in an about 1 m% lower moisture content compared to the samples before drying. The results for a constant pulse width (2200 μs) and varying frequency can be seen in Figure 6a. A constant rise in peel strength with increasing frequency and therefore increasing power can be observed. The power is a function of the frequency, pulse width, and voltage of the setting. An increase in any of those three settings will lead to a higher power and therefore temperature (if all of the other process variables stay the same). The increase in temperature led to a higher bonding strength up to the point where matrix degradation took place or significant matrix squeeze out occurred due to the decrease in matrix viscosity. The lowest value (1.00 N/mm) was at 40 Hz (2153 W with 219 °C process temperature). The highest peel strength value (1.71 N/mm) was achieved with a frequency of 80 Hz and a temperature of 326 °C. However, the variation within the peel strength rose with increasing frequency and, thus, with temperature. Figure 6b shows the results from tests with constant frequency (30 Hz) and varying pulse width. The lowest value (0.94 N/mm) is at 2950 μs (temperature of 212 °C) and the highest value (1.57 N/mm) can be found at 6200 μs with a temperature of 295 °C. The variation within the peel strength increased with pulse width and, thus, again with temperature. In general, the peel strength values found in this study were lower than those reported by Stokes-Griffin et al. [16] for the same material. This could be explained by the lower lay-up speed used in this study (50 mm/s compared to 100 mm/s). This could lead to a higher degree of more ductile γ phases in the tapes placed at 100 mm/s, due to faster cooling. This increases the wedge peel strength. Furthermore the consolidation force was set to 500 N for 25.4 mm wide tapes in this study, whereas Stokes-Griffin et al. [16] used 130 N for 12 mm wide tapes. An analysis of the pressure distribution for the roller used in this study at 500 N showed that an average pressure of 1.57 MPa was applied to the tapes over the area of contact with the roller. The pressure was measured with a Prescale LLW pressure measuring film at a room temperature of 22.8 °C and a humidity of 26.8%. However, the resulting pressure and consolidation tape width were not known for the reference study. There, the highest wedge peel strength was reported with 4.3 N/mm at a process temperature of 260 °C. In general, a higher contact area is favorable to develop higher degrees of intimate contact; this, however, limits the pressure on the tapes, which in turn has negative consequences on the intimate contact [23].

In order to investigate the influence of humidity on the wedge peel strength, three samples for each of the drying conditions were prepared (Table 2).

Table 2. Drying conditions for investigating the moisture influence on the wedge peel strength (ND: nondried, D: dried).

Label	Drying before Placement	Drying before Peel-Test
ND/ND	72 h at 21 °C and 41% humidity	72 h at 21 °C and 41% humidity
ND/D	72 h at 21 °C and 41% humidity	72 h at 60 °C in vacuum oven
D/ND	48 h at 100 °C in vacuum oven	72 h at 21 °C and 41% humidity
D/D	48 h at 100 °C in vacuum oven	72 h at 60 °C in vacuum oven

The results can be seen in Figure 8. The ND/ND samples had the highest peel strength (2.04 N/mm), whereas the samples dried before the placement showed a significant reduction in peel strength resulting in a peel strength of around 0.9 N/mm. This is to be expected, because moisture acts as a plasticizer in the polymer. Plasticizers decrease the T_g and T_m

and provide chain mobility by reducing the secondary forces between the macromolecules. This effect results in a decrease in tensile strength, hardness, or elastic modulus while increasing the elongation at break or toughness [61]. Those effects results in a more ductile crack propagation compared to dry samples [62]. Samples not dried before the placement but dried before the peel test also showed a reduction in peel strength, but it was not as drastica compared to the samples dried before the placement. The peel strength for those samples reached an average level of 1.39 N/mm.

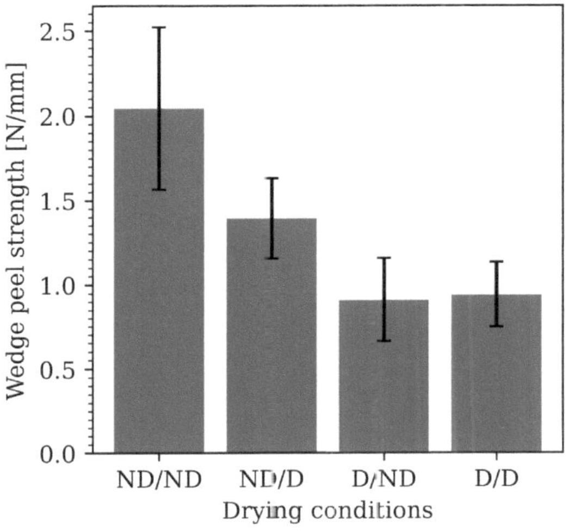

Figure 8. Effect of drying the samples before the placement and before the peel test on the wedge peel strength (ND: nondried, D: dried).

3.3. TGA Measurements

Thermal degradation deteriorates the polymer matrix by cutting the molecular chains by rupture or scission [63]. During ATP, heating rates of 500 °C/s can occur. However, the exposure to high temperature levels is usually very short (<0.2 s) [4,16]. Figure 9 shows the results from TGA tests on five composite material samples. The onset of degradation was annotated at around 362 °C and was set at a weight loss of 3%, because PA6 absorbs around 3 m% of water at standard conditions (23 °C and 50% humidity) [64]. These results were verified by measuring 17 tapes with the moisture analyzer. The tapes have been kept at a humidity of 43% and a temperature of 22 °C. All tapes were dried for 7 min at a temperature of 120 °C with the halogen lamp of the moisture analyzer. The tapes lost 2.73 m% water on average with a standard deviation of 0.11 m%. The degradation temperature for PA6 was found by other researchers at around 350 °C [65] to 400 °C [66], and the material was fully decomposed at around 500 °C [67] to 530 °C [66] in a nitrogen atmosphere. A temperature above 390 °C was shown to lead to a more intense degradation (the degradation took place faster) [68]. The water loss took place from the start of the measurements (25 °C) to the degradation point. A faster mass loss was seen from the start to around 150 °C. After this point to the degradation point, a slower loss was observed (around 1% mass loss from 150 °C to 360 °C). The degradation temperature was determined after the loss of all water (around 3 m%). This yielded a degradation temperature of 361.8 °C with a standard deviation of 8.6 °C. After the initial mass loss due to the absorbed water, the matrix degraded from 361.8 °C to around 500 °C, thus leaving around 56.3 m% (standard deviation of 0.5 m%) carbon fibers behind (at a temperature of 550 °C). The fiber mass content was 2.3 m% higher compared to the data sheet of the material. Slight deviations in the degradation temperature between the samples were observed. The fastest degradation was found at 455.8 °C with a standard deviation of 1.0 °C. The results from Hanna [69] verify the decomposition stages found in this study: At temperature levels up to 150 °C, the matrix was loosing absorbed water, which resulted in a small weight loss of about 2%. The second stage ranged from 280 to 450 °C, where the PA6 was getting decomposed, and above 450 °C, were the impurities and undecomposed PA6 were lost. A higher degree of crystallinity was found to lead to faster decomposition at higher temperature. A higher activation energy is needed at higher crystallinity to break the crystalline parts [69].

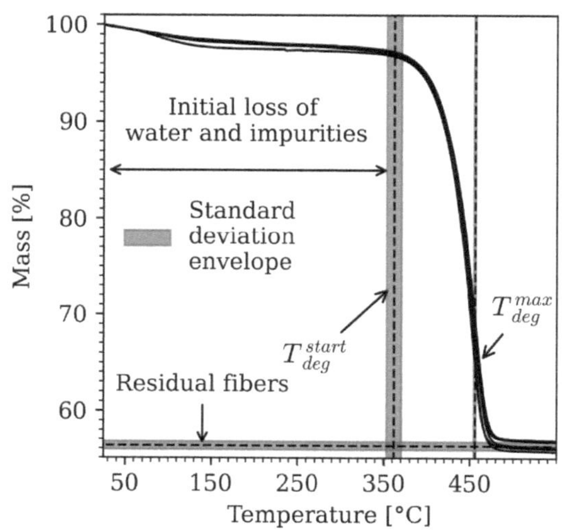

Figure 9. TGA curves of 5 samples with degradation onset annotated at 350 °C.

3.4. DSC Measurements

Figure 10a shows the two heating cycles and the cooling cycle for the sample placed at a frequency of 80 Hz and a pulse width of 2200 µs. The heating cycle 1 was offset by 0.4 W/g for better visualization. The curves would otherwise fall together with the heating cycle 2. Both the melting and crystallization temperature were taken from the endothermic and exothermic peaks. It can be seen that the melting temperature of the as-processed samples (heating cycle 1) was around 2 °C higher compared to the heating cycle after cycling to the infinite polymer melting point T_m^∞. This can also be seen for all other tested samples in Figure 10b, which showed samples ranging from 40 to 80 Hz and a sample at 30 Hz, which showed no bonding. A higher melting point was associated with the α phase crystals, whereas a lower melting point was associated with the γ phase crystals. However, the α phase formed at slow cooling rates, and the γ phase at fast cooling rates. α phase crystals have a melting temperature of around 225 °C and the γ phase crystals at 215 °C [63]. The heating and cooling rates of the ATP process, however, were much higher compared to the 10 °C/min used in the DSC. Therefore, the samples manufactured by ATP would have a higher degree of the γ phase. The formation of the crystalline structure is influenced by thermal conditions, applied stress, additives within the polymer [70], water content [71], or high pressure during crystallization [72]. The α phase is observed to form at temperatures above 190°C, and the γ phase below 130 °C when the polymer is kept at those temperatures for longer periods of time. At temperature levels between 130 °C and 190 °C, a mixture of both phases will be present. Short crystallization times at 200 °C can lead to both crystallization forms [73]. The results from Figure 10a indicate that there was a higher amount of α phase crystals in the first heating cycle resulting in a higher melting temperature. The endothermic heat flow was more narrow compared to the heat flow of the second heating cycle, thus indicating the α phase [74]. A wide heat flow curve indicates either the γ phase or the β phase (also called mesophase). The mesophase, however, was a result of high cooling rates of thin samples and not likely to occur at 10 °C/min cooling rates [75,76]. Both the γ and β phase are hard to distinguish using DSC. Wide-angle X-ray scattering is a method commonly used to distinguish between the different crystalline phases. However, the γ and β phase share a similar scattering angle [74,76]. A clear description of the crystalline phase is not possible with DSC tests. Changes in the crystallinity can also occur during the heating cycle of the DSC tests. The evaluation of the crystalline phase of the processing parameters used in this work needs to be investigated in a future study. The T_g could not be determined in the samples. This is possibly due the moisture in the sample. Studies have found that the time of sample preparation is sufficient to absorb enough moisture for the T_g to drop from the dry value of around 54 °C to a value between 15 °C and 20 °C. Due to the starting point of the DSC in this test (25 °C), the T_g did not appear in the samples. Furthermore, the T_g is likely to be overseen in undried samples, as the moisture loss during the DSC would overlap with the silent exothermic crystallization [77].

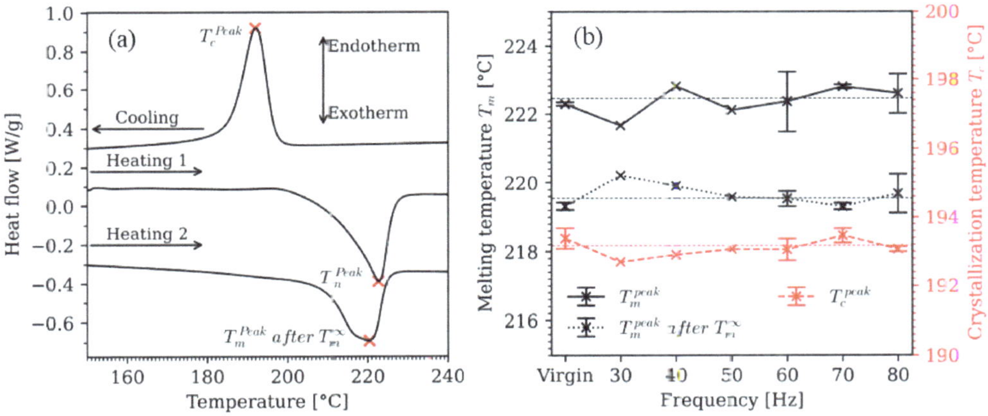

Figure 10. (**a**) DSC heating and cooling cycles for different heating system frequencies (Heating cycle 1 offset by 0.4 W/g for better visualization). (**b**) Melting temperature before and after cycling to the melting point of the infinite crystals, as well as crystallization temperature (all temperatures are peak values).

3.5. Micrographs

Figure 11 shows cross-sections of polished samples from different heating system parameter variations (corresponding to the values in Table 1) around 300 mm from the start of the lay-up. The cross-sections were used to evaluate the homogeneity of the laminates and not to analyze any damages resulting from the placement. All images were taken with a digital microscope from *Keyence* (VHX-7000) at a magnification of 150×. At the lower power levels (to around 3229 W) of the tapes with constant pulse width, the individual tapes of each layer can be clearly distinguished, whereas the layers of the tapes with constant frequency cannot be clearly separated from 2676 W upwards. This corresponds to the peel strength values shown in Figure 6. The clearly separable samples show lower wedge peel strength (<1.3 N/mm). At higher wedge peel strengths, the different states of inhomogeneity of the tape became less visible. The thickness of the samples was measured with ImageJ over 25 points across the tape width. The thickness of the tapes with a constant pulse width showed a linear trend of decreasing thickness with increasing frequency. The tape with 40 Hz had a average thickness of 0.796 mm with a standard deviation of 0.011 mm, whereas the tape placed with 80 Hz had a thickness of 0.651 mm with a standard deviation of 0.015 mm. The thickness of the tapes placed with constant frequency showed more variation and no linear trend. The tape placed with 2950 μs had a thickness of 0.765 mm and a standard deviation of 0.013 mm, whereas the tape placed with 6200 μs had a thickness of 0.675 mm and a standard deviation of 0.025 mm. A full overview of the measured tape thickness is shown in Table 3. The reduction in thickness is attributed to the higher processing temperature and matrix squeeze-out due to lower matrix viscosity at elevated temperature. The fluctuation of the thickness with increasing temperature can be explained with uneven tape thickness and varying temperature during the placement due to the pulsing operation of the system and nonuniform fiber volume fraction along the tape. Another possible explanation for the varying thickness reduction in the samples with constant frequency could be due to a higher matrix viscosity compared to the samples with varying frequency. Higher frequencies could lead to a higher excitement of the molecular chains and thus reducing the London dispersion forces, which reduces the matrix viscosity [78]. Moreover, a shift in frequency could also leads to a shift in wavelength, which could result in a better absorption of the energy emitted from the light source by the molecules.

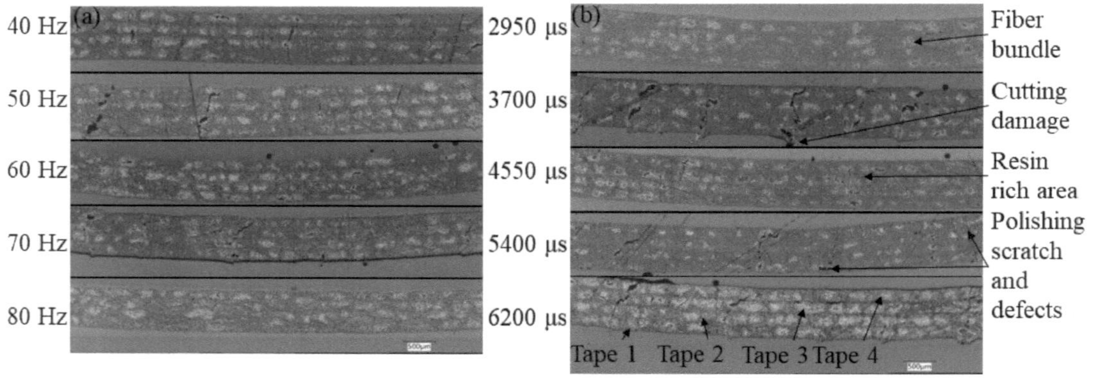

Figure 11. Micrographs of the center of the samples from different heating system parameters. (**a**) Frequencies from 40 Hz to 80 Hz and constant pulse width of 2200 μs. (**b**) Pulse widths from 2950 μs to 6200 μs and constant frequency of 30 Hz.

Table 3. Tape thickness measured from micrographs for all processing conditions.

Frequency [Hz]	Pulse Width [μs]	Mean Thickness [mm]	Standard Deviation [mm]
40	2200	0.796	0.011
50	2200	0.767	0.012
60	2200	0.685	0.010
70	2200	0.667	0.011
80	2200	0.651	0.015
30	2950	0.765	0.013
30	3700	0.729	0.012
30	4550	0.747	0.010
30	5400	0.756	0.014
30	6200	0.675	0.025

4. Conclusions

In this study, the impact of the processing conditions of a flashlamp heating system on the bonding strength and the temperature evolution of the incoming tape was evaluated. Furthermore, the influence of moisture within the CF/PA6 tapes on the bonding strength was investigated. The bonding strength between the layers was investigated using a wedge peel test rig. The frequency and the pulse width of the flashlamp heating system were varied, and their effects on 25.4 mm wide tapes placed at 50 mm/s were analyzed. TGA was used to determine the degradation temperature of the material used in this study. A degradation temperature of around 362 °C was found by analyzing five samples. DSC measurements of samples placed with different frequency settings and constant pulse width were analyzed, as well as micrographs for all parameter combinations. The DSC measurements showed no thermal degradation, which would lead to changes in the melting or crystallization temperature. The melting temperature was found to be around 222 °C right after processing and at 219 °C after cycling to the infinite melting point of the polymer. The crystallization temperature was found to be at around 193 °C. No significant differences could be seen for the tested samples. Micrographs showed a linear thickness reduction of the samples manufactured at a constant pulse width, whereas the thickness of the samples placed with constant frequency showed varying thickness with higher temperature. A decreasing thickness correlated with an increase in wedge peel strength. It has to be noted that a higher matrix squeeze out could lead to a significant loss in wedge peel strength. Matrix squeeze out reduced the resin rich areas between the tapes, which is expected to reduce the bonding strength [16]. More matrix squeeze out was to be expected at higher temperature. It was seen that a higher temperature led to higher residue buildup on the quartz guides used for the flashlamp system. Chamfered guides were more susceptible to residue buildup, as the tape drags over the chamfered radiating surfaces at the end of the lay-up. The quartz guide with only one radiating surface (directed at the nip-point) was seen to be more stable in terms of temperaturem as less residue was building up on this geometry. The humidity was found be an significant factor regarding the bonding. Samples dried before placement had lower peel strength compared to nondried samples. This is the result of the plasticizing effect of moisture within the polymer matrix. The highest peel strength was found to be at a frequency of 80 Hz and a pulse width of 2200 µs. This sample furthermore showed the highest temperature. The variation in the wedge peel strength increased with higher temperature, as well as the variation within the temperature. The results for the DSC and TGA measurements in this study were expected. The increase in peel strength with power is to be expected; however, the differences between a higher frequency and a higher pulse width (while keeping the other parameter low) were not expected at identical power values. More research needs to be conducted to explain these results in more detail. An analysis of the emitted spectral range of the flashlamp system should be conducted to gain a deeper insight into the heat transfer of the heating system.

Author Contributions: Conceptualization, A.L.; methodology, A.L. and E.F.; validation, A.L.; formal analysis, A.L.; investigation, A.L.; resources, E.F.; data curation, A.L.; writing—original draft preparation, A.L.; writing—review and editing, E.F.; visualization, A.L.; supervision, E.F. All authors have read and agreed to the published version of the manuscript.

Funding: This research received no external funding.

Data Availability Statement: The data presented in this study are available upon reasonable request from the corresponding author.

Acknowledgments: Technical support from the Chair of Materials Science and Testing of Polymers and the Chair of Chemistry of Polymeric Materials, Montanuniversität Leoben, is kindly acknowledged.

Conflicts of Interest: The authors declare no conflicts of interest.

References

1. Neitzel, M.; Mitschang, P.; Breuer, U. (Eds.) *Handbuch Verbundwerkstoffe: Werkstoffe, Verarbeitung, Anwendung*; Carl Hanser Verlag GmbH Co KG: München, Germany, 2014.
2. Schledjewski, R. Thermoplastic tape placement process—In situ consolidation is reachable. *Plast. Rubber Compos.* **2009**, *38*, 379–386. [CrossRef]
3. Lukaszewicz, D.H.J.; Ward, C.; Potter, K.D. The engineering aspects of automated prepreg layup: History, present and future. *Compos. Part B Eng.* **2012**, *43*, 997–1009. [CrossRef]

4. Stokes-Griffin, C.M.; Compston, P. The effect of processing temperature and placement rate on the short beam strength of carbon fibre–PEEK manufactured using a laser tape placement process. *Compos. Part A Appl. Sci. Manuf.* **2015**, *78*, 274–283. [CrossRef]
5. Jiang, W.; Chen, C.; Chen, Z.; Huang, Z.; Zhou, H. Effect of crystallinity on optical properties of PEEK prepreg tapes for laser-assisted automated fiber placement. *Compos. Commun.* **2023**, *38*, 101490. [CrossRef]
6. Shafaq; Donough, M.J.; Farnsworth, A.L.; Phillips, A.W.; St John, N.A.; Gangadhara Prusty, B. Influence of deposition rates on the mode I fracture toughness of in-situ consolidated thermoplastic composites. *Compos. Part B Eng.* **2023**, *251*, 110474. [CrossRef]
7. Khodaei, A.; Shadmehri, F. Intimate contact development for automated fiber placement of thermoplastic composites. *Compos. Part C Open Access* **2022**, *8*, 100290. [CrossRef]
8. Zhang, C.; Duan, Y.; Xiao, H.; Wang, B.; Ming, Y.; Zhu, Y.; Zhang, F. Effect of porosity and crystallinity on mechanical properties of laser in-situ consolidation thermoplastic composites. *Polymer* **2022**, *242*, 124573. [CrossRef]
9. Oromiehie, E.; Prusty, B.G.; Compston, P.; Rajan, G. Automated fibre placement based composite structures: Review on the defects, impacts and inspections techniques. *Compos. Struct.* **2019**, *224*, 110987. [CrossRef]
10. Çelik, O.; Bussink, T.; Peeters, D.; Teuwen, J.; Dransfeld, C. The effect of Laser-Induced deconsolidation on the compaction behavior of thermoplastic composite tapes. *Compos. Part A Appl. Sci. Manuf.* **2021**, *151*, 106670. [CrossRef]
11. Miao, Q.; Dai, Z.; Ma, G.; Niu, F.; Wu, D. Effect of consolidation force on interlaminar shear strength of CF/PEEK laminates manufactured by laser-assisted forming. *Compos. Struct.* **2021**, *266*, 113779. [CrossRef]
12. Chanteli, A.; Bandaru, A.K.; Peeters, D.; O'Higgins, R.M.; Weaver, P.M. Influence of repass treatment on carbon fibre-reinforced PEEK composites manufactured using laser-assisted automatic tape placement. *Compos. Struct.* **2020**, *248*, 112539. [CrossRef]
13. Clancy, G.; Peeters, D.; Oliveri, V.; Jones, D.; O'Higgins, R.M.; Weaver, P.M. A study of the influence of processing parameters on steering of carbon Fibre/PEEK tapes using laser-assisted tape placement. *Compos. Part B Eng.* **2019**, *163*, 243–251. [CrossRef]
14. Martín, M.I.; Rodríguez-Lence, F.; Güemes, A.; Fernández-López, A.; Pérez-Maqueda, L.A.; Perejón, A. On the determination of thermal degradation effects and detection techniques for thermoplastic composites obtained by automatic lamination. *Compos. Part A Appl. Sci. Manuf.* **2018**, *111*, 23–32. [CrossRef]
15. Slange, T.K.; Warnet, L.L.; Grouve, W.; Akkerman, R. Deconsolidation of C/PEEK blanks: On the role of prepreg, blank manufacturing method and conditioning. *Compos. Part A Appl. Sci. Manuf.* **2018**, *113*, 189–199. [CrossRef
16. Stokes-Griffin, C.M.; Kollmannsberger, A.; Compston, P.; Drechsler, K. The effect of processing temperature on wedge peel strength of CF/PA 6 laminates manufactured in a laser tape placement process. *Compos. Part A Appl. Sci. Manuf.* **2019**, *121*, 84–91. [CrossRef]
17. Stokes-Griffin, C.M.; Compston, P. An inverse model for optimisation of laser heat flux distributions in an automated laser tape placement process for carbon-fibre/PEEK. *Compos. Part A Appl. Sci. Manuf.* **2016**, *88*, 190–197. [CrossRef]
18. Comer, A.J.; Ray, D.; Obande, W.O.; Jones, D.; Lyons, J.; Rosca, I.; O' Higgins, R.M.; McCarthy, M.A. Mechanical characterisation of carbon fibre–PEEK manufactured by laser-assisted automated-tape-placement and autoclave. *Compos. Part A Appl. Sci. Manuf.* **2015**, *69*, 10–20. [CrossRef]
19. Stokes-Griffin, C.M.; Compston, P. Optical characterisation and modelling for oblique near-infrared laser heating of carbon fibre reinforced thermoplastic composites. *Opt. Lasers Eng.* **2015**, *72*, 1–11. [CrossRef]
20. Khan, M.A.; Mitschang, P.; Schledjewski, R. Parametric study on processing parameters and resulting part quality through thermoplastic tape placement process. *J. Compos. Mater.* **2013**, *47*, 485–499. [CrossRef]
21. Pérez-Martín, H.; Buchalik-Bopp, S.; Guettler, B.E.; Mackenzie, P.; Baidak, A.; Ó Brádaigh, C.M.; Ray, D. Effect of crystallinity and morphology on the mechanical properties of CF/PEKK composites manufactured under compression moulding and automated tape placement. *Mater. Today Commun.* **2023**, *36*, 106442. [CrossRef]
22. Fuessel, L.; Cender, T.A.; Austermann, V.; Gillespie, J.W., Jr.; Heider, D. Tow Steering of stretchable TuFF thermoplastic tape with laser tape placement. In Proceedings of the SAMPE 2022—Charlotte, NC, Proceedings, Charlotte, NC, USA, 23–26 May 2022; pp. 700–711.
23. Çelik, O.; Peeters, D.; Dransfeld, C.; Teuwen, J. Intimate contact development during laser assisted fiber placement: Microstructure and effect of process parameters. *Compos. Part A Appl. Sci. Manuf.* **2020**, *134*, 105888. [CrossRef]
24. Chen, J.; Fu, K.; Li, Y. Understanding processing parameter effects for carbon fibre reinforced thermoplastic composites manufactured by laser-assisted automated fibre placement (AFP). *Compos. Part A Appl. Sci. Manuf.* **2021**, *140*, 106160. [CrossRef]
25. Chadwick, A.R.; Kotzur, K.; Nowotny, S. Moderation of thermoplastic composite crystallinity and mechanical properties through in situ manufacturing and post-manufacturing tempering: Part 1 – Mechanical characterisation. *Compos. Part A Appl. Sci. Manuf.* **2021**, *143*, 106286. [CrossRef]
26. Grouve, W.J.B.; Vanden Poel, G.; Warnet, L.L.; Akkerman, R. On crystallisation and fracture toughness of poly(phenylene sulphide) under tape placement conditions. *Plast. Rubber Compos.* **2013**, *42*, 282–288. [CrossRef]
27. Grouve, W.; Warnet, L.L.; Rietman, B.; Akkerman, R. On the weld strength of in situ tape placed reinforcements on weave reinforced structures. *Compos. Part A Appl. Sci. Manuf.* **2012**, *43*, 1530–1536. [CrossRef]
28. Stokes-Griffin, C.M.; Kollmannsberger, A.; Ehard, S.; Compston, P.; Drechsler, K. Manufacture of steel–CF/PA6 hybrids in a laser tape placement process: Effect of first-ply placement rate on thermal history and lap shear strength. *Compos. Part A Appl. Sci. Manuf.* **2018**, *111*, 42–53. [CrossRef]
29. Schaefer, P.M.; Guglhoer, T.; Sause, M.G.; Drechsler, K. Development of intimate contact during processing of carbon fiber reinforced Polyamide-6 tapes. *J. Reinf. Plast. Compos.* **2017**, *36*, 593–607. [CrossRef]

30. Schaefer, P.M.; Gierszewski, D.; Kollmannsberger, A.; Zaremba, S.; Drechsler, K. Analysis and improved process response prediction of laser-assisted automated tape placement with PA-6/carbon tapes using Design of Experiments and numerical simulations. *Compos. Part A Appl. Sci. Manuf.* **2017**, *96*, 137–146. [CrossRef]
31. Kukla, C.; Peters, T.; Janssen, H.; Brecher, C. Joining of Thermoplastic Tapes with Metal Alloys Utilizing Novel Laser Sources and Enhanced Process Control in a Tape Placement Process. *Procedia CIRP* **2017**, *66*, 85–90. [CrossRef]
32. Steyer, M. Laserunterstütztes Tapelegeverfahren zur Fertigung Endlosfaserverstärkter Thermoplastlaminate. Ph.D. Thesis, RWTH Aachen, Aachen, Germany, 2013.
33. Brecher, C.; Dubratz, M.; Stimpfl, J.; Emonts, M. Innovative manufacturing of 3D-lightweight components. *Laser Tech. J.* **2011**, *8*, 36–40. [CrossRef]
34. Chu, Q.; Li, Y.; Xiao, J.; Huan, D.; Zhang, X.; Chen, X. Processing and characterization of the thermoplastic composites manufactured by ultrasonic vibration–assisted automated fiber placement. *J. Thermoplast. Compos. Mater.* **2018**, *31*, 339–358. [CrossRef]
35. Rizzolo, R.H.; Walczyk, D.F. Ultrasonic consolidation of thermoplastic composite prepreg for automated fiber placement. *J. Thermoplast. Compos. Mater.* **2016**, *29*, 1480–1497. [CrossRef]
36. Baley, C.; Kervoëlen, A.; Lan, M.; Cartié, D.; Le Duigou, A.; Bourmaud, A.; Davies, P. Flax/PP manufacture by automated fibre placement (AFP). *Mater. Design* **2016**, *94*, 207–213. [CrossRef]
37. Brecher, C.; Emonts, M.; Schares, R.L.; Stimpfl, J. CO_2-laser-assisted processing of glass fiber-reinforced thermoplastic composites. In Proceedings of the High-Power Laser Materials Processing: Lasers, Beam Delivery, Diagnostics, and Applications II; SPIE Processings, San Francisco, CA, USA, 22 February 2013; pp. 144–154. [CrossRef]
38. Wazeer, A.; Das, A.; Abeykoon, C.; Sinha, A.; Karmakar, A. Composites for electric vehicles and automotive sector: A review. *Green Energy Intell. Transp.* **2023**, *2*, 100043. [CrossRef]
39. Nishida, H.; Carvelli, V.; Fujii, T.; Okubo, K. Thermoplastic vs. thermoset epoxy carbon textile composites. *IOP Conf. Ser. Mater. Sci. Eng.* **2018**, *406*, 012043. [CrossRef]
40. Moghadamazad, M.; Hoa, S.V. Models for heat transfer in thermoplastic composites made by automated fiber placement using hot gas torch. *Compos. Part C Open Access* **2022**, *7*, 100214. [CrossRef]
41. Gain, A.K.; Oromiehie, E.; Prusty, B.G. Nanomechanical characterisation of CF-PEEK composites manufactured using automated fibre placement (AFP). *Compos. Commun.* **2022**, *31*, 101109. [CrossRef]
42. Oromiehie, E.; Gain, A.K.; Donough, M.J.; Prusty, B.G. Fracture toughness assessment of CF-PEEK composites consolidated using hot gas torch assisted automated fibre placement. *Compos. Struct.* **2022**, *279*, 114762. [CrossRef]
43. Satheesh, B.; Tonejc, M.; Potakowskyj, L.; Pletz, M.; Fauster, E.; Kaynak, B.; Schledjewski, R. Peel strength characterisation on ply/ply interface using wedge and T-peel/pull-type tests. *Polym. Polym. Compos.* **2018**, *26*, 431–445. [CrossRef]
44. Rajasekaran, A.; Shadmehri, F. Steering of carbon fiber/PEEK tapes using Hot Gas Torch-assisted automated fiber placement. *J. Thermoplast. Compos. Mater.* **2022**, *36*, 1651–1679. [CrossRef]
45. Grouve, W. Weld Strength of Laser-Assisted Tape-Placed Thermoplastic Composites. Ph.D. Thesis, University of Twente, Enschede, The Netherlands, 2012.
46. Danezis, A.; Williams, D.; Edwards, M.; Skordos, A.A. Heat transfer modelling of flashlamp heating for automated tape placement of thermoplastic composites. *Compos. Part A Appl. Sci. Manuf.* **2021**, *145*, 106381. [CrossRef]
47. Di Boon, Y.; Joshi, S.C.; Bhudolia, S.K. Review: Filament Winding and Automated Fiber Placement with In Situ Consolidation for Fiber Reinforced Thermoplastic Polymer Composites. *Polymers* **2021**, *13*, 1951. [CrossRef]
48. Meister, S.; Kolbe, A.; Groves, R.M. Reflectivity and emissivity analysis of thermoplastic CFRP for optimising Xenon heating and thermographic measurements. *Compos. Part A Appl. Sci. Manuf.* **2022**, *158*, 106972. [CrossRef]
49. Brandt, L.; Deden, D.; Fischer, F.; Bruckner, F.; Dreher, P.; Williams, D.J.; Engelschall, M.; Nieberl, D.; Nowotny, S. Xenon flashlamp based in-situ automated fiber placement of thermoplastic composites. In Proceedings of the 2019 International Conference on Composite Materials (ICCM), Melbourne, Australia, 11–16 August 2019; pp. 1–8.
50. Monnot, P.; Williams, D.; Di Francesco, M. Power Control of a Flashlamp-based Heating Solution for Automated Dry Fibre Placement. In Proceedings of the Proceedings 18th European Conference on Composite Materials, Athens, Greece, 25–28 June 2018; pp. 1–8.
51. Kollmannsberger, A. Heating Characteristics of Fixed Ofcus Laser Assisted Thermoplastic-Automated Fiber Placement of 2D and 3D Parts. Ph.D. Thesis, Technische Universität München, Munich, Germany, 2019.
52. Dolo, G.; Férec, J.; Cartié, D.; Grohens, Y.; Ausias, G. Model for thermal degradation of carbon fiber filled poly(ether ether ketone). *Polym. Degrad. Stab.* **2017**, *143*, 20–25. [CrossRef]
53. Stokes-Griffin, C.M.; Compston, P. Investigation of sub-melt temperature bonding of carbon-fibre/PEEK in an automated laser tape placement process. *Compos. Part A Appl. Sci. Manuf.* **2016**, *84*, 17–25. [CrossRef]
54. Stokes-Griffin, C.M.; Compston, P. A combined optical-thermal model for near-infrared laser heating of thermoplastic composites in an automated tape placement process. *Compos. Part A Appl. Sci. Manuf.* **2015**, *75*, 104–115. [CrossRef]
55. Schledjewski, R. Mechanical performance of in-situ consolidated thermoplastic fiber reinforced tape materials. In Proceedings of the 11th European Conference on Composite Materials, Rhodes, Greece, 31 May–3 June 2004; ECCM, Ed.; pp. 1–8.
56. Wanigasekara, C.; Oromiehie, E.; Swain, A.; Prusty, B.G.; Nguang, S.K. Machine Learning Based Predictive Model for AFP-Based Unidirectional Composite Laminates. *IEEE Trans. Ind. Inform.* **2020**, *16*, 2315–2324. [CrossRef]

57. Qureshi, Z.; Swait, T.; Scaife, R.; El-Dessouky, H.M. In situ consolidation of thermoplastic prepreg tape using automated tape placement technology: Potential and possibilities. *Compos. Part B Eng.* **2014**, *66*, 255–267. [CrossRef]
58. Schäfer, P.M. Consolidation of Carbon Fiber Reinforced Polyamide 6 Tapes Using Laser-Assisted Tape Placement. Ph.D. Thesis, Technische Universität München, Munich, Germany, 2018.
59. Zhang, C.; Duan, Y.; Xiao, H.; Wang, B.; Ming, Y.; Zhu, Y.; Zhang, F. The effects of processing parameters on the wedge peel strength of CF/PEEK laminates manufactured using a laser tape placement process. *Int. J. Adv. Manuf. Technol.* **2022**, *120*, 7251–7262. [CrossRef]
60. Hulcher, B.; Marchello, J.M.; Hinkley, J.A. Correlation between double cantilever beam and wedge peel tests for automated tow placement. In Proceedings of the 43rd International SAMPE Symposion and Exhibition, Anaheim, CA, USA, 31 May–4 June 1998; SAMPE, Ed.; pp. 1955–1965.
61. Bonifacio, A.; Bonetti, L.; Piantanida, E.; de Nardo, L. Plasticizer design strategies enabling advanced applications of cellulose acetate. *Eur. Polym. J.* **2023**, *197*, 112360. [CrossRef]
62. Piao, H.; Chen, L.; Kiryu, Y.; Ohsawa, I.; Takahashi, J. Influence of Water Absorption and Temperature on the Mechanical Properties of Discontinuous Carbon Fiber Reinforced Polyamide 6. *Fibers Polym.* **2019**, *20*, 611–619. [CrossRef]
63. Liang, J.; Xu, Y.; Wei, Z.; Song, P.; Chen, G.; Zhang, W. Mechanical properties, crystallization and melting behaviors of carbon fiber-reinforced PA6 composites. *J. Therm. Anal. Calorim.* **2014**, *115*, 209–218. [CrossRef]
64. Thirumalai, D.P.R.; Andersen, T.L.; Lystrup, A. Influence of Moisture Absorption on Properties of Fiber Reinforced Polyamide 6 Composites. In Proceedings of the 26th Annual Technical Conference of the American Society for Composites 2011 and the 2nd Joint US-Canada Conference on Composites, Montreal, QC, Canada, 26–28 September 2011; Hyer, M., Ed.; pp. 500–510.
65. Levchik, S.V.; Weil, E.D.; Lewin, M. Thermal decomposition of aliphatic nylons. *Polym. Int.* **1999**, *48*, 532–557. [CrossRef]
66. An, H.J.; Kim, J.S.; Kim, K.Y.; Lim, D.Y.; Kim, D.H. Mechanical and thermal properties of long carbon fiber-reinforced polyamide 6 composites. *Fibers Polym.* **2014**, *15*, 2355–2359. [CrossRef]
67. Tranchard, P.; Duquesne, S.; Samyn, F.; Estèbe, B.; Bourbigot, S. Kinetic analysis of the thermal decomposition of a carbon fibre-reinforced epoxy resin laminate. *J. Anal. Appl. Pyrolysis* **2017**, *126*, 14–21. [CrossRef]
68. Luederwald, I.; Merz, F.; Rothe, M. Ueber den thermischen abbau des poly-e-caprolactams (Nylon-6). *Angew. Makromol. Chem.* **1978**, *67*, 193–202. [CrossRef]
69. Hanna, A.A. Thermal and dielectric properties of nylon 6. *Thermochim. Acta* **1984**, *76*, 97–103. [CrossRef]
70. Fornes, T.D.; Paul, D.R. Crystallization behavior of nylon 6 nanocomposites. *Polymer* **2003**, *44*, 3945–3961. [CrossRef]
71. Campoy, I.; Gómez, M.A.; Marco, C. Structure and thermal properties of blends of nylon 6 and a liquid crystal copolyester1Dedicated to the memory of Prof. J.G. Fatou.1. *Polymer* **1998**, *39*, 6279–6288. [CrossRef]
72. Hiramatsu, N.; Hirakawa, S. Melting and Transformation Behavior of y Form Nylon 6 under High Pressure. *Polym. J.* **1982**, *14*, 165–171. [CrossRef]
73. Kyotani, M.; Mitsuhashi, S. Studies on crystalline forms of nylon 6. II. Crystallization from the melt. *J. Polym. Sci. Part A-2 Polym. Phys.* **1972**, *10*, 1497–1508. [CrossRef]
74. Millot, C.; Fillot, L.A.; Lame, O.; Sotta, P.; Seguela, R. Assessment of polyamide-6 crystallinity by DSC. *J. Therm. Anal. Calorim.* **2015**, *122*, 307–314. [CrossRef]
75. Parodi, E.; Govaert, L.E.; Peters, G. Glass transition temperature versus structure of polyamide 6: A flash-DSC study. *Thermochim. Acta* **2017**, *657*, 110–122. [CrossRef]
76. Seguela, R. Overview and critical survey of polyamide6 structural habits: Misconceptions and controversies. *J. Polym. Sci.* **2020**, *58*, 2971–3003. [CrossRef]
77. Khanna, Y.P.; Kuhn, W.P. Measurement of crystalline index in nylons by DSC: Complexities and recommendations. *J. Polym. Sci. Part B Polym. Phys.* **1997**, *35*, 2219–2231. [CrossRef]
78. Heinrich, M.; Decker, R.; Reindel, P.; Böttcher, K.; Roth-Panke, I.; Kroll, L. Effect of acoustic excitation on fiber-reinforced polypropylene and the influence on melt viscosity. *Int. J. Adv. Manuf. Technol.* **2021**, *117*, 2395–2403. [CrossRef]

materials

MDPI

Article

A Peel Test Method to Characterize the Decay Law of Prepreg Tape Tack at Different Temperatures

Jiaqi Shi, Wang Wang, Yuequan Wang *, Junwei Qi and Jun Xiao *

College of Material Science and Technology, Nanjing University of Aeronautics and Astronautics, Nanjing 210016, China
* Correspondence: yuequanwang@nuaa.edu.cn (Y.W.); j.xiao@nuaa.edu.cn (J.X.)

Abstract: The tack of prepreg is a key factor affecting the automatic tape laying process. During the manufacturing process of large composite parts, prepreg material may be stored at room temperature for several days, resulting in a decrease in its tack. In this study, a new tack test tool was designed, and the decay rate of prepreg tack at different temperatures was tested. We proposed a prepreg tack decay model, which assumes that the main factor in tack decay is the reduction in resin chain activity during storage. The maximum deviation between the model calculation results and the experimental results of the tack decay rate is 9.7%. This study also proposed a new statistical unit for prepreg tack, which can establish the relationship between the tack of prepreg and its remaining storage time and reduce prepreg management costs.

Keywords: handling life; tack; prepreg; composite

1. Introduction

Composite materials can enhance the fuel efficiency of commercial aircrafts; thus, their mass fraction becomes a benchmark for assessing the manufacturing level of commercial aircraft. Currently, the main automated production processes for producing aircraft composite parts are automated fiber placement (AFP) and automatic tape laying (ATL). The ATL process is often used in the production of large wings. Currently, the ATL process for wings employs thermosetting prepreg materials, and the tack of thermosetting prepregs is a critical factor affecting the manufacturing process [1]. Prepregs are stored at room temperature during production, and large parts may cause this storage to approach the handling life limit of the prepreg [2]. Therefore, it is necessary to study the decay of prepreg tack with storage time. However, the tack of prepreg as a material property lacks clear quantification testing standards and mechanistic models [3].

Initial studies on the tack of prepregs used pressure-sensitive adhesive testing methods, such as ASTM D6195 adhesive loop tack strength testing [4] and ASTM D3121 rolling ball test [5]. However, both of these test methods are difficult to accurately measure prepreg tack. Dubois et al. [6] used a probe method to measure the tack of prepreg and analyzed the load curves during bonding and separation. However, this method is essentially a modified version of measuring tack using fingers [7]. Ahn et al. [8] measured the bonding strength between prepregs by compressing and separating multiple layers of prepreg. This test method of removing the prepreg as a whole is very different from the process of peeling the prepreg from the laminate. Nguyen et al. [9] designed a single-lap-based adhesion measurement method that is suitable for evaluating the connection strength of lap joints used for prepreg extension in automated production. Böckl et al. [10] developed a friction-based tack test method that measures the lateral friction of the prepreg tape passing through a measuring roller. The test results can be used to monitor the automatic placement process, but the transferability of the test results to the prepreg bonding strength remains to be verified. Crossley [11] designed a tack-measuring fixture that measures the force required to peel a prepreg from a stainless steel plate. Endruweit et al. [12,13] used the same method to study the effect of tape laying process parameters on prepreg tack. Brooks [14] also used the peeling method to measure the tack of prepreg. His test method had an independent sample preparation device, and he tested the peeling force using a universal testing machine. The peel test method designed by Crossley and Brooks is closer to the reverse process of tape laying and is more suitable as a quantitative test standard for prepreg tack.

Researchers found that the tack of prepregs decreased with aging until it could not be measured [15,16]. The aging process is also accompanied by changes in the properties of the resin. Researchers have found that high-pressure liquid chromatography shows that the resin composition will change during the aging process [17]. Moreover, aging will increase the glass transition temperature (Tg) of the prepreg [18], and Fourier-transform infrared spectroscopy results also prove that the functional groups of the resin will change during the aging process [19]. The researchers not only tested the tack decay of prepregs but also proposed different models for the mechanism of tack decay. The tack decay of prepreg is a change in its surface properties [20]. Unlike adhesives that form chemical bonds with the bonded surfaces [21,22], the prepreg tape laying process only takes a few seconds, which is not enough time for the reaction to form a chemical connection. The adhesive force comes from the van der Waals force and hydrogen bonds generated by the resin chains [23,24]. The changes in the resin during the aging process will lead to a decrease in the van der Waals force and hydrogen bond strength between the prepreg resin molecules, thereby reducing the tack.

This study designed a tack testing device based on the automatic tape laying machine, and measured the tack changes of prepreg during the aging process. Based on the test results, a prepreg tack decay model based on resin molecule movement and the Arrhenius equation was proposed. Finally, this study proposes a new tack measurement unit to reduce the prepreg handling life management cost.

2. Experimental Section

2.1. Experimental Device

The test device of Crossley is a peeling fixture that needs to be operated by a universal testing machine. Therefore, the speed, pressure, and other parameters will be limited by the performance of the testing machine. Brooks used a specialized sample preparation device to overcome the parameter limitation, but there was no fixture constraint during the peeling process in his test method, so the measured load curve fluctuated greatly. This study combines the advantages of the two methods, specialized sampling device that allows the parameters to be set closer to the ATL process, and the design of a peeling fixture to make the load curves less noisy. The test device is shown in Figure 1, which includes an automatic tape laying device and a peeling fixture. The automatic tape laying device was designed based on the basic functional requirements of the ATL machine and consists of a work platform and a laying head. The work platform surface can fix the prepreg by negative pressure, and a weighing sensor is placed at the bottom of the work platform to provide feedback on the laying pressure. The laying head consists of a 60 mm diameter rubber pressure roller, cylinder, and support frame. The hardness value of the pressure roller is 30A–35A. The cylinder provides uniform and controllable pressure to the prepreg tape being laid through the pressure roller. The laying head is installed on the guide rail of the work platform and is driven by a motor, allowing the platform to lay prepreg up to a maximum width of 150 mm. The entire system is controlled by a PLC (Programmable Logic Controller). During the sample preparation process, the prepreg to be bonded is placed on the platform and in the guide groove, and the laying pressure and speed are set. The device then starts to complete the bonding. In this study, the laying speed was set to 50 mm/s, and the laying pressure was set to 500 N.

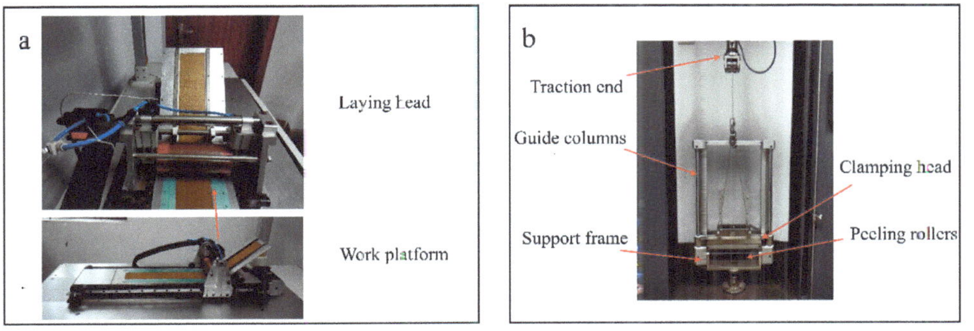

Figure 1. Test device. (**a**) Automatic tape-laying platform; (**b**) peeling test fixture.

To solve the problem of high noise from the force sensor caused by sample oscillation during peeling, this work designed a special peeling fixture, as shown in Figure 1b. The fixture consists of a support frame, clamping head, traction end, guide columns, and peeling rollers. The clamping head is connected to the traction end via a wire rope pulley system. After clamping the prepreg, the universal testing machine pulls the traction end to peel it. During the peeling process, the guide column on the frame can constrain the prepreg to only move vertically, the peeling roller constrains the peeling angle of the prepreg, and the transparent limiting plate constrains the swinging direction of the prepreg.

The aging experiment was conducted using the CH225R-type environmental test chamber from Tuode Environmental Testing Equipment Co., Ltd, Dongguan, China. This chamber has a temperature control accuracy of ±1 °C within the range of −10 °C to 100 °C, and a humidity control accuracy of ±5% RH.

2.2. Experimental Materials

This paper tests three types of carbon fiber epoxy uni-directional prepregs that are widely used in the aerospace industry, namely T, C, and H. The T-type prepreg has a resin content of 35% by weight, while the C-type and H-type prepregs have a resin content of 34% by weight. The manufacturer of T-type prepregs has provided a handling life table, which shows the decay rate of the handling life at different aging temperatures, as presented in Table 1.

Table 1. Manufacturer's recommended handling life table for T-type prepregs.

Temperature	Handling Life Decay Rate Ratio
<26 °C	1
26~32 °C	2
32~37 °C	3
37~43 °C	4.5

2.3. Experiment Procedure

To evaluate the tack decay rate of T-type prepreg at different aging temperatures, an experimental matrix was established, as presented in Table 2. Since the humidity in the production workshop was constant and the manufacturer's data did not include changes in humidity, it was not considered a variable in this study. To ensure significant differences among the test values, the most severe value in Table 1 was chosen as the experimental condition, and the humidity was set at 65% RH. To guarantee uniform changes in the life value between adjacent test points, aging times at various temperatures were calculated based on the life decay rate ratio.

Table 2. Aging test table.

Aging temperature/°C	43	37	32	26
Aging time/Hour	5.3	8	12	24
	21.3	32	48	96
	37.3	56	84	168
	53.3	80	120	240
	69.3	104	156	312

The prepregs were cut to the size shown in Figure 2a and then placed into the CH225R-type environmental test chamber. After aging, the prepregs were transferred to a constant temperature room at 23 °C for testing following sample preparation on the automatic lay-up platform. The samples were clamped on the designed peeling fixture and tested. The load–displacement curve obtained from the experiment includes a low plateau region and a high plateau region, as shown in Figure 2b. The former is due to the frictional force generated by the entire system, and the latter is the result of the former plus the tack of the prepreg. Figure 2c shows that the resin distribution is inhomogeneous, and therefore the loads do not present a straight line. Contamination of the bonding surface was prevented during the aging process and sample preparation. To minimize sample aging during the waiting period, the nine samples were divided into three groups for testing, and each point on the graph was obtained from the average of nine experiments.

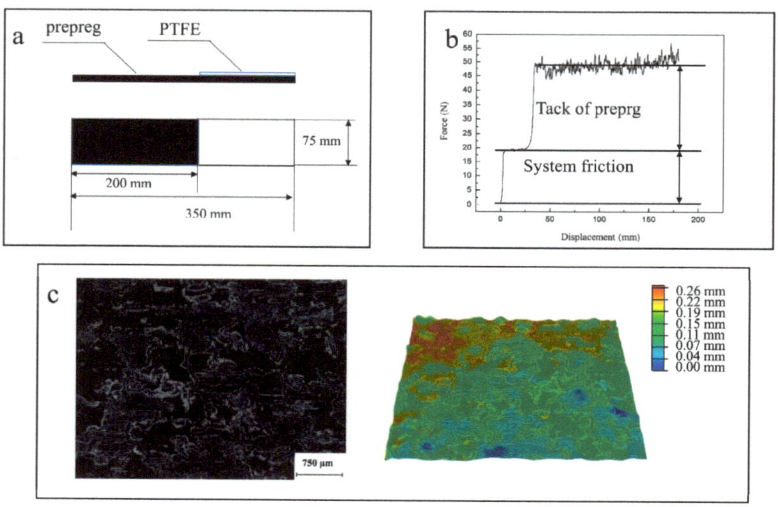

Figure 2. (**a**) Dimensions of the specimen. (**b**) Load–displacement curve for T stripping. (**c**) Inhomogeneous distribution of the resin.

3. Results Analysis and Discussion

3.1. Analysis of Aging Results

The tack decay curve of the T-type prepreg after aging is shown in Figure 3. Due to the long experimental time, it is impossible to use the same batch of prepreg to complete all aging test points within its process life. Therefore, the experiment only ensures that the same batch of prepreg is used at the test points of each curve to ensure the continuity of the aging process. The slope of the tack decay rate after aging at different temperatures was fitted, and it was found that the decay rate increased with increasing temperature. We calculated the ratio of decay rates at different temperatures relative to 26 °C. The ratio of decay speed at different temperatures was 1.0 (26 °C):1.9 (32 °C):3.1 (37 °C):4.9 (43 °C). The Nash–Sutcliffe model efficiency coefficient test formula for the fitting results is as follows:

$$R^2 = 1 - \frac{SS_{res}}{SS_{tot}} \tag{1}$$

$$SS_{res} = \sum_i (y_i - \hat{y}_i)^2 \tag{2}$$

$$SS_{tot} = \sum_i (y_i - \overline{y})^2 \tag{3}$$

where y is the true value, $\hat{y}$ is the predicted value, $\overline{y}$ is the mean value, R^2 is the coefficient of determination, SS_{res} is the sum squared regression, and SS_{tot} is the total sum of squares.

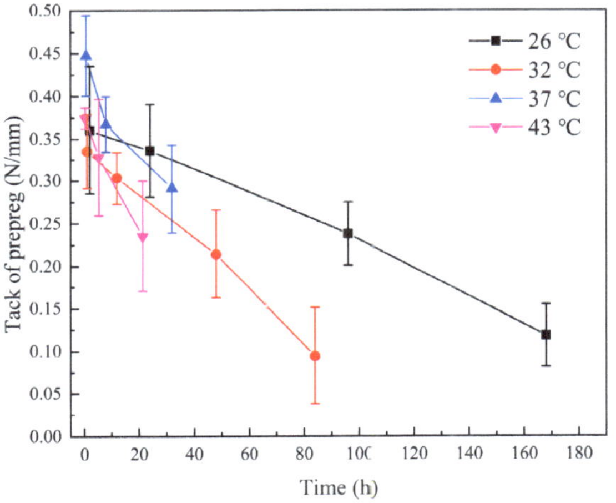

Figure 3. Decay of tacks of T-type prepreg at different aging temperatures.

The results of R^2 are shown in Table 3, which shows a good fit for the decay results. The deviations between the ratio of the fitted results and the manufacturer's results were −5% (32 °C), 3.3% (37 °C), and 8.9% (43 °C), respectively. While it is not possible to know the manner in which the manufacturer evaluates the prepreg decay rate, the methodology used herein yields similar results.

Table 3. R^2 test for T-type prepreg.

Temperature	T-Type Prepreg of R^2
26 °C	0.996
32 °C	0.995
37 °C	0.905
43 °C	0.989

3.2. Tack Decay Model

In this study, a tack decay model of prepreg based on resin chain diffusion and the Arrhenius equation was proposed. Wool [25] proposed a resin movement theory for thermoplastic resin welding. The model assumes that the depth of resin chain diffusion determines the joint strength, and the movement of the resin chain can be equivalent to a random walk chain diffusing in a tube. The resin chain movement ability is affected by the molecular mass of the resin. The prepreg bonding process also involves the formation of van der Waals forces and hydrogen bonds between resin chains after resin diffusion.

Therefore, the model proposed by Wool is also applicable to uncured thermosetting resins. The resin chain movement model is as follows:

$$\sigma_d = q n_0 \chi \tag{4}$$

where σ_d indicates the force between resin chains, q is a constant, n_0 means the total number of constraints per unit volume of the virgin bulk material, and χ means chains self-diffusing across the interface to an interpenetration depth.

The diffusion motion of the molecular chain can be equivalent to the model of a random walk chain diffusing in a tube [26,27], as shown in Figure 4.

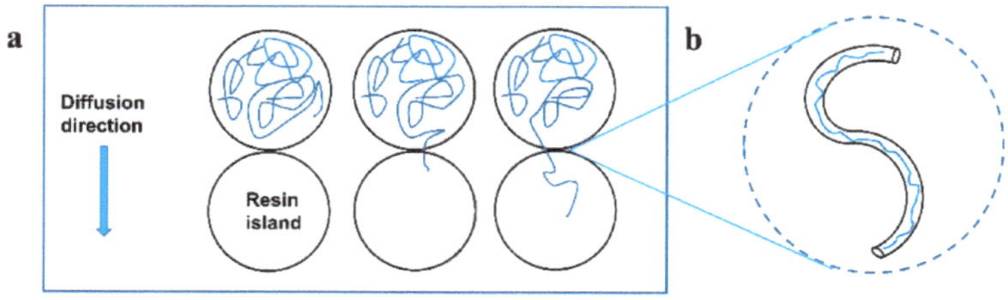

Figure 4. (**a**) Processes of resin movement. (**b**) The model of a random walk chain diffusing in a tube.

The relationship of random penetration depth can be obtained as follows:

$$\left\langle \chi^2 \right\rangle^{1/2} = \alpha(2D_c t)^{1/4} \tag{5}$$

where α is the coefficient, D_c is the diffusion coefficient, and t is the diffusion time.

Since D_c depends on the molar mass M of the resin,

$$D_c \sim 1/M \tag{6}$$

The diffusion capacity of the resin chain depends on its molecular mass, and cross-linking reactions of the resin will increase its molecular mass and thus reduce its diffusion capacity. Therefore, the rate of decay of prepreg tack at different temperatures should be positively correlated with the rate of reaction of the resin. The reaction rate of the resin at different temperatures follows the Arrhenius equation.

$$k = Ae^{-E_a/RT} \tag{7}$$

where k is the reaction rate constant, A is the Arrhenius constant, E_a is the activation energy of the reaction, R is the gas constant, and T is the thermodynamic temperature.

Janković [28] used the invariant kinetic parameters method to obtain a reaction activation energy of 71.4 kJ/mol for the T-type prepreg's resin. The ratios of the reaction rates calculated using the Arrhenius Equations (7) were 1.0 (26 °C), 1.8 (32 °C), 2.8 (37 °C), and 4.7 (43 °C), respectively. This result deviated from the data provided by the manufacturer of the T-prepreg by −10.0% (32 °C), −6.7% (37 °C), and 4.4% (43 °C), and from the experimental results by −5.3% (32 °C), −9.7% (37 °C), and −4.1% (43 °C). This result indicates that the results of tack decay are consistent with the rate of chemical reaction and indirectly demonstrates that the curing reaction of the resin is a factor in the tack decay of the prepreg. The ratios of decay rates are shown in Figure 5.

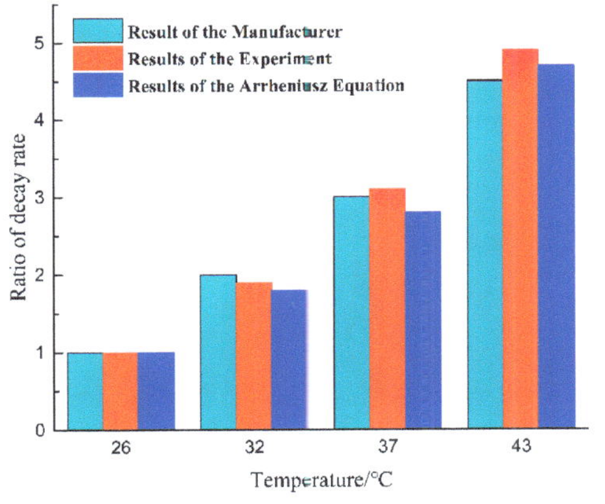

Figure 5. Ratios of tack decay rates obtained by different methods.

3.3. The Unit of Handling Life

In order to establish a relationship between the change in tack value of the prepregs and time, the magnitude of the decrease in tack for one hour of aging at 26 °C/65% RH was recorded as a unit of handling life. A 26 °C/65% RH meets the environmental conditions of most composite material workshops. Using the unit of handling life as the horizontal coordinate, the curve of tack change with the unit of handling life is plotted. Figure 6 shows the change in tack of T-type prepregs with handling life unit. The four curves in the figure show the same slope, which indicates that when the handling life unit is the horizontal coordinate, the tack decay at different temperatures shows the same law. The unit of handling life allows the composite manufacturer to translate the available tack range of the prepreg into the maximum operating downtime at 26 °C/65% RH. Even if the prepreg is subjected to complex temperature aging during downtime or transport, the remaining handling life can be determined by measuring the difference between the tack of the prepreg and the minimum tack permitted for production. The handling life unit establishes a relationship between tack and remaining shelf life and reduces the cost of managing prepregs for manufacturers.

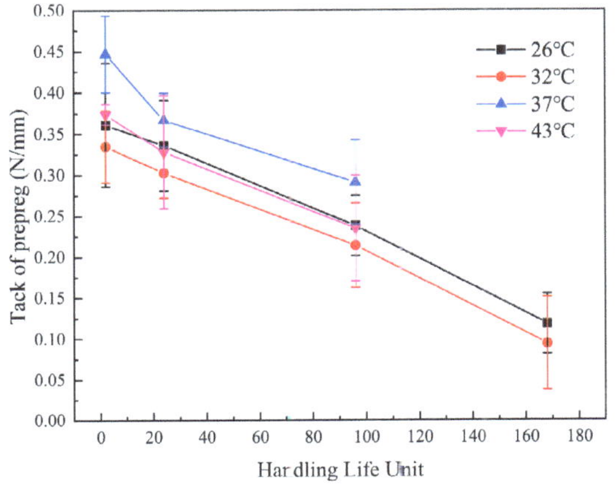

Figure 6. The tack of T-prepregs at different aging temperatures decays with handling life unit.

3.4. Model Adaptability

To evaluate the method's applicability to other products, C-type and H-type epoxy prepregs were also tested using the method, and the results are presented in Figures 7a and 8a. The decay rates of C-type prepreg at different aging temperatures are 1 (26 °C):2 (32 °C):3 (37 °C):12.3 (43 °C), while those of H-type are 1 (26 °C):2 (32 °C):3.6 (37 °C):4.6 (43 °C). The decay curves for each prepreg are parallel when using the handling life unit as the horizontal coordinate, as shown in Figures 7b and 8b.

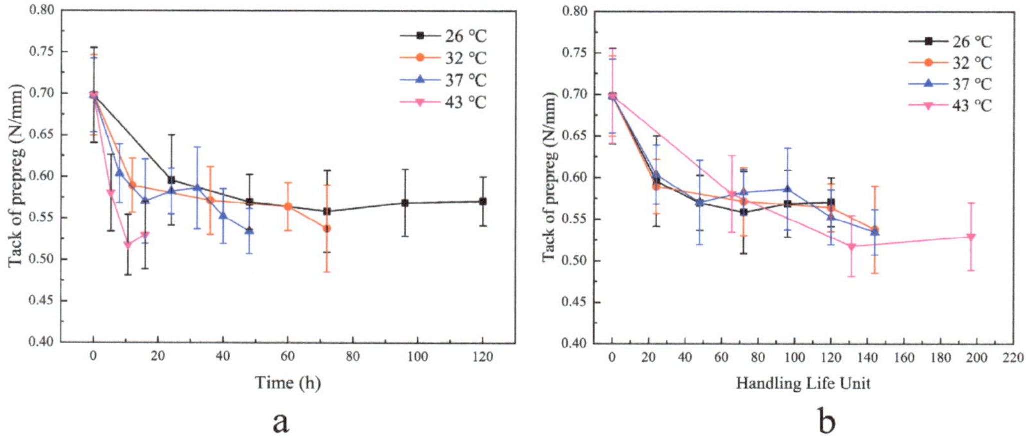

Figure 7. Decay of tack of C-type prepreg at different aging temperatures. (**a**) Change in tack over time. (**b**) Change in tack with handling life unit.

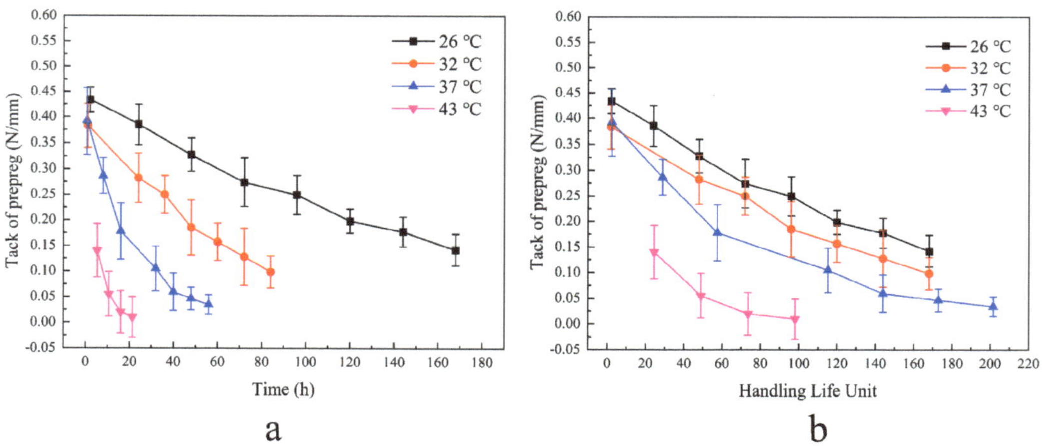

Figure 8. Decay of tack of H-type prepreg at different aging temperatures. (**a**) Change in tack over time. (**b**) Change in tack with handling life unit.

The test method in this paper can also obtain the results of its tack decay rate at different temperatures on H-type and C-type prepregs. It proves that this method has good adaptability between different types of prepregs. The R^2-test results of the above two prepregs are shown in Table 4. It can be seen from the R^2 results that the linear fitting effect of C-type prepreg is worse than that of H-type prepreg. The reason is that with the increase in aging time, the attenuation amplitude of C-type prepreg gradually decreases. It has been found that C-type prepregs add thermoplastic components to the resin in order to improve toughness [29], and therefore may have an effect on the change in tack, which requires further study.

Table 4. R^2 test for H-type prepreg and C-type prepreg.

Temperature	H-Type Prepreg of R^2	C-Type Prepreg of R^2
26 °C	0.981	0.547
32 °C	0.983	0.705
37 °C	0.900	0.707
43 °C	0.862	0.793

4. Conclusions

In this paper, a new prepreg tack test device is designed. The device was used to test the law of prepreg tack decay over time at different temperatures. Based on the experimental results, a prepreg tack decay model was proposed. The model assumes that the resin reaction reduces the mobility of the resin chain, resulting in a decrease in tack. The reaction rates at different temperatures were calculated using the Arrhenius equation, and the maximum deviation from the decay rate measured by the tack experiment was 9.7%. We proposed the concept of a handling life unit, which is used as a life unit by measuring the value of tack lost by aging a prepreg for one hour at the operational ambient temperature. This concept allows for the establishment of a relationship between prepreg tack and remaining shelf life, reducing the administrative costs for manufacturers.

Author Contributions: Data curation, J.S. and W.W.; Writing—original draft, J.S.; Writing—review & editing and Resources, Y.W. and J.Q.; Supervision, J.X. All authors have read and agreed to the published version of the manuscript.

Funding: This research did not receive any specific grant from funding agencies in the public, commercial, or not-for-profit sectors.

Institutional Review Board Statement: Not applicable.

Informed Consent Statement: Not applicable.

Data Availability Statement: Dataset available on request from the authors.

Conflicts of Interest: The authors declare no conflict of interest.

References

1. Putnam, J.W.; Seferis, J.C.; Pelton, T.; Wilhelm, M. Perceptions of Prepreg Tack for Manufacturability in Relation to Experimental Measures. *Sci. Eng. Compos. Mater.* **1995**, *4*, 143–154. [CrossRef]
2. Miller, S.G.; Sutter, J.K.; Scheiman, D.A.; Maryanski, M.; Schlea, M. Study of Out-Time on the Processing and Properties of IM7/977-3 Composites. In Proceedings of the SAMPE, Seattle, WA, USA, 17 May 2010.
3. Budelmann, D.; Schmidt, C.; Meiners, D. Prepreg Tack: A Review of Mechanisms, Measurement, and Manufacturing Implication. *Polym. Compos.* **2020**, *41*, 3440–3458. [CrossRef]
4. Pierik, R. Experimental Setup and Method for the Characterization of Ply-Ply Adhesion for Fiber-Reinforced Thermoplastics in Melt. In Proceedings of the 26th International ESAFORM Conference on Material Forming, ESAFORM 2023, Kraków, Poland, 19–21 April 2023; pp. 267–276.
5. Jois, K.C.; Mölling, T.; Schuster, J.; Grigat, N.; Gries, T. Towpreg Manufacturing and Characterization for Filament Winding Application. *Polym. Compos.* **2024**. [CrossRef]
6. Dubois, O.; Le Cam, J.B.; Béakou, A. Experimental Analysis of Prepreg Tack. *Exp. Mech.* **2010**, *50*, 599–606. [CrossRef]
7. Mohammed, I.K.; Charalambides, M.N.; Kinloch, A. Modelling the Interfacial Peeling of Pressure-Sensitive Adhesives. *J. Nonnewton. Fluid Mech.* **2015**, *222*, 141–150. [CrossRef]
8. Ahn, K.J.; Seferis, J.C.; Pelton, T.; Wilhelm, M. Analysis and Characterization of Prepreg Tack. *Polym. Compos.* **1992**, *13*, 197–206. [CrossRef]
9. Nguyen, C.D.; Krombholz, C. Influence of Process Parameters and Material Aging on the Adhesion of Prepreg in AFP Processes. In Proceedings of the ECCM 2016—Proceeding of the 17th European Conference on Composite Materials, Munich, Germany, 26–30 June 2016; pp. 26–30.

10. Böckl, B.; Jetten, C.; Heller, K.; Ebel, C.; Drechsler, K. Online Monitoring System for the Tack of Prepreg Slit Tapes Used in Automated Fiber Placement. In Proceedings of the ECCM18—18th European Conference on Composite Materials, Athens, Greece, 25–28 June 2018.
11. Crossley, R.J.; Schubel, P.J.; Warrior, N.A. Experimental Determination and Control of Prepreg Tack for Automated Manufacture. *Plast. Rubber Compos.* **2011**, *40*, 363–368. [CrossRef]
12. Endruweit, A.; Choong, G.Y.H.; Ghose, S.; Johnson, B.A.; Younkin, D.R.; Warrior, N.A.; De Focatiis, D.S.A. Characterisation of Tack for Uni-Directional Prepreg Tape Employing a Continuous Application-and-Peel Test Method. *Compos. Part A Appl. Sci. Manuf.* **2018**, *114*, 295–306. [CrossRef]
13. Endruweit, A.; Ghose, S.; Johnson, B.A.; Kelly, S.; De Focatiis, D.S.A.; Warrior, N.A. Tack testing to aid optimisation of process parameters for automated material placement in an industrial environment. In Proceedings of the 21st International Conference on Composite Materials, Xi'an, China, 20–25 August 2017; pp. 20–25.
14. Brooks, J.R.; Platt, P.R. 5513537 Method and Apparatus to Determine Composite Prepreg Tack. *Compos. Part A Appl. Sci. Manuf.* **1996**, *27*, 1114.
15. Cole, K.C.; Noël, D.; Hechler, J.-J.; Cielo, P.; Krapez, J.-C.; Chouliotis, A.; Overbury, K.C. Room-temperature Aging of Narmco 5208 Carbon-epoxy Prepreg. Part II: Physical, Mechanical, and Nondestructive Characterization. *Polym. Compos.* **1991**, *12*, 203–212. [CrossRef]
16. Blass, D.; Kreling, S.; Dilger, K. The Impact of Prepreg Aging on Its Processability and the Postcure Mechanical Properties of Epoxy-Based Carbon-Fiber Reinforced Plastics. *Proc. Inst. Mech. Eng. Part L J. Mater. Des. Appl.* **2017**, *231*, 62–72. [CrossRef]
17. Scola, D.A.; Vontell, J.; Felsen, M. Effects of Ambient Aging of 5245C/Graphite Prepreg on Composition and Mechanical Properties of Fabricated Composites. *Polym. Compos.* **1987**, *8*, 244–252. [CrossRef]
18. Hübner, F.; Meuchelböck, J.; Wolff-Fabris, F.; Mühlbach, M.; Altstädt, V.; Ruckdäschel, H. Fast Curing Unidirectional Carbon Epoxy Prepregs Based on a Semi-Latent Hardener: The Influence of Ambient Aging on the Prepregs Tg0, Processing Behavior and Thus Derived Interlaminar Performance of the Composite. *Compos. Sci. Technol.* **2021**, *216*, 109047. [CrossRef]
19. Jones, R.W.; Ng, Y.; McClelland, J.F. Monitoring Ambient-Temperature Aging of a Carbon-Fiber/Epoxy Composite Prepreg with Photoacoustic Spectroscopy. *Compos. Part A Appl. Sci. Manuf.* **2008**, *39*, 965–971. [CrossRef]
20. Ahn, K.J.; Seferis, J.C.; Pelton, T.; Wilhelm, M. Deformation Parameters Influencing Prepreg Tack. *SAMPE Q. Soc. Aerosp. Mater. Process Eng. States* **1992**, *23*, 2.
21. Shin, Y.; Qiao, Y.; Ni, Y.; Ramos, J.L.; Nickerson, E.K.; Merkel, D.R.; Simmons, K.L. Interfacial Bond Characterization of Epoxy Adhesives to Aluminum Alloy and Carbon Fiber-Reinforced Polyamide by Vibrational Spectroscopy. *Surf. Interfaces* **2023**, *42*, 103346. [CrossRef]
22. Semoto, T.; Tsuji, Y.; Yoshizawa, K. Molecular Understanding of the Adhesive Force between a Metal Oxide Surface and an Epoxy Resin. *J. Phys. Chem. C* **2011**, *115*, 11701–11708. [CrossRef]
23. Voyutskii, S.S.; Vakula, V.L. The Role of Diffusion Phenomena in Polymer-to-polymer Adhesion. *J. Appl. Polym. Sci.* **1963**, *7*, 475–491. [CrossRef]
24. Budelmann, D.; Schmidt, C.; Meiners, D. Adhesion-Cohesion Balance of Prepreg Tack in Thermoset Automated Fiber Placement. Part 1: Adhesion and Surface Wetting. *Compos. Part C Open Access* **2021**, *6*, 100204. [CrossRef]
25. Wool, R.P.; O'Connor, K.M. A Theory of Crack Healing in Polymers. *J. Appl. Phys.* **1981**, *52*, 5953–5963. [CrossRef]
26. De Gennes, P.G. Reptation of a Polymer Chain in the Presence of Fixed Obstacles. *J. Chem. Phys.* **1971**, *55*, 572–579. [CrossRef]
27. Edwards, S.F. The Statistical Mechanics of Polymerized Material. *Proc. Phys. Soc.* **1967**, *92*, 9–16. [CrossRef]
28. Janković, B. Kinetic and Reactivity Distribution Behaviors during Curing Process of Carbon/Epoxy Composite with Thermoplastic Interface Coatings (T800/3900-2 Prepreg) under the Nonisothermal Conditions. *Polym. Compos.* **2018**, *39*, 201–220. [CrossRef]
29. Ma, X.Q.; Gu, Y.Z.; Li, M.; Li, Y.X.; Zhang, D.M.; Jia, L.J.; Zhang, Z.G. Properties of Carbon Fiber Composite Laminates Fabricated by Coresin Film Infusion Process for Different Prepreg Materials. *Polym. Compos.* **2013**, *34*, 2008–2018. [CrossRef]

fibers

MDPI

Article

Analysis and Modeling of the System Boundaries of a High-Speed Direct-Yarn-Placement System for In Situ Impregnation of Carbon Fibre Heavy Tows as Textile Reinforcements for Concrete Parts

Erik Knoch [1,2,*], Steffen Rittner [1] and Klaus Holschemacher [1]

1 Faculty of Civil Engineering, Structural Concrete Institute (IfB), Leipzig University of Applied Sciences, 04722 Leipzig, Germany; steffen.rittner@htwk-leipzig.de (S.R.); klaus.holschemacher@htwk-leipzig.de (K.H.)
2 Chair Additive Manufacturing, Faculty of Mechanical, Process and Energy Engineering, TU Bergakademie Freiberg, 09599 Freiberg, Germany
* Correspondence: erik.knoch@htwk-leipzig.de

Abstract: This study investigates a novel approach in modeling the system limits of a braked, high-speed yarn-laying process with in situ impregnation. Special attention is paid to the investigation of the yarn spool overrun after the robot has come to a standstill. This phenomenon occurs at low yarn tensions in combination with high traversing speed and/or acceleration. The modeling of the yarn spool overrun is carried out using physical equations, taking into account the travel speed, acceleration of the robot, and braking force of the spool brake. Previous research has confirmed various operating points of the yarn-laying process, but a comprehensive and complete analysis of the system limits at different operating points and speeds up to 2 m/s is missing. The result of the study is a novel model that describes the system boundaries of the direct-yarn-placement. Furthermore, models for robot braking time, carbon spool diameter, and spool mass are developed. The proposed models have an $R^2 > 0.9674$. Regarding the system stability boundaries, the calculations reveal that, as acceleration rises, the minimum tension requirement also increases. The same trend is found for system velocity. At $a = 12.5\%$, a minimum tension of 16 N suffices, compared to 23 N and 32 N at $a = 25\%$ and 50%, respectively. The impact on tension of quadrupling the speed outweighs that of acceleration, with tension increasing by factors of up to 22.5 and 2, respectively.

Keywords: direct-yarn-placement; yarn laying; filament winding; fiber placement; pultrusion; pin-assisted impregnation; in situ; carbon fiber heavy tows; carbon fiber composite; robotic filament winding; unwinding; process stability; spinning conditions; spooling

1. Introduction

The current developments of population growth and global warming pose major challenges for the construction industry. To reduce CO_2 emissions in building construction, reinforcements based on textile materials offer possible solutions. Direct-yarn-placement technology using an industrial robot is proving to be a promising method for the automated production of individual reinforcements. Here, typically a carbon yarn is unwound and impregnated in situ with the resin matrix directly within the placement head.

High efficiency and quality are required for an economically efficient automated process. The efficiency of the process can be achieved primarily by increasing the production speed. In addition to good impregnation, a key factor with regard to quality is a continuous yarn tension during yarn placement. Modeling of the yarn-placement system limits is an important challenge, especially for fiber placement with high speed. They require a precise knowledge of the yarn spool overrun after the stoppage of a robot movement.

The focus of this scientific article is the detailed investigation of a specific yarn-placement process, which is characterized by the simultaneous application of low yarn tensions, high travel speeds, and an exclusively braked placement. Previous studies have demonstrated setups for selected operating points of up to 0.5 m/s, whereby systems with braked and driven bobbins were considered. However, to the best of the author's knowledge, a comprehensive analysis of the system boundaries at different operating points in conjunction with a braked spool has not yet been carried out.

Since the fabrication of a single textile reinforcement requires the laying of multiple strands in vertical and horizontal directions, achieving short cycle times is highly dependent on production speed. To be competitive with warp knitting machines, it is essential to reduce production time. Consequently, the authors investigate system limits at speeds of up to 2 m/s, a fourfold increase from the currently proposed 0.5 m/s.

The aim of this research is to address this gap by systematically analyzing the interactions between the aforementioned variables of speed, acceleration, and braking force. Through modeling of the system limits, knowledge can be gained that will contribute to the optimization of the investigated yarn-placement process. These furthermore form the basis for the subsequent design of the in situ impregnation unit in terms of the yarn tension to be tolerated.

2. State of the Research

The method of direct-yarn-placement (DYP) involves the precise deposition of continuous filaments with in situ impregnation onto a base frame. This technique offers advantages such as enhanced control over fiber placement and the ability to create intricate structures with varying fiber orientations.

A key feature is that, unlike textile reinforcements fabricated with conventional textile technologies such as weaving or knitting, no subsequent trimming of the grid reinforcement is necessary. As a result, no waste or scrap is generated, and the necessary finishing steps are eliminated. This makes DYP resource- and cost-efficient compared to traditional manufacturing technologies [1].

Currently, only small literature entries are found for DYP, but the technology is comparable to yarn-laying advancements in additive manufacturing, fiber placement, and filament winding. This paper aims to provide a overview on recently developed technologies, exploring different mechanical builds and their process variables, such as speed, acceleration, and yarn tension.

2.1. Search Criteria

To explore the state of research, a comprehensive search was conducted using Google Scholar, focusing on papers published since 2020. The search encompassed a wide range of topics related to carbon fiber composite production, with specific emphasis on filament winding, fiber placement, and spool overrun. Different combinations of keywords extracted from the paper's keyword section were used (Table A1).

2.2. Additive Manufacturing

2.2.1. Typical Builds of the Laying Head

The research landscape in the realm of additive manufacturing for continuous fiber-reinforced composites (CFRCs) is diverse, reflecting a range of methodologies and designs aimed at optimizing the yarn-laying head. Terekhina et al. [2] explore in-nozzle impregnation during fused filament fabrication (FFF), employing a modified nozzle combining a brake and motor for yarn and polymer filaments. Mosleh et al. [3] also leverage modified FFF nozzles, utilizing a single port nozzle for CFRTC manufacturing. An et al. [4] adopt an in situ pin-assisted melt impregnation approach, utilizing a heating block with pins for impregnation. Wang et al. [5] focus on melt impregnation, employing an impregnation mold designed for optimal impregnation pressure and filament shaping. Additionally, Zhi et al. [6] utilize hot-melt impregnation for direct 3D printing, incorporating heating rods and a shaping die. Elderfield et al. [7] introduce discrete in situ consolidation (DISC) using a conical stainless steel tool, whereas Parmar et al. [8] discuss various designs tailored for prepregs, including preimpregnation with subsequent additive manufacturing. Cheng et al. [9] provide a comprehensive review, encompassing various impregnation methods, such as in-nozzle and in-line impregnation. Furthermore, Tao et al. [10] and Hu et al. [11] offer extensive reviews covering a spectrum of impregnation techniques and designs, including dual-channel extruder heads and specialized FDM nozzles. Kipping et al. [12] focus on path planning for reinforcement rather than impregnation specifics, utilizing multi-axis additive manufacturing for carbon fiber reinforcement. Lastly, Wang et al. [13] employ a specific nozzle design for carbon fiber impregnation in conjunction with low-energy electron beam curing. These studies collectively contribute to advancing the state of the art in yarn-laying head design for CFRC additive manufacturing.

2.2.2. Investigated Speeds, Acceleration, and Tension Settings

In examining the state of research concerning laying speed, acceleration, and tension in additive manufacturing for continuous fiber-reinforced composites (CFRCs), a variety of methodologies and designs have been explored across different studies. Terekhina et al. [2] focused on in-nozzle impregnation during fused filament fabrication (FFF) with a laying speed of 0.01 m/s. Elderfield et al. [7] explored discrete in situ consolidation (DISC), with a laying speed of 0.004 m/s. An et al. [4] adopted an in situ pin-assisted melt impregnation approach with a laying speed of 0.00167 m/s. Wang et al. [5] investigated melt impregnation with an optimal speed of 0.067 m/s. Mosleh et al. [3] studied solution impregnation for fiber prepreg with a laying speed of 0.001 m/s. Chen et al. [14] examined melt impregnation with a speed of 0.0167 m/s. Wang et al. [13] combined EB curing with FFF printing with a speed of 0.01 m/s. Notably, laying acceleration and yarn tension were not explicitly specified in these documents.

2.3. *Fiber Placement*

2.3.1. Typical Builds of the Laying Head

In the landscape of advanced composite manufacturing, the design and functionality of yarn-laying effectors are recognized as critical factors for achieving high-quality output and operational efficiency. Donough et al. ([15]) emphasize the importance of heat sources and roller configurations in their review of in situ consolidation processes, offering valuable insights into effector considerations. Ji et al. ([16]) focus on laser irradiation, utilizing a semiconductor laser with adjustable power settings, showcasing a specific effector configuration tailored to their process. Parmar et al. ([8]) provide a comprehensive overview of various manufacturing techniques, which indirectly underscores the significance of effector design in composite manufacturing. Raps et al. ([17]) employ heated tooling for in situ consolidation, contributing to the understanding of effector functionalities, although specific details about effector design remain undisclosed. These studies collectively highlight the importance of further research to explore and refine effector design, thereby advancing composite manufacturing practices.

2.3.2. Investigated Speeds, Acceleration, and Tension Settings

The yarn-laying speed, acceleration, and tension results from the studies are summarized as follows: Donough et al. [15] did not provide specific values but discussed speeds of up to 0.35 m/s. Ji et al. [16] reported movement speeds ranging from 0.02 m/s to 0.08 m/s during the laser-assisted in situ consolidation additive manufacturing process. Parmar et al. [8] highlighted a maximum yarn-laying speed of 0.1 m/s in their review. Raps et al. [17] specified a layup speed of 0.125 m/s, but detailed information on acceleration and tension was not provided beyond the use of heated tooling for in situ consolidation.

2.4. Filament Winding

2.4.1. Typical Builds of the Laying Head

Research on yarn-laying head design [17–44] showcases a diverse array of effector builds catering to specific impregnation needs. Traditional wet impregnation baths with multiple pins remain prevalent [21,30–33,35,38–41,43,44], offering versatility in impregnation.

Moreover, some studies explore specialized setups optimized for specific processes, including robotic coreless filament winding systems [23,24,41]. These systems often integrate tension control mechanisms within the effector, ensuring uniform impregnation and winding. Another notable trend is the incorporation of tension sensors directly into end-effectors for real-time monitoring [41].

In addition to practical implementations, digital approaches such as Bayesian optimization are gaining traction for fine-tuning impregnation machine parameters [34]. These methods offer increased efficiency and accuracy in impregnation, contributing to advancements in yarn-laying head technology.

2.4.2. Investigated Speeds, Acceleration, and Tension Settings

The laying speed, acceleration, and tension settings in the various papers exhibit a wide range of variables and values, showcasing the diverse approaches in filament winding and related processes [19,20,24,33,35–37]. Usun et al. (2021) focused on 3D printed composites, utilizing a laying speed of 0.01 m/s [19]. Shi et al. (2022) optimized winding parameters with a laying speed of 1 m/s, and tensions varied from 5 N to 25 N [20]. Bodea et al. (2022) developed a cyber-physical manufacturing process for large coreless filament wound composite elements, achieving an average laying speed of up to 0.148 m/s, with tensions ranging from 190 N to 320 N during traveling [24]. Sinha et al. (2024) studied the effect of fiber tension on carbon/epoxy composite specimens, with a processing speed of 0.056 m/s during impregnation [33]. Sieira et al. (2021) explored the impact of spinning conditions on the diameter and tensile properties of mesophase petroleum pitch carbon fibers, achieving a winding speed of up to 0.57 m/s [36]. Hopmann et al. (2021) detected and evaluated fiber deposition parameters during wet filament winding, with winding speeds ranging from 0.28 to 0.96 m/s and resin bath temperatures from 40 to 60 °C [37]. Siegl et al. (2024) examined the winding process of fiber-reinforced thermoplastic tubes, achieving an impregnation line speed of up to 1 m/s [35].

Further information regarding acceleration was not provided in the reviewed papers. The focus primarily remained on laying speed and tension settings, with little mention of acceleration values.

2.5. Winding Models

In the realm of winding dynamics research, several models have been proposed to elucidate various facets of the process. Pušnik et al., in their study [45], explored the effect of winding angle on yarn unwinding, emphasizing the optimization of boundary conditions influencing balloon formation. They developed a mathematical model describing yarn motion during unwinding, incorporating balloon theory, Coriolis force, and the effects of gravity and tangential air drag. Through theoretical modeling and computer simulations, they demonstrated the significance of winding angle and apex angle on yarn angular velocity, thereby impacting tension reduction during unwinding.

Kevac et al. delved into the dynamics of sudden jumps during rope winding processes [46]. Their investigation centered on analyzing the nonlinear and pulsed nature of these dynamics, with a test matrix incorporating variables such as the rotational speed of the winch and rope tension. The employed methods include the analysis of nonlinear dynamics through simulation and experimental validation. The results encompassed the identification of cyclicality in the rope winding (unwinding) process and demonstrated the nonlinear and pulsed nature of the dynamic variables. However, specific details regarding the interaction of tension, velocity, and acceleration on spool overrun were not discussed in their study.

The study by Zhang et al. (2023) endeavored to comprehend yarn tension and vibration dynamics during carbon fiber bobbin unwinding [47]. Employing a test matrix with variables including spring preload and unwinding speed, the researchers developed a dynamic model grounded in an axially moving strings theory, utilizing MATLAB simulations for analysis. Their study showed a close relation between yarn tension during unwinding and spring preload. To avoid damage to the carbon fiber yarn, the speed and tension should be controlled.

2.6. Research Gap

While the given studies contribute valuable insights into winding processes, there remains a gap in the comprehensive examination of acceleration, velocity, and tension dynamics simultaneously, along with the analysis of spool overrun in degrees. Existing literature lacks detailed analyses in these areas, particularly across a broad speed range. While previous studies have explored single work points in additive manufacturing and fiber placement, they often lack specifics on tension, acceleration, and drive-train dynamics. Similarly, in filament winding, research primarily focuses on limited speed ranges and lacks comprehensive tension and acceleration data. In most cases, the system is optimized with regard to a single work point. The proposed model aims to bridge this gap by providing a holistic framework for understanding and optimizing laying processes across a wider range of work points for different combinations of the parameters laying speed, yarn tension, and robot acceleration.

3. Modeling the Carbon Spool Overrun

Efficient filament winding processes rely on precise control of the carbon spool to prevent overrun during robot movements. As the robot accelerates and decelerates while laying yarn, the inertia of the rotating spool fluctuates, potentially leading to overrun if not properly managed. To mitigate this issue, it is essential to accurately model the carbon spool overrun dynamics. The deceleration curve of the robot is modeled as illustrated in Figure 1.

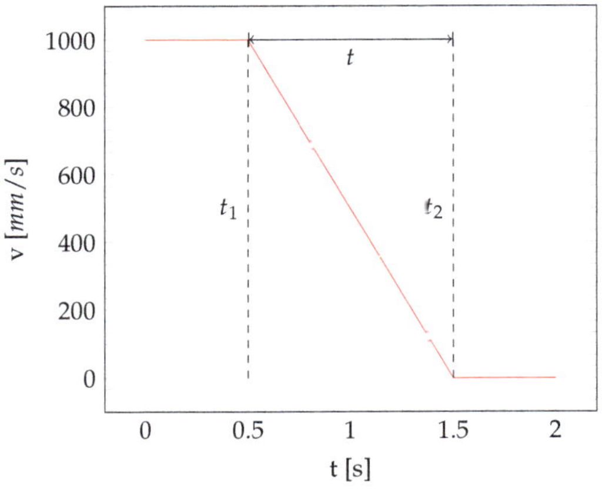

Figure 1. Illustration of the necessary constants for modeling the robot deceleration ramp.

At time t_1, the robot begins decelerating from a constant speed, v. This deceleration is characterized by a linear decrease between the starting time, t_1, and the ending time, t_2. The duration of this deceleration, denoted by the variable t, represents the time delta from start to finish.

The speed progression of the spool is encapsulated within t. The constant speed, n, of the carbon spool before t_1 is assumed to be based on the speed, v, of the robot according to Equation (5). The deceleration behavior of the spool is influenced by the spool brake torque, M, and the combined inertia of the spool and brake shaft, J. The required time for the carbon spool to come to a stop is modeled as follows. Equations (1), (4), (8) and (10) are derived from the work of Heinemann et al. [48].

$$M = J\dot{\omega} = 2\pi J\dot{n} \tag{1}$$

$$\int_{t_1}^{t_2} \dot{n} = \int_{t_1}^{t_2} \frac{M}{J * 2\pi} dt = \int_0^t \frac{M}{J2\pi} dt, \tag{2}$$

$$t = \frac{-nJ2\pi}{M}, \tag{3}$$

where the rotational speed, n, as a factor of the yarn speed, v, is given by:

$$v = \omega * r \tag{4}$$

$$n = \frac{v}{\pi d_o}. \tag{5}$$

To calculate their inertia, the brake shaft and carbon spool are defined as cylindrical tubes with an inner diameter, d_i, outer diameter, d_o, and mass, m.

$$J_{brake,spool} = \frac{1}{2} m \left(\left(\frac{d_o}{2} \right)^2 + \left(\frac{d_i}{2} \right)^2 \right) \tag{6}$$

$$J = k_1 (J_{brake} + J_{spool}) \tag{7}$$

The total inertia is represented by the sum of the two individual components. Additionally, a calibration factor, k_1, is introduced to account for potential variations between the assumed inertia values based on supplier data and the actual system. The torque applied to the carbon spool during unwinding depends on the carbon spool's outer diameter, d_o (Figure 2).

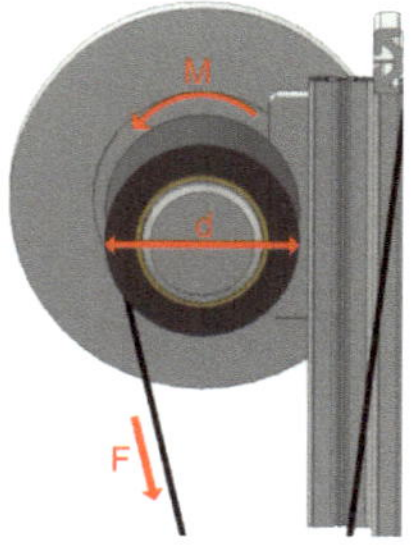

Figure 2. Torque applied to the carbon spool.

$$M = Fd_o/2 \tag{8}$$

Based on the spool stopping time, t, and the robot deceleration time, t_f, the time for spool overrun, Δt, can be determined.

$$\Delta t = t\_t_f \tag{9}$$

In setups where the robot deceleration time is shorter than the necessary spool stopping time, an overrun will occur. The overrun is represented in degrees of spool rotations, denoted as α.

$$\omega = \frac{d\alpha}{dt} \tag{10}$$

$$\alpha = \omega \Delta t = \frac{v * \Delta t}{r}[rad] \tag{11}$$

In the event of a spool overrun, the spool rotation speed is lower than that at the beginning of the deceleration process. The prevailing speed at the time of the overrun, referred to as Δv (immediately after the robot has stopped), is modeled by downscaling the original speed with a constant factor, k_2.

$$\Delta v = v * k_2 \tag{12}$$

4. Materials and Methods

4.1. Investigation of the Spool Overrun

A passive yarn feeding mechanism has been selected for yarn laying, where the yarn is exclusively tensioned by braking forces acting on the spool. This was chosen over an active feed mechanism because it would have required frequency converters combined with high torque motors, which increase the weight and cost of the laying head. Should any slack be present within the setup, the laying head must traverse to eliminate it, as back-winding of the spool is not possible. The presence of slack could result in yarn guiding failures within the feeding apparatus, potentially causing the yarn to slip out of the guiding rollers, leading to severe system damage. To maintain high process stability, it is imperative to ensure that no slack occurs.

During yarn laying, the electro-mechanical systems of the robot and laying head interact. An industrial robot, KUKA KR150, is utilized for yarn placement, programmed in KUKA Robot Language with a soft linear velocity profile (SLIN). The robot's code incorporates velocities and accelerations as per the test matrix.

In the process of yarn laying, a specially developed robot effector (laying head) is utilized (Figure 3a). The carbon filament spool is mounted on a magnetic powder brake (IBD Wickeltechnik GmbH, type B.0351.V24), featuring an expansion clamping shaft of length l = 300 mm and a mass of approximately 6 kg. Multiple sensors on the laying head monitor process variables such as input tension, output tension, filament speed, yarn width, ambient temperature, ambient humidity, and impregnation unit filling degree. These sensors are calibrated and managed by an Adafruit Feather STM32 Express programmed in Python, while yarn tension control is executed by an Industrial Shields ARDBOX 20 I/Os Analog HF Plus programmed in C. Yarn exits the laying effector via a 360-degree rotatable, ball-bearing-mounted deflection head.

The carbon fiber yarn material used is TENAX-E STS40 F13 3200 tex [49]. The spools used weigh approximately 2 kg. The spool clamping diameter is d_i =77 mm. The outer diameter is initially d_o = 145 mm and changes throughout the process.

To monitor spool overrun, a grid pattern with 45° angles has been affixed to the outer end of the brake shaft. Optical tracking of the overrun is conducted using an Apple iPhone SE 2020 using its slow-motion recording capabilities. Subsequently, overrun values are manually extracted from the recorded videos based on the 45° gridding.

The impregnated yarn is positioned by the robot onto a worktable equipped with magnetic deflection points. These points are automatically positioned using a robot equipped with suitable fixtures for the deflection elements, according to the required size and mesh width of the desired textile reinforcement (Figure 3b). For yarn laying, a horizontal grid comprising 20 strands was selected. The laying distance spans 3.3 m, with the overall width of the reinforcement measuring 1.2 m. The diameter of the deflection rollers is 50 mm.

To assess the influence of the robot, brake system, and spool properties on spool overrun, the key variables outlined in Table 1 were identified. The primary variables, actively manipulated during the trials, are the focus of the experiment. Meanwhile, the secondary variables remain unmanaged throughout the experiment. With regard to the spool properties, such as diameter and weight, this implies that, due to carbon yarn consumption throughout the experiment, its values will alter over the course of the trial.

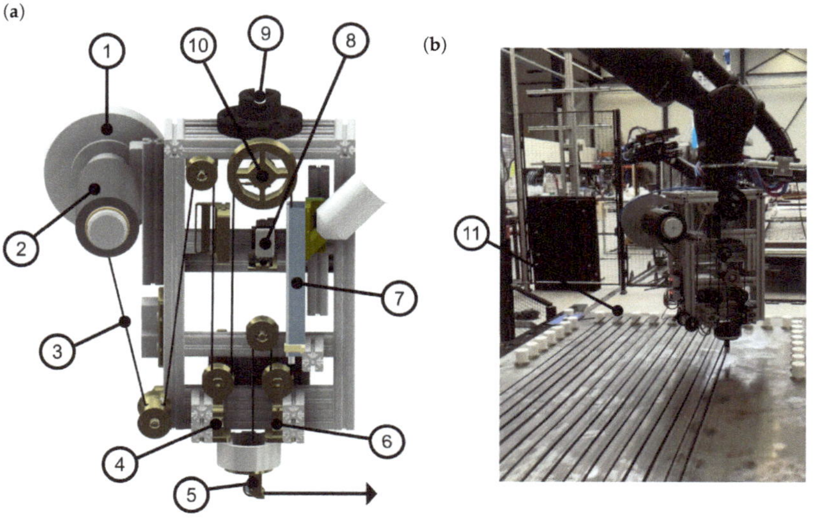

Figure 3. Insights on the direct-yarn-laying process: (**a**) yarn-laying head with sensor: (1) magnetic power brake, (2) carbon spool, (3) unwound carbon filament, (4) input tension sensor, (5) 360-degree rotatable deflection head, (6) output tension sensor, (7) impregnation unit, (8) camera for yarn width sensing, (9) robot mount, (10) unwinding speed sensor; (**b**) yarn-laying head during the DYP process: (11) deflection element used for yarn positioning.

Table 1. Identified variables of the experiment.

Primary Variables	Unmanaged Variables
velocity v acceleration a brake value	spool diameter d_0 spool mass m

The primary variables were tested in a full factorial matrix, as shown in Table 2. Three stages were tested for each variable, resulting in $3 \times 3 \times 3 = 27$ trials. For each trial, six samples were tested, totaling 162 samples in all. Each sample is represented by one horizontal strand (refer to Figure 3b for illustration).

To construct the spool overrun model, the following measurements are required: spool overrun, robot deceleration time, spool properties (diameter and mass), and yarn tension. Spool overrun is measured through video recording. The speed velocity profile is directly logged by the robot output via a Python script. The spool circumference is determined using a tape measure, allowing back calculation of spool diameter and mass for the model construction. Yarn tension is measured by the input force sensor. The data is then extracted using a Python code, which calculates the mean values in the robot's constant speed zone.

Table 2. Overrun test settings to determine system stability.

Laying Acceleration (%)	Brake Value (%)	Laying Speed (mm/s)
12.5, 25, 50	12, 20, 28	250, 1000, 2000

The variable settings are controlled as follows: the acceleration value is set in the KUKA software KSS 8.5.8 as a percentage. The brake value is adjusted to various constant percentages of the maximum brake torque of 35 Nm. Based on the authors previous work, these selected values are known to be compatible with the in situ impregnation device, ranging from approximately 5 N to 25 N. The speed is configured in the KUKA software, utilizing an SLIN movement.

4.2. Carbon Spool Properties

The characteristics of the carbon spool play a critical role in the overrun model. Equations (5) and (8) demonstrate that the outer diameter of the spool serves as an important input parameter. For determining d_o, a methodology relying on circumference measurement was adopted. A tape measure marked at 1 mm intervals was wrapped around the spool to determine its circumference, U.

With the values obtained through circumference measurements and the manufacturer's data, it is possible to first calculate the outer diameter and second determine mass based on the outer diameter. To achieve this, it is essential to compute the outer spool diameter from the measured variable, U. For determining the spool weight, a conversion function based on the manufacturer data is selected (Table 3).

$$d_o = \frac{U}{\pi} \tag{13}$$

Table 3. Manufacturer data of mass to outer diameter ratio [49].

Outer Diameter d_o (mm)	Net Weight m (kg)
105	0.5
120	1
145	2
180	4
205	6
225	8
245	10

4.3. Robot Speed Ramp

Characterizing the robot speed ramp involves key variables: rising time, constant speed, and falling time, all logged in the experiment as previously stated. Data extraction from the KUKA speed log follows a specific procedure. Initially, the maximum value of the robot speed signal, v_{max}, is determined. Subsequently, the timestamps where the signal enters and exits the tolerance goal of 90% v_{max} is ascertained. Based on these values, the rising and falling times of the robot speed curve is calculated. The rising time signifies the duration of the acceleration ramp from $v = 0$ mm/s to 90% v_{max}. In contrast, for the falling time, it represents the period from leaving the 90% v_{max} tolerance band until reaching a minimum speed value. These configurations were found to effectively filter the complex speed behavior at different acceleration settings of the robot, as depicted in Figure 6d. Due to the robot's continuous movement during the laying process, the minimum speed value is 50 mm/s.

4.4. Spool Overrun Model Development

For the development of the spool overrun model, equations from the literature are employed, with solutions calculated separately for each trial. The process involves multiple steps conducted using Excel.

Firstly, rotational speed, n, is computed using Equation (5). Subsequently, the inertias of the carbon spool (J_{spool}) and the magnetic powder brake shaft (J_{brake}) are determined with Equation (6). For the carbon spool, parameters, including a diameter (d_i) of 77 mm and model values for outer diameter (d_o) and mass (m), are employed. For the brake shaft, a mass $m = 6$ kg and an outer diameter $d_o = 77$ mm are employed. The inner diameter is set to $d_i = 50$ mm.

The momentum is then calculated by multiplying the measured tension, T, with the outer diameter of the carbon spool. The estimated time for spool stopping is derived from Equation (2), while spool overrun time is determined using Equation (9), with a measured values of t_f.

Before computing the overrun angle, the remaining overrun speed (Δv) is computed with Equation (12). For determining the values of the constants k_1 and k_2, a python code for a curvefit based on the library scipy/optimize/curve_fit is used. The code numerically solves the overrun model for different values of k_1 and k_2, in combination with the measured variables of each trial. It then returns the values of the constants that showed the strongest model fit to the experimental data.

4.5. Simulation of System Boundaries

In exploring the system boundaries, the previously developed overrun model is tailored to the following framework conditions. To ensure system stability, it is imperative to eliminate any spool overrun, implying that the maximum tolerated overrun time, Δt, must be zero. Consequently, it follows that constant k_2 becomes unnecessary for predicting system stability. Furthermore, from Equations (3) and (9) with $\Delta t = 0$, it can be deduced that $t = t_f$. With this representation, the model for predicting the system boundaries is established.

$$t_f = \frac{-nJ2\pi}{M} \tag{14}$$

In achieving high productivity, production speed plays a key factor and should be as high as possible. Furthermore, different accelerations can be given to the system in order to reach long constant speed times. Assuming the latter are fixed, the remaining factor to transfer the system into a stable state is the yarn tension. This is, however, limited by the requirement of the impregnation unit in respect to a low input tension.

Based on the illustrated conditions, the lowest possible input tension is simulated in respect to different speed and acceleration inputs.

With $M = F * r$:

$$F = \frac{-nJ2\pi}{t_f * r} \tag{15}$$

To calculate the falling time, the regression model derived from the robot speed analysis is utilized. To ensure a conservative estimate of system boundaries, a worst-case scenario approach is adopted. Here, the model prediction $t_{f_{wc}}$ is downscaled based on the identified errors of the falling time model. This is due to the representation of a worst-case scenario by short robot deceleration time, t_f.

$$t_{f_{wc}} = t_f * t_{e2} - t_{e1} \tag{16}$$

The selected error values, based on the results of the previous section, are set to $t_{e1} = 0.218$ s and $t_{e2} \approx 10\%$. Concerning the carbon fiber spool properties, it is assumed that a weight of 2 kg is consistent across all simulations. An unused spool is expected to exhibit the highest overshoot, given that its mass and outer diameter are at maximum compared to an already unwound spool. Utilizing manufacturer data, $d_o = 2r = 145$ mm and $m = 2$ kg are employed for the simulation. The speed range is interpolated from 250 to 2000 mm/s in increments of 50 mm/s, with three accelerations of 12.5%, 25%, and 50% employed for the simulation.

5. Results

5.1. Experiment on the Spool Overrun

The experiment results indicate the significant influences of the primary variables on the system. Both robot acceleration (a) and robot speed (v) demonstrate direct influences, while the magnetic powder brake value exhibits an indirectly proportional effect. Additionally, the secondary variable, spool circumference, U, experiences significant fluctuations throughout the experiment. Raw data are presented in Table 4, detailing the determined target values: carbon spool circumference, U, robot deceleration time, t_f, yarn tension, T_{in}, and spool overrun, α. These values are obtained by calculating the means of the six samples in each trial, with circumference measurements conducted every three trials. The standard deviations are available in Table A2, located in Appendix A.

Table 4. Results of the overrun experiment.

Trial	a (%)	Brake Value (%)	v (mm/s)	U (cm)	t_f (s)	T_{in} (N)	α (°)
1	25	12	250	41.0	2.37	3.5	0
2	25	12	1000		0.34	3.5	285
3	25	12	2000		0.52	5.5	1380
4	25	20	250	39.8	2.39	4.5	0
5	25	20	1000		0.31	5.0	83
6	25	20	2000		0.52	5.5	1110
7	25	27	250	38.6	2.4	17.0	0
8	25	27	1000		0.31	18.0	3
9	25	27	2000		0.6	14.0	183
10	50	12	250	37.2	1.42	5.5	0
11	50	12	1000		0.53	6.0	170
12	50	12	2000		0.35	7.0	1080
13	50	20	250	36.1	1.41	6.5	0
14	50	20	1000		0.55	7.5	113
15	50	20	2000		0.39	8.5	848
16	50	27	250	34.8	1.45	18.0	0
17	50	27	1000		0.53	17.5	0
18	50	27	2000		0.35	15.0	244
19	12.5	12	250	33.5	3.52	7.0	0
20	12.5	12	1000		1.40	7.5	0
21	12.5	12	2000		0.39	8.0	473
22	12.5	20	250	32.0	3.53	8.0	0
23	12.5	20	1000		1.45	8.0	0
24	12.5	20	2000		0.9	10.0	263
25	12.5	27	250	30.6	3.54	20.8	0
26	12.5	27	1000		1.43	20.5	0
27	12.5	27	2000	29.4	0.34	20.0	0

The raw data illustrates the general relationship between the settings. It is observed that, as the speed increases, the spool overshoot rises sharply; for example, from 0 to 1380° from trial 1 to trial 3. Additionally, an increase in the brake value corresponds to a lower spool overshoot. Regarding acceleration, it is observed that only at the lowest setting ($a = 12\%$) is a stable condition of zero overrun consistently maintained across all speeds (trials 25–27).

5.2. Carbon Spool Properties

5.2.1. Spool Diameter d_o

The calculated values are depicted in Figure 4, revealing a linear relationship between the number of trials and the spool diameter. To allow subsequent solving of the overrun model for each trial, a linear regression analysis is performed using the following equation. The resulting linear regression exhibits an R^2 value of 0.9978 and a maximum error of 1.1%. This demonstrates a strong fit to the experimental data and is utilized for further analysis and processing.

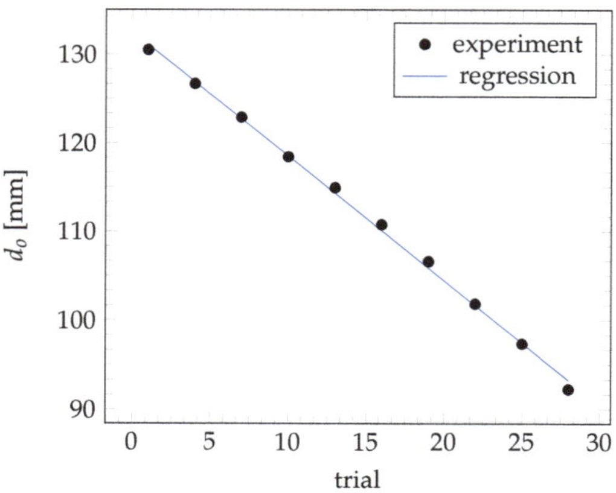

Figure 4. Visualization of the calculated spool outer diameter, d_o, per trial.

5.2.2. Spool Mass m

To obtain a function describing the net weight in relation to d_o, both linear and quadratic regression functions, denoted as m_{regL} and m_{regQ}, respectively, are evaluated. The results, depicted in Figure 5, reveal that, within the available range of spool masses from 0.5 kg to 10 kg, an exponential relationship exists. Consequently, the quadratic function closely aligns with the manufacturer data compared to the linear regression and is thus chosen for further analysis. The R^2 values obtained are 0.9746 and 0.9998, respectively, with maximum errors of 63% and 1.1%.

$$m_{regL} = 0.0720 * d_o - 8.2755 \tag{17}$$

$$m_{regQ} = 3.3266 * 10^{-4} * x^2 - 4.9041 * 10^{-2} * x + 2.0957 \tag{18}$$

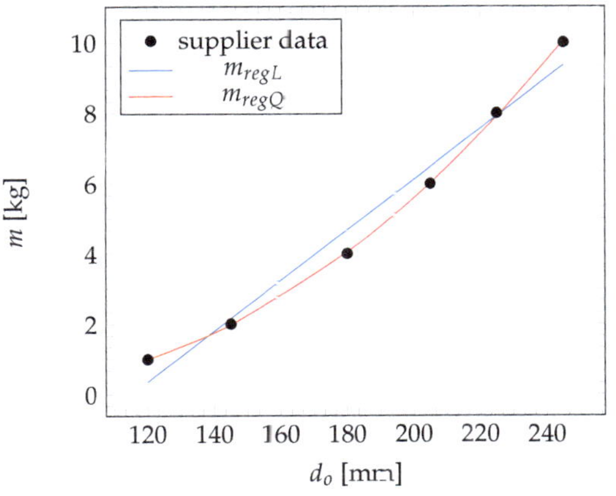

Figure 5. Visualization of the calculated spool outer diameter, d_o, per trial. The supplier data is taken from the data sheet [49].

5.3. Robot Speed Ramp

5.3.1. Data Overview

The analysis indicates that the robot signal times remain consistent across varying brake values. To consolidate common data points for the robot timing (brake values are equal), mean calculation is employed. This process yields values for the robot signal times based on 18 samples each. The summarized results are presented in Table A3 and visually depicted in an overview plot shown in Figure 6.

The experimental data further demonstrates the dependence of robot traveling times on both speed and acceleration. Rising and falling times decrease with increasing speed or acceleration. However, an asymmetry is observed, with rising times being generally longer than falling times. This asymmetry increases with lower acceleration values. This trend is illustrated in Figure 6d through mean curves interpolated from six samples of trials 2, 11, and 20, driven at a constant speed of 1000 mm/s at varying accelerations. The average standard deviation of the robot speed was 2.3%, 1.2%, and 0.6% for 250, 1000, and 2000 m/s, respectively.

Moreover, the size of the constant speed zone is directly proportional to acceleration. The maximum speed of the robot also varies with a, notably reaching a plateau at $a = 25\%$ that is not equal to the target speed of 1000 mm/s.

This behavior suggests potential areas for future research, particularly in addressing the non-stable speed curve behavior observed in trials with SLIN movement.

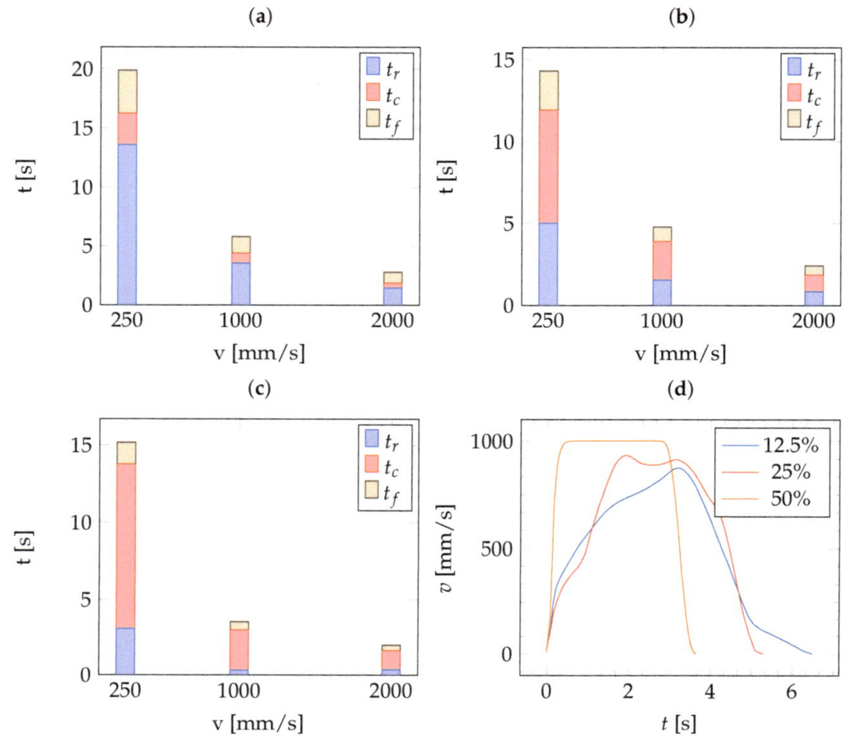

Figure 6. Results of the robot speed ramp analysis: (**a**) rising time, t_r, constant speed time, t_c, and falling speed time, t_f, for an acceleration of $a = 12.5\%$; (**b**) timings for an acceleration of $a = 25\%$; (**c**) timings for an acceleration of $a = 50\%$; (**d**) mean robot speed curves of 6 samples for different acceleration settings at a target speed of 1000 mm/s.

5.3.2. Model Development

Based on the behavior observed in the robot SLIN movement, a model for the falling time, t_f, is developed. This is done due to t_f being one of the main influencing factors for later calculation of the spool overrun. Through trials, it is observed that increasing speed or acceleration leads to a decrease in t_f. An empirical model is thus selected to represent this relationship, with constants c_1 and c_2 calibrated based on experimental data.

$$t_f = \frac{c_1}{v * a} + c_2 \tag{19}$$

The values for acceleration (a) are input as percentages and for velocity (v) in m/s. Utilizing a Python code for curve fitting from the library *scipy/optimize/curve_fit*, the following constants are obtained: $c_1 = 0.1046$ and $c_2 = 0.4618$. The model exhibits a fit to the data with an R^2 value of 0.978 and a maximum time error of 0.22 s. Experimental and model predictions are visually compared in Figure 7. For $t_f > 0.8$ s, the model predicts the experiment with a maximum error of 10.9%, while for values below 0.8 s the maximum error is $\Delta t = 0.218$ s (27%).

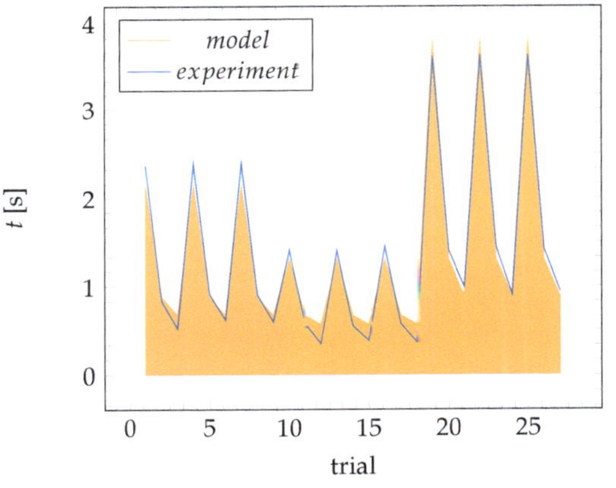

Figure 7. Comparison of the developed falling time model with the experimental results of each trial.

5.4. Modeling the Spool Overrun

Based on the measurements of t_f, T_{in}, and α, combined with the models for the spool outer diameter, d_o, and mass, m, the overrun model is calibrated. This calibration involves calculating values for the constants. The model solutions are calculated for each trial separately and are displayed in Table A4. A comparison of the experimental and model behavior per trial is presented in Figure 8a.

The model demonstrates the ability to predict both large and small overrun angles, α, while also following the reduction of spool weight and outer diameter during the trials. Notably, large overruns ($\alpha > 400°$) are reliably detected, although error percentages increase for smaller angles. The maximum angle errors are 191° and 176° for angles of $\alpha \leq 400°$ and $\alpha > 400°$, respectively. However, for predicting system boundaries, a precise modeling of experimental angles to the degree is unnecessary. Instead, the focus is on detecting the presence or absence of overrun (Boolean True or False detection), where True indicates $\alpha > 0$ and False indicates $\alpha = 0$. Intermediate calculations are summarized in Table A4.

A comparison of Boolean overrun detection is depicted in Figure 8b, demonstrating the model's ability to accurately predict instability due to overrun without false detections across all trials. All instances of overrun are accurately anticipated, demonstrating the model's proficiency in capturing the influence of speed, brake value, acceleration, and spool properties. The rate of detection is 100%.

Parameter values for constants k_1 and k_2 are obtained via curve fitting with Python code, yielding $k_1 = 1.68$ and $k_2 = 0.58$. The model's strong fit to experimental data is reflected in an R^2 value of 0.9674.

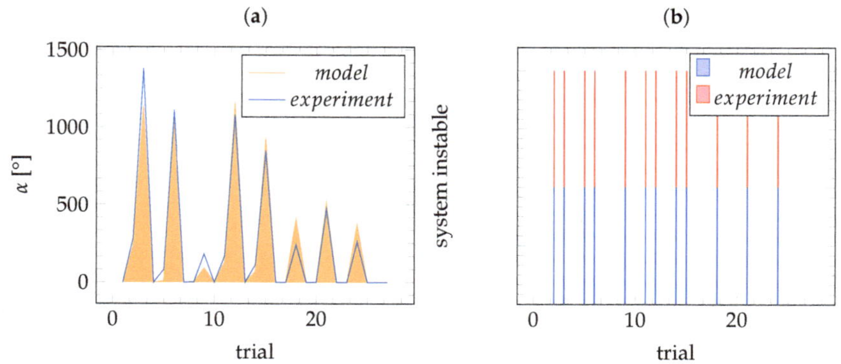

Figure 8. Results of the overrun model development: (**a**) comparison of the predicted model overrun angles with experimental results of each trial; (**b**) comparison of the Boolean system instability detection as a stacked plot (peak present: 1, no peak present: 0).

5.5. Modeling System Boundaries Based on Overrun Model

The adjusted overrun model enables the prediction of crucial minimum tension values for system stability in respect to a worst-case scenario. While there is potential for full overrun compensation at slightly lower values in practical scenarios, the study primarily aims to reliably estimate values where system stability is thought to be assured.

The calculations, illustrated in Figure 9a, show that, with rising acceleration, the necessary minimum tension increases, showing a similar trend concerning system velocity. For high velocities with low tension, a minimum acceleration is recommended. For instance, at $a = 12.5\%$, only 16 N is required, compared to 23 N and 32 N at $a = 25\%$ and 50%, respectively.

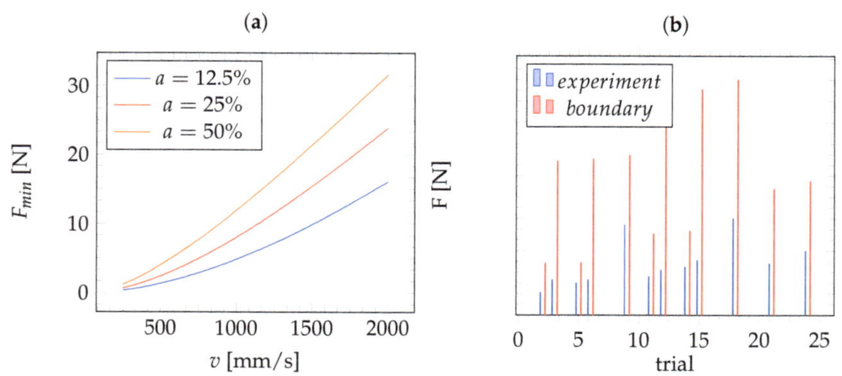

Figure 9. Results of the system boundary simulation: (**a**) expected minimum tension values in respect to laying speed, v, and acceleration, a, for the no-overrun scenario (for values, refer to Table A5); (**b**) comparison of the present tension values of the trials with overrun (blue) with the minimum necessary input tension according to the boundary model (red).

The influence of speed on tension is significantly greater than that of acceleration. This is evident from the observation that the necessary tension increases by factors of 45, 35, and 26 for accelerations of 12.5%, 25%, and 50%, respectively, over a speed range increase of a factor of 8. Conversely, when the acceleration is increased by a factor of 4, the maximum tension increase only rises by a factor of 2.

A comparison of the boundary model results with the experiment are displayed in Figure 9b. It demonstrates that the model predicts higher tension values than those observed in the experiment, suggesting an average tension increase of 2.9 times to avoid overrun.

Based on the evaluated curves, it is possible to predict necessary minimum tensions, which are a key factor for impregnation unit development. In scenarios where the system demands high-speed yarn laying, optimizing the impregnation unit to accommodate greater input forces becomes imperative. Therefore, future research will delve into developing a modeling approach tailored to addressing this aspect.

6. Discussion

6.1. Spool Diameter Model

A linear regression model, developed based on the experiments, demonstrates a good fit with a low maximum error. However, its validity is limited to the current experimental background, as the yarn usage per trial depends on the total laying distance and the number of samples. Further research is necessary to create a more comprehensive model, considering variations in laying length and sample size. Alternatively, a mathematical approach involving back calculation of the spool trial number relative to laying path length and sample number could be explored. Despite its limitations, the model proves valid within the proposed experimental context.

6.2. Spool Mass Model

A model was constructed to establish a relationship between spool diameter and mass using manufacturer data. Both linear and quadratic models were examined, revealing an exponential behavior in the manufacturer data. However, as the spool was not weighed during or after each trial in the experiment, fluctuations to the estimated model may occur in reality. While these influences could not be validated for the selected model, a sufficient fit is estimated based on the manufacturer data. Combining the mass model with the overrun model compensates for fluctuations between reality and experiment by using calibration factors.

6.3. Robot Speed Ramp Model

To the authors' knowledge, recent yarn-laying experiments combined with in situ impregnation have not yielded any models concerning a robot movement model for calculating deceleration time (falling time, t_f). Hence, a model was devised based on the acceleration and velocity settings combined with an SLIN movement type.

Experimental findings showed a complex interplay between acceleration and velocity settings concerning the falling time. Despite minor discrepancies in dimension estimation, the selected model captures the interaction of these settings on falling times. To refine the model, falling times were adjusted downwards using error values derived from maximum fluctuations between reality and model. Consequently, the speed ramp model allows for practical application through worst-case scenario calculations. This worst-case scenario is defined by the possibility that the calculated minimum yarn tension may exceed the actual requirement in practice.

Future research should concentrate on enhancing the model by exploring additional settings and various motion types affecting the resulting robot falling time. Despite its fluctuations, within the chosen experimental parameters the model reliably predicts values with essential accuracy.

6.4. Modeling the Spool Overrun

Although overrun prediction relied on mechanical equations, two fitting constants were required to achieve a good match between reality and experiment. Subsequent calibration improved the model's accuracy in predicting spool overrun across varying speed, acceleration, and brake values, effectively accounting for changes in spool properties. However, the model exhibits weaknesses in precisely calculating the overrun angle, resulting in both over- and underpredictions during experimentation. Nevertheless, for system stability analysis, estimating the occurrence of overrun suffices, obviating the need for precise angle measurements. Analysis of the model in a Boolean manner demonstrated its capability to predict situations of system instability with a 100% success rate, indicating its suitability for delineating system boundaries. While acknowledging the potential for slight deviations in reality, this highlights the potential for further investigation, especially in cases where the model predicts small overrun angles (<15°). Here, predictive accuracy may be refined through targeted research efforts.

6.5. Modeling System Boundaries Based on Overrun Model

The evaluation of the model for system boundaries involves utilizing the calibrated overrun model to predict situations where no overrun occurs initially, representing stable system conditions. By incorporating known errors into the model prediction, they serve as a safety coefficient to ensure stability of the evaluated behavior in reality. The model, acting as an initial prediction tool, can be used to develop subsequent components of the laying head based on worst-case assumptions. Additionally, it is noted that, in reality, the final laying head may potentially be stable at lower values as estimated, although this adjustment is not obligatory if the system is already stable.

Regarding the comparison between model predictions and experimental outcomes, it is evident that the model consistently recommends an increasing tension to compensate the spool overrun that occurred in the experiments. This observation underscores the model's capability to accurately approximate the necessary data points.

6.6. Area of Application of the Models

In the development of a direct-yarn-laying effector, the proposed model serves as the initial step, offering valuable insights. Once the specifics of yarn spool weights, materials, process speeds, and acceleration are determined, it is possible to approximate the minimum yarn tension required based on the model. This estimated tension becomes crucial for designing the impregnation unit. Within the impregnation unit, tension on the yarn increases due to shear and friction, potentially leading to fiber breakage. Therefore, based on the estimated input tension values, the impregnation unit can be optimized to ensure that tension increases remain within the tolerable process window for the textile yarn.

7. Conclusions

Based on the comprehensive analysis conducted, it is evident that this research has made significant strides in understanding the dynamics of the direct-yarn-laying process and its associated parameters. The experiments conducted on spool overrun, carbon spool properties, robot speed ramp modeling, spool overrun modeling, and system boundary modeling have provided valuable insights into the intricate relationships between various process variables.

The results of the experiment on spool overrun highlighted the direct influences of robot acceleration and speed, as well as the indirect influence of the magnetic powder brake value on the system. Additionally, the fluctuations in spool circumference were observed throughout the experiment, further emphasizing the complex nature of the process.

The models developed for spool diameter and mass exhibited strong fits to the experimental data, providing valuable tools for further analysis and processing. Similarly, the robot speed ramp model successfully captured the interplay between acceleration and velocity settings, allowing for practical applications in approximate calculations.

The spool overrun model, despite its reliance on fitting constants, demonstrated a high degree of accuracy in predicting system instability across varying parameters. While precise angle measurements proved challenging, the model's ability to anticipate overrun situations with a 100% success rate underscores its utility in delineating system boundaries. Furthermore, the evaluation of system boundaries based on the overrun model highlighted the model's effectiveness in recommending minimum tension values for stable system conditions.

Overall, the models developed in this research hold immense potential for practical applications in the development of direct-yarn-laying effectors. By providing insights into minimum tension requirements, these models pave the way for optimizing the impregnation unit to ensure stable process conditions while minimizing the risk of fiber breakage. Future research endeavors could focus on refining these models and exploring additional variables to further enhance their predictive capabilities in real-world scenarios.

Author Contributions: Conceptualization, all authors; methodology, all authors; software, E.K.; validation, all authors; formal analysis, E.K.; investigation, E.K.; resources, all authors; data curation, E.K.; writing—original draft preparation, E.K.; writing—review and editing, all authors; visualization, E.K., with review by S.R. and K.H.; supervision, K.H.; project administration, S.R. and K.H.; funding acquisition, S.R. and K.H. All authors have read and agreed to the published version of the manuscript.

Funding: The authors disclosed receipt of the following financial support for the research, authorship, and/or publication of this article: The results described in the article are based on research and development work supported and funded by the German Research Foundation (DFG), the Federal Ministry of Economics and Research (BMWF), and the Federal Ministry of Economics and Climate Protection (BMWK) on the basis of a resolution of the German Bundestag. Furthermore, the author Knoch is partly sponsored by a scholarship from Deutsche Bundesstiftung Umwelt (DBU).

Data Availability Statement: Data are contained within the article.

Acknowledgments: We would like to thank our colleagues Alexander Kahrt, Otto Grauer, Tobias Haft, Tobias Rudloff, Felix Tröger, Lena Völker, and Felix Zakner for their support in preparing and carrying out the research work.

Conflicts of Interest: The authors declare no conflicts of interest. The funders had no role in the design of the study; in the collection, analyses, or interpretation of data; in the writing of the manuscript; or in the decision to publish the results.

Appendix A

Appendix A.1. Search Entries Used for the State of Research

Table A1. Search entries used.

Direct-Yarn-Placement with In Situ Impregnation	Spool Overrun
• carbon textile reinforcement production direct yarn laying • textile reinforcement though carbon impregnation during direct yarn laying • in situ impregnation of textiles • carbon filament winding in situ impregnation • carbon fiber placement in situ impregnation • carbon fiber robotic automated fiber placement in situ impregnation • carbon robotic automated filament winding in situ impregnation • carbon filament winding pin-assisted impregnation	• carbon filament winding spool overrun analysis • carbon filament winding spool tension control • carbon filament winding spool unwinding analysis • carbon filament winding process stability • carbon filament winding spinning conditions • carbon filament winding spooling optimization • carbon filament winding spooling dynamics • carbon filament winding spool overrun compensation • carbon filament winding spool overshoot compensation • carbon yarn placement in situ impregnation

Appendix A.2. Standard Deviations in the Overrun Experiment

Table A2. Standard Deviations in the overrun experiment.

Trial	t_f (s)	T_{in} (N)	α (°)
1	2.37 ± 0.05 s	3.5 ± 0.1 N	0 ± 0°
2	0.84 ± 0.05 s	3.5 ± 0.2 N	285 ± 65°
3	0.52 ± 0.03 s	5.5 ± 0.6 N	1380 ± 90°
4	2.39 ± 0.07 s	4.5 ± 0 N	0 ± 0°
5	0.91 ± 0.05 s	5 ± 0.3 N	83 ± 13°
6	0.62 ± 0.06 s	5.5 ± 0.3 N	1110 ± 40°
7	2.4 ± 0.07 s	17 ± 0.9 N	0 ± 0°
8	0.91 ± 0.06 s	18 ± 1.4 N	3 ± 3°
9	0.6 ± 0.06 s	14 ± 1.1 N	183 ± 3°
10	1.42 ± 0.05 s	5.5 ± 0.3 N	0 ± 0°
11	0.58 ± 0.06 s	6 ± 1 N	170 ± 12°
12	0.35 ± 0.04 s	7 ± 0.8 N	1080 ± 36°
13	1.41 ± 0.06 s	6.5 ± 0.1 N	0 ± 0°
14	0.55 ± 0.02 s	7.5 ± 0.3 N	113 ± 18°
15	0.39 ± 0.05 s	8.5 ± 0.9 N	848 ± 68°
16	1.45 ± 0.05 s	18 ± 0.4 N	0 ± 0°
17	0.58 ± 0.04 s	17.5 ± 0.3 N	0 ± 0°
18	0.36 ± 0.04 s	15 ± 0.9 N	244 ± 41°
19	3.62 ± 0.16 s	7 ± 0.3 N	0 ± 0°
20	1.4 ± 0.08 s	7.5 ± 0.1 N	0 ± 0°
21	0.99 ± 0.05 s	8 ± 0.6 N	473 ± 68°
22	3.63 ± 0.15 s	8 ± 0.2 N	0 ± 0°
23	1.45 ± 0.08 s	8 ± 0.2 N	0 ± 0°
24	0.9 ± 0.04 s	10 ± 0.5 N	263 ± 60°
25	3.64 ± 0.18 s	20.8 ± 0.4 N	0 ± 0°
26	1.43 ± 0.11 s	20.5 ± 0.6 N	0 ± 0°
27	0.94 ± 0.05 s	20 ± 1.3 N	0 ± 0°

Appendix A.3. Robot Speed Ramp Data Overview

Table A3. Results from the robot speed ramp analysis.

a (%)	v (mm/s)	t_{rise} (s)	$t_{constant}$ (s)	t_{fall} (s)
12.5	250	13.58 ± 0.12	2.65 ± 0.01	3.63 ± 0.01
12.5	1000	3.55 ± 0.51	0.83 ± 0.0	1.42 ± 0.02
12.5	2000	1.47 ± 0.03	0.4 ± 0.03	0.94 ± 0.04
25	250	5.01 ± 0.1	6.9 ± 5.13	2.39 ± 0.01
25	1000	1.55 ± 0.09	2.35 ± 0.13	0.89 ± 0.04
25	2000	0.85 ± 0.15	1.01 ± 0.05	0.58 ± 0.05
50	250	3.13 ± 0.03	10.62 ± 0.01	1.42 ± 0.02
50	1000	0.3 ± 0.03	2.72 ± 0.01	0.57 ± 0.02
50	2000	0.34 ± 0.0	1.25 ± 0.03	0.37 ± 0.02

Appendix A.4. Overrun Angle Calculation Overview

Table A4. Overview of the calculated variables for the computation of the overrun per trial.

Trial	a (%)	$brakeVal$ (%)	v (mm/s)	α_{exp} (°)	n (1/s)	J_{spool} (kgm^2)	J_{brake} (kgm^2)	M (Nm)	α_{model} (°)
1	25	12.5	250	0	0.6098	0.0039	0.0077	0.2284	0
2	25	12.5	1000	285	2.4651	0.0037	0.0077	0.2260	232
3	25	12.5	2000	1380	4.9834	0.0035	0.0077	0.3513	1140

Table A4. *Cont.*

Trial	a (%)	$brakeVal$ (%)	v (mm/s)	α_{exp} (°)	n (1/s)	J_{spool} (kgm^2)	J_{brake} (kgm^2)	M (Nm)	α_{model} (°)
4	25	20	250	0	0.6297	0.0033	0.0077	0.2843	0
5	25	20	1000	83	2.5467	0.0031	0.0077	0.3125	10
6	25	20	2000	1110	5.1502	0.0030	0.0077	0.3399	1107
7	25	27	250	0	0.6510	0.0028	0.0077	1.0390	0
8	25	27	1000	0	2.6339	0.0026	0.0077	1.0877	0
9	25	27	2000	183	5.3286	0.0025	0.0077	0.8363	93
10	50	12.5	250	0	0.6739	0.0023	0.0077	0.3248	0
11	50	12.5	1000	170	2.7273	0.0022	0.0077	0.3501	126
12	50	12.5	2000	1080	5.5198	0.0021	0.0077	0.4037	1163
13	50	20	250	0	0.6983	0.0020	0.0077	0.3704	0
14	50	20	1000	113	2.8275	0.0018	0.0077	0.4222	70
15	50	20	2000	848	5.7252	0.0017	0.0077	0.4726	930
16	50	27	250	0	0.7246	0.0016	0.0077	0.9884	0
17	50	27	1000	0	2.9354	0.0015	0.0077	0.9488	0
18	50	27	2000	244	5.9465	0.0014	0.0077	0.8029	411
19	12.5	12.5	250	0	0.7530	0.0013	0.0077	0.3699	0
20	12.5	12.5	1000	0	3.0519	0.0013	0.0077	0.3911	0
21	12.5	12.5	2000	473	6.1856	0.0012	0.0077	0.4117	517
22	12.5	20	250	0	0.7837	0.0011	0.0077	0.4062	0
23	12.5	20	1000	0	3.1780	0.0010	0.0077	0.4006	0
24	12.5	20	2000	263	6.4447	0.0010	0.0077	0.4939	374
25	12.5	27	250	0	0.8170	0.0009	0.0077	1.0106	0
26	12.5	27	1000	0	3.3149	0.0009	0.0077	0.9842	0
27	12.5	27	2000	0	6.7265	0.0008	0.0077	0.9464	0

Appendix A.5. Simulation Results of the System Boundaries

Table A5. Detailed overview of the simulation results for the necessary minimum tension values in respect to selected speed and acceleration settings.

v (m/s)	T (N), a = 12.5%	T (N), a = 25%	T (N), a = 50%
250	0.359741	0.677758	1.214638
300	0.511727	0.953846	1.679269
350	0.688147	1.269509	2.197946
400	0.888132	1.622174	2.764676
450	1.110853	2.009483	3.374321
500	1.355517	2.429277	4.02245
550	1.62137	2.87957	4.705221
600	1.907692	3.358538	5.419289
650	2.213794	3.864496	6.161722
700	2.539019	4.395892	6.929949
750	2.882737	4.951287	7.721695
800	3.244348	5.529352	8.534949
850	3.623274	6.128853	9.36792
900	4.018966	6.748643	10.21901
950	4.430894	7.387656	11.0868
1000	4.858553	8.0449	11.96999
1050	5.301457	8.719451	12.86745
1100	5.75914	9.410443	13.77811
1150	6.231156	10.11707	14.70105
1200	6.717076	10.83858	15.63541
1250	7.216486	11.57426	16.5804
1300	7.728992	12.32344	17.53533

Table A5. *Cont.*

v (m/s)	T (N), a = 12.5%	T (N), a = 25%	T (N), a = 50%
1350	8.254214	13.08552	18.49954
1400	8.791783	13.8599	19.47244
1450	9.34135	14.64603	20.45349
1500	9.902575	15.44339	21.4422
1550	10.47513	16.2515	22.43809
1600	11.0587	17.0699	23.44076
1650	11.65299	17.89815	24.4498
1700	12.25771	18.73584	25.46487
1750	12.87256	19.58259	26.48561
1800	13.49729	20.43803	27.51174
1850	14.13162	21.3018	28.54295
1900	14.77531	22.1736	29.57898
1950	15.42812	23.05309	30.61958
2000	16.0898	23.93999	31.66453

References

1. Schladitz, F.; Curbach, M.; Rittner, S.; Kahnt, A.; Tietze, M. Textile Bewehrung und Deren Herstellung. German Patent No.: DE102016100455B4, 19 November 2020. Available online: https://register.dpma.de/DPMAregister/pat/register?AKZ=1020161004553 (accessed on 31 January 2024).
2. Terekhina, S.; Egorov, S.; Tarasova, T.; Skornyakov, I.; Guillaumat, L.; Hattali, M.L. In-nozzle impregnation of continuous textile flax fiber/polyamide 6 composite during FFF process. *Compos. Part Appl. Sci. Manuf.* **2022**, *153*, 106725. [CrossRef]
3. Mosleh, N.; Rezadoust, A.M.; Dariushi, S. Determining process-window for manufacturing of continuous carbon fiber-reinforced composite Using 3D-printing. *Mater. Manuf. Process.* **2020**, *36*, 409–418. [CrossRef]
4. An, Y.; Myung, J.H.; Yoon, J.; Yu, W. Three-dimensional printing of continuous carbon fiber-reinforced polymer composites via in situ pin-assisted melt impregnation. *Addit. Manuf.* **2022**, *55*, 102860. [CrossRef]
5. Wang, Q.; Zhang, Q.; Kang, Y.; Wang, Y.; Liu, J. An investigation of preparation of continuous carbon fiber reinforced PLA prepreg filament. *Compos. Commun.* **2023**, *39*, 101530. [CrossRef]
6. Zhi, Q.; Li, D.; Zhang, Z.; Zhu, W. High-content continuous carbon fiber reinforced multifunctional prepreg filaments suitable for direct 3D-printing. *Compos. Commun.* **2023**, *44*, 101726. [CrossRef]
7. Nicholas, E.; Wong, J.C.H. Discrete in-situ consolidation of additively manufactured continuous fiber-reinforced polymer composites. *Compos. Part Appl. Sci. Manuf.* **2023**, *171*, 107562. [CrossRef]
8. Parmar, H.; Khan, T.; Tucci, F.; Umer, R.; Carlone, P. Advanced robotics and additive manufacturing of composites: towards a new era in Industry 4.0. *Mater. Manuf. Process.* **2021**, *37*, 483–517. [CrossRef]
9. Cheng, P.; Peng, Y.; Li, S.; Rao, Y.; Le Duigou, A.; Wang, K.; Ahzi, S. 3D printed continuous fiber reinforced composite lightweight structures: A review and outlook. *Compos. Part Eng.* **2023**, *250*, 110450. [CrossRef]
10. Tao, Y.; Li, P.; Zhang, J.; Wang, S.; Shi, S.Q.; Kong, F. A review of fused filament fabrication of continuous natural fiber reinforced thermoplastic composites: Techniques and materials. *Polym. Compos.* **2023**, *44*, 8200–8222 . [CrossRef]
11. Hu, Y.; Lin, Y.; Yang, L.; Wu, S.; Tang, D.; Yan, C.; Shi, Y. Additive Manufacturing of Carbon Fiber-reinforced Composites: A Review. *Appl. Compos. Mater.* **2023**, *31*, 353–398. [CrossRef]
12. Kipping, J.; Kállai, Z.; Schüppstuhl, T. A Set of Novel Procedures for Carbon Fiber Reinforcement on Complex Curved Surfaces Using Multi Axis Additive Manufacturing. *Appl. Sci.* **2022**, *12*, 5819. [CrossRef]
13. Wang, B.; Ming, Y.; Zhu, Y.; Yao, X.; Ziegmann, G.; Xiao, H; Zhang, X.; Zhang, J.; Duan, Y.; Sun, J. Fabrication of continuous carbon fiber mesh for lightning protection of large-scale wind-turbine blade by electron beam cured printing. *Addit. Manuf.* **2020**, *31*, 100967. [CrossRef]
14. Chen, X.; Wang, Y.; Liu, M.; Qu, S.; Zhang, Q.; Chen, S. Preparation and Process Parameter Optimization of Continuous Carbon Fiber-Reinforced Polycarbonate Prepreg Filament. *Polymers* **2023**, *15*, 607. [CrossRef]
15. Donough, M.J.; Shafaq; St John, N.A.; Philips, A.W.; Prusty, B.G. Process modelling of In-situ consolidated thermoplastic composite by automated fibre placement A review. *Compos. Part Appl. Sci. Manuf.* **2022**, *163*, 107179. [CrossRef]
16. Ji, Y.; Luan, C.; Yao, X.; Ding, Z.; Niu, C.; Dong, N.; Fu, J. Mechanism and behavior of laser irradiation on carbon fiber reinforced polyetheretherketone (CF/PEEK) during the laser-assisted in-situ consolidation additive manufacturing process. *Addit. Manuf.* **2023**, *74*, 103713. [CrossRef]
17. Raps, L.; Chadwick, A.R.; Mössinger, I.; Vinot, M.; Behling, T.; Schaefer, Y. Characteristics of in-situ automated fiber placement carbon-fiber-reinforced low-melt polyaryl ether ketone laminates part 2: Effect of prepreg composition. *J. Compos. Mater.* **2024**, *58*, 1523–1535. [CrossRef]
18. Lunetto, V.; Galati, M.; Settineri, L.; Iuliano, L. Sustainability in the manufacturing of composite materials: A literature review and directions for future research. *J. Manuf. Process.* **2023**, *85*, 858–874. [CrossRef]

19. Usun, A.; Gümrük, R. The mechanical performance of the 3D printed composites produced with continuous carbon fiber reinforced filaments obtained via melt impregnation. *Addit. Manuf.* **2021**, *46*, 102112. [CrossRef]
20. Shi, J.; Yuan, S.; Zhang, W.; Wang, G.; Zhang, J.; Chen, H.; Cheng, H. Jute yarn-wound composites: optimization of methods for evaluating mechanical properties and improvement of mechanical properties. *J. Mater. Res. Technol.* **2022**, *21*, 827–840. [CrossRef]
21. Nikravesh, Y.; Muralidharan, K.; Frantziskonis, G. Techno-economic assessment and design optimization of compressed air energy storage using filament wound carbon fiber reinforced plastic pressure vessels. *J. Energy Storage* **2021**, *40*, 102754. [CrossRef]
22. Liu, J.; Phoenix, S.L. Analytical Prediction and Numerical Verification of Stress Concentration Profile Around an In-situ Tow Break in Resin-impregnated Filament-wound Composites. *arXiv* **2023**, arXiv.2311.01736.
23. Mindermann, P.; Rongen, B.; Gubetini, D.; Knippers, J.; Gresser, G.T. Material Monitoring of a Composite Dome Pavilion Made by Robotic Coreless Filament Winding. *Materials* **2021**, *14*, 5509. [CrossRef]
24. Bodea, S.; Mindermann, P.; Gresser, G.T.; Menges, A. Additive Manufacturing of Large Coreless Filament Wound Composite Elements for Building Construction. *3D Print. Addit. Manuf.* **2022**, *9*, 145–160. [CrossRef]
25. Esfandiari, P.; Silva, J.F.; Novo, P.J.; Nunes, J.P.; Marques, A.T. Production and processing of pre-impregnated thermoplastic tapes by pultrusion and compression moulding. *J. Compos. Mater.* **2022**, *56*, 1667–1676. [CrossRef]
26. Friese, D.; Scheurer, M.; Hahn, L.; Gries, T.; Cherif, C. Textile reinforcement structures for concrete construction applications—A review. *J. Compos. Mater.* **2022**, *56*, 4041–4064. [CrossRef]
27. Ferreira, F.; Fernandes, P.; Correia, N.; Marques, A.T. Development of a Pultrusion Die for the Production of Thermoplastic Composite Filaments to Be Used in Additive Manufacture. *J. Compos. Sci.* **2021**, *5*, 120. [CrossRef]
28. Çelik, M.; Noble, T.; Jorge, F.; Jian, R.; Ó Brádaigh, C.M.; Robert, C. Influence of Line Processing Parameters on Properties of Carbon Fibre Epoxy Towpreg. *J. Compos. Sci.* **2022**, *6*, 75. [CrossRef]
29. Budiyantoro, C.; Rochardjo, H.S.B.; Nugroho, G. Design, Manufacture, and Performance Testing of Extrusion—Pultrusion Machine for Fiber-Reinforced Thermoplastic Pellet Production. *Machines* **2021**, *9*, 42. [CrossRef]
30. Mörl, S.; Knorr, D.; Streinz, M.; Mörl, M.; Altstädt, V. Melt impregnation of woven glass fabric reinforced composites in situ modified with short glass fibers in the interlaminar free spacing: Morphology, microstructure and static mechanical properties. *Polym. Compos.* **2020**, *41*, 4117–4129. [CrossRef]
31. Jois, K.C.; Mölling, T.; Schuster, J.; Grigat, N.; Gries, T. Towpreg manufacturing and characterization for filament winding application. *Polym. Compos.* **2024**, 1–13. [CrossRef]
32. Chukov, D.I.; Tcherdyntsev, V.V.; Stepashkin, A.A.; Zadorozhnyy, M.Y. Structure, Thermal, and Mechanical Behavior of the Polysulfone Solution Impregnated Unidirectional Carbon Fiber Yarns. *Polymers* **2023**, *15*, 4601. [CrossRef]
33. Sinha, S.K.; Buragohain, M.K.; Bose, P.S.C. Effect of fiber tension on physical and mechanical properties of filament wound carbon/epoxy composite specimen. *Sadhana* **2024**, *49*, 116. [CrossRef]
34. Schönl, F.; Hübner, F.; Luik, M.; Thomas, J.; de Albuquerque, R.; Ruckdäschel, H. *Digital Approaches for Optimization of Composite Processing: Bayesian Optimization for Impregnation and Fibre Spreading In-Situ Monitoring*; Springer International Publishing: Cham, Switzerland, 2023. [CrossRef]
35. Siegl, M.; Jungbauer, B.; Gebhardt, J.; Judenmann, A.; Ehrlich, I. *Winding Process of Fibre-Reinforced Thermoplastic Tubes with Integrated Tape Production through In-Situ Roving Impregnation and Infrared Consolidation*; Research Square Platform LLC: Durham, NC, USA, 2024. [CrossRef]
36. Sieira, P.; de Souza Mendes, P.R.; de Castro, A.; Pradelle, F. Impact of spinning conditions on the diameter and tensile properties of mesophase petroleum pitch carbon fibers using design of experiments. *Mater. Lett.* **2021**, *285*, 129110. [CrossRef]
37. Hopmann, C.; Magura, N.; Lopez, N.R.; Schneider, D.; Fischer, K. Detection and evaluation of the fibers' deposition parameters during wet filament winding. *Polym. Eng. Sci.* **2021**, *61*, 1353–1367. [CrossRef]
38. Schneider, K.; Michel, A.; Liebscher, M.; Mechtcherine, V. Verbundverhalten mineralisch gebundener und polymergebundener Bewehrungsstrukturen aus Carbonfasern bei Temperaturen bis 500 °C. *Beton-Und Stahlbetonbau* **2018**, *113*, 886–894. [CrossRef]
39. Zhao, J.; Liebscher, M.; Michel, A.; Junger, D.; Trindade, A.; Silva, F.; Mechtcherine, V. Development and testing of fast curing, mineral-impregnated carbon fiber (MCF) reinforcements based on metakaolin-made geopolymers. *Cem. Concr. Compos.* **2021**, *116*, 103898. [CrossRef]
40. Liebscher, M.; Zhao, J.; Wilms, G.; Michel, A.; Wilhelm, K.; Mechtcherine, V. Influence of Roller Configuration on the Fiber-Matrix Distribution and Mechanical Properties of Continuously Produced, Mineral-Impregnated Carbon Fibers (MCFs). *Fibers* **2022**, *10*, 42. [CrossRef]
41. Mindermann, P.; Bodea, S.; Menges, A.; Gresser, G.T. Development of an Impregnation End-Effector with Fiber Tension Monitoring for Robotic Coreless Filament Winding. *Processes* **2021**, *9*, 806. [CrossRef]
42. Mindermann, P.; Gil Pérez, M.; Knippers, J.; Gresser, G.T. Investigation of the Fabrication Suitability, Structural Performance, and Sustainability of Natural Fibers in Coreless Filament Winding. *Materials* **2022**, *15*, 3260. [CrossRef]
43. Friese, D.; Mersch, J.; Hahn, L.; Cherif, C. Development of a yarn guiding and impregnation technology for robot-assisted fiber manufacturing of 3D textile reinforcement structures. In Proceedings of the 11th International Conference on Fiber-Reinforced Polymer (FRP) Composites in Civil Engineering (CICE 2023), Rio de Janeiro, Brazil, 23–26 July 2023 . [CrossRef]
44. Marissen, R.; van der Drift, L.; Sterk, J. *Technology for Rapid Impregnation of Fibre Bundles with a Molten Thermoplastic Polymer*; Elsevier: Amsterdam, The Netherlands, 2000. [CrossRef]
45. Pusnik, N.; Pracek, S. The Effect of Winding Angle on Unwinding Yarn. *Trans. Famena* **2016**, *40*, 29–42. [CrossRef]

46. Kevac, L.; Filipovic, M.; Rakic, A. Dynamics of the process of the rope winding (unwinding) on the winch. *Appl. Math. Model.* **2017**, *48*, 821–843. [CrossRef]
47. Zhang, Y.; Yang, Z.; Wang, Z.; Li, X.; Meng, Z. Yarn tension model and vibration analysis during unwinding of carbon fiber bobbins. *J. Ind. Text.* **2023**, *53*, 152808372311789. [CrossRef]
48. Heinemann, H.; Krämer, H.; Zimmer, H.; Martin, R. *Kleine Formelsammlung PHYSIK*, 8th ed.; Carl Hanser Verlag: Munich, Germany, 2023; pp. 230–245. [CrossRef]
49. Teijin Carbon Europe GmbH. Product Data Sheet. (2022, 1st March). Version 1.4. Available online: https://www.teijincarbon.com/fileadmin/user_upload/Datenblätter/Filament_Yarn/Product_Data_Sheet_TSG01en__EU_Filament_.pdf (accessed on 31 January 2024).

polymers

MDPI

Article

Influence of Extrusion Parameters on the Mechanical Properties of Slow Crystallizing Carbon Fiber-Reinforced PAEK in Large Format Additive Manufacturing

Patrick Consul *, Matthias Feuchtgruber, Bernhard Bauer and Klaus Drechsler

Chair of Carbon Composites, Department of Aerospace and Geodesy, Technical University of Munich, 85748 Garching, Germany; matthias.feuchtgruber@tum.de (M.F.); bernhard.georg.bauer@tum.de (B.B.); klaus.drechsler@tum.de (K.D.)

* Correspondence: patrick.consul@tum.de

Abstract: Additive Manufacturing (AM) enables the automated production of complex geometries with low waste and lead time, notably through Material Extrusion (MEX). This study explores Large Format Additive Manufacturing (LFAM) with carbon fiber-reinforced polyaryletherketones (PAEK), particularly a slow crystallizing grade by Victrex. The research investigates how extrusion parameters affect the mechanical properties of the printed parts. Key parameters include line width, layer height, layer time, and extrusion temperature, analyzed through a series of controlled experiments. Thermal history during printing, including cooling rates and substrate temperatures, was monitored using thermocouples and infrared cameras. The crystallization behavior of PAEK was replicated in a Differential Scanning Calorimetry (DSC) setup. Mechanical properties were evaluated using three-point bending tests to analyze the impact of thermal conditions at the deposition interface on interlayer bonding and overall part strength. The study suggests aggregated metrics, enthalpy deposition rate and shear rate under the nozzle, that should be maximized to enhance mechanical performance. The findings show that the common practice of setting fixed layer times falls short of ensuring repeatable part quality.

Keywords: material extrusion; additive manufacturing; high-performance polymers; PAEK

1. Introduction

AM is a family of production techniques that enables the automated production of complex geometries with low material waste and lead time without needing tooling [1]. Among the AM technologies, MEX is the most common [2]. In desktop-sized printers, it is known as Fused Filament Fabrication (FFF) or Fused Deposition Modeling, but it can also be scaled to large formats. These machines often use single screw extruders for material plasticization, and the process is referred to as LFAM or Extrusion Deposition Additive Manufacturing (EDAM) [3,4].

MEX has thus far primarily been used for prototyping and to produce tooling and molds in more advanced applications, especially as LFAM. However, using technical or high-performance polymers with fiber reinforcement, the technology can also address the production of end-use parts replacing both metals and other manufacturing techniques. The granulate fed single screw extruders of LFAM systems are capable of processing materials with high fiber content from a cost-effective feedstock and at high material output. This makes LFAM both technologically and economically attractive.

The implementation of AM by MEX, both on small and large scales, is challenging due to the quality of mechanical properties in the produced material. MEX deposits material along lines in stacked layers to form the part. The processing conditions of each line, as well as the conditions of the layer being built upon, affect how well these bond together and can significantly affect the quality of the part [5–7]. Therefore, a comprehensive understanding of the interactions between the process parameters is the foundation for optimizing the quality towards the requirements of end-use parts.

Material selection for MEX often has relied on polymers originally developed for injection molding, which had been adapted for AM using additives and fillers during the compounding process. This generally results in a compromise between the needed properties in the final part and the processability. Adding fiber reinforcement, particularly carbon fibers, offers benefits both to the final part and processing. It improves the material's strength and stiffness and, particularly relevant to molds, limits the coefficient of thermal expansion (CTE). The low CTE reduces warping and stabilizes the part at higher temperatures during processing [8].

Initially, the polymers used in MEX were predominantly amorphous grades, as crystallization hinders the bonding of the printed lines and layers in the printing process. Crystal formation stops chain mobility and, thus, bonding of the layers by diffusion across the interface. Crystallization introduces a time dependence of layer bondability in addition to the temperature dependence and elevates the bondable temperature. Most materials developed for injection or compression molding have been optimized to accelerate crystallization and reduce the cycle time in these processes, resulting in inferior performance in MEX, particularly in LFAM [9].

In recent years, new grades of semi-crystalline polymers have been specifically developed for AM. These materials possess tailored crystallization kinetics, with some of these materials crystallizing slow enough to be printed fully amorphous [9]. This enables the deposition of lines on amorphous layers, followed by gradual crystallization as the following layers are built, thereby showing amorphous behavior during the critical stages of processing, but then developing crystallinity and the associated favorable properties, while avoiding additional annealing steps in post-processing [10].

Polyaryletherketones (PAEKs), such as Polyetheretherketone (PEEK) and Polyetherketoneketone (PEKK), have particularly benefited from these developments, with several manufacturers introducing new grades. These polymers are used in industries striving for the highest performance, like aerospace, oil and gas, and medical, which have already been early adopters of AM as an innovative technology. PAEK materials, while costly, are valued by these industries for their superior thermal stability, chemical resistance, and mechanical properties, making them suitable for high-performance applications [11–13].

The objective of this study is to investigate the effect of extrusion parameters on the mechanical properties of the final parts in LFAM using a slow crystallizing, short carbon fiber-reinforced PAEK material. A set of experiments is conducted varying the line width, layer height, layer time, and extrusion temperatures [14]. The resulting layer contact temperature is measured during processing. We attempted to determine fiber misalignment in the lines after printing from cross-section microscopy, and crystallization is reproduced by replicating the thermal history in differential scanning calorimetry (DSC) measurements. Through this intermediate step of dynamic in-process temperatures and material microstructure, the observed effects on material strength and stiffness are expected to be better explained. The printed parts are evaluated by three-point bending to assess the mechanical properties along the extrusion lines and between the layers. The understanding of the impact of basic extrusion parameters and their interactions on the resulting material quality, including macroscopic mechanics and microstructural features like crystallinity and fiber alignment, can then be used to optimize processes for future high-performing applications for AM, transitioning further from the original use in prototyping towards direct part manufacturing.

2. Materials and Methods

2.1. Materials

The material used is a 30 wt.% short carbon fiber-reinforced PAEK developed for AM by Victrex plc (Thornton Cleveleys, UK). The recently commercialized AM200 stems from the same approach and is comparable, yet this material has an even slower crystallization behavior. DSC measurements showed that for constant cooling rates above 20 K/min, the crystallization is completely suppressed. The onset of cold crystallization is at 193 °C and it has three melting peaks at 288 °C, 316 °C, and 328 °C, and the glass transition is at 154 °C. An exemplary DSC curve is shown in Figure 1. The material's base polymer has previously been investigated and has shown significant potential for LFAM with a 650% improvement compared to an injection molding grade [15]. The material was compounded and provided directly by Victrex.

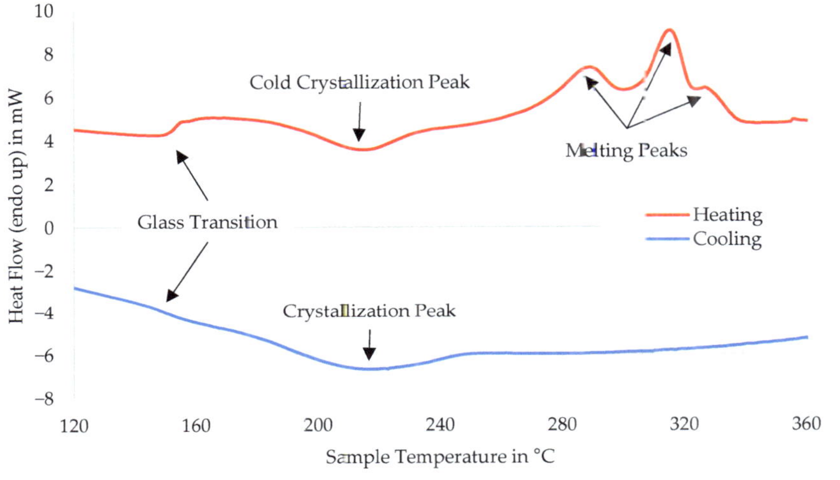

Figure 1. DSC measurement at 10 K/min.

2.2. Sample Preparation

2.2.1. Large Format Additive Manufacturing

The material was dried for at least eight hours at 120 °C in a Vismec (Padua, Italy) DP80 dryer and fed directly to the single screw extruder by an air conveyor. An AM Flexbot by CEAD (Delft, The Netherlands) with a G25 extrusion unit, consisting of a single screw extruder with 25 mm screw diameter and a 5 cc volumetric gear pump for accurate material dispensing, was used to manufacture the samples. An 8 mm brass nozzle was used for printing. The heat zones were set to 330 °C and 350 °C for the first two, with the remaining three zones (extruder zone 3, gear pump, and nozzle) set to the values of the design of experiments. The print bed was heated to 100 °C and the machine was enclosed in a protective cell which maintained an average ambient temperature of 31 °C (minimum 25 °C, maximum 33 °C) during printing; the ambient temperature for each print was recorded.

The temperature of the part during the printing process was recorded in two ways. An infrared camera, a Teledyne FLIR (Wilsonville, OR, USA) a325sc was positioned so that it could observe the printed specimens horizontally. Additionally, thermocouples were introduced into the print by placing them on top of a printed line and holding them in place while the next layer was deposited on top and solidified, fixing the thermocouple in place. The measurements yielded a surface and core temperature for the printed lines and were used subsequently for the analysis of assumed surface-temperature-based bonding between layers and core-temperature-based crystallization of the polymer by replicating the measured thermal history.

2.2.2. Sample Geometry and Parameter Settings

Sample geometry was chosen as cubes with 300 mm edge length and corners rounded to 50 mm radii in the printing plane, printed in a spiral mode. This yielded 200 mm by 300 mm flat plates on each side for sample preparation and allowed for the extrusion to be under almost stationary conditions. Coupons for subsequent testing were then milled from the plane faces, an approach which has already been validated on a smaller scale to yield more repeatable results than direct printing of the coupons [16].

For this, the cubes were separated into the four side planes. One was used for coupons for analysis of fiber orientation and two were used for coupons for mechanical testing by 3-point bending. The coupons for mechanical testing were cut on one along X and the other along Z. This allowed for testing of fiber-dominated behavior along X, and matrix-dominated behavior along Z. The specimens for the analysis of fiber alignment and porosity were cut so the XY and YZ planes were visible with fiber orientation determined in XY and porosity in YZ. The remaining plate had the thermocouples inside and was used for DSC analysis samples. Crystallization was then reproduced with the material and approximate thermal history of the respective cube.

The experimental set was based on parameter variation in line width, layer height, layer time, and extrusion temperature using a D-optimal design of experiments in Modde 13. The set consisted of 30 experiments including three center point repetitions. The observed responses and method of measurement were as follows in Table 1.

Table 1. Responses measured in the experimental design.

Response	Measurement Method	Unit
Material core temperature at deposition (Temperature the material is quench-cooled to)	Thermocouple overprinted	°C
Material core temperature at second layer (Target temperature for fast cooling in DSC cycles)	Thermocouple overprinted	°C
Cooling rate (high): Topmost layer	Thermocouple overprinted	K/min
Cooling rate (low): After coverage with additional layers	Thermocouple overprinted	K/min
Layer contact temperature: surface just before nozzle	IR Camera	°C
Fiber misalignment	Image analysis on cross-section micrographs in XY plane	Standard deviation of orientation distribution in °
Bending strength along extrusion lines (fiber-dominated)	3-point bending	MPa
Bending strength across layers (matrix-dominated)	3-point bending	MPa
Elastic modulus along extrusion lines (fiber-dominated)	3-point bending	MPa
Elastic modulus across layers (matrix-dominated)	3-point bending	MPa
Strain at break along extrusion lines (fiber-dominated)	3-point bending	%
Strain at break across layers (matrix-dominated)	3-point bending	%
Crystallization onset (temperature and time)	DSC of feedstock	°C/s
Crystallization peak (temperature and time)	DSC of feedstock	°C/s
Crystallization enthalpy	DSC of feedstock	J/g
Relative crystallinity	DSC of feedstock	%

The factor ranges were as follows in Table 2.

Table 2. Factor ranges set in the experimental design.

Parameter	Lower Limit	Upper Limit
Line width	8 mm (100% of nozzle diameter)	12 mm (150% of nozzle diameter)
Layer height	2 mm	4 mm
Extrusion temperature	350 °C	400 °C
Layer time	40 s	120 s

2.2.3. Machining of Coupons and Polishing of Cross-Sections

Coupons for mechanical testing were machined from the plates by first machining both sides of the plate flat to remove layer lines and then cutting the specimens by contour milling. The resulting coupons were smooth on all sides to avoid notch effects of the layers influencing mechanical characterization and provide homogeneous specimen shape throughout extrusion width and height settings. Flood cooling was used to minimize the temperature increase during milling, which could potentially weld the layers and affect the measured layer bonding. Samples were dried and then conditioned at 23 °C and 50% humidity for at least 48 h and kept in the climate chamber until their testing.

For cross-section microscopy, samples were embedded in transparent epoxy resin for stabilization prior to polishing. Polishing was conducted in steps on a water-lubricated disc grinder using silicon carbide sandpaper in grit steps of 180, 320, 800, 1200, 2400, and 4000 to achieve scratch-free surfaces for analysis. The cross-sections were recorded on an Olympus BX41M incident light optical microscope (Tokyo, Japan) with a 10× magnification and automatic multiple image alignment to record the entire extrusion line width in one image. Five specimens were imaged per parameter set. Analysis was conducted with a greyscale analyzing program previously used for the study of smaller specimens from fused filament fabrication [17].

2.3. Material Characterization

2.3.1. DSC Measurement for Recreating Thermal History

DSC measurements were performed to accurately recreate the LFAM process's thermal history. The DSC measurements were performed to investigate the occurrence of crystallization during the cooling cycle and determine if it was suppressed due to the fast cooling rates present in the additive manufacturing process.

The thermocouples introduced during printing provided time-continuous temperature data that allowed the replication of the three distinct cooling phases experienced by the material during printing.

The three cooling phases consisted of the following:

1. Quench-Cooling: After extrusion, the fresh molten material immediately comes into contact with the print surface, leading to rapid quench-cooling. During this phase, the material was rapidly cooled to an elevated temperature slightly below the extrusion temperature. In Figure 2, this corresponds to the drop in the extrusion temperature of 350 °C to the maximum recorded temperature of 260 °C. This occurs within the 15 s from the first onset temperature increase at −15s and corresponds to a cooling rate of 360 K/min in this example.
2. Fast Cooling: Once the layer was deposited and free on three sides, fast cooling occurred to a second, lower temperature at time 0 to 90 s in Figure 2 to 170 °C. Cooling rates during this phase were between 60 and 120 K/min, in the example shown, at a relatively low rate of 60 K/min compared to the other experiments.
3. Slow Cooling: After the next layer was deposited, after a short increase in temperature, the cooling rate of the material, now only free on its sides, was reduced. This slower cooling phase could provide sufficient time for the material to achieve crystallization, if transition to the slow phase occurred at a high enough temperature level. Cooling rates in this phase were lower than 25 K/min, in the example, from 120 s onwards at on average 15 K/min until 360 s.

To recreate the three observed cooling phases, a Perkin Elmer (Shelton, CT, USA) DSC8500 was used. This DSC enabled cooling rates of up to 750 K/min, providing the capability to recreate the rapid cooling rates and measure the effect on material crystallization.

The cooling profiles were set to begin after a five-minute isothermal step at the extrusion temperature with a so-called "ballistic cooling" step in the instrument settings, which is uncontrolled cooling at the maximum achievable rate of the instrument to the first target temperature. From there, constant cooling occurred at the average rate of phase 2 to the temperature at which the next layer was deposited, determined by the lowest temperature measured by the thermocouple before the spike in temperature from the deposition. The last cooling step was conducted with the cooling rate determined by the pre-deposition temperature and the temperature recorded three minutes later. The sample was cooled at this rate until 120 °C, which is below the glass transition temperature. At that point, no more crystallization was expected, and it was held there in a two-minute isothermal step, allowing the sample and instrument to settle, before beginning a heating cycle at a constant 20 K/min to the extrusion temperature.

During cooling, the temperatures of onset, peak, and end of crystallization were observed. During heating, the melt enthalpy of crystals formed was determined. As the DSC is only calibrated for one heating rate, the temperatures determined during the cooling cycles may be less accurate than the values specified during heating. However, as crystallization generally occurred during the last stage of cooling, which was at rates lower than 25 K/min, and the instrument was calibrated for the 20 K/min of the heating cycle, the expected error is smaller than 1 K, which was accepted.

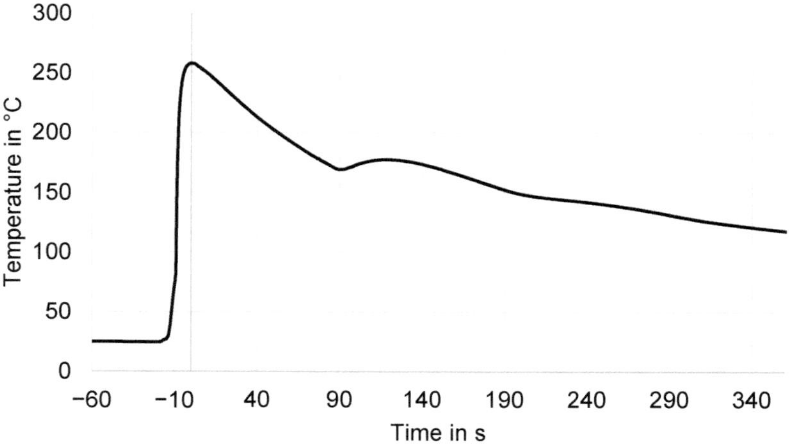

Figure 2. Exemplary cooling curve as recorded by thermocouple during experiment #9. The time zero point is set at the maximum recorded temperature at the deposition.

Crystallization of the material is important as it both hinders bonding of the layers if it occurs too early in processing and improves the material's properties in the finished print if completed [18].

2.3.2. Fiber Alignment

The alignment of fibers inside the printed lines through the squeeze flow underneath the nozzle has a strong effect on the mechanical properties [19–21]. Accurate assessment of the fiber orientation within the prints is crucial for understanding the resulting mechanical properties. A proven precise method for determining the fiber orientation is computed tomography; however, to achieve a resolution high enough to identify the individual fibers, specimens must be small and substantial amounts of data are generated. This makes it unpractical for the screening of sets with a high number of parameters and large specimens, as in this study. For this reason, for this study, we attempted to use image analysis of cross-section micrographs with the help of a previously developed Python script. The core assumption is that fibers are mostly oriented in the XY plane and cross-sections can be made parallel. This had previously proven accurate for fused filament fabrication printed specimens with low layer heights [17]. However, in this study, it became evident that the approach is less effective for LFAM, as the significantly higher layer heights result in more out-of-plane fiber orientation. With regions within the extruded line showing significantly different fiber alignment, highly aligned around the edges of the line and highly randomized in the core, the resulting distribution of the measurements is strongly affected by the position of the cross-section inside the line. This resulted in a very high variance in measurement results of repetitions on the same specimen and would only have been possible to remedy with a significant additional effort to ensure precise positioning of the cross-section, defeating the purpose of being able to analyze a large number of samples quickly. For interested readers, the results are still included in the tables in Appendix A but will not be included in the article. The measurement is performed by creating a histogram of all fiber orientations, which follows a normal distribution, and then determining the standard deviation of the distribution, effectively measuring the fiber misalignment rather than alignment.

2.3.3. Mechanical Testing

Mechanical testing was performed by 3-point bending on a universal test machine, a UPM250 by Hegewald&Peschke (Neu-Ulm, Germany). The loads were measured using a 5 kN load cell, and deflection was measured using a DK830SLR extensometer by Magnescale (Tokyo, Japan). Conditioned specimens were removed from the climate chamber as a set and immediately tested. Six specimens were evaluated per orientation for each parameter set.

2.3.4. Data Analysis

In addition to the original four factors, extrusion width, extrusion height, extrusion temperature, and layer time, two calculated metrics were formed that consider the interactions of these factors. For both of these, the cross-section area of the extrusion line is used and approximated as a stadium or discorectangle shape, as shown in Figure 3, with the layer height w corresponding to the radius and the extrusion line width w to the line width, which is the length of the straight sides a and twice the radius. The extrusion line cross-section area A_{EL} can then be calculated as:

$$A_{EL} = \pi \times \left(\frac{h}{2}\right)^2 + h \times (w - h)$$

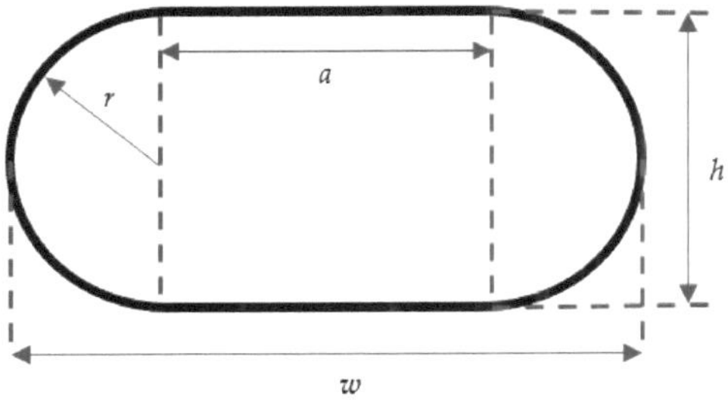

Figure 3. Approximated extrusion line cross-section as stadium shape.

With this approximation, the following metrics can be calculated.

Enthalpy Deposition Rate $\dot{H}$: As the combination of extrusion temperature and material output, this metric is the enthalpy of the material deposition rate. It indicates the power at which thermal energy is introduced into the part. Output is calculated from the line cross-section, simplified as a rectangle of extrusion width by layer height, with Tool Center Point (TCP) speed indirectly set in the experiments by the layer time.

$$\dot{H} = \left(c \times (T_{Ex} - T_A) + H_{Crys_{Max}}\right)/\rho \times (A \times v_{TCP})$$

with specific heat c, extrusion temperature T_{Ex}, ambient temperature T_A, maximum crystallization enthalpy H_{Crys_Max}, density ρ, and Tool Center Point Speed v_{TCP}.

This assumes that cooling of the part and the enthalpy deposition rate must form an equilibrium and the layer temperature will be related to this aggregate.

Shear Rate $\dot{\gamma}$: Pibulchinda et al. [19] have previously proposed using Jeffery's equation [22] using empirical modifications to calculate the shear rate as a metric to investigate how shear and flow affect fiber orientation. This approach was also mentioned in previous publications [23–29]. The equation is modified from the formulation by Pibulchinda et al. using the extrudate and bed velocity, or in this case, nozzle velocity, by using the relation of nozzle orifice area A_N to extrusion line cross-section A_{EL}. The extrudate velocity is calculated by assuming that the volume flow through the nozzle must be equal to that flowing into the extrusion line, with the ratio of flow velocities then equal to the inverse ratio of areas. By simplifying the line cross-section in the same manner as before, the shear rate can then be calculated from the factors of the experiment using the TCP speed of the machine, which depends on the set layer time, as follows:

$$\dot{\gamma} = \frac{v_{TCP}}{h} \times \left(1 - \frac{A_{EL}}{A_N}\right)$$

The nozzle orifice area is assumed to be a circle with nozzle diameter D_{Nozzle} and $A_N = \pi + \left(\frac{D_{Nozzle}}{2}\right)$.

The effect of process responses is visualized by linking these metrics to mechanical properties as desired part responses, under the assumption that the enthalpy deposition rate will affect the properties dominated by welding processes between layers and shear rate the properties dominated by fiber orientation in the squeeze flow under the nozzle. This aims to improve the understanding of the material responses by linking them to process conditions resulting from the process parameters, which only indirectly affect the material responses.

3. Results and Discussion

A complete table of results for the experiments is included in Appendix A. The key results are summarized in graphs and interpreted in the following sections.

3.1. Thermal History and Crystallization

The basic parameters showed no correlation to only a very weak correlation and high variance between the resulting layer temperatures during processing. Only the layer time was found to have a strong negative correlation with the substrate temperature. However, the substrate temperature varied by almost 100 °C for identical layer times, depending on the combination with the other parameters of extrusion height, extrusion width, and extrusion temperature, as can be seen in Figure 4.

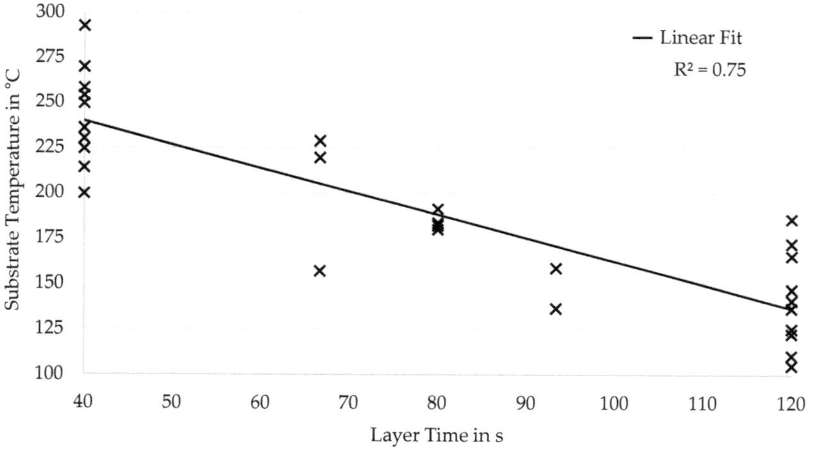

Figure 4. Correlation of substrate temperature over layer time.

The previously introduced enthalpy deposition rate showed much higher correlation with the resulting layer temperature just before deposition of the next layer, as can be seen in Figure 5. A comparison of temperature measurements during printing by thermography and introduced thermocouples is included in the same figure. As expected, the thermocouple readings show slightly higher temperature than the thermography, since they measure the inside of the deposition rather than the free surface. The difference between readings is demonstrated by the one-side error bars in Figure 5, with the marker indicating the reading of the IR camera and the end of the error bar the reading of the thermocouple. This difference is most prominent in the range between about 180 and 260 °C, which is close to the crystallization window of the material of about 180 to 245 °C. While the difference indicates that the temperature gradient within a single layer is small, mostly less than 10 °C, outside of the crystallization window, it can reach up to 35 °C when the layer is completed just as the crystallization reaches its peak.

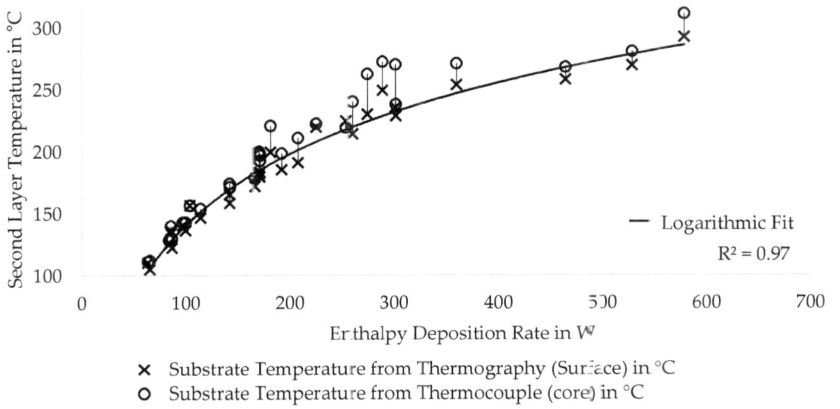

Figure 5. Measured substrate temperature by thermography and overprinted thermocouple, showing higher internal temperature in crystallization window between 180° and 260 °C.

Substrate surface temperature correlates with the enthalpy deposition rate, and experiments with very different process parameters achieve similar layer temperature if the aggregate is similar, as can be seen in the pairwise examples of Table 3.

Table 3. Pairwise comparison of experiments with similar deposition powers but different extrusion parameters.

Number in Figure 6	Extrusion Temperature in °C	Extrusion Width in mm	Layer Height in mm	Layer Time in s	Enthalpy Deposition Rate in W	Surface Temperature in °C
1	350	8	4	93	142.1	170.3
2	400	9	4	120	141.9	172
3	400	12	2	40	301.6	257.7
4	350	12	4	67	301.8	236.9

Figure 6 shows the side IR image of the prints, showing that for the observed temperature, it is similar not just for the last layer, but also for the material below, independent of the extrusion height, width, and temperature, as long as the enthalpy deposition rate is similar. Higher enthalpy deposition rate leads to a higher substrate temperature at contact with the deposition but also shifts the thermal profile along Z to a higher level, as can be seen in Figure 7. But while the spatial temperature distribution is very similar, due to the different build up rates, the cooling rates can be significantly different.

The difference of the readings of the surface by IR camera and internal thermocouple, as shown in Figure 5, increases when transitioning below about 260 °C, at which the internal thermocouple reading seems to plateau in the 300 to 400 W range, before achieving higher temperatures at higher enthalpy deposition rates. This matches the measured onset of crystallization during DSC. This could explain the relative increase in temperature of the thermocouple reading inside the material against the IR reading of the surface.

The DSC measurement also showed that the achieved crystallinity was highest for experiments in which the consecutive layers were deposited when the layer had cooled to temperatures slightly above the window of crystallization onset. A higher layer temperature showed lower crystallinity, as shown by the histogram in Figure 8. This is likely due to the two-staged nature of crystal growth, with nucleation preceding crystal growth. Supercooling the melt causes a higher nucleation rate but decreases crystal growth rate. An optimum for crystallization seems to exist at which a supercooling to the crystallization onset temperature causes a high nucleation rate and seeds crystals which can grow in the subsequent slower cooling phase. Higher layer temperature may either result in less nuclei being formed or nuclei being dissolved by the temperature increase caused by the deposition of the next layer, as can be seen in Figure 9. Another plausible explanation could be the increased formation of a higher enthalpy crystal phase, as indicated by the multiple melting peaks [30–32]. If the layer had cooled below ~200 °C, the crystallization was almost suppressed.

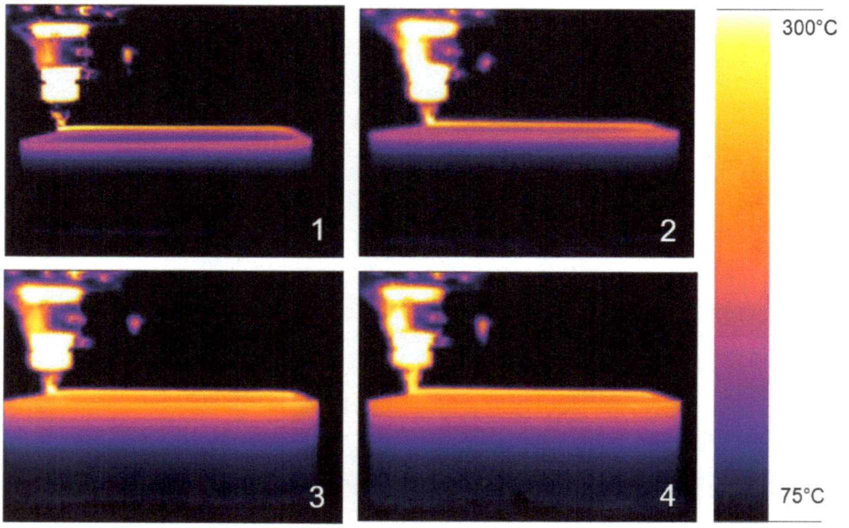

Figure 6. IR images of examples of Table 1, numbers 1–4 indicating the corresponding row of Table 3.

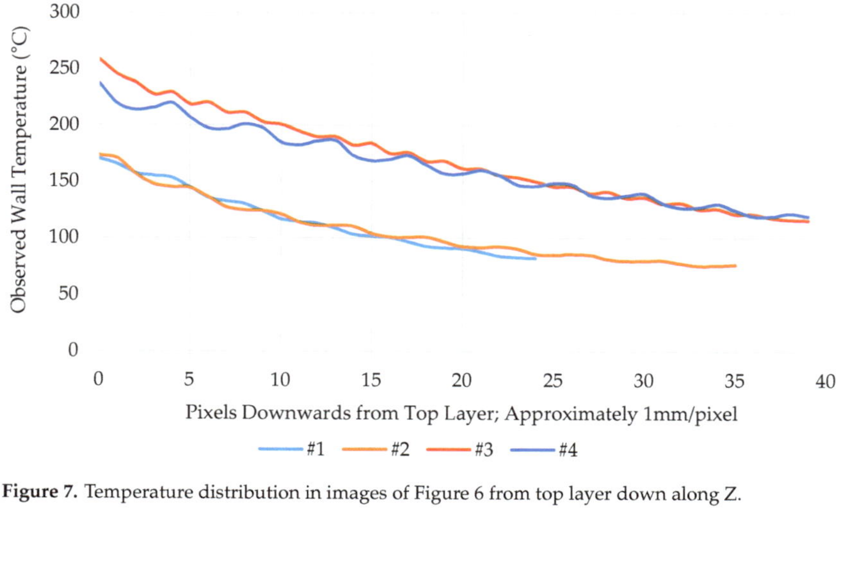

Figure 7. Temperature distribution in images of Figure 6 from top layer down along Z.

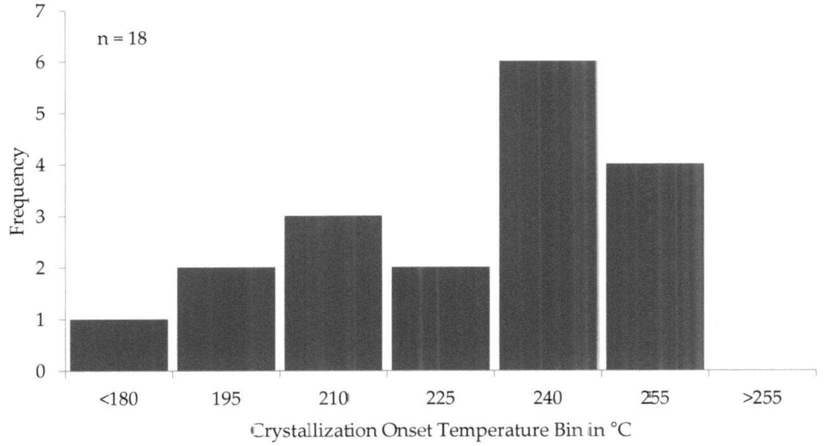

Figure 8. Crystallization onset histogram in DSC measurements recreating experiment cooling rates. Of the experiments, 18 showed crystallization, and for the remainder, it was fully suppressed.

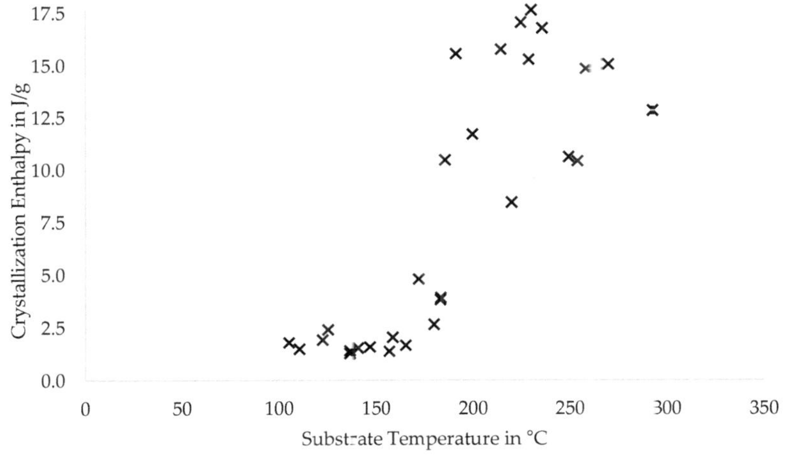

Figure 9. Crystallization enthalpy over substrate temperature.

3.2. Fiber Misalignment

Determining the fiber orientation through micrographs proved susceptible to measurement errors. Therefore, the results should be seen as indicative values and are only included for completeness. A trend to more aligned fibers with shear rate was noticeable and agrees with the determined fiber-dominated mechanical properties in X. For completeness, this trend is shown in Figure 10 over the shear rate.

Overall, the experiments agree with the expectation that a higher shear rate leads to lower fiber misalignment. However, because of the flawed measurement technique, no conclusion should be drawn due to the high variance of measurement results. Logarithmic trendlines have been added for the highest and lowest extrusion temperature experiments, respectively, 400 and 350 °C, which have a difference in viscosity. The trend for lower temperatures points towards better alignment at high temperatures, which is plausible, as the higher viscosity would lead to higher stress and larger aligned edge regions in the extrusion line.

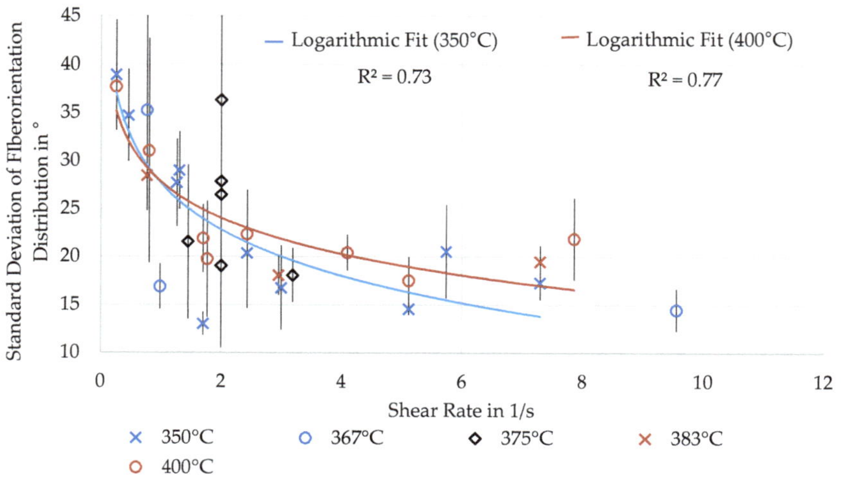

Figure 10. Fiber misalignment over shear rate.

3.3. Mechanical Properties

The measured thermal history and microstructure of fiber alignment inside the lines during processing were expected to have a strong effect on the mechanical properties of the resulting part, which three-point bending tests could confirm. The results showed that the intermediate step of thermal history has a clear correlation with the resulting matrix-dominated mechanical properties. In contrast, the impact of the material flow on fiber-dominated properties along the extrusion lines could only show a tendency without a clear correlation.

3.3.1. Matrix and Thermal History Dominated in Z

The matrix dominates the mechanical properties of the part perpendicular to the layer plane. Strength is mostly determined by the bonding of the layers through welding by chain diffusion. This is a strongly temperature-driven process, and an increase in substrate layer temperature at the time of deposition leads to an increase in strength, as shown in Figure 11. As the substrate temperature depends strongly on the enthalpy deposition rate, as previously described, and substrate temperature influences strength and stiffness; a correlation between deposition power and mechanical properties was also found, as seen in the figures. However, as fusion bonding is also a time-dependent process, cooling rates from the substrate temperatures will also have an effect that is not captured by this metric.

A similar dependence was found for the determined modulus of the specimens. As fibers align mostly in the layer plane, this increase is expected to be due to a higher crystallinity of the matrix and a more difficult crack propagation along layer interfaces [33].

Figure 12 shows that samples with high crystallinity have high mechanical properties; however, the correlation is weak. Results above 70 MPa strength and 3000 MPa stiffness occurred only at high crystallinity. However, the presence of results with lower mechanical properties at high crystallinity indicate that other influencing factors are present.

Overall, it can be concluded that higher deposition rates leading to higher processing temperatures improve the matrix-dominated material properties. However, there is a natural limitation to this. One of the experiments, with an extrusion temperature of 400 °C, line width of 12 mm, layer height of 4 mm, and layer time of 40 s, resulting in a deposition power of 579 W and a substrate layer temperature of 311 °C, was on the limit of collapse and substrate layers were pushed aside during printing. Testing was only possible as line width was high enough to machine deep enough into an uncompromised layer stack.

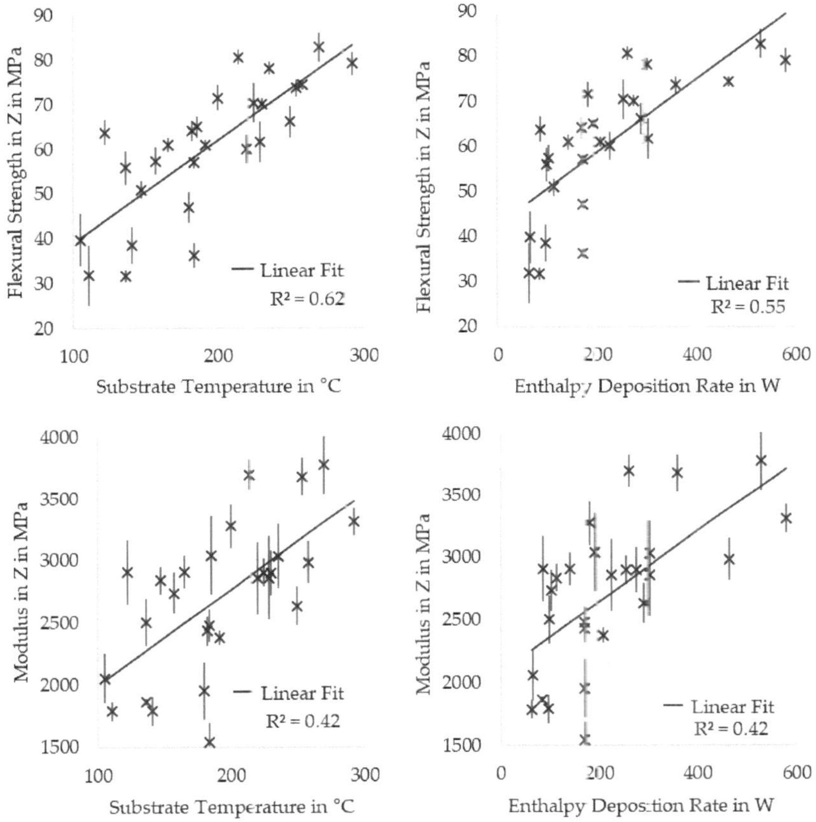

Figure 11. Three-point bending strength and modulus in Z over substrate temperature and enthalpy deposition rate.

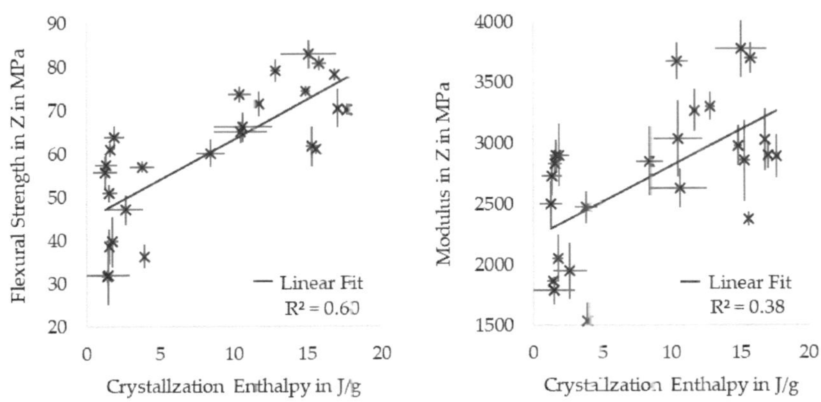

Figure 12. Strength and stiffness over crystallization enthalpy.

3.3.2. Fiber Alignment and Material Flow Dominated in X

To investigate the effect of extrusion parameters on mechanical properties affected by fiber alignment, the shear rate is used. As the measurement of fiber alignment was inaccurate and unreliable, the mechanical properties were not correlated to these measurements. Figures 13 and 14 illustrate that a correlation between the shear rate and the mechanical properties exists, with high shear rates leading to high strength and stiffness. This is assumed to be due to a higher fiber alignment. Linear trendlines have been added for the experiments at 350 and 400 °C again to give an indication of the effect of lower viscosity. The trendline for the cooler and, thereby, more viscous material shows higher material strength and stiffness, possibly due to more fibers being aligned through thicker edge regions with high alignment. These experimental results agree with the simulated results of Pibulchinda et al. [19].

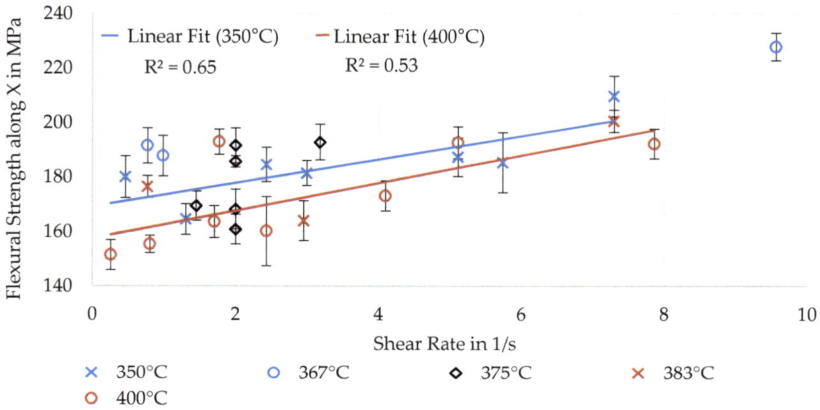

Figure 13. Flexural strength determined by 3-point bending along X.

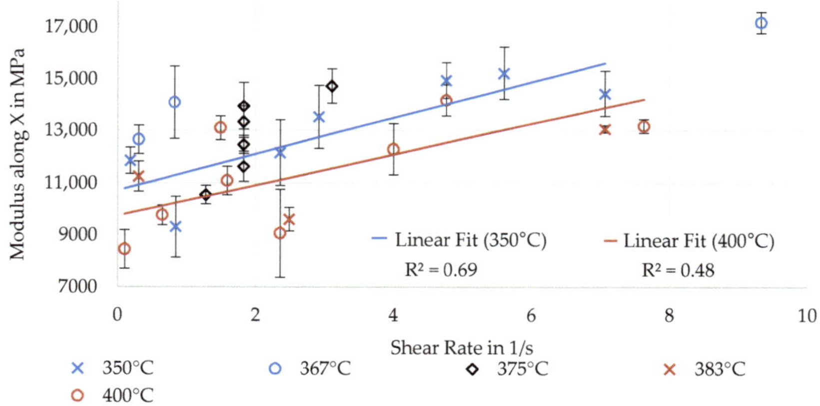

Figure 14. Modulus determined through 3-point bending along X.

Strain at break is low for all specimens with only weak correlation to the shear rate, shown in Figure 15. A higher strain at break at higher temperatures is visible for higher extrusion temperatures.

3.4. Effects of Extrusion Parameters

The results show indirect correlations between the settable extrusion parameters and the resulting material properties, with their interactions having a more direct impact, as seen from the calculated metrics deposition power and shear rate. While the shear rate affects mostly fiber-dominated properties of the material, the deposition affects mostly the matrix-dominated properties.

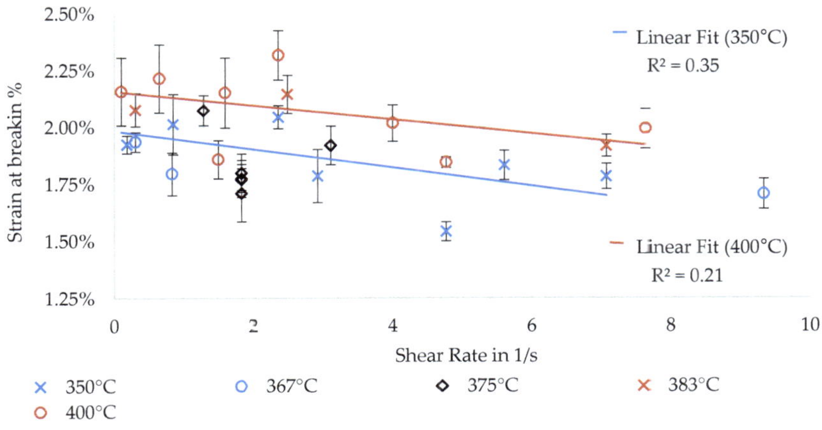

Figure 15. Strain at break determined through 3-point bending along X.

Fiber misalignment was expected to strongly affect the mechanical properties of the material, but it could not be accurately measured using only microscopy. However, a high degree of fiber misalignment leads to lower stiffness and strength along the extrusion lines. High values of shear rate were shown to improve these, suggesting that it also helps in aligning the fibers inside the extrusion lines [24,34,35]. This trend fits well with the mean values of the observed fiber misalignment. Large layer heights resulted in more inaccurate measurements of the fiber misalignment and a larger portion of the line core having high out-of-plane fiber misalignment. This behavior is similar to injection molding, where the fiber misalignment in the core increases [36,37]. A design guideline like a maximum wall thickness in injection molding for a maximum extrusion line height and possibly also for the ratio of width to height in MEX should be considered.

The thermal history of the part during printing has a strong influence on the matrix material, affecting the welding process between the layers, the crystallization, and warping of the part. This temperature strongly correlates with the deposition power of the process, with different extrusion settings resulting in similar thermal conditions when the aggregates are similar. Higher deposition powers lead to higher substrate temperatures during deposition. This also suggests that the common approach of using a set layer time is not valid if the layer height or line width is changed.

Perpendicular to the layer plane, the matrix dominates the mechanical properties. The strength developed during bonding of the layers through chain diffusion is a temperature-driven process. The substrate layer temperature during deposition is a key factor affecting the final part strength. Higher substrate temperatures strengthen bonding and improve mechanical performance [38,39].

Layer temperature and cooling rates are also critical for the crystallization behavior, and the experiments demonstrate that higher layer temperatures lead to higher crystallinity. However, there is an optimum temperature range for crystallization, and excessively high layer temperatures can result in lower degrees of crystallinity. Highly crystalline samples showed high stiffness compared to samples with low crystallinity.

AM materials with a slower crystallization kinetic, like the investigated PAEK material, show significantly enhanced strength between layers compared to materials designed for injection molding with faster crystallization kinetics [12,40]. For end-use applications, crystallization must be completed to ensure the desired material properties. Yet, the experiments have shown that for a wide range of layer temperatures, a high degree of crystallinity can be reached [41,42]. This would make post-processing annealing necessary. The transition to full suppression is fast when the material cools fast enough to affect the crystallinity developed.

4. Conclusions

This experimental study emphasizes the importance of extrusion parameters in LFAM to optimize the mechanical properties of printed parts [43]. The results provide insights for designing end-use parts with LFAM, particularly when using high-performance polymers with fiber reinforcement. As the investigated aggregates affecting the material properties are influenced by the same extrusion parameters, such as layer height, line width, and print speed, not all can be adjusted independently to achieve a desired outcome of material properties, and they need to be considered with their interactions.

The enthalpy deposition rate should always be maximized to ensure that the layers are bonded together; this results in high crystallinity and productivity. This can be achieved both through small lines at high print speeds or large lines and small print speeds; however, material degradation and part slumping set limits on how far this can be increased. The shear rate can be adjusted to achieve a desired fiber alignment, stiffness, and strength along X.

If the highest strength and stiffness are desired, the shear rate should be high by using small extrusion line dimensions and high print speed, ideally with extrusion of the same width as the nozzle orifice diameter to avoid material flow and fiber alignment in the Y direction, in the print plane but perpendicular to the movement direction. On the other hand, if a more isotropic behavior with more random fiber orientation is desired, shear rates should be lower, by using large lines printed at low speeds, ideally overexpanding the line to widths larger than the nozzle orifice diameter to achieve a squeeze flow under the nozzle with the resulting fiber misalignment from the movement direction.

The shear rate depends only on the squeeze flow under the nozzle and can be predicted with little effort to design the process according to the desired material morphology. Layer temperature depends on the balance of introduced energy through the enthalpy deposition rate and the dissipated heat from the part that is dependent on the geometry and ambient conditions and therefore much more difficult to predict. It would be beneficial to develop a system to monitor and adjust this in real time during the process. It is likely easier to control in a closed loop by adjusting cooling or introducing energy as needed, or if possible, without compromising the shear rate and affecting the properties along X, by adjusting the machine speed.

Author Contributions: Conceptualization, P.C. and M.F.; methodology, P.C.; formal analysis, P.C.; investigation, P.C., M.F. and B.B.; resources, K.D.; data curation, P.C.; writing—original draft preparation, P.C.; writing—review and editing, P.C. and M.F; visualization, P.C.; supervision, K.D.; project administration, P.C.; funding acquisition, P.C. All authors have read and agreed to the published version of the manuscript.

Funding: This research received no external funding.

Institutional Review Board Statement: Not applicable.

Data Availability Statement: The raw data supporting the conclusions of this article will be made available by the authors upon request.

Acknowledgments: The authors would like to thank Victrex for the materials provided.

Conflicts of Interest: The authors declare no conflicts of interest.

Appendix A

Table A1. Full list of factor settings.

	Set Factors					Calculated Metrics		
Number	Temperature in °C	Extrusion Width in mm	Extrusion Height in mm	Layer Time in s	TCP Speed in mm/s	Shear Rate in 1/s	Enthalpy Deposition Rate in W
1	350	12	2	40	27.05	7.298	260.30
2	400	12	2	40	27.05	7.298	301.58

Table A1. *Cont.*

	Set Factors				Calculated Metrics		
Number	Temperature in °C	Extrusion Width in mm	Extrusion Height in mm	Layer Time in s	TCF Speed in mm/s	Shear Rate in 1/s	Enthalpy Deposition Rate in W
3	400	12	4	40	27.05	0.767	578.87
4	350	12	2	120	9.02	2.433	86.67
5	400	12	2	120	9.02	2.433	99.79
6	350	12	4	120	9.02	0.256	166.59
7	400	12	4	120	9.02	0.256	192.17
8	350	8	2	67	16.43	5.742	103.87
9	350	8	4	93	11.73	1.266	142.12
10	350	8	3	40	27.37	5.117	253.65
11	350	8	3	120	9.12	1.705	83.87
12	350	12	4	67	16.25	0.461	301.84
13	350	9	2	120	9.10	2.998	65.25
14	350	11	4	40	27.13	1.309	464.80
15	400	8	2	93	11.73	4.099	85.49
16	400	8	4	67	16.43	1.774	225.31
17	400	8	3	40	27.37	5.117	289.09
18	400	8	3	120	9.12	1.705	96.91
19	400	11	2	40	27.13	7.861	274.31
20	400	9	4	120	9.10	0.801	141.90
21	366.7	8	2	40	27.37	9.561	181.54
22	383.3	8	2	120	9.12	3.185	63.45
23	383.3	8	4	40	27.37	2.953	360.03
24	366.7	8	4	120	9.12	0.984	114.01
25	366.7	12	4	40	27.05	0.767	528.51
26	375	12	3	80	13.53	1.454	207.82
27	375	10	3	80	13.60	2.002	171.48
28	375	10	3	80	13.60	2.002	171.30
29	375	10	3	80	13.60	2.002	170.91
30	375	10	3	80	13.60	2.002	170.58

Table A2. Responses of the experiments, part 1 of 2.

	Responses Part 1							
Number	Material Core Temperature at Deposition in °C	Material Core Temperature at Second Layer in °C	Cooling Rate (High): Topmost Layer in °C/min	Cooling Rate (Low): After Coverage with Additional Layers in °C/min	Layer Contact Temperature: Surface Just before Nozzle in °C	Fiber Misalignment in ° with Standard Dev.	Bending Strength along Extrusion Lines (Fiber-Dominated) in MPa with Standard Dev.	Bending Strength across Layers (Matrix-Dominated) in MPa with Standard Dev.
1	275.60	240.00	60	13	214.20	17.30 ± 1.75	209.71 ± 7.52	80.60 ± 1.66
2	315.86	270.12	92	13	235.60	19.50 ± 1.61	200.57 ± 4.07	78.08± 1.41
3	357.32	311.10	67	18	292.50	23.43 ± 3.63	176.42 ± 4.06	79.03 ± 2.70
4	264.95	129.56	69	−1	122.40	20.35 ± 5.71	184.63 ± 6.31	53.68 ± 2.84
5	289.64	142.84	77	−2	136.50	22.33 ± 4.57	160.17 ± 12.70	55.82 ± 3.74
6	292.78	179.00	56	6	172.00	33.95 ± 5.57	NA	NA
7	349.14	198.38	73	6	185.50	37.70 ± 4.53	151.44 ± 5.53	54.88 ± 2.45
8	248.70	156.60	90	7	156.80	20.55 ± 4.86	185.38 ± 11.00	57.29 ± 3.03
9	260.38	171.33	61	11	158.60	27.70 ± 4.55	NA	NA
10	277.50	219.25	104	16	224.60	14.58 ± 0.62	187.37 ± 7.06	70.31 ± 4.41
11	266.70	128.48	70	3	125.20	13.03 ± 1.19	NA	NA
12	282.10	237.85	46	13	228.80	34.7 ± 4.79	180.03 ± 7.66	51.54 ± 4.57
13	252.99	111.97	73	−2	105.10	15.75 ± 4.37	181.46 ± 4.58	39.74 ± 5.79

Table A2. *Cont.*

Number	Material Core Temperature at Deposition in °C	Material Core Temperature at Second Layer in °C	Cooling Rate (High): Topmost Layer in °C/min	Cooling Rate (Low): After Coverage with Additional Layers in °C/min	Layer Contact Temperature: Surface Just before Nozzle in °C	Fiber Misalignment in ° with Standard Dev.	Bending Strength along Extrusion Lines (Fiber-Dominated) in MPa with Standard Dev.	Bending Strength across Layers (Matrix-Dominated) in MPa with Standard Dev.
	Responses Part 1							
14	308.83	268.01	48	17	258.20	28.98 ± 4.03	164.59 ± 5.63	74.27 ± 1.14
15	325.98	139.81	123	2	136.50	20.40 ± 1.86	173.13 ± 5.52	31.68 ± 1.36
16	292.26	222.46	65	17	219.80	19.73 ± 6.03	192.92 ± 4.62	60.03 ± 3.22
17	343.44	272.37	105	22	249.30	17.50 ± 2.44	192.66 ± 5.89	66.06 ± 3.55
18	328.86	142.65	99	1	140.70	21.88 ± 3.51	163.62 ± 5.90	38.50 ± 4.03
19	314.73	262.62	95	14	230.10	21.85 ± 4.25	192.25 ± 5.45	70.06 ± 1.29
20	344.27	174.25	84	5	165.30	31.00 ± 11.67	155.40 ± 3.24	60.83 ± 1.63
21	276.93	220.79	94	16	199.70	14.48 ± 2.19	227.92 ± 5.04	71.45 ± 2.87
22	265.19	111.23	80	−2	110.50	18.05 ± 2.78	192.85 ± 6.55	31.85 ± 6.74
23	307.53	270.98	82	21	253.80	18.07 ± 2.06	164.03 ± 7.32	73.54 ± 1.96
24	297.00	153.83	80	3	146.70	16.88 ± 2.33	187.80 ± 7.39	50.85 ± 1.94
25	302.90	280.78	44	18	269.70	35.25 ± 9.88	191.57 ± 6.48	82.75 ± 3.28
26	305.55	210.99	76	8	191.20	21.53 ± 8.01	169.49 ± 5.31	60.82 ± 1.13
27	277.01	196.72	67	10	183.30	36.33 ± 8.77	160.87 ± 5.48	36.25 ± 2.73
28	289.63	192.61	76	9	179.80	27.83 ± 6.94	168.29 ± 7.26	47.06 ± 3.43
29	292.92	198.71	74	10	183.20	26.45 ± 8.59	185.73 ± 2.10	57.01 ± 1.29
30	296.36	199.83	79	10	182.00	19.03 ± 8.49	191.63 ± 6.43	64.00 ± 2.31

Table A3. Responses of the experiments, part 2 of 2.

Number	Elastic Modulus along Extrusion Lines (Fiber-Dominated) in MPa with Standard Dev.	Elastic Modulus across Layers (Matrix-Dominated) in MPa with Standard Dev.	Strain at Break along Extrusion Lines (Fiber-Dominated) in % with Standard Dev.	Strain at Break across Layers (Matrix-Dominated) in % with Standard Dev.	Crystallization Onset (Temperature and Time)	Crystallization Peak (Temperature and Time)	Crystallization Enthalpy in J/g	Relative Crystallinity
	Responses Part 2							
1	14,442.09 ± 882.07	3699.49 ± 128.40	1.78 ± 0.06	2.27 ± 0.06	232.36 °C/89.7 s	219.82 °C/146.72 s	15.79	89%
2	13,056.70 ± 139.78	3030.32 ± 258.25	1.92 ± 0.05	2.90 ± 0.11	240.13 °C/137.26 s	214.87 °C/246.4 s	16.80	95%
3	11,260.26 ± 580.24	3304.45 ± 116.92	2.08 ± 0.07	2.62 ± 0.08	242.03 °C/305.84 s	220 °C/391.5 s	12.84	73%
4	12,162.61 ± 1246.39	2905.32 ± 257.26	2.04 ± 0.05	2.40 ± 0.08	NA	NA	1.92	11%
5	9066.43 ± 1696.87	2502.00 ± 191.26	2.32 ± 0.11	2.62 ± 0.18	NA	NA	1.28	7%
6	NA	NA	NA	NA	NA	NA	4.81	27%
7	8449.26 ± 742.96	3039.19 ± 311.83	2.16 ± 0.15	2.24 ± 0.13	195.01 °C/128.846 s	185.15 °C/199.04 s	10.48	59%
8	15,237.24 ± 1015.82	2735.67 ± 166.47	1.83 ± 0.06	2.47 ± 0.17	NA	NA	1.37	8%
9	NA	NA	NA	NA	163.31 °C/216.3 s	50.29 °C/284.04 s	2.04	12%
10	14,957.735 ± 690.65	2898.58 ± 110.57	1.54 ± 0.04	2.71 ± 0.20	231.95 °C/78.54 s	217.92 °C/131.22 s	17.07	97%
11	NA	NA	NA	NA	219 °C/82.17 s	124.09 °C/100.17 s	2.41	14%
12	11,860.29 ± 493.62	2858.21 ± 334.94	1.92 ± 0.04	1.82 ± 0.11	228.43 °C/111.04 s	214.28 °C/174.98 s	15.31	87%
13	13,539.99 ± 1219.94	2054.35 ± 195.74	1.79 ± 0.12	2.06 ± 0.36	NA	NA	1.80	10%
14	9321.80 ± 1171.16	2982.68 ± 167.64	2.01 ± 0.13	2.55 ± 0.11	243.21 °C/131.36 s	225.57 °C/189.6 s	14.86	84%
15	12,292.56 ± 969.49	1864.83 ± 24.20	2.02 ± 0.08	2.34 ± 0.20	NA	NA	1.40	8%
16	13,099.85 ± 458.59	2854.93 ± 287.18	1.86 ± 0.08	2.29 ± 0.14	214.05 °C/161.52 s	198.14 °C/213 s	8.44	48%
17	14,182.86 ± 607.97	2632.01 ± 159.46	1.84 ± 0.02	2.73 ± 0.14	226.98 °C/160.54 s	221.62 °C/217.26 s	10.63	60%
18	11,098.65 ± 543.68	1787.53 ± 116.11	2.15 ± 0.15	2.71 ± 0.32	NA	NA	1.53	9%
19	13,170.27 ± 271.35	2896.77 ± 177.64	1.99 ± 0.09	2.46 ± 0.14	229.28 °C/137.44 s	204.3 °C/233.7 s	17.66	100%
20	9767.92 ± 381.98	2904.11 ± 129.96	2.21 ± 0.15	2.40 ± 0.11	NA	NA	1.66	9%
21	17,182.23 ± 412.13	3271.46 ± 176.69	1.70 ± 0.07	2.34 ± 0.18	208.51 °C/97.08 s	194.57 °C/147.66 s	11.71	66%

Table A3. *Cont.*

Responses Part 2								
Number	Elastic Modulus along Extrusion Lines (Fiber-Dominated) in MPa with Standard Dev.	Elastic Modulus across Layers (Matrix-Dominated) in MPa with Standard Dev.	Strain at Break along Extrusion Lines (Fiber-Dominated) in % with Standard Dev.	Strain at Break across Layers (Matrix-Dominated) in % with Standard Dev.	Crystallization Onset (Temperature and Time)	Crystallization Peak (Temperature and Time)	Crystallization Enthalpy in J/g	Relative Crystallinity
22	14,738.96 ± 668.09	1786.57 ± 78.67	1.92 ± 0.09	2.33 ± 0.77	NA	NA	1.49	8%
23	9605.70 ± 456.95	3679.72 ± 152.73	2.14 ± 0.08	2.31 ± 0.09	227 °C/125.37 s	199.80 °C/206.04 s	10.42	59%
24	14,095.51 ± 1410.61	2838.41 ± 110.80	1.79 ± 0.09	2.21 ± 0.12	NA	NA	1.59	9%
25	12,653.63 ± 537.95	3781.61 ± 245.08	1.93 ± 0.04	2.46 ± 0.09	241.25 °C/305.34 s	209.94 °C/399.58 s	15.08	85%
26	10,558.26 ± 359.14	2376.69 ± 60.07	2.07 ± 0.07	2.08 ± 0.08	201.84 °C/129.6 s	189.60 °C/208.77 s	15.58	88%
27	11,640.79 ± 566.11	1537.75 ± 151.19	1.80 ± 0.08	2.33 ± 0.31	180.61 °C/192.3 s	90.19 °C/233.04 s	3.92	22%
28	12,463.96 ± 340.84	1950.33 ± 234.21	1.71 ± 0.13	2.10 ± 0.14	NA	NA	2.66	15%
29	13,342.17 ± 619.77	2475.26 ± 130.41	1.77 ± 0.08	1.98 ± 0.10	192.54 °C/189.87 s	183.09 °C/235.02 s	3.82	22%
30	13,958.45 ± 915.31	2430.76 ± 117.29	1.77 ± 0.05	2.23 ± 0.08	NA	NA	0.00	0%

References

1. Saleh Alghamdi, S.; John, S.; Roy Choudhury, N.; Dutta, N.K. Additive Manufacturing of Polymer Materials: Progress, Promise and Challenges. *Polymers* **2021**, *13*, 753. [CrossRef] [PubMed]
2. Sarabia-Vallejos, M.A.; Rodríguez-Umanzor, F.E.; González-Henríquez, C.M.; Rodríguez-Hernández, J. Innovation in Additive Manufacturing Using Polymers: A Survey on the Technological and Material Developments. *Polymers* **2022**, *14*, 1351. [CrossRef]
3. Justino Netto, J.M.; Idogava, H.T.; Frezzatto Santos, L.E.; Silveira, Z.d.C.; Romio, P.; Alves, J.L. Screw-assisted 3D printing with granulated materials: A systematic review. *Int. J. Adv. Manuf. Technol.* **2021**, *115*, 2711–2727. [CrossRef]
4. Barocio, E.; Brenken, B.; Favaloro, A.; Bogdanor, M.; Pipes, R.B. Extrusion deposition additive manufacturing with fiber-reinforced thermoplastic polymers. In *Structure and Properties of Additive Manufactured Polymer Components*; Elsevier: Amsterdam, The Netherlands, 2020; pp. 191–219, ISBN 9780128195352.
5. Bouzaglou, O.; Golan, O.; Lachman, N. Process Design and Parameters Interaction in Material Extrusion 3D Printing: A Review. *Polymers* **2023**, *15*, 2280. [CrossRef] [PubMed]
6. Nowka, M.; Hilbig, K.; Schulze, L.; Jung, E. Vietor, T. Influence of Process Parameters in Material Extrusion on Product Properties Using the Example of the Electrical Resistivity of Conductive Polymer Composites. *Polymers* **2023**, *15*, 4452. [CrossRef] [PubMed]
7. Sharafi, S.; Santare, M.H.; Gerdes, J.; Advani, S.G. Extrusion-Based Additively Manufactured PAEK and PAEK/CF Polymer Composites Performance: Role of Process Parameters on Strength, Toughness and Deflection at Failure. *J. Compos. Sci.* **2023**, *7*, 157. [CrossRef]
8. Feuchtgruber, M.; Holmes, J.; Sommacal, S.; Strobel, M.; Gehringer, F.; Consul, P.; Bublitz, D.; Weghorst, J.; Thomson, R.; Strachauer, F.; et al. Using carbon fiber tape to tailor the coefficient of thermal expansion in 3D-Printed composite tooling. *J. Thermoplast. Compos. Mater.* **2024**. [CrossRef]
9. Yi, N.; Davies, R.; Chaplin, A.; McCutchion, P.; Ghita, O. Slow and fast crystallising poly aryl ether ketones (PAEKs) in 3D printing: Crystallisation kinetics, morphology, and mechanical properties. *Addit. Manuf.* **2021**, *39*, 101843. [CrossRef]
10. Shang, Y.; Xu, Q.; Jiang, B.; Yang, Y.; Liu, X.; Jiang, Z.; Yu, C.; Li, X.; Zhang, H. Slowing crystallization to enhance interlayer strength of 3D printed poly (ether ether ketone) parts by molecular design. *Addit. Manuf.* **2022**, *59*, 103104. [CrossRef]
11. Santiago, C.C.; Yelamanchi, B.; La Diosdado De Peña, J.A.; Lamb, J.; Roguski, K.; Turzyński, F.; Faruqui, R.; Choo, K.; Du Plessis, A.; Sillani, F.; et al. Thermoplastic Extrusion Additive Manufacturing of High-Performance Carbon Fiber PEEK Lattices. *Crystals* **2021**, *11*, 1453. [CrossRef]
12. Liaw, C.-Y.; Tolbert, J.W.; Chow, L.W.; Guvendiren, M. Interlayer bonding strength of 3D printed PEEK specimens. *Soft Matter* **2021**, *17*, 4775–4789. [CrossRef]
13. Yi, N.; Davies, R.; McBean, M.; Chaplin, A.; Ghita, O. Mobile amorphous fraction in polyaryletherketones and its influence on interlayer bonding in material extrusion. In Proceedings of the 38th International Conference of the Polymer Processing Society (PPS-38), St. Gallen, Switzerland, 22–26 May 2023; AIP Publishing: Melville, NY, USA, 2024; p. 150002.
14. Dou, H.; Cheng, Y.; Ye, W.; Zhang, D.; Li, J.; Miao, Z.; Rudykh, S. Effect of Process Parameters on Tensile Mechanical Properties of 3D Printing Continuous Carbon Fiber-Reinforced PLA Composites. *Materials* **2020**, *13*, 3850. [CrossRef] [PubMed]
15. Consul, P.; Chaplin, A.; Tagscherer, N.; Zaremba, S.; Drechsler, K. Interlaminar strength in large-scale additive manufacturing of slow crystallizing polyaryletherketone carbon composites. *Polym. Int.* **2020**, *70*, 1099–1108 [CrossRef]
16. Davies, R.; Yi, N.; McCutchion, P.; Ghita, O. Mechanical property variance amongst vertical fused filament fabricated specimens via four different printing methods. *Polym. Int.* **2021**, *70*, 1073–1079. [CrossRef]

17. Consul, P.; Beuerlein, K.-U.; Luzha, G.; Drechsler, K. Effect of Extrusion Parameters on Short Fiber Alignment in Fused Filament Fabrication. *Polymers* **2021**, *13*, 2443. [CrossRef]
18. Androsch, R.; Schick, C.; Di Lorenzo, M.L. Kinetics of Nucleation and Growth of Crystals of Poly(l-lactic acid). In *Synthesis, Structure and Properties of Poly(lactic acid)*; Advances in Polymer Science; Springer: Cham, Switzerland, 2017. [CrossRef]
19. Pibulchinda, P.; Barocio, E.; Favaloro, A.J.; Pipes, R.B. Influence of printing conditions on the extrudate shape and fiber orientation in extrusion deposition additive manufacturing. *Compos. Part B Eng.* **2023**, *261*, 110793. [CrossRef]
20. Matschinski, A.; Ziegler, P.; Abstreiter, T.; Wolf, T.; Drechsler, K. Fiber Formation of Printed Carbon Fiber/Poly (Ether Ether Ketone) with Different Nozzle Shapes. *Polym. Int.* **2021**, *70*, 1109–1117. [CrossRef]
21. Huang, C.-T.; Chen, X.-W.; Fu, W.-W. Investigation on the Fiber Orientation Distributions and Their Influence on the Mechanical Property of the Co-Injection Molding Products. *Polymers* **2019**, *12*, 24. [CrossRef] [PubMed]
22. Jeffery, G.B. The motion of ellipsoidal particles immersed in a viscous fluid. *Proc. R. Soc. Lond. A* **1922**, *102*, 161–179. [CrossRef]
23. Heller, B.P.; Smith, D.E.; Jack, D.A. Planar deposition flow modeling of fiber filled composites in large area additive manufacturing. *Addit. Manuf.* **2019**, *25*, 227–238. [CrossRef]
24. Heller, B.P.; Smith, D.E.; Jack, D.A. Effects of extrudate swell and nozzle geometry on fiber orientation in Fused Filament Fabrication nozzle flow. *Addit. Manuf.* **2016**, *12*, 252–264. [CrossRef]
25. Wang, Z.; Smith, D.E. Finite element modelling of fully-coupled flow/fiber-orientation effects in polymer composite deposition additive manufacturing nozzle-extrudate flow. *Compos. Part B Eng.* **2021**, *219*, 108811. [CrossRef]
26. Wang, Z.; Smith, D.E. Numerical analysis of screw swirling effects on fiber orientation in large area additive manufacturing polymer composite deposition. *Compos. Part B Eng.* **2019**, *177*, 107284. [CrossRef]
27. Wang, Z.; Smith, D.E. A Fully Coupled Simulation of Planar Deposition Flow and Fiber Orientation in Polymer Composites Additive Manufacturing. *Materials* **2021**, *14*, 2596. [CrossRef] [PubMed]
28. Wang, Z. A numerical study on the predicted fiber orientation of large area extrusion deposition additive manufactured composites. *Polym. Compos.* **2022**, *43*, 6862–6876. [CrossRef]
29. Behdani, B.; Senter, M.; Mason, L.; Leu, M.; Park, J. Numerical Study on the Temperature-Dependent Viscosity Effect on the Strand Shape in Extrusion-Based Additive Manufacturing. *J. Manuf. Mater. Process.* **2020**, *4*, 46. [CrossRef]
30. Wang, Y.; Beard, J.D.; Evans, K.E.; Ghita, O. Unusual crystalline morphology of Poly Aryl Ether Ketones (PAEKs). *RSC Adv.* **2016**, *6*, 3198–3209. [CrossRef]
31. Wang, Y.; Chen, B.; Evans, K.E.; Ghita, O. Novel fibre-like crystals in thin films of Poly Ether Ether Ketone (PEEK). *Mater. Lett.* **2016**, *184*, 112–118. [CrossRef]
32. Comelli, C.A.; Yi, N.; Davies, R.; van der Pol, H.; Ghita, O. Material extrusion thermal model mapped across polyetheretherketone isothermal and continuous cooling transformation charts. *Addit. Manuf.* **2022**, *59*, 103129. [CrossRef]
33. Ma, X.; Wen, L.; Wang, S.; Xiao, J.; Li, W.; Hou, X. Inherent relationship between process parameters, crystallization and mechanical properties of continuous carbon fiber reinforced PEEK composites. *Def. Technol.* **2023**, *24*, 269–284. [CrossRef]
34. Spoerk, M.; Savandaiah, C.; Arbeiter, F.; Traxler, G.; Cardon, L.; Holzer, C.; Sapkota, J. Anisotropic properties of oriented short carbon fibre filled polypropylene parts fabricated by extrusion-based additive manufacturing. *Compos. Part A Appl. Sci. Manuf.* **2018**, *113*, 95–104. [CrossRef]
35. Yang, D.; Zhang, H.; Wu, J.; McCarthy, E.D. Fibre flow and void formation in 3D printing of short-fibre reinforced thermoplastic composites: An experimental benchmark exercise. *Addit. Manuf.* **2021**, *37*, 101686. [CrossRef]
36. Kim, E.G.; Park, J.K.; Jo, S.H. A study on fiber orientation during the injection molding of fiber-reinforced polymeric composites. *J. Mater. Process. Technol.* **2001**, *111*, 225–232. [CrossRef]
37. Tseng, H.-C.; Chang, R.-Y.; Hsu, C.-H. Improved fiber orientation predictions for injection molded fiber composites. *Compos. Part A Appl. Sci. Manuf.* **2017**, *99*, 65–75. [CrossRef]
38. Coogan, T.J.; Kazmer, D.O. Prediction of interlayer strength in material extrusion additive manufacturing. *Addit. Manuf.* **2020**, *35*, 101368. [CrossRef]
39. Tagscherer, N.; Osswald, T.A.; Drechsler, K. Targeted Temperature Manipulation and Analysis of the Influence on Mechanical Properties in Large-Scale Extrusion Additive Manufacturing. *Appl. Sci.* **2022**, *12*, 2998. [CrossRef]
40. Wu, W.; Geng, P.; Li, G.; Zhao, D.; Zhang, H.; Zhao, J. Influence of Layer Thickness and Raster Angle on the Mechanical Properties of 3D-Printed PEEK and a Comparative Mechanical Study between PEEK and ABS. *Materials* **2015**, *8*, 5834–5846. [CrossRef]
41. Pastorek, M.; Kovalcik, A. Effects of thermal annealing as polymer processing step on poly(lactic acid). *Mater. Manuf. Process.* **2018**, *33*, 1674–1680. [CrossRef]
42. Yeh, G.; Hosemann, R.; Loboda-Čačković, J.; Čačković, H. Annealing effects of polymers and their underlying molecular mechanisms. *Polymer* **1976**, *17*, 309–318. [CrossRef]
43. Gao, G.; Xu, F.; Xu, J.; Tang, G.; Liu, Z. A Survey of the Influence of Process Parameters on Mechanical Properties of Fused Deposition Modeling Parts. *Micromachines* **2022**, *13*, 553. [CrossRef]

materials

Article

Designing Prepregnation and Fused Filament Fabrication Parameters for Recycled PP- and PA-Based Continuous Carbon Fiber Composites

Marah Baddour [1], Ruth Garcia-Campà [2], Pablo Reyes [1,3], Dagmar R. D'hooge [3,4], Ludwig Cardon [1] and Mariya Edeleva [1,*]

1 Centre for Polymer and Material Technologies, Department of Materials, Textiles and Chemical Engineering, Ghent University, Technologiepark 130, 9052 Zwijnaarde, Belgium; marah.baddour@ugent.be (M.B.); pablo.reyesisaacura@ugent.be (P.R.); ludwig.cardon@ugent.be (L.C.)
2 Applied Chemistry and Materials Department, Leitat Technological Center, C/Innovació 2, 08225 Terrassa, Spain; rgarcia@leitat.org
3 Laboratory for Chemical Technology, Department of Materials, Textiles and Chemical Engineering, Ghent University, Technologiepark 125, 9052 Zwijnaarde, Belgium; dagmar.dhooge@ugent.be
4 Centre for Textile Science and Engineering, Department of Materials, Textiles and Chemical Engineering, Ghent University, Technologiepark 70a, 9052 Zwijnaarde, Belgium
* Correspondence: mariya.edeleva@ugent.be

Abstract: Continuous carbon fiber (cCF)-based 3D-printed polymer composites are known for their excellent flexural properties; however, the optimization of the overall process is still desired, depending on the material types involved. Here, the improved manufacturing of cCF-based composites is reported, considering virgin polyamide (PA) and postindustrial waste polypropylene (PP), and the parameters affecting the material properties are evaluated. Firstly, the prepregnation technique was optimized to manufacture cCF polymer filaments with various fiber-to-polymer ratios. Secondly, the fused filament fabrication (FFF) technique was optimized. It was observed that the layer height needs to be sufficiently low for proper interlayer adhesion. The influence of the printing temperature is more complicated, with filaments characterized by a lower fiber-to-polymer ratio requiring a higher nozzle diameter and higher temperatures for efficient printing; and for lower diameters, the best flexural properties are observed for parts printed at lower temperatures, maintaining a high interspace distance. Plasma treatment of the cCF was also explored, as was annealing of the produced parts to enhance the flexural properties, the latter being specifically interesting for the PP-based composite due to a lower wetting caused by a higher viscosity, despite supportive interfacial interactions. Eventually, overall guidelines were formulated for the successful production of cCF-based composites.

Keywords: 3D printing; polyamide (PA); polypropylene (PP); composite; continuous carbon fiber (cCF); enhanced flexural properties

1. Introduction

Additive manufacturing (AM) is a processing technique to fabricate parts with basic and complex shapes via layer-by-layer deposition of materials such as metals, polymers, and ceramics [1–3]. The advantages of AM, i.e., short design cycles, the possibility to manufacture complex structures, and high material utilization as compared to traditional manufacturing, have facilitated its wide application [4–13]. AM is specifically used for the processing of (thermoplastic) polymers and polymer-based composites, the latter being the focus of the present work, considering fused deposition modeling (FDM), which is also known as fused filament fabrication (FFF).

FDM/FFF is one of the leading AM processes based on extrusion and relies on the melting of a polymer filament in the extrusion nozzle and its subsequent deposition on a moving bed to create the final product layer by layer [5–7,14]. FFF has already established itself within the manufacturing industry [8,9]; nevertheless, the main limitation remains the relatively poor tensile properties of the printed products due to void formation [6,15,16]. Hence, it is worthwhile to, for instance, reinforce the printed parts, explaining why the production of 3D-printed polymer matrix composites with short or long continuous carbon fibers (cCFs) has grown [17,18]. It can be expected that such modification enables enhancement of the tensile properties and achievement of a high performance while maintaining the lightweight nature of the produced components in diverse industries.

cCF is often used as reinforcement for polymeric composites due to its low density, high strength, and high elastic modulus with applications for aerospace, windmills, the automotive industry, and sports appliances [15,16]. In general, cCF significantly enhances the tensile and flexural properties of the polymer composite parts [19]. As, e.g., shown by Isobe et al. [20], the utilization of cCF allowed for a seven-times increase in tensile strength and a five-times increase in elastic modulus. More recently recycled products could be made as well, broadening the market potential [19]. For instance, Alarifi et al. [21] showed that cCF-based composites from virgin and recycled polyethylene terephthalate glycol (PETG) have comparable thermal, mechanical, and rheological properties. The authors also showed that recycled PETG has better interfacial interactions with cCF, due to a lower viscosity. Consistently, Tian et al. [22] reported the production of recyclable cCF-poly(lactic acid) (PLA) composites with 100% cCF and 73% PLA recovery potential.

Another important characteristic of cCF, highlighted by Li et al. [23,24], is its superior fatigue and corrosion resistance when compared to traditional metal materials. This quality explains why cCF-based composites are widely used in various anchoring technologies.

As shown in Figure 1, two techniques are typically used to exploit the FFF principle to produce cCF-reinforced composites [22,25], i.e., a technique based on in situ (in-printhead) melt impregnation [26] and a technique based on pre-impregnation followed by 3D printing [27]. During in situ melt impregnation (Figure 1A), the cCF is dragged through the die of an extruder, which melts the matrix material typically in filament form, toward simultaneous composite production and printing. For the pre-impregnation technique, the cCF polymer filament is first manufactured in a separate device and then the filament is used for the actual 3D printing [27,28]. Note that both techniques are two-stage, as highlighted in Figure 1B. However, the pre-impregnation-based process gives more control over the quality of the produced filament, as the so-called prepreg parameters can be varied, ensuring a good wetting of cCF with the polymer [29,30], playing with different polymer-to-cCF ratios [31].

One of the ways to enhance interfacial interactions during pre-impregnation is cCF plasma treatment [32]. For instance, Yuan et al. [33,34] showed that plasma treatment increased the surface roughness of the carbon fibers and changed the functional groups on the surface. Cho et al. [35] additionally showed that plasma treatment significantly increased the storage modulus and impact strength of CF–polycarbonate composites. Furthermore, Ma et al. [36] studied the effect of oxygen plasma treatment on the interface of CF–epoxy composites. The authors highlighted that the treatment enhanced the surface roughness and increased the surface free energy by 45% after 3 min of plasma treatment. Another parameter that can influence the wetting of cCF by the matrix is the viscosity of the polymer. If the viscosity is too high, the wetting is rather ineffective, even in the case of favorable interfacial interactions from a functional group point of view [37,38].

It should be admitted that it remains a challenge to develop an impregnation device that allows the production of prepreg filaments at high speeds and low cost [39]. The prepregnation system should allow sufficient contact of the cCF with the polymer melt, knowing that the nozzle diameter of the prepregnation chamber influences the ideal polymer-to-cCF ratio. Installing a good winding system is also important, as the diameter of the produced prepreg filament is small, making it very brittle.

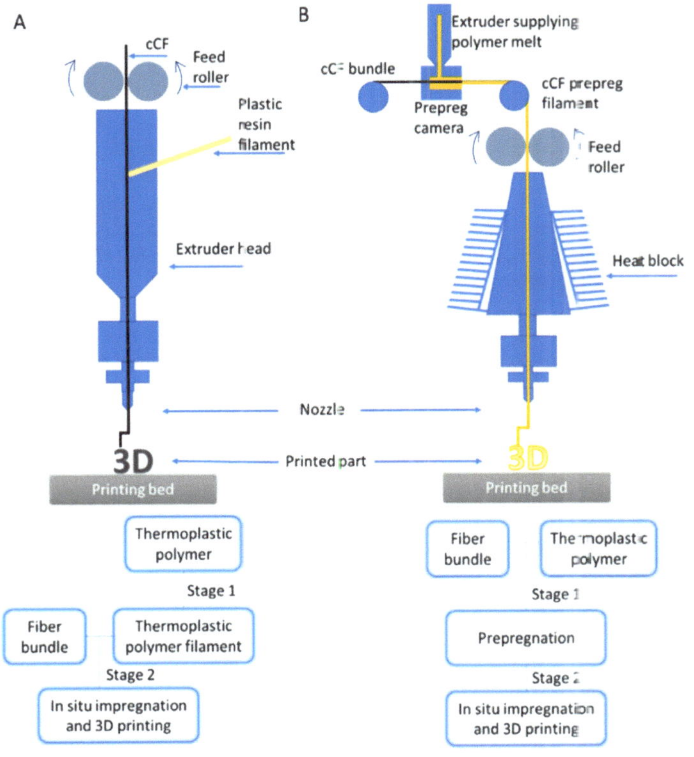

Figure 1. Two techniques were used to print continuous carbon fiber (cCF)-reinforced polymers. (**A**) Set-ups. (**B**) General descriptions.

To engineer a proper contact of the cCF with the polymer melt, three main techniques exist, namely, the pultrusion, passive pin, and active pin techniques [40]. The pultrusion technique is applied in a liquid matrix medium which can be a monomer precursor, a solution, or a polymer melt, whereas the passive pin technique uses the generated pressure between the spreader pin and the fiber bundle in the melt chamber as a driving force for the proper cCF impregnation. The active pin technique involves the injection of a liquid/molten polymer between a bundle of fibers and a cylindrical friction surface to ensure that the entire volume of the matrix material is impregnated with cCF.

It should be further realized that the prepregnation is only one step in the right-hand part of Figure 1, and one should also put emphasis on the optimization of the FFF process as such, i.e., the tuning of the layer height (h), the nozzle temperature (T_n), and the printing pattern, to minimize voids during printing. In this framework, Kuznetsov et al. [41] investigated the effect of h on the flexural strength (σ_{flex}) of PLA printed parts, showing that a smaller h provides a higher σ_{flex} and implicitly lesser voids. Similarly, Wang et al. [42] highlighted the better tensile, flexural, and impact strength of 3D-printed polyetheretherketone (PEEK), CF/PEEK, and glass fiber/PEEK parts. Hu et al. [27] showed that h significantly influences the final strength and modulus of PLA-based parts, while the printing temperature and speed have only a minor effect. Li et al. [43], in turn, demonstrated that when the printing temperature increased from 180 °C to 230 °C, σ_{flex} and the flexural modulus (E_{flex}) of PLA printed parts increased by 46% and 32%, but as soon as h increased from 0.35 mm to 0.55 mm, the flexural properties of the specimens decreased dramatically. In agreement with the above findings, the model of Garzon-Hernandes et al. [44] predicts an increase in tensile strength in acrylonitrile–butadiene–styrene (ABS) polymer parts with a lower h.

Complementary material design can also be performed after the FFF or 3D printing step, as the 3D-printed parts can be subjected to a post-treatment, e.g., annealing, to induce closure of the voids, adhesion of the layers, and recrystallization of semi-crystalline polymers and residual stress release of amorphous ones [45,46]. By annealing above the glass transition temperature (T_g) and below the melt temperature (T_m) for amorphous polymers, or below the recrystallization temperature (T_{cc}) for semi-crystalline thermoplastics, the void content can be significantly decreased [47]. Notably, Bhandari et al. [48] explained that the main mechanism of tensile property increase by annealing is caused by the better incorporation of polymer molecules between layers, providing adhesion, rather than by increasing the crystallinity. Typically, annealing has only a positive effect on the mechanical properties if a slow cooling is applied [48]. Hence, to adjust cCF-reinforced polymeric parameters, several design steps can be conducted, as summarized in Figure 2.

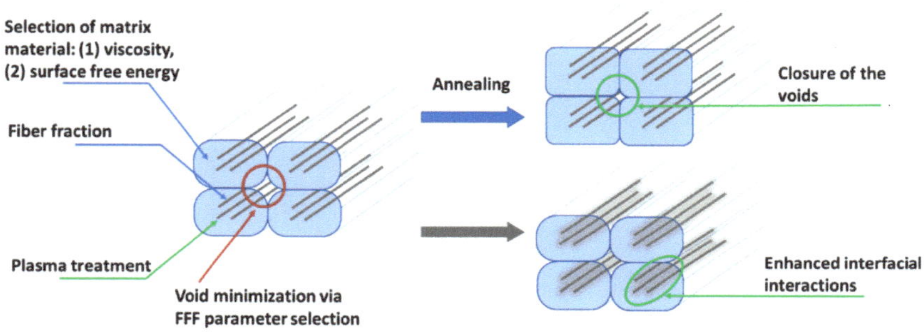

Figure 2. Methods for the enhancement of the material properties of cCF-based printed materials during different steps in Figure 1 or via post-treatment.

In this contribution, cCF-based composite filaments were manufactured with a recycled polypropylene (PP) or a virgin polyamide (PA) matrix. The prepregnation filament fabrication technique was employed to produce the parts, employing an in-house machine for the prepregnation process with an enhanced active pin system. The processing parameters for prepregnation of the filament were systematically varied (e.g., nozzle diameter (D)), as were the conditions for FFF (e.g., layer height (h), interpath distance (s), and nozzle temperature (T_n)), to gain an understanding of their influence on the flexural properties of the resulting parts. To enhance the quality of the filaments, the influence of cCF plasma treatment was also studied, along with an exploration of the impact of annealing on the morphology of the printed parts and their flexural properties. In other words, all optimization methods shown in Figure 2 were addressed, making the current work detailed and complementary to the state of the art.

Based on the overall findings, general guidelines for 3D printing-based manufacturing of cCF-reinforced composite parts for both virgin and recycled polymers were formulated.

2. Materials and Methods

2.1. Materials

Sabic PP 108MF10 postindustrial clean waste polypropylene (PP) from car parts supplied by Maier s.coop group (Bizkaia, Spain), and Rilsamid® AMNO TLD–PA12 from Arkema (Colombes, France) were used as polymer matrix materials. Torayca® T300B-3000 yarn from Toray Composite Materials America (Inc, Tacoma, WA, USA), containing an average of 3000 individual fibers, was used as the cCF. Table S1 in the Supplementary Materials shows an overview of the most relevant material properties for the matrix materials and cCF.

2.2. Manufacturing cCF Prepreg Filaments

cCF prepreg film manufacturing was conducted on an in-house machine, with the main components displayed in Figure 3 [49]. The machine consists of a single screw micro-extruder (element 2 in Figure 3) with two temperature zones (T_{s1} and T_{s2}) [50] which delivers molten polymer to a heated mold (element 3 in Figure 3). To further optimize the prepreg process, three active temperature-controlled pins were introduced (temperatures T_{p1}, T_{p2}, and T_{p3}), which are equipped with a slit to inject the liquid polymer between the cylindrical contact surface and the fiber bundle to avert the dry contact between the fiber and the spreader pin, as shown in Figure S1 in the Supplementary Materials. The pulling system (elements 1 and 4 in Figure 3) pulls the cCF through the polymer melt in the impregnation mold (element 3 in Figure 3). The produced filament is then mounted on a spool by a winding system (element 5 in Figure 3) in view of subsequent FFF. The system is controlled by a Programmable Logic Controller (PLC) through TwinCAT3 software (Version 3.1, Build 4024.56), which gives real-time data and feedback through sensors.

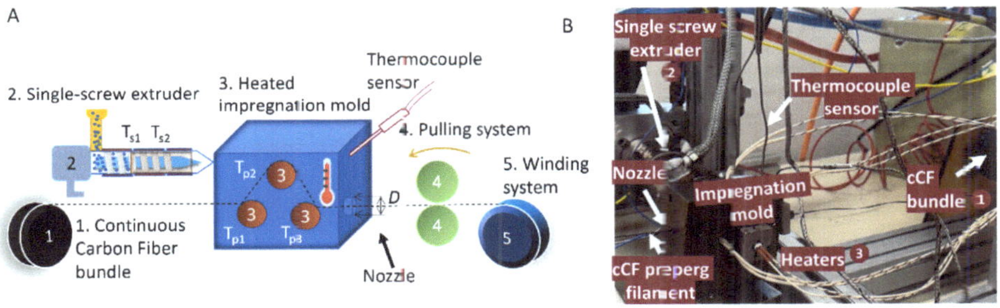

Figure 3. (**A**) Concept of the pre-impregnation process, starting from the principles introduced in Figure 1. (**B**) Actual implementation. 1: Continuous carbon fiber (cCF) bundle, 2: single screw extruder with two heating zones (T_{s1} and T_{s2}), 3: heated impregnation mold with 3 impregnation pins (T_{p1}, T_{p2}, and T_{p3}), 4: pulling system which controls the pultrusion speed ($V_{pultrusion}$), 5: winding system.

For the filament production, the extruder screw temperatures were optimized (T_{s1} and T_{s2}, element 2 in Figure 3), as were the rotational speed of the extruder (N, element 2 in Figure 3), the temperature of the impregnation chamber (T_C, element 3 in Figure 3), the temperature of the impregnation pins (T_{p1}, T_{p2}, and T_{p3}; element 3 in Figure 3), and the pultrusion speed ($V_{pultrusion}$, elements 1 and 4 in Figure 3).

The heated impregnation chamber is also equipped with nozzles with variable diameters, which control the diameter of the prepreg filament and consequently the polymer-to-fiber fraction. Four nozzle diameters were used (D, element 3 in Figure 3) for PA- and PP-based prepreg filaments, namely, 1.1 mm, 0.9 mm, 0.7 mm, and 0.6 mm. The PA fiber volume fractions (V_f) were, e.g., 13.5%, 18.7%, 30%, and 37.6%, respectively. These fiber volume fractions were measured experimentally by thermal degradation of the polymer matrix and weighing the remaining fibers using sample sizes ranging from 500 mg to 5 g of the initial composite material and performing the degradation below 500 °C in a vacuum oven for 2 h. The following equations were used to calculate the filler volume fraction:

$$M_f = \frac{m_f}{m_c} \quad (1)$$

$$V_f = \frac{\rho_m M_f}{\rho_f + M_f\left(\rho_m - \rho_f\right)} \quad (2)$$

where m_c is the composite mass [kg], m_f is the filler mass [kg], M_f is the filler mass fraction [-], ρ_f is the filler density [kg m^{-3}], ρ_m is the matrix density [kg m^{-3}], and V_f is the filler volume fraction [-] [49].

Table 1 lists the specifications of the 17 filaments produced with the optimized prepreg parameters. The variables were determined through trial and error by testing various parameters to identify the optimal ones.

Table 1. Parameters of the filament production from polypropylene PP108MF10 (PP), polyamide Rilsamid AMN O TLD (PA12), and Torayca T300B-3000 continuous carbon fiber (cCF) potentially treated with plasma. V_f—fiber volume fraction; $V_{pultrusion}$—pultrusion speed; T_{S1} and T_{S2}—extruder screw temperature; T_C—temperature of impregnation chamber; T_{P1}, T_{P2}, and T_{P3}—temperature of impregnation pins; *D*—nozzle diameter; and *N*—rotational speed of the extruder.

Matrix	Filament	V_f (%)	$V_{pultrusion}$ [mm/min]	T_{S1} [°C]	T_{S2} [°C]	T_C [°C]	T_{P1} [°C]	T_{P2} [°C]	T_{P3} [°C]	*D* [mm]	*N* [rpm]
PP	F001	16.43	300	180	230	230	230	230	230	1.1	0.68–0.8
	F002 [1]	14.84	300	180	230	230	230	230	230	1.1	2.05
	F003 [2]	15.23	300	180	230	230	230	230	230	1.1	2.05–2.2
	F004	16.37	300	180	230	230	230	230	230	1.1	2.05–2.2
	F005	11.27	300	185	225	215	240	240	240	1.1	0.65
PA	F006	13.50	300	185	225	225	240	240	240	1.1	1.15
	F007	12.63	300	185	225	225	240	240	240	1.1	1.1
	F008	14.91	300	185	225	225	240	240	240	1.1	1.1
	F009	14.73	550	185	225	225	240	240	240	1.1	2
	F010	14.02	425	185	225	225	240	240	240	1.1	1.8
	F011	14.39	675	185	225	225	240	240	240	1.1	2.3
	F012	14.33	800	185	225	225	240	240	240	1.1	2.3
	F013	18.70	300	185	225	225	240	240	240	0.9	0.95
	F014	30	300	185	225	225	240	240	240	0.7	0.47
	F015	30	300	185	225	225	240	240	240	0.7	0.47
	F016	37.60	300	185	225	225	240	240	240	0.6	0.35
	F017	37.60	300	185	225	225	240	240	240	0.6	0.35

[1] Carbon fiber was treated with O_2 + He plasma. [2] Carbon fiber was treated with Ar + He plasma.

2.3. Fused Filament Fabrication of cCF Prepreg Filaments

For FFF, an in-house developed machine was used (Figure 4A) consisting of an aluminum printing bed coated with polyetherimide moving in the X, Y, and Z directions, with a printing volume of 250 × 200 × 145 mm^3. The printing chamber is heated by two infrared radiators mounted below the nozzle. The machine is insulated by using an insulator wall of 5 cm thickness from Recticel (Wetteren, Belgium) to minimize warping and improve layer adhesion. During FFF, the layer height (*h*), interpath distance (*s*), and nozzle temperature (T_n) were controlled. The nozzle temperature was chosen between 200 and 235 °C. The bed temperature was set at 110 °C for cCF-PA and ranged from 60 to 110 °C for cCF-PP (see Table S8 in the Supplementary Materials).

It should be further noted that the printing of cCF prepreg filaments could not be performed with a zig-zag pattern to obtain a complex shape (e.g., a dog-bone shape for the tensile test) because of the high stiffness of cCF. Hence, rectangular specimens were printed, as shown in Figure 4B. The individual tracks of the test specimens were deposited in the same direction due to the spiraling pattern. Upon adding more than one layer, a small loop was made to start a new layer using the same X and Y pattern.

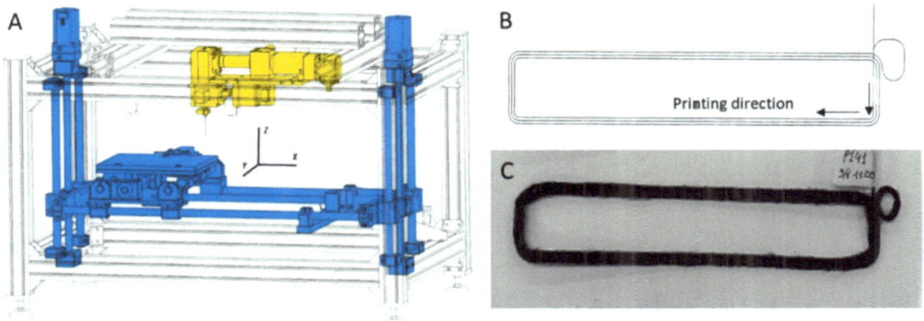

Figure 4. (**A**) Sketch of in-house FFF printer. The bed and driving system are in blue, and the extruder is in yellow. (**B**) Top-view sketch of the toolpath of the printed specimens. (**C**) Actual printed part (printing condition P141 in the Supplementary Materials).

To optimize h and minimize voids, the long straight paths were printed at 120 mm/min, and the curves and short sides were printed between 30 and 60 mm/min. The actual view of the printed specimen is shown in Figure 4C. All printing conditions are summarized in Table S8 in the Supplementary Materials, while Table 2 lists examples of key printing parameters that were used for the production of the parts (notation with P), starting from filaments listed in Table 1 (notation with F).

Table 2. Layer height (h) and interpath distances (theoretical s and actual s, s_{actual}) of printed parts, starting from 4 filaments with specifications in Table 1.

D (mm)	Filament	Sample	T_n (°C)	h (mm)	s (mm)	s_{actual} ($3 \times s_{actual}$) (mm)
1.1	F001	P002	215–235	0.35	1.67	7.02
	F001	P004	215–235	0.45	1.30	5.46
	F001	P005	215–235	0.55	1.06	4.46
0.9	F013	P126	215–235	0.29	1.37	5.50
	F013	P135	215–235	0.37	1.06	4.50
	F013	P145	215–235	0.45	0.87	3.85
0.7	F014	P154	205–235	0.22	1.06	4.85
	F014	P164	205–235	0.29	0.83	4.10
	F014	P172	205–235	0.35	0.68	3.75
0.6	F016	P184	200–225	0.19	0.91	4.40
	F017	P185	200–225	0.25	0.71	3.80
	F017	P195	200–225	0.30	0.58	3.35

2.4. Plasma Treatment

Certain cCF materials were treated in continuous mode using an Atmospheric Pressure Glow Discharge (APGD) device, model PLATEX 600–LAB VERSION, obtained from Grinp, S.r.l. (Settimo Torinese, Italy). The two-planar electrode equipment operated with a frequency range from 20 to 45 kHz to partially ionize gases and/or vapors of precursors. Two types of plasma treatment were used, namely, oxidation with He/O_2 and etching with He/Ar gases. The flow was kept at 2.5/1.5 L/min in both cases, and the power was set to 700 W. The speed of the cCF pultrusion was 1 m/min.

After plasma treatment of cCF, the pre-impregnation process was run with PP as a matrix material to produce the prepreg filaments F002 and F003 in Table 1 with oxygen–helium plasma- and argon–helium plasma-treated cCF, respectively. Conditions P093 and

P095 were used for FFF, incorporating oxygen–helium plasma and argon–helium plasma treatments for cCF, respectively, as shown in Table 3, which also includes the reference untreated case.

Table 3. Printing parameters of untreated and treated CF with PP. Specification of original filaments in Table 2.

Filament	Printing Conditions	cCF	h (mm)	s (mm)	T_n (°C)	T_{bed} (°C)
F001	P004	untreated	0.45	1.30	230	60
F002	P093	He/O_2	0.45	1.30	235	80
F003	P095	He/Ar	0.45	1.30	235	80

2.5. Annealing

The printed specimens for both polymer types were annealed in a vacuum oven for 3 h (excluding heating time). All annealed selected specimens were compared to the unannealed specimens with the same printing parameters, as shown in Table 4.

Table 4. The annealing and printing parameters. Specification of original filaments in Table 1.

Matrix	Printing Conditions	Filament	D (mm)	h (mm)	s (mm)	T_n (°C)	T_b (°C)	Annealing Temperature (°C)
PP	P088	F004	0.9	0.45	1.30	235	80	90, 120, and 140
	P089							
	P092							
PA12	P113–P115	F009	0.9	0.35	1.67	215	110	140, and 165
	P122–P125	F010						
	P116–P119	F011						
	P120–P121	F012						

2.6. Rheological, Morphological, and Thermal Property Characterization

Rheological measurements for the PP matrix material were performed for the compression-molded disk-shaped specimens with a 25 mm diameter and a 1 mm thickness manufactured in a hot press (Fontijne Holland, Vlaardingen, The Netherlands) at 200 and 230 °C. The frequency sweep tests were performed in an MCR 702 rheometer (Anton-Paar, Graz, Austria), using the parallel plate configuration with a 25 mm diameter and a 1 mm gap. The complex viscosity (η^*) was monitored as a function of the angular frequency (from 600 to 0.1 rad/s), with a strain amplitude of 0.1%, under a nitrogen atmosphere. The strain amplitude was defined employing amplitude sweep tests, and all the materials were assumed to be tested in the linear viscoelastic regime. All materials were dried at 60 °C in a vacuum dryer overnight prior to molding and rheological testing. The rheological data for PA12 were obtained from the datasheet.

Optical microscopy for the printed parts was performed on a VHX-7000 Keyence OM (Keyence International NV/SA, Mechelen, Belgium) and applying its accompanying software. The obtained images from the OM were subsequently subjected to processing and analysis through the ImageJ software (Version 1.51), to evaluate the voids volume. Scanning electron microscopy (SEM) images were obtained on a Phenom Pro electron microscope (Benelux Scientific, Ede, The Netherlands). The samples were coated with a 10 nm gold layer, and images were taken at 5 kV.

The thermal properties of the polymer matrices and the prepreg filaments were studied via differential scanning calorimetry (DSC) with a DSC 214 Polyma device (NETZSCH-Gerätebau GmbH, Selb, Germany). All the materials were heated up from 25 to 250 °C with a 10 °C/min ramp considering two heating–cooling cycles. Thermographic analysis (TGA) was performed on a Netsch STA 449 F3 (NETZSCH-Gerätebau GmbH, Selb, Germany). Tests were performed under a nitrogen (N_2) atmosphere, with a flow rate of 50 mL·s^{-1} and a heating rate of 10 °C·min^{-1}, according to the standard ISO 11358-1 [51]. All samples had an initial mass of ca. 10 mg.

2.7. Flexural Property Measurements

The test samples were cut from the printed specimens (Figure 4B,C) according to ISO 14125 [52] (Table S2). All flexural tests were performed for samples that were conditioned for 48 h in the lab before conducting the "climatized samples" test, using a three-point flexural testing set-up and a 2 kN load cell on an Instron 4464 machine at a rate of 1 mm/min. The flexural modulus (E_{flex}) was calculated to be between 0.15% and 0.20% strain. The values mentioned for σ_{flex} are the maximal stress values of the flexural curves, which are typically reported for continuous fiber composites [49].

3. Results and Discussion

To gain an understanding of the influence of the composition on the manufacturing process, 17 filament formulations were prepared (see Table 1) with varying matrix materials (PP or PA12) and fiber volume fractions (V_f). These formulations were used for printing according to 215 conditions (see Table S8 in the Supplementary Materials) to identify relations between printing and overall process parameters and morphology, including plasma treatment and annealing. In what follows, the main results are discussed, focusing first on the wetting potential, both theoretically and by morphological analysis, and then on the void minimization by tuning the process parameters, and finally on the tuning of flexural properties.

3.1. Wetting Degree after Impregnation

The interfacial interactions can be theoretically assessed based on the surface free energies of the matrix and filler, γ_A and γ_B, respectively. In general, the surface energy of the filler should be greater than that of the matrix for proper wetting to take place. For the chosen cCF and the selected matrix polymers, data in the literature suggest that this is the case, with the PA12 γ_{PA} being equal to 40 mN/m [53] and the PP γ_{PP} being equal to 30.1 mN/m [54], both lower than the γ_{CF} of 53 mN/m [55]. Thus, from a theoretical standpoint, both polymers should demonstrate good adhesion to cCF.

However, the SEM analysis in Figure 5A,B shows deviation from this theoretical insight, as the compression fracture of CF-PP displays a clear separation of the individual fibers and matrix, indicative of poor wetting. In more detail, a delamination zone inside the fiber bundles was observed, which was due to inefficient wetting causing the fibers to break during printing (Figure S2b in the Supplementary Materials). On the contrary, PA12 followed the theoretical insights and showed much better adhesion to cCF (Figure 5C,D).

A good dispersion of cCF in the matrix can be observed here and a very flat fracture surface, where the fiber and matrix broke as a whole (Figure 5D), which shows that individual fibers are (close to) fully coated with PA12, which was not observed for PP. Additional SEM images (Figure S3 in the Supplementary Materials) for the PA12-based composite further support the good wetting of the fibers with the matrix material. Evidently, on the microscopic level, PA12-based composites possess superior properties due to the proper wetting of the fibers by the matrix.

Another parameter to consider for the optimization of the wetting is the viscosity of the polymer melt, with a lower viscosity facilitating the wetting. Figure S6 in the Supplementary Materials shows that the viscosity of PP is significantly higher than that of PA, disfavoring the wetting of the cCF with PP, as was observed via SEM. Note that in the theoretical calculations, such viscosity variation is not included, at least partially explaining the differences with the SEM images. For further investigation, the estimated shear rate ($\gamma.w$) for PP during the printing process was calculated at three different temperatures, 215 °C, 225 °C, and 235 °C, employing the following equations [56]:

$$\gamma\,\dot{}\mathrm{a} = \frac{4Q}{\pi R^3}\ [\mathrm{s}^{-1}] \quad (3)$$

$$\gamma\dot{}w = \frac{\gamma\dot{}a}{4}\,(3 + \frac{1}{n})\;[\mathrm{s}^{-1}] \qquad (4)$$

where Q represents the volumetric flow rate, which is derived from the melt density (ρ) and the weight per unit time discharged from the printer at the used extrusion speed. The melt density (ρ) was obtained from the melt flow index test, the nozzle diameter was $2R = 2$ mm, and n was the power law index determined from rheological data, taking on a value of $n = 0.2$ for PP. Table S3 in the Supplementary Materials shows low values for the estimated shear rate ($\gamma\dot{}w$). Therefore, the lack of wetting in the cCF-PP specimens was due to high viscosity. Additionally, the voids analysis showed a slight reduction in void content for samples at higher temperatures (235 °C) compared to samples printed at 225 °C (see Table S4 in the Supplementary Materials). This decrease is attributed to a lower viscosity, facilitating improved wetting between the matrix and the filler.

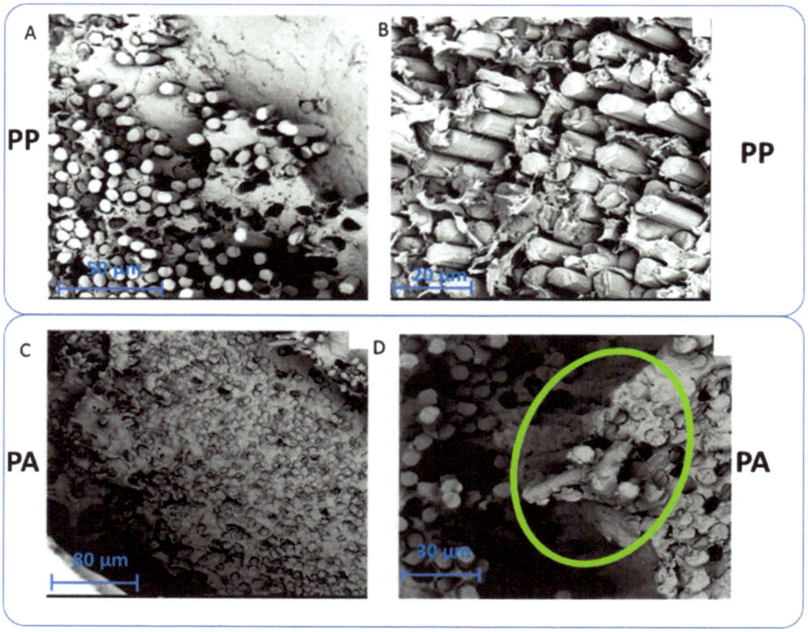

Figure 5. SEM images of the compression fracture surface of cCF/polypropylene (PP) (**A**,**B**) and cCF/polyamide (PA) (**C**,**D**) printed parts. (**A**) Filament F001, printing condition P002. (**B**) Higher magnification of image (**A**). (**C**) Filament F013, printing condition P126. (**D**) Higher magnification of image (**C**). The green oval shows the (almost) complete wetting of the fiber with the polymer matrix.

3.2. Optimization of FFF Parameters to Decrease Voids

Due to the layer-by-layer nature of FFF with imperfect interlayer adhesion, the 3D-printed part may contain voids, which can affect the material properties, justifying the control of printing parameters to reduce the void content and sizes. The filaments were first theoretically considered to be incompressible solids with an elliptical cross-section so that the void cross-section would have a four-arm star shape (Figure 6A). These macro-void volume fractions ($Vmac$) exist between the individual deposited tracks and are determined using Equation (5). By reducing the ratio of the layer thickness (h) to the individual strand width (w), $Vmac$ can be reduced. Additionally, by adjusting s, which is the distance between two tracks, the shape and the content of these voids can change. In particular, Equation (6) can be used for the connection of s and w:

$$Vmac = \frac{h}{w}(1 - \frac{\pi}{4}) \quad (5)$$

$$s = w - \xi \cdot h \cdot (1 - \frac{\pi}{4}) \quad (6)$$

where ξ is the void-filling factor, which ranges from 0 to 1. The complete derivation of Equation (6) is included in the Supplementary Materials.

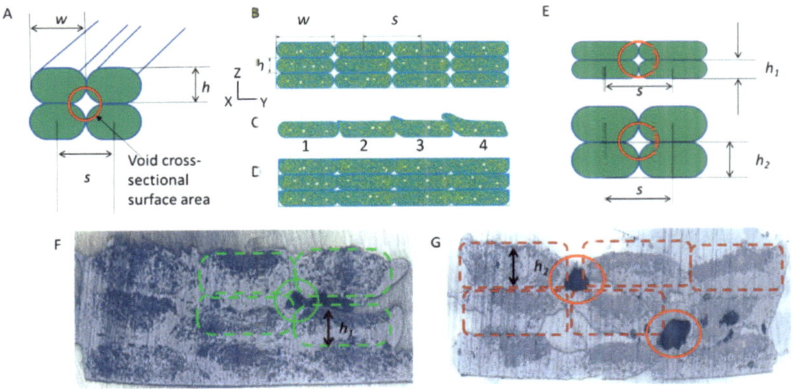

Figure 6. (**A**) Theoretical cross-section of the 3D-printed incompressible printed specimen. w—the width of an individual strand, s—the distance between two tracks, h—the layer height; the void cross-sectional surface area (highlighted with a red circle) is star-shaped. (**B**–**D**) Influence of the void filling factor on the cross-section of the 3D-printed parts: (**B**) $s = w$, $\xi = 0$; (**C**) $s << w$, $\xi = 1$; (**D**) $s < w$, $\xi = 0.5$. (**E**) Dependence of the void cross-sectional surface area on the layer's height for smaller (**upper image**) and larger (**lower image**) h values. Optical images of (**F**) cross-section of CF-PA12-based part, with s = 1.37 mm and h = 0.29 mm (printing condition P126), and of (**G**) cross-section of CF-PA12 specimen, with s = 1.06 mm and h = 0.22 mm (printing condition P154). Colors used to facilitate identification of printing quality, with green indicating better quality.

Figure 6B shows a theoretical situation in which filament strands are deposited next to each other, with $w = s$ and $\xi = 0$, which is the case for an incompressible filament with star-shaped voids. If $\xi = 1$ (Figure 6C), the nearby tracks overlap too much, decreasing the quality of the printed part. If $\xi = 0.5$ (Figure 6D), a good balance between void filling and printability can be achieved [49]. Additional discussion of the influence of ξ on the void content is provided in the Supplementary Materials. Notably, h affects the void size as well, with Figure 6E making it theoretically evident that the voids between layers with different thicknesses diminish if the layer thickness becomes smaller. Figure S9 in the Supplementary Materials shows a comparison between the calculated and experimental impact of h on the void content. In general, h is determined by the filament diameter (D), by the distance between the printing head and the bed, and by the deformation of the filament during printing. Furthermore, with a larger s, the filament has more room to deform during melting, which results in smaller voids as well.

In view of the theory mentioned above, four nozzle diameters were selected (D = 1.1 mm, D = 0.9 mm, D = 0.7 mm, and D = 0.6 mm) to study the influence of h. Optical microscopy was used to study the effect of the printing conditions on the quality of the part and voids content. Figure 6F,G, for example, show optical microscopy images of a three-layered CF-PA12 specimen with s = 1.37 mm and s = 1.06 (specimen F013, printing conditions P126 and P154; Table 2). Indeed, a denser composite with a lower void content was observed when h was smaller (Figure 6F). Upon exploring the range of printing conditions, s could be gradually brought to 1.2 mm, at which point both cCF-PP and cCF-PA12 experienced complete breakage of the filament during printing. A complete overview of the results for all printing conditions is given in Table S8 in the Supplementary Materials. More optimal parameters always include a larger s and a smaller h.

3.3. Optimal FFF Parameters to Tune Flexural Properties

To evaluate the material properties, a comparison was made between the flexural strength, the flexural modulus, and the failure mode of PP- and PA-based printed parts. This failure mode strongly differed for cCF-PP and cCF-PA12, consistent with the already discussed wetting results. The cCF-PA12 part was always totally fractured after delamination or partially cracked (Figure S10B in the Supplementary Materials), unlike the cCF-PP samples, which never broke as a whole (Figure S10A in the Supplementary Materials).

A typical flexural curve of a cCF-PP sample exhibits, first, a linear slope, and then multiple smaller stress drops, followed by a larger drop, as shown in Figure S10A in the Supplementary Materials. The small drops can be attributed to partial delamination and breakage of fiber bundles. During the last drop, the remaining polymer broke but the specimens were still held together by an outer PP layer and a cCF intact bundle. The flexural test was manually stopped at one point as the specimens never broke completely, and the stress level remained constant. The convex side of the sample, i.e., the side in which the tensile stress is created, showed PP stress whitening. There are two potential explanations for discoloration, strain-induced crystallization or formation of cavities [57]. For more insights, the DSC test was conducted on samples from a broken area and samples that did not undergo the flexural test to assess the degree of crystallinity using the following equation [58]:

$$Xc(\%) = \frac{\Delta H_m - \Delta H_{cc}}{\Delta H_m^{\infty}.w} \times 100\% \quad (7)$$

where ΔH_m is the melting enthalpy, ΔH_{cc} is the cold crystallization enthalpy, ΔH_m^{∞} is the melting enthalpy, and w is the weight fraction of the sample in grams.

The results indicated a significant reduction in crystallinity, dropping from 23.7% to 6.69% for the samples that were not subjected to the flexural test and those from the broken area, respectively. This suggests that the likely cause of the observed stress whitening is the formation of cavities (see Figures S11 and S12 in the Supplementary Materials).

Figure S10B in the Supplementary Materials displays two typical flexural curves of cCF-PA12 samples. The dotted curve shows a typical first type of failure mode, which indicates that the composite breaks as one homogeneous material. This mode of failure suggests that the interface between the CF and the PA12 polymer matrix is optimal. The second failure mode, which is displayed as a full line in Figure S10B, exhibits several drops in flexural stress after the linear increase. Unlike the failure mode in cCF-PP parts, the final drop of stress in the second failure mode of cCF-PA parts indicates a complete fracture of the specimen. It should be noted that only in rare cases did cCF-PA12 samples not show the second type of failure mode.

To obtain a deeper insight into the relation between printing conditions and the material properties of the samples, the response surface analysis approach was additionally applied [59]. The experimental data that were used to construct the response surfaces can be found in the Supplementary Materials. Table S5 in the Supplementary Materials summarizes the flexural measurement results, considering both the flexural modulus (E_{flex}) and the flexural strength (σ_{flex}), while Figures S16–S22 in the Supplementary Materials show the strength–strain curves for the measured samples and the bar charts used to analyze the results.

Selecting a D of 0.6 mm, Figure 7 shows for the PA-based composite the response surfaces for σ_{flex} and E_{flex} depending on s, h, and the printing temperature (T_n). A major influence of the printing temperature can be observed in Figure 7A on σ_{flex}, due to the need for sufficient melting of the polymer and thus better interlayer adhesion. It is interesting to note that the best flexural strength results were observed when the printing temperature was at the lower limit for a smaller D, which evidently minimizes the degradation of the polymer material. Increasing the value of s from 3.35 to 4.4 mm has a notable effect on the material properties, as shown in Figure 7A. A synergy is even observed when the material is printed at lower temperatures and a higher s, as the (molten) filament has enough room to flow during melting, resulting in smaller voids and enhanced material properties.

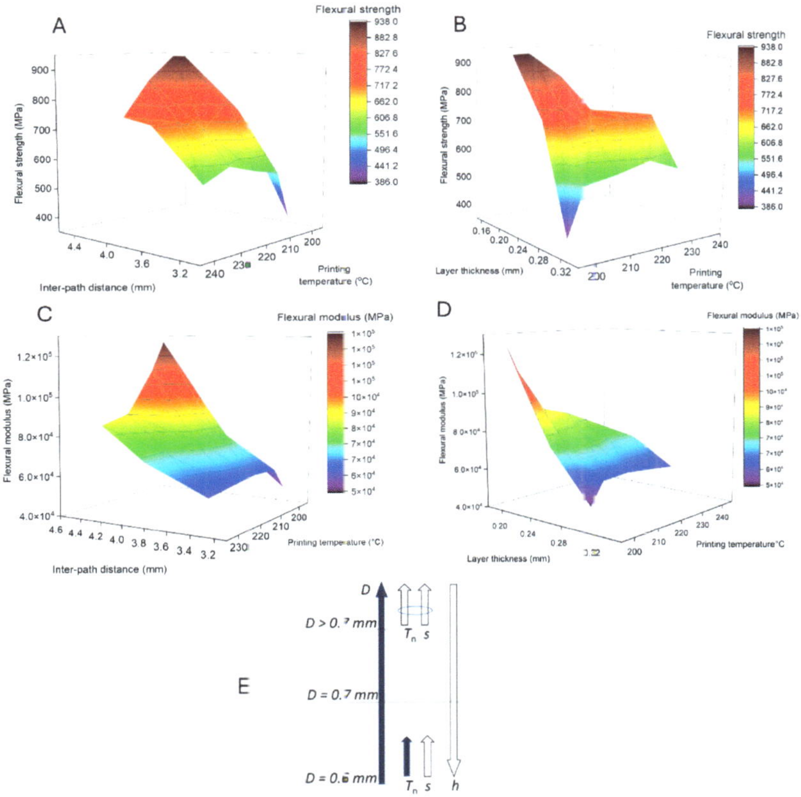

Figure 7. For the PA-based composite, response surface plots of the relationship between the (**A**) flexural strength (σ_{flex}), interpath distance, and printing temperature; (**B**) flexural strength (σ_{flex}), layer thickness, and printing temperature; (**C**) E_{flex} modulus, interpath distance, and printing temperature; and (**D**) E_{flex} modulus, layer thickness, and printing temperature of the samples. Specimens from F016 and F017, printing conditions P184–P215, with a D of 0.6 mm (Table S8). (**E**) Schematic representation of how the filament and printing parameters influence the flexural properties: the direction of the arrows indicates an increase/decrease in the printing parameter; the color reflects a positive (white) or negative (dark blue) impact on the flexural properties, and the blue ellipse highlights that the combination of the parameters has an influence.

Figure 7B further shows that h significantly contributes to the σ_{flex} of the printed parts. σ_{flex} increases more than two times, from ca. 400 MPa to 900 MPa, when h decreases from ca. 0.3 mm to ca. 0.2 mm. However, the influence of temperature is more complicated in comparison with Figure 7A. A positive effect on σ_{flex} is observed with a higher temperature, but the best results are only obtained when h and the temperature are smallest. In general, a small h ensures more efficient melting and interlayer adhesion, even at the lowest printing temperature, whereas a higher temperature may induce polymer degradation and overflow of the printing stands, as schematically shown in Figure 6C, leading to poor material properties.

The dependence of E_{flex} on temperature and s (Figure 7C), as well as on temperature and h (Figure 7D), follows the same trend as for σ_{flex}, already discussed in Figure 7A,B. In particular, the highest E_{flex} is obtained when the part is printed at the lowest temperature with the highest s (Figure 7C). It should be noted that, although the dependence is more pronounced, the modulus values decrease more rapidly if s deviates from the maximum and the temperature increases. In particular, E_{flex} decreases by 35% upon going from the lowest printing temperature to the highest, whereas only a 20% variation upon changing the printing temperature was observed for σ_{flex} (Figure 7A). A similar evolution of modulus vs. printing temperature can be observed when h changes (Figure 7D). The synergy of a low printing temperature and sufficient s is thus more important for the modulus value than for the strength.

A similar dependence of the flexural properties on the temperature and h was observed for the filament with D = 0.7 mm (instead of D = 0.6 mm), still focusing on the PA-based composite, as shown in Figure S13 in the Supplementary Materials. However, it is interesting to note that for the filaments with an even higher D = 0.9 mm, the influence of the temperature becomes more crucial. As Figure S14 in the Supplementary Materials shows, only the highest printing temperature leads to the best material performance independently of s and h. Evidently, a large filament diameter and a higher fraction of polymer (a higher diameter and a lower fraction of cCF) require more energy to enable sufficient melting, resulting in a greater influence on the printing temperature.

Figure 7E summarizes conceptually the influence of s, h, and printing temperature (T_n) on the flexural properties (assuming a good wetting, as for the PA-based composite), with the direction of the arrows indicating an increase/decrease in the respective printing parameter, whereas the color variation reflects the positive/negative influence on the flexural properties.

3.4. *Effect of Plasma Treatment and Annealing*

As observed from the SEM images, the interfacial interactions of the cCF and PP matrix are poor. Hence, the attempt to improve the flexural properties was initiated with plasma treatment of the cCF before pre-impregnation. The results of the flexural tests of CF-PP with plasma-treated CF are summarized in Figure S16 in the Supplementary Materials. No significant difference can, however, be observed between E_{flex} and σ_{flex} for the sample prepared with the plasma treatment (printing conditions P093 and P095, Table S8 in the Supplementary Materials) and the non-treated cCF sample (printing condition P004, Table S8 in the Supplementary Materials). Thus, the main factor hindering the efficient wetting of the cCF with the PP matrix can be considered to be the high melting viscosity of the PP matrix.

The printed samples were alternatively annealed to close the voids and enhance the flexural properties. Prior to annealing, DSC analysis was performed to select the annealing conditions for PP and PA12, as the annealing temperature for semi-crystalline polymers should be selected between T_g and T_m, and ideally it should be close to T_{cc}. To exclude the negative influence of the possible crystallinity increase after annealing, the DSC data were compared before and after annealing for both polymer matrix materials, as shown in Figures S24–S29 in the Supplementary Materials. The results show that the crystallinity percentage and the shape of the melting peak of the annealed parts are almost the same as those of the unannealed ones.

For the recycled PP (Figure S24 in the Supplementary Materials), no exothermal cold crystallization peak was observed during the first and the second heating DSC cycle. The small exothermal peak at approximately 120 °C in the heating stage was possibly due to a small contamination by (low-density) polyethylene ((LD)PE). The melting temperature was T_m = 167 °C. Consequently, an annealing temperature should be selected between T_g = −20 °C and this melting temperature. Three temperatures were selected, namely, 90 °C, 120 °C, and 140 °C, for the annealing of PP samples. The DSC curve of PA12 (the right-hand side of Figure S24 in the Supplementary Materials), by contrast, shows in its first run a cold-crystallization peak at around 165 °C. In this context, 140 °C and 165 °C were selected to anneal the cCF-PA12 parts.

Analysis of the flexural curves shows that the annealed cCF-PP exhibits a slight change-over in the failure mechanism, as shown in Figure 8A. The annealed samples showed larger and more uniform drops in stress levels compared to the unannealed samples, which showed multiple smaller drops. Such a flexural curve profile is typical in case of increased interlayer strength and/or increased strength between the paths, which is achieved due to (partial) void closure. ImageJ analysis results demonstrated that the void volume in an annealed sample at 140 °C is reduced compared to the unannealed sample, as illustrated in Figure S32 in the Supplementary Materials. The small drops in the flexural strength (σ_{flex}) of the unannealed samples are due to more local delamination or fractures, whereas the layers of the annealed material are more uniformly delaminated.

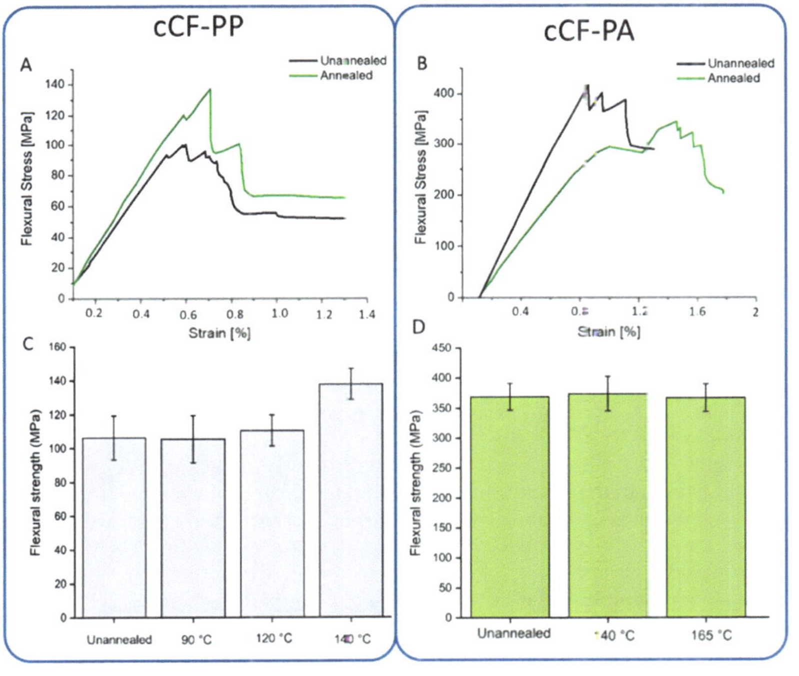

Figure 8. Representative flexural curves of (**A**) cCF-PP parts unannealed and annealed at 140 °C (specimen F004, printing conditions P002 and P088, respectively) and (**B**) cCF-PA12 parts unannealed and annealed at 165 °C (specimen F007, printing condition P107; specimen F009, printing condition P114, respectively). Effect of annealing at different temperatures on (**C**) cCF-PP parts at 90 °C, 120 °C, and 140 °C and (**D**) cCF-PA12 parts at 140 °C and 165 °C.

Note that a change-over of failure mode for the cCF-PP could result in better interlayer bonding in the annealed samples. No significant increase in flexural strength (σ_{flex}), however, was noticed for the annealed specimens at 90 °C and 120 °C (Figure 8C). In contrast, at 140 °C, which allows for PP melting, an increase of ca. 30% in the flexural strength (σ_{flex}) was noticed compared to the unannealed samples and the samples that were annealed at lower temperatures. The flexural modulus (E_{flex}), however, remained unaffected (Figure 8C). Hence, only a careful selection of the annealing conditions allowed us to improve the mechanical properties of the cCF-PP parts.

For cCF-PA, no changes in the failure mode were observed, as the annealed parts broke similarly to the unannealed parts (Figure 8B). This suggests that the interlaminar strength did not improve significantly with the annealing process. Indeed, no increase in flexural strength (σ_{flex}) was observed for cCF-PA12 (Figure 8D) after annealing at all the selected temperatures. This result implies that the printed (unannealed) cCF-PA12 already exhibited optimal mechanical properties, and additional thermal treatment could even degrade PA12. Figure S33 in the Supplementary Materials displays unannealed and annealed samples of cCF-PA12 exhibiting effective closure of voids. Annealing is therefore likely unnecessary for improving the mechanical properties of the cCF-PA12 composite. In other words, annealing seems to be only necessary when the wetting is less.

4. Conclusions

This research focused on the manufacturing of polymer-based composites reinforced with continuous carbon fibers for utilization in 3D-printing applications. Initially, 17 filament formulations were developed using different matrix materials (recycled polypropylene and virgin polyamide 12) via a prepregnation process. Subsequently, bending bars were 3D-printed via FFF using the prepared prepreg filaments.

The study examined the influence of various 3D printing parameters, including the layer height (h), interpath distance (s), printing temperature (T_n), and nozzle diameter of the impregnation mold (D), on the void formation and the flexural properties of the printed components. Explorations involved a range of h values from 0.19 to 0.55 mm, s values from 0.58 to 1.67 mm, T_n values from 200 to 235 °C, and D values from 0.6 to 1.1 mm. Additionally, plasma treatment of cCF was assessed, and optimization of prepregnation parameters, such as cCF fractions ranging from 11.27% to 37.60%, was performed.

The cCF-PP specimens exhibited poor wetting characteristics. Analysis of the estimated shear rate ($\gamma.w$) demonstrated low values, suggesting that the inadequate wetting observed in cCF-PP specimens can be attributed to high viscosity levels. Moreover, the void analysis indicated a slight reduction of 8.16% in void content for samples printed at higher temperatures of 235 °C compared to those printed at 225 °C. This decline is associated with reduced viscosity, facilitating improved wetting between the matrix and the filler components.

DSC results of cCF-PP specimens exhibited a substantial reduction in crystallinity of 72%. This reduction indicates stress whitening, attributed to cavitation formation. However, plasma treatment showed no significant difference and did not result in improvement. On the contrary, the cCF-PA specimens exhibited proper wetting of the fibers by the matrix, resulting in superior properties.

Both a high printing temperature (T_n) and a larger interpath distance (s) had a positive impact on flexural properties, especially when combined with larger nozzle diameters (D), leading to a notable 20% increase in flexural strength. However, specimens with a high cCF fraction of 37.60% and a smaller nozzle diameter of 0.6 mm experienced negative effects from high temperatures, where a synergistic effect occurred with higher interpath distances and lower printing temperatures. In any case, reducing the layer height (h) from 0.3 mm to 0.2 mm resulted in a significant enhancement in flexural strength, demonstrating a remarkable increase of 125%.

In summary, this research offers guidelines for enhancing the flexural properties of cCF-based parts. For instance, annealing at sufficiently high temperatures proves particularly beneficial for PP-based composites, potentially increasing flexural strength by up to 30%. This strategy compensates for the reduced wetting observed in PP-based composites due to higher viscosity. Contrarily, for PA-based composites, wetting does not need to be further optimized and the focus can be on the optimal printing settings alone.

Supplementary Materials: The following supporting information can be downloaded at: https://www.mdpi.com/article/10.3390/ma17081788/s1, Figure S1: Cross-section of a spreader pin and fiber bundle (dashed line) forces; Table S1: The properties of the cCF and the polymers matrices; Table S2: Dimensions of ISO-14125 flexural specimens (ISO-14125); Figure S2: SEM results of PP composite obtained from printing F001: (a) compression fracture, (b) delamination zone, (c) poor interfacial interaction between the matrix and the fiber, and (d) broken fibers inside the matrix; Figure S3: SEM results for PA12 composite obtained from F013: (a) CF dispersion in the matrix PA12, (b) compressive fracture for composites, (c) interfacial interaction between the matrix and the fiber, and (d) individual fiber that is fully coated with PA12; Figure S4: Cross-section of the annealed sample made with CF-PA12; Figure S5: Cross-section of the annealed sample made with CF-PP; Figure S6: Viscosity vs. shear rate temperature dependence for (a) PA12 and (b) PP; Table S3: The estimated shear rate for PP during the printing process; Table S4: The results of the void analysis of PP printed samples; theoretical support for Equation (6) in the main text; Figure S7: (A) Deformation of the printing filament during the FFF process. (B) Cross-section of the printed strand; Figure S8: (A) The effect of partially overlapping composite tracks, where $s \ll w$ and $\xi \approx 1$. The order of strand deposition is indicated by increasing numbers. Cross-section of a single-layer composite perpendicular to the fiber direction. (B) Reduction in the macro-void content by slightly overlapping the composite tracks, where ξ = 0.5. Cross-section of a triple-layer composite perpendicular to the fiber direction; Figure S9: The relationship between h and the voids showcases a comparison between calculated and actual values; Figure S10: Averaged flexural stress curves for (A) cCF-PP (specimen F001, printing condition P002; Table S8) and (B) cCF-PA12 (specimens F007 and F009, printing conditions P107–P113; Table S8); Figure S11: DSC results for PP-cCF samples taken from a broken area; Figure S12: DSC results for PP-cCF samples taken from samples that did not undergo the flexural test; Figure S13: Response surface graphs of the relationship between the (A) flexural strength (σ_{flex}), interpath distance, and printing temperature; (B) flexural strength (σ_{flex}), layer thickness, and printing temperature; (C) E_{flex} modulus, interpath distance, and printing temperature; and (D) E_{flex} modulus, layer thickness, and printing temperature of the samples (specimen F014, printing conditions P154–P183, with D = 0.7 mm); Figure S14: Response surface graphs of the relationship between (A) flexural strength (σ_{flex}), interpath distance, and printing temperature; (B) flexural strength (σ_{flex}), layer thickness, and printing temperature; (C) E_{flex} modulus, interpath distance, and printing temperature; and (D) E_{flex} modulus, layer thickness, and printing temperature of the samples (specimen F013, printing conditions P126–P153, with D = 0.9 mm); Figure S15: Response surface graphs of the relationship between (A) flexural strength (σ_{flex}), interpath distance, and printing temperature; (B) flexural strength (σ_{flex}), layer thickness, and printing temperature; (C) E_{flex} modulus, interpath distance, and printing temperature; and (D) E_{flex} modulus, layer thickness, and printing temperature of the samples (specimen F001, printing conditions P001–P066, with D = 1.1 mm); Figure S16: Flexural strength (σ_{flex}) and flexural modulus (E_{flex}) of plasma-treated and untreated cCF-PP (specimens F001, F002, and F003). The printing conditions are highlighted in the plots; Figure S17: The sample failure mode of F013, printing condition P126, with D = 0.9 mm (PA12 as a matrix); Figure S18: The sample failure mode of F014, printing condition P164, with D = 0.7 mm (PA12 as a matrix); Figure S19: The sample failure mode of F017, printing condition P190, with D = 0.6 mm (PA12 as a matrix); Figure S20: Flexural strength and modulus of the samples with filament diameter D = 1.1 mm; Figure S21: Flexural strength and modulus of the samples with D = 0.9 mm; Figure S22: Flexural strength and modulus of the samples with D = 0.7 mm; Figure S23: Flexural strength and modulus of the samples with D = 0.6 mm; Table S5: Flexural strength and modulus of the samples with different D, T, s, and h values; Table S6: DSC test parameters; Figure S24: DSC results for PP; Figure S25: DSC results for PA12; Figure S26: DSC results for unannealed cCF-PP; Figure S27: DSC results for annealed cCF-PP; Figure S28: DSC results for unannealed cCF-PA12; Figure S29: DSC results for annealed cCF-PA12; Figure S30: TGA results for PP; Figure S31: TGA results for PA; Table S7: Estimated flexural properties of the cCF composites for varying fiber fractions; Figure S32: ImageJ results for (A) a non-annealed cCF-PP (P62) sample and (B) an annealed cCF-PP sample at 140 °C; Figure S33: ImageJ results for (A) a non-annealed cCF-PA12 (P142) sample and (B) an annealed cCF-PA12 sample at 165 °C; Table S8: Printing conditions for the test specimens.

Author Contributions: Conceptualization, L.C. and M.E.; software, M.B., D.R.D. and L.C.; investigation, M.B., L.C., D.R.D. and M.E.; writing—original draft preparation M.B. and M.E.; writing—review and editing, M.B., R.G.C., P.R., D.R.D., L.C. and M.E.; supervision, D.R.D., L.C. and M.E.; funding acquisition, D.R.D. and L.C. All authors have read and agreed to the published version of the manuscript.

Funding: This work was funded by the European Union's Horizon 2020 research and innovation programs entitled: "Recycling and Repurposing of Plastic Waste for Advanced 3D Printing Applications" (Repair 3D) under GA No. 814588. The Commission is not responsible for any use that may be made of the information it contains.

Institutional Review Board Statement: Not applicable.

Informed Consent Statement: Not applicable.

Data Availability Statement: Data are contained within the article and Supplementary Materials.

Conflicts of Interest: The authors declare no conflicts of interest.

References

1. Li, N.; Huang, S.; Zhang, G.; Qin, R.; Liu, W.; Xiong, H.; Shi, G.; Blackburn, J. Progress in Additive Manufacturing on New Materials: A Review. *J. Mater. Sci. Technol.* **2019**, *35*, 242–269. [CrossRef]
2. Ngo, T.D.; Kashani, A.; Imbalzano, G.; Nguyen, K.T.Q.; Hui, D. Additive Manufacturing (3D Printing): A Review of Materials, Methods, Applications and Challenges. *Compos. Part B Eng.* **2018**, *143*, 172–196. [CrossRef]
3. Zhai, X.; Jin, L.; Jiang, J. A Survey of Additive Manufacturing Reviews. *Mater. Sci. Addit. Manuf.* **2022**, *1*, 21. [CrossRef]
4. Joshi, S.C.; Sheikh, A.A. 3D Printing in Aerospace and Its Long-Term Sustainability. *Virtual Phys. Prototyp.* **2015**, *10*, 175–185. [CrossRef]
5. Fico, D.; Rizzo, D.; Casciaro, R.; Corcione, C.E. A Review of Polymer-Based Materials for Fused Filament Recycled Materials. *Polymers* **2022**, *14*, 465. [CrossRef] [PubMed]
6. Cuan-Urquizo, E.; Barocio, E.; Tejada-Ortigoza, V.; Pipes, R.; Rodriguez, C.; Roman-Flores, A. Characterization of the Mechanical Properties of FFF Structures and Materials: A Review on the Experimental, Computational and Theoretical Approaches. *Materials* **2019**, *12*, 895. [CrossRef] [PubMed]
7. Shanmugam, V.; Pavan, M.V.; Babu, K.; Karnan, B. Fused Deposition Modeling Based Polymeric Materials and Their Performance: A Review. *Polym. Compos.* **2021**, *42*, 5656–5677. [CrossRef]
8. Patel, A.; Taufik, M. Nanocomposite Materials for Fused Filament Fabrication. *Mater. Today Proc.* **2021**, *47*, 5142–5150. [CrossRef]
9. Lalehpour, A.; Barari, A. Post Processing for Fused Deposition Modeling Parts with Acetone Vapour Bath. *IFAC-PapersOnLine* **2016**, *49*, 42–48. [CrossRef]
10. Wang, Y.; Chen, T.; Yeh, Y. Advanced 3D Printing Technologies for the Aircraft Industry: A Fuzzy Systematic Approach for Assessing the Critical Factors. *Int. J. Adv. Manuf. Technol.* **2019**, *105*, 4059–4069. [CrossRef]
11. Pei, E.; Shen, J.; Watling, J. Direct 3D Printing of Polymers onto Textiles: Experimental Studies and Applications. *Rapid Prototyp. J.* **2015**, *21*, 556–571. [CrossRef]
12. Vithani, K.; Goyanes, A.; Jannin, V.; Basit, A.W.; Gaisford, S.; Boyd, B.J. A Proof of Concept for 3D Printing of Solid Lipid-Based Formulations of Poorly Water-Soluble Drugs to Control Formulation Dispersion Kinetics. *Pharm. Res.* **2019**, *36*, 102. [CrossRef] [PubMed]
13. Revilla-León, M.; Özcan, M. Additive Manufacturing Technologies Used for Processing Polymers: Current Status and Potential Application in Prosthetic Dentistry. *J. Prosthodont.* **2019**, *28*, 146–158. [CrossRef] [PubMed]
14. Van Waeleghem, T.; Marchesini, F.H.; Cardon, L.; D'hooge, D.R. In Silico Quantification of 3D Thermal Gradients and Voids during Fused Filament Fabrication Deposition to Enhance Mechanical and Dimensional Stability. *Addit. Manuf.* **2023**, *72*, 103624. [CrossRef]
15. Ćwikła, G.; Grabowik, C.; Kalinowski, K.; Paprocka, I.; Ociepka, P. The Influence of Printing Parameters on Selected Mechanical Properties of FDM/FFF 3D-Printed Parts. *IOP Conf. Ser. Mater. Sci. Eng.* **2017**, *227*, 012033. [CrossRef]
16. Rajpurohit, S.R.; Dave, H.K.; Rajurkar, K.P. Prediction of Tensile Strength of Fused Deposition Modeling (FDM) Printed PLA Using Classic Laminate Theory. *Eng. Solid Mech.* **2022**, *10*, 13–24. [CrossRef]
17. Ali, Z.; Gao, Y.; Tang, B.; Wu, X.; Wang, Y.; Li, M.; Hou, X.; Li, L.; Jiang, N.; Yu, J. Preparation, Properties and Mechanisms of Carbon Fiber/Polymer Composites for Thermal Management Applications. *Polymers* **2021**, *13*, 169. [CrossRef]
18. Das, T.K.; Ghosh, P.; Das, N.C. Preparation, Development, Outcomes, and Application Versatility of Carbon Fiber-Based Polymer Composites: A Review. *Adv. Compos. Hybrid Mater.* **2019**, *2*, 214–233. [CrossRef]
19. Czyżewski, P.; Bieliński, M.; Sykutera, D.; Jurek, M.; Gronowski, M.; Ryl, Ł.; Hoppe, H. Secondary Use of ABS Co-Polymer Recyclates for the Manufacture of Structural Elements Using the FFF Technology. *Rapid Prototyp. J.* **2018**, *24*, 1447–1454. [CrossRef]
20. Isobe, T.; Tanaka, T.; Nomura, T.; Yuasa, R. Comparison of Strength of 3D Printing Objects Using Short Fiber and Continuous Long Fiber. *IOP Conf. Ser. Mater. Sci. Eng.* **2018**, *406*, 012042. [CrossRef]
21. Alarifi, I.M. PETG/Carbon Fiber Composites with Different Structures Produced by 3D Printing. *Polym. Test.* **2023**, *120*, 107949. [CrossRef]
22. Tian, X.; Todoroki, A.; Liu, T.; Wu, L.; Hou, Z.; Ueda, M.; Hirano, Y.; Matsuzaki, R.; Mizukami, K.; Iizuka, K.; et al. 3D Printing of Continuous Fiber Reinforced Polymer Composites: Development, Application, and Prospective. *Chin. J. Mech. Eng. Addit. Manuf. Front.* **2022**, *1*, 100016. [CrossRef]
23. Li, C.; Xian, G.; Li, H. Effect of Postcuring Immersed in Water under Hydraulic Pressure on Fatigue Performance of Large-diameter Pultruded Carbon/Glass Hybrid Rod. *Fatigue Fract. Eng. Mater. Struct.* **2019**, *42*, 1148–1160. [CrossRef]

24. Wu, J.; Zhu, Y.; Li, C. Experimental Investigation of Fatigue Capacity of Bending-Anchored CFRP Cables. *Polymers* **2023**, *15*, 2483. [CrossRef] [PubMed]
25. Blyweert, P.; Nicolas, V.; Fierro, V.; Celzard, A. 3D Printing of Carbon-Based Materials: A Review. *Carbon* **2021**, *183*, 449–485. [CrossRef]
26. Liu, T.; Tian, X.; Zhang, Y.; Cao, Y.; Li, D. High-Pressure Interfacial Impregnation by Micro-Screw in-Situ Extrusion for 3D Printed Continuous Carbon Fiber Reinforced Nylon Composites. *Compos. Part A Appl. Sci. Manuf.* **2020**, *130*, 105770. [CrossRef]
27. Hu, Q.; Duan, Y.; Zhang, H.; Liu, D.; Yan, B.; Peng, F. Manufacturing and 3D Printing of Continuous Carbon Fiber Prepreg Filament. *J. Mater. Sci.* **2018**, *53*, 1887–1898. [CrossRef]
28. Ligon, S.C.; Liska, R.; Stampfl, J.; Gurr, M.; Mülhaupt, R. Polymers for 3D Printing and Customized Additive Manufacturing. *Chem. Rev.* **2017**, *117*, 10212–10290. [CrossRef] [PubMed]
29. Huang, H.; Talreja, R. Effects of Void Geometry on Elastic Properties of Unidirectional Fiber Reinforced Composites. *Compos. Sci. Technol.* **2005**, *65*, 1964–1981. [CrossRef]
30. Bijsterbosch, H.; Gaymans, R.J. Polyamide 6—Long Glass Fiber Injection Moldings. *Polym. Compos.* **1995**, *16*, 363–369. [CrossRef]
31. Zhang, H.; Huang, T.; Jiang, Q.; He, L.; Bismarck, A.; Hu, Q. Recent Progress of 3D Printed Continuous Fiber Reinforced Polymer Composites Based on Fused Deposition Modeling: A Review. *J. Mater. Sci.* **2021**, *56*, 12999–13022. [CrossRef]
32. Shih, C.-C.; Burnette, M.; Staack, D.; Wang, J.; Tai, B.L. Effects of Cold Plasma Treatment on Interlayer Bonding Strength in FFF Process. *Addit. Manuf.* **2019**, *25*, 104–111. [CrossRef]
33. Yuan, L.Y.; Shyu, S.S.; Lai, J.Y. Plasma Surface Treatments of Carbon Fibers. Part 2: Interfacial Adhesion with Poly(Phenylene Sulfide). *Compos. Sci. Technol.* **1992**, *45*, 9–16. [CrossRef]
34. Yuan, L.Y.; Chen, C.S.; Shyu, S.S.; Lai, J.Y. Plasma Surface Treatment on Carbon Fibers. Part 1: Morphology and Surface Analysis of Plasma Etched Fibers. *Compos. Sci. Technol.* **1992**, *45*, 1–7. [CrossRef]
35. Cho, B.-G.; Hwang, S.-H.; Park, M.; Park, J.K.; Park, Y.-B.; Chae, H.G. The Effects of Plasma Surface Treatment on the Mechanical Properties of Polycarbonate/Carbon Nanotube/Carbon Fiber Composites. *Compos. Part B Eng.* **2019**, *160*, 436–445. [CrossRef]
36. Ma, K.; Chen, P.; Wang, B.; Cui, G.; Xu, X. A Study of the Effect of Oxygen Plasma Treatment on the Interfacial Properties of Carbon Fiber/Epoxy Composites. *J. Appl. Polym. Sci.* **2010**, *116*, 2658–2667. [CrossRef]
37. Giannotta, G.; Morra, M.; Occhiello, E.; Garbassi, F.; Nicolais, L.; D'Amore, A. Dynamic Wetting of Carbon Fibers by Viscous Fluids. *J. Colloid Interface Sci.* **1992**, *148*, 571–578. [CrossRef]
38. Park, S.-J.; Kim, M.-H.; Lee, J.-R.; Choi, S. Effect of Fiber–Polymer Interactions on Fracture Toughness Behavior of Carbon Fiber-Reinforced Epoxy Matrix Composites. *J. Colloid Interface Sci.* **2000**, *228*, 287–291. [CrossRef] [PubMed]
39. Marissen, R.; van der Drift, L.T.; Sterk, J. Technology for Rapid Impregnation of Fibre Bundles with a Molten Thermoplastic Polymer. *Compos. Sci. Technol.* **2000**, *60*, 2029–2034. [CrossRef]
40. Van De Steene, W. A Generic Process for Impregnation and Additive Manufacturing of Endless Fibre-Reinforced Thermoplastic Composites. Ph.D. Thesis, Ghent University, Ghent, Belgium, 2020.
41. Kuznetsov, V.; Solonin, A.; Urzhumtsev, O.; Schilling, R.; Tavitov, A. Strength of PLA Components Fabricated with Fused Deposition Technology Using a Desktop 3D Printer as a Function of Geometrical Parameters of the Process. *Polymers* **2018**, *10*, 313. [CrossRef]
42. Wang, P.; Zou, B.; Ding, S.; Li, L.; Huang, C. Effects of FDM-3D Printing Parameters on Mechanical Properties and Microstructure of CF/PEEK and GF/PEEK. *Chin. J. Aeronaut.* **2021**, *34*, 236–246. [CrossRef]
43. Li, H.; Liu, B.; Ge, L.; Chen, Y.; Zheng, H.; Fang, D. Mechanical Performances of Continuous Carbon Fiber Reinforced PLA Composites Printed in Vacuum. *Compos. Part B Eng.* **2021**, *225*, 109277. [CrossRef]
44. Garzon-Hernandez, S.; Garcia-Gonzalez, D.; Jérusalem, A.; Arias, A. Design of FDM 3D Printed Polymers: An Experimental-Modelling Methodology for the Prediction of Mechanical Properties. *Mater. Des.* **2020**, *188*, 108414. [CrossRef]
45. Papon, E.A.; Haque, A.; Spear, S.K. Effects of Functionalization and Annealing in Enhancing the Interfacial Bonding and Mechanical Properties of 3D Printed Fiber-Reinforced Composites. *Mater. Today Commun.* **2020**, *25*, 101365. [CrossRef]
46. Wang, K.; Long, H.; Chen, Y.; Baniassadi, M.; Rao, Y.; Peng, Y. Heat-Treatment Effects on Dimensional Stability and Mechanical Properties of 3D Printed Continuous Carbon Fiber-Reinforced Composites. *Compos. Part A Appl. Sci. Manuf.* **2021**, *147*, 106460. [CrossRef]
47. Nassar, A.; Younis, M.; Elzareef, M.; Nassar, E. Effects of Heat-Treatment on Tensile Behavior and Dimension Stability of 3D Printed Carbon Fiber Reinforced Composites. *Polymers* **2021**, *13*, 4305. [CrossRef] [PubMed]
48. Bhandari, S.; Lopez-Anido, R.A.; Gardner, D.J. Enhancing the Interlayer Tensile Strength of 3D Printed Short Carbon Fiber Reinforced PETG and PLA Composites via Annealing. *Addit. Manuf.* **2019**, *30*, 100922. [CrossRef]
49. Van De Steene, W.; Verstockt, J.; Degrieck, J.; Ragaert, K.; Cardon, L. An Evaluation of Three Different Techniques for Melt Impregnation of Glass Fiber Bundles with Polyamide 12. *Polym. Eng. Sci.* **2018**, *58*, 601–608. [CrossRef]
50. La Gala, A.; Fiorio, R.; Erkoç, M.; Cardon, L.; D'hooge, D.R. Theoretical Evaluation of the Melting Efficiency for the Single-Screw Micro-Extrusion Process: The Case of 3D Printing of ABS. *Processes* **2020**, *8*, 1522. [CrossRef]
51. *ISO 11358-1*; Plastics—Thermogravimetry (TG) of Polymers—General Principles. ISO: Geneva, Switzerland, 2022.
52. *ISO 14125*; Fibre-Reinforced Plastic Composites—Determination of Flexural Properties. ISO: Geneva, Switzerland, 2015.
53. Stachewicz, U.; Barber, A.H. Enhanced Wetting Behavior at Electrospun Polyamide Nanofiber Surfaces. *Langmuir* **2011**, *27*, 3024–3029. [CrossRef]

54. Cheng, S.; Janimak, J.; Zhang, A. Regime Transitions in Fractions of Isotatic Polypropylene. *Macromolecules* **1995**, *34*, 1.
55. Tsutsumi, K.; Ishida, S.; Shibata, K. Determination of the Surface Free Energy of Modified Carbon Fibers and Its Relation to the Work of Adhesion. *Colloid Polym. Sci.* **1990**, *268*, 31–37. [CrossRef]
56. Feng, K.-C.; Pinkas-Sarafova, A.; Ricotta, V.; Cuiffo, M.; Zhang, L.; Guo, Y.; Chang, C.-C.; Halada, G.P.; Simon, M.; Rafailovich, M. The Influence of Roughness on Stem Cell Differentiation Using 3D Printed Polylactic Acid Scaffolds. *Soft Matter* **2018**, *14*, 9838–9846. [CrossRef] [PubMed]
57. Pawlak, A.; Galeski, A.; Rozanski, A. Cavitation during Deformation of Semicrystalline Polymers. *Prog. Polym. Sci.* **2014**, *39*, 921–958. [CrossRef]
58. Herrera, N.; Roch, H.; Salaberria, A.M.; Pino-Orellana, M.A.; Labidi, J.; Fernandes, S.C.M.; Radic, D.; Leiva, A.; Oksman, K. Functionalized Blown Films of Plasticized Polylactic Acid/Chitin Nanocomposite: Preparation and Characterization. *Mater. Des.* **2016**, *92*, 846–852. [CrossRef]
59. Ohnmacht, H.; Fiorio, R.; Wieme, T.; D'hooge, D.R.; Cardon, L.; Edeleva, M. Response-Surface-Methodology-Based Increasing of the Isotropic Thermal Conductivity of Polyethylene Composites Containing Multiple Fillers. *Polymers* **2022**, *15*, 39. [CrossRef]

Journal of
Manufacturing and Materials Processing

MDPI

Article

Experimental Characterization of Screw-Extruded Carbon Fibre-Reinforced Polyamide: Design for Aeronautical Mould Preforms with Multiphysics Computational Guidance

Juan Carlos Antolin-Urbaneja [1,*], Haritz Vallejo Artola [1], Eduard Bellvert Rios [1], Jorge Gayoso Lopez [1], Jose Ignacio Hernández Vicente [1] and Ana Isabel Luengo Pizarro [2]

1 TECNALIA, Basque Research and Technology Alliance (BRTA), 20009 Donostia-San Sebastian, Spain; haritz.vallejo@tecnalia.com (H.V.A.); eduard.bellvert@tecnalia.com (E.B.R.); jorge.gayoso@tecnalia.com (J.G.L.); jose.hernandez@tecnalia.com (J.I.H.V.)
2 Internacional de Composites S.A., 45007 Toledo, Spain; anaisabel.luengo@aernnova.com
* Correspondence: juancarlos.antolin@tecnalia.com

Abstract: In this research work, the suitability of short carbon fibre-reinforced polyamide 6 in pellet form for printing an aeronautical mould preform with specific thermomechanical requirements is investigated. This research study is based on an extensive experimental characterization campaign, in which the principal mechanical properties of the printed material are determined. Furthermore, the temperature dependency of the material properties is characterized by testing samples at different temperatures for bead printing and stacking directions. Additionally, the thermal properties of the material are characterized, including the coefficient of thermal expansion. Moreover, the influence of printing machine parameters is evaluated by comparing the obtained tensile moduli and strengths of several manufactured samples at room temperature. The results show that the moduli and strengths can vary from 78% to 112% and from 55% to 87%, respectively. Based on a real case study of its aeronautical use and on the experimental data from the characterization stage, a new mould design is iteratively developed with multiphysics computational guidance, considering 3D printing features and limitations. Specific design drivers are identified from the observed material's thermomechanical performance. The designed mould, whose mass is reduced around 90% in comparison to that of the original invar design, is numerically proven to fulfil thermal and mechanical requirements with a high performance.

Keywords: 3D printing; material extrusion; carbon fibre-reinforced polyamide; mechanical properties; thermal characterization; design and numerical simulation

1. Introduction

Traditionally, moulds for manufacturing fibre-reinforced polymer matrix composite components for the aeronautical sector include low-cost metals such as aluminium or plain carbon steel; high-end nickel–steel alloys such as invar; and composites [1–3]. Invar is usually the preferred material due to its unique dimensional stability and proper material properties, which facilitate the fulfilment of the strict tolerances required by end users. Aluminium is also frequently used because it is lighter and easier to machine than invar. Mechanical and thermal requirements are commonly specified, which aim at ensuring the parts' dimensional tolerances and the needed material properties, respectively. The thermomechanical properties of metallic materials are commonly well known, so moulds can be designed based on accurate simulations that allow designers to numerically validate their behaviour under operating conditions. However, the prices and lead times of large metallic moulds are high. Therefore, alternative materials, structures and manufacturing technologies are being investigated to provide lightweight, low-cost, durable and functional tools for manufacturing high-performance composite structures [2,4].

Moulds made by tooling prepreg laminates from a machined master model are also being developed. In this regard, Galiana et al. [5] conducted a cost analysis of the engineering, management and production of an autoclave curing mould when using a carbon fibre-reinforced polymer (CFRP) or invar, providing guidelines for their selection. CFRP properties have been extensively analysed; therefore, composite mould designs with multi-physics computation guidance are regularly used [6,7]. In fact, CFRPs usually have a low coefficient of thermal expansion (CTE), thus minimizing problems related to shrinkage and spring-back. Nevertheless, metal moulds used and maintained properly can have a long service life for the mass production of thousands of composite parts. On the other hand, tooling prepreg moulds are limited to service in relatively low-volume production areas [3], mainly due to the surface wear at the lamination area and to the development of interlaminar vacuum leaks, which are difficult to repair. In addition, the manufacturing of moulds with tooling prepreg is tedious due to the bonding of the master boards together, their subsequent machining and the application of coatings, and the need to reinforce the laminated skin with a reinforced backing structure [8].

Regarding new manufacturing technologies, additive manufacturing (AM) described in the UNE-EN ISO/ASTM 52900:2021 standard [9] has become a very interesting alternative technology to commonly used machinery for the development of moulds because of associated reductions in lead time, material waste and manufacturing costs [10–13]. Moreover, some investigations have focused on additive manufacturing's impact in the supply chain due to the fact that customers can turn into manufacturers by printing their products on demand and reducing their stock [14]. One of these additive manufacturing technologies is focused on material extrusion, in filament or pellet form. In the past few years, there has been an increasing amount of interest in assessing this technology, i.e., developing parts for different applications [15,16]. However, the development of large moulds made of thermoplastics to be used for the production of aeronautical composite parts using additive manufacturing technology is not easy, since there are very few technical materials that can withstand the curing conditions [17]. Based on research conducted by Olusanmi Adeniran et al. [6], in which a review of the different thermoplastic matrices for additive manufacturing is carried out, these can be classified by commodity, engineering and high-performance grades. As a general rule, aeronautical requirements imply the use of high-performance grade materials, which can generally be found commercially in filament form for FDM (fused deposition modelling) technology [18] at high prices with critical processing parameters [19,20]. Another more economical alternative is to use engineering-grade thermoplastics with short-fibre reinforcements [17,21,22]. In this regard, Ahmed Arabi Hassen et al. [23] describe the development of autoclave moulds that can be used to fabricate aerospace composite parts with materials that are capable of withstanding elevated temperatures, such as polyphenylene sulphide (PPS) with variations in high carbon fibre loading (40%, 50% and 60% by weight). In the same line, Park et al. developed and tested plastic injection moulds using polyetheretherketone (PEEK) after manufacturing samples at different high temperatures [24].

Most research studies about the additive manufacturing of thermoplastic polymers are focused on the analysis of FDM producibility parameters [25–27]. However, additive manufacturing technology presents some limitations for the fabrication of large moulds [28]. Consequently, in recent years, there has been growing interest in the aeronautical industry in concepts such as big area additive manufacturing (BAAM) [29–31], wide and high additive manufacturing (WHAM) [32] and large-format additive manufacturing (LFAM) [33] for producing large moulds using extrusion-based additive manufacturing (EAM) technology with thermoplastic materials in pellet form. Some of these forms of thermoplastic materials reinforced with chopped fibre increase the deposition rate using direct extrusion, being much more economical in terms of energy intensity consumption [30]. Pignatelli [15] reviewed the responses of large-format pellet-based AM systems introduced in the last decade to different real-world market applications. Among these different uses, the production of tooling for the automotive sector [34,35], moulds for the naval sector [36,37] and moulds for the manufacture of wind turbine blades for the wind energy sector [30,38,39] were highlighted.

Generally, suppliers of the materials to be printed using extrusion AM technology provide a basic characterization of the printed samples under specific parameters related to the temperature or printing speed, among others [40]. Most studies regarding engineering-grade polyamide 6 reinforced with carbon fibre (PA6/CF) are focused on the analysis of FDM producibility parameters. In this sense, the material properties of PA6/CF highly depend on the material form, CF content, manufacturing process and corresponding parameters [26,41–44] being available at room temperature upon request. For example, Table 1 shows the relevant mechanical properties at room temperature of two different polyamides reinforced with 20% carbon fibre after different processing methods. It is remarkable that the mechanical properties of the Bergamid™ B70 KF20 Black do not depend on the material orientation in this case, as they have been obtained from an injection process. In addition, the values published for the tensile modulus and strength are significantly higher than the ones obtained via FDM. The density of the material is also higher.

Table 1. Mechanical properties of two different polyamides reinforced with 20% CF.

	Product #1 [45]	**Product #2** [46]
Material	PolyMide™ PA6/20CF (Supplementary Material S1)	Bergamid™ B70 KF20 Black (Supplementary Material S2)
Manufacturer	Polymaker (Shangai, China)	Avient Corporation (Barbastro, Spain)
Form of material	Filament	Pellet
Processing method	FDM	Injection Molding
Tensile moduli (X-Y)—GPa	7.5	13.8
Tensile moduli (Z)—GPa	4.4	-
Tensile strength (X-Y)—MPa	105	220
Tensile strength (Z)—MPa	68	-
Density—kg/m^3	1170	1210–1250

In case of PA6/CF pellet extrusion-based AM, the data available are very limited and experimental works are being carried out currently to better understand their properties. For example, Xiping Li et al. [43] carried out a detailed material characterization regarding their mechanical properties by assessing the tensile and flexural moduli and strengths depending on the carbon fibre content. They determined that the highest tensile properties at room temperature are obtained for a fibre content of 25%. In this case, a maximum Young's modulus of 6.9 GPa and a maximum tensile strength of 169.7 MPa were obtained. Moreover, they analysed how the operating temperature affects the tensile properties, determining that, in the case of a polyamide reinforced with 35% carbon fibre, the modulus and strength at 150 °C are 75% and 70% lower, respectively, than those at room temperature. Furthermore, Nevine Tagscherer et al. [47] characterized 40% chopped carbon fibre-reinforced polyamide 6 at room temperature with production parameter variations, obtaining a maximum longitudinal modulus of 12.8 GPa. In addition, it was determined that the transversal tensile strength is up to 27% that of the longitudinal, and the transverse tensile modulus is up to 20% that of the longitudinal.

Despite the number of works published in regards to PA6/CF, there is very little information about the dependency of the material properties on the temperature and process parameters, such as printing paths, the nozzle diameter and bead dimensions, which are all considered to be cornerstones. Moreover, the thermal diffusion and thermal expansion of the material are not well known either. Further, the porosity of printed PA/CF has rarely been studied, though the ASTM standard defines two types of porosities generated during AM processes. In addition, pores have an effect on the mechanical properties of manufactured parts in AM [48]. These key deficiencies in our knowledge about this material prevent designers, engineers and manufacturers from accurately analysing how the material behaves under operating conditions (i.e., due to the temperature and pressure inside an autoclave during curing processes) and consequently prevents the development of proper aeronautical mould designs for the manufacturing of composite parts. On the whole, further research is required to improve the knowledge on the actual thermomechanical performance of aeronautical moulds printed through pellet-based PA6/CF.

In the present paper, an extensive mechanical characterization of pellet-based PA6 reinforced with 20% short carbon fibre that is randomly distributed in the polymer is completed for the development of 3D-printed aeronautic mould preforms for composite manufacturing. Furthermore, a comparison between printed tensile test probes' performance manufactured by three different additive manufacturing machines with different parameters is included. Also, the porosity via optical microscopy and the density of printed specimens have been evaluated as these can critically affect the mechanical strength and stiffness, among other parameters [48]. Based on the experimental data obtained, a detailed design process is carried out with finite element analysis (FEA) guidance to accurately predict the thermoelastic behaviour of an aeronautical mould under realistic operating conditions. According to end-user needs, particular thermal and mechanical requirements are specified to guarantee the manufacturing of composite parts with adequate dimensions and material properties. Through these computer modelling activities, several potential problems related to the material behaviour of moulds printed through this technology are identified, and design drivers to overcome them are applied; thus, they are capable of fulfilling the requirements of end users. Apart from this introduction, the paper is divided into four main sections. Section 2 includes a description of a use case in which an aeronautical mould and its requirements are presented. Section 3 describes our methodology, including the working procedure, the material characterization and the computer modelling stages. Section 4 shows the results of the experimental characterization campaign, along with the numerical results for validating the thermoelastic performance of the designed mould. Also, potential problems when using this material for the development of aeronautical moulds are outlined. Finally, in Section 5, the conclusions and prospects for future work are summarized.

2. Description of the Use Case

Our current research is focused on an application in aeronautics, as specified by the end user, in which a mould, also called tool, is required for manufacturing inside an autoclave an aeronautical composite prepreg part, which is marked by the engineering edge of part (EEOP) and the manufacturing edge of part (MEOP) area as seen in Figure 1. The EEOP indicates the dimensions of the finished part, while the MEOP denotes the boundary to be used for manufacturing the laminate. Currently, the end user relies on an invar mould. Because of the complex geometry of the composite part, the mould must be formed by several semi-moulds to ensure the part can be suitably extracted after the curing process. The overall dimensions of the tool are approximately 400 mm × 300 mm × 250 mm.

Regarding the working conditions corresponding to this mould, the end user indicates that the manufacturing process is conducted in an autoclave. In this regard, a particular thermal cycle, characterized by a dwell time of 1 to 5 h at 80 °C and a curing phase of 2 to 6 h at 180 °C, is specified. Heating rates are required to be in the 1.5 to 3 °C/min range. Besides the pressure exerted by the compaction of the vacuum bag, the mould is stated to be subject to an additional pressure of 7 bar which is applied by the autoclave.

Concerning the final properties required for the mould, mechanical and thermal requirements are specified by the end user. On the one hand, it is required that the maximum deflection at the lamination surface must be lower than 0.1 mm during the curing stage. On the other, it is required that the temperature at the lamination surface must be able to properly follow the thermal cycle specified above, with the maximum delay between the maximum and minimum temperatures along the lamination surface below 2 h.

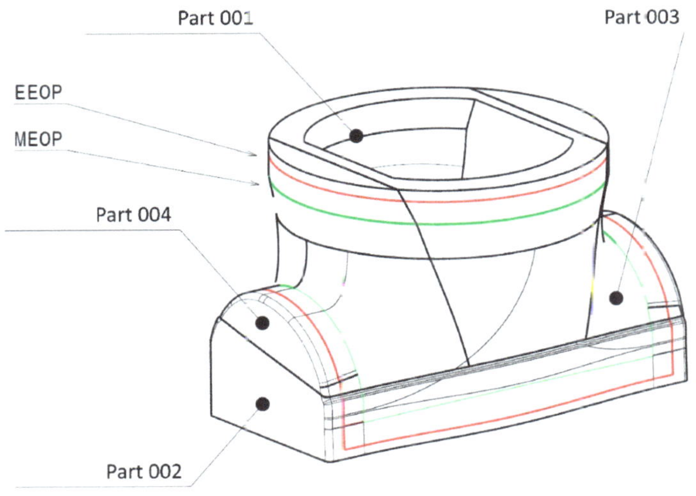

Figure 1. Complete designed mould with finite element analysis guidance.

3. Methodology

The main steps of our methodology for the current research study to design the aeronautical mould are shown in Figure 2. Firstly, the material that best fits the end-user requirements is assessed based on properties found in the literature.

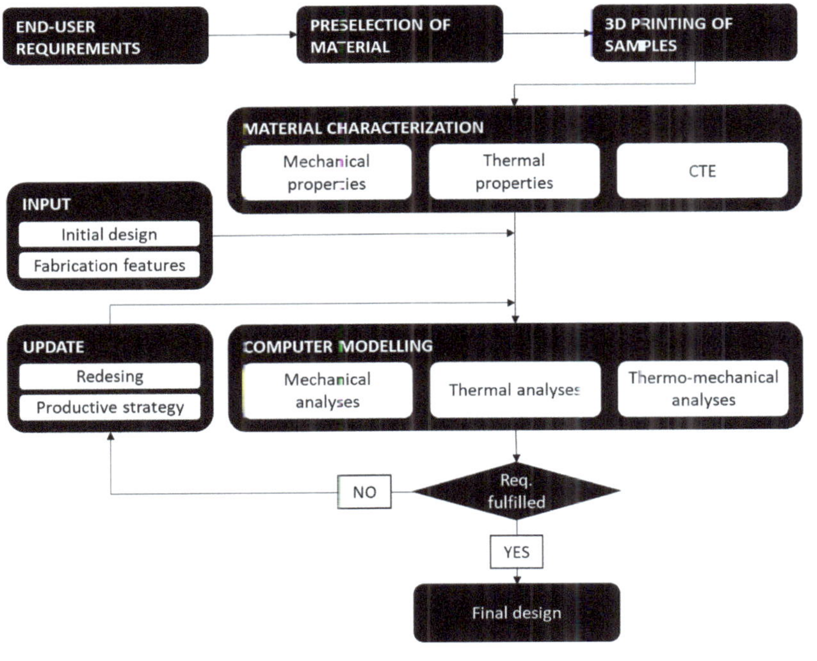

Figure 2. Working methodology for the development of the 3D-printed mould preform for the manufacturing of aeronautical composite parts using pellets of carbon fibre-reinforced PA6.

The chosen thermoplastic material must be printable without using a heating chamber. Providing high-performance-grade polymers require highly controlled conditions to be printed [49], a preselection of engineering thermoplastic materials (allowing broaden range for printing conditions) was conducted [6]. As the tool has to support up to 180 °C at 7 bar during the autoclave process, a preselection of engineering thermoplastic materials was carried out, taking into account the available data provided by the potential suppliers. The preselected material for manufacturing the moulds was polyamide 6 reinforced with short carbon fibre (20%) that was randomly distributed in the polymer, which is included in Table 1 and was supplied by Avient [46]. This thermoplastic material is classified as semi-crystalline [6,19]; therefore, the service temperature is between the glass transition temperature, Tg (60 °C), and the melting temperature, Tm (222 °C), according to the results of the differential scanning calorimetry (DSC) technique carried out using the printed material and the Q100 model (TA Instruments, New Castle, UK). Also, one of the properties able to indicate the stiffness for thermoplastics in service is the heat deflection temperature (HDT). The HDT values provided of the filament form in the datasheets of [45] are within the range of the thermal specifications. However, the performance of the thermoplastic polymer at specific required conditions (in filament or pellet form) is not available in the datasheet. Therefore, an extensive experimental characterization campaign, including tensile, compression, flexural and shear tests, has been completed to enable a detailed assessment of the performance of PA6/CF 20% under operating conditions. The material's thermal conductivity, specific heat at room temperature and CTE in the 25 °C to 180 °C range are also characterized. This extensive experimental characterization campaign was performed using specimens obtained after being printed by a research additive manufacturing system (RAMS). Taking into account that the mould preforms will be printed using another additive manufacturing systems (AMS), some similar specimens were characterized to compare the quality between the printed beads in different AM machines.

3.1. Material Characterization

In order to characterize the printed material, a part was first designed to contain all the envisaged test probes. As an illustration of the types of specimens included, Figure 3 shows the design of a cube in which the type of test probe is represented according to the standards indicated in Table 2. The original part was printed by depositing theoretical 5 mm thick and 1.5 mm high beads using a robotic system (RAMS) similar to the one illustrated in Figure 4 (right), in which a 20XD extruder is attached [50]. The different test specimens have been machined from a square printed cube of 300 × 300 × 290 mm^3 (Figure 4 (left)).

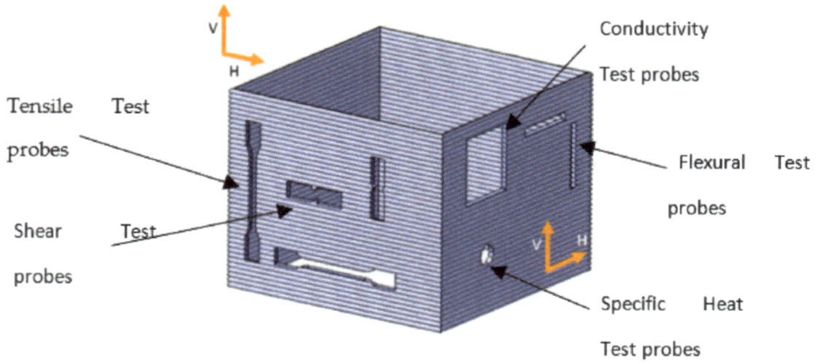

Figure 3. Illustration of the designed square cube to obtain the test specimens for characterization. Example of test probes.

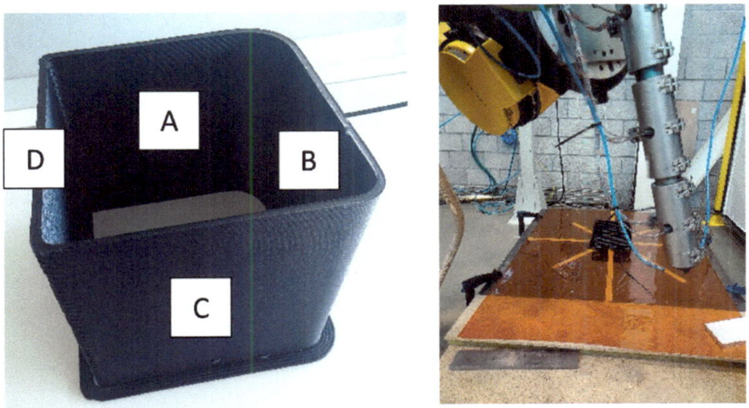

Figure 4. Square cube manufactured with RAMS to obtain the test specimens from named walls (**left**) and MOLDAM additive manufacturing system for printing parts (**right**).

Recently, a study evaluating the printability and mechanical properties of regrinding material previously used in large-format additive manufacturing was conducted. It included an assessment of similar printed parts for the machining of tensile and flexural samples in a similar way [51]. It is noteworthy that the process parameters were iteratively optimized by means of several tests in order to ensure proper manufacturing. Being an orthotropic material, the testing samples are horizontally (H) or vertically (V) machined to obtain the corresponding properties in the printing (1) and stacking (2) directions, respectively. It is assumed that properties in the printing and stacking directions are constant for each wall.

Samples for the tensile and shear tests in horizontal and vertical orientations were machined from wall A and B, respectively. Samples for the flexural test in both directions were obtained from wall C. The machined specimens for the tensile and flexural tests are shown in Figure 5. Finally, samples for the compression, conductivity and specific heat tests were machined from wall D. As one of the requirements is related to rugosity, three additional samples were machined under the same conditions to evaluate the rugosity under the ISO 4287:1997 standard [52] using a Surftest SJ-210 (Mitutoyo, Kawasaki, Japan). The mean values were 3.013 ± 0.568 µm, showing good machinability. As presented in Table 2, the mechanical tests were repeated at three temperatures (i.e., room temperature (RT), 80 °C and 180 °C) to investigate the temperature dependency of the properties. A total of 98 samples were tested during this characterization stage. Figure 6 shows some of the specimens under the tests. Also, some additional samples were machined to assess the difference between specimens printed by three similar machines to compare the behaviour of the same material while applying other process parameters. For the mechanical characterization of the specimens, a Model Instron 5500 R6025 (Instron, Norwood, MA, UK) universal testing machine applying a 100 kN load cell was used. The thermal conductivity and specific heat were determined utilizing a hot disk unit in combination with the measurement unit transient plane surface (TPS) 2500S (Hot Disk AB, Göteborg, Sweden). The CTE was experimentally calculated using the dynamic mechanical thermal analysis (DMTA) technique by means of a Tritec 2000 DMA device (Triton Technology Ltd., Nottinghamshire, UK).

Additionally, the tensile properties at room temperature (RT) are compared with those published in the Technical Data Sheet of Bergamid^{TM} B70 KF20 Black processed by injection moulding by Avient [46]. Moreover, additional tensile samples are manufactured using two different 3D printing machines (AMS I—Additive Manufacturing System I and AMS II—Additive Manufacturing System II) to investigate if similar material properties are obtained. Furthermore, in case of AMS II, samples are printed by using two bead heights, AMS II (1) and AMS II (2), to compare how this printing parameter affects the tensile strength and modulus. The main characteristics and parameters used with the different AM machines are shown in Table 3. As the average dispersion of the measurements (the standard deviation divided by the mean value) was about 5% and the budget was limited, the quantity of samples in each test was restricted to three samples.

In order to determine the quality of the bead deposition compared to theoretical ones, an evaluation of the bead height and width was performed using image analysis techniques. Hence, some micro- and macro-photographs of the bead structure and profile were obtained for each case in our study to evaluate the porosity and the maximum length of the short carbon fibres in the printed beads. Moreover, some additional specimens were machined to calculate the density of the printed material in each case in our study, geometrically and experimentally, by means of the Archimedes method.

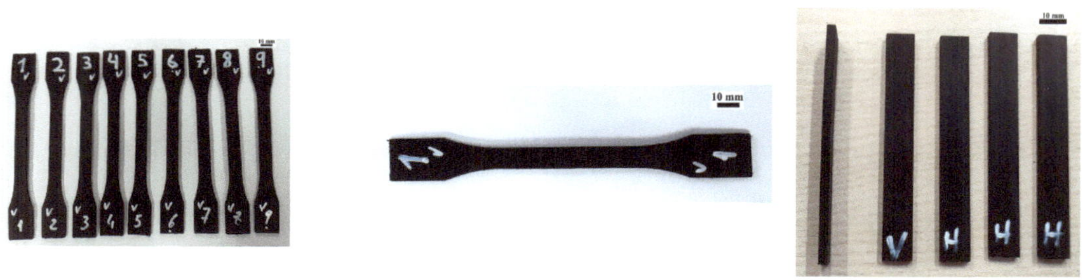

Figure 5. Machined specimens for tensile (**left**) and flexural tests (**right**).

(a) (b) (c) (d) (e) (f)

Figure 6. Specimens manufactured with RAMS during (**a**) tensile, (**b**) compression, (**c**) flexural, (**d**) shear and host-disc tests for the characterization of (**e**) specific heat and (**f**) thermal conductivity.

Table 2. Tests conducted with the samples manufactured with RAMS, corresponding standards, obtained properties and testing conditions.

Test	Standard	Obtained Properties	Material Orientation	Temperature [°C]	Number of Samples
Tensile	ISO 527-1 [53] ISO 527-2 1B Specimens [54]	Tensile Strength Tensile Modulus Nominal Strain at Break	1	RT	3
				80	3
				180	3
			2	RT	3
				80	3
				180	3
Flexural	ISO 14125 3-point test [55]	Flexural Strength Flexural Modulus Flexural Strain at Max Load	1	RT	3
				80	3
				180	3
			2	RT	3
				80	3
				180	3
Compression–Strength	ISO 604 Standard Specimens [56]	Compression Strength Compression Strength (5%)	1	RT	3
				80	3
				180	3
			2	RT	3
				80	3
				180	3
Compression–Modulus	ISO 604 Standard Specimens [56]	Compression Modulus	1	RT	3
				80	3
				180	3
			2	RT	3
				80	3
				180	3
Shear	ASTM D5379 (Iosipescu Fixture) [57]	Ultimate Strength Shear Cord Modulus	1	RT	3
				80	3
				180	3
			2	RT	3
				80	3
				180	3
DMTA	ISO 6721 [58]	CTE	2	−25 ÷ 180	1
Hot disc	ISO 22007-2.2 [59]	Thermal conductivity Specific heat	1	RT	1
			2	RT	2
Rugosity	ISO 4287:1997 [52]	Rugosity, Ra	-	RT	3
Density	UNE-EN ISO 1183-1:2019 [60]	Density	-	RT	1

Table 3. Main characteristics of the machines and parameters used to manufacture samples for the experimental characterization.

Parameter	RAMS	AMS I	AMS II (1)	AMSII (2)
Extruder Type	20XD	Pulsar	20XD	20XD
Nozzle Diameter (mm)	4	3	8	8
Theoretical Bead Width (mm)	5	3.6	8	8
Theoretical Bead Height (mm)	1.5	1.5	4	2
Maximum Mass Flow (kg/h)	30	2.5	30	30
Extruder Temperature (°C) 1. Feed 2. Zone 2 (Middle) 3. Nozzle	 210 230 240	 220 235 250	 190 200 235	 190 200 235
Bed Temperature (°C)	RT	110	90	90
Barrel Length (mm)	720	300	720	720
Print Speed (mm/s)	35	35	35	35

3.2. *Preliminary Computer Modelling of the Mould*

In a parallel effort, an initial design is proposed for each semi-mould out of the entire mould, keeping in mind how to assemble and dissemble them to ensure successful composite part extraction once they are manufactured. Specific fabrication features and limitations are also taken into account in this initial step. Subsequently, the mould design is iteratively developed, based on computer-aided engineering (CAE), and taking into account the material properties obtained in the characterization stage, until a lightweight mould that fulfils the specified requirements is obtained. The production strategy is also optimized, and a better mould position is proposed to improve its thermal response. Computer modelling activities are carried out with Simcenter 3D 2021.2. (1998). Regarding the developed models, solid meshes are used for both the thermal and mechanical simulations as seen in Figure 7. Specific load cases that digitally reproduce the loads and the boundary conditions under the mould are included in the numerical models. For the thermal simulations, a semi-circular autoclave (Ø700 × 1400 mm) is modelled. The analytical estimation of representative convection coefficients along geometrically complex moulds inside an autoclave is not easy, so the results of single thermal analyses with imposed heat transfer coefficients might not be accurate enough for this application. Therefore, coupled thermal-flow analyses, in which the convection coefficients are calculated based on representative operating conditions, are performed instead. To that end, the air volume in contact with the mould is included in the model, and specific air inlets and openings are applied. In particular, an air velocity of 1.75 m/s is considered. The mathematical model developed for this activity is formed by around 42,000 linear tetrahedral elements and 11,000 nodes. A general thermal impedance of 2000 W/m^2 K is modelled among the solid components. For the mechanical analyses, a finite element model formed by around 76,000 parabolic tetrahedral elements and 115,000 nodes is built. Rigid joints are considered among the semi-moulds. The transient temperature contours from the thermal analysis are defined as thermal loads, assuming a stress-free temperature of 20 °C. The operating pressure and gravity are also applied. A rigid joining is considered among the semi-moulds.

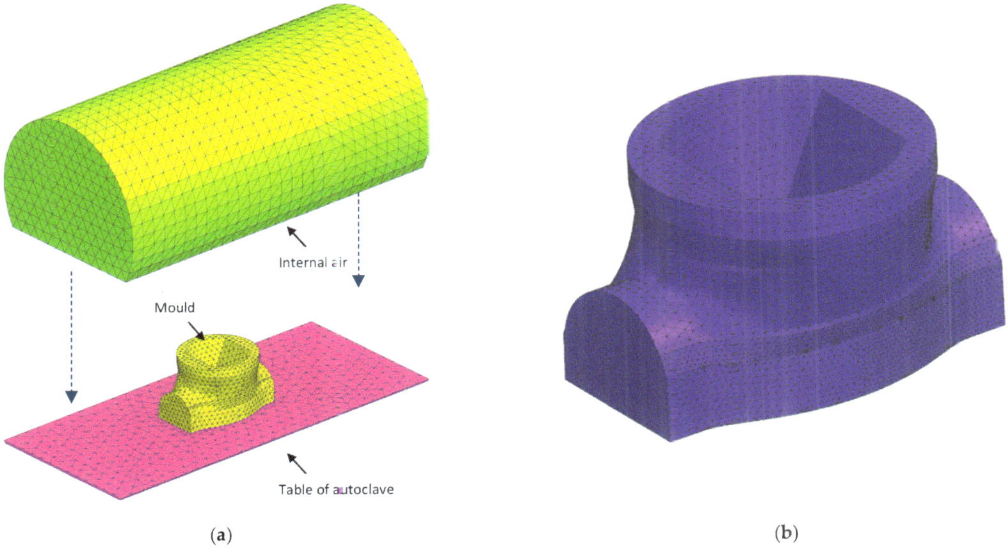

Figure 7. Meshes for the thermal flow (**a**) and mechanical (**b**) simulations. Note that the mesh regarding the internal air is moved upwards to improve the visualization of the internal meshes.

4. Results and Discussion

4.1. Tensile Moduli and Strength Results

The printed material is highly machinable; therefore, testing samples can be easily obtained. Moreover, a good level of machinability is obtained for each machining process performed, regardless of the printer or the process parameters used in the manufacturing of the part from which the sample is extracted. In this regard, the surface roughness, Ra, was measured under the UNE-EN ISO 4287 [52] and UNE-EN ISO 4288 standard [61] as 2.388 to 3.496 µm, according to the tests conducted.

Concerning the experimental results obtained from the extensive characterization stage conducted for the samples manufactured with RAMS, it is noteworthy that the ratio between the longitudinal and transversal moduli and strengths at room temperature is in line with the values published by Nevine Tagscherer et al. [47]. For convenience, only the results of the tensile moduli and strength at three temperatures, RT, 80 °C and 180 °C, are provided. As derived from Figure 8, the tensile strength and modulus decrease dramatically. In this sense, the tensile strength in the printing bead direction is reduced by about 37% and 70% at 80 °C and 180 °C, respectively; meanwhile, the tensile modulus is reduced up to 56% and 71%. In the stacking direction, the tensile strength is reduced by 10% (80 °C) and 73% (180 °C). The tensile modulus is reduced by up to 73% (80 °C) and 85% (180 °C). The reduction in these mechanical properties at a high temperature is very typical in polymers, as explained in [62]. The results in the bead direction depend on the material matrix—in this case, reinforced with 20% carbon fibres. However, the results in the stacking direction depend on the layer union interface and the effect of the adhesion of the different layers. The rest of the properties generated in our research are available upon request. They are included in registered software as part of TECNALIA's intellectual property.

Taking into account that the mould preforms will be printed using AMS machines, some similar specimens are characterized to compare the quality between printed beads. As a consequence, the tensile properties at room temperature are assumed to be a good characteristic to use in such a comparison. The experimental results show a relatively low level of dispersion depending on the machine and the parameters used for the manufacturing of the sample, as depicted in Figure 9. In this sense, taking as a reference the properties of injected Bergamid^{TM} B70 KF20 Black processed by injection moulding, the material tensile modulus along the bead printing direction can decrease up to 78% or increase up to 112%. The lowest and highest obtained values for the material tensile strength are 55% and 87%, respectively. It is noteworthy that the most extreme values concerning the modulus are obtained for the same additive manufacturing machine. Accordingly, it is observed that the lower the bead height, the higher the tensile modulus. It is plausible that this effect is due to the layers being more compacted, consequently reducing the porosity, as estimated in Section 4.3. This parameter has a similar effect on the tensile strength of the material, but its influence is not so notable.

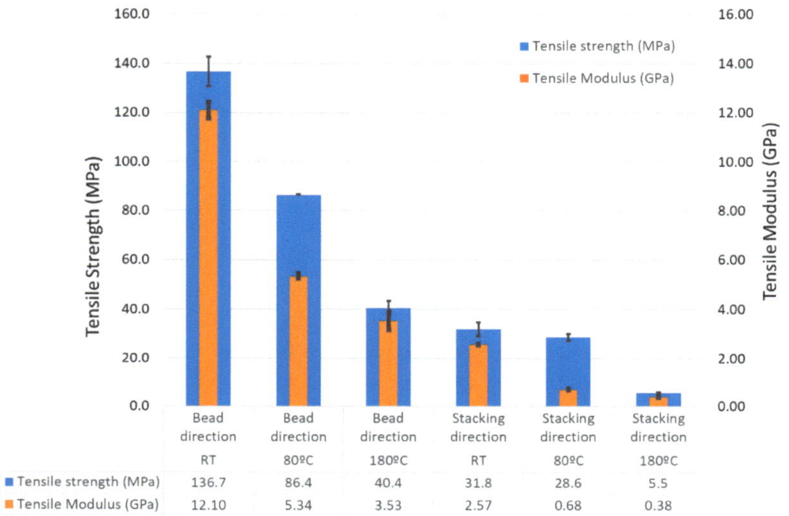

Figure 8. Comparison of the tensile moduli and strengths in the printing bead direction and stacking direction depending on the temperature using printed Bergamid B70 KF20 Black.

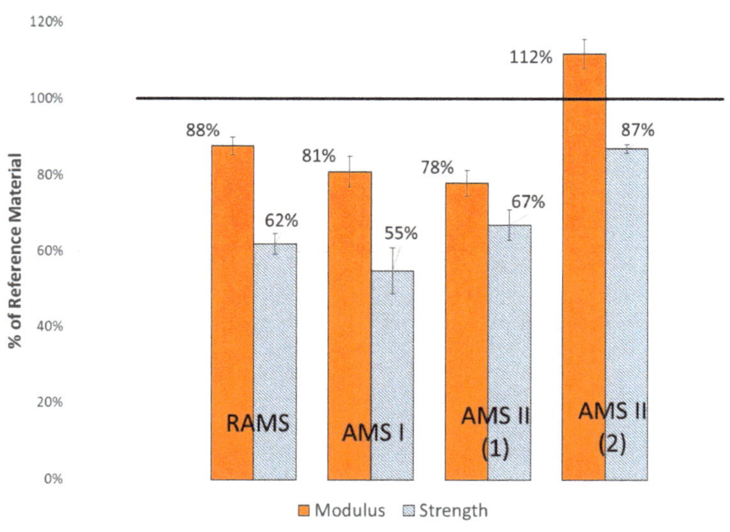

Figure 9. Comparison of the tensile moduli and strengths in the printing bead direction depending on the machines and/or parameters used for the manufacturing of samples with those for Bergamid B70 KF20 Black.

4.2. Bead Height and Width Assessment

At the beginning of our research, it was found that we needed to compare the bead height and width between the theoretical and the real deposited ones after the extrusion process. In this sense, some specimens of each cube were cut and prepared to measure them by means of image analysis software (Photoshop V11.0). In each macrograph, some measurements were selected under the area of the interest, determining the bead height (m) and width (n). Figure 10 illustrates an example of the measurement identification. The mean, standard deviation and error percentage between the theoretical and measured values of the printed cubes from each machine are included. This analysis found that the deposited bead size was quite well adjusted in each machine although some additional adjustments are needed in the so-called AMS I machine. The compared values are included in Table 4.

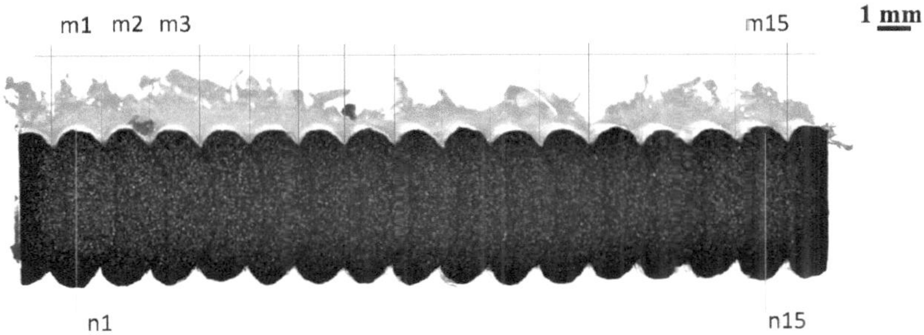

Figure 10. Macrograph of the deposited profile obtained via 3D printer (RAMS).

Table 4. Theoretical and measured bead height and width versus additive manufacturing.

Bead Height and Width	RAMS	AMS I	AMS II (1)	AMSII (2)
Theoretical Bead Width (mm)	5	3.6	8	8
Bead Width Mean, m (mm)	4.70	4.16	8.09	8.01
Bead Width Deviation (mm)	0.08	0.37	0.21	0.17
Theoretical Bead Height (mm)	1.5	1.5	4	2
Bead Height Mean, n (mm)	1.48	1.47	3.87	1.92
Bead Height Deviation (mm)	0.04	0.15	0.07	0.15
B. Height Error vs. Theoretical (%)	1.33	2.00	3.25	4.00
B. Width Error vs. Theoretical (%)	5.96	15.44	1.08	0.12

4.3. Porosity Evaluation

Another aim of this research is the evaluation of the porosity during the fabrication process before any post processing (i.e., thermal treatment) as the porosity can influence the mechanical characteristics of the tensile strength, as reviewed by Al-Maharma [48]. Moreover, the formation of voids is a common problem when fillers are introduced into a thermoplastic material [63]. In order to evaluate the porosity percentage of each case in our study, some specimens were cut and treated to be embedded in bakelite pieces. These parts were photographed and evaluated via inverted microscopy Eclipse MA200 (Nikon, Tokyo, Japan) in combination with properly calibrated analysis software.

To calculate the percentage of the porosity of the printed specimens, an analysis of the volumetric fraction of each case in our study was performed following an internal procedure based on the ASTM E562 standard [64]. Therefore, after the test specimen preparation, five representative and homogeneous areas of the total area illustrated in each macrograph of Figure 11 were selected, as shown in Figure 11a, which analyses the percentage in the area of interest.

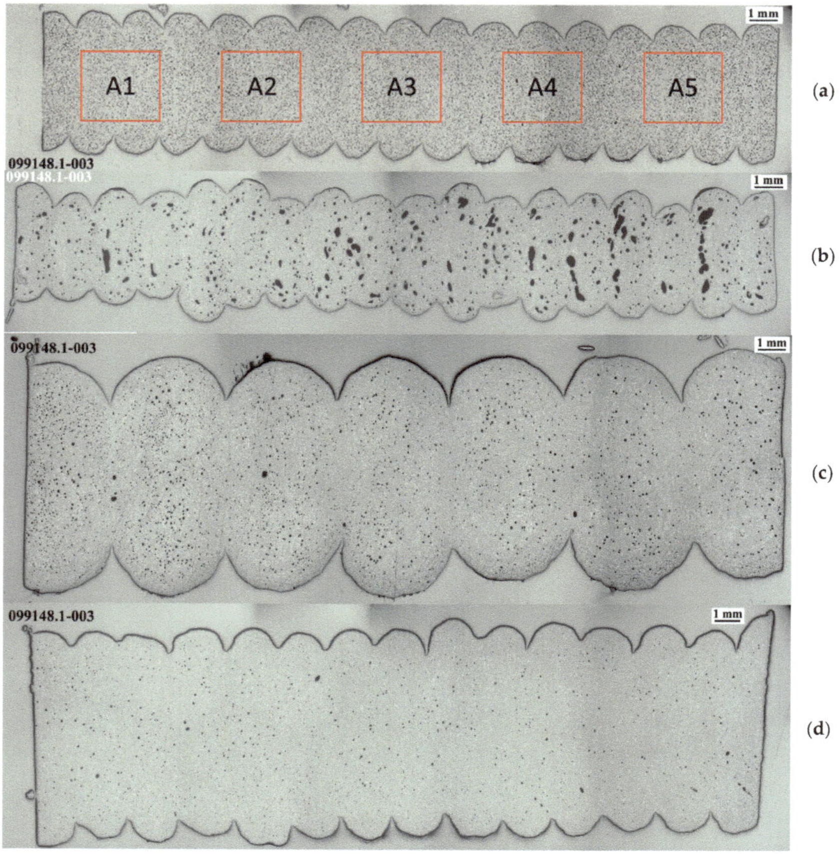

Figure 11. Macrographs of the each case in our study: RAMS (**a**), AMSI (**b**), AMSII (1) (**c**) and AMSII (2) (**d**), showing the profile of the printed beads and the porosity.

For simplicity, only the first micrograph of each case in our study is included in Figure 12. In the images (with ×50 magnification), the polymer matrix (grey area) with short carbon fibres (white marks) and pores (black dots) can be seen. In this study, the longest short carbon fibres and the biggest pores were identified in each photograph. According to the magnified images, the pores are quite uniform (spherical) in each case, apart from AMS I, in which the pores are bigger and irregular. In fact, looking at the micrographs, there is a lower number of pores than in the other cases.

Separately, the porosity percentage was estimated in contrast with the polymer matrix and fibres. A summary of the results can be seen in Table 5, showing that the specimens printed using the AMSII machine with a lower bead height achieve the lowest porosity.

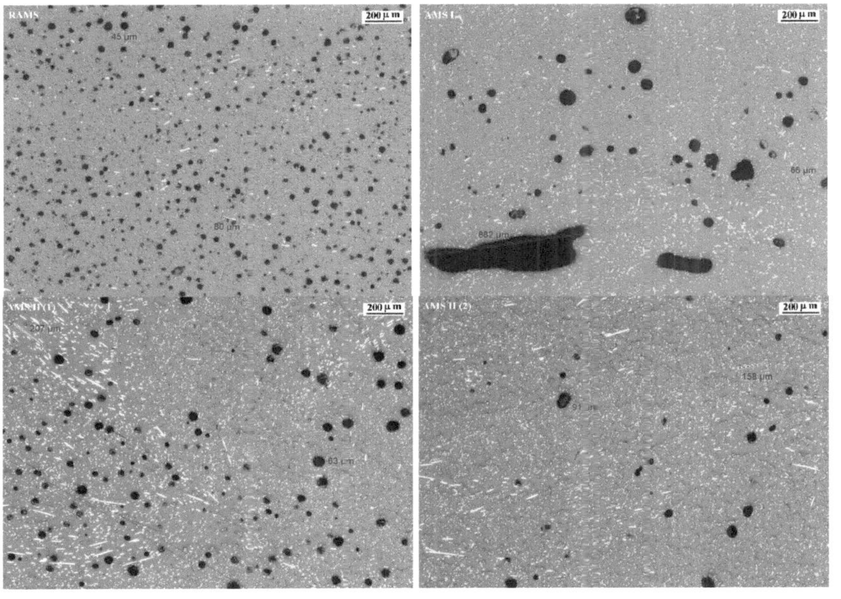

Figure 12. Micrographs of the first representative field for each case in our study: RAMS (**upper left**), AMS I (**upper right**), AMS II (1) (**bottom left**) and AMS II (2) (**bottom right**), showing the greatest carbon fibre length and porosity.

Table 5. Results of the estimated porosity for each type of specimen.

Porosity	RAMS	AMS I	AMS II (1)	AMS II (2)
(A1) Area #1 (%)	4.5	6.73	1.57	1.09
(A2) Area #2 (%)	4.95	3.21	2.29	1.35
(A3) Area #3 (%)	4.47	8.55	2.63	0.43
(A4) Area #4 (%)	4.3	9.77	1.38	0.5
(A5) Area #5 (%)	4.57	3.7	3.84	0.34
Mean Value (%)	4.6	6.4	2.3	0.7
Standard Deviation	0.2	2.9	1	0.4
Confidence interval 95%	0.3	3.6	1.2	0.6
Relative Accuracy (RA %)	6.6	56.3	52	75.2

4.4. Density

An additional evaluation was performed in order to verify the density of the printed beads based on UNE-EN ISO 1183-1:2019 [60]. Therefore, a square plate approximately 30 mm × 30 mm and 3 mm thick was machined. The aim was to calculate the density experimentally through geometrical measurements and the Archimedes method using a Mettler AE 240 (Mettler-Toledo, Cornellá de Llobregat, Spain) weight balance and isopropanol-2 as a liquid medium. The results are included in Table 6, showing that all the calculated values are lower than those provided in the datasheet of the specimens processed by injection moulding (between 1.21 and 1.25 g/cm^3).

Table 6. Density of the printed material versus machine.

Density	RAMS	AMS I	AMS II (1)	AMSII (2)
X length (mm)	30.13	30.17	30.22	30.14
Y width (mm)	29.98	30.15	30.07	30.09
Z height (mm)	2.78	2.81	3.42	3.97
Weight (g)	2.842	2.991	3.643	4.334
Geometrical- ρ (g/cm^3)	1.132	1.170	1.172	1.204
Archimedes- ρ (g/cm^3)	1.12	1.17	1.2	1.19

4.5. Observed Potential Problems

During the computer modelling activities, some potential problems related to the thermoelastic performance of the carbon fibre-reinforced PA6 (manufactured via RAMS) were found. The authors believe that these issues, described in detail hereafter, may occur with any aeronautic mould made of this material and printed through this technology.

1. Severe CTE mismatch-related problems (critically high tensile states along the mould and/or highly heterogeneous deformations that cannot be compensated in practice) occur under operating conditions when using dissimilar materials (metallic screws/positioners to join the semi-moulds, for instance).
2. These problems might also occur in moulds that are entirely made of this material because its CTE is highly orthotropic.
3. Temperature-induced deformations along the fibre-perpendicular direction are large, much greater than the maximum deflection required, as a result of the material's CTE.
4. Large displacements are caused in service along the fibre-perpendicular direction, greater than the maximum deflection required, due to the relatively low stiffness of the material at the operating temperature.
5. Because of the relatively low thermal diffusivity of the material, the thermal response of the mould is not valid (excessive thermal gradients on the lamination surface) if a light enough mould is not designed.

In our research, a specific design strategy is followed to overcome these issues by applying the design drivers described below. Our numerical results prove that this strategy allows us to meet the particular thermal and mechanical requirements specified by the end user for this work.

1. The use of dissimilar materials should be avoided.
2. It is recommended to follow strategic material deposition, thus printing each layer perpendicularly to the previous one (for example 0/90/0/90/etc.). Such a bidirectional printing strategy results in a mould with homogeneous orthotropic properties. Therefore, the deformations due to the thermal expansion become uniform along each direction, and direction-dependent scale factors can be applied to the original mould geometry to compensate for the large temperature-induced displacements.
3. In the case of moulds formed by several semi-moulds, it should be ensured that the printing and stacking directions of every semi-mould matches once the mould is assembled. Positioners made of the same material and placed so that the printing directions match with the ones of the mould should be used to join the semi-moulds.
4. From a mechanical point of view, solid moulds should be developed as a general rule. This way, in combination with the printing strategy described in 2, the deformations due to the operating pressure (which are stiffness dependent) become uniform, and direction-dependent scale factors to compensate for the pressure-induced displacements can be applied to the original mould geometry.
5. In terms of thermal performance, a thermally light design is required. Moreover, the design must enable similar thermal impedances at ambient temperature and the

thermal capacitance should be as uniform as possible along the lamination surfaces to homogenise the temperature contour.

6. For the cases in which an entirely solid mould does not allow us to ensure proper thermal behaviour, infill volumes with a particular percentage of material should be designed instead of totally hollow parts. This way, a high enough stiffness can be kept without excessively enlarging the thermal capacitance of the component. The deformations due to the pressure might be uniform enough in these cases. Moreover, the scale factors for compensating for the deformations due to the autoclave process could be applied to result in a proper thermal response.

Table 7 identifies the potential problems related to the material performance under operating conditions with design drivers to overcome them.

Table 7. Identification of the potential problems versus design drivers.

Potential Problem	Design Strategy
Severe problems of CTE mismatch.	Avoid the use of dissimilar materials. Strategically print the semi-moulds and the joining elements to have uniform thermal expansion in each direction.
Large temperature-induced displacements.	Accurately determine and apply specific scale factors to compensate for the displacements.
Large pressure-induced displacements.	Manufacture solid semi-moulds as a general rule. Accurately determine and apply specific scale factors to compensate for the displacements.
Excessive thermal gradients on the lamination surface.	In case of a poor thermal performance, print partially hollow semi-moulds with infill volumes.

4.6 Result of the Computer Modelling of the Mould

As a result of several iterations, the design of the mould finally developed in this research paper is formed by four semi-moulds. The final mechanical design is shown in Figures 1 and 13. Such a configuration ensures proper part extraction after the curing process. Moreover, the design drivers described in previous sections were applied for the development of this design. It is also remarkable that the simulations conducted during the iterative design process prove that some of the semi-moulds cannot be entirely solid to have a proper thermal response. The reasons for this are the quite heterogeneous distribution of the convection coefficients along the mould and the relatively low thermal diffusivity of the material. Therefore, a combination of solid and partially hollow (30% infill) semi-moulds is proposed instead. Concerning the assembly method, the central semi-mould is joined to the base part by means of two positioners, printed with the same material, while the external ones are joined through two rectangular pins. Regarding the printing strategy, every component is strategically printed to ensure a mould with homogeneous orthotropic properties once the semi-moulds are assembled. Scale factors of 98.2%, 98.2% and 96.6% are applied for the X, Y and Z directions, respectively. These factors are calculated bearing in mind the temperature- and pressure-induced deformations. The mass of the whole mould is lightened by around 90% with respect to the original invar design.

According to the numerical results, the streamlines predicted inside the autoclave (Figure 14a) involve a relatively heterogenous distribution of convection coefficients (Figure 14b). It is noticeable that the values obtained are in line with those published in the literature [65]. The heat exchange with the ambient temperature along the mould's external faces is not uniform enough and might involve a heterogenous temperature contour. However, our results indicate that the strategic design of lateral semi-moulds, with internal infill volumes, enables proper thermal behaviour. Actually, even for the most demanding heating rates of 3 °C/min, an admissible delay of 2 h is foreseen between the hottest and coldest points to reach the curing temperature if a minimum dwell time of 3 h is completed (Figure 14c). The raw displacements, that is, the actual displacements that take place along the scaled mould due to the operating loads, are proven to mainly be a consequence of the temperature increase, as expected due to the high CTE of the material in the fibre-perpendicular direction. The scale factors applied to the mould are, however, able to compensate for the deformations well enough. Consequently, the determined displacement with respect the target geometry (net displacements) is proven to be low enough, lower than 0.1 mm, thus confirming that the mechanical requirements are fulfilled (Figure 14d). Our numerical results show a very low tensile state once the steady-state condition is reached (i.e., once the curing temperature is reached in the whole mould). However, quite high stresses are identified in the transient processes during the heating stages. It is numerically determined that the most unfavourable condition takes place at 4 h, when the maximum amounts of stress are reached. Based on the worst principal stresses in the printing direction and the material strength at corresponding temperatures (≈150 °C on average), an FoS of 2.8 is predicted. It can therefore be confirmed that the material can withstand the operating conditions with no risks.

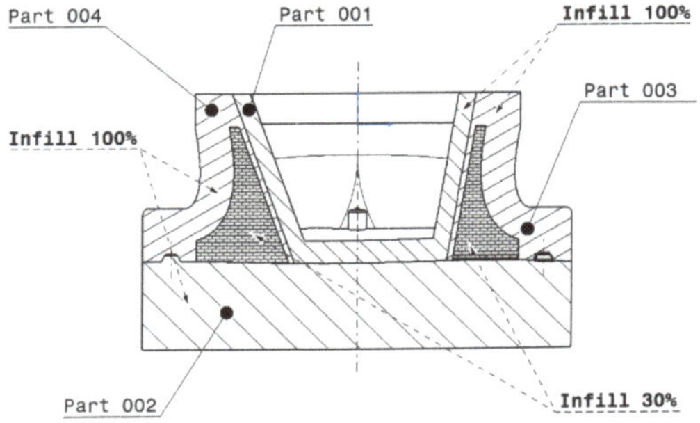

Figure 13. Cut view of the complete designed mould according to Figure 1.

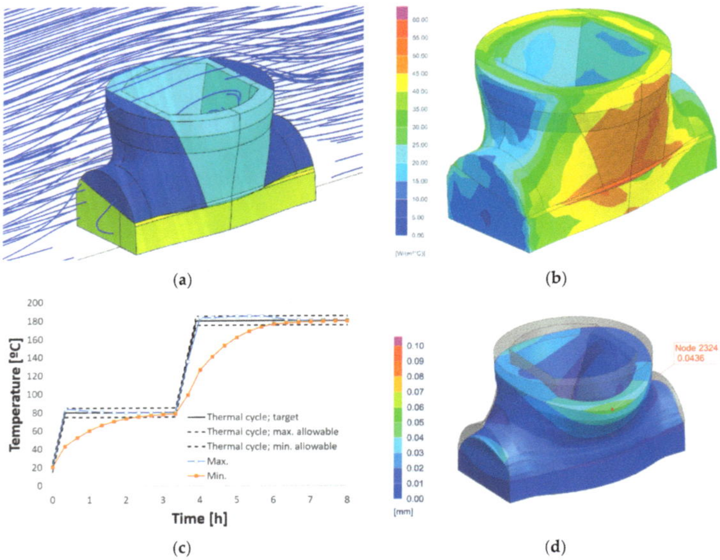

Figure 14. (**a**) Air streamlines around the mould, (**b**) resulting convection coefficients at 180 °C, (**c**) evolution of the maximum and minimum temperatures at the lamination surface (3 °C/min, 3 h at 80 °C) and (**d**) net displacement contour under maximum operating conditions.

5. Conclusions and Further Work

The printed pellet-based PA6 reinforced (20%) with short carbon fibre used in this research paper is found to be a suitable material to print preforms for the aeronautic moulding of aeronautic parts of composites, considering the design drivers established in the research.

The completed full experimental characterization allows the authors to determine the main properties concerning the thermoelastic performance of the material under operating conditions. The temperature dependency of the mechanical properties was characterized, and the complete results are available upon request. The experimental results include the tensile, compression, flexural and shear moduli and strength, both in the printing and the stacking directions and at room temperature, 80 °C and 180 °C. The influence of the temperature reduces the tensile strength in the printing bead direction from 37% to 70%; meanwhile, the tensile modulus decreases down to 56% and 71% at 80 °C and 180 °C, respectively. In the stacking direction, the tensile strength and modulus also decrease significantly. The effects of the printing machine and/or process parameters on the tensile moduli and strength, the porosity and the density are also analysed. Regarding the results of this study, the lower the porosity, the higher the tensile modulus and the tensile strength, whose values approach the tensile values of specimens processed by injection moulding. Comparing the tensile properties of injected Bergamid™ B70 KF20 Black at room temperature, the moduli and strength can vary from 78% to 112% and from 55% to 87%, depending on the parameters and type of machine. The main properties regarding thermal response are characterized as well, and we found that the printed material is highly orthotropic, with different CTE values depending on whether the stacking or printing direction is used. In addition, due to the fact that the test specimens have been machined after the printing process, a good level of machinability was found, with surface roughness, Ra, values between 2.3388 and 3.496 µm. This work contributes to the literature by increasing our knowledge about the thermoelastic behaviour of printed carbon fibre-reinforced PA6.

Based on a real aeronautical application specified by the end user and thanks to its characterized properties, simulations that accurately reproduce the thermoelastic performance of the material under representative operating conditions are completed. In this sense, the conducted modelling activities allow the authors to iteratively design a 3D-printed mould for composite parts using pellets of carbon fibre-reinforced PA6. Moreover, the full design of the mould's behaviour has been numerically validated in service. The mould is provn to be able to properly withstand both the curing temperature (180 °C) and the operating pressure (7 bar). A maximum net displacement of 0.04 mm is numerically determined, which is clearly lower than the maximum distortion required by the end user (0.10 mm). Moreover, it can ensure good temperature evolution and thermal uniformity along the lamination surface during the curing cycle. This suitable thermal performance is numerically predicted to continue even for the most demanding heating rates, in which the delay between the hottest and coldest points to reach the curing temperature matches the maximum delay required by the end user (2 h), if a minimum dwell time of 3 h is completed. Thanks to the advantages of this technology and the proposed material, the developed solution is 90% lighter than the original invar design. However, some potential problems that might occur in service for aeronautical moulds printed with this material are identified as well. These are basically related to the CTE, as the material is extremely orthotropic and very high in the fibre-perpendicular direction. In this work, the authors propose and successfully apply some design drivers to overcome these issues, which could be extended to different aeronautical moulds and sectors.

Taking into account that the mould will be fabricated using AMS II, a balance between the required time for printing the parts and the highest quality regarding the mechanical behaviour will determine the parameters for the final process.

Further work is planned concerning the research of moulds printed with PA6/CF6 in pellet form. Currently, semi-moulds based on the design presented in this paper are being printed through the AMS II extruder system, fed by pellets and installed at Tecnalia facilities. Once properly manufactured, it is planned to investigate methods for post-processing the laminate surface to ensure a smooth finish that is well-adjusted to the strict dimensional requirements before assembly. Afterwards, we plan to test the moulds under operating conditions to validate the proposed technology and material for the development of moulds for composite parts.

Supplementary Materials: The following supporting information can be downloaded at: https://www.mdpi.com/article/10.3390/jmmp8010034/s1, Supplementary Material S1: PolyMide^{TM} PA6/CF. Supplementary Material S2: Bergamid^{TM} B70 KF20 Black.

Author Contributions: Conceptualization, J.C.A.-U. and E.B.R.; Data curation, J.G.L.; Formal analysis, H.V.A. and J.G.L.; Investigation, J.C.A.-U. and H.V.A.; Methodology, H.V.A.; Project administration, J.C.A.-U.; Software, H.V.A. and J.I.H.V.; Supervision, J.C.A.-U., H.V.A. and E.B.R.; Validation, E.B.R. and A.I.L.P.; Visualization, H.V.A., J.G.L. and J.I.H.V.; Writing—original draft, J.C.A.-U. and E.B.R.; Writing—review and editing, J.C.A.-U., H.V.A. and A.I.L.P. All authors have read and agreed to the published version of the manuscript.

Funding: This research was co-funded by EIT Manufacturing, code 22017.

Data Availability Statement: The data are available upon request due to restrictions, e.g., privacy or ethical reasons.

Acknowledgments: The work leading to this publication was co-funded by EIT Manufacturing under the code 22017. EIT Manufacturing is supported by the European Institute of Innovation and Technology (EIT), a body of the European Union. The views and opinions expressed herein are those of the author(s) only and do not necessarily reflect those of the European Union or EIT Manufacturing. Neither the European Union nor the EIT Manufacturing can be held responsible for them. Also, the authors would like to thank TECNALIA LabServices-Industry and other departments members, in particular, Urko Larramendi, David Villalta, Egoi Garmendia, Aitor Urbistondo, Xabier Cruz, Mariasun Mendizabal, Naiara Azurmendi and Gorka Imbuluzqueta, who provided support and input throughout this research project.

Conflicts of Interest: Ana Isabel Luengo Pizarro was employed by Internacional de Composites S.A. The remaining authors declare that the research was conducted in the absence of any commercial or financial relationships that could be construed as a potential conflict of interest. The Internacional de Composites S.A. had no role in the design of the study; in the collection, analyses, or interpretation of data; in the writing of the manuscript, or in the decision to publish the results.

References

1. Grankäll, T.; Hallander, P.; Petersson, M.; Åkermo, M. The true shape of composite cure tools. *J. Manuf. Process.* **2020**, *59*, 279–286. [CrossRef]
2. Stewart, R. New mould technologies and tooling materials promise advances for composites. *Reinf. Plast.* **2010**, *54*, 30–36. [CrossRef]
3. Hasan, Z. Chapter 1—Introduction. In *Tooling for Composite Aerospace Structures*; Hasan, Z., Ed.; Butterworth-Heinemann: Oxford, UK, 2020; pp. 1–19, ISBN 978-0-12-819957-2.
4. Li, Y.; Xiao, Y.; Yu, L.; Ji, K.; Li, D. A review on the tooling technologies for composites manufacturing of aerospace structures: Materials, structures and processes. *Compos. Part A Appl. Sci. Manuf.* **2022**, *154*, 106762. [CrossRef]
5. Galiana, J. Guías básicas para elección de material de útil de curado en autoclave para fabricar piezas de composite. CFRP vs. INVAR36. In Proceedings of the XII Congreso Nacional de Materiales Compuestos MATCOMP 2017, Donostia-San Sebastian, Spain, 21–23 June 2017; Asociación Española de Materiales Compuestos (AEMAC): Getafe, Spain, 2017.
6. Adeniran, O.; Cong, W.; Aremu, A. Material design factors in the additive manufacturing of Carbon Fiber Reinforced Plastic Composites: A state-of-the-art review. *Adv. Ind. Manuf. Eng.* **2022**, *5*, 100100. [CrossRef]
7. Garmendia, I.; Vallejo, H.; Osés, U. Composite Mould Design with Multiphysics FEM Computations Guidance. *Computation* **2023**, *11*, 41. [CrossRef]
8. Materials & Processes: Tooling for Composites. Available online: https://www.compositesworld.com/articles/tooling (accessed on 4 August 2023).
9. *UNE-EN ISO/ASTM 52900:2021*; Additive Manufacturing- General Principles- Fundamentals and Vocabulary. UNE-EN (Una Norma Española sobre Norma Europea): Madrid, Spain, 2021.
10. Frazier, W.E. Metal Additive Manufacturing: A Review. *J. Mater. Eng. Perform.* **2014**, *23*, 1917–1928. [CrossRef]

11. Lee, J.-Y.; An, J.; Chua, C.K. Fundamentals and applications of 3D printing for novel materials. *Appl. Mater. Today* **2017**, *7*, 120–133. [CrossRef]
12. Tofail, S.A.M.; Koumoulos, E.P.; Bandyopadhyay, A.; Bose, S.; O'Donoghue, L.; Charitidis, C. Additive manufacturing: Scientific and technological challenges, market uptake and opportunities. *Mater. Today* **2018**, *21*, 22–37. [CrossRef]
13. Yan, L. Wire and Arc Additive Manufacture (WAAM) Reusable Tooling Investigation. Master's Thesis, Cranfield University, Cranfield, UK, 2013.
14. Calignano, F.; Mercurio, V. An overview of the impact of additive manufacturing on supply chain, reshoring, and sustainability. *Clean. Logist. Supply Chain* **2023**, *7*, 100103. [CrossRef]
15. Pignatelli, F.; Percoco, G. An application- and market-oriented review on large format additive manufacturing, focusing on polymer pellet-based 3D printing. *Prog. Addit. Manuf.* **2022**, *7*, 1363–1377. [CrossRef]
16. Bi, X.; Huang, R. 3D printing of natural fiber and composites: A state-of-the-art review. *Mater. Des.* **2022**, *222*, 111065. [CrossRef]
17. Abderrafai, Y.; Diouf-Lewis, A.; Sosa-Rey, F.; Farahani, R.D.; Piccirelli, N.; Lévesque, M.; Therriault, D. Additive manufacturing and characterization of high temperature thermoplastic blends for potential aerospace applications. *Compos. Sci. Technol.* **2023**, *231*, 109839. [CrossRef]
18. Dey, A.; Eagle, I.N.R.; Yodo, N. A review on filament materials for fused filament fabrication. *J. Manuf. Mater. Process.* **2021**, *5*, 69. [CrossRef]
19. Spoerk, M.; Holzer, C.; Gonzalez-Gutierrez, J. Material extrusion-based additive manufacturing of polypropylene: A review on how to improve dimensional inaccuracy and warpage. *J. Appl. Polym. Sci.* **2020**, *137*, 48545. [CrossRef]
20. Saroia, J.; Wang, Y.; Wei, Q.; Lei, M.; Li, X.; Guo, Y.; Zhang, K. A review on 3D printed matrix polymer composites: Its potential and future challenges. *Int. J. Adv. Manuf. Technol.* **2020**, *106*, 1695–1721. [CrossRef]
21. Zhuo, P.; Li, S.; Ashcroft, I.A.; Jones, A.I. Material extrusion additive manufacturing of continuous fibre reinforced polymer matrix composites: A review and outlook. *Compos. Part B Eng.* **2021**, *224*, 109143. [CrossRef]
22. Patel, A.; Taufik, M. Extrusion-Based Technology in Additive Manufacturing: A Comprehensive Review. *Arab. J. Sci. Eng.* **2022**, *225*, 111505. [CrossRef]
23. Hassen, A.A.; Lindahl, J.; Chen, X.; Post, B.; Love, L.; Kunc, V. Additive Manufacturing of Composite Tooling using High Temperature Thermoplastic Materials. In Proceedings of the Society for the Advancement of Material and Process Engineering, Long Beach, CA, USA, 23–26 May 2016; North America Society for the Advancement of Material and Process Engineering (SAMPE): Diamond Bar, CA, USA.
24. Park, S.J.; Lee, J.E.; Park, J.; Lee, N.K.; Son, Y.; Park, S.H. High-temperature 3D printing of polyetheretherketone products: Perspective on industrial manufacturing applications of super engineering plastics. *Mater. Des.* **2021**, *211*, 110163. [CrossRef]
25. Ajinjeru, C.; Kishore, V.; Liu, P.; Lindahl, J.; Hassen, A.A.; Kunc, V.; Post, B.; Love, L.; Duty, C. Determination of melt processing conditions for high performance amorphous thermoplastics for large format additive manufacturing. *Addit. Manuf.* **2018**, *21*, 125–132. [CrossRef]
26. Oberlercher, H.; Heim, R.; Laux, M.; Berndt, A.; Becker, C.; Amancio-Filho, S.T.; Riemelmoser, F.O. Additive manufacturing of continuous carbon fiber reinforced polyamide 6: The effect of process parameters on the microstructure and mechanical properties. *Procedia Struct. Integr.* **2021**, *34*, 111–120. [CrossRef]
27. Shashikumar, S.; Sreekanth, M.S. The effect of printing parameters on tensile properties of thermoplastics prepared by fused deposition modeling (FDM) based additive manufacturing technique. *Mater. Today Proc.* **2023**, *90*, 256–261. [CrossRef]
28. Chesser, P.; Post, B.; Roschli, A.; Carnal, C.; Lind, R.; Borish, M.; Love, L. Extrusion control for high quality printing on Big Area Additive Manufacturing (BAAM) systems. *Addit. Manuf.* **2019**, *28*, 445–455. [CrossRef]
29. Love, L.J.; Duty, C.E.; Post, B.K.; Lind, R.F.; Lloyd, P.D.; Kunc, V.; Peter, W.H.; Blue, C.A. Breaking Barriers in Polymer Additive Manufacturing. In *SAMPE 2015*; SAMPE North America: Baltimore, MD, USA, 2015.
30. Post, B.K.; Richardson, B.; Lind, R.; Love, L.J.; Lloyd, P.; Kunc, V.; Rhyne, B.J.; Roschli, A.; Hannan, J.; Nolet, S.; et al. Big Area Additive Manufacturing application in wind turbine molds. In Proceedings of the 28th Annual International Solid Freeform Fabrication Symposium—An Additive Manufacturing Conference, SFF 2017, Austin, TX, USA, 7–9 August 2017; pp. 2430–2446.
31. Roschli, A.; Gaul, K.T.; Boulger, A.M.; Post, B.K.; Chesser, P.C.; Love, L.J.; Blue, F.; Borish, M. Designing for Big Area Additive Manufacturing. *Addit. Manuf.* **2019**, *25*, 275–285. [CrossRef]
32. Sher, D. WHAM, There's a New "Largest 3D Printer in the World" in Town. Available online: https://www.voxelmatters.com/wham-theres-a-new-largest-composite-3d-printer-in-the-world-in-town/ (accessed on 7 August 2023).
33. Thompson, W.; Huelskamp, S.R.; Allessio, T.; Ly, K. Large-Format Additive Manufacturing Viable for Autoclave Tooling? Available online: https://www.additivemanufacturing.media/articles/large-format-additive-manufacturing-viable-for-autoclave-tooling (accessed on 7 August 2023).
34. Chambon, P.; Curran, S.; Huff, S.; Love, L.; Post, B.; Wagner, R.; Jackson, R.; Green, J. Development of a range-extended electric vehicle powertrain for an integrated energy systems research printed utility vehicle. *Appl. Energy* **2017**, *191*, 99–110. [CrossRef]
35. Curran, S.; Chambon, P.; Lind, R.; Love, L.; Wagner, R.; Whitted, S.; Smith, D.; Post, B.; Graves, R.; Blue, C.; et al. Big Area Additive Manufacturing and Hardware-in-the-Loop for Rapid Vehicle Powertrain Prototyping: A Case Study on the Development of a 3-D-Printed Shelby Cobra. In *SAE 2016 World Congress and Exhibition*; SAE International: Warrendale, PA, USA, 2016.
36. Moreno Nieto, D.; Casal López, V.; Molina, S.I. Large-format polymeric pellet-based additive manufacturing for the naval industry. *Addit. Manuf.* **2018**, *23*, 79–85. [CrossRef]

37. Post, B.; Chesser, P.; Lind, R.; Sallas, M.; Love, L.J. *Feasibility of Using Big Area Additive Manufacturing to Directly Manufacture Boat Molds;* Final Report NN-17-1062; Oak Ridge National Laboratory: Oak Ridge, TN, USA, 2018.
38. Wahlström, N.; Gabrielsson, O. Additive Manufacturing Applications for Wind Turbines. Master of Science Thesis, KTH Industrial Engineering and Management, Stockholm, Sweden, 2017.
39. Post, B.K.; Richardson, B.; Lloyd, P.; Love, L.J.; Nolet, S.; Hannan, J. *Additive Manufacturing of Wind Turbine Molds;* CRADA Final Report NFE-16-06051; Oak Ridge National Labnoratory: Oak Ridge, TN, USA, 2017; ISBN 1800553684.
40. Silva, M.R.; Pereira, A.M.; Alves, N.; Mateus, G.; Mateus, A.; Malça, C. Development of an additive manufacturing system for the deposition of thermoplastics impregnated with carbon fibers. *J. Manuf. Mater. Process.* **2019**, *3*, 35. [CrossRef]
41. Peng, X.; Zhang, M.; Guo, Z.; Sang, L.; Hou, W. Investigation of processing parameters on tensile performance for FDM-printed carbon fiber reinforced polyamide 6 composites. *Compos. Commun.* **2020**, *22*, 100478. [CrossRef]
42. Zhuo, P.; Li, S.; Ashcroft, I.A.; Jones, A.I. Continuous fibre composite 3D printing with pultruded carbon/PA6 commingled fibres: Processing and mechanical properties. *Compos. Sci. Technol.* **2022**, *221*, 109341. [CrossRef]
43. Li, X.; He, J.; Hu, Z.; Ye, X.; Wang, S.; Zhao, Y.; Wang, B.; Ou, Y.; Zhang, J. High strength carbon-fiber reinforced polyamide 6 composites additively manufactured by screw-based extrusion. *Compos. Sci. Technol.* **2022**, *229*, 109707. [CrossRef]
44. Li, L.; Liu, W.; Sun, L. Mechanical characterization of 3D printed continuous carbon fiber reinforced thermoplastic composites. *Compos. Sci. Technol.* **2022**, *227*, 109618. [CrossRef]
45. Polymaker Technical Data Sheet—PolyMideTM PA6-CF. Available online: https://filament2print.com/es/index.php?controller=attachment&id_attachment=414 (accessed on 13 April 2023).
46. Avient Technical Data Sheets. Available online: https://www.avient.com/resources/technical-data-sheets (accessed on 13 April 2023).
47. Tagscherer, N.; Bär, A.M.; Zaremba, S.; Drechsler, K. Mechanical Analysis of Parameter Variations in Large-Scale Extrusion Additive Manufacturing of Thermoplastic Composites. *J. Manuf. Mater. Process.* **2022**, *6*, 36. [CrossRef]
48. Al-Maharma, A.Y.; Patil, S.P.; Markert, B. Effects of porosity on the mechanical properties of additively manufactured components: A critical review. *Mater. Res. Express* **2020**, *7*, 122001. [CrossRef]
49. Yang, C.; Tian, X.; Li, D.; Cao, Y.; Zhao, F.; Shi, C. Influence of thermal processing conditions in 3D printing on the crystallinity and mechanical properties of PEEK material. *J. Mater. Process. Technol.* **2017**, *248*, 1–7. [CrossRef]
50. Antolín-Urbaneja, J.C.; Bengoa Ganado, P.; Mateu, A.; Fernández Valares, J.B.; Hernandez Vicente, J.; Bellvert Rios, E.; Vallejo Artola, H.; Alberdi Olaizola, N.; Pacheco Goñi, R.; Luengo Pizarro, A.I. Automated MOLDAM Robotic System for 3D printing: Manufacturing aeronautical mould preforms. In *V International Congress on Computer Science, Electronic and Industrail Engineering (CSEI);* Universidad Técnica de Ambato, Facultad de Ingeniería en Sistemas, Electrónica e Industrial: Ambato, Ecuador, 2023.
51. Ly, K.; Thompson, W.; Voorde, D. Evaluating the printability and Mechanical Properties of LFAM Regrind. Available online: https://www.additivemanufacturing.media/articles/evaluating-the-printability-and-mechanical-properties-of-lfam-regrind (accessed on 1 February 2023).
52. *ISO 4287:1997;* Geometrical Product Specifications (GPS). Surface Texture: Profile Method. ISO (International Organization for Standardization): Geneva, Switzerland, 1997.
53. *ISO 527-1;* Plastics—Determination of Tensile Properties—Part 1: General Principles. ISO (International Organization for Standardization): Geneva, Switzerland, 2019.
54. *ISO 527-2:2012;* Plastics—Determination of Tensile Properties—Part 2: Test Conditions for Moulding and Extrusion Plastics. ISO (International Organization for Standardization): Geneva, Switzerland, 2012.
55. *ISO 14125:1998/Amd 1:2011;* Fibre-Reinforced Plastic Composites—Determination of Flexural Properties (Amendment 1). ISO (International Organization for Standardization): Geneva, Switzerland, 2011.
56. *ISO 604:2022;* Plastics—Determination of Compressive Properties. ISO (International Organization for Standardization): Geneva, Switzerland, 2022.
57. *ASTM D5379/D5379M-19e1;* Standard Test Method for Shear Properties of Composite Materials by the V-Notched Beam Method. ASTM International (American Society for Testing and Materials): West Conshohocken, PA, USA, 2021.
58. *ISO6721-1:2019;* Plastics. Determination of Dynamic Mechanical Properties. Part1: General Principles. ISO (International Organization for Standardization): Geneva, Switzerland, 2019.
59. *ISO 22007-2:2016;* Plastics—Determination of Thermal Conductivity and Thermal Diffusivity—Part 2: Transient Plane Heat Source (Hot Disc) Method. ISO (International Organization for Standardization): Geneva, Switzerland, 2016.
60. *UNE-EN ISO 1183-1:2019;* Plastics—Methods for Determining the Density of Non-Cellular Plastics—Part 1: Immersion Method, Liquid Pycnometer Method and Titration Method (ISO 1183-1:2019, Corrected Version 2019-05). UNE-EN (Una Norma Española sobre Norma Europea): Madrid, Spain, 2019.
61. *UNE-EN ISO 4288:1998;* Geometrical Product Specifications (GPS). Surface Texture: Profile Method- Rules and Procedures for the Assessment of Surface Texture. UNE-EN (Una Norma Española sobre Norma Europea): Madrid, Spain, 1998.
62. Brinson, H.F.; Brinson, L.C. Time and Temperature Behavior of Polymers. In *Polymer Engineering Science and Viscoelasticity: An Introduction;* Springer: Boston, MA, USA, 2008; pp. 221–274, ISBN 978-0-387-73861-1.
63. Valino, A.D.; Dizon, J.R.C.; Espera, A.H.; Chen, Q.; Messman, J.; Advincula, R.C. Advances in 3D printing of thermoplastic polymer composites and nanocomposites. *Prog. Polym. Sci.* **2019**, *98*, 101162. [CrossRef]

64. Park, J.; Zobeiry, N.; Poursartip, A. Tooling materials and their effect on surface thermal gradients. In Proceedings of the International SAMPE Technical Conference, Seattle, WA, USA, 22–26 May 2017; pp. 2554–2568.
65. *ASTM E562-19e1*; Standard Test Method for Determining Volume Fraction by Systematic Manual Point Count. ASTM International (American Society for Testing and Materials): West Conshohocken, PA, USA, 2020.

Journal of
Manufacturing and Materials Processing

MDPI

Article

The Impact of PP-g-MAH on Mechanical Properties of Injection Molding of Long Glass Fiber/Polypropylene Pellets from Thermoplastic Pultrusion Process

Ponlapath Tipboonsri and Anin Memon *

Department of Industrial Engineering, Faculty of Engineering, Rajamangala University of Technology Thanyaburi (RMUTT), Klong Luang 12110, Thailand; ponlapath_t@mail.rmutt.ac.th
* Correspondence: anin.m@en.rmutt.ac.th

Abstract: Long fiber thermoplastic pellets are pellets containing discontinuous reinforced fibers and a matrix, offering excellent mechanical properties, good processability, recyclability, and low cost. Typically, commercial LFTP is manufactured through the hot melt impregnation process, combining extrusion and pultrusion. Although there is a thermoplastic pultrusion process for LFTP production, characterized by a simple machine and an easy method, its mechanical properties have not yet approached those of commercial LFTP. In improving the mechanical characteristics of LFTP manufactured via thermoplastic pultrusion, this research employed polypropylene-graft-maleic anhydride as a coupling agent during the injection molding procedure. The LFTP is composed of polypropylene material reinforced with glass fiber. Mechanical and physical properties of the LFTP were investigated by introducing PP-g-MAH at concentrations of 4, 8, and 12 wt% through injection molding. The results revealed that, at a 4 wt% concentration of PP-g-MAH, the LFTP composites exhibited heightened tensile, flexural and impact strengths. However, these properties began to decrease upon exceeding 4 wt% PP-g-MAH. The enhanced interfacial adhesion among glass fibers, induced by PP-g-MAH, contributed to this improvement. Nonetheless, excessive amounts of PP-g-MAH led to a reduction in molecular weight, subsequently diminishing the impact strength, tensile modulus, and flexural modulus. In LFTP composites, both tensile and flexural strengths exhibited a positive correlation with the PP-g-MAH concentration, attributed to improved interfacial adhesion between glass fibers and polypropylene, coupled with a reduction in fiber pull-out. Based on morphological analysis by SEM, the incorporation of PP-g-MAH improved interfacial bonding and decreased fiber pull-out. The presence of maleic anhydride in the LFTPc was confirmed through the utilization of FTIR spectroscopy. Mechanical properties of LFTP containing 4 wt% PP-g-MAH were found to be equivalent to or superior to those of commercial LFTP, according to the results of a comparative analysis.

Keywords: LFTP composite; thermoplastic pultrusion; long glass fiber reinforced polypropylene; PP-g-MAH

1. Introduction

Long fiber thermoplastic pellets (LFTPs) are plastic pellets composed of reinforced fibers and thermoplastic. The fibers embedded in the pellets are discontinuous and possess a length-to-diameter aspect ratio exceeding the critical aspect ratio [1–3]. Normally, LFTPs are utilized in compression and injection molding processes. Various industries, including automobile, sports, wind energy, military, aerospace, and electronics [4–6], employ LFTPs in manufacturing due to their superior mechanical properties, weight savings, improved damping, good processability, recyclability, corrosion resistance, and low cost [1,6,7]. Several types of thermoplastic matrices are used in the fabrication of LFTP, such as polyamide (PA), polymethyl methacrylate (PMMA), polyacrylonitrile–butadiene–styrene (ABS), and polypropylene (PP), among others [8–11]. These matrices are reinforced with glass fiber and carbon fiber [12]. The popular type of LFTP incorporates glass fiber (GF) reinforcement in polypropylene (PP) due to its low density, low molding temperature, and cost-effectiveness [13–15]. Glass fiber provides high tensile stress and is also cost-effective [16,17]. The combination of glass fiber and polypropylene results in a composite with excellent mechanical properties, low density, and cost-effectiveness. The typical manufacturing process for LFTP involves a hot melt impregnation process, combining extrusion and pultrusion processes. Continuous reinforced fibers are drawn into the hot melt impregnation die, where molten thermoplastic from the extruder impregnates the reinforcing fibers. Subsequently, the material undergoes cooling and is subsequently cut into the desired length. LFTP products usually demonstrate dimensions of 6 to 25 mm in length and 2 to 4 mm in diameter [1,18,19].

Currently, achievements have been made in producing LFTP using the thermoplastic pultrusion process [20,21]. This method demonstrates comparable mechanical properties to the conventional process but incurs lower costs due to the absence of an extruder requirement in the machinery setup. Generally, pultrusion processes are used for the fabrication of constant cross-section and continuous composites. The advantages of this process include high mechanical properties of products, a high production rate, and low cost [22]. Pultrusion processes can fabricate both thermoplastic and thermoset composites [23]. Thermoplastics are gaining interest due to their recyclability, contributing to the reduction in environmental pollution problems [24]. Thermoplastic composites are highly ductile, tough to break, impact resistant, recyclable, and quickly processed [25,26]. Therefore, the thermoplastic pultrusion process is interesting to develop for adding new products, with the capability of fabricating LFTP through this method. In the process of manufacturing LFTP through thermoplastic pultrusion, continuous thermoplastic fibers and continuous fibers are drawn into the heated die. The thermoplastic fibers undergo melting at elevated temperatures, impregnating the fibers, followed by cooling and subsequent cutting into the required length [21,23]. Although LFTP is fabricated using thermoplastic pultrusion, which imparts good mechanical properties, its mechanical characteristics may not be sufficient when compared to some commercial LFTP products. Surface treatment is one among numerous techniques that can be implemented to boost the mechanical properties of composites. Interfacial adhesion is improved as a result of surface treatment of the fiber, which in turn leads to improvements in mechanical properties [27]. Using a coupling agent [28,29] is a popular method for surface treatment. GF and PP are treated with polypropylene-graft-maleic anhydride (PP-g-MAH), one such agent [30]. The incorporation of PP-g-MAH improves the mechanical properties of the GF/PP composite by enhancing interfacial adhesion [31,32].

The aim of this study is to improve the interfacial adhesion between GF and PP in the LFTP composite (LFTPc) by utilizing PP-g-MAH as a coupling agent. The preparation of LFTPc, which is made up of GF/PP, was carried out through the use of the injection molding process, alongside the incorporation of PP-g-MAH into this process. The LFTP was produced through the thermoplastic pultrusion process. The literature review indicates that PP-g-MAH has been widely used as a coupling agent to enhance interfacial adhesion, leading to a notable growth of mechanical properties by 10–40% [27,30,33–35]. In the literature review of PP composites reinforced with long glass fiber (LGF), GF is typically used at a volume fraction of 10–40 wt% in the composite [1,35,36], with most LFTPs in the literature being fabricated using the hot melt impregnation process. Therefore, this work differs in the manufacturing process of LFTP, which is fabricated using the thermoplastic pultrusion process: a method known for its cost-effectiveness and high productivity.

2. Materials and Methods

2.1. Materials

The thermoplastic pultrusion method was utilized to produce LFTP in this study. PP yarn with a density of 0.946 g/cm^3 (133 tex or 1200 denier) was purchased from Praditkorn Co., Ltd., Bangkok, Thailand, and GF yarn with a density of 2.620 g/cm^3 (EDR171200386) was purchased from China Jushi Co., Ltd., Tongxiang, China. The diameter and length of the LFTP are 3 mm and 10–12 mm, respectively. For injection, LFTP was combined with PP pellets (POLIMAXX 1100NK) that were purchased from Irpc Polyol Co., Ltd. in Bangkok, Thailand. As a coupling agent, PP-g-MAH was purchased from Merick Polymer Co., Ltd., Bangkok, Thailand.

2.2. Methods

The thermoplastic pultrusion technique was employed in this experiment for manufacturing LFTP, with the molding temperature ranging from 160 to 230 °C and the pulling speed set at 10 cm/min. In the process of manufacturing LFTP through thermoplastic pultrusion, continuous PP fibers and continuous glass fibers are drawn into the heated die. The PP fibers undergo melting at elevated temperatures, impregnating the glass fibers. This is followed by fan cooling and subsequent cutting into the required length using a plastic pellet cutting machine. The preheat die and heating die for the lab-scale thermoplastic pultrusion machine employed for manufacturing LFTP are both 645 mm in length. The cross-sectional length of both dies is 300 mm, while their taper length is 345 mm and their radius is 1 degree. A characteristic of the pultrusion die is its 3 mm diameter.

The LFTP consists of a combination of PP and GF with volume fractions of 87.03 and 12.97%, respectively. The determination of the volume fraction of LFTP employed the filling ratio equation outlined in the literature [20,21]. LFTP, characterized by a specified diameter of 3 mm and length ranging from 10 to 12 mm, as indicated in the literature [18] and aligned with commercial LFTP standards, was utilized. LFTP and virgin PP were mixed together in dry mixer, with a volume fraction of 15% by weight of GF. Subsequently, PP-g-MAH was added at weight concentrations of 4%, 8%, and 12% during the injection molding process. Intermediate materials were subjected to overnight drying at 100 °C in an oven before injection to eliminate moisture from PP-g-MAH, given that MAH is hygroscopic. The injection molding machine (LG model LGH-50N) was used to prepare LFTPc specimens for property testing. Molding temperature ranges were set between 170 and 190 °C, with an injection speed set at 100 mm/sec, injection pressure at 90 MPa, and holding pressure at 30 MPa. These parameters were adjusted according to the datasheet of virgin PP. Figure 1 displays the schematic of the experiment. Specimens were evaluated for their mechanical and physical properties.

2.3. Characterization and Measurement

2.3.1. Mechanical Properties

Mechanical testing, including impact, flexural, and tensile tests, was conducted to investigate the properties of LFTPc. A tensile test was performed in accordance with ASTM D638 [37] standards using a Hounsfield universal testing machine (Load 25 kN) at a testing speed of 50 mm/min. The dog bone type I structure was used for fabricating the tensile specimens. A flexural test was carried out, corresponding with the standards of ASTM D790 [38]. A Hounsfield universal testing machine was utilized, which consisted of a three-point bending model (load of 25 kN), a test speed of 1.30 mm/min, and a span length of 48 mm. The head speed was set according to ASTM D790 [38] standards. The impact properties of LFTPc were carried out using an impact Izod tester fitted with a 2J hammer in accordance with ASTM D256 [39] guidelines.

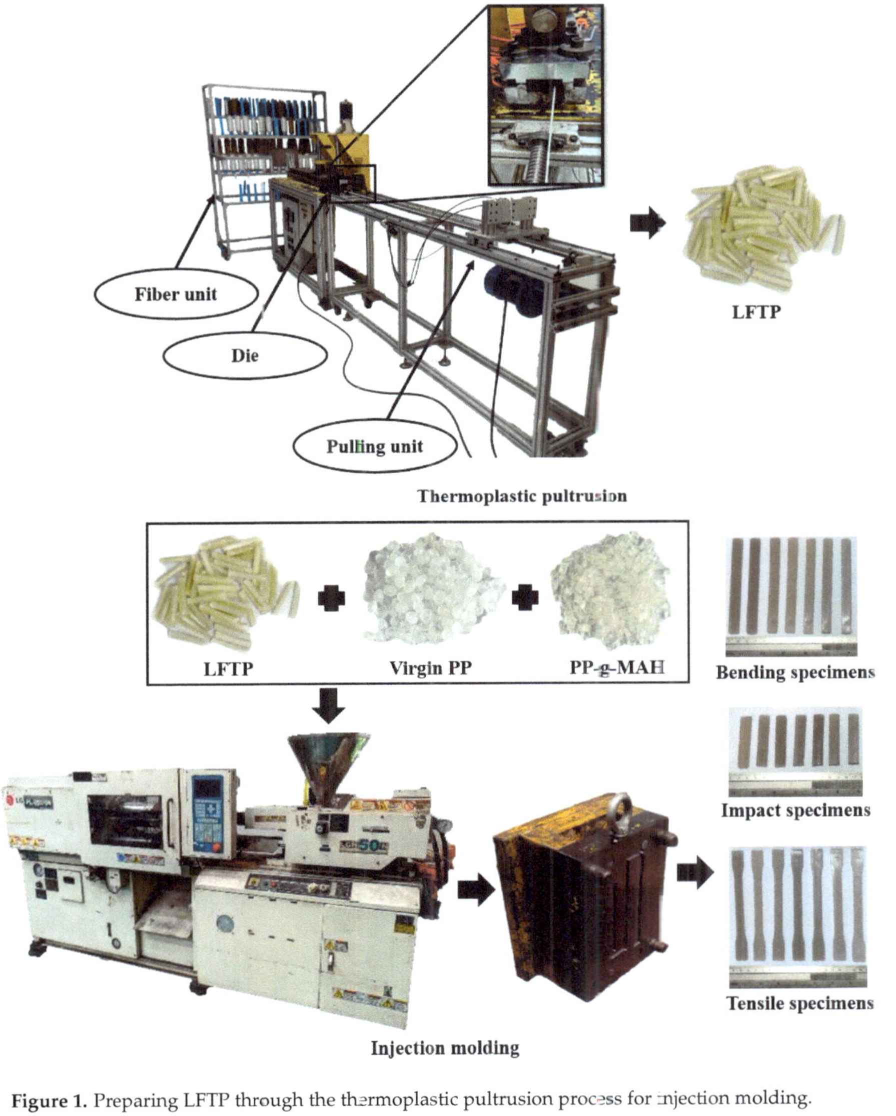

Figure 1. Preparing LFTP through the thermoplastic pultrusion process for injection molding.

2.3.2. Fiber Length Measurement

The fiber length of LFTPc was measured using specimens from injection molding. These specimens were placed on the black substrate of ceramic tiles and burned at a temperature of 500 °C for 2 h using a Nabertherm furnace. The decomposition of PP in LFTPc occurred due to the higher degradation temperature of PP, leaving only GF and ash. The GF was distributed, and the fiber length was measured using a microscope (Motic model SMZ-171) at 10× magnification. At least 1000 fibers were sampled and measured using the ImageJ program. The fiber count results were calculated by determining the number of fibers in each range of fiber length per total number of fibers. Figure 2 illustrates the fiber length measuring process of LFTPc.

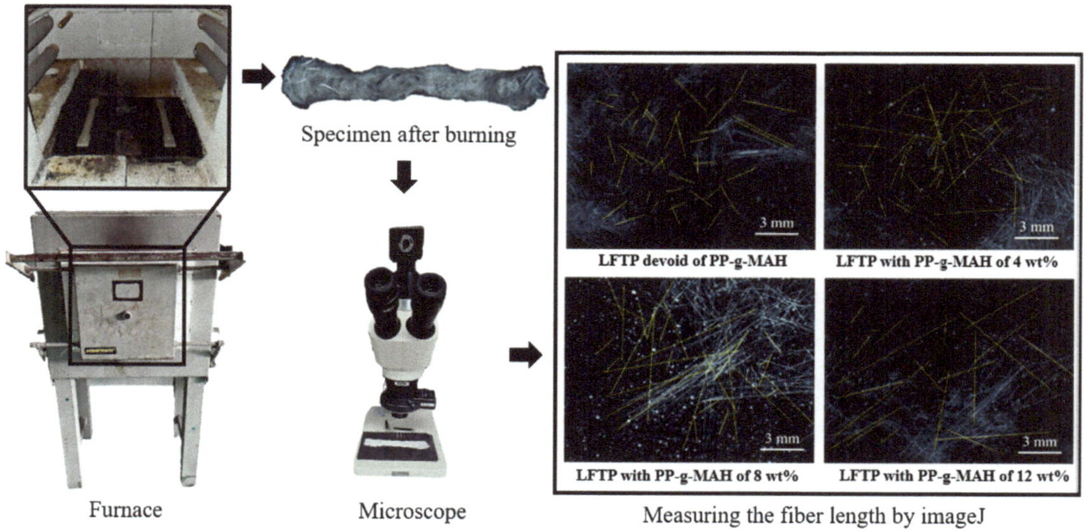

Figure 2. Fiber length measuring process of LFTPc.

2.3.3. FTIR Spectroscopy

The LFTPc specimens were examined for chemical composition using Fourier transform infrared spectroscopy (FTIR) measurement (using the Thermo Scientific Nicolet iS5, Waltham, MA, USA) to characterize their functional groups. The specimens were analyzed over a range of 4000 to 400 cm^{-1}.

2.3.4. Morphology

The surface morphology of LFTPc specimens was examined utilizing a scanning electron microscope (SEM), examining both the fiber pullout from the specimens and the interfacial adhesion between GF and PP (using the JEOL model JSM-5410LV, Tokyo, Japan). Specimens after the impact test were examined at 50× and 500× magnifications.

3. Results

3.1. The Influence of the Coupling Agent on the Mechanical Properties of the LFTPc

Both tensile modulus and tensile strength values are determined through tensile tests. The tensile modulus of LFTPc is illustrated in Figure 3 at varying concentrations of PP-g-MAH. Additionally, the characteristics of specimens during testing are depicted. It was found that the tensile modulus increased while the PP-g-MAH concentration was at 4 wt%, but it decreased when the PP-g-MAH concentration grew higher. For PP-g-MAH concentrations of 0, 4, 8, and 12 wt%, the tensile modulus values for the LFTPc were 2710, 3183, 2954, and 2830 MPa, respectively. The LFTPc, containing 4 wt% PP-g-MAH, exhibited a tensile modulus approximately 17.45% higher than that of the LFTPc devoid of PP-g-MAH, reaching its maximum tensile strength. Despite the decrease in tensile modulus beyond 4 wt% PP-g-MAH, these values remained greater than those of LFTP lacking PP-g-MAH.

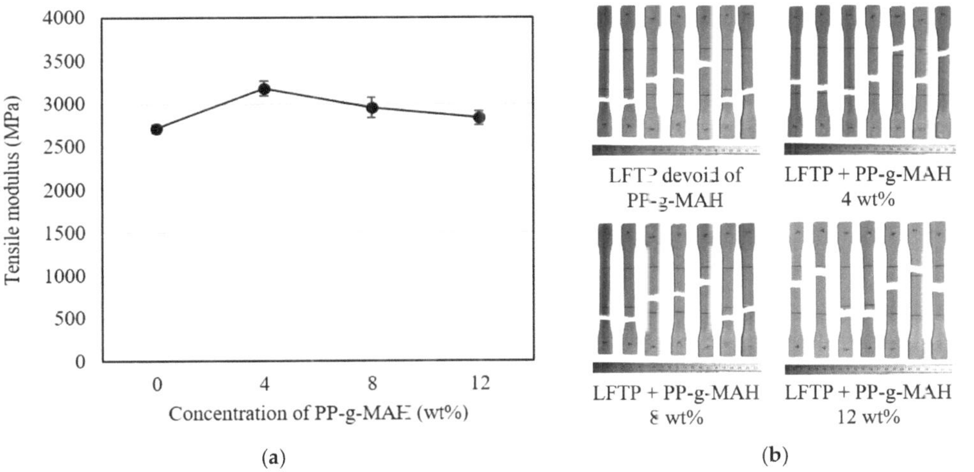

Figure 3. Correlation between (**a**) tensile modulus and the concentration of PP-g-MAH in LFTPc and (**b**) testing specimens.

Figure 4, which exhibits the results of employing various concentrations of PP-g-MAH, depicts the tensile strength of LFTPc. The results indicate that an increase in the amount of PP-g-MAH correlates with a related increase in the tensile strength. At PP-g-MAH concentrations of 0, 4, 8, and 12 wt%, the tensile strength values for the LFTPc were 38.92, 55.37, 59.84, and 61.98 MPa, respectively. The LFTPc with a 12 wt% PP-g-MAH component exhibited a tensile strength increase of approximately 59.24% compared to LFTPc devoid of PP-g-MAH.

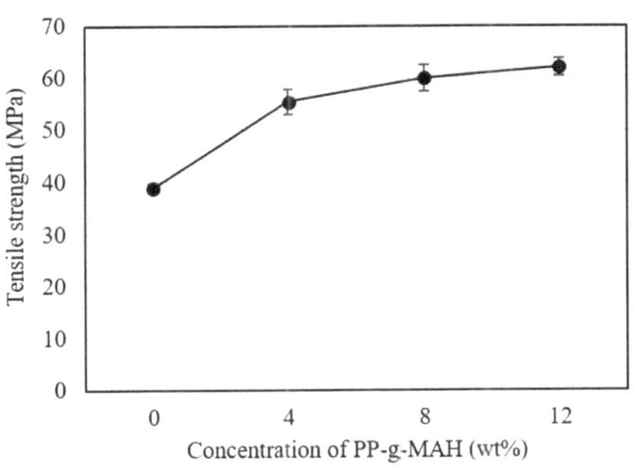

Figure 4. Correlation between tensile strength and the concentration of PP-g-MAH in LFTPc.

The results of testing specimens and the flexural modulus of the LFTPc with varying concentrations of PP-g-MAH are shown in Figure 5. Similar to the tensile modulus, the results indicate that the flexural modulus increased at a PP-g-MAH concentration of 4 wt% and reduced when the concentrations of PP-g-MAH surpassed 4 wt%. The flexural modulus values of LFTPc, with concentrations of PP-g-MAH at 0, 4, 8, and 12 wt%, were 1968, 2376, 2244, and 2146 MPa, respectively. The LFTPc containing a concentration of 4 wt% PP-g-MAH exhibited the greatest flexural modulus, showing an increase of approximately 20.73% compared to the LFTPc devoid of PP-g-MAH. In the same way that the flexural modulus values exhibited a reduction when the concentrations of PP-g-MAH surpassed 4 wt%, they consistently remained at levels superior to those of the LFTPc devoid of PP-g-MAH.

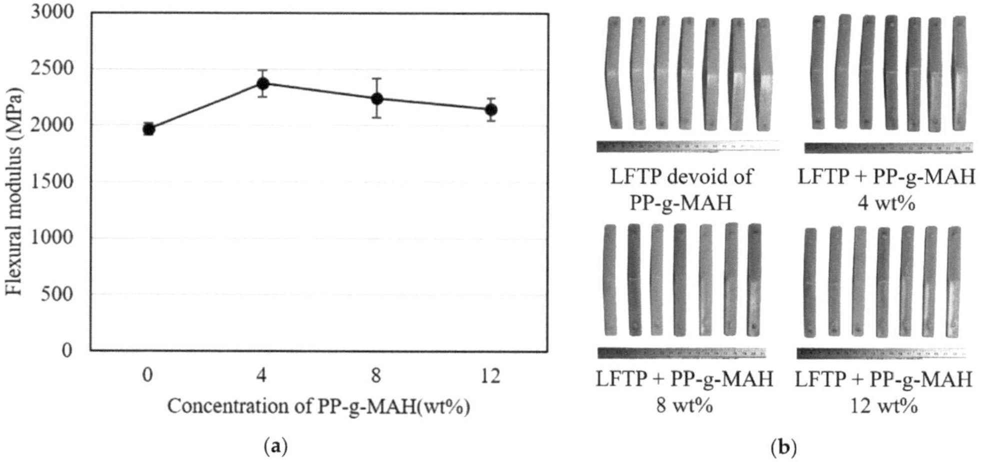

Figure 5. Correlation between (**a**) flexural modulus and the concentration of PP-g-MAH in LFTPc and (**b**) testing specimens.

Figure 6 presents an illustration of the flexural strength values of the LFTPc using various concentrations of PP-g-MAH. The results indicate that the flexural strength exhibited an increase in correlation with the concentration of PP-g-MAH. The respective flexural strengths of the LFTPc were obtained at 47.27, 68.16, 77.35 and 80.02 MPa, when the concentrations of PP-g-MAH were 0, 4, 8, and 12 wt%. The LFTPc containing 12 wt% PP-g-MAH demonstrated the maximum flexural strength values, an increase of approximately 69.66% in comparison to the LFTPc devoid of PP-g-MAH.

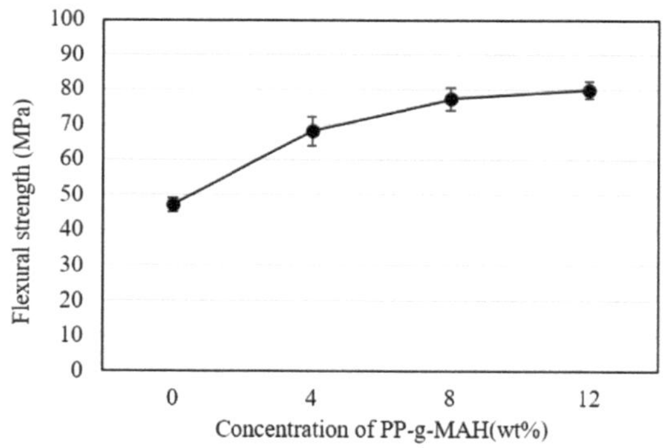

Figure 6. Correlation between flexural strength and the concentration of PP-g-MAH in LFTPc.

Figure 7 shows the impact strength of the LFTPc with varying concentrations of PP-g-MAH, along with features of the test specimens. Similar trends were observed in the impact strength data and the tensile and flexural modulus values. The impact strength increased at a concentration of 4 wt% PP-g-MAH and decreased beyond that point. The LFTPc exhibited impact strength values of 54.92, 82.57, 67.43, and 64.90 J/m, respectively, when PP-g-MAH was present in concentrations of 0, 4, 8, and 12 wt%. The LFTPc that incorporated 4 wt%

PP-g-MAH demonstrated the most significant increase in impact strength, approximately 50.34%, in comparison to the LFTPc that did not contain PP-g-MAH. While the impact property decreased with the concentration of PP-g-MAH surpassing 4 wt%, it continued to surpass that of the LFTPc devoid of PP-g-MAH. These differences corresponded with the tensile and flexural modulus values.

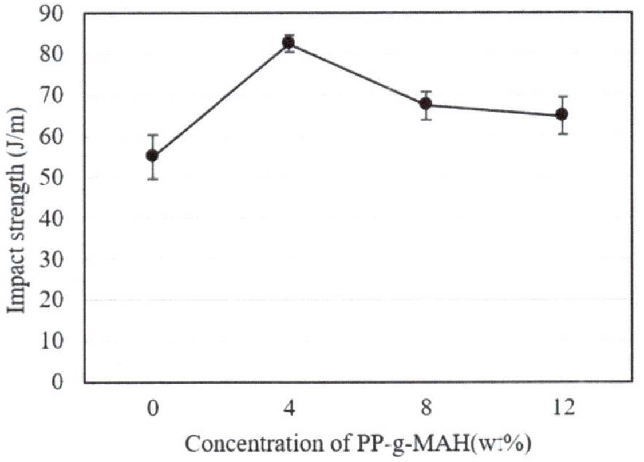

(a)

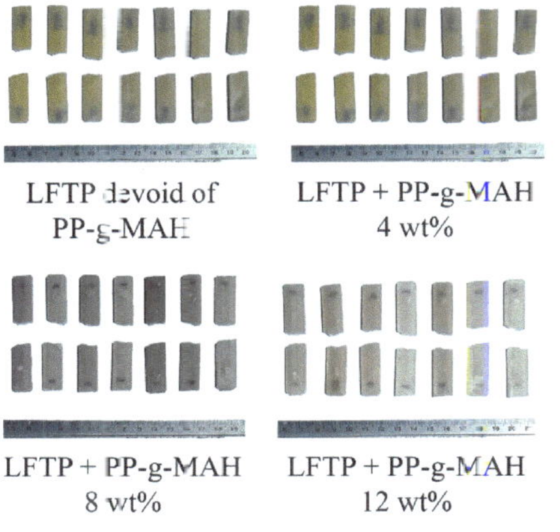

(b)

Figure 7. Correlation between (**a**) impact strength and the concentration of PP-g-MAH in LFTPc and (**b**) testing specimens.

The mechanical properties of the LFTPc were enhanced by the improved contact adhesion between PP and GF, which was caused by the incorporation of PP-g-MAH. Lin et al. and Fu et al. [27,36] elucidated the interaction between the hydroxyl groups of GF and the maleic anhydride of PP-g-MAH. The results align with the literature review [27,35,36,40], confirming the increased tensile modulus of the LFTPc when incorporated with PP-g-MAH.

The tensile modulus, flexural modulus, and impact strength values of the LFTPc, however, decreased at concentrations of PP-g-MAH beyond 4 wt% because, as explained by Lin et al. [27], PP-g-MAH can only enhance the interfacial adhesion between GF and PP; it cannot increase deformation resistance. Furthermore, a rise in the PP-g-MAH concentration results in a reduction in molecular weight [28]. The decrease in molecular weight affected the tensile strength and flexural strength, causing them to remain constant or decrease. In accordance with the literature, the elongation at break exhibited a rise as the molecular weight decreased [41]. Thus, when the amount of PP-g-MAH exceeded 4 wt%, the tensile and flexural modulus values decreased. Gumus [34] described the negative effects of PP-g-MAH on notched impact properties when the LFTPc incorporated a concentration of PP-g-MAH exceeding 4 wt%, resulting in decreased impact properties. In addition, the impact strength has a result in accordance with another study in the literature [36].

3.2. Fiber Length of LFTPc

The fiber length of GF decreased during the injection molding process due to the screw, cylinder, nozzle, and runner [35]. Figure 8 shows the residual fiber length of LFTPc with varying concentrations of PP-g-MAH. The results indicate that the fiber length of GF is distributed across the entire range, from 0.1 to 12 mm. The fiber length ranging from 0 to 2 mm exhibited the highest values, while the range of fiber length from 3 to 4 mm showed a decrease of only 5–15%. This finding suggests that LFTPc is reinforced by long fibers. Additionally, the percentage of fiber length in the range of 5 to 6 mm was 9–18%, providing clear evidence of reinforcement by long fibers. This is supported by the fact that short fiber-reinforced composites typically have fiber lengths in the range of 0.1–0.5 mm. [18]. This evidence confirms the reinforcement by long fibers.

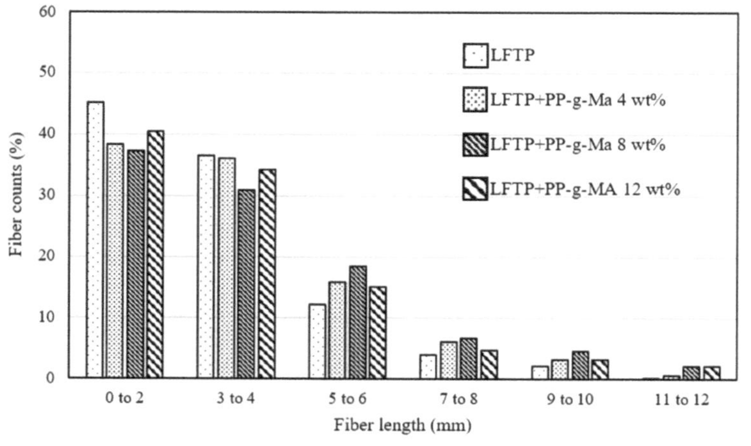

Figure 8. Residual fiber length of LFTP during the injection molding.

3.3. The Influence of the Coupling Agent on the Morphology of the LFTPc

Specimens were examined using SEM after the Izod impact test to investigate the surface morphology of the LFTPc with varying concentrations of PP-g-MAH, and the findings are shown in Figure 9. It was evident that the LFTPc's pull-out of GF and the spaces between them and PP was improved by the incorporation of PP-g-MAH. Figure 9b–e illustrates that the pull-out of fibers decreased as the concentration of PP-g-MAH increased. Consequently, the fiber pull-out and the interface between GF and PP were effectively improved by PP-g-MAH. Additionally, SEM images of the LFTPc containing varying concentrations of PP-g-MAH are displayed in Figure 10 at a 500× magnification. It was evident that as the amount of PP-g-MAH increased, the interstices between the GF and PP decreased. Furthermore, the incorporation of PP-g-MAH to the GF surface of the LFTPc resulted in an appearance of cohesive resin, which indicated a great interface between the GF and resin. Consequently, an improvement in the GF-PP interface and pull-out of fibers leads to an increase in mechanical characteristics. The images of the LFTPc with varying concentrations of PP-g-MAH corresponded with the literature review [27,36,40]. It effectively communicates that the mechanical properties of LFTPc were enhanced through the enhancement of interfacial adhesion between GF and PP, a result of the incorporation of PP-g-MAH.

3.4. FTIR Characterization

The spectra of virgin PP and the LFTPc with a varying PP-g-MAH content were examined at wave numbers ranging between 4000 and 400 cm^{-1}; these spectra are displayed in Figure 11. The FTIR spectra exhibit consistent peaks at 2949 and 2915 cm^{-1}, corresponding to aliphatic C-H stretching, and at 1457 and 1375 cm^{-1}, indicating C-H bending, which, respectively, demonstrate the presence of PP in the LFTPc [34,42–44]. The carbonyl group (C=O) of the anhydride and the acid groups, respectively, is represented by peaks at 1740 and 1714 cm^{-1} in the FTIR spectra of the LFTPc with varying concentrations of PP-g-MAH [44–46]. It is confirmed that maleic anhydride is present in the LFTPc containing PP-g-MAH. Therefore, the presence of PP-g-MAH in LFTPc improved interfacial adhesion, as observed in SEM micrographs. This effect was explained by Tselios et al. [47], with the reaction of MAH groups of PP-g-MAH with the hydroxyl groups of GF. The generated carboxylic groups have the ability to form hydrogen bonds with the hydroxyl groups on GF.

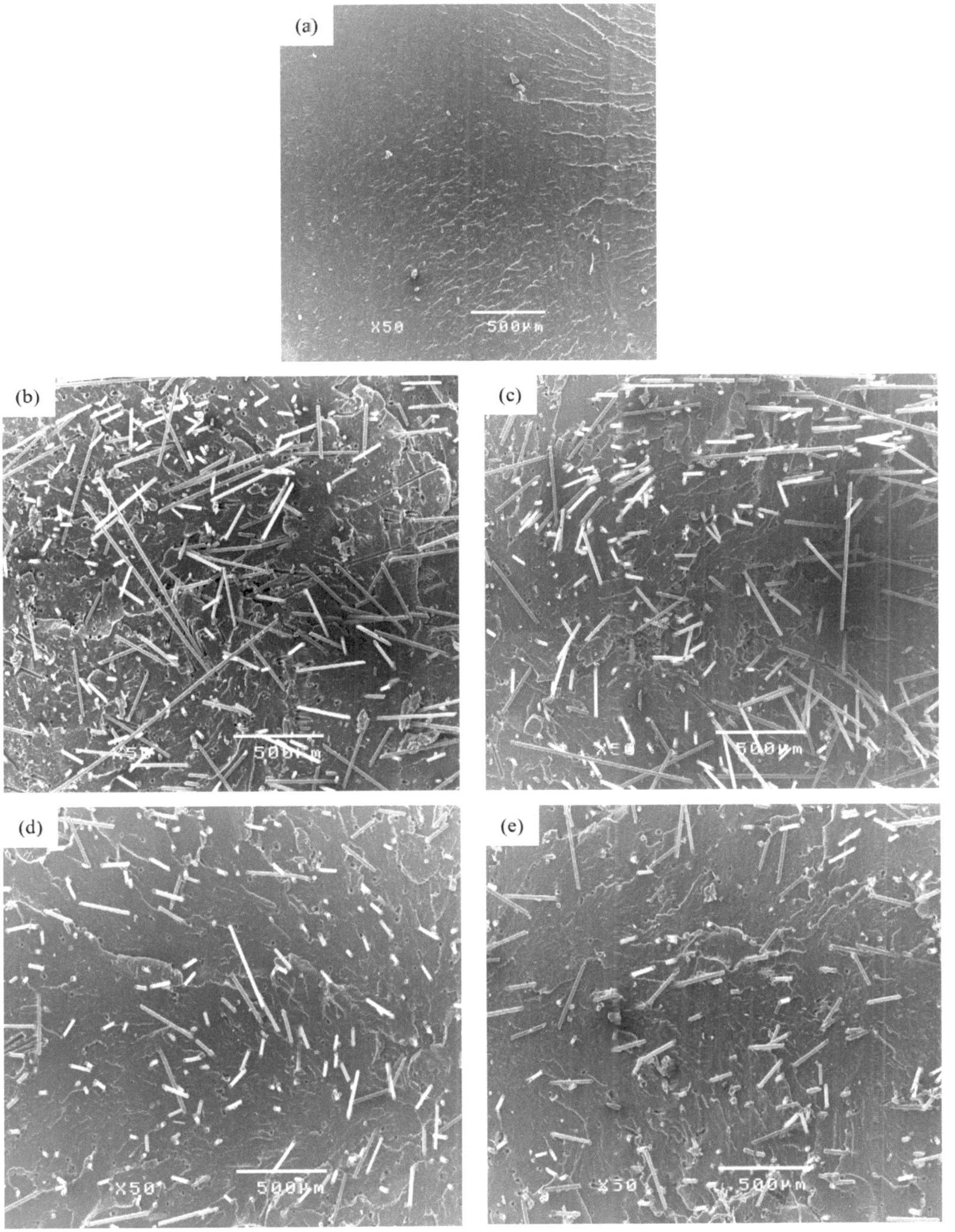

Figure 9. SEM micrographs at a magnification of 50× of (**a**) virgin PP, (**b**) LFTP devoid of PP-g-MAH, (**c**) LFTP with PP-g-MAH of 4 wt%, (**d**) LFTP with PP-g-MAH of 8 wt%, and (**e**) LFTP with PP-g-MAH of 12 wt%.

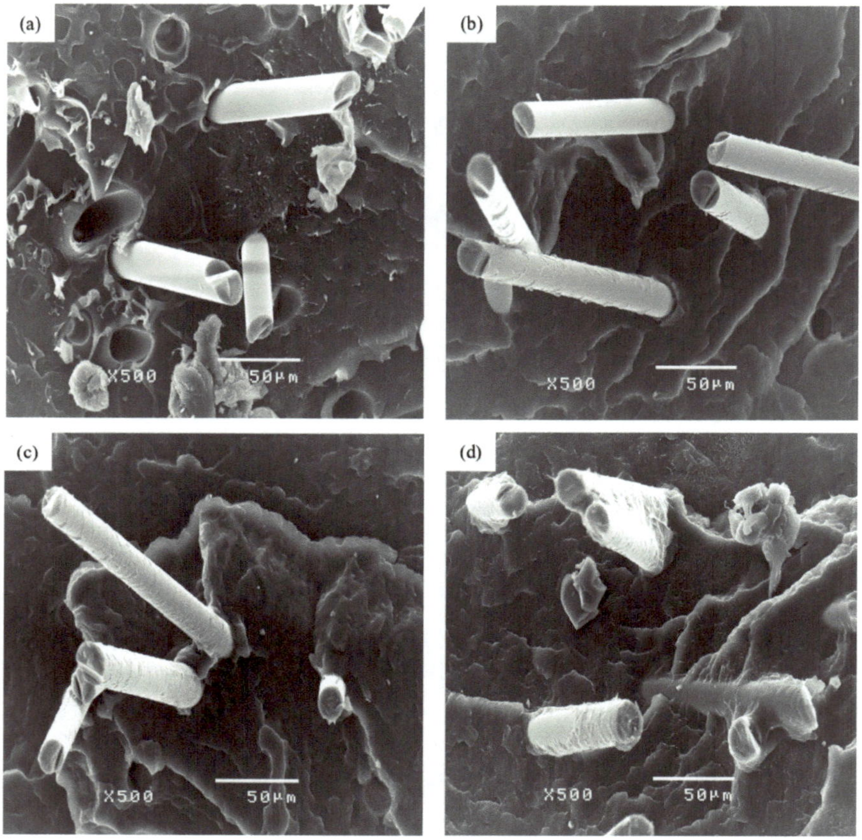

Figure 10. SEM micrographs at a magnification of 500× of (**a**) LFTP devoid of PP-g-MAH, (**b**) LFTP with PP-g-MAH of 4 wt%, (**c**) LFTP with PP-g-MAH of 8 wt%, and (**d**) LFTP with PP-g-MAH of 12 wt%.

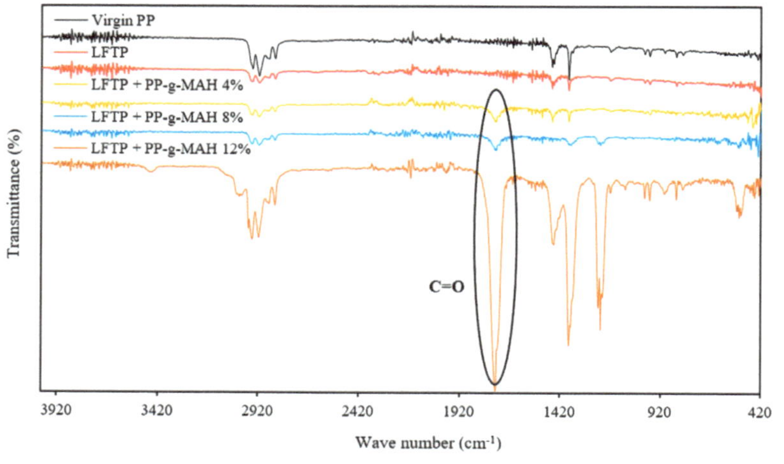

Figure 11. FTIR spectra of LFTPc with varying concentrations of PP-g-MAH.

3.5. A Comparison of Properties between LFTPc and Commercial LFT

The mechanical properties were conducted to compare with commercial LFTP at the same volume fraction of GF, which is 15 wt%, as depicted in Figure 12. The results clearly demonstrate that the mechanical properties of the LFTPc were better with the addition of PP-g-MAH. There was an approximate increase of 4.44–17.44% in the tensile modulus values, 42.25–59.26% in the tensile strength values, 9.07–20.72% in the flexural modulus values, 44.19–69.29% in the flexural strength values, and 18.16–50.33% in the impact property data. While the addition of PP-g-MAH into LFTP led to positive mechanical properties, exceeding the limit of PP-g-MAH addition will result in a reduction in impact strength, tensile modulus, and flexural modulus. Furthermore, the cost of PP-g-MAH incorporation into LFTPcs is taken into consideration. In comparison to the commercial LFTP product, the mechanical properties of the LFTPc devoid of PP-g-MAH have been found to be inferior. All mechanical property values in the LFTPc increased to levels that were comparable to commercial LFTP after PP-g-MAH was incorporated. The LFTPc containing filled PP-g-MAH component had a slightly greater tensile modulus, tensile strength, and flexural modulus values than commercial LFTP, by approximately 1–16.76%. The flexural strength values of the LFTPc incorporating PP-g-MAH content of 8 and 12 wt% were slightly higher than those of commercial LFTP, by approximately 7.17–10.87%. The impact strength values of the LFTPc containing a 4 wt% PP-g-MAH content were slightly greater compared to those of commercial LFTP by approximately 12.36%. Therefore, the LFTPc incorporated with a concentration of 4 wt% PP-g-MAH was considered optimal for manufacturing LFTP, as it exhibits nearly equivalent or higher mechanical properties compared to commercial LFTP. Although the flexural strength value was lower than those of commercial LFTP, it was only 5.57%.

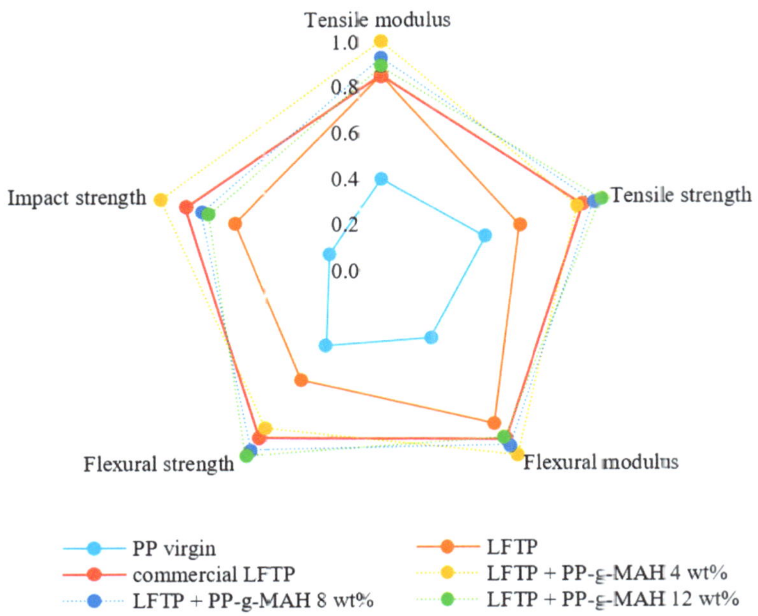

Figure 12. Comparison of properties between LFTPc and commercial LFT.

4. Conclusions

This study involved the incorporation of PP and GF into LFTP, a material produced using the thermoplastic pultrusion process. LFTP was incorporated with a varying PP-g-MAH content through injection molding for the preparation of specimens. The mechanical properties of injection specimens were investigated, including tensile, flexural, and impact tests. The fiber length of LFTPc was measured to confirm the long fiber reinforcement. The morphology was examined to analyze the interfacial interaction between GF and PP, while FTIR was employed to study the differences in the chemical structure after incorporating PP-g-MAH into LFTPc. The study can be summarized as follows:

- Significant improvements were observed in the tensile modulus, flexural modulus, and impact strength of the LFTPc when an amount of 4 wt% PP-g-MAH was incorporated. Nevertheless, with the PP-g-MAH concentration exceeding 4 wt%, a decline in these properties was observed.
- The amount of PP-g-MAH added to LFTPc resulted in increases in both tensile and flexural strength. The enhanced interfacial bonding between GF and PP, along with the reduction in fiber pull-out, is the reason for this improvement. Therefore, the optimal concentration of PP-g-MAH for the LFTPc was found to be 4 wt%. This choice resulted in mechanical properties nearly equivalent to those of commercial products.
- The fiber length results indicate long fiber reinforcement, as there is a significant amount of fiber in the range of 3 to 6 mm of fiber length.

This research is significant because it contributes to the advancement of LFTP product development employing the thermoplastic pultrusion technique. It presents an alternative approach distinguished by affordable machinery costs and a straightforward process for LFTP production. In addition, the mechanical properties of LFTP from thermoplastic pultrusion are nearly equivalent to or higher than those of commercial LFTP. Thermoplastic pultrusion is an advanced process that should be further developed. In future research, thermoplastic pultrusion will be used for manufacturing hybrid long fiber pellets to enhance mechanical properties and improve product performance.

Author Contributions: Conceptualization, A.M. and P.T.; methodology, A.M. and P.T.; software, A.M. and P.T.; validation, A.M. and P.T.; formal analysis, A.M. and P.T.; investigation, A.M. and P.T.; resources, A.M. and P.T.; data curation, A.M. and P.T.; writing—original draft preparation, A.M. and P.T.; writing—review and editing, A.M. and P.T.; visualization, A.M. and P.T.; supervision, A.M. and P.T.; project administration, A.M. and P.T.; funding acquisition, A.M. and P.T.; All authors have read and agreed to the published version of the manuscript.

Funding: This research received no external funding.

Data Availability Statement: The data presented in this study are available on request from the corresponding author.

Acknowledgments: The Department of Industrial Engineering, Faculty of Engineering, Rajamangala University of Technology Thanyaburi, Thailand, has been acknowledged by the authors for allowing permission to utilize its laboratory equipment for the purpose of this research.

Conflicts of Interest: The authors declare no conflicts of interest. The funders had no role in the design of the study; in the collection, analyses, or interpretation of data; in the writing of the manuscript; or in the decision to publish the results.

References

1. Ning, H.; Lu, N.; Hassen, A.A.; Chawla, K.; Selim, M.; Pillay, S. A review of Long fibre-reinforced thermoplastic or long fibre thermoplastic (LFT) composites. *Int. Mater. Rev.* **2019**, *65*, 164–188. [CrossRef]
2. Duncan, R.K.; Chen, X.G.; Bult, J.B.; Brinson, L.C.; Schadler, L.S. Measurement of the critical aspect ratio and interfacial shear strength in MWNT/polymer composites. *Compos. Sci. Technol.* **2010**, *70*, 599–605. [CrossRef]
3. Migneault, S.; Koubaa, A.; Erchiqui, F.; Chaala, A.; Englund, K.; Wolcott, M.P. Application of micromechanical models to tensile properties of wood–plastic composites. *Wood Sci. Technol.* **2011**, *45*, 521–532. [CrossRef]
4. Kim, S.-E.; Ahn, J.-G.; Ahn, S.; Park, D.-H.; Choi, D.-H.; Lee, J.-C.; Yang, H.-I.; Kim, K.-Y. Development of PA6/GF Long-Fiber-Reinforced Thermoplastic Composites Using Pultrusion and Direct Extrusion Manufacturing Processes. *Appl. Sci.* **2022**, *12*, 4838. [CrossRef]
5. Alwekar, S.; Ogle, R.; Kim, S.; Vaidya, U. Manufacturing and characterization of continuous fiber-reinforced thermoplastic tape overmolded long fiber thermoplastic. *Compos. Part B Eng.* **2021**, *207*, 108597. [CrossRef]
6. Markarian, J. Long fibre reinforced thermoplastics continue growth in automotive. *Plast. Addit. Compd.* **2007**, *9*, 20–24. [CrossRef]

7. Balaji Thattaiparthasarathy, K.; Pillay, S.; Ning, H.; Vaidya, U.K. Process simulation, design and manufacturing of a long fiber thermoplastic composite for mass transit application. *Compos. Part A Appl. Sci. Manuf.* **2008**, *39*, 1512–1521. [CrossRef]
8. Xie, T.; Yang, G. Interface and mechanical properties of poly(methyl methacrylate)-fiber composites. *J. Appl. Polym. Sci.* **2004**, *93*, 2478–2483. [CrossRef]
9. Krause, W.; Henning, F.; Tröster, S.; Geiger, O.; Eyerer, P. LFT-D—A Process Technology for Large Scale Production of Fiber Reinforced Thermoplastic Components. *J. Thermoplast. Compos. Mater.* **2003**, *16*, 289–302 [CrossRef]
10. Sheikh, M.R.; Hassan, A.; Yahya, R.; Mohd Isa, M.R.; Hussin, A.; Hornsby, P.R. Interfacial shear strength and tensile properties of injection-molded, short- and long-glass fiber-reinforced polyamide 6,6 composites. *J. Reinf. Plast. Compos.* **2011**, *30*, 1233–1242 [CrossRef]
11. Inoue, A.; Morita, K.; Tanaka, T.; Arao, Y.; Sawada, Y. Effect of screw design on fiber breakage and dispersion in injection-molded long glass-fiber-reinforced polypropylene. *J. Compos. Mater.* **2015**, *49*, 75–84. [CrossRef]
12. Kunc, V.; Frame, B.; Nguyen, B.; Tucker, C.; Velez-Garcia, G. Fiber length distribution measurement for long glass and carbon fiber reinforced injection molded thermoplastics. In Proceedings of the Automotive Composite Conference & Exhibition, Society of Plastics Engineers, Troy, MI, USA, 11–13 September 2007.
13. Van de Velde, K.; Kiekens, P. Thermoplastic pultrusion of natural fibre reinforced composites. *Compos. Struct.* **2001**, *54*, 355–360. [CrossRef]
14. Yang, C.; Wang, G.; Zhao, J.; Zhao, G.; Zhang, A. Lightweight and strong glass fiber reinforced polypropylene composite foams achieved by mold-opening microcellular injection molding. *J. Mater. Res. Technol.* **2021**, *14*, 2920–2931. [CrossRef]
15. Junaedi, H.; Baig, M.; Dawood, A.; Albahkali, E.; Almajid, A. Modeling analysis of the tensile strength of polypropylene base Short Carbon Fiber reinforced composites. *J. Mater. Res. Technol.* **2021**, *11*, 1611–1621. [CrossRef]
16. Rajak, D.K.; Wagh, P.H.; Linul, E. Manufacturing Technologies of Carbon/Glass Fiber-Reinforced Polymer Composites and Their Properties: A Review. *Polymers* **2021**, *13*, 3721. [CrossRef] [PubMed]
17. El-Ghaoui, K.; Chatelain, J.-F.; Ouellet-Plamondon, C. Effect of Graphene on Machinability of Glass Fiber Reinforced Polymer (GFRP). *J. Manuf. Mater. Process.* **2019**, *3*, 78. [CrossRef]
18. Tan, Y.; Wang, X.; Wu, D. Preparation, microstructures, and properties of long-glass-fiber-reinforced thermoplastic composites based on polycarbonate/poly(butylene terephthalate) alloys. *J. Reinf. Plast. Compos.* **2015**, *34*, 1804–1820. [CrossRef]
19. Hwang, D.; Cho, D. Fiber aspect ratio effect on mechanical and thermal properties of carbon fiber/ABS composites via extrusion and long fiber thermoplastic processes. *J. Ind. Eng. Chem.* **2019**, *80*, 335–344. [CrossRef]
20. Tipboonsri, P.; Memon, A. Optimizing thermoplastic pultrusion parameters for quality long fiber thermoplastic pellets in glass fiber-reinforced polypropylene. *Polym. Polym. Compos.* **2023**, *31*, 1–11. [CrossRef]
21. Tipboonsri, P.; Wattanahitsiri, V.; Memon, A. Long fiber thermoplastic pellets of glass fiber/polypropylene from pultrusion process. *J. Phys. Conf. Ser.* **2021**, *1719*, 012066. [CrossRef]
22. Minchenkov, K.; Vedernikov, A.; Safonov, A.; Akhatov, I. Thermoplastic Pultrusion: A Review. *Polymers* **2021**, *13*, 180. [CrossRef]
23. Tipboonsri, P.; Pramoonmak, S.; Uawongsuwan, P.; Memon, A. Optimization of Thermoplastic Pultrusion Parameters of Jute and Glass Fiber-Reinforced Polypropylene Composite. *Polymers* **2024**, *16*, 83. [CrossRef]
24. Memon, A.; Nakai, A. Mechanical Properties of Jute Spun Yarn/PLA Tubular Braided Composite by Pultrusion Molding. *Energy Procedia* **2013**, *34*, 818–829. [CrossRef]
25. Nordin, N.A.; Yussof, F.M.; Kasolang, S.; Salleh, Z.; Ahmad, M.A. Wear Rate of Natural Fibre: Long Kenaf Composite. *Procedia Eng.* **2013**, *68*, 145–151. [CrossRef]
26. Yadav, R.; Tirumali, M.; Wang, X.; Naebe, M.; Kandasubramanian, B. Polymer composite for antistatic application in aerospace. *Def. Technol.* **2020**, *16*, 107–118. [CrossRef]
27. Lin, J.-H.; Huang, C.-L.; Liu, C.-F.; Chen, C.-K.; Lin, Z.-I.; Lou, c.-w. Polypropylene/Short Glass Fibers Composites: Effects of Coupling Agents on Mechanical Properties, Thermal Behaviors, and Morphology. *Materials* **2015**, *8*, 8279–8291. [CrossRef]
28. Wenzhong, N. The effect of coupling agents on the mechanical properties of carbon fiber-reinforced polyimide composites. *J. Thermoplast. Compos. Mater.* **2015**, *28*, 1572–1582. [CrossRef]
29. Denault, J.; Vu-Khanh, T. Fiber/Matrix Interaction in Carbon/PEEK Composites. *J. Thermoplast. Compos. Mater.* **1993**, *6*, 190–204. [CrossRef]
30. Chen, M.; Wan, C.; Zhang, Y.; Zhang, Y. Fibre Orientation and Mechanical Properties of Short Glass Fibre Reinforced PP Composites. *Polym. Polym. Compos.* **2005**, *13*, 253–262. [CrossRef]
31. Nayak, S.K.; Mohanty, S. Sisal Glass Fiber Reinforced PP Hybrid Composites: Effect of MAPP on the Dynamic Mechanical and Thermal Properties. *J. Reinf. Plast. Compos.* **2010**, *29*, 1551–1568. [CrossRef]
32. Abd Rahman, N.M.M.; Hassan, A.; Yahya, R. Extrusion and injection-molding of glass fiber/MAPP/polypropylene: Effect of coupling agent on DSC, DMA, and mechanical properties. *J. Reinf. Plast. Compos.* **2011**, *30*, 215–224. [CrossRef]
33. Kim, H.-S.; Lee, B.-H.; Choi, S.-W.; Kim, S.; Kim, H.-J. The effect of types of maleic anhydride-grafted polypropylene (MAPP) on the interfacial adhesion properties of bio-flour-filled polypropylene composites. *Compos. Part A Appl. Sci. Manuf.* **2007**, *38*, 1473–1482. [CrossRef]
34. Gümüş, B. Effect of MA-g-PP addition on mechanical properties of polypropylene/hollow glass spheres/nanoclay composites. *Polym. Bull.* **2022**, *80*, 3405–3422. [CrossRef]

35. Kim, Y.; Park, O.O. Effect of Fiber Length on Mechanical Properties of Injection Molded Long-Fiber-Reinforced Thermoplastics. *Macromol. Res.* **2020**, *28*, 433–444. [CrossRef]
36. Fu, X.; He, B.; Chen, X. Effects of Compatibilizers on Mechanical Properties of Long Glass Fiber-Reinforced Polypropylene. *J. Reinf. Plast. Compos.* **2009**, *29*, 936–949. [CrossRef]
37. *ASTM D638-14*; Standard Test Method for Tensile Properties of Plastics. ASTM International: West Conshohocken, PA, USA, 2020.
38. *ASTM D790-10*; Standard Test Method for Flexural Properties of Unreinforced and Reinforced Plastics and Electrical Insulating Materials. ASTM International: West Conshohocken, PA, USA, 2016.
39. *ASTM D256-10*; Standard Test Methods for Determining the Izod Pendulum Impact Resistance of Plastics. ASTM International: West Conshohocken, PA, USA, 2015.
40. Wong, K.H.; Syed Mohammed, D.; Pickering, S.J.; Brooks, R. Effect of coupling agents on reinforcing potential of recycled carbon fibre for polypropylene composite. *Compos. Sci. Technol.* **2012**, *72*, 835–844. [CrossRef]
41. Tharanikkarasu, K.; Kim, B. Modification of aqueous polyurethane dispersions via tetraphenylethane iniferters. *J. Appl. Polym. Sci.* **1999**, *73*, 2993–3000. [CrossRef]
42. Abbasian, M.; Ghaeminia, H.; Jaymand, M. A facile and efficient strategy for the functionalization of multiple-walled carbon nanotubes using well-defined polypropylene-grafted polystyrene. *Appl. Phys. A* **2018**, *124*, 522. [CrossRef]
43. Lian, Z.; Xu, Y.; Zuo, J.; Qian, H.; Luo, Z.; Wei, W. Preparation of PP-g-(AA-MAH) Fibers Using Suspension Grafting and Melt-Blown Spinning and its Adsorption for Aniline. *Polymers* **2020**, *12*, 2157. [CrossRef]
44. Burgada, F.; Fages, E.; Quiles-Carrillo, L.; Lascano, D.; Ivorra-Martinez, J.; Arrieta, M.P.; Fenollar, O. Upgrading Recycled Polypropylene from Textile Wastes in Wood Plastic Composites with Short Hemp Fiber. *Polymers* **2021**, *13*, 1248. [CrossRef]
45. Fan, M.; Zhang, Y.; Li, X.; Zeng, B.; Chen, S.; Zhu, W.; Wang, S.; Xu, J.; Feng, N. Facile synthesis and applications of polypropylene/polydimethylsiloxane graft copolymer. *Polym. Adv. Technol.* **2019**, *30*, 1226–1233. [CrossRef]
46. Martinez, G.; Sanchez, S.; Ramos, L.; Perez, O.; Ramírez, E.; Benavides-Cantú, R.; Avila-Orta, C.; Cruz, V.; Mata, J.; Lozano-Ramirez, T.; et al. Aniline-Modified Polypropylene as a Compatibilizer in Polypropylene Carbon Nanotube Composites. *Polym.-Plast. Technol. Eng.* **2017**, *57*, 1360–1366. [CrossRef]
47. Tselios, C.; Bikiaris, D.; Savidis, P.; Panayiotou, C.; Larena, A. Glass-fiber reinforcement of in situ compatibilized polypropylene/polyethylene blends. *J. Mater. Sci.* **1999**, *34*, 385–394. [CrossRef]

fibers

MDPI

Article

Nanomaterial-Enhanced Sizings: Design and Optimisation of a Pilot-Scale Fibre Sizing Line

Dionisis Semitekolos [1], Ioannis Papadopoulos [1], Stavros Anagnou [1], Behnam Dashtbozorg [2], Xiaoying Li [2], Hanshan Dong [2] and Costas A. Charitidis [1,*]

1 Research Lab of Advanced, Composite, Nano-Materials and Nanotechnology (R-NanoLab), School of Chemical Engineering, National Technical University of Athens, 9 Heroon Polytechniou, 15773 Athens, Greece; diosemi@chemeng.ntua.gr (D.S.); papad67@chemeng.ntua.gr (I.P.); sanagnou@chemeng.ntua.gr (S.A.)

2 School of Metallurgy & Materials, University of Birmingham, Birmingham B15 2TT, UK; b.dashtbozorg@bham.ac.uk (B.D.)

* Correspondence: charitidis@chemeng.ntua.gr; Tel.: +30-210-7724046

Abstract: This study focuses on the development of a pilot-scale sizing line, including its initial design and installation, operational phases, and optimization of key process parameters. The primary objective is the identification of critical parameters for achieving a uniform sizing onto the fibres and the determination of optimal conditions for maximum production efficiency. This investigation focused on adjusting the furnace desizing temperature for the removal of commercial sizing, adjusting the drying temperature, as well as optimizing the corresponding residence time of carbon fibres passing through the furnaces. The highest production rate, reaching 1 m sized carbon fibres per minute, was achieved by employing a desizing temperature of 550 °C, a drying temperature of 250 °C, and a residence time of 1 min. Furthermore, a range of sizing solutions was investigated and formulated, exploring carbon-based nanomaterial types with different surface functionalizations and concentrations, to evaluate their impact on the surface morphology and mechanical properties of carbon fibres. In-depth analyses, including scanning electron microscopy and contact angle goniometry, revealed the achievement of a uniform coating on the carbon fibre surface, leading to an enhanced affinity between fibres and the polymeric epoxy matrix. The incorporation of nanomaterials, specifically N_2-plasma-functionalized carbon nanotubes and few-layer graphene, demonstrated notable improvements in the interfacial shear properties (90% increase), verified by mechanical and push-out tests

Keywords: sizing; carbon fibre; fibre matrix interface; push-out test

1. Introduction

Carbon-fibre-reinforced polymers (CFRPs) are an essential part of present and future materials. They are among the most widely employed materials in several sectors (e.g., aerospace engineering, automotive industry, leisure and sports industry) due to their enhanced mechanical and physical properties. This can be mainly attributed to carbon fibres' (CFs) highly ordered graphite crystal structure and their ability to absorb and redistribute mechanical loads imposed on the composites [1–3]. Some unique properties of CFRPs include their high specific tensile strength and stiffness, high strength-to-weight ratio, corrosion and thermal resistance [4,5]. The properties of these composite materials depend also on the chosen matrix, as well as on the degree of interaction at the interface between the fibre and matrix [6]. Nowadays, research is focused on the development of novel CFRPs with improved functionalities and the investigation of matrix-CF-tailored interactions for optimum adhesion [7].

Despite the outstanding individual characteristics of CFs, several problems arise during the manufacturing process of CFRPs (e.g., brittle nature of CFs can lead to difficulties or handling concerns). One main issue is the lack of reactive functional groups on the surface of CFs and the subsequent poor adhesion to the polymer matrix, which negatively affects the overall performance of composites (e.g., displaying of relatively poor interlaminar shear strength) and sets limitations on their applications [3,6,8].

Various studies have focused on modifying the CFs' surface to enhance and improve the adhesion between CFs and the polymer matrix. These include electrochemical treatment, plasma treatment, chemical grafting, surface roughness adjustment, the addition of carbon nanotubes (e.g., CVD/microwave-produced CTNs) or nanoparticles for chemical activation of the surface, and sizing [4,8,9]. However, not all the above methods can improve the interface bonding effectively and, in addition, some of these cannot be practically applied in industrial production, due to many problems such as fibre damage or excessive cost [10].

Sizing treatment is considered to be a simple cost-effective process and, during this process, a thin, homogenous polymeric layer is formed on the surface of the CFs (in most cases of commercial carbon fibres, the thickness of this layer is about 100 nm) [11]. The sizing formula includes one or several polymeric compounds, a coupling agent, a lubricant, and several additives (plasticizers, adhesion promoters, rheology modifiers, etc.). The mixture is commonly diluted in water to form an aqueous solution, which can be either an emulsion or a dispersion, in which CFs are immersed for a short period. Afterwards, CFs are subjected to mild heat treatment to dry and remove excess solvent [12]. The result of this process is the formation of a coating on the surface of CFs and thus the modification of the CF's surface in terms of physical and chemical characteristics, without causing damage [4]. Lastly, a new trend is observed in current research, which involves the incorporation of nanomaterials in the sizing agent, such as carbon nanotubes (CNTs) and nanoparticles of various morphologies and dimensions (e.g., zirconium dioxide, graphene oxide/CNTs), due to their prominent functional, physicochemical, and mechanical properties [2,13,14].

This study focuses on the design, installation, and operation of a pilot-scale CF sizing line. The main goal is to study all parameters for applying a uniform coating on the CFs' surface under optimum conditions, and subsequently the production of fibres with enhanced mechanical properties and multifunctionality, through the incorporation of nanomaterials in sizing solutions.

2. Materials and Methods

2.1. Materials

For this study, a 6k CF (Toray, New York, NY, USA) was used. Physical and mechanical properties of CFs are presented in Table 1 below:

Table 1. Technical information of Torayca T700S.

	TORAYCA T700S
Tensile Strength (MPa)	4.900
Tensile Modulus (GPa)	230
Strain (%)	2.1
Density (g/cm^3)	1.80
Filament Diameter (μm)	7 μm

Commercial sizing solution, Hydrosize® HP2-06 (Michelman, Aubange, Belgium), was used to size the CFs (Table 2), with or without the addition of functionalized nanoparticles (multiwalled carbon nanotubes—MWCNTs; few-layer graphene—FLG). HP2-06 is an anionic/nonionic phenoxy aqueous dispersion designed for use as a fibre sizing agent that enhances compatibility between fibres and matrix, resulting in better mechanical performance of composites. The addition of nanomaterials can have a positive effect on those properties, due to incorporation of stiff nanoparticles that create stress concentration areas and lead to additional energy absorption [15].

Table 2. Datasheet of Michelman's Hydrosize® HP2-06.

Physical Properties	
pH	6.5–8.5
Emulsifier Charge	Amine-dispersed
Percent Non-Volatile (%)	24.5–26.5
Recommended pH Range	6.5–8.5
Brookfield Viscosity Range (cps)	<2000
Appearance	White emulsion

Functionalized CNTs and FLGs were incorporated in the sizing solutions, which were provided by Haydale Ltd. (UK) [16]. Haydale performs functionalization of nanomaterials by utilizing their patented HDPlas plasma functionalization process. For this case study, O_2, N_2, NH_3 process gases were dissociated into their compartment parts by applying an electrical potential that subsequently bombards the nanomaterials, producing chemical groups at their surface. The effect of plasma functionalization is analysed in the Supplementary File [17–20]

To assess the fibre–resin affinity, the SR1710/SD8822 structural epoxy system from Fibremax Composites (Greece) was used (Table 3). SR1710/SD8822 is a two-component epoxy system that cures at 25 °C for 24 h and post-cures at 40 °C for an additional 24 h.

Table 3. Datasheet of SR1710 injection/SD8822.

Technical Properties	
Modulus of elasticity (GPa)	3.65
Elongation at break (%)	2.2
Flexural Strength (MPa)	115
Charpy impact strength (KJ/m^2)	17
Shear Strength (MPa)	53
Glass Transition Temperature (°C)	67
Tensile Strength (MPa)	70

2.2. Design and Description of Pilot Line

In industry, CF sizing is a continuous process; thus, a pilot-scale continuous sizing line had to be designed for this study. The core compartments of the line were identified, and an initial sketch was drawn. IZUMI International, a company with experience in fibre line manufacturing, was selected for the line construction. The compartments were digitalized with use of SolidWorks, a CAD desktop system that provides product-level automated design tools [21]. A model was proposed that included all the core compartments and extra parts that compose the fibre sizing line (Figure 1). The latter are analysed below.

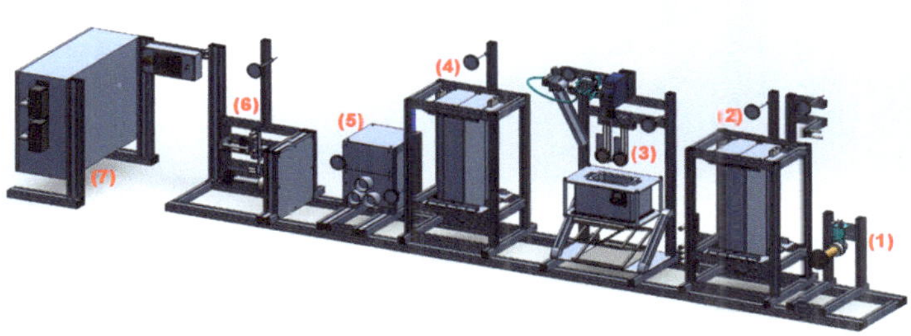

Figure 1. Fibre sizing line designed at SolidWorks.

For the support frame, on which these compartments were mounted, aluminium was selected. Aluminium frame, known for its cost-effectiveness, durability, and lightweight properties, provides an optimal structural foundation for the sizing line. For the line compartments, stainless steel was the material of choice, serving multiple purposes. Its corrosion-resistant properties are particularly crucial in areas exposed to the aqueous sizing solution and traversed by the CFs. Additionally, stainless steel facilitates the smooth movement of CFs through the rollers without causing damage. Inner parts of furnaces were made of ceramic material. This material was selected for its ability to endure high temperatures and exhibit a low thermal expansion coefficient, making it well suited for the demanding conditions inherent in the sizing process. In terms of general design considerations, efforts were made to maintain adequate distances between each compartment. This design choice was driven by the aim of facilitating easy handling and maintenance for operators, emphasizing user-friendly principles throughout the engineering of fibre sizing line.

1. The let-off tension creel's role is to feed the mounted fibre into the desizing unit and, in cooperation with the take-up winder, to provide the required tension (analysed below).
2. The desizer furnace is responsible for removal of the already existing sizing on a commercial fibre. The temperature range is wide and covers temperatures from 300 °C up to 600 °C, enabling the user to remove every unnecessary coating.
3. The fibre sizing bath consists of bath rollers, sizing bath, squeeze rollers, and an overhead stirring system. Bath rollers guide the desized fibre through the sizing bath where a solution (aqueous is the most used for CFs [22]) with users' desired composition (solids in various concentrations) coats the fibre. The overhead steering rotor is submerged in the solution to avoid sedimentation. Subsequently, squeeze rollers remove the excess solution from the fibre so that a uniform coating is achieved.
4. As the fibre passes through the squeeze rollers and the excess is removed, fibre drying heater evaporates the remaining solvent and slightly solidifies the coating around the fibre. The temperature does not exceed 300 °C, to ensure that the coating remains intact. This process ensures that the through-thickness fibre entanglements during winding are avoided and the fibre preform is diminished (fibre needs to be fully dried before winding so that it does not retain its cylindrical shape during its unwinding).
5. The feed roller system is used to pull the fibre from the let-off creel. Its speed is adjusted by the user at the control panel (min 0.2 m/min–max 2 m/min). It is a key parameter since it affects the production rate and controls the residence time of fibre in the desizing and drying unit.
6. The take-up winder collects the fibre with a mechanical traverse system, in which the spindle is driven by a constant-torque motor. High tension is not required for fibre winding, but a specific ratio between the tension of the feed roller inlet and outlet needs to be followed. Tension is adjusted by changing the torque on control panel.
7. The power control unit supplies electric current to the whole line.

2.3. Run and Evaluation of Parameters

One of the most important aspects of this study was to pinpoint the parameters that optimize the process. For this purpose, excessive runs were completed to determine these optimum conditions (Table 4).

- Desizer framework

Every commercially available carbon fibre is coated with a sizing agent that protects the fibre from being damaged during transport, facilitates handling during operation, and has a small effect on the fibre–resin adhesion. To assess the sizing solutions developed within this work, the already existing sizing had to be removed. The desizing furnace temperature in combination with the line's operation speed were investigated. Six different temperatures for the desizer were chosen for the complete removal of commercial carbon fibre sizing, starting from 350 °C to 600 °C with a step of 50 °C, while the line's operation speed varied from 0.20 to 1 and 2 m/min.

Table 4. Parameter investigation per station.

	Furnaces		Sizing Solution		Winding System
	Desizer	Dryer	Solid Concentration	Nanomaterial Concentration	Operation speed
Parameter Investigation	Temperature (°C)	Temperature (°C)	Nanomaterial/Solid content (% wt)	Weight ratio (% wt)	Residence time in furnaces (min)

- Dryer framework

Another important aspect as described in Section 2.2 is the effective drying of resized fibre. Similarly with the aforementioned method, three different temperatures (200, 250, 300 °C) and operation speeds (0.2, 1, 2 m/min) were tested.

- Solid content concentration

The sizing solutions developed within this work were water dispersions with solids (HP2-06 solids and CNTs and/or FLGs). Based on previous results and the literature review [23–25], a nanomaterial content of 0.05, 0.1, and 0.25% wt with total solid contents (HP2-06 solids + CNTs/FLGs) of 1, 2.5, and 5% wt were examined. To produce steady aqueous solutions with HP2-06 and CNTs/FLGs, the addition of surfactants along with a sonication process had to take place.

Initially, the goal was to identify the solid concentration (without nanomaterials) that had the most uniform distribution on the fibre's surface without any excess.

- Nano-enhanced sizings

As mentioned above, two different nanomaterials were examined, FLGs and CNTs in different weight ratios (0.05, 0.1, and 0.25% wt) with various modifications (plasma modification with O_2, N_2, and NH_3), resulting in 18 unique cases. The selection criteria focused on surface morphology, fibre–resin affinity, and the resulting mechanical properties.

2.4. Characterization Methods

2.4.1. Scanning Electron Microscopy (SEM) and Thermogravimetric Analysis (TGA)

The evaluation of different sizing solution parameters (solid content/nanomaterial type) on the carbon fibre's surface was studied using an FEI Quanta 650 FEG SEM (FEI, Hillsboro, OR, USA), in magnifications up to ×10,000.

The commercial sizing that was removed during the desizing stage was evaluated also by thermogravimetric analysis using a TGA apparatus (STA 449 F5 Jupiter, Hamburg, Germany). Instrument calibration for temperature and sensitivity was performed prior to testing. Two sets of tests were performed with 16 and 24 mg of carbon fibres at a heating rate of 20 °C/min in N_2 atmosphere (50 mL/min) from room temperature to 750 °C, and mass change was measured as a function of temperature. The analysis was performed in accordance with ISO 11358 [26].

2.4.2. Contact Angle Goniometry (CAG)

The modified fibre–matrix affinity was assessed by testing single fibres, aiming to evaluate the modification's effectiveness at microscopic level. The fibres were coated with epoxy resin type SR1710/SD8822. Each isolated fibre was placed on a Plexiglas holder. Once positioned, the fibres were coated with resin droplets using a micro-pipette and then examined by optical microscopy through a Zeiss Axio Imager A2 microscope (Carl Zeiss Microscopy, Ostfildern, Germany). For each specimen, 10 microdroplets were measured to calculate the average contact angle. Zeiss Axiovision imaging processing program was utilized to measure the inner contact angles formed between droplets and fibres. The measured angle was used as an indicator of improved wettability, with larger angles indicating enhanced wetting capabilities [27,28].

2.4.3. Mechanical Tests

For the mechanical properties, 3-point bend (3PB), tensile, and push-out tests were carried out. Regarding 3PB, the specimen was mounted between two supporting pins as loading pin force was increased progressively in the middle of the specimen, thus causing it to break. The used device was the SAUTER FH 500 with maximum load of 500 N and minimum load of 0.1 N. The loading pin's speed was set at 3 mm/min.

To calculate flexural strength, the following formula was used:

$$\sigma_f = (1.5 \times F \times L)/(b \times d^2) \tag{1}$$

σ_f: flexural strength (MPa);
F: load at a given point on the load deflection curve (N);
L: support span (mm);
b: width of test beam (mm);
d: depth of tested beam (mm).

The tensile strength test was performed according to ASTM D 4018 [29]. The used specimens were tabbed resin-impregnated and consolidated fibre bundles. A universal tensile machine TE Forcespeed/Jinan WDW Series (TE Forcespeed Corporation, building 4-B-3, No.5577, Industrial North Road, Jinan, 250109, China) was used with a load cell of 5 kN.

To calculate tensile strength, the following formulas were used:

$$MUL = W_1/L \tag{2}$$

MUL: mass per unit length (g/m);
W_1: mass of the specimen (g);
L: length of the specimen (m).

$$UTS = P \times \rho_f/MUL \tag{3}$$

UTS: ultimate tensile strength (MPa);
P: maximum load measured in tensile test (N);
ρ_f: fibre density (g/cm^3);
MUL: mass per unit length (g/m).

2.4.4. Push-Out Test

The interfacial shear strength (IFSS) of resin-embedded fibres was assessed using single-fibre push-out testing, where the load–displacement curve was measured when a single fibre was successfully pushed out from a resin matrix disc (of known thickness). To accurately and repeatably perform single-fibre push-out testing, thin (<55 μm) composite samples (with fibres embedded perpendicular to the polished surface) were prepared. The composite's matrix consisted of a mixture of EPIKOTE™ Resin MGS RIMR 135 resin and 4-aminophenyl disulfide hardener in a ratio of 100 g (resin) to 55.051 g (hardener). The hardener was initially melted by heating at 80–85 °C (SNOL 3/1100 muffle furnace) before the resin was added. The fibres were then held under tension within silicone rubber moulds (1-inch diameter, Agar Scientific Ltd., Stansted, UK) before being embedded in the resin and hardener mixture. The composite mixture was then cured by heating at 130 °C for 1.5 h (SNOL 3/1100 muffle furnace). Thin discs (~1 mm) were abrasion-cut from initial composite and then ground to reduce disc thickness to a few hundred microns, using #4000 grit SiC abrasive paper. The surface of the thin slices was polished to a mirror finish (Struers OP-S) and then attached (using wax) onto a GATAN TEM sample disc grinder. The other surface was used to thin the composite discs to an approximate thickness of 30–50 μm (using #4000 grit SiC paper and 1 µm diamond suspension). A final polish (Struers OP-S) of both surfaces of the composite discs was carried out and then assessed using optical imaging (Figure 2). The final composite discs were then glued (Loctite Superglue Precision) to a metal support with 20 µm wide grooves (which provide the necessary depth needed for pushing out the fibres) produced by femtosecond-laser fabrication (Figure 3).

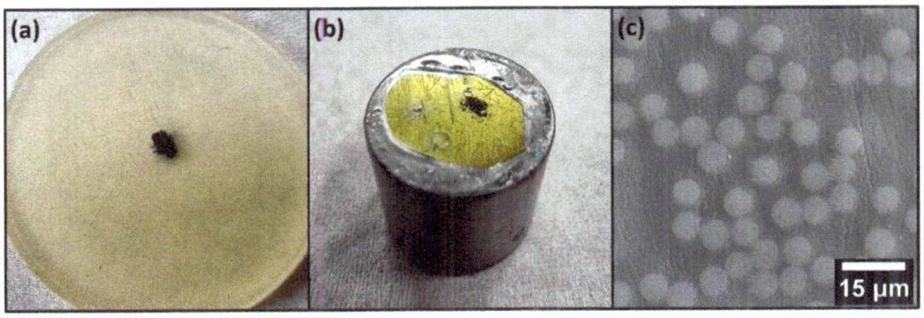

Figure 2. Sample preparation steps for single-fibre push-out testing of composite discs: (**a**) thin slice cut from original composite with CFs in middle, (**b**) specimen mounted to stub during final thinning stage, (**c**) electron microscope image of composite surface after final polishing.

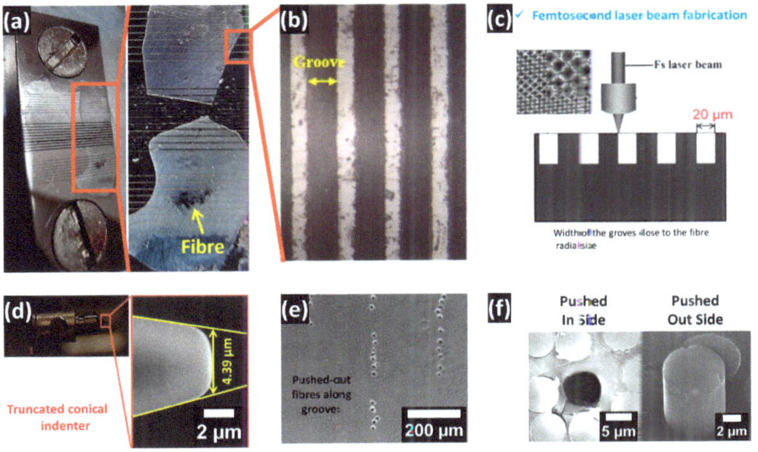

Figure 3. Push-out test sample holder with attached composite slices (**a**), optical microscope image of the fabricated grooves on the sample holder (**b**), overview of the femtosecond laser fabrication of the grooves (**c**), overview of the shape of the push-out test conical indenter (**d**), electron microscope image of pushed-out fibres on thinned composite slices along the fabricated grooves (**e**), and magnified view of the pushed-in and pushed-out sides of a typical carbon fibre following single-fibre push-out testing (**f**). Magnified images of specific regions are indicated by red frames and lines.

Single-fibre push-out tests were performed using NanoTest Vantage (Micro Materials Ltd., Wrexham, UK) nanomechanical instrument fitted with a conical diamond indenter with a maximum tip diameter of 4.39 µm (Figure 3). To more accurately record the applied load and to precisely aim the indenter tip, both load and cross-hair calibration of the instrument were performed on regions of the composite disc above grooves and with no fibres present. Testing was performed by manually choosing ideal single fibres (i.e., with no damage and within the grooves) using a 400× optical microscope attached to NanoTest device. Disc thickness was locally measured at each site by recording physical travel distance from the focus point on the metal support to the surface point on the composite. Testing was performed under loading and unloading rates of 0.5 mN/s, with a dwell time of 5 s at the final load.

The load–displacement curves for each push-out test were analysed to confirm the successful pushing-out of fibre without significant prior crack formation or damage to the fibre or disc. Finally, the composite discs were removed from the metal support (using ethanol) and imaged (JEOL 7000F FEG SEM). Only push-out data corresponding to a well-centred contact and with intact fibres (i.e., not crushed) were used for subsequent IFSS calculations (Figure 3).

The IFSS was calculated by examining the load for complete debonding of the fibre–resin matrix interface (sudden change in displacement with no increase in load) using the following equation:

$$\tau = F/(\pi \times d \times l) \tag{4}$$

τ: IFSS between fibre and resin (Pa);
F: push-out load (N);
d: diameter of the pushed-out fibre (m);
l: length of the fibre being pushed out (m).

3. Results

3.1. De-Sizing/Drying Temperatures

As mentioned before, the temperature of the furnaces and the residence time of CFs inside these furnaces are two of the most important factors that the affect efficiency and productivity of the sizing line, as both parameters impact the effective removal of sizing. Higher temperatures facilitate the process, but they might have a negative impact on the fibre's properties. On the other hand, a longer residence time can accelerate sizing removal, though it reduces the production rate (lower-speed line results in fewer meters treated per hour). The parametric study that follows assesses the parameters that ensure effective desizing temperatures with a higher production rate.

Operation speeds of 0.2, 1, and 2 m/min correspond to 18 s and 1 and 2 min of residence time. For this purpose, six different temperatures for the desizer were tested, starting from 350 °C to 600 °C with a step of 50 and for the three different dryer temperatures (200/250/300 °C). The line's operation speed was set to the lowest possible, 0.2 m/min, that correlates with 18 s for the residence time of CFs inside the furnaces. The results from SEM analysis of these runs are depicted in Figure 4.

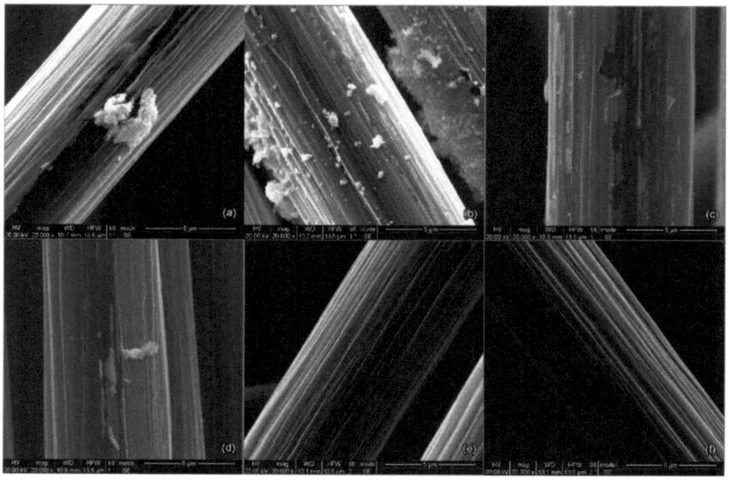

Figure 4. Temperature parametric study of sizing removal at operation speed of 0.2 m/min: (**a**) 350 °C; (**b**) 400 °C; (**c**) 450 °C; (**d**) 500 °C; (**e**) 550 °C and (**f**) 600 °C.

Evidently, temperatures of 350 and 400 °C have an effect on the CF's surface, as can be observed from Figure 4a,b. The pre-existing polymeric sizing starts to degrade and debond from the fibre surface. This indicates that the residence time is not long enough or there is not enough energy derived from heat (due to low temperature) to entirely remove the coating. The desizing phenomenon intensifies at temperatures of 450 and 500 °C, where it is obvious from Figure 4c,d that a large amount of the thin polymeric film is detached from the surface, leaving the exposed CF area available for new sizing. However, it is only at temperatures of 550 and 600 °C that the CF surface area is totally exposed, as the preexisting sizing is completely removed.

To accelerate the desizing process and increase the production rate, lower but efficient values of residence time were put to test. Additional runs were executed at 550 and 600 °C (Figure 5) in which two values of residence time were tested (18 s and 1 min).

Figure 5. Temperature and residence parametric study of sizing removal: (**a**) 550 °C—18 s; (**b**) 550 °C—1 min; (**c**) 600 °C—18 s; and (**d**) 600 °C—1 min.

Regardless of residence time, 600 °C is sufficient to efficiently remove all the pre-existing sizing. At 550 °C, 1 min of heating is required, due to some polymeric residues that may be observed on the CF surface, as proved from the tests performed with 18 s of residence time. As a conclusion, to achieve the highest production rate of the sizing line, a desizing temperature of 600 °C is required, as the line can operate in 2 m/min. The results from mechanical tests would determine if 600 °C affects the properties of the fibres. If the mechanical properties decrease, then an operational speed of 1 m/min must be implemented using 550 °C as the desizing temperature.

TGA was used to study the thermal decomposition of the commercially sized and desized CFs to validate the SEM results.

In graphs like those presented in Figure 6, where mass change differences are considerably low, it is quite challenging to extract solid quantitative conclusions. In this case, the graph form is of great interest and is the one that provides all necessary information. As highlighted in the temperature ranges of 231.7 to 342.3 °C for the reference fibres, there is a steep change in the form of the graph, a sudden drop that is attributed to decomposition, evaporation, or other chemical reactions. In our case, it is decomposition of the pre-existing sizing. On the contrary, for the desized fibre, there is no similar phenomenon, besides the normal negligible mass loss over time, indicating that the pre-existing sizing is removed during the desizing process, confirming what was already assessed from SEM analysis.

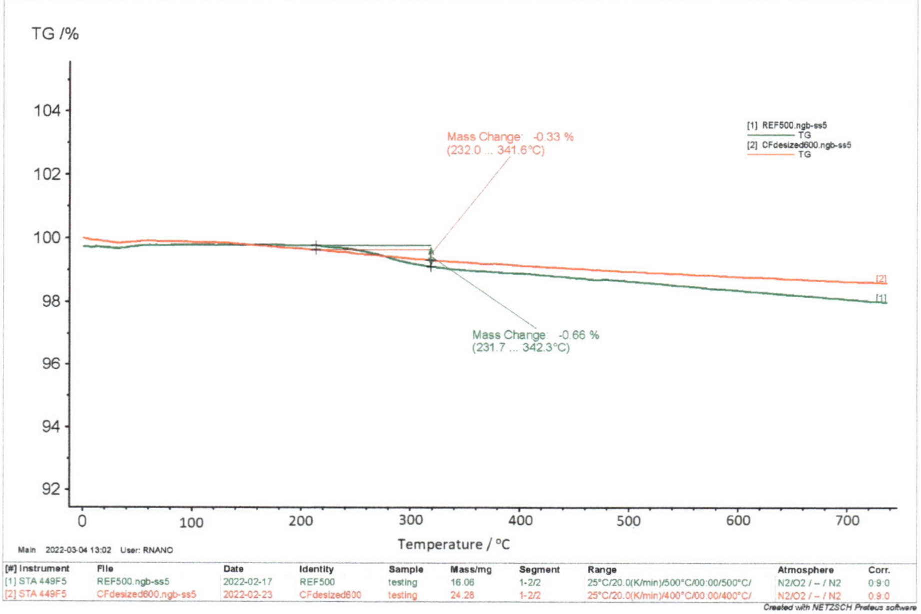

Figure 6. TGA results from desizing temperature at 600 °C.

Finally, an operation speed of 1 or 2 m/min must be taken into consideration for the effective drying of CFs. As indicated by the desizing investigation, these two operation speeds will give the highest productivity. The drying furnace can reach temperatures up to 300 °C, so test runs at 200; 250 and 300 °C at operational speeds of 1 and 2 m/min were executed. As the fibre exited the furnace, filter papers were used to detect the moisture level. At 200 °C and regardless of operation speed, the filter paper was wet. At 250 °C, and at an operation speed of 1 m/min, the filter paper was dry. Evidently, at 300 °C, the carbon fibre was also completely dry for every operation speed.

To conclude, for a desizing temperature of 550 °C and drying temperature of 250 °C, the maximum operation speed was 1 m/min, whereas for a desizing temperature of 600 °C and drying temperature of 300 °C, the sizing line could work at full speed, 2 m/min, without any drawbacks.

3.2. Solid Content Concentration

A uniform distribution of sizing solids onto the CF surface ensures the functionality of this polymeric layer, which facilitates fibre handling and affects mechanical properties. Solid excess could potentially hinder the solution's adhesion to the CF. A thicker layer would act as an extra layer instead of an interlayer, which would improve the affinity of the CF with the resin matrix system. Alternatively, a thinner layer might fail to provide adequate protection during handling and mechanical reinforcement. Achieving the ideal solid content concentration in the sizing solution and subsequently on the CF surface is essential for the utilization of the sizing system. SEM analysis provided information regarding solid content percentages of 1, 2.5, and 5% wt, aiding in the identification of the optimal ratio (Figure 7).

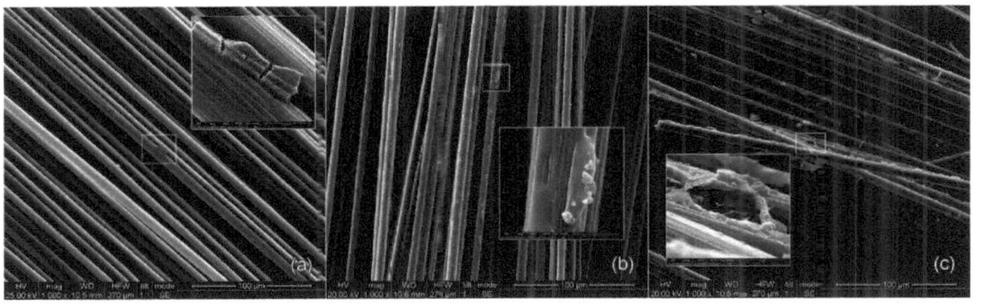

Figure 7. Solid content concentration study of (**a**) 1; (**b**) 2.5 and (**c**) 5% wt at 1000× magnification and 20,000× magnification.

At the 1% wt solid content level, the coating's performance displayed uniformity across the total carbon fibre surface. CF stripes were visible, without any local solid excess spots. In contrast, when the solid content was elevated to 2.5% and 5% wt, it was noticeable that the desired uniformity was nonexistent. The increase in solid content led to an uneven distribution and the creation of excess mass, impacting the coating's ability to uniformly adhere to the carbon fibre surface. As a result, the 1% wt solid content coating stood out as the optimal choice, ensuring even distribution and thickness.

3.3. *Nanoenhanced Sizings*

Based on the findings from Section 3.2, a total solid content concentration of 1% wt was identified as the optimal choice, demonstrating an ideal layer thickness and uniform coverage across the CF surface. To study the potential use of nanomaterials in sizing solutions, CNT- and FLG-sized fibres were examined in terms of surface morphology through SEM. To define their optimum concentration in sizing solutions, the affinity of CFs with resin was assessed through CAG and the mechanical performance was assessed through tensile, 3PB, and push-out testing.

3.3.1. Surface Morphology Assessment

In the case of CNTs, this investigation focused on assessing layer uniformity and nanomaterial agglomeration and distribution across the CF's surface. On the other hand, due to the unique dimensions of FLGs, SEM analysis was primarily focused on uniformity, as conventional metrics of agglomeration and distribution cannot be applied. Analysis was performed on samples with 0.25% wt CNTs or FLGs out of the three weight ratios under investigation (0.05, 0.1, 0.25% wt) and all three possible functionalizations (O_2, NH_3, N_2), resulting in studying six different cases.

Taking into consideration Figure 8, the addition of CNTs in the sizing solutions does not appear to have a negative impact on the sizing layer uniformity, with no observation of sizing pile-up, the presence of boundaries, or delamination of the sizing layer following the CNT and FLG sizing treatments. This suggests that the sizing layer is uniformly distributed across the surface with no unsized or excessively sized spots. However, some potential drawbacks that can be derived from SEM analysis are related to the formation of agglomerates, especially for the cases of O_2 and NH_3 plasma-functionalized CNTs deposited on fibres. The nanomaterial distribution on the fibre surface is inconsistent, displaying specific localized areas with varying nanomaterial concentrations, alongside spots with minimal or absent nanomaterial presence.

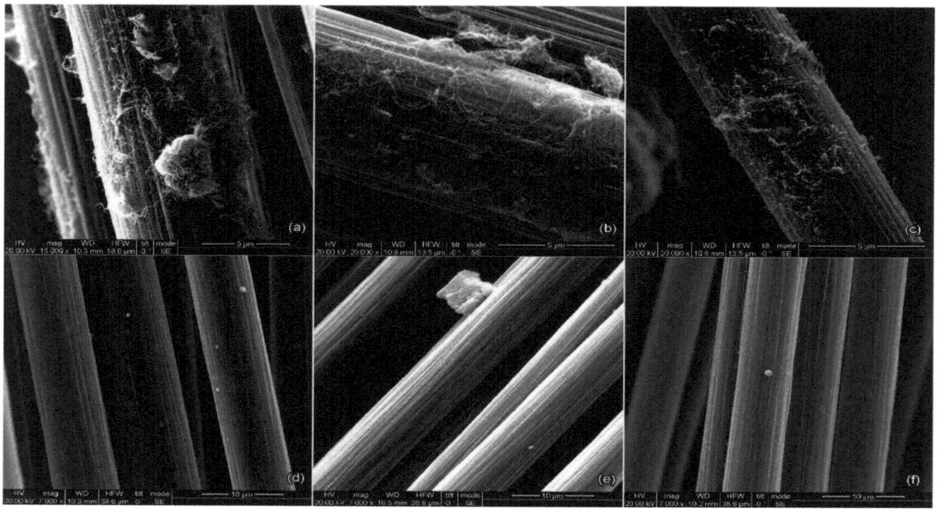

Figure 8. Surface morphology study of (**a**) CNTs O_2; (**b**) CNTs N_2; (**c**) CNTs NH_3; (**d**) FLG O_2; (**e**) FLG N_2 and (**f**) FLG NH_3.

Addressing these drawbacks demands extensive further study. Specifically, understanding the interactions among the three different plasma-functionalized particles and the sizing solution is crucial, offering insights into mitigating the inconsistent distribution on the fibre surface and the formation of agglomerates. Initial enhancements could be achieved by investigating the duration for which a stable solution can be maintained during the preparation of the sizing solution, without the occurrence of agglomerates or sediment formation. Mechanical tests would determine whether these drawbacks have a significant effect on the properties of the fibres. FLGs appeared to exhibit a highly uniform distribution without spots of inconsistency.

3.3.2. Fibre–Resin Affinity

The importance of affinity in composites made of carbon fibre and resin is critical as it signifies the strength of the bond between them, significantly impacting the material's performance and durability. Good affinity facilitates load transfer and enhances mechanical properties. It also contributes to the material's resistance to environmental factors, thermal cycling, and fatigue [26].

Table 5 represents the average contact angle as measured in the Zeiss Axiovision 4.9 imaging processing software (Figure 9). It can be observed that the desized fibre exhibits a contact angle of 46.1°, while the contact angle for all the different functionalizations seems to be smaller, resulting in better wettability. The greatest decrease is observed for samples with N_2 functionalization for both CNTs and FLGs, with contact angle reductions of 7.6% and 13.4%, respectively.

Table 5. Affinity results.

	Desized	**Michelman**	**CNT-Sized Fibres**			**FLG-Sized Fibres**		
			O_2	N_2	NH_3	O_2	N_2	NH_3
Contact Angle (°)	46.1 ± 2.3	43.2 ± 2.1	43 ± 4.5	42.6 ± 1.8	43.8 ± 0.8	43 ± 2.3	39.9 ± 2.8	44.9 ± 1.4

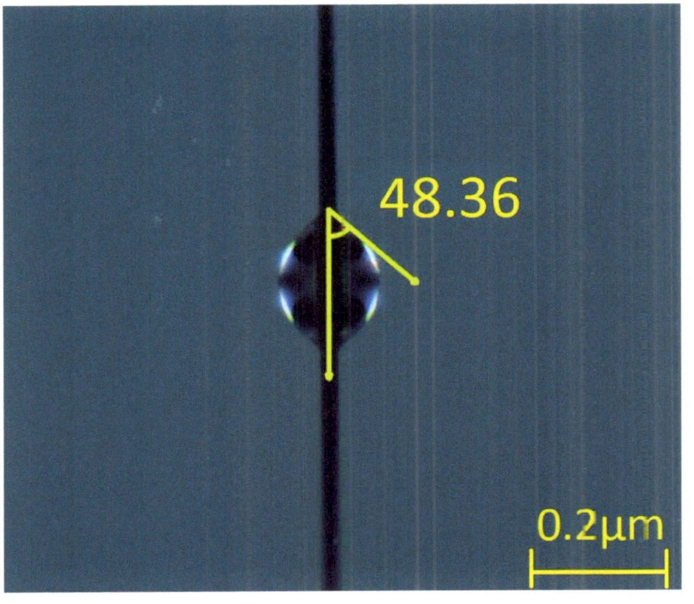

Figure 9. Indicative contact angle measurement.

The contact angle results for CNT-sized fibres are in agreement with Section 3.3.2, where fewer agglomerates are observed for the case of CNT-N_2-sized fibres, indicating a potential correlation between fibre–resin affinity and nanomaterial agglomerates.

3.3.3. Mechanical Performance

Three-Point-Bend and Tensile Results

Initially, samples were subjected to tensile and 3PB tests to assess any potential impact of the sizing process and its steps (e.g., desizing, drying, winding) on the mechanical properties of fibres. For example, desizing is a process step performed at relatively high temperatures (600 °C); thus, it is important to investigate its effect on fibre performance and compare it with data retrieved from the literature. Tensile strength is a property that is mostly dependent on the Young's modulus of the material. Despite sizing not altering the Young's modulus of the fibre, it can promote the fibre–matrix adhesion, which is another parameter that contributes to a material's tensile strength [28,30]. A potential increase in the tensile strength could be related to the latter, especially in the case with nano-enhanced sizing, indicating some cases with improved properties.

A correlation of the results from mechanical tests with those from morphology and affinity assessments is the validation mechanism and decision tool to promote the cases for push-out testing. The push-out test is the optimal method to define the interfacial shear strength of sized fibres, a property which is highly related to the fibre–matrix adhesion.

Upon comparing the results presented in Table 6 for desized fibres and Michelman-sized fibres, it is evident that the latter outperformed in both tensile and 3PB strength. This improvement can be attributed to the sizing's ability to neatly rearrange the fibres and reduce disorientation [31,32].

Table 6. Results regarding the mechanical properties of the different sizing solutions.

	Desized Fibres	Michelman-Sized Fibres	CNT-Sized Fibres								
			O_2			N_2			NH_3		
Weight ratio (%)			0.05	0.1	0.25	0.05	0.1	0.25	0.05	0.1	0.25
3PB (Mpa)	31.3 ± 3	37.1 ± 4.3	35.6 ± 4.5	36.1 ± 5	37.9 ± 3.8	41.8 ± 3.7	44.5 ± 3	34.2 ± 3.5	36.6 ± 2.4	39.1 ± 3.3	37.9 ± 2
Tensile (Mpa)	262 ± 31	391 ± 49	317 ± 34	271 ± 24	334 ± 33	430 ± 27	434 ± 12	376 ± 44	333 ± 27	380 ± 32	388 ± 47
	FLG-sized Fibres										
	O_2			N_2			NH_3				
Weight ratio (%)	0.05	0.1	0.25	0.05	0.1	0.25	0.05	0.1	0.25		
3PB (Mpa)	41.3 ± 2.7	43.1 ± 6.6	41.9 ± 3.6	43 ± 4.1	45.1 ± 2.8	41.3 ± 3.1	33.6 ± 3	42.2 ± 3.9	34.2 ± 2.6		
Tensile (Mpa)	343 ± 27	391 ± 49	421 ± 61	391 ± 34	444 ± 39	354 ± 40	441 ± 156	336 ± 41	351 ± 15		

In the 3PB test among the three different plasma functionalizations and weight ratios, 0.1% N_2-CNT-sized fibres and 0.1% N_2-FLG-sized fibres stood out with the best results, 44.5 ± 3 MPa and 45.1 ± 2.8 MPa accordingly. On the other hand, the desized fibres exhibited the lowest strength, 31.3 ± 3 MPa, as expected.

In the tensile test, and in agreement with the 3PB data, 0.1% N_2-CNT-sized fibres and 0.1% N_2-FLG-sized fibres exhibited the highest tensile strengths of 434 ± 12 MPa and 444 ± 39 MPa, respectively. The desized fibres showed again the lowest tensile strength, measuring at 262 ± 31 MPa.

O_2-CNT-sized fibres and NH_3-CNT-sized fibres exhibited less favourable or similar results to Michelman-sized fibres. This can be attributed to the presence of CNT agglomerates on the fibre surface which can suspend the beneficial effect of CNTs on the fibre–matrix adhesion. These agglomerates pose challenges for interfacial properties and may potentially introduce stress concentration points during loading, which can impact the mechanical performance [33].

Evidently, 0.1% N_2-CNT-sized fibres and 0.1% N_2-FLG-sized fibres consistently yielded superior results in both the 3-point bend and tensile tests, while maintaining a uniform coating. These results are in agreement with those from SEM analysis where agglomeration was observed for O_2-CNT-sized fibres and NH_3-CNT-sized fibres, and the CAG findings where the optimum affinity was with N_2-CNT-sized fibres and N_2-FLG-sized fibres.

The enhanced performance of N_2-functionalized nanoparticles can be attributed to their chemical affinity with the sizing solution. As shown in Table 2, HP206 is amine-dispersed, while N_2 functionalization introduces C-N bonds, as confirmed by XPS results (see Supplementary File). The incorporation of C-N bonds contributes to a stronger chemical affinity compared to O_2 functionalization, which introduces C-O bonds. Although NH_3 functionalization also introduces C-N bonds, the quantity is considerably lower than in N_2 functionalization (area of C-N bond in C1s scan: 4.1% for NH_3 FLG compared to 4.7% for N_2 FLGs and 0.49% for NH_3 CNTs compared to 1.78% for N_2 CNTs). This variation accounts for the superior overall properties associated with N_2 functionalization.

Based on these results, 0.1% N_2-CNT-sized fibres and 0.1% N_2-FLG-sized fibres were examined under push-out tests to determine their influence on the interfacial shear properties of composites.

Push-Out Test Results

Three groups of composites were produced for IFSS evaluation, consisting of pristine, 0.1% N_2-CNT-sized fibres, and 0.1% N_2-FLG-sized fibres embedded within resin matrices. The reported IFSS measurements in Table 7 correspond to data from undisturbed push-out tests (i.e., fibres pushed out smoothly in one go and without damage to the fibres). A clear increase in IFSS can be seen following the FLG N_2 and CNT N_2 sizing treatments of the fibres, with an increase from 50.3 ± 5.5 MPa (pristine) to 70.7 ± 2.9 MPa and 95.4 ± 7.6 MPa, respectively. As compared with the pristine fibres, this constitutes an IFSS increase of 40.6%

and 89.7% for the respective treated fibres. Figure 10 illustrates example hysteresis data of each composite type during the single-fibre push-out testing. Both treated fibres are found to exhibit significantly greater loading rates (gradient before the push-out event), thus reaching the push-out load at approximately half the normalised depth as the pristine fibre. This demonstrates the more efficient transfer of load between the indenter and the fibre interface, leading to a more minimal deflection of the resin matrix during loading. The most significant enhancement is found with the CNT N_2 treatment, which is observed to have the best structural rigidity and IFSS within this study.

Table 7. Average IFSS measurements of the pristine, FLG-N_2-sized, and CNT-N_2-sized carbon fibres.

Treatment	IFSS (MPa)
Pristine	50.3 ± 5.5
FLG-N_2-Sized	70.7 ± 2.9
CNT-N_2-Sized	95.4 ± 7.6

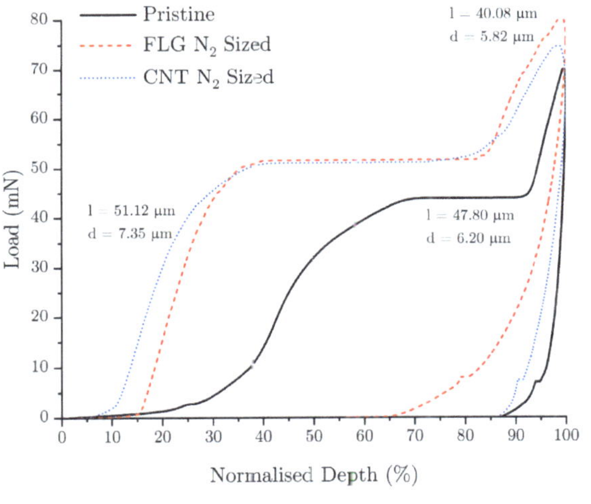

Figure 10. Normalised depth against indentation load during single-fibre push-out testing of pristine, FLG-N_2-sized, and CNT-N_2-sized fibres embedded in resin. The length of the fibre (l) and the average diameter of the fibre (d) are also provided for each sample.

4. Conclusions

In this study, a pilot-scale sizing line for carbon fibres was designed and installed, serving as an established tool for extensive investigation. Various operational parameters were thoroughly explored to optimize the line's performance and productivity. A key parameter was the necessity of desizing temperature for commercial fibres. This research demonstrated that a desizing temperature of 600 °C was essential to achieve maximum productivity, ensuring complete removal of the pre-existing sizing agents and enabling the line to operate at its peak efficiency.

Detailed surface morphology analysis, conducted through SEM, confirmed the significance of maintaining a 1% solid content in the coating solution. This specific concentration was identified as the optimum choice, ensuring uniformity and appropriate film thickness on the carbon fibre surface. This study then focused on the integration of nanomaterials, such as carbon nanotubes and few-layer graphene, aiming to enhance mechanical performance. This research highlighted the importance of good affinity between these nanomaterials and the fibres through contact angle tests, a critical factor influencing the strength and durability of the resulting composites.

The mechanical tests indicated improved and desirable properties for the treated fibres, emphasizing the effectiveness of optimized parameters. This study further employed push-out tests, revealing substantial improvements in the interfacial shear strength of 89.7% for 0.1% N_2-CNT-sized fibres and 40.6% for 0.1% N_2-FLG-sized fibres, as compared to the reference fibres, thereby demonstrating enhanced bonding between fibres and the surrounding matrix.

The improved chemical affinity of the sizing solution (amine-dispersed) with N_2-functionalized CNTs/FLGs is attributed to the addition of C-N bonds, surpassing both (a) the C-O bonds introduced by O_2 plasma functionalization and (b) the quantity of C-N bonds in NH_3 plasma functionalization.

In conclusion, the outcomes of this research underline the significant potential for tailoring sizing solutions to match specific requirements. By utilizing the pilot-scale sizing line, besides the scientific finding this study reveals, a practical tool for developing customized sizing solutions is also provided. These findings encourage the exploration of tailored sizing solutions, contributing to the advancement of understanding in this complex domain, enabling innovative applications in the field of composite materials.

Supplementary Materials: The following supporting information can be downloaded at https://www.mdpi.com/article/10.3390/fib12020016/s1, Figure S1: FLG 0.1% wt water dispersions; Figure S2: CNTs 0.1% wt water dispersions; Figure S3: 3510 BRANSON sonication bath (left) and Thielsen ultrasonic processor (right); Figure S4: Samples 24 h after sonication for 1 h in 3510 BRANSON sonication bath; Figure S5: Samples after ultrasonication with Thielsen sonication probe at 72 h; Figure S6: FLG unfunctionalized; Figure S7: FLG NH3; Figure S8: FLG N_2; Figure S9: FLG O_2; Figure S10: CNT unfunctionalised; Figure S11: CNT NH3; Figure S12: CNT N_2; Figure S13: CNT O_2; Table S1: Table of surfactants that were tested; Table S2: Samples prepared for step 1; Table S3: Samples prepared for step 2; Table S4: XPS element survey results of unfunctionalised and functionalised nanomaterials; Table S5: XPS high-resolution data peak assignment FLG materials; Table S6: XPS high-resolution data peak assignment CNT materials.

Author Contributions: Conceptualization, D.S.; methodology, D.S., I.P., S.A., B.D. and X.L.; investigation, D.S., I.P., S.A., B.D. and X.L.; writing—original draft preparation, D.S., I.P., B.D. and X.L.; writing—review and editing, D.S., H.D. and C.A.C.; visualization, D.S.; supervision, H.D. and C.A.C.; funding acquisition, C.A.C. All authors have read and agreed to the published version of the manuscript.

Funding: This research was funded by the EU H2020 project "New generation of offshore turbine blades with intelligent architectures of hybrid, nano-enabled multi-materials via advanced manufacturing" (Carbo4Power), under Grant Agreement no. 953192.

Data Availability Statement: Data will be available on the Zenodo platform upon request.

Acknowledgments: The authors would like to thank Haydale Ltd. for the supply of functionalized nanomaterials and XPS measurements and Maria Modestou for the preparation of graphical abstract.

Conflicts of Interest: The authors declare no conflicts of interest.

References

1. Becker-Staines, A.; Bremser, W.; Tröster, T. Cyclodextrin as sizing for carbon fibers: New bonding mechanism improves adhesion in carbon fiber reinforced epoxy resin. *Heliyon* **2020**, *6*, e03766. [CrossRef]
2. Wu, Q.; Zhao, R.; Xi, T.; Yang, X.; Zhu, J. Comparative study on effects of epoxy sizing involving ZrO_2 and GO on interfacial shear strength of carbon fiber/epoxy composites through one and two steps dipping routes. *Compos. Part A Appl. Sci. Manuf.* **2020**, *134*, 105909. [CrossRef]
3. Rankin, S.M.; Moody, M.K.; Naskar, A.K.; Bowland, C.C. Enhancing functionalities in carbon fiber composites by titanium dioxide nanoparticles. *Compos. Sci. Technol.* **2021**, *201*, 108491. [CrossRef]
4. Lim, S.H.; On, S.Y.; Kim, H.; Bang, Y.H.; Kim, S.S. Resin impregnation and interfacial adhesion behaviors in carbon fiber/epoxy composites: Effects of polymer slip and normalized surface free energy with respect to the sizing agents. *Compos. Part A Appl. Sci. Manuf.* **2021**, *146*, 106424. [CrossRef]

5. Eyckens, D.J.; Arnold, C.L.; Simon, Ž.; Gengenbach, T.R.; Pinson, J.; Wickramasingha, Y.A.; Henderson, L.C. Covalent sizing surface modification as a route to improved interfacial adhesion in carbon fibre-epoxy composites. *Compos. Part A Appl. Sci. Manuf.* **2021**, *140*, 106147. [CrossRef]
6. Chu, C.; Ge, H.; Gu, N.; Zhang, K.; Jin, C. Interfacial microstructure and mechanical properties of carbon fiber composite modified with carbon dots. *Compos. Sci. Technol.* **2019**, *184*, 107856. [CrossRef]
7. Wang, S.; Yang, Y.; Mu, Y.; Shi, J.; Cong, X.; Luan, J.; Wang, G. Synergy of electrochemical grafting and crosslinkable crystalline sizing agent to enhance the interfacial strength of carbon fiber/PEEK composites. *Compos. Sci. Technol.* **2021**, *203*, 108562. [CrossRef]
8. Ali, I.; Shchegolkov, A.; Shchegolkov, A.; Chumak, M.A.; Viktorovich, N.A.; Vasilievich, L.K.; Imanova, G.; Kurniawan, T.A.; Habila, M.A. Facile microwave synthesis of multi—Walled carbon nanotubes for modification of elastomer used as heaters. *Polym. Eng. Sci.* **2023**, *63*, 3975–3985. [CrossRef]
9. Zhang, Z.; Li, X.; Jestin, S.; Termine, S.; Trompeta, A.-F.; Araújo, A.; Santos, R.M. Charitidis, C.; Dong, H. The Impact of Carbon Nanofibres on the Interfacial Properties of CFRPs Produced with Sized Carbon Fibres. *Polymers* **2021**, *13*, 3457. [CrossRef]
10. Yuan, C.; Li, D.; Yuan, X.; Liu, L.; Huang, Y. Preparation of semi-aliphatic polyimide for organic-solvent-free sizing agent in CF/PEEK composites. *Compos. Sci. Technol.* **2021**, *201*, 108490. [CrossRef]
11. Yuan, X.; Jiang, J.; Wei, H.; Yuan, C.; Wang, M.; Zhang, D.; Liu, L.; Huang, Y.; Gao, G.-L.; Jiang, Z. PAI/MXene sizing-based dual functional coating for carbon fiber/PEEK composite. *Compos. Sci. Technol.* **2021**, *201*, 108496. [CrossRef]
12. Michelman.com. Reinforced Plastic Composites-Fibre SizingReinforced Plastic Composites-Fibre Sizing. Available online: https://www.michelman.com/markets/reinforced-plastic-composites/fibre-sizing/ (accessed on 3 November 2023).
13. Li, M.; Gu, Y.; Liu, Y.; Li, Y.; Zhang, Z. Interfacial improvement of carbon fiber/epoxy composites using a simple process for depositing commercially functionalized carbon nanotubes on the fibers. *Carbon* **2013**, *52*, 109–121. [CrossRef]
14. Sharma, M.; Gao, S.; Mäder, E.; Sharma, H.; Wei, L.Y.; Bijwe, J. Carbon fiber surfaces and composite interphases. *Compos. Sci. Technol.* **2014**, *102*, 35–50. [CrossRef]
15. Knoll, J.; Riecken, B.; Kosmann, N.; Chandrasekaran, S.; Schulte, K.; Fiedler, B. The effect of carbon nanoparticles on the fatigue performance of carbon fibre reinforced epoxy. *Compos. Part A Appl. Sci. Manuf.* **2014**, *67*, 233–240. [CrossRef]
16. Haydale Ltd. Creating Material Chang. 2019. Available online: https://haydale.com/wp-content/uploads/2023/03/Haydale_Brochure_Functionalised_Graphene.pdf (accessed on 11 November 2023).
17. Evgeniy, T.E.; Marcos, G.; Gijsbertus, D.W.; Cor, K.E. The use of surfactants for dispersing carbon nanotubes and graphene to make conductive nanocomposites. *Curr. Opin. Colloid Interface Sci.* **2012**, *17*, 225–232.
18. Sezer, N.; Koc, M. Stabilization of the aqueous dispersion of carbon nanotubes using different approaches. *Therm. Sci. Eng. Prog.* **2018**, *8*, 411–417. [CrossRef]
19. Rastogi, R.; Kaushal, R.; Tripathi, S.K.; Sharma, A.L.; Kaur, I.; Bharadwaj, L.M. Comparative study of carbon nanotube dispersion using surfactants. *J. Colloid Interface Sci.* **2008**, *328*, 421–428. [CrossRef] [PubMed]
20. Keinänen, P.; Siljander, S.; Koivula, M.; Sethi, J.; Sarlin, E.; Vuorinen, J.; Kanerva, M. Optimized dispersion quality of aqueous carbon nanotube colloids as a function of sonochemical yield and surfactant/CNT ratio. *Heliyon* **2018**, *4*, e00787. [CrossRef] [PubMed]
21. Zhang, H.-Y.; Li, T.-C.; Li, Z.-K. Modeling in SolidWorks and analysis of temperature and thermal stress during construction of intake tower. *Water Sci. Eng.* **2009**, *2*, 95–102. [CrossRef]
22. Tiwari, S.; Bijwe, J. Surface Treatment of Carbon Fibers—A Review. *Procedia Technol.* **2014**, *14*, 505–512. [CrossRef]
23. Körbelin, J.; Kötter, B.; Voormann, H.; Brandenburg, L.; Selz, S.; Fiedler, B. Damage tolerance of few-layer graphene modified CFRP: From thin-to thick-ply laminates. *Compos. Sci. Technol.* **2021**, *209*, 108765. [CrossRef]
24. Hao, B.; Ma, Q.; Yang, S.; Mäder, E.; Ma, P.-C. Comparative study on monitoring structural damage in fiber-reinforced polymers using glass fibers with carbon nanotubes and graphene coating. *Compos. Sci. Technol.* **2016**, *129*, 38–45. [CrossRef]
25. Lai, M.; Jiang, L.; Wang, X.; Zhou, H.; Huang, Z.; Zhou, H. Effects of multi-walled carbon nanotube/graphene oxide-based sizing on interfacial and tribological properties of continuous carbon fiber/poly(ether ether ketone) composites. *Mater. Chem. Phys.* **2022**, *276*, 125344. [CrossRef]
26. ISO. 2022. Available online: https://www.iso.org/standard/79999.html (accessed on 11 November 2023).
27. Semitekolos, D.; Kainourgios, P.; Jones, C.; Rana, A.; Koumoulos, E.P.; Charitidis, C.A. Advanced carbon fibre composites via poly methacrylic acid surface treatment; surface analysis and mechanical properties investigation. *Compos. Part B Eng.* **2018**, *155*, 237–243. [CrossRef]
28. Termine, S.; Naxaki, V.; Semitekolos, D.; Trompeta, A.-F.; Rovere, M.; Tagliaferro, A.; Charitidis, C. Investigation of Carbon Fibres Reclamation by Pyrolysis Process for Their Reuse Potential. *Polymers* **2023**, *15*, 768. [CrossRef]
29. ASTM. ASTM International, ASTM, 22 September 2023. Available online: https://www.astm.org/d4018-17.html (accessed on 28 September 2023).
30. Zhang, J. Different Surface Treatments of Carbon Fibres and Their Influence on the Interfacial Properties of Carbon Fibre/Epoxy Composites. Ph.D. Thesis, University of Paris-Saclay, Gif-sur-Yvette, France, 2012.
31. Pozegic, T.; Jayawardena, K.D.G.I.; Chen, J.-S.; Anguita, J.V.; Ballocchi, P.; Stolojan, V.; Silva, S.; Hamerton, I. Development of sizing-free multi-functional carbon fibre nanocomposites. *Compos. Part A Appl. Sci. Manuf.* **2016**, *90*, 306–319. [CrossRef]

32. Yao, H.; Sui, X.; Zhao, Z.; Xu, Z.; Chen, L.; Deng, H.; Liu, Y.; Qian, X. Optimization of interfacial microstructure and mechanical properties of carbon fiber/epoxy composites via carbon nanotube sizing. *Appl. Surf. Sci.* **2015**, *347*, 583–590. [CrossRef]
33. Godara, A.; Mezzo, L.; Luizi, F.; Warrier, A.; Lomov, S.V.; van Vuure, A.W.; Gorbatikh, L.; Moldenaers, P.; Verpoest, I. Influence of carbon nanotube reinforcement on the processing and the mechanical behaviour of carbon fiber/epoxy composites. *Carbon* **2009**, *47*, 2914–2923. [CrossRef]

materials

MDPI

Article

Influence of the Thermoplastic Fiber Ratio on the Mechanical Properties of Recycled Carbon Fibers During the Carding Process

Jean Ivars, Ahmad Rashed Labanieh and Damien Soulat *

ULR 2461-GEMTEX-Génie et Matériaux Textiles-Gemtex, Ensait, University of Lille, F-59000 Lille, France; jean.ivars@ensait.fr (J.I.); ahmad.labanieh@ensait.fr (A.R.L.)
* Correspondence: damien.soulat@ensait.fr

Abstract: This study investigates the impact of carding and blending recycled carbon fibers (rCF) with crimped thermoplastic polypropylene (PP) fibers on the mechanical properties of rCF, using a Weibull statistical approach. Tensile properties of rCF were evaluated before and after carding with varying rCF/PP blend ratios (100/0%, 85/15%, 70/30%, and 50/50%). A comparison between the two-parameter and three-parameter Weibull models showed that the two-parameter model provided a better fit for rCF properties before carding. The results show that adding crimped PP fibers during carding helps to decrease the stress-at-break disparity and move their distribution to higher values. Furthermore, a slight increase in tensile modulus was observed in carded rCF, with higher PP ratios associated with smaller scatter modulus distributions. Elongation at break remained consistent, with the Weibull modulus increasing slightly with carding and the inclusion of PP fibers, indicating improved consistency. Overall, carding rCF with PP fibers helped in the mechanical property uniformity of the resulting carded webs without compromising tensile performance. This work shows the potential of the carding process with or without thermoplastic fibers to efficiently realign and give continuity to discontinuous recycled carbon fibers.

Keywords: recycled carbon fiber; carding; tensile properties; statistical distribution; Weibull

1. Introduction

The reuse of recycled carbon fibers (rCF) is a response to growing environmental concerns associated with the composites industry. Carbon fiber-reinforced polymers (CFRPs) are widely used in high-performance applications, such as aerospace, automotive, and renewable energy [1], due to their high strength-to-weight ratios and durability. Recycling and reusing carbon fibers represents a more sustainable alternative by reducing waste at the end of the life cycle of composite materials and decreasing dependency on virgin raw materials. The quality of the initial fiber source and the recycling method used [2] can have an impact on the variability of the tensile properties of rCF, such as breaking stress, tensile modulus, and elongation at break. The process used to realign those fibers can also have an influence on their properties.

While the potential environmental benefits of recycling carbon fibers are significant, the process also presents technical challenges. Recycling methods, such as pyrolysis [3] and solvolysis [4,5], are widely used to recover carbon fibers from end-of-life composite products or manufacturing scrap. However, these recycling processes often result in fibers that are shorter than their virgin precursors and stripped of their original surface treatments, or sizings [6], which were applied to improve fiber–matrix bonding and handling. The absence of sizing, combined with the inherent brittleness of carbon fibers, makes handling and reprocessing more difficult, particularly when recycled fibers need to be aligned or integrated into high-performance applications. Despite these challenges, strategies and techniques have been developed to reuse recycled carbon fibers in new materials: depending on the structure and the length, several processing methods can effectively reuse rCFs. For long rCFs, nonwoven technologies, for instance, involve combining multiple layers of randomly oriented fibers to create mats. This technique is simple and cost-effective, but results in limited strength and stiffness when a cross-lapping and needle-punching manufacturing route is applied [7]. Nonwoven carbon fiber structures are not only cost-effective but also versatile for a variety of applications beyond semi-structural reinforcements [8,9]. Recent advancements have demonstrated the feasibility of using hybrid nonwovens made of rCFs for electromagnetic interference shielding and sound absorption [10].

The carding process is a preparatory step in the manufacturing of these materials. Carding disentangles and cleans fibers while forming a continuous web, which serves as a base for further processing. Depending on the intended application, this process can lead to different pathways: the production of nonwoven materials, or the preparation of webs with doubling and drawing steps, which improve fiber alignment and enhance structural uniformity. Another advantage of the carding process is its ability to create hybrid structures by blending rCFs with thermoplastic fibers, making it a suitable option for producing thermoplastic composites.

Various studies have already demonstrated the feasibility of carding rCFs with thermoplastic fibers. Hengstermann et al. [11] were able to card 40 mm and 60 mm long rCFs, and highlighted the influence of fiber length at the card infeed. In another study, a core-sheath yarn was produced, also with 60 mm carded rCFs in the core, with a co-polyamide as a sheath [12]. Also, it has been shown that crimped polypropylene helped with mixing and carrying rCFs, diminishing fiber breakage through the carding process [13]. This approach aligns with previous works where polyamide and polyester were used as carrier fibers in blends with rCFs [14]. The morphological and mechanical properties of the fibers and the resulting web are influenced by both the material properties and the machine parameters during carding. For example, Manis et al. [15] were able to demonstrate that a higher number of worker/stripper pairs led to a greater degradation of 60 mm fiber length. Another study looked at the influence of the distance between the worker and the stripper on the length of the fibers leaving the carding machine [12]. A higher clearance between the worker and stripper led to lower fiber length degradation.

Statistical methods help with analyzing and interpreting the variability of the mechanical properties of individual fibers. Among these, the Weibull distribution is the most commonly used in the literature to study the tensile behavior of brittle fibers, such as carbon fiber [16–19], but also for natural fibers [20,21], as it accounts for variability in fiber strength caused by defects.

This study focuses on evaluating the effect of the carding process and blending rCFs with crimped thermoplastic fibers on the mechanical properties of rCFs. Polypropylene (PP) is incorporated in the carding process of rCFs in the present study. First, the mechanical performance of rCF is assessed before and after carding across different rCF/PP blend ratios. A Weibull statistical approach is then used to analyze the variability in tensile properties, such as stress at break, tensile modulus, and elongation at break. The comparative analysis uses a two-parameter Weibull model, which is found to fit the experimental data more effectively, particularly for rCF properties before carding when fitted with the Maximum Likelihood Estimation (MLE) technique. This study provides new insights into the mechanical behavior of rCF, a topic that remains underexplored in the literature. Few studies have specifically investigated the individual behavior of fibers after processes like carding or the influence of blending ratios with thermoplastic fibers.

2. Materials and Methods

The carbon fibers used in this study are Standard Modulus (SM) carbon fibers reclaimed by steam thermolysis. The fibers are unsized and cut to a length of 60 mm. rCF recovered using this technique shows a slight degradation in its initial tensile properties, about 10%, a good surface condition [22], and well-preserved diameter [23]. rCF before carding's properties are listed in Table 1.

Table 1. Fibre properties before and after carding according to their ratio in the webs.

	Data Points	Diameter (μm)		Stress at Break (MPa)		Tensile Modulus (GPa)		Elong. at Break (%)	
	Numb.	Mean	Std.	Mean	Std.	Mean	Std.	Mean	Std.
100% rCF	45	7.36	0.45	3221	769	192.8	13.7	1.57	0.36
15–85% rCF	49	7.21	0.46	3406	741	195.5	12.1	1.57	0.29
30–70% rCF	47	7.18	0.34	3563	656	198.5	13.6	1.68	0.30
50–50% rCF	25	7.10	0.48	3413	685	206.4	15.9	1.53	0.27
rCF bef. Card.	51	7.41	0.25	3143	773	186.2	20.8	1.57	0.37

These values are extracted from 51 tensile tests performed using the protocol detailed in Section 2.2. at a gauge length of 25 mm. These values enabled us to classify these recycled carbon fibers as low-SM-type fibers [24]. The modulus and stress values were calculated under the assumption of a circular fiber cross-section.

2.1. Web Manufacturing with a Carding Machine

For carding, individualize the staple fibers provided in entangled pocket, called tuft, by successive passes between toothed cylinders (Figure 1). The relative orientation of the faced cylinders permits the opening and transmission of the fibers from the feeder to the doffer and transforms the fiber in a low-inter-fiber-cohesion web. The fibers in the manufactured web are distributed randomly on the web surface. However, a preferred orientation for card manufacturing is reported for the nonwoven rCF needle punch [6].

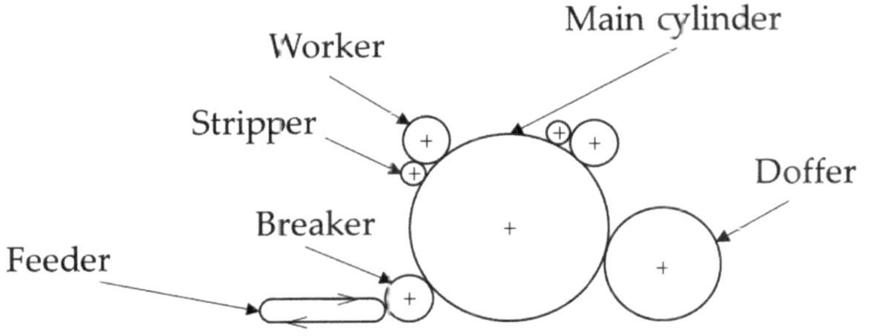

Figure 1. Carding machine schema.

To manufacture the web, rCF reclaimed from the steam thermolysis process are used alongside 60 mm long crimped thermoplastic polypropylene fibers at different ratios as shown in Figure 2. Polypropylene fibers are widely used in the textile industry due to its low-cost and processability, frequently used as a matrix in thermoplastic composites.

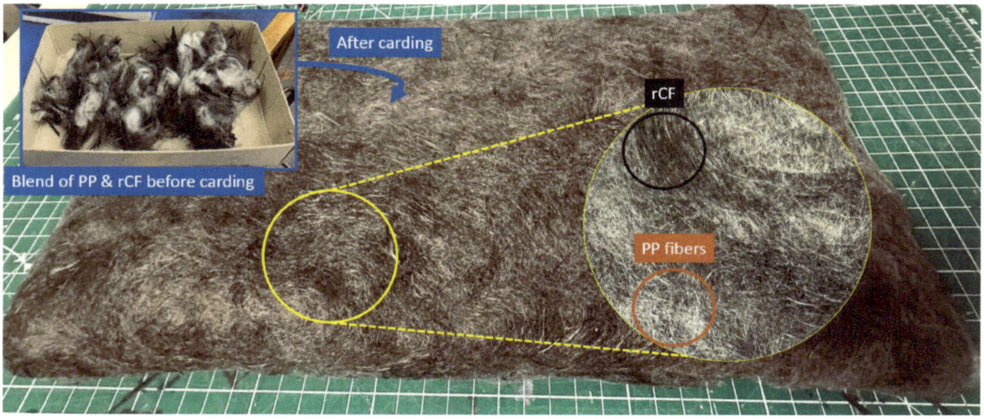

Figure 2. Carded web with a mix of PP and rCF.

In this study, four webs are manufactured with only changing the rCF/PP blend's ratio with, respectively, 50/50%, 70/30%, 85/15%, and finally 100/0%. Other carding parameters such as feeding density and production speed are kept constant. The physical (diameter) and mechanical tensile properties of individual fibers extracted from each manufactured web are then identified.

2.2. Characterization Method and Data Extraction

2.2.1. Fiber Preparation and Testing

The tensile test for individual carbon fibers is performed using NF ISO 11566 standard [25], which is specific to single carbon fibers. After individualization, a single fiber is fixed to a cardboard frame using cyanoacrylate glue. The cardboard frame is mounted on an INSTRON MTS tensile tester (Norwood, MA, USA) mounted with a 10 N load cell, and the sides of the frame are cut to ensure that the load applied by the tensile tester is transmitted only to the fiber. The test speed is set at 2 mm/min and the gauge length is set to 25 mm. Cardboard frames were cut on a laser cutting machine to avoid errors in the initial length of the fibers. Doing so helps to reduce errors due to misalignment of the fiber on the cardboard frame, which can introduce measurement errors at shorter gauge lengths [26]. The compliance values were not considered for study [27–29].

The NF ISO 11566 standard recommends twenty measurements per fiber type. However, the inherent fragility of carbon fibers makes preparing the test difficult, potentially leading to measuring errors. Furthermore, a minimum number of tensile tests must be carried out to obtain robust data and representative results for a batch of fibers. For 50 tensile tests on carbon fiber, the Weibull modulus could vary by +/−1 [30]. Beyond 80 tests, there is no significant difference in this Weibull modulus value [18], so a balance must be struck between the time spent on the test preparation and the accuracy of the value.

Considering the number of webs and the preparation time for tensile test at fiber scale, it was considered necessary to carry out a more thorough test procedure of at least sixty tests to ensure more sufficient and reliable data.

2.2.2. Diameter Measurements and Circularity Assumption

For each tensile sample, diameter measurements are taken at five points along the fiber using an OLYMPUS DSX1000 microscope (Waltham, MA, USA, Figure 3, right). A circular cross-section is assumed for the carbon fibers, as it is supposed that non-circularity is not significant for the carbon fiber [26].

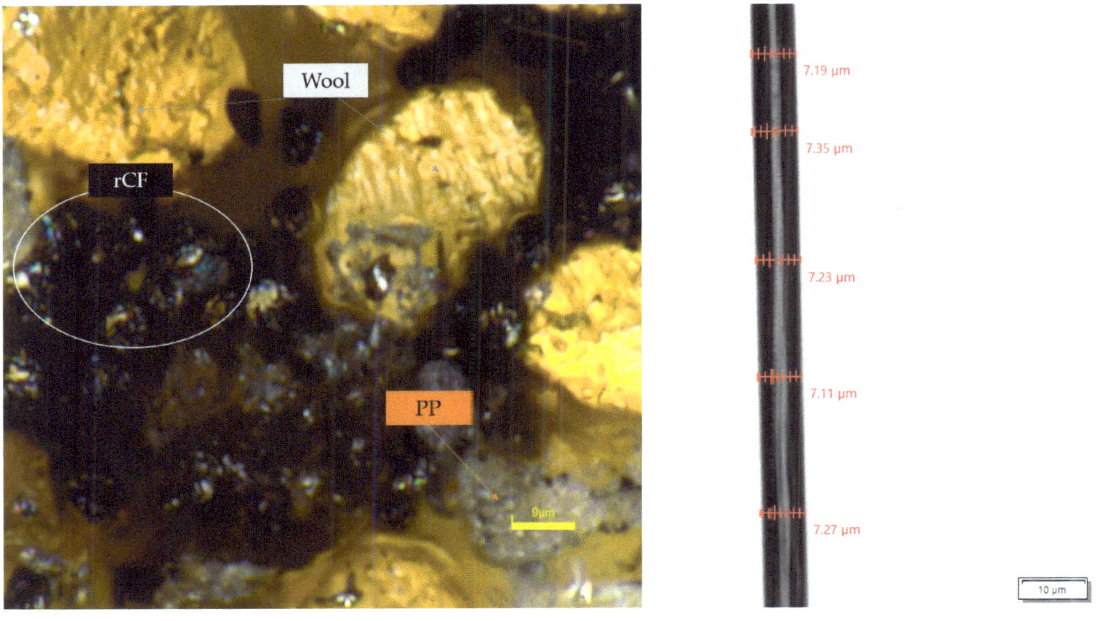

Figure 3. Cross-section of rCF before carding (**left**), diameter measurement of rCF before tensile testing (**right**).

To support this assumption, cross-sectional measurements were performed using a microtome (Figure 3, left). The wool is added during the sample preparation step for cross-section measurements to improve the visualization of the carbon fibers. As carbon fibers are black and do not allow light to pass through, the wool provides contrast, making the fibers easier to observe. The circularity coefficient, calculated from the cross-sectional area and perimeter, reflects how close the fibers are to being perfectly circular, with a value closer to 1 indicating greater circularity [31]. This relationship is defined in Equation (1):

$$Circularity = \frac{4 * \pi * Area}{Perimeter^2} \quad (1)$$

Measurements of the recycled carbon fibers (rCF) before carding, carried out with ImageJ (version 1.54f), gave a circularity coefficient of 0.956 ± 0.014 across 11 samples, confirming that the fibers are highly circular prior to carding.

2.2.3. Data Extraction and Analysis

Post-processing on the collected raw data, including apparent diameter, load and displacement, is performed using a Python script. The start of fiber loading is determined by identifying the second inflection point on stress–elongation curve. This eliminates the foot of the curve and ensures that all curves have the same starting point for comparison. The fiber gauge length is corrected with identified slack distance on the recorded displacement. The stress and elongation at break are determined for each test and collected for each fiber batch, per the design of the experiments. The tensile modulus is calculated between 0.1% and 0.6% of elongation [25] using a regression on the linear part of the stress/strain curve in order to avoid introducing errors due to data resolution.

These data can be used to conduct a classical comparative statistical study using mean and standard deviation for each of tensile characteristics mentioned (tensile modulus, stress and elongation at break). Additionally, it enables the carrying out of Weibull statistical modeling, suitable for the brittle behavior of fibers. Comparing Weibull parameters provides information about the variability of data distribution [17,20,30,32] and enables us to differentiate between the different batches of fibers from the webs produced. This information on the disparity in the fiber properties cannot be estimated based on the average values and standard deviations of the tensile characteristics. This provides a better understanding of how the crimped thermoplastic fibers can affect the tensile properties of the recycled carbon fiber during the carding process.

2.3. Statistical Analysis Using Weibull Distribution

2.3.1. Introduction and Model

The elastic–brittle behavior of carbon fiber can be depicted by the Weibull's statistic, which assumes that carbon fiber is an assembly of links. Tensile failure of the fiber occurs when the weakest link in the assembly breaks [16]. Although the Weibull distribution, used widely in carbon fiber mechanical analysis, is not an intrinsic characteristic of carbon fiber, it is a statistical model applied to assess the strength variability in carbon fiber composites which can facilitate the prediction of failure probabilities under varying stress conditions. The cumulative distribution function (CDF) is used to represent the probability of failure P at a given value of σ, the applied stress on the fiber following Equation (2), also known as the 3-parameter Weibull CDF:

$$P(\sigma) = 1 - e^{-\left(\frac{\sigma-\sigma_u}{\sigma_0}\right)^m} \tag{2}$$

The 3-parameter Weibull model uses the following key parameters: σ_0—scale parameter (indicating the average lifetime under stress at which 63.2% of fiber are broken), m—shape of the distribution, or Weibull modulus (measuring the variability, the spread or scatter of fiber strength), σ_u— location parameter (reflecting the threshold for fiber failure initiation; below this value, the failure probability is null) and σ— applied stress. A Simpler form of the Weibull CDF is also considered to simulate the brittle fracture of fiber by setting the location parameter σ_u in Equation (2) to zero, known as 2-parameter Weibull CDF. The location parameter in the 3-parameter Weibull distribution represents a threshold stress below which failures are unlikely to occur, enhancing the model's accuracy, particularly when data exhibit an initial delay in failure observations. Studies on natural fibers [20,21] have demonstrated differences between the two- and three-parameter Weibull models in predicting strength, elongation at break, and tensile modulus, depending on the gauge length.

According to the different manufactured webs, these statistical models are compared with each other, but also with recycled carbon fiber tensile data before carding.

2.3.2. Model Fitting on Experimental Data

One of the statistical methods used to fit a distribution to experimental data is Maximum Likelihood Estimation (MLE). It is the maximization of a likelihood function that quantifies the probability of the observed data under a particular statistical model. In the case of the 3-parameter Weibull distribution, MLE is used to determine the shape m, scale σ_0, and location σ_u parameters, which are the most likely to produce the observed dataset [33]. Nketiah [34] reported the reliability of MLE in comparison with the Least Squares method. The likelihood function is expressed in Equation (3) with a set of points $x_1,\ x_2,\ \ldots,\ x_n$.

$$L(m,\ \sigma_0,\ \sigma_u;\ x_i) = \prod_{i=1}^{n}\left(\frac{m}{\sigma_0}\left(\frac{x_i-\sigma_u}{\sigma_0}\right)^{m-1} e^{-\left(\frac{x_i-\sigma_u}{\sigma_0}\right)^m}\right) \tag{3}$$

Once each parameter in the Weibull distribution is determined with MLE, the next step involves the computation of the Hessian Matrix [34]. This square matrix contains the second derivatives of the log-likelihood function and is essential for establishing confidence intervals for the estimated parameters. By taking the inverse of the Hessian Matrix, the covariance matrix is obtained. This matrix contains the variances of the parameters along the diagonal and the covariances of the parameters off the diagonal. The size of the covariance matrix corresponds to the number of parameters.

For instance, in case of a 2-parameter Weibull model fitted on experimental data, the covariance matrix would take the following form, given in Equation (4).

$$\begin{bmatrix} Var(m) & Cov(m, \sigma_0) \\ Cov(m, \sigma_0) & Var(\sigma_0) \end{bmatrix} \tag{4}$$

By using this 2 × 2 matrix and taking the square root of the variance of m and σ_0, it is possible to determine the standard errors of each parameter at a certain confidence interval and be able to explain the difference between the different Weibull moduli calculated for each batch of fibers before carding or extracted from webs manufactured by carding. This method was used to fit experimental tensile data for carbon fibers [19] and also natural fibers [21] with the Weibull model.

2.3.3. Fitting Performance Assessment

To compare statistical distributions, one non-parametric statistical method is the Kolmogorov–Smirnoff (K-S) test. This test assesses the fit between a sample of data and a reference distribution by measuring the maximum distance between their respective cumulative distribution functions [19,21]. A higher relative value indicates a divergence between the distribution tested and the experimental data. On the other hand, a lower value suggests a higher degree of similarity between the data and the computed distribution. In addition, the p-value can be used to assess the significance of the model in relation to the data. A p-value greater than 0.05 indicates compatibility between the data and the proposed distribution. With these two metrics, it is possible to determine whether the Weibull distribution is a good fit for the experimental data. Additionally, to determine the statistical model best suited to the theoretical data, the Sum Square Error (SSE) can be calculated. This metric aims to achieve convergence by minimizing the sum of the squares of the differences between the observed and predicted values. The higher the value, the greater the disparity between the theoretical and actual values. The K-S test was conducted using a 2-parameter Weibull distribution model to analyze the static strength and fatigue life of T300 carbon fiber [35]. Similarly, Rao et al. applied the K-S test to evaluate the strength probability distribution of virgin carbon fibers, modeled with a two-parameter Weibull CDF [36].

The K-S test and the SSE values will be used simultaneously to determine the best model based on the data and the rCF ratios in the web.

3. Results

3.1. Weibull Model Selection

The tensile properties of recycled carbon fibers are experimentally identified, as described in Section 2.2, and the data are fitted to two Weibull models, the two-parameter model and the three-parameter model. The parameters estimated by MLE are then compared with the actual distribution using the K-S test. Figure 4 gives an example of the K-S test p-values calculated on recycled fibers before carding, for the three tensile parameters under study, and for the Weibull distributions with two and three parameters.

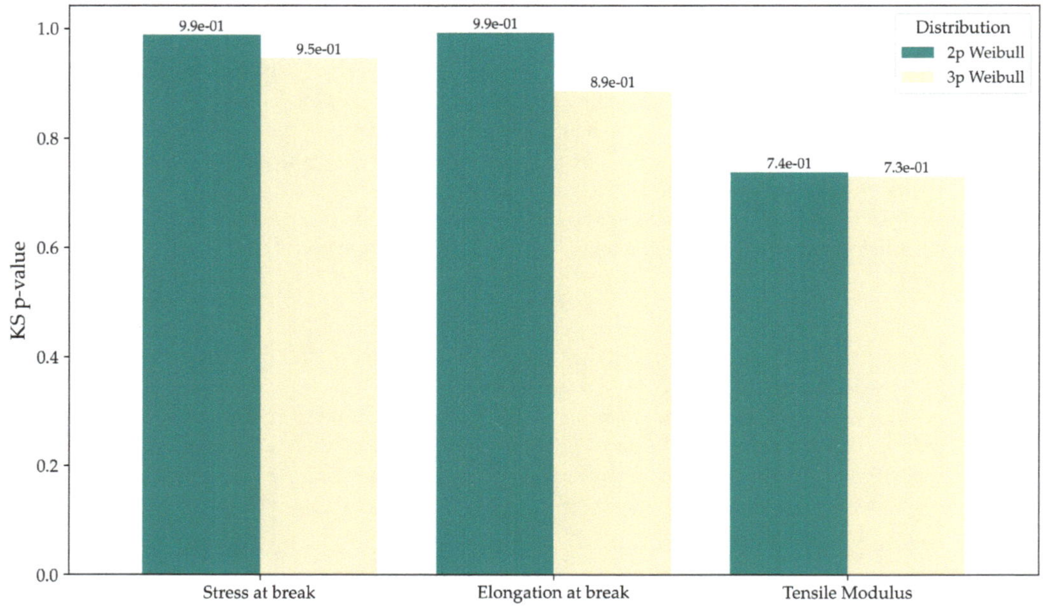

Figure 4. Weibull model evaluation for rCF before carding.

For fibers before carding, the highest K-S test p-value for tensile modulus, stress at break, and elongation at break corresponds to the two-parameter Weibull model. The p-values, which reflect the fit between the theoretical and experimental models, measured for the two-parameter Weibull model, are higher than for the three-parameter model for rCF before carding. The fitting between theoretical and experimental values gave a p-value for stress at break and elongation at break of 0.988 and 0.992, respectively, for the two-parameter distribution, which suggests a high degree of fit with the experimental model, with the value being close to 1. For the tensile modulus, the value of the p-value is lower, at 0.737, but still slightly higher than the one obtained with the three-parameter model for fibers before carding, also suggesting a reasonable fit.

Table A1 (in Appendix A) gives the different values computed with Weibull with two and three parameters for the different rCF/PP blend ratios. Using the K-S test value, the K-S test p-value, and SSE calculations to choose the best model, it was decided that the two-parameter Weibull distribution be used.

Having these values, for generalization purposes and to avoid overfitting for each of the cases, the two-parameter Weibull (Equation (1), with $\sigma_u = 0$) model was used throughout this study. Additionally, it is worth noting that the choice of the two-parameter Weibull model offers a balance between model complexity and goodness of fit, enhancing the interpretability and robustness of the analysis.

3.2. Tensile Properties

Before each tensile strength test, the fiber diameter is measured using an OLYMPUS DSX1000 optical microscope. The diameters of the recycled carbon fibers measured for each batch (Table 1) showed similar variability to virgin carbon fibers, with CV% ranging from 3.37 to 6.11% [30]. After carding, the average diameter of the fibers decreases very slightly, while the standard deviation increases. However, all values remain within the standard deviations of one another, indicating a limited impact of the carding process on fiber diameter. Additionally, it was observed that the transverse section shape of the fibers was not modified by the carding process and remains circular. Figure 5 provides further insights into the variability of fiber properties depending on the fiber ratio. The dispersion observed in these tests highlights significant variability, supported by the standard deviations reported in Table 1.

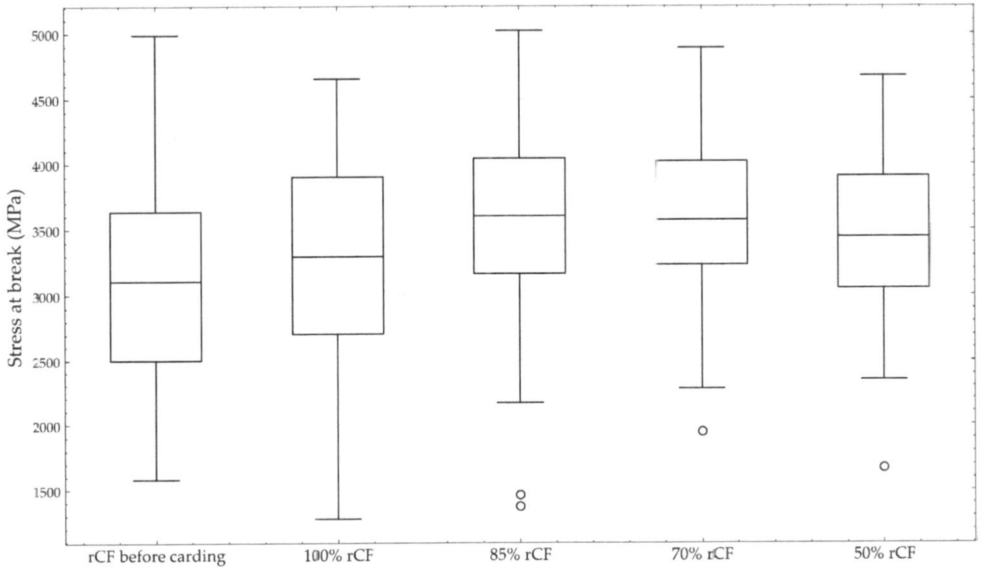

Figure 5. Stress-at-break boxplots for different blending ratios.

The coefficient of variation (CV%) for the tensile modulus is approximately 11%, while for stress and strain at break, it reaches 23%. This level of dispersion is typical for single-fiber tensile tests and underscores the necessity of employing statistical models, particularly probabilistic approaches, to characterize the behavior of these fibers at failure. The variability shown in Table 1 for tensile modulus, stress, and elongation at break highlights the need to use probabilistic models to better analyze and understand the dispersion in these mechanical properties. For the 50–50% ratio, a sample size of 25 tests was retained following post-processing of the tensile data. This number was confirmed to be sufficient through complementary statistical analyses. Levene's test for homogeneity of variance [37] ($p = 0.7144$) indicated no significant differences in variance across all batches, and a cumulative mean analysis demonstrated that 25 tests were sufficient to achieve result stability, ensuring the sample size reliably represented the fiber properties.

3.2.1. Stress at Break

From the values in Table 1, the average stress at break for each batch of fiber showed no significant change in terms of standard deviations. Before carding, the average stress at break of the rCF was 3143 ± 773 MPa. All the stress-at-break values after carding, regardless of the proportion of polypropylene fiber in the blend, are within this standard deviation value.

In order to determine the differences in distributions of the stress at break according to the PP fiber ratio, these values are also studied using a Weibull modulus from the two-parameter model in Figure 6.

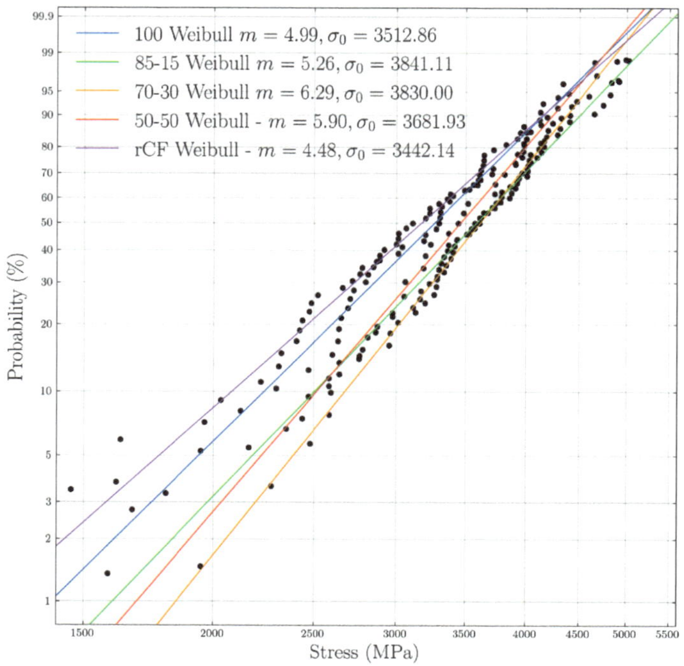

Figure 6. Weibull two-parameter probability plot—stress at break.

The MLE technique is used to obtain a mean value and a standard deviation on the calculated Weibull modulus values. It can already be noted that the number of tests has a significant impact on the standard deviation of the Weibull modulus. In fact, for the 50–50 blend, whose dataset contains 20 fewer usable trials than the 100% rCF batch, the standard deviation rises from 0.59 to 0.92 (Table 2).

Table 2. Two-parameter Weibull modulus values of stress at break for each fiber ratio.

	Data Points	Two-Parameter Weibull Modulus		Kol. Smirn.
		Mean	Std.	p-Value
100% rCF	45	4.99	0.59	0.89
15–85% rCF	49	5.26	0.59	0.99
30–70% rCF	47	6.29	0.71	0.99
50–50% rCF	25	5.90	0.92	0.99
rCF bef. carding	51	4.48	0.47	0.99

The p-values provided in Table 2, particularly those close to 1, indicate a good fit of the experimental data to the Weibull model, apart from the 100% rCF batch where the p-value is slightly lower at 0.89, suggesting a slightly less consistent fitting for this set.

The Weibull modulus values increase with the carding process (100% rCF) compared with fibers recycled before carding (Figure 6), indicating a reduced spread in stress at break data. In relation to the average values and Weibull moduli for each of the fibers extracted from the webs, it can be observed that on the criterion of stress at break, carding has no significantly damaging impact on the carbon fibers.

The slight increase in (m, σ_0) can be attributed to the selection of fibers with a length longer than 30 mm for the test. Carding eliminates the weakest fibers, which are broken during the process. This reduces the probability of these weak fibers being collected during the tensile test. Otherwise, weak fibers present in the initial batch of fibers prior to carding might be selected and included in the tensile test, leading to lower measured strength values.

Adding the crimped PP fiber leads to decreased data scattering regarding the stress at break with a higher Weibull's scale parameter (σ_0). This effect varies with the PP ratio. The highest impact is noted around 30% for a PP–fiber ratio (from 4.99 to 6.29). This leads to larger increases in the scale parameter in comparison with raw rCF before carding, which can be attributed to advantage gained in regards to the elimination of weak fibers, which drafts the distribution to a higher scale parameter. Further, it can be supposed that the crimped PP fiber (with higher inter-fiber entanglement) induces a higher pull-out force exerted to individualize a fiber from a fiber tuft.

3.2.2. Tensile Modulus

To avoid variations in stress due to the resolution of the force sensor, the tensile modulus is calculated by linear regression in the deformation range between 0.1 and 0.6%, in accordance with standard NF ISO 11566. The average tensile modulus of recycled fibers calculated before carding is equal to 186.2 ± 20.8 GPa (Table 1) which places them at the lower end of the range compared with standard modulus (SM) virgin carbon fiber [24].

After carding (Table 1) the tensile modulus of rCFs extracted from webs increases, first comparatively to the tensile modulus of rCF before carding, also and in relation to the ratio of PP in webs. This modulus reaches 206.4 GPa for the fibers extracted from the webs with a ratio of 50–50.

In parallel with the increase in the tensile modulus values for carded fibers compared to fibers before carding, a corresponding increase in the Weibull modulus on the tensile modulus value is observed (Figure 7 and Table 3). For the 50–50 ratio, with the lower *p*-value, the higher standard deviation on the Weibull modulus can be explained with the lower number of experimental data points.

Table 3. Two-parameter Weibull modulus values of tensile modulus for each fiber ratio.

	Data Points	Two-Parameter Weibull Modulus		Kol. Smirn.
		Mean	Std.	*p*-Value
100% rCF	45	16.33	1.84	0.83
15–85% rCF	49	17.85	1.89	0.82
30–70% rCF	47	17.67	1.99	0.84
50–50% rCF	25	14.40	2.18	0.76
rCF bef. carding	51	10.99	1.18	0.74

The heatmap shown in Figure 8 can be used to describe the relationships between two variables by calculating the Pearson coefficient. An absolute value close to one indicates a strong linear correlation between two variables. The "Fiber_pct" variable corresponding to the fiber content in the web shows a strong and positive correlation (0.89) with the value of the Weibull modulus in the tensile modulus, denoted "Point Estimate" in Figure 8.

This indicates a homogenization of the tensile modulus values with carding, on the one hand, but also with the increase in the PP ratio in the web where the tested fibers are extracted. The Weibull modulus values reach a peak for blends 70–30% and 85–15%, suggesting an optimum value of PP ratio in the webs produced.

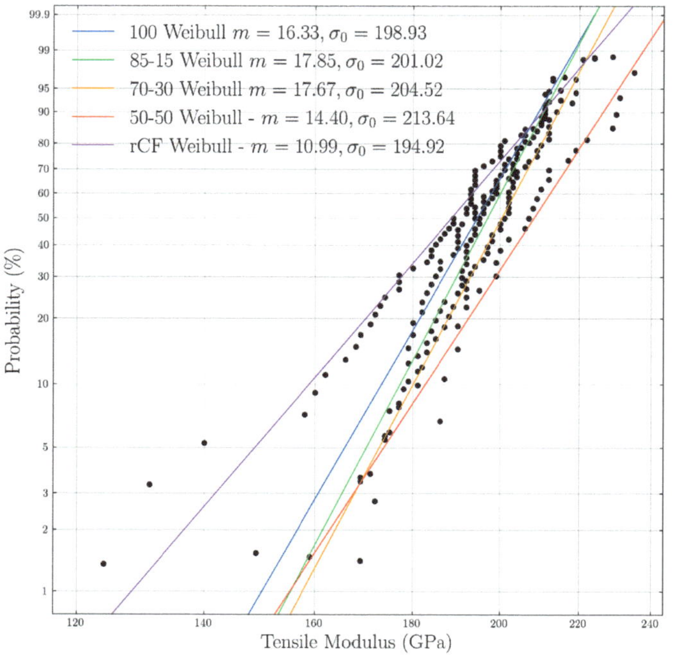

Figure 7. Weibull two-parameter probability plot—tensile modulus.

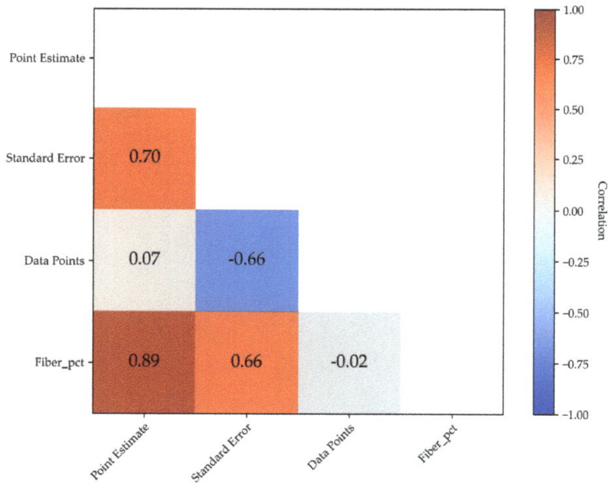

Figure 8. Heatmap for Weibull modulus on the tensile modulus.

This increase in tensile modulus values is highlighted by scientific studies describing the phenomenon of the stiffening of the carbon fibers either during the tensile test with the increase in deformation [38] or during cyclic tensile tests where a slight increase in the tensile modulus of the fibers was observed due to a realignment of the graphite crystal microstructure along the fiber axis [39]. In the fiber extraction process using the carding machine, mechanisms of fiber rupture can occur. When a carding tooth catches a fiber loop, as shown in Figure 9, if the fiber is trapped at both ends within the fiber tuft, it breaks.

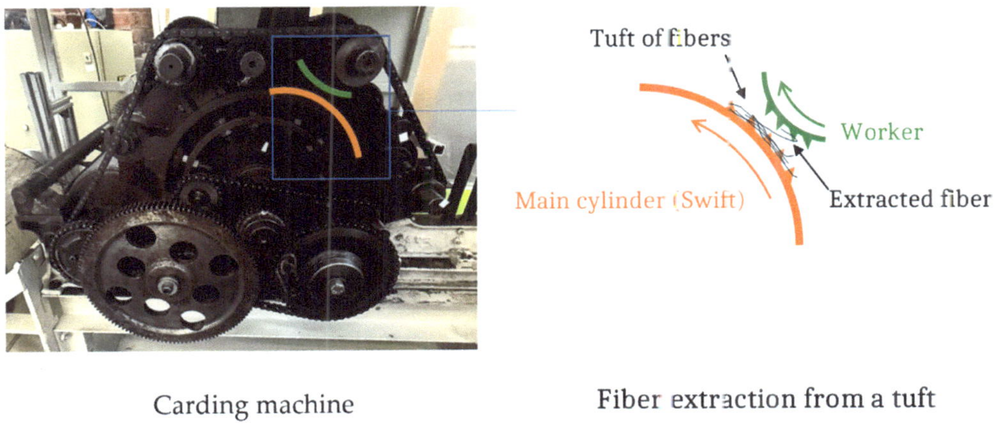

Figure 9. Fiber pull-out from tuft during carding.

Also, if the force applied to extract the fiber exceeds its resistance, the fiber will also break. The extracted fiber is subjected to successive tensile forces, which can either cause it to break or to be extracted from a tuft of fibers. Linked to the observation made by Bunsell et al. [39] on cyclic tensile tests for carbon fibers, this mechanism within the carding machine provides the beginnings of an explanation for the increase in tensile modulus values and stiffening, but also for the homogenization of values resulting in an increase in Weibull modulus values.

3.2.3. Elongation at Break

To complete this investigation, a study of elongation at break values was also carried out. Before carding, the average elongation at break value of the rCF was 1.57 ± 0.37% (Table 1).

Since carbon fiber is a rigid fiber, the trends shown for stress at break also apply to elongation at break. The average values of strain at break of carded rCF, regardless of the level of PP fiber in the blend, remain stable and within the standard deviation of fibers recycled before carding (Table 1).

The Weibull modulus values (Table 4 and Figure 10) show the same trend as that seen for the stress-at-break values. A slight increase in Weibull modulus values was observed with carding and the addition of PP fiber to the blend. For instance, the modulus increased from 4.79 for non-carded fibers to 6.94 for carded carbon fibers in a 50–50 mix, reflecting a homogenization of values as a result of carding and the elimination of weak fibers.

Table 4. Two-parameter Weibull modulus values of elongation at break for each fiber ratio.

	Data Points	Two-Parameter Weibull Modulus		Kol. Smirn.
		Mean	Std.	p-Value
100% rCF	45	5.24	0.62	0.99
15–85% rCF	49	5.87	0.64	0.43
30–70% rCF	47	6.36	0.69	0.96
50–50% rCF	25	6.94	1.10	0.96
rCF bef. carding	51	4.79	0.52	0.99

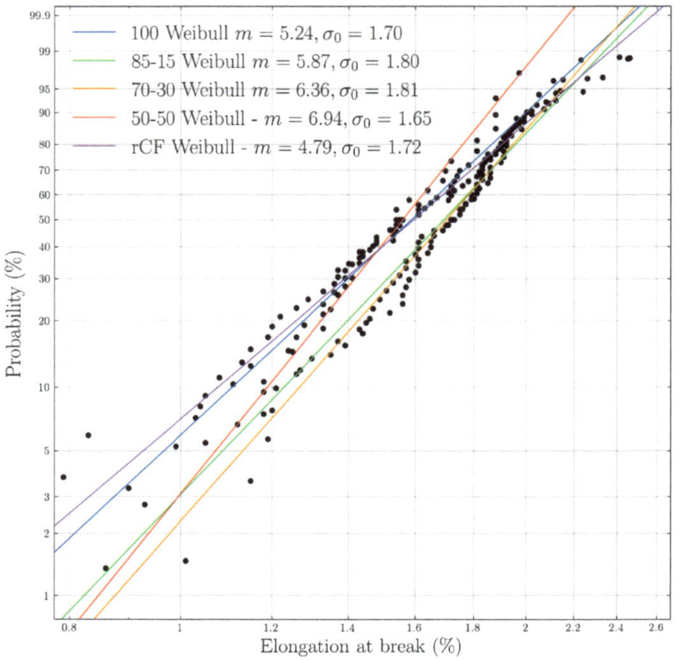

Figure 10. Weibull two-parameter probability plot—elongation at break.

The *p*-values in Table 4. further confirm the good fit of the Weibull two-parameter model to the experimental data, except for the 15–85 rCF blend, where the *p*-value is noticeably lower. This deviation suggests that the two-parameter Weibull model may not fully capture the distribution of elongation-at-break data for this specific blend. While no definitive explanation is available for this anomaly, it is worth noting that when a three-parameter Weibull model is applied (Table A1), the *p*-value for the 15–85 rCF blend remains similarly low.

4. Discussion

The influence of the proportion of thermoplastic polypropylene (PP) fibers on the mechanical properties of recycled carded carbon fibers (rCF) was investigated using a statistical Weibull approach. The comparison between two-parameter and three-parameter Weibull models highlights the better fitting of the two-parameter Weibull model for describing the tensile properties of rCF before carding, leading to its use for the study. Four different web compositions were prepared with varying rCF/PP blend ratios (100/0%, 85/15%, 70/30%, and 50/50%). The experimental results lead to the following conclusions:

- Impact of the proportion of PP fibers on mechanical properties:
  The addition of crimped PP fibers influences the mechanical properties of rCF. Regarding stress at break, the presence of PP fibers decreases the disparity among fibers and shifts the stress distribution to higher values. This improvement is attributed to the removal of the weakest fibers during the blending process. For the tensile modulus, the inclusion of PP fibers further amplifies the slight increases observed after carding alone, particularly in blends with 85/15% and 70/30% rCF/PP ratios. The addition of PP fibers ensures a more consistent distribution of tensile modulus values, as reflected by the increased Weibull modulus. In contrast, elongation at break values remain

stable across all PP blend ratios, with minor improvements in uniformity, as shown by slight increases in the Weibull modulus.

- Impact of the carding process on mechanical properties:
  Carding improves the mechanical consistency of rCF, particularly in webs composed of 100% rCF. The process leads to a slight increase in tensile modulus values for fibers extracted from these webs, indicating improved alignment and structural uniformity. Carding increased the Weilbull modulus values of stress at break, tensile modulus, and elongation at break across all rCF/PP blends. Moreover, carding facilitates the intimate integration of PP fibers with rCF, creating hybrid blends.

Overall, the study demonstrates the potential of carding rCF in conjunction with PP to produce carded webs. A major innovation of this work lies in the integration of a mechanical and statistical approach to evaluate the effects of carding and PP fiber blending. The application of the Weibull statistical model provided insights into the mechanical behavior of rCF. However, the limitations of the Weibull model in exploring mechanical variations have been acknowledged. An upcoming publication will investigate alternative statistical models, including normal, lognormal, and three-parameter Weibull distributions.

Future research will also focus on studying the impact of varying carding parameters on fiber quality and mechanical performance.

In conclusion, the incorporation of thermoplastic fibers in the production of hybrid composites leads to modifications in the mechanical properties of recycled carbon fibers. This impact can be effectively analyzed through a combination of processing techniques and statistical analysis, as demonstrated in this study. Furthermore, the proposed approach can be extended to other carding parameters and textile processing techniques, providing deeper insights into the mechanical properties and ensuring greater consistency in the processing of recycled carbon fibers.

Author Contributions: J.I.: Conceptualization, Formal analysis, Investigation, Methodology, Software, Writing—original draft. A.R.L.: Supervision, Writing—review and editing. D.S.: Project administration, Supervision. Writing—review and editing. All authors have read and agreed to the published version of the manuscript.

Funding: This work was supported by the "Hauts-de-France" region.

Institutional Review Board Statement: Not applicable.

Informed Consent Statement: Not applicable.

Data Availability Statement: The original contributions presented in the study are included in the article, further inquiries can be directed to the corresponding author.

Acknowledgments: The authors would like to thank Kim-Phuc TRAN for his assistance with numerical and statistical methods.

Conflicts of Interest: The authors declare no conflicts of interest.

Appendix A

Table A1. K-S test and SSE parameter values for 2 and 3-parameter Weibull model for each rCF/PP ratio.

rCF Ratio	Two-Parameter Weibull									Three-Parameter Weibull								
	Tensile Modulus			Stress at Break			Elongation at Break			Tensile Modulus			Stress at Break			Elongation at Break		
	SSE	KS	*p*-Val	SSE	KS	*p*-Val	SSE	KS	*p*-Val	SSE	KS	*p*-Val	SSE	KS	*p*-Val	SSE	KS	*p*-Val
50–50	2.31×10^{-4}	1.28×10^{-1}	7.56×10^{-1}	5.23×10^{-9}	6.76×10^{-2}	9.99×10^{-1}	9.45×10^{-1}	9.64×10^{-2}	9.57×10^{-1}	1.40×10^{-4}	8.56×10^{-2}	9.85×10^{-1}	5.06×10^{-9}	6.80×10^{-2}	9.99×10^{-1}	9.11×10^{-1}	9.67×10^{-2}	9.56×10^{-1}
30–70	7.90×10^{-5}	8.64×10^{-2}	8.44×10^{-1}	3.18×10^{-8}	5.38×10^{-2}	9.98×10^{-1}	4.85×10^{-1}	7.12×10^{-2}	9.57×10^{-1}	8.68×10^{-5}	7.75×10^{-2}	9.19×10^{-1}	3.18×10^{-8}	6.09×10^{-2}	9.90×10^{-1}	5.32×10^{-1}	6.91×10^{-1}	9.67×10^{-1}
85–15	4.22×10^{-4}	8.69×10^{-2}	8.22×10^{-1}	2.42×10^{-8}	5.71×10^{-2}	9.94×10^{-1}	1.49×10^{0}	1.22×10^{-1}	4.27×10^{-1}	3.48×10^{-4}	8.28×10^{-2}	8.62×10^{-1}	4.21×10^{-4}	8.69×10^{-2}	8.22×10^{-1}	1.48×10^{0}	1.16×10^{-1}	4.80×10^{-1}
100	2.43×10^{-7}	9.01×10^{-1}	8.26×10^{-1}	3.63×10^{-8}	8.32×10^{-2}	8.88×10^{-1}	9.79×10^{-2}	6.36×10^{-2}	9.87×10^{-1}	1.94×10^{-4}	7.87×10^{-2}	9.22×10^{-1}	3.65×10^{-8}	8.19×10^{-2}	8.98×10^{-1}	1.03×10^{-1}	7.03×10^{-2}	9.68×10^{-1}
rCf b. c.	4.63×10^{-5}	9.27×10^{-2}	7.37×10^{-1}	3.90×10^{-8}	5.97×10^{-2}	9.88×10^{-1}	1.30×10^{-1}	5.74×10^{-2}	9.92×10^{-1}	4.61×10^{-5}	9.28×10^{-2}	7.29×10^{-1}	4.43×10^{-8}	7.05×10^{-2}	9.46×10^{-1}	2.25×10^{-1}	7.87×10^{-1}	8.86×10^{-1}

References

1. Zhang, J.; Chevali, V.S.; Wang, H.; Wang, C.-H. Current Status of Carbon Fibre and Carbon Fibre Composites Recycling. *Compos. Part B Eng.* **2020**, *193*, 108053. [CrossRef]
2. Pimenta, S.; Pinho, S.T. Recycling of Carbon Fibers. In *Handbook of Recycling*; Elsevier: Amsterdam, The Netherlands, 2014; pp. 269–283, ISBN 978-0-12-396459-5.
3. Fernández, A.; Lopes, C.S.; González, C.; López, F.A. Characterization of Carbon Fibers Recovered by Pyrolysis of Cured Prepregs and Their Reuse in New Composites. In *Recent Developments in the Field of Carbon Fibers*; Khanna, R., Cayumil, R., Eds.; InTech: London, UK, 2018; ISBN 978-1-78923-518-0.
4. Oliveux, G.; Bailleul, J.-L.; Gillet, A.; Mantaux, O.; Leeke, G.A. Recovery and Reuse of Discontinuous Carbon Fibres by Solvolysis: Realignment and Properties of Remanufactured Materials. *Compos. Sci. Technol.* **2017**, *139*, 99–108. [CrossRef]
5. Oliveux, G.; Dandy, L.O.; Leeke, G.A. Current Status of Recycling of Fibre Reinforced Polymers: Review of Technologies, Reuse and Resulting Properties. *Prog. Mater. Sci.* **2015**, *72*, 61–99. [CrossRef]
6. Goethals, F.; Demeyer, E.; De Schrijver, I.; Vanneste, M. Pretreating Recycled Carbon Fiber Nonwoven with a Sizing Formulation to Improve the Performance of Thermoplastic Recycled Fiber-Reinforced Composites. *Polymers* **2024**, *16*, 561. [CrossRef] [PubMed]
7. Ivars, J.; Labanieh, A.R.; Soulat, D. Effect of the Fibre Orientation Distribution on the Mechanical and Preforming Behaviour of Nonwoven Preform Made of Recycled Carbon Fibres. *Fibers* **2021**, *9*, 82. [CrossRef]
8. Pimenta, S.; Pinho, S.T. Recycling Carbon Fibre Reinforced Polymers for Structural Applications: Technology Review and Market Outlook. *Waste Manag.* **2011**, *31*, 378–392. [CrossRef]
9. Souppez, J.-B.; Pavar, G. Recycled Carbon Fibre Composites in Automotive Manufacturing. *Int. J. Automot. Manuf. Mater.* **2023**, *2*, 12. [CrossRef]
10. Pakdel, E.; Kashi, S.; Baum, T.; Usman, K.A.S.; Razal, J.M.; Varley, R.; Wang, X. Carbon Fibre Waste Recycling into Hybrid Nonwovens for Electromagnetic Interference Shielding and Sound Absorption. *J. Clean. Prod.* **2021**, *315*, 128196. [CrossRef]
11. Hengstermann, M.; Raithel, N.; Abdkader, A.; Hasan, M.; Cherif, C. Development of New Hybrid Yarn Construction from Recycled Carbon Fibers for High Performance Composites. Part-I: Basic Processing of Hybrid Carbon Fiber/Polyamide 6 Yarn Spinning from Virgin Carbon Fiber Staple Fibers. *Text. Res. J.* **2016**, *86*, 1307–1317. [CrossRef]
12. Abdkader, A.; Bachor, S.; Hasan, M.M.B.; Cherif, C. Development of Yarns from Recycled Carbon Fiber Based on Friction Spinning Technology with Specific Properties for Thermoset Composites. *Text. Res. J.* **2024**, *94*, 12–23. [CrossRef]
13. Akonda, M.H.; Lawrence, C.A.; Weager, B.M. Recycled Carbon Fibre-Reinforced Polypropylene Thermoplastic Composites. *Compos. Part A Appl. Sci. Manuf.* **2012**, *43*, 79–86. [CrossRef]
14. Colombo, B.; Gaiardelli, P.; Dotti, S.; Caretto, F. An Innovative Spinning Process for Production and Characterisation of Ring-Spun Hybrid Yarns from Recycled Carbon Fibre. *J. Clean. Prod.* **2023**, *406*, 137069. [CrossRef]
15. Manis, F.; Stegschuster, G.; Wölling, J.; Schlichter, S. Influences on Textile and Mechanical Properties of Recycled Carbon Fiber Nonwovens Produced by Carding. *J. Compos. Sci.* **2021**, *5*, 209. [CrossRef]
16. Thionnet, A.; Chou, H.Y.; Bunsell, A. Fibre Break Failure Processes in Unidirectional Composites. Part 2: Failure and Critical Damage State Induced by Sustained Tensile Loading. *Appl. Compos. Mater.* **2015**, *22*, 141–155. [CrossRef]
17. Islam, F.; Joannès, S.; Bunsell, A.; Laiarinandrasana, L. Adaptation of Weibull Analysis to Represent Strength Behaviour of Brittle Fibres. In Proceedings of the 22th International Conference on Composite Materials (ICCM22) 2019, Melbourne, Australia, 11–16 August 2019.
18. Thomason, J.L. On the Application of Weibull Analysis to Experimentally Determined Single Fibre Strength Distributions. *Compos. Sci. Technol.* **2013**, *77*, 74–80. [CrossRef]

19. Harikrishnan, R.; Mohite, P.M.; Upadhyay, C.S. Generalized Weibull Model-Based Statistical Tensile Strength of Carbon Fibres. *Arch. Appl. Mech.* **2018**, *88*, 1617–1636. [CrossRef]
20. Amroune, S.; Belaadi, A.; Bourchak, M.; Makhlouf, A.; Satha, H. Statistical and Experimental Analysis of the Mechanical Properties of Flax Fibers. *J. Nat. Fibers* **2022**, *19*, 1387–1401. [CrossRef]
21. Belaadi, A.; Bourchak, M.; Aouici, H. Mechanical Properties of Vegetal Yarn: Statistical Approach. *Compos. Part B Eng.* **2016**, *106*, 139–153. [CrossRef]
22. Cai, G.; Wada, M.; Ohsawa, I.; Kitaoka, S.; Takahashi, J. Tensile Properties of Recycled Carbon Fibers Subjected to Superheated Steam Treatment under Various Conditions. *Compos. Part A Appl. Sci. Manuf.* **2020**, *133*, 105869. [CrossRef]
23. Ateeq, M. A State of Art Review on Recycling and Remanufacturing of the Carbon Fiber from Carbon Fiber Polymer Composite. *Compos. Part C Open Access* **2023**, *12*, 100412. [CrossRef]
24. Matsuhisa, Y.; Bunsell, A.R. Tensile Failure of Carbon Fibers. In *Handbook of Tensile Properties of Textile and Technical Fibres*; Elsevier: Amsterdam, The Netherlands, 2009; pp. 574–602, ISBN 978-1-84569-387-9.
25. *NF ISO 11566*; AFNOR Détermination Des Propriétés En Traction Sur Éprouvette Monofilament. European Union: Brussels, Belgium, 1997.
26. Islam, F.; Joannès, S.; Laiarinandrasana, L. Evaluation of Critical Parameters in Tensile Strength Measurement of Single Fibres. *J. Compos. Sci.* **2019**, *3*, 69. [CrossRef]
27. Engelbrecht-Wiggans, A.E.; Forster, A.L. Analysis of Strain Correction Procedures for Single Fiber Tensile Testing. *Compos. Part A Appl. Sci. Manuf.* **2023**, *167*, 107411. [CrossRef]
28. Jeannin, T.; Arnold, G.; Bourmaud, A.; Corn, S.; De Luycker, E.; Dumont, P.J.J.; Ferreira, M.; François, C.; Grégoire, M.; Harzallah, O.; et al. A Round-Robin Study on the Tensile Characterization of Single Fibres: A Multifactorial Analysis and Recommendations for More Reliable Results. *Compos. Part A Appl. Sci. Manuf.* **2024**, *185*, 108323. [CrossRef]
29. Huguet, E.; Corn, S.; Le Moigne, N.; Ienny, P. Single-Fibre Tensile Testing of Plant Fibres: Set-up Compliance as a Key Parameter for Reliable Assessment of Their Mechanical Behaviour. *Ind. Crops Prod.* **2024**, *222*, 119762. [CrossRef]
30. Mesquita, F.; Bucknell, S.; Leray, Y.; Lomov, S.V.; Swolfs, Y. Single Carbon and Glass Fibre Properties Characterised Using Large Data Sets Obtained through Automated Single Fibre Tensile Testing. *Compos. Part A Appl. Sci. Manuf.* **2021**, *145*, 106389. [CrossRef]
31. Cox, E.P. A Method of Assigning Numerical and Percentage Values to the Degree of Roundness of Sand Grains. *J. Paleontol.* **1927**, *1*, 179–183.
32. Atoui, S.; Belaadi, A.; Boumaaza, M.; Ibrahim, M.H.A.; Ghernaout, D. Weibull Statistics and ANOVA Analysis of the Tensile Mechanical Characteristics of Chamaerops Humilis Cellulose Fibers and Sustainable Twisted Yarns. *J. Nat. Fibers* **2024**, *21*, 2431677. [CrossRef]
33. Zanakis, S.H.; Kyparisis, J. A Review of Maximum Likelihood Estimation Methods for the Three-Parameter Weibull Distribution. *J. Stat. Comput. Simul.* **1986**, *25*, 53–73. [CrossRef]
34. Nketiah, E.A. Parameter Estimation of the Weibull Distribution; Comparison of the Least-Squares Method and the Maximum Likelihood Estimation. *Int. J. Adv. Eng. Res. Sci.* **2021**, *8*, 210–224. [CrossRef]
35. Song, J.; Wen, W.; Cui, H.; Zhao, S. Study on Static and Fatigue Behaviors of Carbon Fiber Bundle and the Statistical Distribution by Experiments. *J. Compos. Mater.* **2015**, *49*, 3157–3168. [CrossRef]
36. Rao, G.S.; Albassam, M.; Aslam, M. Evaluation of Bootstrap Confidence Intervals Using a New Non-Normal Process Capability Index. *Symmetry* **2019**, *11*, 484. [CrossRef]
37. Gastwirth, J.L.; Gel, Y.R.; Miao, W. The Impact of Levene's Test of Equality of Variances on Statistical Theory and Practice. *Stat. Sci.* **2009**, *24*, 343–360. [CrossRef]
38. Kant, M.E.; Crabtree, J.D.; Young, S.; Penumadu, D. Concept of Limit Stress for the Tensile Behavior of Carbon Fiber Composite Tows. *Compos. Part B Eng.* **2020**, *201*, 108384. [CrossRef]
39. Bunsell, A.R.; Somer, A. The Tensile and Fatigue Behaviour of Carbon Fibres. *Plast. Rubber Compos. Process. Appl.* **1992**, *18*, 263–267.

polymers

MDPI

Article

Influence of Cooling Rate on the Flexural and Impact Properties of Compression Molded Non-Woven Flax/PLA Biocomposites

Anurag Pisupati [1,†], Marco Curto [2,†], Thomas Laurent [3], Benoit Cosson [1], Chung Hae Park [1,*] and Hom Nath Dhakal [4,*]

1 Center for Materials and Processes, IMT Nord Europe, Institut Mines-Télécom, Université de Lille, 59000 Lille, France; anurag.pisupati@ct-ipc.com (A.P.); benoit.cosson@uttop.fr (B.C.)
2 Wärtsilä Defence Solutions UK Limited, 4 Marples Way, Havant PO9 1NX, UK; marco.curto@wartsiladefence.com
3 Centre Européen des Textiles Innovants, 59355 Tourcoing, France; t.laurent@team2.fr
4 Advanced Polymers and Composites (APC) Research Group, School of Mechanical and Design Engineering, University of Portsmouth, Portsmouth PO1 3DJ, UK
* Correspondence: chung-hae.park@imt-nord-europe.fr (C.H.P.); hom.dhakal@port.ac.uk (H.N.D.)
† These authors contributed equally to this work.

Abstract: This work investigates the influence of crystallinity on the mechanical properties of needle-punched non-woven flax/polylactic acid (PLA) biocomposites with different flax fiber contents. Biocomposites were fabricated by a compression molding adopting different cooling rates to understand the mechanism of crystallinity and their contribution to the mechanical properties. Image-based analysis of the fiber distribution in non-woven preform indicates the probable origins of the residual porosities and the potential nucleation sites for crystal formation within the composites. The improvement of 25% and 100% in flexural modulus is observed for the composites with 40% and 50% of flax fiber mass fractions, respectively, when subjected to a lower cooling rate, which implies the significant influence of the void content on the brittleness of composites. The impact properties of the composites decrease from 11% to 18% according to the flax fiber mass fraction when the cooling rate decreases to 1 °C/min, and the composites become more brittle. The induced impact and flexural properties of the composites are compared with those of other composites in the literature to emphasize their applicability to semi-structural applications.

Keywords: flax fiber; polylactic acid (PLA); crystallinity; flexural properties; process cycle time; thermal degradation

1. Introduction

Flax fibers as composite reinforcement are of great interest owing to their reasonable cost, low density, and high specific mechanical properties, and they are a promising replacement for glass fibers in composite materials. In many industrial parts, however, the applicability of flax fiber thermoplastic composites is limited to semi-structural applications or non-structural applications. These composites are manufactured using woven or non-woven (oriented or random) fabrics and mats. Although woven reinforcements are of great interest owing to high mechanical performance and drapability, their cost is also relatively higher than their counterparts, i.e., non-wovens. The non-woven fabric reinforcements are becoming popular in the automotive industry for semi-structural or non-structural components since the product cost becomes lower than that of woven fabric parts, which is one of the key driving factors in the automotive sector. Moreover, these non-woven fabrics are easy to handle and provide greater formability. Additionally, the

https://doi.org/10.3390/polym17040493

non-wovens with random fiber arrangement provide quasi-isotropic behavior, unlike oriented reinforcements [1], which can be tailored according to the required loading direction and the application. Furthermore, these random non-woven composites exhibit mechanical performance higher than the injection-molded composites [2] and can be molded into large complex shapes without drapability issues.

Most thermoplastic flax composites adopt polyolefins matrices [3–7]. Using thermoplastics such as polylactic acid (PLA) or other bio-based thermoplastics can be interesting, owing to their easier end-of-life treatment [3,5,8]. Several studies have investigated the feasibility of flax/PLA composites in the past two decades [9–18]. Most of these studies employed injection or compression molding to manufacture flax/PLA composites. The limitation of injection molding is that one cannot achieve high fiber volume fractions (V_f), whereas, with compression molding, high fiber volume fractions such as over 30% can easily be achieved. Furthermore, the fiber length can be much shorter in the case of injection molded composites leading to low mechanical performance.

Compression molding is one of the main manufacturing processes used to produce woven and non-woven polymeric composite parts. Compression molding of composites can be sub-categorized into several manufacturing routes depending on the state of reinforcements. In the case of prepregs, the process is simple and straightforward: the prepregs are cut to the mold shape and consolidated under pressure and heat [19]. In the case of film stacking, the influence of through-thickness permeability comes into play, which significantly affects the impregnation quality [20,21]. A higher processing temperature can be imposed to avoid poor impregnation issues and achieve composite parts with low void content. Temperatures higher than 200 °C are not advised for flax fibers, however, because they may lead to the thermal degradation of flax fibers [7]. Another solution is using commingled fabrics since they reduce the resin flow path and process cycle time, which is a prerequisite for high production volume manufacturing. Commingled fabrics can also be transformed into tapes, which can be used in laser-assisted tape placement for manufacturing complex-shaped parts [22,23] or can be directly used for other manufacturing processes such as pultrusion [24–26] and additive manufacturing [27]. Commingled fabrics can be in the form of fabrics with hybrid yarns or needle-punched non-wovens. In the case of hybrid yarns, thermoplastic and natural fiber yarns are twisted to maintain the form and facilitate weaving, but this leads to poor impregnation and low mechanical performance. As the natural fiber yarns are twisted, the resin penetration can be difficult, leading to high void content. This arrangement also leads to a significant drop in the mechanical properties of the composites [28–31].

Polylactic acid (PLA) is a semi-crystalline polymer whose properties are highly sensitive to the cooling rate imposed during composite processing. Rapid cooling results in reduced crystallinity by limiting the time available for crystal growth at nucleation sites, such as flax fiber surfaces, whereas slower cooling promotes crystallization, allowing for the development of more ordered crystal structures [32,33]. Although rapid cooling is often favored to decrease production time, it can detriment mechanical properties if crystallization is insufficient. This relationship between cooling rate and mechanical performance has been well characterized for glass fiber-reinforced composites [34], yet remains underexplored for flax fiber-reinforced PLA composites.

The relationship between the degree of crystallinity and flax fiber content in PLA composites has been sparsely studied [33,35]. While several investigations in the literature have detailed the crystallization behavior of PLA in the presence of natural fibers [36–38], the scope of this current work is specifically limited to flax fibers at high mass fractions. Xia et al. [33] demonstrated that increasing flax fiber content enhances the transcrystallinity of the PLA matrix, attributing this to the nucleating effect of the fibers. Similarly, Aliotta et al. [35] reported that a higher fiber content leads to increased PLA crystallinity, which improves the mechanical properties of the resulting composites. Furthermore, Bayart et al. [32] improved the crystallization and interfacial adhesion of flax/PLA composites by incorporating titanium dioxide. However, these studies lack comprehensive analyses of the interplay between the crystallization kinetics and the mechanical performance of flax/PLA composites. This study aims to bridge these gaps by systematically investigating the contribution of crystallization kinetics to the mechanical properties of non-woven flax/PLA composites with high fiber mass fractions ($M_f > 30\%$). The research involves characterizing the mechanical and thermal properties of the composites and comparing the results with those in the existing literature. Furthermore, the crystallization kinetics are modeled using analytical approaches. The non-woven composites are fabricated under consistent thermal processing conditions to ensure reproducibility, and their microstructures are examined to establish correlations between the process cycle, crystallinity, and resultant mechanical properties. This comprehensive approach provides a deeper understanding of how thermal processing affects the structural integrity and applicability of flax/PLA biocomposites in semi-structural applications.

2. Materials and Methods

2.1. Reinforcement and Matrix Materials

In this study, needle-punched non-wovens were prepared using commercial PLA filaments from Max model, France, and flax fibers from Procotex SA, Mouscron, Belgium. The preparation steps of dry-laid fabrics are shown in Figure 1. These fibers were delivered in an opening Laroche system using two Laroche feeding rooms. The fibers were opened and weighed to provide an excellent blend with the expected mixture rate to a carding system. The fibers were laid down on a conveying belt and superposed to be fed to the opening system. The feeding system was composed of one pin cylinder that nibbled two layers of fibers and fed a constant and open fiber blend. The fibers were then fed to a carding machine (Excelle Andritz, Graz, Austria) composed of a pre-working unit with three carding groups, a main cylinder with five carding cylinders, and two doffers. The veils coming from these cards were then cross-lapped to form non-woven mats that were bonded through needle punching. The target areal weight of the fabrics was set to 175 g/m^2 with a nominal thickness of 1.8 mm. Three different configurations were manufactured in this study with varying flax mass fractions. The following nomenclature was adopted for non-wovens and their composites: F-XX, where F denotes flax and XX refers to the mass fraction of flax fibers. The actual areal weight of the fabric was measured by weighing at least three large fabric samples with dimensions of 500×500 mm^2. The variation in the areal weights could be caused by the variability in the carding process of both fibers. Furthermore, to investigate the spatial distribution of flax and PLA fibers within the non-woven, the optical measurements of fiber distribution were carried out according to the method suggested by Cosson [39]. This approach provides the information about the distribution of fibers and the fiber orientations within the fabric, which can then be used to predict the degree of anisotropy of the composite properties.

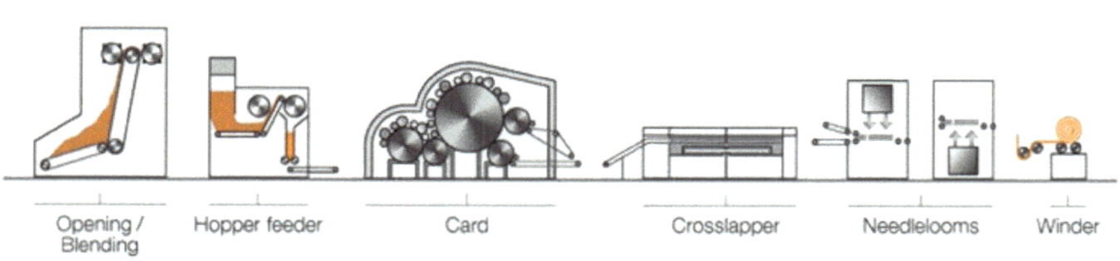

Figure 1. Schematic representation of the carding process.

2.2. Fabrication of Composite Laminates

In this study, the composite laminates were manufactured using the compression molding technique. The non-woven fabrics were cut to the mold dimensions, i.e., 250 × 150 mm^2, and were stored in a conditioning room (23 °C and RH 50%) for at least 24 h and were not dried before the manufacturing. Owing to the varied areal weight of fabrics (see Table 1), the number of layers in each composite plate was not maintained constant. Nevertheless, the total weight of the stack was maintained constant for each plate. The weight of the stack was calculated based on the fiber densities of flax and PLA mentioned in Section 2.1. The cavity height of the mold was 2 mm. The non-woven fabrics were stacked and transferred onto a preheated hydraulic press (PEI, Charlon sur Saône, France). For rapid cooling, the mold was immediately transferred to another hydraulic press (Douloets, Soustons, France) maintained at 23 °C. All the composites were demolded when the mold temperature reached about 40 °C. The adopted consolidation cycle is shown in Figure 2. In this study, a stepped consolidation cycle was selected based on the recommendations of a previous study [7,40]. A stepped consolidation cycle is advantageous for two reasons: since the preform is compacted sequentially, the polymer flow will induce lower drag forces on the fibers, avoiding significant movement during the consolidation step. This approach also ensures that the entrapped air moves out easily. Furthermore, if a sudden pressure is applied on a smaller surface area, there could be significant squeeze flow, which would induce defective parts. At least three plates were manufactured for each composition.

Table 1. Properties of non-wovens.

Non-Woven	Areal Weight (g/m^2)	Thickness (mm)	Strength (N/50 mm)		Elongation (%)	
			MD	CD	MD	CD
F40	161.67 ± 4.19	1.80 ± 0.045	64.0	117.9	73.0	43.5
F50	183.16 ± 0.86	1.88 ± 0.01	52.0	85.9	70.7	43.1
F60	171.67 ± 1.24	1.78 ± 0.02	41.2	63.9	72.1	44.3

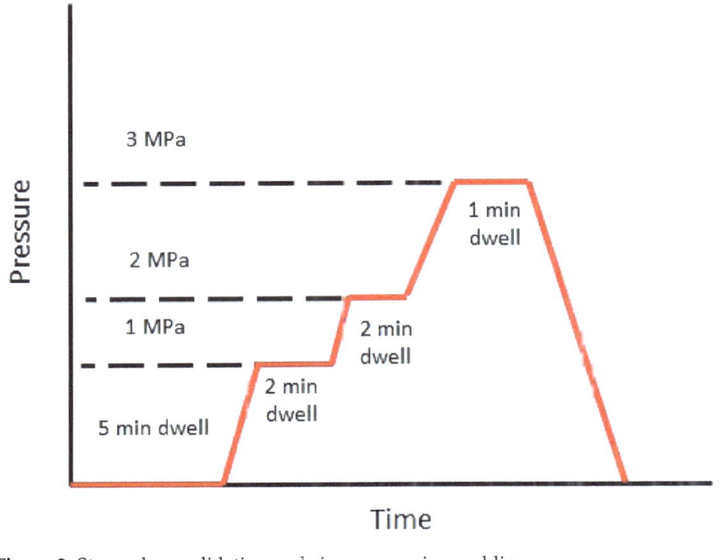

Figure 2. Stepped consolidation cycle in compression molding.

2.3. Flax Fiber Distribution

To understand the variability of the distribution of flax fibers, the optical approach suggested by Cosson was adopted in this study [39]. The method employs a bracketing approach that facilitates the calculation of energy received by the camera and facilitates the calculation of the mean difference to estimate the fiber distributions [41,42]. The approach is macroscopic and ensures faster analysis compared to microscopy or other advanced techniques. Single-layer preforms of dimensions 250 × 150 mm^2 were consolidated at 170 °C and rapidly cooled to ensure no crystals were formed. These single-layer composites were placed against a light source (a 17-inch LED laptop screen) to capture the images. These images were further processed using MATLAB to report the local variations. The image treatment steps are described elsewhere [39].

2.4. Void Content Measurement

The density of the non-woven composites was measured using the Archimedes principle employing ethanol as the test liquid. The samples with dimensions of 20 × 20 mm^2 were cut from the plates randomly to report the average values. Using the apparent density of composites, the void volume fraction of the composite was estimated according to the CRAG report [43]. The density of flax and PLA was assumed to be 1.5 g/cm^3 [44] and 1.24 g/cm^3 [40], respectively.

2.5. Thermal Characterization

2.5.1. Differential Scanning Calorimetry (DSC)

Differential scanning calorimetry (DSC) was used to identify the crystallization kinetics of the pure polymer (i.e., PLA) and flax/PLA non-wovens. The DSC test samples were hermetically sealed in an aluminum crucible and weighed about 6–8 mg. The samples were subjected to a sequential heating and cooling cycle. The heating rate for all the tests was maintained constant at 5 °C/min. To analyze the nonisothermal crystallization, the samples were subjected to three different cooling rates, viz., 1 °C/min, 5 °C/min, and 10 °C/min. On the other hand, isothermal crystallization was investigated at three different temperatures, viz., 90 °C, 100 °C, and 120 °C. These temperature values were selected based on the literature review to analyze the different crystallization behaviors of PLA [45]. The crystallinity (χ) was calculated based on the melting (ΔH_m) and cold crystallization enthalpy (ΔH_{cc}) of the polymer using the following equation (Equation (1)).

$$\chi = \frac{\Delta H_m - \Delta H_{cc}}{(1-\phi)\Delta H_{100\%PLA}} \times 100 \quad (1)$$

where $\Delta H_{100\%PLA}$ corresponds to the complete melting enthalpy of PLA, equivalent to 93.2 J/g, and ϕ corresponds to the flax fiber mass fraction. Generally, the mathematical model suggested by Avrami [46] should be used to model the crystallization kinetics of flax/PLA composites. Avrami suggested a mathematical relation between the polymer's relative crystallinity (χr) and the crystallization time (t). The Avrami equation does not apply to cases with primary and secondary crystallization. This equation only applies to the linear part of the Avrami plot [47,48]. Owing to the nonlinearity of relative crystallinity with time, however, a parallel Avrami model that describes both primary and secondary crystallization was considered in this study [46]. Equation (2) describes the primary and secondary crystallization of the polymer and is generally used for PEEK [46]

$$\chi_r(t) = w_1(1-\exp[-k_1 t^{n_1}]) + w_2(1-\exp[-k_2 t^{n_2}]) \quad (2)$$

where χ_r is the relative crystallinity of the polymer at time t, w_1, and w_2 are weight factors corresponding to primary and secondary crystallizations, respectively, $k_{i=1,2}$ is the crystallization rate constant, and $n_{i=1,2}$ is the Avrami exponent. The weight factors are defined as $w_1 + w_2 = 1$, indicating that in the absence of secondary crystallization, Equation (2) will transform into the conventional Avrami equation [46]. The coefficients in Equation (2) were identified using curve fitting in Python, Version 3.11.

The Avrami equation is applicable for isothermal conditions, and the equation in its current form cannot be used for nonisothermal conditions. Generally, in order to reduce the processing time, the isothermal steps are avoided. Hence, the crystallization occurs during the nonisothermal step, i.e., the cooling step of the process cycle. In order to analyze the nonisothermal crystallization kinetics, Ozawa [49] suggested a model for the nonisothermal step by dividing it into infinitesimally small isothermal steps. This was further modified by Liu et al. [49] by combining Ozawa and Avrami equations, (a.k.a. Mo's model) and is written as follows.

$$\log\beta = \frac{1}{m}\log\left(\frac{K(T)}{Z}\right) - \frac{n}{m}logt \tag{3}$$

$$log\beta = \log\left(\mathrm{F}(T)\right) - alog(T) \tag{4}$$

where β is the cooling rate, n and m are Avrami and Ozawa coefficients, respectively, $K(T)$ is the crystallinity rate, Z is the rate constant considering both nucleation and growth. Equation (3) can be further simplified into Equation (4), where $F(T)$ refers to the value of the cooling rate at a unit crystallization time at a given degree of crystallinity.

2.5.2. Thermogravimetry (TGA) Analysis

To study the thermal stability of the flax/PLA composites, thermogravimetry analysis (TGA) was performed using a Mettler Toledo TGA/DSC1, Viroflay, France. The test samples had a mass between 6 and 9 mg. The samples were placed in a ceramic crucible and were heated at a constant rate of 10 °C/min from 25 °C to 500 °C under an inert atmosphere. At least three samples were tested for each composition to report the average values.

2.6. Mechanical Tests

2.6.1. Flexural Test

The three-point flexural tests were performed at room temperature in a Zwick/Roell Z030 Universal Testing Mleaachine (ZwickRoell ltd., Hertfordshire, UK) fitted with a 30 kN static load cell. The flexural properties (flexural strength and modulus) of compression-molded materials were retrieved according to sample dimensions and support span set respecting the British standard BS EN ISO 178:2003 [50]. Samples were loaded under displacement control at 5 mm/min. Flexural strength and moduli were evaluated according to Equations (5) and (6).

$$E = \frac{L^3m}{4bd^3} \tag{5}$$

$$\sigma = \frac{3FL}{2bd^2} \tag{6}$$

where L is the length of the specimen, b and d are the width and thickness of the specimen, respectively, m is the ratio between the change in force F and the change in displacement s.

2.6.2. Impact Test

The impact characteristics of composite samples were determined with a Charpy impact pendulum device type 5102 (Zwick GmbH, Ulm, Germany) according to ISO 179-2:2020 [51], at a room temperature of 21 °C. Flatwise positioning of the compression-molded un-notched samples was adopted, and a strike energy of 1 Joule was used to test five specimens (80 mm × 10 mm) for each composite type at a bearing distance of 48 mm.

3. Results and Discussion

3.1. Properties of Flax/PLA Non-Wovens

This section presents the properties of non-wovens, such as the areal weight, thickness, and tensile strength (see Table 1). As mentioned in Section 2.1, the target areal weight of the fabrics has been set to 175 g/m^2. However, deviations occurred due to the intrinsic variability of the carding process, which can lead to non-uniform distribution of flax and PLA fibers. Such variability is particularly distinguished in non-woven systems, where the fiber orientation and dispersion directly influence the mechanical performance of the composite, which is a phenomenon previously documented in the literature [15,32,40,52]. These variations can be caused by the non-uniformity of flax, PLA, or both fibers. Using the methodology described in Section 2.2, an initial trial with unconsolidated preform was made, but the distribution of fibers in the thickness direction made it difficult to quantify the variations. In Figure 3, the distribution of flax fibers is presented. The contour plot represents the normalized fiber distribution of the flax fibers within the preform. In the case of F40, the fibers are not well distributed, thus resulting in matrix-rich zones (indicated by dark blue zones), whereas F60 has rather uniform distributions with relatively low matrix-rich zones. These differences in fiber distribution can significantly affect the local impregnation and the subsequent change of mechanical properties [40,53]. Owing to these fiber clusters, the local rigidity of the preform during compaction can increase, leading to low compaction ratios. This local rigidity can be demonstrated by the void volume fraction variations within the composite plate, which will be discussed in Section 3.2.1.

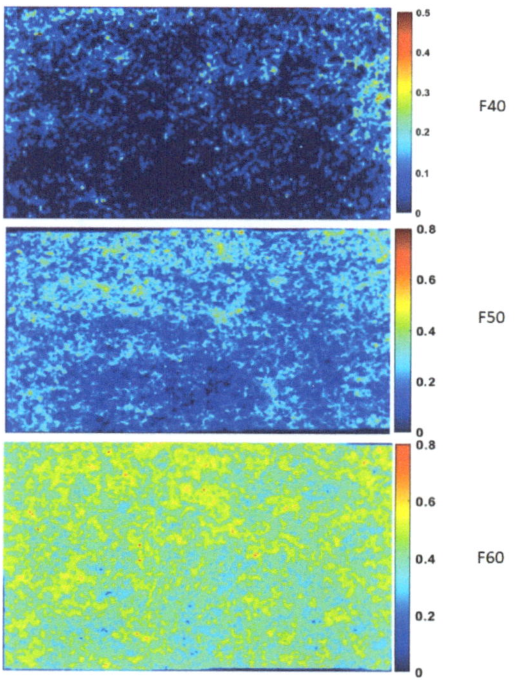

Figure 3. Contour plots of fiber distributions in different flax/PLA non-wovens.

3.1.1. Crystallization Kinetics

In order to analyze the crystallization kinetics of flax/PLA composites, as mentioned in Section 2.5.1, Avrami and Mo's equations were used. For the Avrami model, the relative crystallinity of the material was calculated using Equation (7)

$$X_r(t) = \frac{\int_0^t Q(t)dt}{\int_0^{t_\infty} Q(t)dt} \quad (7)$$

where $Q(t)$ is the heat flow at time t and t_∞ is the total isothermal duration. The relative crystallinity was then used to analyze the different crystallization kinetics of flax/PLA composites.

Table 2 shows the fitting parameters for the parallel Avrami equation, as presented in Section 2.5.2. It can be noticed that the crystallization of PLA occurs in two stages. The primary crystallization refers to the linear part of the curve, which is induced by the growth of crystal lamellas from nucleation sites. In contrast, secondary crystallization is linked to the completion of spherulite formation [54]. The linear part of the Avrami plot can be directly linked to the accelerated initial crystal growth. In works related to the study of PEEK crystallization, it was suggested to use the linear part of the curve to identify the Avrami exponent (n_1 and n_2) and use them to identify the other unknowns in Equation (2). In this study, we adopted a universal approach, where the model was fit to the whole curve without any presumptions concerning the Avrami exponents. Generally, the growth and number of crystals are governed by the primary crystallization. The molecular arrangement is more significant at this phase than at the secondary crystallization phase. The crystals would gradually grow in local zones until the maximum extent is reached, and the completion of this crystallization can be classified as secondary crystallization [55]. In Equation (2), the Avrami exponent n provides information concerning the nature of the nucleation and growth process of the crystals during the cooling phase. It is generally accepted that if the n value is close to 3, the growth of crystal structures is three dimensional, whereas if the value is between 2 and 3, the growth is two dimensional, such as circular lamellas. In Table 2, the n_1 values indicate that the initial growth of crystals was two dimensional. Since most nucleating sites are on the surface of flax fibers, crystals tend to form 2D structures quickly. The n_2 values indicate that the nucleation is sporadic in nature, and the crystals grow into 2D or 1D lamellar aggregates. The addition of flax fibers to PLA significantly improved the crystallization kinetics; however, no further improvement in crystallinity can be noticed. It has been previously reported in the literature that a threshold of flax content exists after which no significant improvement in crystallinity can be observed [56]. In this study, we can notice a similar phenomenon where the crystallization kinetics seem to be the same for all three flax/PLA non-wovens. The study of the growth rate of spherulites or crystal lamellas can be of interest since they can affect the local mechanical properties as well as the microstructure of the materials [32,54].

Nonisothermal crystallization kinetics (Figure 4) were modeled using the approach of Mo's model. This model has been identified to be more accurate than conventional Avrami and Ozawa models in representing secondary crystallization kinetics (see Equation (4)). $F(T)$ is the cooling rate parameter, which can be physically described as the cooling rate required to achieve a relative crystallinity at unit crystallization time. In other words, a higher value of $F(T)$ indicates that a slower cooling rate is needed to achieve crystallization. In Table 3, the coefficients of Equation (4) are presented. The $F(T)$ values for the composites are much smaller than those of the pure PLA matrix, indicating that the crystallization process is faster for a given cooling rate. This is expected since fibers act as nucleating sites and increase the crystallization rate, as observed in the case of isothermal crystallization.

Table 2. Parameters for isothermal crystallization model.

Non-Woven	Temperature (°C)	w_1	k_1 (10^{-7} s^{-1})	n_1	w_2	k_2(10^{-7} s^{-1})	n_2
F40	80	0.65	1.083	2.182	0.344	2.179	1.369
	100	0.63	1.285	2.311	0.363	2.415	1.457
	120	0.62	1.251	2.314	0.372	3.343	1.413
F50	80	0.64	0.961	2.19	0.353	2.496	1.346
	100	0.62	1.284	2.307	0.371	3.071	1.414
	120	0.62	1.253	2.311	0.376	3.975	1.382
F60	80	0.64	0.965	2.195	0.353	2.336	1.357
	100	0.62	1.272	2.311	0.371	2.702	1.437
	120	0.62	1.264	2.312	0.378	3.801	1.389

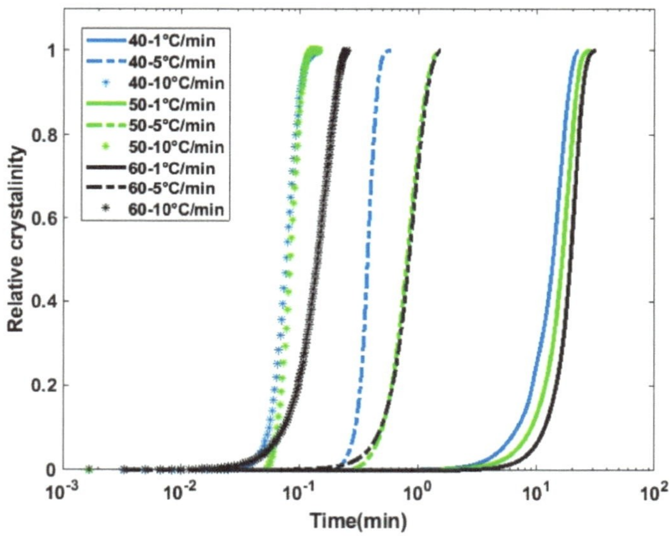

Figure 4. Nonisothermal crystallization kinetics of flax/PLA non-woven composites.

Table 3. Coefficients of Mo's model.

	F40		F50		F60		PLA	
χ (%)	F (T)	a	F (T)	a	F (T)	a	F (T)	a
20	1.042	0.46	0.9328	0.3723	1.01	0.3618	1.409	0.4873
40	1.138	0.448	1.289	0.444	1.395	0.4704	1.526	0.4794
60	1.193	0.4419	1.348	0.44	1.47	0.4969	1.606	0.4765
80	1.242	0.4395	1.393	0.4303	1.54	0.4835	1.681	0.4814

3.1.2. Thermal Degradation Analysis

Figure 5 illustrates the thermograms from thermogravimetric analysis (TGA) of various flax/PLA composites, including pure flax and pure PLA for comparative analysis.

These thermograms reveal distinct thermal degradation patterns, showing that the majority of mass loss occurs around 300 °C for all specimens, indicative of thermal decomposition stages for both the flax and PLA components. Initial degradation begins at approximately 220 °C, aligning with the known thermal instability of flax fibers at elevated temperatures [7]. As flax fiber mass fraction increases, the thermal stability of the composites improves slightly, attributed to the higher residual char content, which acts as a thermal barrier, slowing down further decomposition. The initial mass reduction below 100 °C is likely due to the evaporation of inherent moisture within the specimens, which is a common characteristic in natural fibers. The mass loss observed between 100 °C and 200 °C can be linked to the decomposition of organic compounds such as pectin, lignin, and waxes within the flax fibers, contributing to a gradual weight decrease in the flax and its composites [57]. This breakdown of organic matter occurs prior to the onset of main degradation, which involves the degradation of the PLA matrix and further breakdown of cellulose. Interestingly, the thermal responses of F50 and F60 composites are nearly identical, implying two possible scenarios: a non-uniform distribution of flax and PLA in the preform or a threshold beyond which additional flax integration no longer improves the thermal stability significantly. Higher flax fiber mass fraction may provide enhanced stability by increasing the amount of char, which acts as an insulating layer, slowing the rate of thermal degradation in the composite. Additionally, the increased char residue at high flax mass fractions could imply a protective effect, further enhancing the thermal stability and integrity of the composite materials during prolonged thermal exposure. These findings emphasize the role of flax mass fraction in improving composite performance under heat, which is crucial for applications where thermal endurance is required at elevated temperatures.

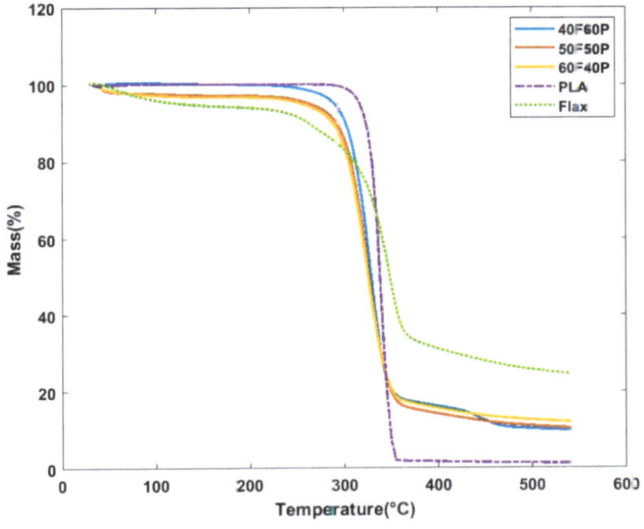

Figure 5. Thermal behavior of flax/PLA composites.

3.2. Mechanical Properties of Flax/PLA Composites

3.2.1. Void Volume Fraction and Crystallinity of Composites

In Figure 6, the void volume fraction and crystallinity of flax/PLA composites are presented, highlighting critical structural characteristics influenced by the flax fiber mass fraction within the preforms. As hypothesized in Section 3.1, the void volume fraction increases as the flax fiber mass fraction increases. This rise in void volume fraction can be attributed to two primary factors: the spatial distribution of flax fibers and the mechanical behavior of the preform under compaction. Firstly, the non-uniform distribution of fibers within the composite can lead to local agglomerations of flax fibers, forming a more tortuous path for the PLA melt to penetrate and fully impregnate the reinforcement. This phenomenon not only hinders uniform PLA distribution but also creates micro-regions within the composite where resin infiltration is incomplete, leading to air entrapment. Secondly, the accumulation of fiber clusters increases the local rigidity of the preform, which restricts compaction and can result in uneven density across the composite. This variability in compaction can further contribute to the development of discrete compaction zones, with varying degrees of fiber packing and air retention. The void volume fraction in these composites can be classified into two primary categories: air-entrapment voids and fiber–matrix interfacial voids. Air-entrapment voids are primarily due to the incomplete displacement of air pockets during the impregnation phase. Meanwhile, voids at the fiber–matrix interface arise due to insufficient bonding between flax fibers and PLA. These interfacial voids can significantly influence the composite's mechanical properties, as they weaken the stress transfer between the fibers and the polymer matrix, reducing the overall structural integrity [58]. Additionally, the degree of crystallinity of PLA is sensitive to flax fiber mass fraction and distribution. A higher flax fiber mass fraction can increase the number of nucleation sites for the crystallization of the PLA matrix, potentially enhancing its thermal stability and stiffness. An excess of voids due to poor impregnation and interfacial gaps may disrupt the crystallinity, however, as air pockets interrupt the molecular alignment needed for uniform crystallite formation.

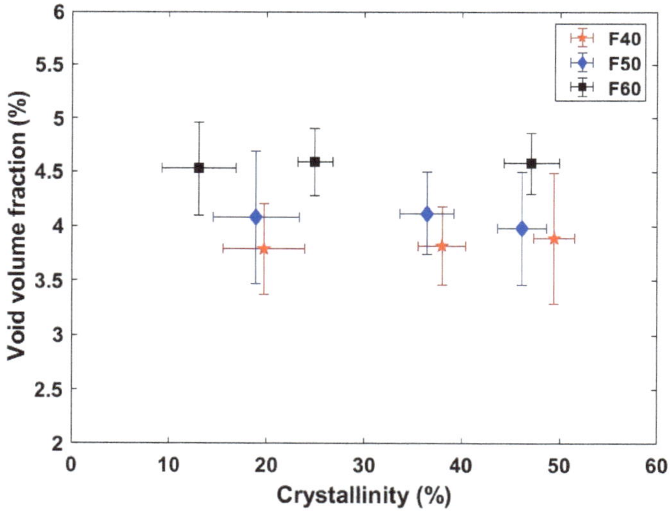

Figure 6. Residual void content and crystallinity of flax/PLA composites.

3.2.2. Flexural Properties

In this study, flexural tests were carried out using a three-point bending method as shown in Figure 7. The influence of cooling rate on the flexural properties of flax/PLA composites can be observed in Figure 8. F40 and F50 composites exhibited a trend that was well reported in the literature [34] (see Figure 8). With an increase in the cooling rates, the F40 and F50 composites showed a drop in flexural modulus, which is coherent with the cooling rate and degree of crystallinity. Conversely, the trend for the other F60 flax/PLA composites was different. This can be caused by the lack of impregnation and increased void volume fraction, as stated in Section 2.2. It can thus be inferred that the distribution of PLA and flax fibers in the non-woven fabric was non-uniform, leading to fiber-rich and matrix zones. However, it has been observed that the flexural moduli obtained in the current study are slightly higher than the values reported in the literature [59,60].

Figure 7. Experimental setups for mechanical characterization: flexural tests (**left**) and impact tests (**right**).

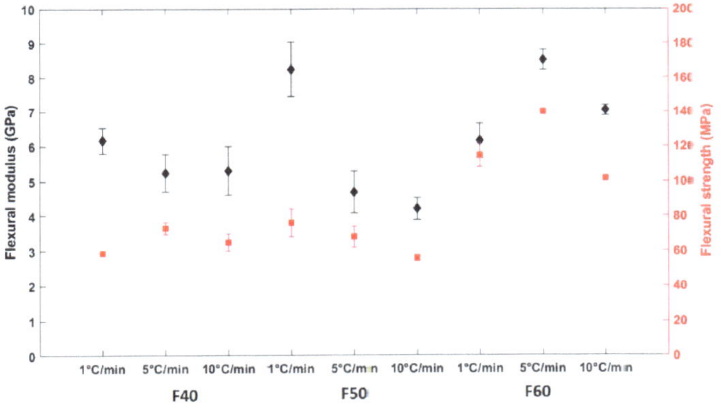

Figure 8. Flexural properties of flax/PLA composites subjected to different cooling rates.

It is difficult to find a global trend in the flexural strength against the cooling rate. We can also see that the flexural strength of F60 is significantly greater than that of F40. It can be hypothesized that this difference can be linked to the flax fiber mass fraction. With the increase in the flax fiber mass fraction, the reinforcing effect of flax fibers becomes greater. At the same time, the probability of available nucleating zones increases significantly, enabling easier crystallization of PLA and resulting in a higher degree of crystallization. For F60 composites, the flexural strength increases with a decrease in the cooling rate (from 10 to 5 °C/min), indicating the transition of the properties of composites into a relatively brittle behavior. A similar trend can also be noticed for F50 composites. In the case of F40 composites, a drop in the flexural strength was noticed with a decrease in the crystallinity.

To evaluate the performance of flax/PLA composites, their specific flexural modulus and strength are compared with those of commercial short glass fiber thermoplastic composites designed for semi-structural applications [61]. In our work, flax/PLA composites with a 40% fiber mass fraction exhibit average specific flexural modulus and strength of

4.11 GPa/g/cm^3 and 55 MPa/g/cm^3, respectively. These values are comparable to the specific flexural modulus and strength of short glass fiber composites (30% fiber mass volume fraction), which are 3.5 GPa/g/cm^3 and 100 MPa/g/cm^3, respectively. Although the flexural strength of flax/PLA composites is lower, it is expected that increasing the flax fiber mass fraction can reduce this difference. When F60 composites are analyzed, they demonstrate superior average specific flexural modulus and strength of 6.03 GPa/g/cm^3 and 100.45 MPa/g/cm^3, respectively. This implies that composites manufactured under optimal process conditions and material configurations can offer a promising alternative to short glass fiber composites for such applications.

3.2.3. Impact Properties

In Figure 9, the results of the impact resistance of flax/PLA composites under varying cooling rates reveal significant differences, highlighting the influence of the cooling rate on the material brittleness. As the cooling rate decreases and crystallinity increases, the composites exhibit a more brittle behavior, leading to reduced impact resistance, which is a phenomenon well reported in the literature [9,62]. Specifically, the impact resistance of F40 composites decreased by approximately 11% under a slow cooling rate of 1 °C/min, while F60 composites exhibited a more pronounced reduction of 18% at the same cooling rate. Interestingly, F50 composites showed minimal variation in impact resistance across different cooling rates, implying that the fiber and matrix distribution may play a stabilizing role in this configuration.

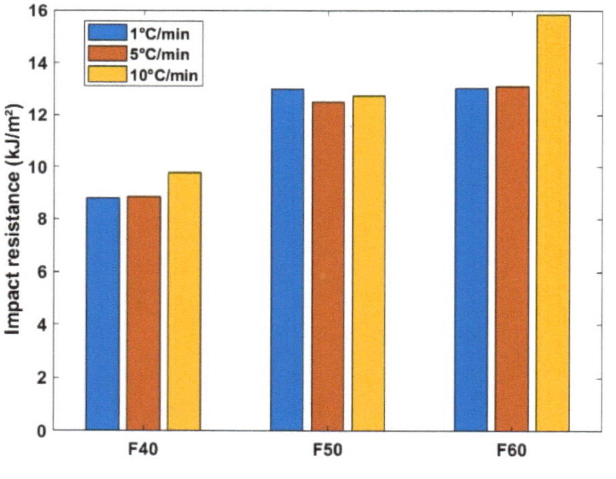

Figure 9. Impact properties of flax/PLA composites subjected to different cooling rates.

Bax and Mussig reported 11.13 kJ/m^2 impact strength for random flax/PLA composites with 30% of fiber mass fraction, whereas Foruzanmehr et al. [63] reported an impact strength of 18 kJ/m^2 for UD flax/PLA composites with 34% of fiber mass fraction. Compared to the results in the literature [64], the current composites exhibit better impact properties, and the distribution of flax fibers is likely to have a substantial influence on the composite's toughness. Higher flax mass fraction can increase the formation of fiber-rich zones, which may lead to uneven crystallization and localized brittleness, reducing impact performance. Furthermore, the differences in void volume fraction have a critical influence on the impact resistance. Composites with high void volume fraction are less capable of withstanding impact loads due to insufficient energy absorption and accelerated crack propagation. The voids, especially those at the fiber–matrix interface, act as stress concentrators, which weaken the material and lead to faster fracture under impact [10]. Thus, the cooling rate and fiber distribution both play crucial roles in determining the impact properties of flax/PLA composites. A slower cooling rate generally promotes higher crystallinity, which can compromise impact resistance by increasing brittleness, even if it is beneficial to improve the stiffness.

4. Conclusions

This study systematically investigated the crystallization kinetics and mechanical properties of flax/PLA non-woven composites under different thermal processing conditions. The composites were fabricated using a dry layup and needle-punching approach, achieving flax fiber mass fractions up to 60%. The results showed that a 40% flax fiber mass fraction led to a 25% improvement in flexural modulus, while a 50% flax fiber mass fraction increased it by 100% under slow cooling conditions. Impact properties, however, decreased by approximately 11% for composites subjected to a cooling rate of 1 °C/min, implying increased brittleness. Void content analysis indicated that higher fiber loading led to localized matrix-rich and fiber-rich zones, influencing structural uniformity. Isothermal crystallization kinetics were accurately modeled using a parallel Avrami approach, revealing non-linear growth behaviors, while non-isothermal kinetics demonstrated significantly enhanced crystallization rates due to the nucleating effect of flax fibers, with F (T) values for flax/PLA composites that are 20–25% lower than those of pure PLA. The specific mechanical properties were excellent, with flexural modulus and strength reaching up to 6.03 GPa/g/cm^3 and 100.45 MPa/g/cm^3, respectively, comparable to those of short glass fiber composites with a 30% fiber mass fraction. These results suggest that flax/PLA composites are promising for semi-structural applications, providing a sustainable alternative with enhanced lightweight performance. Furthermore, in addition to their improved mechanical performance, it would be interesting to carry out a detailed LCA analysis to understand the influence of process parameters on the environmental indicators.

Author Contributions: A.P.: conceptualization, investigation, methodology, data curation, formal analysis, visualization, writing—original draft; M.C.: conceptualization, investigation, methodology, resources, formal analysis, writing—original draft; T.L.: investigation, resources, formal analysis; B.C.: formal analysis, resources, software, visualization; C.H.P.: methodology, writing—review and editing, funding acquisition; H.N.D.: methodology, writing—review and editing, funding acquisition. All authors have read and agreed to the published version of the manuscript.

Funding: European Regional Development Fund (ERDF), INTERREG 2SEAS. Contract No 2S06-006.

Data Availability Statement: Data is contained within the article. The original contributions presented in this study are included in the article. Further inquiries can be directed to the corresponding author(s).

Acknowledgments: The authors would like to thank the European Regional Development Fund (ERDF) for the financial support to the SEABIOCOMP project within the framework of INTERREG2SEAS under subsidy contract No 2S06-006.

Conflicts of Interest: Marco Curto was employed by the company Wärtsilä Defence Solutions UK Limited. This work was conducted before Marco Curto joined Wärtsilä Defence Solutions UK Limited. Thomas Laurent was employed by Centre Européen des Textiles Innovants. The remaining authors declare that the research was conducted in the absence of any commercial or financial.

References

1. Yan, L.; Chouw, N.; Jayaraman, K. Flax fibre and its composites—A review. *Compos. Part B Eng.* **2014**, *56*, 296–317. [CrossRef]
2. Shah, D.U. Natural fibre composites: Comprehensive Ashby-type materials selection charts. *Mater. Des. 1980–2015* **2014**, *62*, 21–31. [CrossRef]

3. Pantaloni, D.; Shah, D.; Baley, C.; Bourmaud, A. Monitoring of mechanical performances of flax non-woven biocomposites during a home compost degradation. *Polym. Degrad. Stab.* **2020**, *177*, 109166. [CrossRef]
4. Martin, N.; Davies, P.; Baley, C. Comparison of the properties of scutched flax and flax tow for composite material reinforcement. *Ind. Crops Prod.* **2014**, *61*, 284–292. [CrossRef]
5. Bourmaud, A.; Le Duigou, A.; Baley, C. What is the technical and environmental interest in reusing a recycled polypropylene–hemp fibre composite? *Polym. Degrad. Stab.* **2011**, *96*, 1732–1739. [CrossRef]
6. Bar, M.; Alagirusamy, R.; Das, A.; Ouagne, P. Low velocity impact response of flax/polypropylene hybrid roving based woven fabric composites: Where does it stand with respect to GRPC? *Polym. Test.* **2020**, *89*, 106565. [CrossRef]
7. Kim, S.H.; Park, C.H. Direct impregnation of thermoplastic melt into flax textile reinforcement for semi-structural composite parts. *Ind. Crops Prod.* **2017**, *95*, 651–663. [CrossRef]
8. Azka, M.A.; Sapuan, S.M.; Abral, H.; Zainudin, E.S.; Aziz, F.A. An examination of recent research of water absorption behavior of natural fiber reinforced polylactic acid (PLA) composites: A review. *Int. J. Biol. Macromol.* **2024**, *268*, 131845. [CrossRef] [PubMed]
9. Oksman, K.; Skrifvars, M.; Selin, J.F. Natural fibres as reinforcement in polylactic acid (PLA) composites. *Compos. Sci. Technol.* **2003**, *63*, 1317–1324. [CrossRef]
10. Bax, B.; Müssig, J. Impact and tensile properties of PLA/Cordenka and PLA/flax composites. *Compos. Sci. Technol.* **2008**, *68*, 1601–1607. [CrossRef]
11. Bodros, E.; Pillin, I.; Montrelay, N.; Baley, C. Could biopolymers reinforced by randomly scattered flax fibre be used in structural applications? *Compos. Sci. Technol.* **2007**, *67*, 462–470. [CrossRef]
12. Nassiopoulos, E.; Njuguna, J. Thermo-mechanical performance of poly(lactic acid)/flax fibre-reinforced biocomposites. *Mater. Des.* **2015**, *66*, 473–485. [CrossRef]
13. Rubio-López, A.; Artero-Guerrero, J.; Pernas-Sánchez, J.; Santiuste, C. Compression after impact of flax/PLA biodegradable composites. *Polym. Test.* **2017**, *59*, 127–135. [CrossRef]
14. Le Duigou, A.; Bourmaud, A.; Davies, P.; Baley, C. Long term immersion in natural seawater of Flax/PLA biocomposite. *Ocean Eng.* **2014**, *90*, 140–148. [CrossRef]
15. Gautreau, M.; Kervoelen, A.; Barteau, G.; Delattre, F.; Colinart, T.; Pierre, F.; Hauguel, M.; Le Moigne, N.; Guillon, F.; Bourmaud, A.; et al. Fibre individualisation and mechanical properties of a flax-PLA non-woven composite following physical pre-treatments. *Coatings* **2021**, *11*, 846. [CrossRef]
16. Wang, X.-Y.; Wang, Q.-H.; Gu, H. Research on Mechanical Behavior of the Flax/Polyactic Acid Composites. *J. Reinf. Plast. Compos.* **2010**, *29*, 2561–2567. [CrossRef]
17. Khanlou, H.M.; Woodfield, P.; Summerscales, J.; Hall, W. Consolidation process boundaries of the degradation of mechanical properties in compression moulding of natural-fibre bio-polymer composites. *Polym. Degrad. Stab.* **2017**, *138*, 115–125. [CrossRef]
18. Couture, A.; Lebrun, G.; Laperrière, L. Mechanical properties of polylactic acid (PLA) composites reinforced with unidirectional flax and flax-paper layers. *Compos. Struct.* **2016**, *154*, 286–295. [CrossRef]
19. Dhakal, H.N.; Zhang, Z.Y.; Guthrie, R.; MacMullen, J.; Bennett, N. Development of flax/carbon fibre hybrid composites for enhanced properties. *Carbohydr. Polym.* **2013**, *96*, 1–8. [CrossRef] [PubMed]
20. Woigk, W.; Fuentes, C.A.; Rion, J.; Hegemann, D.; van Vuure, A.W.; Kramer, E.; Dransfeld, C.; Masania, K. Fabrication of flax fibre-reinforced cellulose propionate thermoplastic composites. *Compos. Sci. Technol.* **2019**, *183*, 107791. [CrossRef]
21. Ouagne, P.; Bizet, L.; Baley, C.; Bréard, J. Analysis of the Film-stacking Processing Parameters for PLLA/Flax Fiber Biocomposites. *J. Compos. Mater.* **2010**, *44*, 1201–1215. [CrossRef]
22. Zhang, Z.; Wang, S.; Ma, Y.; Pan, B.; Sun, M.; Zhang, G.; Chai, H.; Li, J.; Jiang, S. Laser-assisted thermoplastic composite automated fiber placement robot for bonding GF/PP unidirectional composites and braided composites. *Compos. Part B Eng.* **2024**, *287*, 111798. [CrossRef]
23. Islam, M.A.; Islam, M.M.; Yang, C.; Wodag, A.F.; Wang, R.; Chen, W.; Zhou, B.; Gao, S.; Xu, F. Development of three dimensional (3D) woven flax/PLA composites with high mechanical and thermal properties using braided yarns. *Ind. Crops Prod.* **2024**, *222*, 119580. [CrossRef]
24. Pisupati, A.; Boivin, A.; Beigbeder, A.; Méndez-Rial, R.; Le Goff, R. Applicability of circularity protocols to extend the lifetime of a thermoplastic pultrusion line: A case study. *Manuf. Lett.* **2024**, *42*, 56–60. [CrossRef]
25. Linganiso, L.Z.; Bezerra, R.; Bhat, S.; John, M.; Braeuning, R.; Anandjiwala, R.D. Pultrusion of flax/poly(lactic acid) commingled yarns and nonwoven fabrics. *J. Thermoplast. Compos.* **2013**, *27*, 1553–1572. [CrossRef]
26. Minchenkov, K.; Vedernikov, A.; Kuzminova, Y.; Gusev, S.; Sulimov, A.; Gulyaev, A.; Kreslavskaya, A.; Prosyanoy, I.; Xian, G.; Akhatov, I.; et al. Effects of the quality of pre-consolidated materials on the mechanical properties and morphology of thermoplastic pultruded flat laminates. *Compos. Commun.* **2022**, *35*, 101281. [CrossRef]
27. Bi, Z.; Li, Q.; Zhang, Z.; Zhang, Z.; Yang, W.; Li, Y. Experimental and numerical evaluation of the influence of voids on sound absorption behaviors of 3D printed continuous flax fiber reinforced PLA composites. *Compos. Sci. Technol.* **2024**, *255*, 110720. [CrossRef]

28. Bar, M.; Alagirusamy, R.; Das, A. Properties of flax-polypropylene composites made through hybrid yarn and film stacking methods. *Compos. Struct.* **2018**, *197*, 63–71. [CrossRef]
29. Shah, D.U.; Schubel, P.J.; Licence, P.; Clifford, M.J. Determining the minimum, critical and maximum fibre content for twisted yarn reinforced plant fibre composites. *Compos. Sci. Technol.* **2012**, *77*, 1909–1917. [CrossRef]
30. Zhang, L.; Miao, M. Commingled natural fibre/polypropylene wrap spun yarns for structured thermoplastic composites. *Compos. Sci. Technol.* **2010**, *70*, 130–135. [CrossRef]
31. Guillou, E.; King, A.; Perrin, J.; Proudhon, H.; Weitkamp, T.; Shah, D.U.; Beigbeder, A.; Ouagne, P.; Bourmaud, A. Impact of flax fibre micro-structural features on composite damage observed through micro-CT characterisation. *Compos. Part Appl. Sci. Manuf.* **2024**, *181*, 108118. [CrossRef]
32. Bayart, M.; Foruzanmehr, M.R.; Vuillaume, P.Y.; Ovlaque, P.; Robert, M.; Elkoun, S. Poly(lactic acid)/flax composites: Effect of surface modification and thermal treatment on interfacial adhesion, crystallization, microstructure, and mechanical properties. *Compos. Interfaces* **2021**, *29*, 17–36. [CrossRef]
33. Xia, X.; Shi, X.; Liu, W.; Zhao, H.; Li, H.; Zhang, Y. Effect of flax fiber content on polylactic acid (PLA) crystallization in PLA/flax fiber composites. *Iran. Polym. J.* **2017**, *26*, 693–702. [CrossRef]
34. Mahato, B.; Babarinde, V.O.; Abaimov, S.G.; Lomov, S.V.; Akhatov, I. Interface strength of glass fibers in polypropylene: Dependence on the cooling rate and the degree of crystallinity. *Polym. Compos.* **2020**, *41*, 1310–1322. [CrossRef]
35. Aliotta, L.; Gigante, V.; Coltelli, M.B.; Cinelli, P.; Lazzeri, A.; Seggiani, M. Thermo-mechanical properties of PLA/short flax fiber biocomposites. *Appl. Sci.* **2019**, *9*, 3797. [CrossRef]
36. Jin, X.; Chen, X.; Cheng, Q.; Zhang, N.; Cai, S.; Ren, J. Non-isothermal crystallization kinetics of ramie fiber-reinforced polylactic acid biocomposite. *RSC Adv.* **2017**, *7*, 46014–46021. [CrossRef]
37. Lee, S.H.; Wang, S.; Teramoto, Y. Isothermal crystallization behavior of hybrid biocomposite consisting of regenerated cellulose fiber, clay, and poly(lactic acid). *J. Appl. Polym. Sci.* **2008**, *108*, 870–875. [CrossRef]
38. Cai, J.; Liu, M.; Wang, L.; Yao, K.; Li, S.; Xiong, H. Isothermal crystallization kinetics of thermoplastic starch/poly(lactic acid) composites. *Carbohydr. Polym.* **2011**, *86*, 941–947. [CrossRef]
39. Cosson, B. Optical measurement of local permeability of flax fiber fabrics before liquid composite molding. *J. Compos. Mater.* **2018**, *52*, 3289–3297. [CrossRef]
40. Pisupati, A.; Leroy, M.; Laurent, T.; Park, C.H. Development and investigation of nonwoven preforms for self-reinforced polylactic acid composites. *J. Reinf. Plast. Compos.* **2024**, *43*, 72–83. [CrossRef]
41. Jabeen, R.; Cosson, B.; Akué Asséko, A.C.; Verstraete, S.; Desplentere, F.; Park, C.H. Effect of fibre orientation on the light scattering during laser transmission welding. *J. Manuf. Process.* **2023**, *86*, 1–9. [CrossRef]
42. Cosson, B.; Akué Asséko, A.C.; Dauphin, M. A non-destructive optical experimental method to predict extinction coefficient of glass fibre-reinforced thermoplastic composites. *Opt. Laser Technol.* **2018**, *106*, 215–221. [CrossRef]
43. Kretsis, G.; He, Y.; Makeev, A.; Concrete, R.; Goh, G.D.L.; Dikshit, V.; Nagalingam, A.P.; Goh, G.D.L.; Agarwala, S.; Sing, S.L.; et al. CRAG (Composite Research Advisory Group) Test Methods for the Measurement of the Engineering Properties of Fibre Reinforced Plastics. *Fifth Conf. Compos. Mater. Test. Des.* **1988**, *18*, 631–738.
44. Gall, M.L.; Davies, P.; Martin, N.; Baley, C. Recommended flax fibre density values for composite property predictions. *Ind. Crops Prod.* **2018**, *114*, 52–58. [CrossRef]
45. Foglia, F.; De Meo, A.; Iozzino, V.; Volpe, V.; Pantani, R. Isothermal crystallization of PLA: Nucleation density and growth rates of α and α′ phases. *Can. J. Chem. Eng.* **2020**, *98*, 1998–2007. [CrossRef]
46. Velisaris, C.N.; Seferis, J.C. Crystallization kinetics of polyetheretherketone (peek) matrices. *Polym. Eng. Sci.* **1986**, *26*, 1574–1581. [CrossRef]
47. Liao, R.; Yang, B.; Yu, W.; Zhou, C. Isothermal cold crystallization kinetics of polylactide/nucleating agents. *J. Appl. Polym. Sci.* **2007**, *104*, 310–317. [CrossRef]
48. Yu, L.; Liu, H.; Dean, K.; Chen, L. Cold crystallization and postmelting crystallization of PLA plasticized by compressed carbon dioxide. *J. Polym. Sci. Part B Polym. Phys.* **2008**, *46*, 2630–2636. [CrossRef]
49. Liu, T.; Mo, Z.; Wang, S.; Zhang, H. Nonisothermal melt and cold crystallization kinetics of poly(Aryl Ether Ether Ketone Ketone). *Polym. Eng. Sci.* **1997**, *37*, 568–575. [CrossRef]
50. *BS EN ISO 178:2003*; Plastics. Determination of Flexural Properties. BSI: London, UK, 2003.
51. *ISO 179-2.2020*; Plastics—Determination of Charpy Impact properties. Part 2: Instrumented Impact Test. ISO: Geneva, Switzerland, 2020.
52. Pisupati, A.; Deléglise Lagardère, M. Mechanical Performance of Flax Fiber Composites with Waste Glass Fibers as a Core Structure. *Materials* **2022**, *15*, 9017. [CrossRef]
53. Boztepe, S.; Šimáček, P.; Labastie, K.; Chevalier, M.; Sandre, P.; Des, J.-M.; Advani, S.G. Effect of the initial resin distribution in partially impregnated thermoplastic prepregs on consolidation. *Compos. Sci. Technol.* **2022**, *225*, 109488. [CrossRef]
54. Saeidlou, S.; Huneault, M.A.; Li, H.; Park, C.B. Poly(lactic acid) crystallization. *Prog. Polym. Sci.* **2012**, *37*, 1657–1677. [CrossRef]

55. Gumus, S.; Ozkoc, G.; Aytac, A. Plasticized and unplasticized PLA/organoclay nanocomposites: Short- and long-term thermal properties, morphology, and nonisothermal crystallization behavior. *J. Appl. Polym. Sci.* **2012**, *123*, 2837–2848. [CrossRef]
56. Pickering, K.L.; Sawpan, M.A.; Jayaraman, J.; Fernyhough, A. Influence of loading rate, alkali fibre treatment and crystallinity on fracture toughness of random short hemp fibre reinforced polylactide bio-composites. *Compos. Part Appl. Sci. Manuf.* **2011**, *42*, 1148–1156. [CrossRef]
57. Waters, C.L.; Janupala, R.R.; Mallinson, R.G.; Lobban, L.L. Staged thermal fractionation for segregation of lignin and cellulose pyrolysis products: An experimental study of residence time and temperature effects. *J. Anal. Appl. Pyrolysis* **2017**, *126*, 380–389. [CrossRef]
58. Pisupati, A.; Ayadi, A.; Deléglise-Lagardère, M.; Park, C.H. Influence of resin curing cycle on the characterization of the tensile properties of flax fibers by impregnated fiber bundle test. *Compos. Part Appl. Sci. Manuf.* **2019**, *126*, 105572. [CrossRef]
59. O'Donnell, A.; Dweib, M.A.; Wool, R.P. Natural fiber composites with plant oil-based resin. *Compos. Sci. Technol.* **2004**, *64*, 1135–1145. [CrossRef]
60. Pervaiz, M.; Sain, M.M. Sheet-Molded Polyolefin Natural Fiber Composites for Automotive Applications. *Macromol. Mater. Eng.* **2003**, *288*, 553–557. [CrossRef]
61. Glass Fibre Reinforced PA 66—TECAMID 66 GF30 Black | Ensinger. Available online: https://www.ensingerplastics.com/en/shapes/pa66-tecamid-66-gf30-black (accessed on 12 January 2025).
62. Ejaz, M.; Azad, M.M.; Shah, A.U.R.; Afaq, S.K.; Song, J. Mechanical and Biodegradable Properties of Jute/Flax Reinforced PLA Composites. *Fibers Polym.* **2020**, *21*, 2635–2641. [CrossRef]
63. Foruzanmehr, M.; Vuillaume, P.Y.; Elkoun, S.; Robert, M. Physical and mechanical properties of PLA composites reinforced by TiO_2 grafted flax fibers. *Mater. Des.* **2016**, *106*, 295–304. [CrossRef]
64. Sanivada, U.K.; Mármol, G.; Brito, F.P.; Fangueiro, R. Pla composites reinforced with flax and jute fibers—A review of recent trends, processing parameters and mechanical properties. *Polymers* **2020**, *12*, 2373. [CrossRef] [PubMed]

Journal of
Composites Science

MDPI

Technical Note

A Comparative Study of Airbag Covers for Automotive Safety Using Coconut Shell Fiber/PP Composite Materials

Jinsong Li *, You Zhou, Jiatao Chen, Hongtao Hu and Mingze Sun

Mechanical and Electrical Engineering College, Hainan University, Haikou 570228, China;
22220951360017@hainanu.edu.cn (Y.Z.); jiatao.chen@drinda.com.cn (J.C.);
23220855000028@hainanu.edu.cn (H.H.); 99993977@caeri.com.cn (M.S.)
* Correspondence: bmwjsli@163.com; Tel.: +86-13648672888

Abstract: In this study, we compared the physical properties of coconut fiber/polypropylene (PP) composite materials with coconut fiber as a reinforcing agent, produced through a hybrid injection molding process and a layered hot-pressing process. Through comparative experiments, the mechanical properties of both the hybrid injection-molded and layered hot-pressed materials were validated. The results indicated that, when using a coconut fiber content of 5%, the layered hot-pressed composite material exhibited optimal comprehensive performance. Specifically, its tensile strength reached 25.12 MPa, showing a 37.6% increase over that of pure PP materials of the same brand and batch. Its tensile modulus was 1.17 GPa, representing an 11.4% decrease. Additionally, its bending strength was 35.94 MPa, marking a 49.8% increase, and its bending modulus was 2.69 GPa, which is nearly double that of pure PP materials. Furthermore, through Creo modeling and an ANSYS simulation analysis, it was verified that this material could be applied to airbag covers in the field of automotive safety. This study confirmed that layered hot-pressed coconut fiber/PP composite materials exhibit superior mechanical properties to traditional materials and injection-molded composite materials, making them more suitable for airbag covers.

Keywords: automotive safety airbags; coconut shell fiber; polypropylene composite material; stacked compression molding

1. Introduction

With the rapid development of the automotive industry, there is an increasing demand for lightweight and safe vehicles [1,2]. In this context, fiber-reinforced materials have become indispensable in the automotive industry due to their excellent mechanical properties and lightweight characteristics. Notably, natural plant fibers have attracted significant attention as reinforcing materials in automotive components due to their renewable and environmentally friendly characteristics, [3–8]. Traditionally, synthetic fibers such as glass fiber and aramid fiber have been widely used in automotive manufacturing. However, these materials do have disadvantages, such as high costs, associated environmental pollution, and limited raw material resources for their production and recycling. To address these issues, researchers have begun exploring the use of natural plant fibers, such as coconut shell fiber, as alternative materials. Coconut shell fiber is not only abundant and cost-effective but also possesses excellent mechanical properties and thermal stability, making it an ideal choice for reinforcing automotive interior components [9–16].

Coconut shell fiber, a type of cellulose fiber, has a multi-cellular aggregation structure and contains a high proportion of lignin and cellulose. These components confer unique chemical and physical properties to coconut shell fiber, including good heat resistance and high elongation at break [17–21]. These characteristics make coconut shell fiber a highly promising material for the reinforcement of polypropylene (PP) materials, particularly in applications such as automotive airbag covers [22]. Despite the broad prospects of coconut shell fiber for application in composite materials, challenges remain in terms of interface bonding with the PP matrix, uniform dispersion, and processing techniques for the composites [23–30]. Traditional injection molding and emerging layer-by-layer hot-pressing methods each have their advantages in preparing coconut shell fiber/PP composites. This study aimed to explore the impact of different processing methods on the physical properties of composites by comparing the two techniques and evaluating their feasibility in the production of automotive airbag covers.

The purpose of this study was to present safer, lighter, and more cost-effective alternative materials for the production of automotive airbag covers through comprehensively considering their environmental friendliness, cost-effectiveness, and performance. Through in-depth research on coconut shell fiber/PP composites, we hope to contribute to the sustainable development of the automotive industry and promote the application of environmentally friendly materials in a wider range of fields.

2. Materials and Methods

2.1. Primary Materials

Coconut shell fibers, either with lengths of 120–140 mm or 3–5 mm, or in powder form, were sourced from Hainan, China. Polypropylene (PP), grade PP-BG2017F, was supplied by Nanjing Julong Technology Co., Ltd., Nanjing, China. Different chemical reagents were used to treat the coconut fibers: sodium hydroxide, of analytical grade, supplied by China National Pharmaceutical Group Corporation (Beijing, China); anhydrous ethanol, of analytical grade, supplied by China National Pharmaceutical Group Corporation; and coupling agent KH550, of industrial grade, supplied by Shandong Jinan Guanhe Chemical Co., Ltd., Jinan, China.

2.2. Instruments and Equipment

A stainless-steel crusher ZG-J210H (Ningbo Zhaoji Electric Appliance Co., Ltd., Ningbo, China.) was used to cut coconut shell fibers into lengths of 3–5 mm. The treated fibers were dried at a constant temperature in an electric blast-drying oven (101-00B, Shaoxing Huyue Instrument Equipment Co., Ltd., Shaoxing, China.). We used a plastic injection molding machine CJ150M3V, supplied by Dongguan Aiyufa Automation Machinery Co., Ltd., Dongguan, China. which was further equipped with an injection mold provided by Kunshan Yushan Town Oubaijia Testing Instrument Business Department (Kunshan, China) and used to manufacture the short fiber-reinforced composite samples. A hot press-molding machine, HH-100A is provided by Huahui Hydraulic Machinery Factory in Dongguan, China, and is equipped with a hot pressing mold provided by the Oubaijia Testing Instrument Division in Yushan Town, Kunshan City, China, for the manufacture of long fiber reinforced composite material samples. The mechanical properties of the manufactured composite samples were measured using an electronic universal testing machine, KQL WD7-5, from Shenzhen Kaiqiangli Experimental Instrument Co., Ltd., Shenzhen, China. which was equipped with an extensometer to record strain during tensile tests.

2.3. Sample Preparation

2.3.1. Alkali Treatment of Coconut Shell Fiber

For the surface treatment of coconut shell fibers, we pretreated fibers with lengths of 120–140 mm and 3–5 mm and fibers in powder form by washing away impurities and air-drying them at room temperature. We then submerged the cleaned and dried coconut shell fibers in a 5% concentration (mass fraction) NaOH solution for 15 h before filtering. We thoroughly rinsed the alkali-treated coconut shell fibers with distilled water to ensure that no NaOH residue remained before air-drying at room temperature and further drying at a constant temperature of 90 °C in an electric blast-drying oven for 5 h.

2.3.2. Coupling Agent Treatment of Coconut Shell Fibers

For the preparation of the coupling agent, we placed the alkali-treated coconut shell fiber in a 5% concentration KH550 ethanol solution, let it stand for 1 h, and then filtered it.

We air-dried the mixture at room temperature and further dried it in an electric blast-drying oven at a constant temperature of 90 °C for 5 h. The pretreatment of coconut shell fiber was thus completed.

2.3.3. Preparation of Coconut Shell Fiber/PP Composite Material Specimens

During this experiment, we prepared two types of coconut shell fiber/PP composite material specimens with different fiber lengths and arrangements and one specimen of pure PP material. They may be described as follows:

1. Cross-arranged stacked hot-pressed coconut shell fiber/PP composite-material specimens with lengths of 120–140 mm were processed using a hot press-molding machine to form 1 mm thick sheets of polypropylene granules. The pretreated 120–140 mm coconut shell fibers were then vertically interwoven. These fibers were stacked in layers on polypropylene resin sheets in a layering ratio of 2:3 and mass ratios of 5:95, 10:90, and 15:85. The stacked layers were placed in a 4 mm mold. Hot pressing was carried out using a flat vulcanizing machine at a temperature of 200 °C, a molding pressure of 6 MPa, and a holding time of 20 min. The specimens were then allowed to cool gradually after molding.
2. Samples of 3–5 mm mixed injection-molded coconut shell fiber/PP composite material were prepared by uniformly mixing pretreated 3–5 mm coconut shell fibers and PP in mass ratios of 5:95, 10:90, and 15:85. The mixture was then injected into a plastic injection-molding machine, with a temperature of 200 °C, a filling time of 5 s, a cycle time of 15.0 s, a sol time of 4.47 s, a cooling time of 16.1 s, and a shot end point at 2.1 mm. After sample preparation, the specimens were allowed to cool gradually after molding.
3. Pure PP material specimens were prepared using the same method as that used for the 3–5 mm mixed injection-molded coconut shell fiber/PP composite material specimens.

2.4. Mechanical Properties Test

Tensile Test

The specimens of pure PP material and injection-molded composite material in this experiment were directly molded using an injection-molding machine and mold. The tests were conducted according to the GBT1040.1-2018 standard (for Plastics—Determination of Tensile Properties) [31]. The dimensions of the specimen were 170 mm × 10 mm × 4 mm, and the speed of the universal testing machine was set to 2 mm/min.

The specimens of laminated compression-molded composite material were made using a plate-vulcanizing machine for laminated compression molding. The tests were conducted according to the ASTM D3039-507 standard (the Standard Test Method for Tensile Properties of Polymer Matrix Composite Materials) [32]. The dimensions of the specimen were the same as those of the pure PP material and injection-molded composite material specimens, set to 170 mm × 10 mm × 4 mm, with the speed of the universal testing machine set to 2 mm/min. All the tests were carried out at room temperature (23 °C ± 2 °C), and the humidity was controlled at 60% ± 5%.

3. Results and Discussion

3.1. The Influence of Processing and Arrangement on the Tensile and Bending Properties of Composite Materials

The materials required for tensile and bending tests include pure PP material; mixed injection-molded composite materials with 5%, 10%, and 15% coconut shell-fiber contents; and stacked compression-molded composite materials with 5%, 10%, and 15% coconut shell-fiber content. There were a total of seven sets of specimens to be tested, with five specimens in each set. Our results were calculated using formulas based on the measured data and dimensions of the specimens, and averages were taken.

3.1.1. Analysis of the Tensile Properties of Composite Materials

Based on the results of the tensile tests described above, including pure PP material, coconut shell-fiber mixed injection-molded composite materials (with 5%, 10%, and 15% contents), and coconut shell-fiber stacked compression-molded composite materials (with 5%, 10%, and 15% contents), a total of seven sets of specimens were tested for tensile strength and modulus. These results are plotted in Figures 1 and 2.

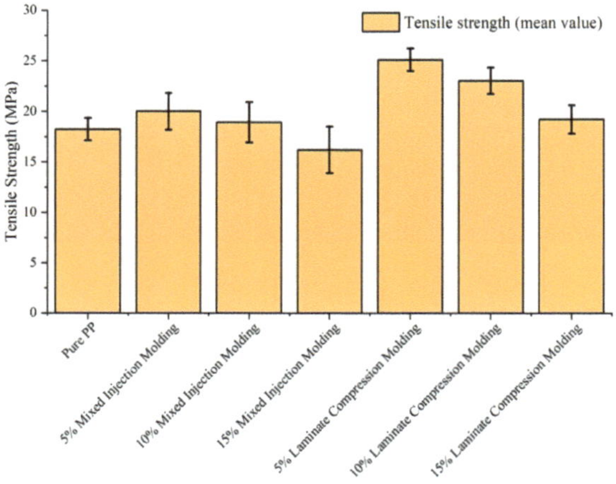

Figure 1. Tensile strength versus coconut shell-fiber content of injection-molded and layered hot-pressed coconut fiber-reinforced polypropylene composites.

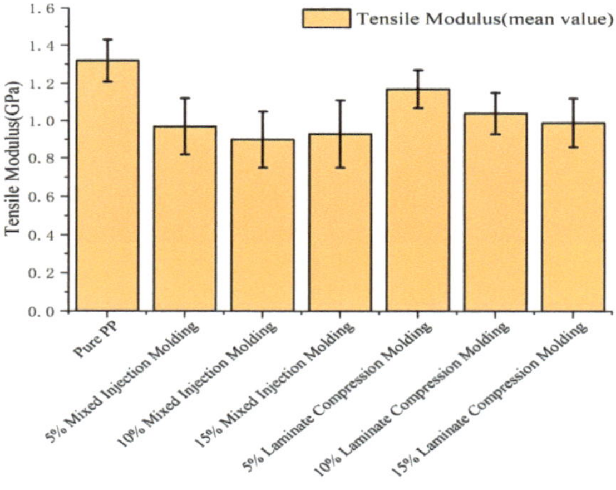

Figure 2. Tensile modulus versus coconut shell-fiber contents of injection-molded and layered hot-pressed coconut fiber-reinforced polypropylene composites.

1. Tensile strength

Figure 1 shows the tensile strength of composite materials with different coconut shell-fiber contents, which were subjected to mixed injection molding and stacked compression molding. The x-axis represents the coconut shell-fiber content in the composite material, and the y-axis represents the tensile strength of the material.

Both mixed injection-molded and stacked compression-molded composite materials achieved maximum tensile strength when the coconut shell-fiber content was 5%. The maximum tensile strength of the stacked compression-molded composite material was

25.12 MPa, representing a 37.6% increase compared to the tensile strength of pure PP material of the same grade and batch. The maximum tensile strength of the mixed injection-molded composite material was 20.23 MPa, representing a 9.8% increase compared to the tensile strength of pure PP material of the same grade and batch.

The reason for this is that, when the composite material is subjected to external forces and undergoes deformation under tension, partial separation of the matrix materials occurs. In this state of separation, the fibers can act as a buffer for the transfer of stress between the PP and fibers, thereby increasing the overall strength of the material.

As the coconut shell-fiber content gradually increased, both the mixed injection-molded and stacked compression-molded composite materials exhibited varying degrees of decrease in their tensile strength. Concerning the stacked compression-molded composite materials, the tensile strength of the materials with 10% and 15% coconut shell-fiber contents decreased by 8.3% and 23.4%, respectively, relative to the maximum value. The materials' tensile strength at first decreased gradually before decreasing abruptly, reaching a minimum of 19.24 MPa in the material with 15% coconut shell-fiber content, slightly higher than that of the pure PP material. Additionally, in the mixed injection-molded composite materials with 10% and 15% coconut shell-fiber content, the tensile strength decreased by 5.4% and 19.1%, respectively, relative to the maximum value. The trend in the decreasing of tensile strength decrease was similar to that of the stacked compression-molded composite materials, transitioning from gradual to abrupt. However, their overall decreasing trend was slower than that of the former. The tensile strength of this material reached a minimum of 16.21 MPa in the material with 15% coconut shell-fiber content, which is slightly lower than that of the pure PP material.

The reason for this is that, as the coconut shell-fiber content gradually increased, the proportion of coconut shell fibers, which have a much lower density than PP, increased to nearly 30% in the composite material. This led to the inability of the coconut shell fibers, as a form of reinforcement, to uniformly mix into the matrix (or be uniformly interlaid in the case of stacked compression-molded composite materials). Consequently, the individual coconut shell fibers came into contact with each other without being bound, thus weakening the overall strength of the material.

2. Tensile Modulus

Figure 2 depicts the tensile moduli of composite materials with different coconut shell-fiber contents that were subjected to mixed injection molding and stacked compression molding. The x-axis shows the coconut shell-fiber content in the composite materials, while the y-axis shows the tensile moduli of the materials.

Both mixed injection-molded and stacked compression-molded composite materials exhibited lower tensile moduli than that of pure PP material of the same grade and batch, at 1.32 GPa. In the stacked compression-molded composite materials, a maximum tensile modulus of 1.17 GPa was achieved when the coconut shell-fiber content was 5%, representing an 11.4% decrease in the tensile modulus of pure PP material. Similarly, in the mixed injection-molded composite materials, a maximum tensile modulus of 0.97 GPa was achieved when the coconut shell-fiber content was 5%, representing a 26.5% decrease in the tensile modulus of PP material. The tensile modulus of both mixed injection-molded and stacked compression-molded composite materials decreased with the increasing coconut shell-fiber content, but this decrease was non-significant. Moreover, the tensile modulus of the 15% mixed injection-molded composite material slightly increased compared to that of the 10% mixed injection-molded composite material, but the overall change was non-significant and did not exceed 5%.

The tensile modulus of the composite materials was lower than that of the pure PP material for the following reasons. First, the high proportion of lignin in coconut shell fibers, which can reach up to 40%, contributes to the decrease in the tensile modulus of coconut shell-fiber composite materials because lignin has higher strength but lower toughness than cellulose. Second, the cellulose and lignin surfaces in coconut shell fibers contain a large number of hydroxyl groups, making them hydrophilic, while the material matrix of PP exhibits hydrophobicity. Although the coconut shell fibers were pretreated with NaOH alkaline solution and coupling agent KH550 solution to significantly alleviate this problem before composite processing, there may still be insufficient bonding between the reinforcement and matrix materials, which could potentially affect the tensile modulus of the composite materials.

3. Sample curves from tensile testing

Based on our data analysis of the tensile strength and modulus of the two different composite materials with varying coconut shell-fiber contents, we determined that the composite material with 5% coconut shell-fiber content exhibited an optimal performance. To provide a more intuitive analysis of the tensile performance of the composite material with 5% coconut shell-fiber content, we produced tensile curves of the two types of composite materials with 5% coconut shell-fiber content and the pure PP-material specimens, as shown in Figure 3.

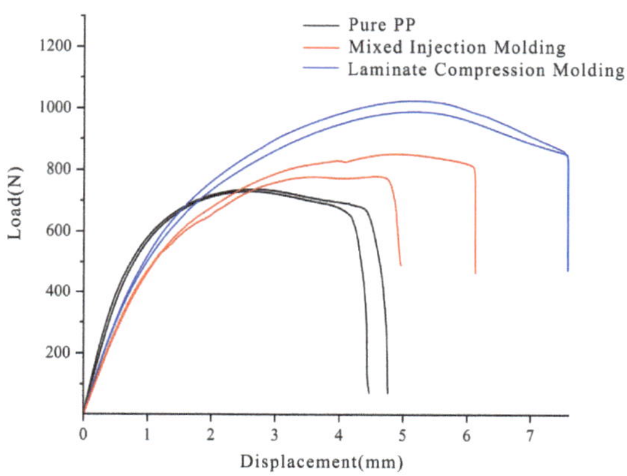

Figure 3. Tensile load–displacement curves of injection-molded and layered hot-pressed 5%wt. coconut fiber-reinforced polypropylene composites, as compared to pure polypropylene.

To provide a clear and comprehensive display of the tensile process of each material specimen, two specimens from each of the three types of materials were selected and plotted. This process facilitated a more intuitive comparison and analysis. Different specimens from the same material showed consistent trends in their load–displacement curves, indicating reproducibility within the same batch of material specimens. The differences between specimens of the stacked compression-molded composite material and pure PP material were relatively small, whereas the differences between specimens of the injection-molded composite material were relatively large. We can mainly attribute this to the difficulty we encountered in achieving homogeneous mixing during the process of blending the short coconut shell-fiber reinforcement with the injection-molded composite material. Consequently, it is possible for local fiber clustering and uneven distribution of fibers to occur, leading to variations in the material properties of different specimens.

Of the three materials, the performance of the stacked compression-molded composite material is superior. The maximum load borne by the standard specimen can reach around 1000 N, with a deformation of approximately 7.5 at the point of fracture. In comparison, although the performance of the injection-molded composite material also represents an improvement upon pure PP material, it is still inferior to the stacked compression-molded composite material.

In conclusion, among the seven different materials tested, the stacked compression-molded composite material with a 5% coconut fiber content exhibited the best overall tensile performance.

3.1.2. Analysis of the Bending Performance of Composite Materials

Based on the bending-test results mentioned above, including pure PP material, coconut-fiber mixed injection-molded composite materials (with 5%, 10%, and 15% content), and coconut-fiber laminated compression-molded composite materials (with 5%, 10%, and 15% content), a total of seven groups of tested materials were used to plot bending strength and bending modulus, as shown in Figures 4 and 5.

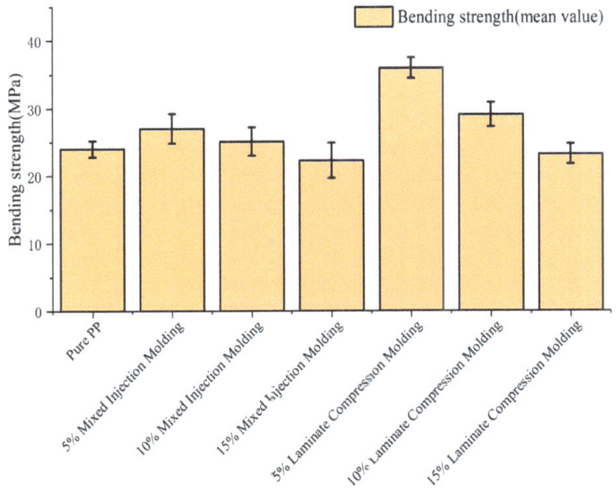

Figure 4. Bending strength versus coconut shell-fiber content of injection-molded and layered hot-pressed coconut fiber-reinforced polypropylene composites.

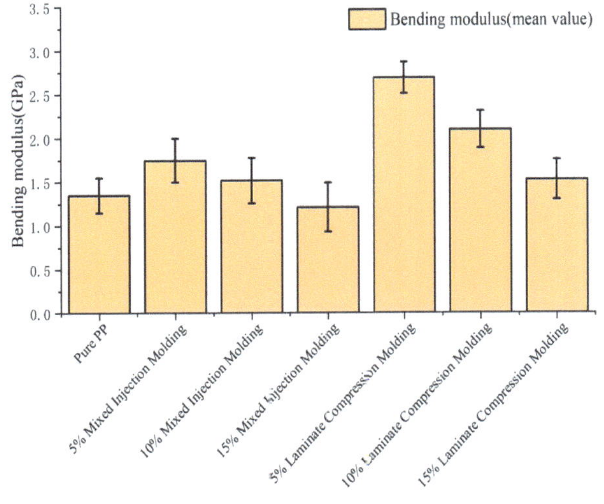

Figure 5. Bending modulus versus coconut shell-fiber content of injection-molded and layered hot-pressed coconut fiber-reinforced polypropylene composites.

1. Bending strength

Figure 4 illustrates the bending strength of composite materials with different coconut fiber contents that were subjected to injection molding and laminate compression molding. The x-axis represents the coconut fiber content in the composite material, while the y-axis represents the bending strength of the material.

Both injection-molded and laminate compression-molded composite materials exhibited their maximum bending strength when a coconut fiber content of 5% was present. The maximum bending strength of laminate compression-molded composite materials was 35.94 MPa, representing a 49.8% increase compared to that of pure PP material of the same brand and batch. On the other hand, the maximum bending strength of injection-molded composite materials was 27.03 MPa, showing a 12.6% increase compared to that of pure PP material of the same brand and batch.

When the coconut fiber content exceeded 5%, both injection-molded and laminate compression-molded composite materials experienced a gradual decrease in bending strength. For laminate compression-molded composite materials, the bending strength decreased by 19.1% and 35.4%, respectively, in materials with 10% and 15% coconut fiber contents, relative to the maximum value. The overall rate of decrease was consistent, but the magnitude of decrease was significant. When the coconut fiber content was 15%, the bending strength reached a minimum of 23.22 MPa, which is slightly lower than that of the pure PP material. Additionally, in injection-molded composite materials with 10% and 15% coconut fiber contents, the bending strength decreased by 7.1% and 17.6%, respectively, relative to the maximum value. Bending strength decreased in a similar manner to the laminate compression-molded composite materials, but the overall decreasing trend was relatively gentle. The material with a coconut fiber content of 15% had a minimum bending strength of 22.26 MPa, which is slightly lower than that of the pure PP material and laminate compression-molded composite materials with the same coconut fiber content.

The significant increase in the bending strength of laminate compression-molded composite materials compared to pure PP material can mainly be attributed to the composite process, within which the coconut fiber reinforcement and PP matrix are compressed under a high temperature and pressure. Additionally, coconut fibers undergo alkali treatment and coupling agent treatment, which removes impurities such as pectin from the surface. The bonding mechanism involves the hydrolysis of alkoxy groups in the silane coupling agent to form trihydroxy silane, in which the hydroxyl groups react with the -OH groups in coconut fibers to form a stable structure [33]. Moreover, the organic functional groups (-NH_2) on the coupling agent react with PP to form a covalently bonded cross-linked structure, significantly enhancing the interfacial bonding between the fiber and PP material. This enhancement leads to improved bending strength in laminate compression-molded composite materials. This principle may be demonstrated as follows:

The ethoxy (-OEt) groups in KH550 hydrolyze to generate silanol (Si-OH) in the presence of water:

$$(CH_3CH_2O)_3Si(CH_2)_3NH_2 + 3H_2O \rightarrow (HO)_3Si(CH_2)_3NH_2 + 3CH_3CH_2OH \quad (1)$$

After hydrolysis, the silanol groups can further condense to form a siloxane network, which reacts with the hydroxyl (-OH) groups on the fiber surface to form a strong chemical bond:

$$(HO)_3Si(CH_2)_3NH_2 + \text{Fiber-OH} \rightarrow \text{Fiber-O-Si}(CH_2)_3NH_2 + 3H_2O \quad (2)$$

The amino (-NH_2) groups of KH550 can react with some active groups (such as carbonyl or carboxyl groups) on the surface of the polypropylene matrix to further enhance the bonding between the fiber and the matrix:

$$\text{PP-COOH} + NH_2\text{-(CH2)}_3\text{Si-O-Fiber} \rightarrow \text{PP-CONH-(CH2)}_3\text{Si-O-Fiber} + H_2O \quad (3)$$

2. Bending modulus

The graph shown in Figure 5 represents the bending modulus of composite materials with different coconut fiber contents subjected to both injection-molding and laminated compression-molding processes. The horizontal axis represents the coconut fiber content in the composite material, while the vertical axis represents the bending modulus of the material.

Both the bending modulus of the injection-molded and laminated compression-molded composite materials reached their maximum values when the coconut fiber content was 5%. The maximum bending modulus of the laminated compression-molded composite material was 2.69 GPa, nearly double that of the pure PP material of the same brand and batch. For the injection-molded composite material, the maximum bending modulus was 1.75 GPa, representing a 29.6% increase compared to that of the pure PP material of the same brand and batch.

When the coconut fiber content exceeded 5%, both the injection-molded and laminated compression-molded composite materials experienced varying degrees of decrease in their bending modulus as the coconut fiber content gradually increased. Specifically, for the laminated compression-molded composite material, the bending modulus in materials with 10% and 15% coconut fiber contents decreased by 21.9% and 43.1%, respectively, relative to the maximum value. The overall rate of decrease in bending modulus was similar, but the magnitude of this decrease was larger. When the coconut fiber content was 15%, the bending modulus reached its lowest point, at 1.53 GPa, which is slightly higher than that of the pure PP material. Additionally, for injection-molded composite materials with 10% and 15% coconut fiber contents, the bending modulus decreased by 13.1% and 30.9%, respectively, compared to the maximum value. The decreasing trend in bending moduli was similar to that of the laminated compression-molded composite material, but the overall rate of decrease was slower. The bending modulus of this material reached its lowest point, 1.21 GPa, when the coconut fiber content was 15%; this is slightly lower than that of the pure PP material and laminated compression-molded composite material with the same coconut fiber content.

After the coconut fiber content exceeded 5%, the decrease in the bending moduli of the composite materials with greater coconut fiber contents could be attributed to the same factors causing a decrease in bending strength. Therefore, further elaboration on this point is unnecessary.

3. Bending test curves of specimens

Based on the analysis of the data above and considering the bending strength and modulus of elasticity of the two different composite materials with varying coconut fiber contents, it can be concluded that the composite material with 5% coconut fiber content exhibited the best performance. In order to provide a clearer and more intuitive analysis of the bending performance of the composite material with 5% coconut fiber content, Figure 6 shows bending curves of the two composite materials with 5% coconut fiber content alongside the curve of the pure PP material.

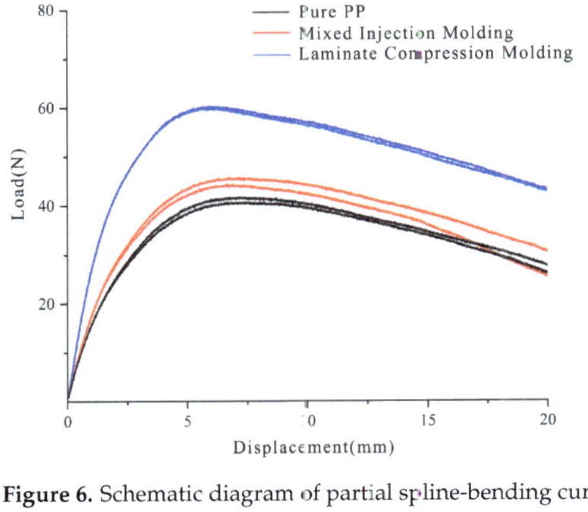

Figure 6. Schematic diagram of partial spline-bending curve.

The graph presented in Figure 6 depicts the load–displacement curves obtained during bending tests of selected samples of pure PP material, the laminated compression-molded composite material with 5% coconut fiber content, and the mixed injection-molded composite material with 5% coconut fiber content. In order to comprehensively illustrate the bending of each material sample and provide a more intuitive comparative analysis, two samples of each of the three tested materials were selected and plotted on the graph.

In conclusion, among the seven tested material groups, the laminated compression-molded composite material with 5% coconut fiber content exhibited the best overall bending performance.

3.2. *Scanning Electron Microscopy (SEM) Cross-Section Characterization Analysis*

In order to effectively observe and visually demonstrate formation of a composite by the coconut fiber reinforcement and the PP matrix, we selected tensile specimens of the laminated compression-molded and injection-molded composite materials with the optimal 5% coconut fiber content (which performed best in the tensile and bending tests) for cross-section slicing. As the bending specimens reached a preset deflection during testing but did not fracture, only tensile specimens were selected. We conducted a physical performance-oriented empirical analysis via a scanning electron microscopy (SEM) scanning analysis and by taking photographs for characterization.

3.2.1. SEM Characterization Analysis of the Tensile Fracture Surface of Laminated Compression-Molded Composite Materials with 5% Coconut Fiber Content

Figure 7 shows partial SEM images of the tensile fracture surface of the laminated compression-molded composite material with 5% coconut fiber content. The bright columnar structures in the image represent exposed coconut fibers on the fracture surface of the composite material, while the areas with lower brightness and surface irregularities correspond to the PP material matrix on the fracture surface of the composite material.

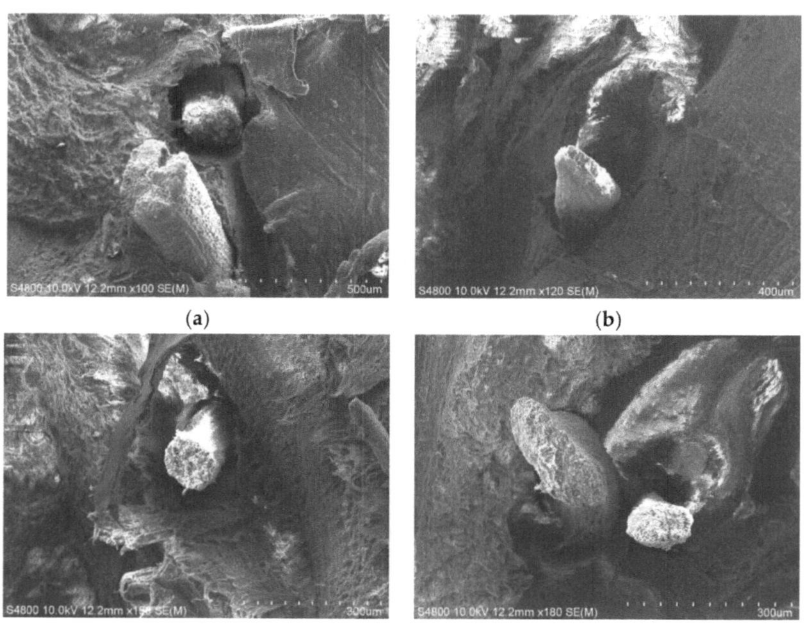

(a) (b) (c) (d)

Figure 7. SEM characterization of tensile sections of laminated hot-pressed composites with 5% coconut fiber content, where images (**a–d**) represent different tensile sections.

The coconut fiber reinforcement bonded tightly to the matrix material, with no signs of intact fibers being pulled out. Additionally, no significant voids could be observed between the fibers and the matrix. However, in the images shown in Figure 7a,c,d, some minor voids are visible between the fibers and the matrix material. This phenomenon may be attributed to the fibers continuing to carry the mechanical load even after the matrix material fractured under tension. As the fibers remained engaged in supporting the load, the applied force may have caused partial separation between the fibers and the matrix near the fracture surface.

3.2.2. SEM Characterization Analysis of the Tensile Fracture Surface of Injection-Molded Composite Materials with 5% Coconut Fiber Content

Figure 8 shows partial SEM images of the tensile fracture surface of injection-molded composite materials with 5% coconut fiber content. The cylindrical and filamentous structures visible in the images are the coconut fibers exposed on the fracture surface of the composite material, while the remaining uneven surfaces feature the polypropylene (PP) matrix of the composite material.

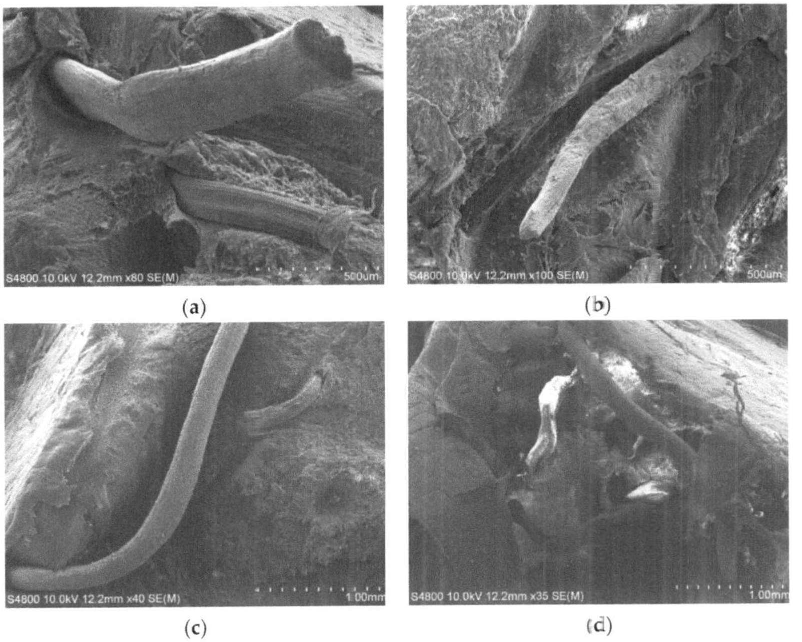

(a) (b) (c) (d)

Figure 8. SEM characterization of tensile sections of mixed injection-molded composites with 5% coconut fiber content, where images (**a–d**) represent different tensile sections.

Partial coconut fibers inside the injection-molded composite materials did not fracture but were fully pulled when the ultimate load was applied during tension. In Figure 8a, the fracture surface of the coconut fiber appears smooth and regular, indicating that it is not a fracture generated by the tensile stress but, rather, a smooth cut produced during fiber processing.

The coconut fibers were distributed unevenly within the matrix, as shown in Figure 8d. In the same region, the number of fibers varies from 1 to 5, indicating uneven distribution. This uneven distribution of fibers leads to the increased instability of the composite material. The primary reason for this uneven distribution is the light weight and large volume of coconut fibers, thus making it challenging to thoroughly mix them during the blending process. Consequently, this uneven distribution can have a certain impact on the performance of the injection-molded composite material.

The coconut fibers, acting as reinforcement, exhibited a relatively loose bonding with the matrix material, as evidenced by the frequent occurrence of fibers being pulled out intact, as seen in Figure 7a,c. Additionally, clear signs of separation between the fibers and the matrix can be observed in Figure 8b. The primary reason for this phenomenon is that the weaker bonding force between the short coconut fibers and the matrix makes them more prone to being pulled out intact, thereby affecting the overall performance of the injection-molded composite material. In the SEM characterization analysis, this manifested as the fibers being pulled out intact and a separation between the fibers and the matrix. Conversely, in the compression-molded composite material, the long fibers had a larger total contact area with the matrix. Taking into account the same binding force per unit contact area (due to frictional force and chemical bonding, among other factors), the total binding force was greater, resulting in a better performance.

The 3–5 mm short fibers in the injection-molded composite material have various advantages over the long fibers in the compression-molded composite material. The advantage of short fibers, serving as reinforcement in the injection-molded composite material, lies in their homogeneity, which results in isotropic properties [34]. On the other hand, although the compression-molded composite material is heterogeneous and exhibits anisotropic behavior, its performance is superior within a single plane, making it more suitable for sheet-like components (e.g., test specimens). Therefore, this compression-molded composite material is more suitable for the production of panel components of airbag covers.

4. Finite-Element Analysis

In the current international automotive market, the cover panel of an airbag is typically integrated with its base, as shown in Figure 9. In actual use, the cover panel is secured to the car dashboard through slots on the base, ensuring that the airbag can be quickly deployed through the weakened slots on the cover panel. To verify the feasibility of using cross-laminated compression-molded coconut fiber/PP composite materials for the cover panels of airbags for automotive safety, this study utilized Creo for model creation and the finite-element software ANSYS for simulation and validation.

Figure 9. Panel of an airbag cover (backside).

4.1. Airbag Cover Panel Modeling

The types of weakening grooves commonly found in airbag cover panels in currently available domestic and international car models include the H-type, U-type, and double Y-type. Among them, the H-type is easy to process and has shown stable performance, making it the most universal. Therefore, we selected an H-type weakening groove airbag cover panel from a Haima automobile, and its actual dimensional data are shown in Table 1.

Based on the dimensions of the H-type weakened groove airbag cover panel, a three-dimensional model was established, as shown in Figure 10, in which two layers of meshed coconut fiber were uniformly embedded in the cover panel. One layer was interrupted due to the H-type weakened groove, while the other layer remained intact.

Table 1. Size of the selected airbag cover panel.

gth of Panel	The Width of the Cover Panel	The Length of the Airbag Compartment	The Width of the Airbag Compartment	The Height of the Highest Point of the Airbag Compartment	The Height of the Lowest Point of the Airbag Compartment	The Thickness of the Airbag Cover Panel
nm	158.46 mm	210.02 mm	112.08 mm	65.39 mm	45.93 mm	4 mm

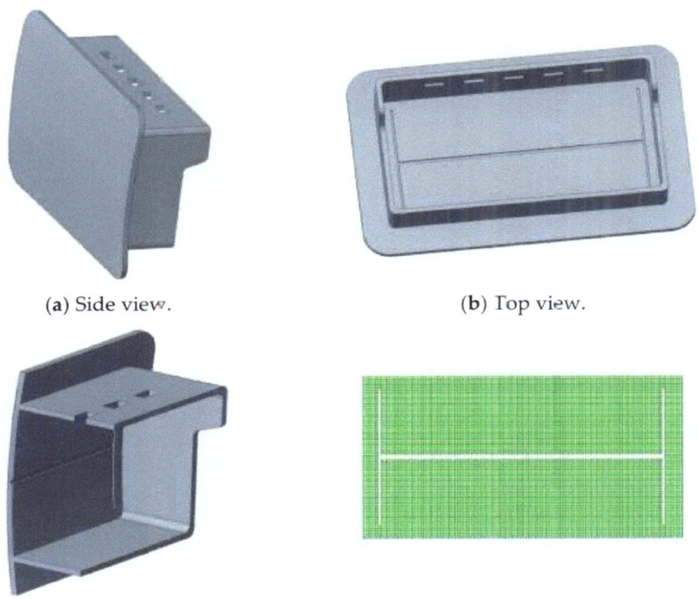

(**a**) Side view. (**b**) Top view.

(**c**) Cross-sectional view. (**d**) Fiber layer.

Figure 10. Model of a laminated hot-pressed composite airbag cover panel.

As shown in Figure 10, the model consists of two layers: one of coconut fiber mesh and the other the main body of the cover panel. The diameter of the mesh fibers ranges from 0.3 to 0.5 mm, and the fibers are stacked and distributed to form the mesh structure. The mesh is evenly distributed within the cover panel, making full contact with the matrix material.

4.2. Simplified Airbag Cover Model

The airbag cover panel is made of laminated compression-molded composite material, and its specific structure is shown in Figure 10. However, due to factors such as the thickness, mechanical strength, and pre-forming condition of the fibrous network, it cannot be quantified as precisely as conventionally cast parts. Moreover, there are numerous contacts between the fibrous network and the matrix, meaning there are high demands in terms of computational resources for finite-element analysis. Therefore, in order to simplify the model of the cover panel's fibrous network layer, the mesh layer model was simplified to a single thin plate [35]. The mechanical properties and stacking direction of the fibers in the mesh were defined using Hypermesh and ANSYS simulation analysis to obtain the final results. The resulting simplified model of the cover panel assembly is shown in Figure 11.

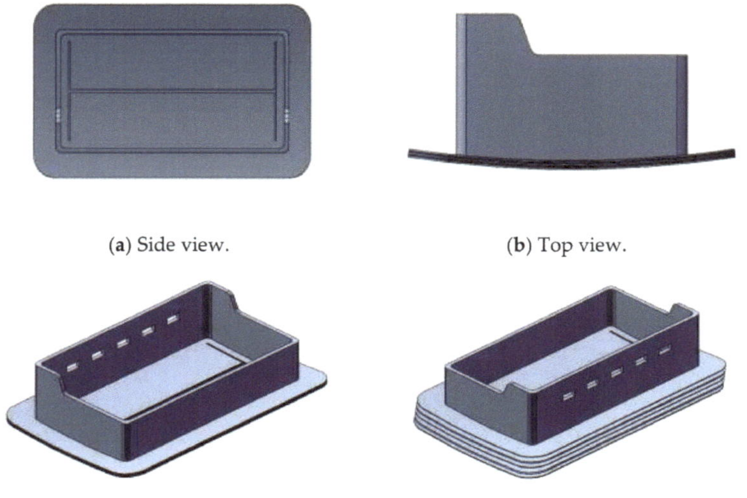

(**a**) Side view. (**b**) Top view.

(**c**) Assembled specimen. (**d**) Pre-assembly specimen.

Figure 11. Simplified model of a laminated hot-pressed composite airbag cover.

4.3. The Selection of Material Models and Element Types

In this study, we utilized cross-laminated hot-pressed coconut fiber/PP composite materials, which generally exhibit anisotropic behavior. However, for the purpose of finite element modeling in ANSYS, the composite was simplified by defining each material as a separate component, assuming isotropic properties for each component [36]. This common simplification in finite-element modeling makes computational analysis more manageable, while still providing useful insights. Defining different materials as separate components, we may acknowledge the distinct mechanical properties of the coconut fiber and PP matrix, which influence the overall behavior of the composite. Consequently, data for the 5% coconut fiber material from Table 2 and data for the pure PP material were individually imported into ANSYS for analysis, allowing for an accurate reflection of the properties of each component and leading to more reliable simulation results.

Table 2. Experimental data of composite materials.

	Tensile Strength/MPa	Tensile Modulus/GPa	Bending Strength/MPa	Bending Modulus/GPa
Pure PP material	18.25	1.32	24	1.35
5% Stacked Compression Molding	25.12	1.17	35.94	2.69

In this study, to select the element type, we utilized the SOLID186 element to define and simulate the performance of the airbag cover panel. SOLID186 is a high-order 3D 20-node solid element, which is defined by 20 nodes, each with three translational degrees of freedom in the x, y, and z directions. SOLID186 exhibits arbitrary spatial anisotropy and supports hyperelasticity, plasticity, stress stiffening, creep, large strain, and large deformation [37]. Therefore, the SOLID186 element is well-suited to this study.

4.4. Mesh Partitioning

Solid mesh partitioning is typically carried out in one of two ways: mapped mesh partitioning and free mesh partitioning. Compared to mapped mesh partitioning, free mesh partitioning is more widely applicable, as it imposes fewer restrictions on element shapes and features, thus offering greater flexibility and convenience. However, free mesh partitioning requires more extensive hardware. Considering the relatively simple nature of our research model, we chose free mesh partitioning. The mesh partitioning of the model is shown in Figure 12.

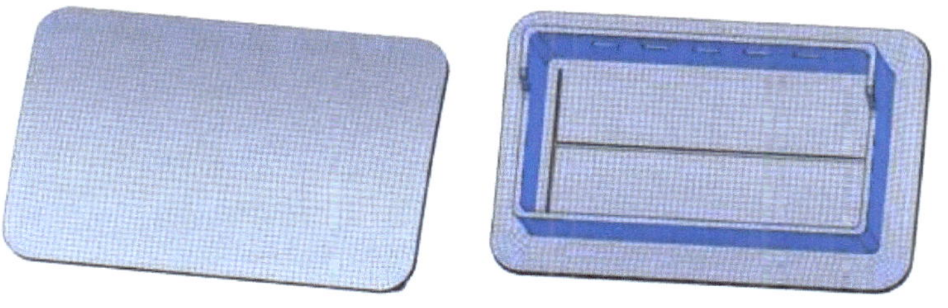

Figure 12. Grid partitioning.

4.5. Setting Up Contact, Constraints, and Applying Loads

When a car airbag is activated, the detonation impact force is approximately 1764 N, resulting in a pressure of about 70 kPa [38]. In actual conditions, the deployment of the airbag is initiated when it receives signals from sensors, which ignite the explosive device, and then fill the airbag inside the cover plate. Subsequently, the pressure gradually increases on the inner wall of the cover plate until it reaches a critical point [39]. At this point, the cover plate ruptures along the weakening groove, releasing the pressure from the airbag. This entire process occurs within a very short span of time; therefore, we can consider the application of the uniform load to the inner wall of the cover plate to be instantaneous. As the cover plate is a single entity and its base is fixed to the car dashboard, fixed constraints can be added to various locations on the airbag base, as shown in Figure 13.

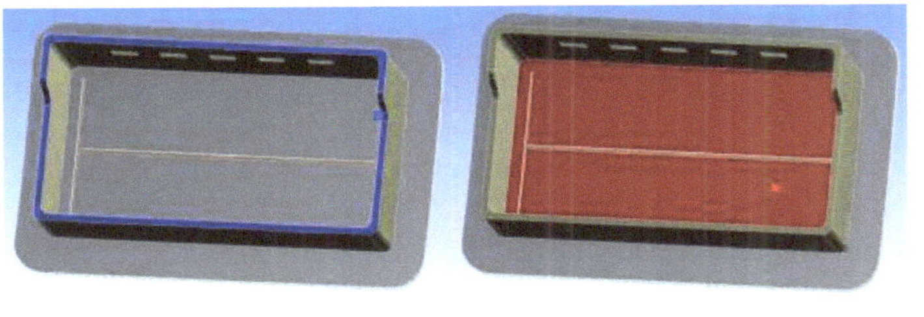

Figure 13. Load and constraints of the model.

4.6. Resolution and Analysis

Once the model's contact, loads, and constraints were set up, the model was solved to obtain the deformation plot and stress contour of the cross-laminated compression-molded coconut fiber/PP composite car airbag cover at the moment of ignition, as shown in Figures 14 and 15.

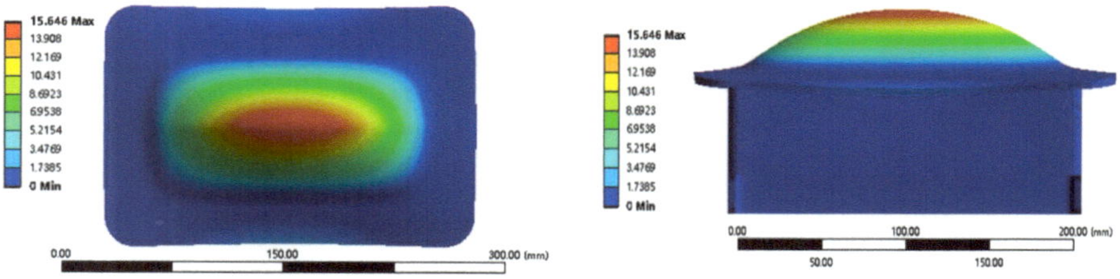

Figure 14. Schematic diagram of model deformation.

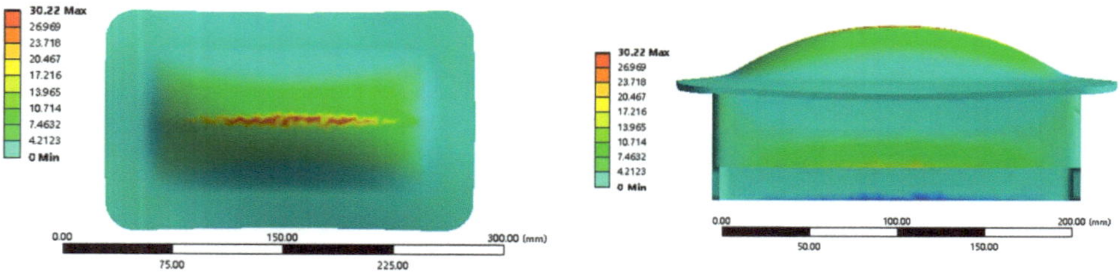

Figure 15. Cloud map of modeled stress.

From Figures 14 and 15, it can be observed that, at the moment of airbag deployment, the stress on the airbag cover is primarily concentrated on the H-shaped weakening groove. This is mainly because the weakening groove is positioned at a relatively weak spot and located at the center of the cover. Among the weakening grooves, the stress is predominantly borne by the central long groove, while the stresses on the shorter grooves on both sides are relatively smaller. Therefore, it can be concluded that, upon airbag deployment, the airbag cover starts to fracture from the central long groove, gradually tearing towards the shorter grooves on both sides until the airbag is fully deployed. Thus, it can be concluded that the studied cross-laminated compression-molded coconut fiber/PP composite material is suitable for application in automotive airbag covers.

5. Conclusions

This study compared the physical properties of coconut shell fiber/polypropylene (PP) composites prepared using different processes, showing that layered hot-pressed materials demonstrated the best comprehensive performance when the coconut fiber content was 5%. In particular, these materials showed enhanced tensile and flexural strength and an almost-doubled flexural modulus. Even with a 15% fiber content, while the tensile strength and modulus decreased, layered hot-pressed materials still outperformed mixed injection-molded materials in terms of mechanical properties. Creo modeling and ANSYS simulation confirmed the potential of this composite for use in automotive airbag covers. This material is lightweight, safe, and cost-effective, and is a great example of an eco-friendly material to be used in the automotive industry which can, therefore, contribute to sustainable development. Future research will focus on optimizing the parameters of its processing to enhance its market competitiveness and potential applications.

Author Contributions: Conceptualization, J.L. and J.C.; methodology, Y.Z.; software, Y.Z.; validation, Y.Z., H.H. and M.S.; formal analysis, J.L.; investigation, M.S.; resources, Y.Z.; data curation, Y.Z.; writing—original draft preparation, Y.Z.; writing—review and editing, Y.Z.; visualization, H.H.; supervision, J.L.; project administration, J.L.; All authors have read and agreed to the published version of the manuscript.

Funding: This research was funded by the National Natural Science Foundation of Hainan Province, grant number 521RC497.

Conflicts of Interest: The authors declare no conflict of interest.

es

g, W.; Xu, J. Advanced lightweight materials for Automobiles: A review. *Mater. Des.* **2022**, *221*, 110994. [CrossRef]
vinski, F. Current trends in automotive lightweighting strategies and materials. *Materials* **2021**, *14*, 6631. [CrossRef]
Med]
, D.K.; Pagar, D.D.; Menezes, P.L.; Linul, E. Fiber-reinforced polymer composites: Manufacturing, properties, and applica-
Polymers **2019**, *11*, 1667. [CrossRef] [PubMed]
, S.; Islam, R.; Uddin, M.A.; Afroj, S.; Eichhorn, S.J.; Karim, N. Sustainable fiber-reinforced composites: A Review. *Adv.*
n. Syst. **2022**, *6*, 2200258. [CrossRef]
mmadi, H.; Ahmad, Z.; Mazlan, S.A.; Johari, M.A.F.; Siebert, G.; Petrů, M.; Koloor, S.S.R. Lightweight glass fiber-reinforced
ner composite for automotive bumper applications: A review. *Polymers* **2022**, *15*, 193. [CrossRef] [PubMed]
ra, F.V.; Pinheiro, I.F.; de Souza, S.F.; Mei, L.H.I.; Lona, L.M.F. Polymer composites reinforced with natural fibers and
cellulose in the automotive industry: A short review. *J. Compos. Sci.* **2019**, *3*, 51. [CrossRef]
na, S.J.; Khoathane, C.; Malwela, T. Driving towards sustainability: A review of natural fiber reinforced polymer composites
o-friendly automotive light-weighting. *J. Thermoplast. Compos. Mater.* **2024**. *online first*. [CrossRef]
, S.; Khatkar, V.; Behera, B.K. Textile-based natural fibre-reinforced polymeric composites in automotive lightweighting.
er. Sci. **2021**, *56*, 18867–18910. [CrossRef]
ar, S.; Saha, A. Utilization of coconut shell biomass residue to develop sustainable biocomposites and characterize the
cal, mechanical, thermal, and water absorption properties. *Biomass Convers. Biorefinery* **2022**, *14*, 12315–12831. [CrossRef]
rudin, S.H.; Basri, M.S.M.; Rayung, M.; Abu, F.; Ahmad, S.; Norizan, M.N.; Osman, S.; Sarifuddin, N.; Desa, M.S.Z.M.;
llah, U.H.; et al. A review on natural fiber reinforced polymer composites (NFRPC) for sustainable industrial applications.
ers **2022**, *14*, 3698. [CrossRef]
le, I.O.; Adelani, S.O.; Makinde-Isola, B.A.; Omotosho, T.F. Coconut/coir fibers, their composites and applications. In *Plant*
, Their Composites, and Applications; Woodhead Publishing: Cambridge, UK, 2022; pp. 181–208.
wal, J.; Sahoo, S.; Mohanty, S.; Nayak, S.K. Progress of novel techniques for lightweight automobile applications through
ative eco-friendly composite materials: A review. *J. Thermoplast. Compos. Mater.* **2020**, *33*, 978–1013. [CrossRef]
ik, K.; Rajamanikkam, R.K.; Venkatesan, E.P.; Bishwakarma, S.; Krishnaiah, R.; Saleel, C.A.; Soudagar, M.E.M.; Kalam, M.;
M.M.; Bashir, M.N. State of the Art: Natural Fibre-Reinforced Composites in Advanced Development and their Physi-
chemical/Mechanical Properties. *Chin. J. Anal. Chem.* **2024**, *52*, 100415. [CrossRef]
B.; Rahman, A.; Alam, R.; Repon, R.; Nawab, Y. Mapping the progress in natural fiber reinforced composites: Preparation,
anical properties, and applications. *Polym. Compos.* **2023**, *44*, 3748–3788. [CrossRef]
A.R.; Kumar, R.; Mural, P.K.S. Natural fiber reinforced polymer composites: A comprehensive review of tribo-mechanical
rties. *Tribol. Int.* **2023**, *189*, 108978. [CrossRef]
S.; Khatri, H.; Naveen, J.; Jawaid, M.; Jayakrishna, K.; Norrrahim, M.; Rashedi, A. Potential of natural fiber based polymeric
osites for cleaner automotive component production—A comprehensive review. *J. Mater. Res. Technol.* **2023**, *25*, 1086–1104.
sRef]
mez, M.A.; Özer, K.; Ozer, D. Biofibers and Their Composites for Industrial Applications. In *Handbook of Bioplastics and*
nposites Engineering Applications; Wiley-Scrivener Publishing LLC: Beverly, MA, USA, 2023; pp. 513–538.
nelli, F.R.B.; Ribeiro, F.R.C.; Marvila, M.T.; Monteiro, S.N.; Filho, F.d.C.G.; de Azevedo, A.R.G. A review of the use of coconut
in cement composites. *Polymers* **2023**, *15*, 1309. [CrossRef]
edion, F.; Inegbedion, F.; Osasona, I.E. Coconut Fibre (Coir) Composites: A Review. *J. Mater. Eng. Struct. Comput.* **2024**,
30.
, W.; Reddy, N.; Barsberg, S.; Sanadi, A.R. Coir from Coconut Processing Waste as a Raw Material for Applications Beyond
tional Uses. *BioResources* **2023**, *18*, 49. [CrossRef]
R.V.; Yadav, A.; Winczek, J. Physical, mechanical, and thermal properties of natural fiber-reinforced epoxy composites for
ruction and automotive applications. *Appl. Sci.* **2023**, *13*, 5126. [CrossRef]
Y.; Li, J.; Li, J.; Hong, S.; Ji, C. Feasibility study on the application of coconut shell fiber/PP composite material based on
us in automotive airbag cover plates. *Plast. Technol.* **2020**, *48*, 94–97.
G.; Ikua, B.; Kabini, S.; Mwangi, J. Characterization and Modeling of Mechanical Properties of Additively Manufactured
nut Fiber-Reinforced Polypropylene Composites. *Adv. Mater. Phys. Chem.* **2024**, *14*, 95–112. [CrossRef]
mora, P.; Simanjuntak, J.; Sinulingga, K.; Laksono, A.D. Mechanical Properties of Polypropylene Composites with Different
orced Natural Fibers—A Comparative Study. *J. Ecol. Eng.* **2023**, *24*, 311–317. [CrossRef] [PubMed]
ascimento, E.P.; Luna, C.B.B.; Ferreira, E.D.S.B.; dos Santos Filho, E.A.; Siqueira, D.D.; Weller, R.M.F.; Araújo, E.M. Extraction
tural fibers of Catole coconut (*Syagrus Cearensis*, application as reinforcing filler in polypropylene-based composites. *Polym.*
os. **2023**, *44*, 5891–5909. [CrossRef]
, Y.; Mustafa, N.; Mastura, M.T.; Latip, M.A.M.; Hadi, S.D. Development of 3D Printing Filament Material Using Recycled
ropylene (rPP) Reinforced with Coconut Fiber. In *Additive Manufacturing for Biocomposites and Synthetic Composites*; CRC
: Boca Raton, FL, USA, 2023; pp. 59–75.

aurya, A.K.; Manik, G. Advances towards development of industrially relevant short natural fiber reinforced and hybridized ›lypropylene composites for various industrial applications: A review. *J. Polym. Res.* **2023**, *30*, 47. [CrossRef]
mes, R.M.S.; Gisip, J.; Mohd Yusof, N. Effect of chemical treatment on physical and mechanical properties of coir fibre- ›lypropylene composites. *Sci. Res. J.* **2023**, *20*, 145–157.
ılogun, O.A.; Daramola, O.O.; Adediran, A.A.; Akinwande, A.A.; Bello, O.S. reinforced polypropylene composites for ıtomobile application: Mechanical, wear and flow properties. *Alex. Eng. J.* **2023**, *65*, 327–341. [CrossRef]
aliappan, S.; Natrayan, L. *Polypropylene Composite Materials with Natural Fiber Reinforcement: An Acoustic and Mechanical Analysis r Automotive Implementations*; SAE Technical Paper No. 2023-01-5130; SAE International: Warrendale, PA, USA, 2024.
B/T 1040.1-2018; Determination of Tensile Properties of Plastics Part 1: General Principles. China National Standardization dministration: Beijing, China, 2008.
STM D3039; Standard Test Method for Tensile Properties of Polymer Matrix Composite Materials. ASTM: West Conshohohoken, A, USA, 1995; pp. 99–109.
ıng, Y.; Zhang, H.; Liu, X.; Deng, Y.; Sun, M.; Wang, J.; Cui, Y.; Pan, L.; Chen, Z. Hierarchical interface design of jute ›ers/polypropylene composites for enhanced interfacial and mechanical properties. *J. Clean. Prod.* **2024**, *450*, 141966. [CrossRef]
üller, V.; Brylka, B.; Dillenberger, F.; Glöckner, R.; Kolling, S.; Böhlke, T. Homogenization of elastic properties of short-fiber inforced composites based on measured microstructure data. *J. Compos. Mater.* **2016**, *50*, 297–312. [CrossRef]
:nek, L.T.; Argyris, J. *Finite Element Analysis for Composite Structures*; Springer Science & Business Media: New York, NY, SA, 2013.
allick, P.K. *Fiber-Reinforced Composites: Materials, Manufacturing, and Design*; CRC Press: Boca Raton, FL, USA, 2007.
ao, W.; Liu, C.; Shen, H. *Application Examples of ANSYS in Civil Engineering*; China Water & Power Press: Beijing, China, 2005; ›. 14–23.
aurya, A. Study of Mechanical Properties of Coconut Coir Fiber Reinforced Epoxy Biocomposite. Master's Thesis, Integral niversity, Lucknow, India, 2021.
heng, Y.; Li, Y.; Yang, C. The design of an airbag automatic inflator and the simulation analysis of airbag in the unfolding ocess. In Proceedings of the 2020 5th International Conference on Mechanical, Control and Computer Engineering (ICMCCE), arbin, China, 25–27 December 2020; IEEE: Toulouse, France, 2020; pp. 961–965.

polymers

MDPI

Article

Development of an Equivalent Analysis Model of PVB Laminated Glass for TRAM Crash Safety Analysis

Yuhyeong Jeong [1], Youngjin Jeon [1,2], Wonjoo Lee [1] and Jonghun Yoon [1,2,*]

[1] Department of Mechanical Engineering, Hanyang University ERICA, 55, Hanyangdaehak-ro, Sangnok-gu, Ansan-si 15588, Gyeonggi-do, Republic of Korea; jtpye9402@gmail.com (Y.J.); jj981226@hanyang.ac.kr (Y.J.); wj6478@gmail.com (W.L.)

[2] Department of Mechanical Engineering, BK21 FOUR ERICA-ACE Center, Hanyang University ERICA, Ansan-si 15588, Gyeonggi-do, Republic of Korea

* Correspondence: yooncsmd@gmail.com; Tel.: +82-31-400-5255

Abstract: This study focuses on an equivalent model of Polyvinyl Butyral (PVB) laminated glass to simulate the Head Injury Criterion (HIC) when a pedestrian collides with a TRAM. To simulate the collision behavior that occurs when a pedestrian's head collides with PVB laminated glass, a comparison was made between the results of the widely used PLC model for PVB laminated glass modeling and an actual dynamic head impact test. The material properties of the tempered glass and PVB film used in the PLC and equivalent models were obtained via four-point bending tests and tensile tests, respectively. The proposed equivalent model is developed by assigning the thickness, material properties, and positional information of each layer in the multilayer PLC model to the integration points of the shell element. The results of the equivalent analysis model were found to accurately simulate the collision behavior when compared with the results of both the dynamic head impact test and the PLC model. Moreover, the analysis cost improved to approximately 15% of that of the traditional PLC model.

Keywords: head injury criterion; equivalent model; PVB laminated glass; numerical analysis; dynamic head impact test

1. Introduction

The TRAM is a type of rail vehicle, consisting of either individual railcars or self-propelled trains coupled into multiple units that run on tramway tracks on public roads. There is an increasing trend in the introduction of trams, driven by the growing demand for eco-friendly energy. Unlike conventional rail vehicles that run on designated railroads, trams operate on public roads, raising significant safety concerns not only for the vehicle itself but also for the objects it may collide with, particularly pedestrians [1–6]. Additionally, due to the vertically designed windshield shape [7,8], the collision behavior between a tram and a pedestrian shows a different pattern compared to that of a typical automobile and pedestrian collision. Therefore, the design of the windshield for trams requires a unique impact analysis model specifically for tram windshields, rather than relying on the collision behavior models of conventional rail vehicles or automobiles.

The front glasses of most automobiles, including trams, are primarily composed of Polyvinyl Butyral (PVB) laminated glass [9–12]. This material combines tempered glass with a PVB film to achieve the high strength of tempered glass while the PVB film prevents glass fragments from scattering during collisions. Due to these characteristics, PVB laminated glass not only ensures sufficient strength for vehicle safety but also prevents

 https://doi.org/10.3390/polym17010025

secondary damage caused by glass fragments during collisions. For this reason, it is widely used in vehicle windshields. Consequently, analysis methods that consider collisions, particularly those involving pedestrians, are well-established. For analyzing the effects of pedestrian-vehicle collisions, dynamic head impact tests are widely utilized. These tests simulate a pedestrian's head using a dummy head, which is impacted against actual vehicles or automotive glass. By analyzing pre- and post-impact velocity and acceleration, the test evaluates the force and impulse exerted on the dummy head. This makes it an essential consideration in designing vehicles to minimize risks in pedestrian collisions.

Henn [13] analyzed the head injury criterion (HIC), which measures the likelihood of head injury from an impact, by comparing real crash tests using Mercedes–Benz cars. They calculated HIC parameters with and without airbags to compare injury risks, and the HIC became the standard by which to judge critical injury to a pedestrian's head during a collision. Bois et al. [14] proposed a numerical analysis to simulate the crash behaviors of laminated safety glasses via computational analysis. They presented a laminated safety structure consisting of glass layers modelled with shell elements and PVB interlayers with membrane elements. In addition, they conducted spherical impact simulations and validated their model by comparing their results with those of a single-material model. Timmel et al. [15] modelled PVB laminated glass as a multilayer structure with solid and shell elements, conducted ball drop analyses, and compared the results with those of actual windshield drop tests. They also verified the consistency of the time–acceleration graphs, which are crucial for measuring the HIC parameters, between the experiments and analyses. Most studies model PVB laminated glass as multiple layers, considering glass layers as shell elements and PVB layers as solid elements, and apply the respective material properties to each layer. They predominantly verified the analysis models by comparing the fracture patterns and HIC parameter-influencing time–acceleration graphs resulting from the impacts of spherical rigid bodies simulating human heads against PVB laminated glass. However, from the perspective of TRAM design, glass often does not break upon collision with pedestrians, owing to the low operational speed of TRAMs. Therefore, front glass designs are required to minimize pedestrian injuries in noncracking situations. Although existing multilayer analysis models effectively simulate the acceleration values applied to a pedestrian's head during a collision and the occurrences of cracks in glass, they require relatively long analysis times and high complexity in re-modeling every layer when modifications are needed. Consequently, such models are unsuitable for applications in the initial design step. Therefore, a new analysis model that is easier to modify, has lower analysis costs, and does not heavily consider glass breakage is needed.

Hence, this study proposes an equivalent analysis model that simplifies the multilayer structure of PVB laminated glass into a single-layer model. The proposed model maintains predictive accuracy comparable to experimental results and traditional multilayer models while dramatically reducing analysis costs to 15%, achieving approximately seven times the efficiency. This efficiency enables the rapid generation of pedestrian collision data specific to TRAMs, whose windshield is nearly vertical, aiding safety standard development. Additionally, it significantly reduces time and costs during frequent design iterations in the early design stages.

2. Experiments and Methodologies

2.1. Acquiring Material Properties of PVB Laminated Glass

The material properties of the tempered glass and PVB film were experimentally obtained to construct an analytical model of the PVB-laminated glass. In addition, 4-point bending tests were conducted on tempered glass to investigate the effect of the strain rate. Experiments were conducted at three different strain rates (0.001, 0.01, and 0.1/s). Tempered glass cannot undergo additional post-processing because of its brittleness. Therefore, the experimental die set was produced by scaling the ISO 11228-3 standard [16] 4-point bending die set to fit the size of the prepared tempered glass, as shown in Figure 1. The PVB film material properties were obtained via quasi-static and high-speed tensile tests that were conducted at various strain rates (0.004, 1, 10, 100, 200, and 400/s) to consider the high strain rate during impact. Figure 2 shows a schematic of the tensile test specimen, following the ISO 573-3 standard. Quasi-static tensile tests were performed using a universal testing machine (UTM), as shown in Figure 3a, whereas high-speed tensile tests were conducted using a high-speed teTnsile testing machine available at the Korea Institute of Materials Science (KIMS), as depicted in Figure 3b. The strain was measured via the 2D digital image correlation (DIC) method. The material properties of the PVB film were obtained based on the strain rate, and fitting was performed using the Mooney–Rivlin fitting equation.

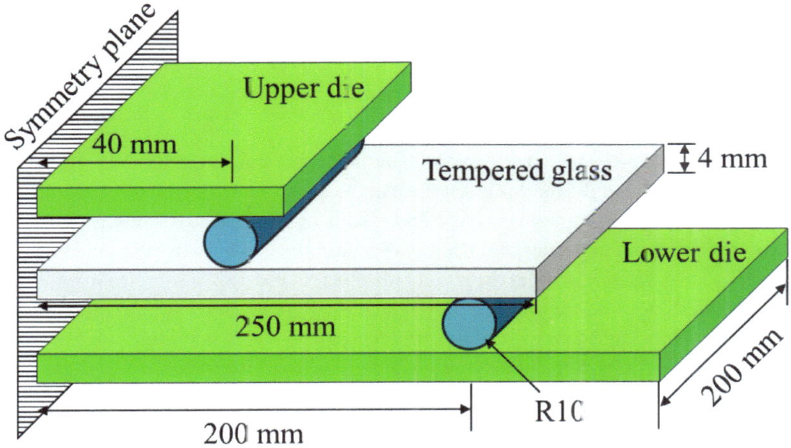

Figure 1. Schematic of the 4-point bending test, following the scaled ISO11228-3 standard.

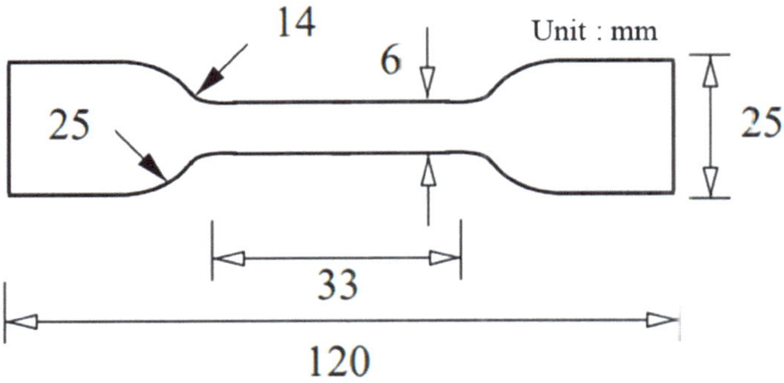

Figure 2. Schematic of a tensile test specimen for a PVB film, following the ISO573-3 standard.

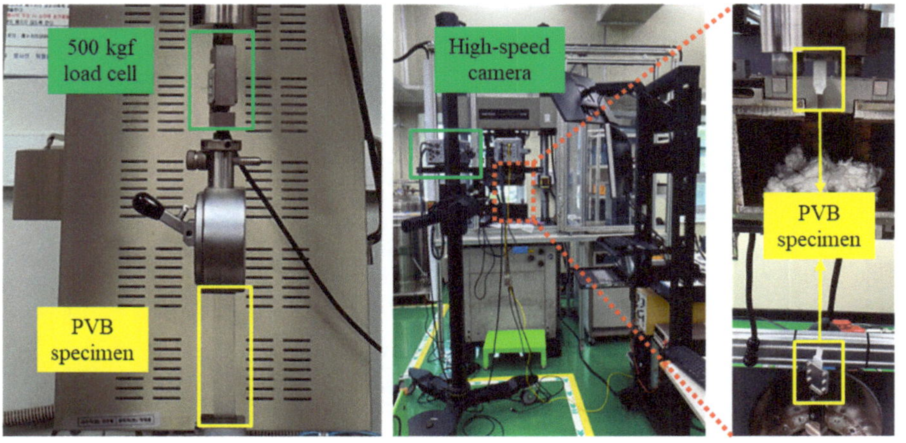

(a) (b)

Figure 3. Experimental setup for a PVB film tensile test: (**a**) quasi-static strain rate; (**b**) high strain rate.

Figure 4 shows the results of the 4-point bending test of the tempered glass. It appears that there is no change in the modulus with respect to the strain rate, which is attributed to the highly brittle nature of tempered glass and results in minimal strain rate dependency. Therefore, the material properties of the tempered glass applied in both the multilayer and equivalent models did not consider the strain rate effect. In contrast, a significant strain rate dependency was observed for the PVB film, which is a hyperelastic material that displays nonlinear behavior. The Mooney–Rivlin fitting equation, which is widely used to fit the stress–strain curves of hyperelastic materials, was used to fit the stress–strain data of the PVB film. The Mooney–Rivlin fitting equation is given by Equation (1):

$$\sigma = 2\lambda\left(1-\lambda^{-3}\right)\left(A_1\lambda + A_2 + A_3\left(\lambda^2 + 2\lambda^{-1} - 3 + \lambda\left(\lambda^{-2} + 2\lambda - 3\right)\right)\right) \quad (1)$$

$$\lambda = 1 + \varepsilon \text{ (stretch ratio)}$$

$$A_1, A_2, A_3 : \text{ Material constants}$$

The material constants for each strain rate were derived by fitting the stress–strain curve of the PVB film obtained from the tensile tests at different strain rates, as shown in Table 1. Figure 5 presents both the raw stress–strain data from the tensile test and data fitted using the Mooney–Rivlin equation. The fitted data were applied as material properties that accounted for the strain–rate dependency in the numerical analysis model.

Table 1. Material constants of the Mooney–Rivlin fitting equation with respect to the strain rate.

	1/s	10/s	100/s	200/s	400/s
A_1	−0.295	−1.027	−4.935	−4.962	−10.984
A_2	1.157	3.451	10.793	11.359	20.835
A_3	0.073	0.216	0.614	0.562	1.212

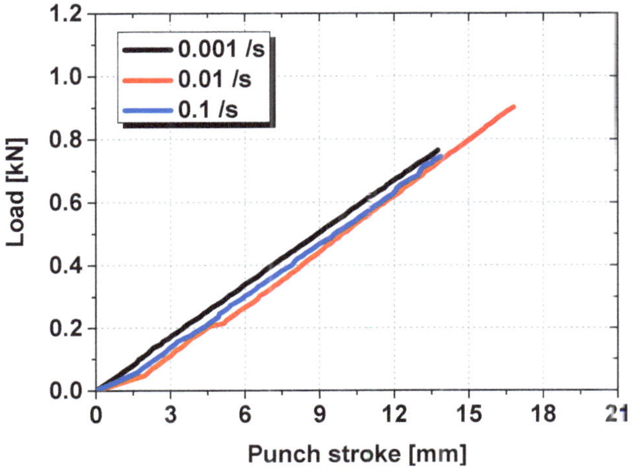

Figure 4. Punch stroke–load graph of tempered glass with respect to the strain rate.

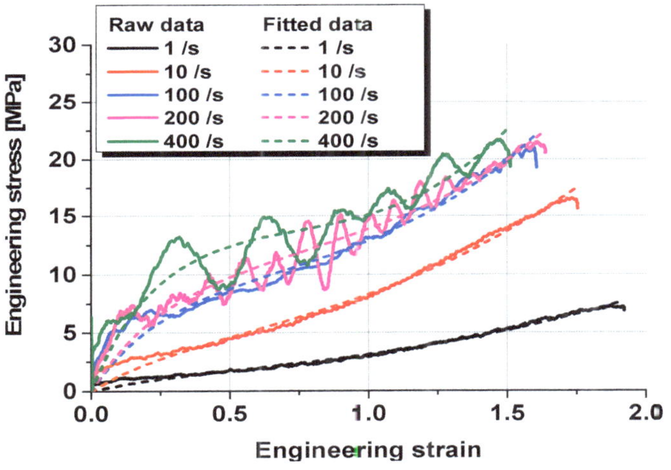

Figure 5. Stress–strain curves of PVB film with raw and fitted data obtained via the Mooney–Rivlin equation.

2.2. Dynamic Head Impact Test

A dynamic head impact test was conducted using the free-fall method with a ball equipped with an internal accelerometer. PVB-laminated glass measuring 1000 mm in width, 700 mm in height, and 9.22 mm in thickness was manufactured for the experiment. The glass was attached to 30 mm wide and 4 mm thick rubber frame packing and fixed to a test jig using bolts. Figure 6 shows the PVB-laminated glass attached to the dynamic head impact test jig and ball for the free-fall. The experiment was conducted for six cases, using two different impact locations and three impact speeds. The impact locations were the center impact condition, for which the ball was dropped on the exact center of the PVB-laminated glass, and the side impact condition, for which the ball was shifted 325 mm from the center along the longer direction of the glass. The impact speeds for the center impact were set at 20, 21, and 22 km/h, and those for the side impact were 20, 22, and 24 km/h. These speeds were determined by converting the drop heights of the Dynamic Head Impact Test equipment into equivalent velocities. The experiments were conducted by adjusting the drop height to control the corresponding impact speed. After the free-fall of the ball, the accelerometer was removed from the ball, and the time-acceleration data were extracted at the moment of impact between the ball and PVB laminated glass. Figure 7 presents the time–acceleration graphs obtained from the center impact and side impact experiments. The detailed experimental conditions are also described in Ahn et al. [17].

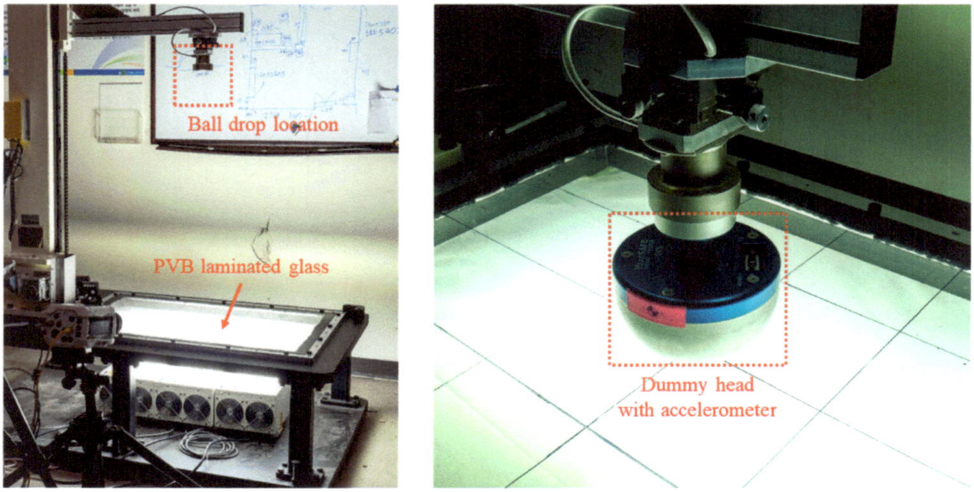

(a) (b)

Figure 6. Experimental settings for the dynamic head impact test: (**a**) test jig for the dynamic impact test; (**b**) dropped ball, including the accelerometer.

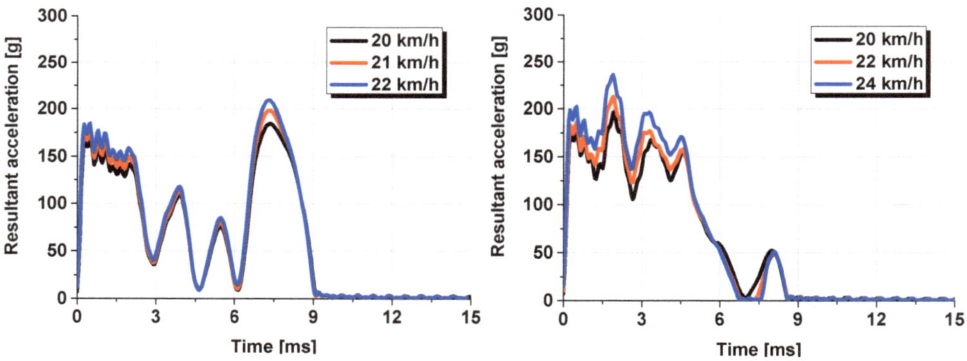

(a) (b)

Figure 7. Time–acceleration results from the dynamic head impact test: (**a**) center impact results; (**b**) side impact results.

3. Numerical Model for PVB-Laminated Glass

3.1. PVB-Laminated Glass Modeling with a Nonlocal Failure Criterion

The PLC criterion, which Pyttel et al. [18] called the nonlocal criterion, is an effective model for simulating the fracture behavior of glass caused by the impact between PVB-laminated glass and a pedestrian's head. Owing to its highly brittle nature, tempered glass exhibits nonlinear fracture characteristics in which cracks propagate at a high speed once a certain point of external impact is reached. Consequently, accurate simulations using general fracture methods are impossible. Hence, an energy-based threshold criterion was added to the main criterion. If this criterion is not met, no fractures occur. Pyttel et al. determined the fracture occurrence threshold based on the energy accumulation of elements within a certain radius from the point where the initial impact occurs on the glass. Figure 8 illustrates the initial impact point on the glass and the radius of the circle used to calculate the energy accumulation around that point. The first impact between the ball and tempered glass follows the major stress condition represented in Equation (2):

$$\sigma_{fail} = max(|\sigma_1|, |\sigma_2|) \quad (2)$$

The first impact area then becomes the center of the circle, using a predefined radius (R_C). The strain energy inside the finite region is calculated at each time step for each element.

The strain energy for each time step (i) and each element (e) can be represented as shown in Equation (3):

$$\Delta E_i^e = \iint \sigma : d\dot{\varepsilon} dV^e \Delta t_i \quad (3)$$

The total strain energy in the impact area at time t_n is represented by Equation (4):

$$E^e(t_n) = \sum_{i=0}^{i=n} \Delta E_i^e \quad (4)$$

If the accumulated strain energy in the region exceeds the critical value, then the failure criterion in Equation (2) is activated for each element of the entire glass area. The critical value (E_C) was predefined based on the experiments:

$$E > E_c \quad (5)$$

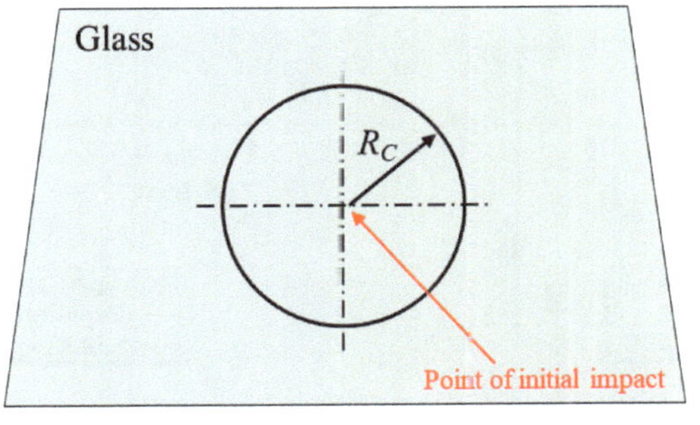

Figure 8. General parameters of the nonlocal approach.

Using the above method, a ball-drop analysis was applied to the PLC fracture model using LS-DYNA 2021 R1 version, based on the widely used multilayer modeling method. The model was constructed under the same experimental conditions as those used in the dynamic head impact test. The tempered glass and PVB film layers of the PVB-laminated glass were modeled separately and combined by applying node–node tie conditions. For each gripping rubber, hard contact conditions were applied to the parts in contact with the glass, using a dynamic friction coefficient of 0.2 and a static friction coefficient of 0.3, whereas the nodes on the opposite side were fixed using a fixed boundary condition. Figure 9 shows a model of the dynamic head impact test. For the glass, a simple linear elastic model was used based on the experimental results in Section 2.1, and mechanical properties from the European standard DIN EN 572-1 [19] were applied. These values included a Young's modulus of 70 GPa, Poisson's ratio of 0.23, and density of 2.53 kg/m^3. The material properties of the PVB film were determined by tabulating the stress–strain curves based on the strain rate. Figure 10 shows the analytical results of the dynamic head impact test and the experimental results. The analytical results indicate that the acceleration trend increases and decreases, closely following the experimental data.

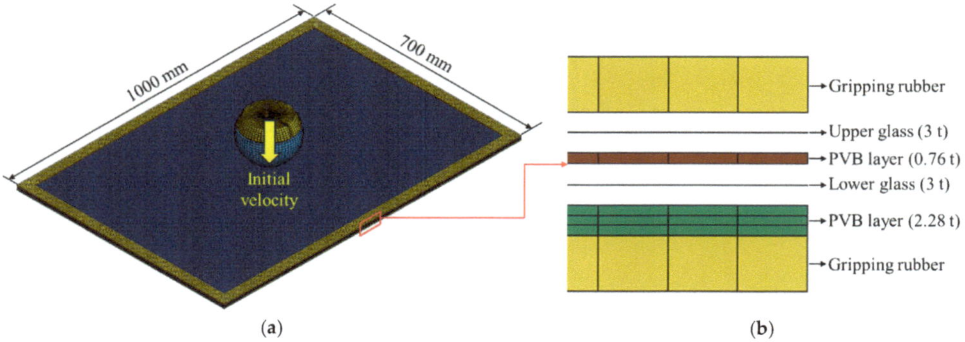

Figure 9. Numerical analysis model for the dynamic head impact test: (**a**) overall modeling; (**b**) structure of PVB-laminated glass in the direction of thickness.

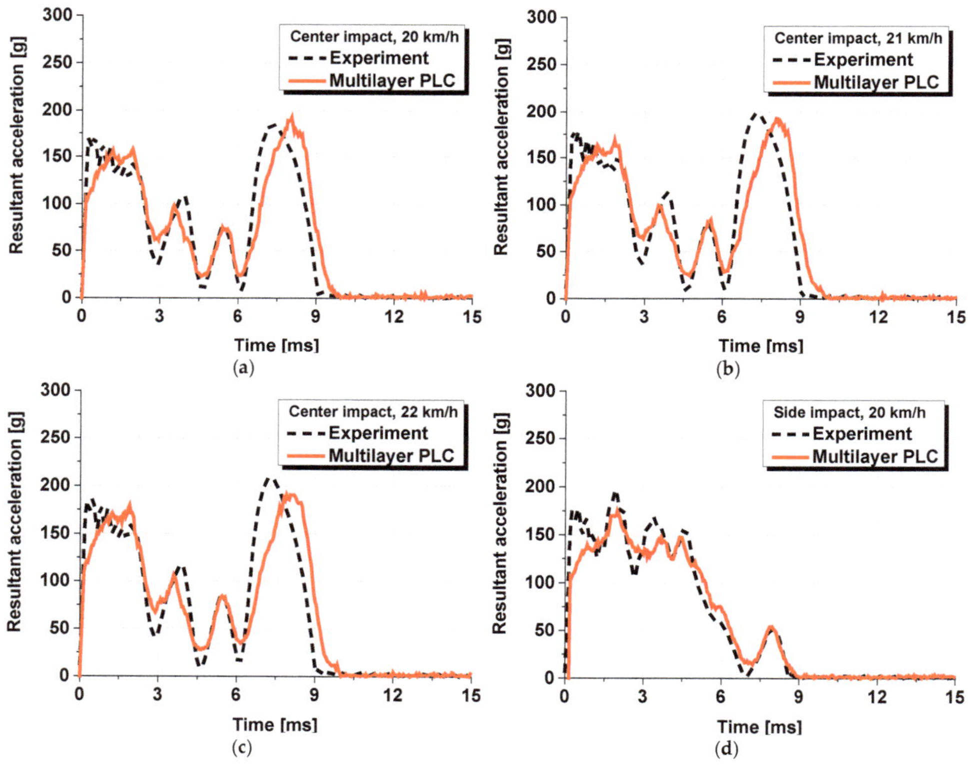

Figure 10. *Cont.*

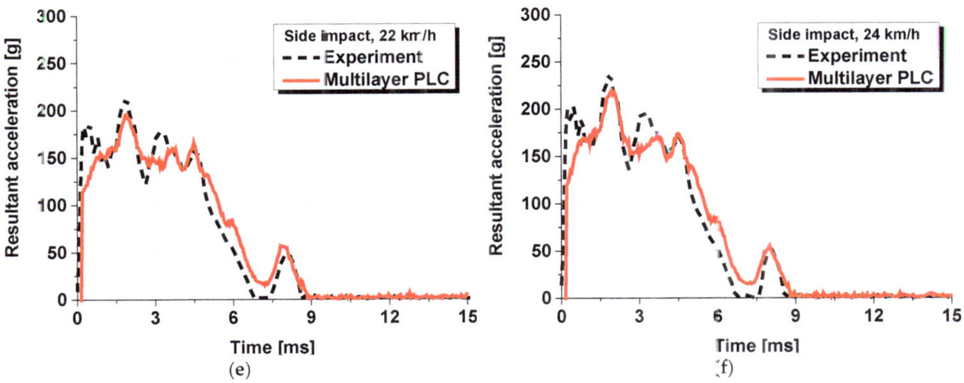

Figure 10. Analytical results of the dynamic head impact test with respect to the impact location and speed: (**a**) center impact at a speed of 20 km/h; (**b**) center impact at 21 km/h; (**c**) center impact at 22 km/h; (**d**) side impact at 20 km/h; (**e**) side impact at 22 km/h; (**f**) side impact at 24 km/h.

3.2. Equivalent Model for PVB-Laminated Glass

A single-layer structure was modelled for the equivalent model, including the material properties of the tempered glass and PVB film. In the proposed model, the integration points used in the shell elements were used to simulate each layer of the multilayer model. The location, thickness, and material properties of each integration point were controlled to simulate the multilayer structure; therefore, it played a role similar to that of the multilayer model. In LS-DYNA, the positions and thicknesses of each integration point should be converted to scaled values, with the total thickness of the structure being normalized to a range from −1 to 1. Thus, the total thickness of the PVB-laminated glass, that is, 9.22 mm, was normalized to 1, and 0 mm was normalized to −1. Then, the center points of each tempered glass and PVB film layer were calculated and applied. However, due to the single-layer structure of the proposed equivalent model, it is difficult to directly apply the failure criteria specific to the glass part in the PLC model. As a result, the equivalent model cannot simulate glass fracture behavior. Nonetheless, dynamic head impact tests considering TRAM operating speeds demonstrated no fracture of the PVB laminated glass, validating the equivalent model's applicability. Figure 11 illustrates the structure of the PVB-laminated glass and numerical structure of the equivalent model simulated using integration points.

A numerical analysis identical to the one conducted in Section 3.1 was conducted by applying the equivalent model proposed in this study. Figure 12 shows the numerical analysis model with the equivalent model applied, in which the PVB-laminated glass was modelled as a single layer. All other analysis conditions remained the same, and six cases from the previous analysis were conducted repeatedly. Figure 13 presents the time–acceleration graphs measured from the head dummy to depict the results of the actual dynamic head impact test, PLC analysis model, and equivalent model. The results demonstrated that the acceleration tendency of the equivalent model closely matched the experimental results and those of the existing PLC model. The trends aligned under various experimental conditions. Moreover, in terms of analysis cost, the PLC model required approximately 8 h to complete the analysis, whereas the equivalent model reduced the analysis time to less than 1 h for the same case, thus, demonstrating a dramatic decrease in the analysis cost.

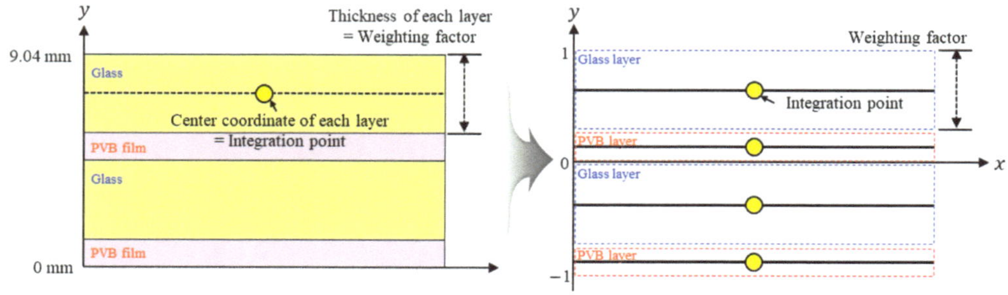

(**a**) (**b**)

Figure 11. (**a**) Structure of PVB-laminated glass; (**b**) numerical structure of the equivalent model.

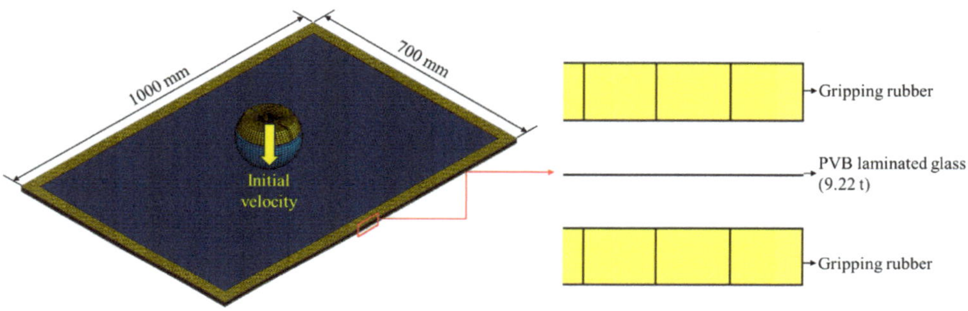

(**a**) (**b**)

Figure 12. Numerical analysis model for the dynamic head impact test using the equivalent model: (**a**) overall modeling; (**b**) equivalent modeling in the direction of thickness.

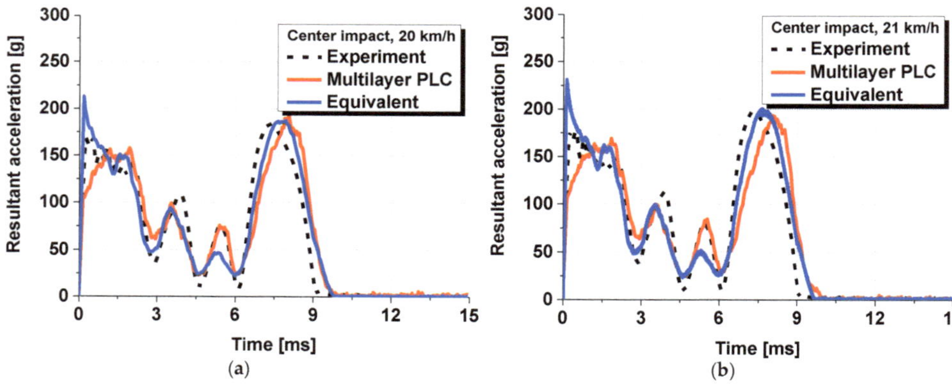

(**a**) (**b**)

Figure 13. *Cont.*

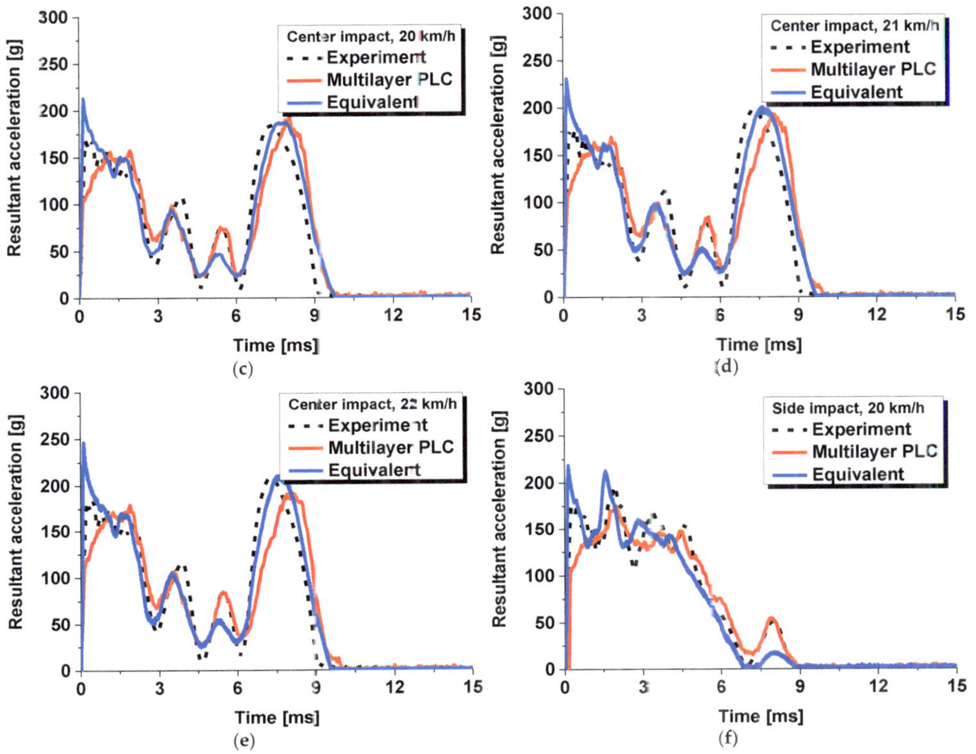

Figure 13. Analytical results of the dynamic head impact test using the equivalent model: (**a**) center impact at a speed of 20 km/h; (**b**) center impact at 21 km/h; (**c**) center impact at 22 km/h; (**d**) side impact at 20 km/h; (**e**) side impact at 22 km/h; (**f**) side impact at 24 km/h.

4. Conclusions

In this study, an equivalent model of PVB-laminated glass is proposed to simulate collision behavior. The PLC model replicates the multi-layer structure of PVB-laminated glass, including tempered glass and PVB film. However, the proposed equivalent model simplifies this by modeling tempered glass and PVB film as a single shell element layer, incorporating the multi-layer characteristics via integration points. This simplification maintains analytical equivalency with the PLC model. Consequently, the equivalent model produces results comparable to the PLC model while significantly reducing analysis time from 8 h to 1 h under the same hardware and software conditions. Although the proposed equivalent model cannot analyze the fracture pattern of PVB-laminated glass, it demonstrates high computational efficiency at impact speeds that do not cause glass fracture. Thus, it is suitable as an analytical model for low-speed vehicles such as TRAMs. Particularly for TRAMs, which are currently in the introduction and operational stages, the demand for computational analysis for front glass design is expected to be significant. By applying the proposed method to the design of TRAMs' front windshields, which are increasingly being adopted, the behaviors of pedestrian impacts can be reliably and quickly analyzed to significantly reduce the time required for front-end design in the initial design phase of a TRAM.

Author Contributions: Conceptualization, J.Y.; Methodology, Y.J. (Yuhyeong Jeong); Software, W.L.; Validation, Y.J. (Youngjin Jeon); Formal analysis, Y.J. (Yuhyeong Jeong) and Y.J. (Youngjin Jeon); Data curation, Y.J. (Youngjin Jeon); Investigation, W.L.; Visualization, W.L.; Writing—original draft, Y.J. (Yuhyeong Jeong); Writing—review & editing, J.Y. Project administration, J.Y.; Funding acquisition, J.Y. All authors have read and agreed to the published version of the manuscript.

Funding: This work was supported by the Industrial Strategic Technology Development Program-A program for win-win type innovation leap between middle market enterprise and small and medium-sized enterprise (P0024516, Development and commercialization of a customized dental solution with intelligent automated diagnosis technology based on virtual patient data), funded by the Ministry of Trade, Industry and Energy (MOTIE, Republic of Korea) and the Korea Institute for Advancement of Technology (KIAT). This research was financially supported by the Ministry of Trade, Industry, and Energy (MOTIE), Korea, under the "170k closed section roll forming and free curvature bending technology development for electric vehicle body" (reference number 20022814) supervised by the Korea Institute for Advancement of Technology (KIAT). This work was also supported by This research was financially supported by the Ministry of Trade, Industry, and Energy (MOTIE), Korea, under the "Innovative Digital Manufacturing Platform" (reference number P00223311) supervised by the Korea Institute for Advancement of Technology (KIAT).

Institutional Review Board Statement: Not applicable.

Data Availability Statement: Data are contained within the article.

Conflicts of Interest: The authors declare no conflict of interest.

References

1. Špirk, S.; Špička, J.; Vychytil, J.; Křížek, M.; Stehlík, A. Utilization of the validated windshield material model in simulation of tram to pedestrian collision. *Materials* **2021**, *14*, 265. [CrossRef] [PubMed]
2. Marti, C.M.; Kupferschmid, J.; Schwertner, M.; Nash, A.; Weidmann, U. Tram safety in mixed traffic: Best practices from Switzerland. *Transp. Res. Rec.* **2016**, *2540*, 125–137. [CrossRef]
3. Guerrieri, M. Tramways in urban areas: An overview on safety at road intersections. *Urban Rail Transit* **2018**, *4*, 223–233. [CrossRef]
4. Hedelin, A.; Björnstig, U.; Brismar, B. Trams—A risk factor for pedestrians. *Accid. Anal. Prev.* **1996**, *28*, 733–738. [CrossRef]
5. Gaca, S.; Franek, L. Pedestrian fatality risk as a function of tram impact speed. *Open Eng.* **2021**, *11*, 1105–1113. [CrossRef]
6. Peng, Y.; Hu, Z.; Liu, Z.; Che, Q.; Deng, G. Assessment of pedestrians' head and lower limb injuries in tram–pedestrian collisions. *Biomimetics* **2024**, *9*, 17. [CrossRef] [PubMed]
7. Lopot, F.; Tomsovsky, L.; Marsik, F.; Masek, J.; Kubovy, P.; Jezdik, R.; Sorfova, M.; Hajkova, B.; Hylmarova, D.; Havlicek, M.; et al. Pedestrian safety in frontal tram collision, part 1: Historical overview and experimental-data-based biomechanical study of head clashing in frontal and side impacts. *Sensors* **2023**, *23*, 8819. [CrossRef]
8. Tomsovsky, L.; Lopot, F.; Kubovy, P.; Jezdik, R.; Hajkova, B.; Rulc, V.; Jelen, K. Tram-pedestrian collisions: The severity of head injuries due to secondary impact with the surrounding infrastructure (ground). In Proceedings of the 2022 International Conference on Electrical, Computer, Communications and Mechatronics Engineering (ICECCME), Maldives, Maldives, 16–18 November 2022; pp. 1–6.
9. Shahriari, M.; Saeidi Googarchin, H. Prediction of vehicle impact speed based on the post-cracking behavior of automotive PVB laminated glass: Analytical modeling and numerical cohesive zone modelling. *Eng. Fract. Mech.* **2020**, *240*, 107352. [CrossRef]
10. Molnár, G.; Vigh, L.G.; Stocker, G.; Dunai, L. Finite element analysis of laminated structural glass plates with polyvinyl butyral (PVB) interlayer. *Period. Polytech. Civ. Eng.* **2012**, *56*, 35–42. [CrossRef]
11. Müller-Braun, S.; Brokmann, C.; Schneider, J.; Kolling, S. Strength of the individual glasses of curved, annealed and laminated glass used in automotive windscreens. *Eng. Fail. Anal.* **2021**, *123*, 105281. [CrossRef]
12. Valera, T.S.; Demarquette, N.R. Polymer toughening using residue of recycled windshields: PVB film as impact modifier. *Eur. Polym. J.* **2008**, *44*, 755–768. [CrossRef]
13. Henn, H.W. Crach tests and the head injury criterion. *Teach. Math. Appl.* **1998**, *17*, 162–170.
14. Du Bois, P.A.; Kolling, S.; Fassnacht, W. Modelling of safety glass for crash simulation. *Comput. Mater. Sci.* **2003**, *28*, 675–683. [CrossRef]
15. Timmel, M.; Kolling, S.; Osterrieder, P.; Du Bois, P.A. A finite element model for impact simulation with laminated glass. *Int. J. Impact Eng.* **2007**, *34*, 1465–1478. [CrossRef]

16. *ISO 11228-3:2007*; Ergonomics—Manual Handling—Part 3: Handling of Low Loads at High Frequency. ISO: Geneva, Switzerland, 2007.
17. Ahn, S.H.; Jung, H.S.; Kim, J.S.; Hwang, J.H. Finite Element Simulation of Windshield for Tram-pedestrian Collision *Int. J. Crashworthiness*, 2024; *preprint*.
18. Pyttel, T.; Liebertz, H.; Kai, J. Failure criterion for laminated glass under impact loading and its application in finite element simulation. *Int. J. Impact Eng.* **2011**, *38*, 252–263. [CrossRef]
19. *EN 572-1:2012*; Glass in Building—Basic Soda Lime Silicate Glass Products—Part 1: Definitions and General Physical and Mechanical Properties. CEN: Brussels, Belgium, 2012

Journal of
Manufacturing and Materials Processing

MDPI

Article

A Comprehensive Study on the Optimization of Drilling Performance in Hybrid Nano-Composites and Neat CFRP Composites Using Statistical and Machine Learning Approaches

Tanzila Nargis [1], S. M. Shahabaz [2], Subash Acharya [2,*], Nagaraja Shetty [2,*], Rashmi Laxmikant Malghan [3] and S. Divakara Shetty [4]

1 Department of Information Science and Engineering, NMAM Institute of Technology, NITTE (Deemed to be University), Karkala 574110, India; tanzilanargis@nitte.edu.in
2 Department of Mechanical & Industrial Engineering, Manipal Institute of Technology, Manipal Academy of Higher Education, Manipal 576104, India; shahabaz.sm@learner.manipal.edu
3 Department of Data Science and Computer Applications, Manipal Institute of Technology, Manipal Academy of Higher Education, Manipal 576104, India; rashmi.malghan@manipal.edu
4 Alva's Institute of Engineering and Technology, Mijar, Moodabidri, Mangalore 574225, India
* Correspondence: subash.acharya@manipal.edu (S.A.); nagaraj.shetty@manipal.edu (N.S.)

Abstract: Carbon fiber-reinforced polymer (CFRP) composites have gradually replaced metals due to their exceptional strength-to-weight ratio compared to metallic materials. However, the drilling process often reveals various defects, such as surface roughness, influenced by different drilling parameters. This study explores the drilling quality of uni-directional CFRP composites, as well as hybrid Al_2O_3 alumina and hybrid SiC silicon carbide nano-composites, through experimental exploration using step, core, and twist drills. Response surface methodology (RSM) and statistical tools, including main effect plots, ANOVA, contour plots, and optimization techniques, were used to analyze the surface roughness of the hole. Optimization plots were drawn for optimal conditions, suggesting a spindle speed of 1500 rpm, feed of 0.01 mm/rev, and a 4 mm drill diameter for achieving minimum surface roughness. Furthermore, two machine learning models, artificial neural network (ANN) and random forest (RF), were used for predictive analysis. The findings revealed the robust predictive capabilities of both models, with RF demonstrating superior performance over ANN and RSM. Through visual comparisons and error analyses, more insights were gained into model accuracy and potential avenues for improvement.

Keywords: carbon fiber-reinforced polymer; drilling; response surface methodology; artificial neural network; random forest

1. Introduction

Among the various classifications of composites, carbon fiber-reinforced polymer (CFRP) has garnered considerable attention in aerospace circles owing to its exceptional property of offering a high strength-to-weight ratio [1]. The substantial weight of engine components in automobiles increases fuel consumption, diminishing efficiency. Consequently, replacing these components with composite materials has been shown to enhance overall performance [2]. Similarly, weight reductions of 25% have been observed for both commercial and military aircraft structures, respectively. Notably, in the case of the Boeing 777, a prominent passenger aircraft constructed primarily of carbon fiber epoxy, weight savings of 15–20% have been achieved. Components fabricated from composites in the Boeing 777 include flaperons, ailerons, inboard and outboard flaps, landing gear doors, and engine cowlings [3].

There has been a notable surge in demand for advanced composites with enhanced performance characteristics in recent years. Nano-composites represent a cutting-edge category of materials wherein one or more distinct nanoparticles are incorporated into the matrix material to further augment performance [4]. Epoxy resin, derived from the interaction of bisphenol A and epichlorohydrin, has served as a primary matrix material for the fabrication of these nano-composites for many years [5]. Nano-composites comprising epoxy and nanoparticle reinforcements have exhibited exceptional mechanical, thermal, and electrical properties. Key factors such as molecular structure, curing conditions, and the ratio of epoxy resin to curing agent(s) significantly influence their performance. Nanoparticles are characterized by a high surface area, rendering them invaluable across numerous applications [6]. Inorganic particles such as alumina (Al_2O_3) and silicon carbide (SiC) show high hardness, excellent wear, and temperature resistance, as well as satisfactory chemical inertness, making them extensively utilized in metallurgical components, composite fabrication, and electronic industries [7,8].

In a study conducted by Mohanty et al. [9], it was found that incorporating Al_2O_3 nanoparticles into hybrid carbon and glass fiber-reinforced composites significantly enhanced both tensile strength and modulus. Similarly, Priyadarshi et al. [10] investigated the mechanical characteristics of Al_2O_3-filled jute epoxy composite under various conditions, revealing notable improvements such as heightened impact strength (1.902 Joules), augmented flexural strength (72.94 MPa), and a maximum hardness reaching 29.9 Vickers hardness number. Additionally, the authors, in their prior work, reported experimental results on the enhancement of mechanical properties at different nano-filler loadings (Al_2O_3 and SiC) in neat CFRP composite compared to hybrid nano-composites. The maximumHighest mechanical properties were observed at 1.75 wt.% filler loading for Al_2O_3 hybrid nano-composites and at 1.25 wt.% filler loading for SiC hybrid nano-composites [11,12].

Defects such as fiber pull-out, debonding, microcracking, surface roughness, and delamination occur in the drilled holes after drilling. To determine the extent of drilling-induced damage, researchers have used a wide range of methods, such as digital image processing, optical microscopy, C-scan, X-ray, and laser-based imaging, to scan electron microscopic images. The Taguchi technique and analysis of variance (ANOVA) have been used in experimental studies, including those conducted by Davim and Reis, [13] to establish a correlation between surface roughness, feed rate, and cutting speed. Tsao et al. [14] also suggested that the right tool geometry and cutting parameters might decrease surface roughness in CFRP composite drilling. Response surface methodology (RSM) is a straightforward and efficient method for establishing a relationship between output variables and machining parameters [15]. RSM was utilized by Palanikumar and Davim [16] to forecast surface roughness in glass fiber-reinforced polymer (GFRP) composite drilling.

Soft computing techniques serve as valuable supplements to conventional statistical methods in the analysis of composite drilling processes. Their utilization in this context aims to tackle the intricate, nonlinear and ambiguous aspects inherent in process variables. Through soft computing modeling, researchers can rely on a robust approach that consistently yields thorough, accurate, and dependable results. Recently, researchers have been using various soft computing techniques, including response surface methodology (RSM) [17], random forest technique (RF) [18], artificial neural network (ANN) [19], and design of experiments (DOE) [20]. These methods provide researchers with versatile tools to navigate the complexities of the drilling process in composites.

Given the typically extensive and high-cost nature of experiments required to assess the machinability of metals or materials, an effective approach lies in the utilization of statistically or numerically designed tests, commonly known as design of experiments (DOE). This methodology enables researchers to strategically plan experimental set-ups, evaluate the impact of each process parameter, and ultimately minimize the total number of tests needed to achieve optimal conditions [21].

Jayabal and Natarajan [22] conducted a study to investigate the influence of process parameters, including drill diameter, spindle speed, and feed rate, on thrust force, torque, and tool wear during the drilling of coir fiber-reinforced composite materials. Utilizing the Box–Behnken design and genetic algorithm (GA) techniques, the researchers were able to identify the optimal process parameters. Their findings indicated that this approach was effective in forecasting both main and interaction effects and drilling process output variables. Moreover, it facilitated the determination of optimal values for drilling parameters. Therefore, Jayabal and Natarajan concluded that the employed technique demonstrated practicality in optimizing the drilling process. To predict the delamination and surface roughness during the drilling of CFRP composite, Enemuoh et al. [23] employed a multi-layered perceptron neural network (MLPNN) model. They discovered that the predicted and experimental results for delamination and surface roughness agreed well. Similarly, to optimize the drilling process output variables during the drilling of the Al/SiCp composite, Karthikaya et al. [24] applied fuzzy logic and genetic algorithm techniques. The experimental data were trained and simulated using fuzzy logic, and the GA model was utilized to optimize the process parameters.

It is evident from the literature that the appropriate machining parameters and tool geometry can lessen drilling-induced damage to CFRP composites. In order to measure the output variable surface roughness (R_a), this study examines the drilling of hybrid nano-composites composed of Al_2O_3, as well as SiC at different cutting parameters (spindle speed, feed, drill diameter, and drill type). Surface roughness is a crucial factor in machining processes, influencing product quality and performance of composites. Accurate prediction models can aid in optimizing machining parameters for desired surface finish. In this study, along with the experimental and RSM results, two popular machine learning models are implemented—artificial neural network (ANN) and random forest (RF)—for predicting R_a in different materials, like Al_2O_3, SiC, and neat CFRP composites. The implemented models (ANN, RF) are evaluated based on prediction accuracy, that is, relative error attained for R_a. Later, a comparative analysis is made against the R_a values obtained from response surface methodology (RSM) with ANN and RF. RSM is employed to design experiments with different machine parameters, and main effects plots, contour plots, and optimization plots aregenerated using experimental results.

2. Material and Methods

2.1. Materials

In this investigation, an unidirectional carbon fiber-reinforced polymer (CFRP) material has been employed as a reinforcing agent, along with the utilization of bisphenol-A epoxy resin and an amine-based hardener serving as the polymer matrix. To enhance the structural attributes, distinct inorganic nano-fillers, namely Al_2O_3 and SiC, were carefully incorporated at varying filler loadings (1, 1.5, 1.75, 2 wt% for Al_2O_3 and 1, 1.25, 1.5, 2 wt% for SiC), yielding hybrid nano-composites. The achievement of a homogenous dispersion of nano-fillers was methodically accomplished through a combined application of sonication and magnetic stirring methods.

The manufacturing process encompassed the fabrication of both the CFRP and hybrid nano-composites through a hand lay-up method, succeeded by a compression molding technique with a curing duration of 24 h at room temperature. The drilling was performed utilizing a computer numerical control vertical machining center, as represented in Figure 1. The maximum properties obtained for hybrid nano-composites were at 1.75 wt% for Al_2O_3 hybrid nano-composites and at 1.25 wt% for SiC hybrid nano-composites, as reported by authors in their previous work. Drilling was performed for the above hybrid nano-composites, and their surface roughness was measured.

2.2. Machining Parameters

Table 1 illustrates the selected machining parameters chosen for the investigation. An experimental design incorporating response surface methodology (RSM) through Minitab V15 software was employed, utilizing the following machining parameters. Composite strips measuring 250 mm × 25 mm were precisely cut using an abrasive water jet cutting machine. A total of 180 holes were drilled, with 60 holes dedicated to each composite type. Three types of drill geometry were selected, i.e., twist drill, step drill, and core drill, as shown in Figure 2. The drill bits were made of solid carbide material, and a diamond coating was applied using the physical vapor deposition (PVD) method. The PVD diamond coating provides a higher tool life and improved quality of drilled holes. Also, it provides resistance to oxidation, corrosion, and wearing of tools. PVD diamond coating was performed by exposing the cutting tool to the vapor of coating material at higher temperatures, up to 1000 °C, and then allowing it to adhere to the cutting tool.

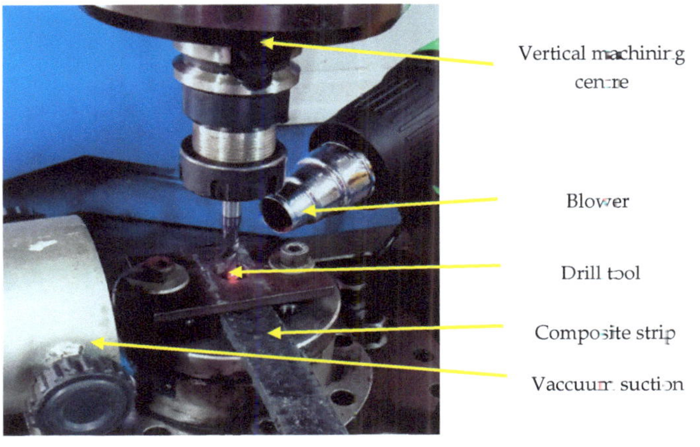

Figure 1. Schematic representation of drilling set-up.

Table 1. Various machining parameters selected for drilling.

Sl. No.	Parameters			
	Spindle Speed (rpm)	Feed (mm/rev)	Drill Diameter (mm)	Drill Type
1	1500	0.01	4	Twist
2	3500	0.02	6	Step
3	5500	0.03	8	Core

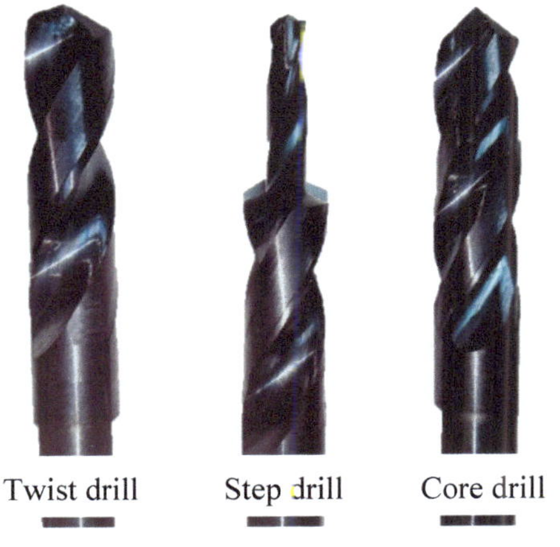

Figure 2. Drill types with PVD coating.

2.3. Surface Roughness Measurement

The surface roughness (R_a) of the drilled hole was measured by using the Taylor-Hobson Surtronic 3+ instrument (Model: Surtronic 3+, Taylor Hobson Ltd., Leicester, UK), as shown in Figure 3. Surface roughness was measured at six different locations with a probe speed of 0.5 mm/s in the transverse direction, up to 4 mm in length. The average surface roughness value was noted to determine the surface quality of the hole wall.

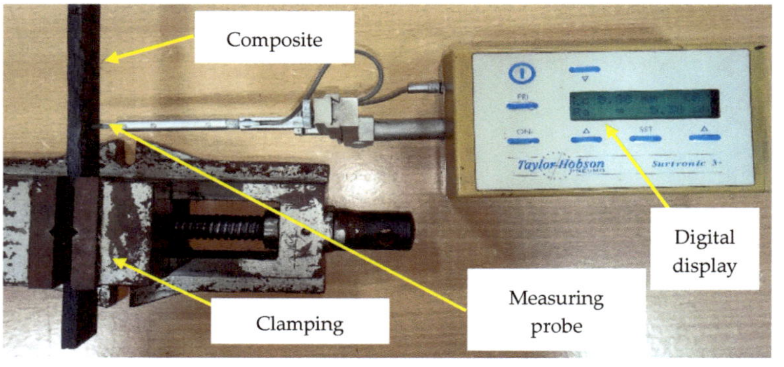

Figure 3. Measurement of surface roughness of the drilled hole.

2.4. Statistical Tools

RSM serves as a predictive and optimization tool for output variables affected by multiple factors. The modeling and analysis of data involved the application of RSM employing a full factorial method. An empirical connection between process parameters was established using RSM's central composite design (CCD). Equation (1) illustrates the correlation between various output parameters ('y'), such as delamination factor, burr area, surface roughness, and hole temperature, and the corresponding process parameters (spindle speed (A), feed (B), drill diameter (C), and type of drill (D).

$$y = \beta_o + \beta_1 A + \beta_2 B + \beta_3 C + \beta_4 D + \beta_5 A^2 + \beta_6 B^2 + \beta_7 C^2 + \beta_8 D^2 + \beta_9 AB + \beta_{10} AC + \beta_{11} AD + \beta_{12} BC + \beta_{13} BD + \beta_{14} CD + \varepsilon \quad (1)$$

where, $\beta_0, \beta_1, \beta_2, \ldots, \beta_{14}$ are the regression coefficients and ε is the random error.

This study also employed analysis of variance (ANOVA) to assess the impact of process parameters on the output variables of the drilling process in both the hybrid nano-composites and the neat CFRP composite. The objective of ANOVA is to identify the factors and their combinations that have a substantial influence on the machining operation. The analysis was conducted with a significance level set at $\alpha = 0.05$, corresponding to a confidence level of 95%. Similarly, the impact of different process parameters on surface roughness values was assessed using main effects plots and contour plots. A main effects plot illustrates the mean response values for each level of a process parameter, allowing for a clear understanding of its influence. Comparatively, the main effects plot enables the assessment of the relative intensity of the impacts of various factors. Likewise, contour plots offer a 2D surface projection to depict the variation in response. These plots are generated by simultaneously considering two input factors while maintaining the remaining factors at their median values. The resulting contour lines connect data points with identical response values, providing a comprehensive visualization of the surface's behavior.

2.5. Optimization of Process Parameters Using RSM

The process parameters were optimized through response surface methodology (RSM). In RSM, the desirability function method is commonly employed for the optimization of multi-output variables. According to this methodology, any process exhibiting a quality characteristic beyond the acceptable limits is considered entirely unacceptable. The primary objective was to identify process parameters that yield the most favorable output variables. In this investigation, the desirability function approach was utilized to optimize the output variable within the framework of RSM. The desirability function transforms the output variable, specifically the surface roughness value, into a dimensionless variable known as the desirability index (d_i), as defined in Equation (2). The desirability index ranges within the closed interval of (0, 1). A higher desirability index value for output variable indicates a greater contribution to the overall performance by the respective output variable. The desirability index (d_i) is a function of the output variables (y_i), and each individual desirability function assigns a number between 0 and 1, where 0 signifies undesirability and 1 signifies a desirable or ideal value for the output variable.

$$d_i = \left(\frac{y - y_{min}}{y_{target} - y_{min}}\right)^q \quad y_{min} < y < y_{target} \tag{2}$$

In Equation (2), the output variable y is represented, where y_{min} corresponds to the lowest value, y_{target} denotes the target values, and q stands for the weight. The weight q can assume either low value ($0 < q < 1$) or high value ($q > 1$), depending on the desired effect. A value of one produces a linear ramp function between the low, target, and high values. Increasing the weight facilitates moving the result closer to the desired goal. The comprehensive assessment of product performance involves calculating the geometric mean of all desirability indices to derive an aggregate (global or composite) desirability index D, as expressed in Equation (3).

$$D = (d_1 \times d_2 \times d_3 \ldots\ldots d_m)^{1/m} \tag{3}$$

In Equation (3), the desired output variable is denoted as 'm'. An association is established between the output responses and the various process or machining factors through the incorporation of a second-degree polynomial regression equation into the experimental data. Subsequently, the determination of the output variable's desirability is executed to gauge the overall desirability. To identify the optimal overall desirability, the univariate search method is applied, systematically exploring diverse combinations of machining factors within the specified experimental range

2.6. Prediction Methods

2.6.1. Artificial Neural Network (ANN)

ANN is a brain-structured computation model. The structure of neurons provides the computing capability to identify the relations between the input and the output parameters through the mapping model. The ANN structure consists of 3 layers:

- Input layer: number of input parameters = number of neurons placed in the input layer.
- Hidden layer: a greater number of neurons placed compared to the input layer.
- Output layer: number of output parameters = number of neurons.

A relation will be created by flowing the information between the input layer to the hidden layer and from the hidden to the output layer. The forward and back propagation techniques can be implemented for each iteration, that is, for the input–output pair. In the case of backpropagation, initially, the result "Y" will be calculated. Then, if the desired results are not met, the weight adjustment needs to be performed backward from the output layer to the hidden layer and from the hidden layer to the input layer.

2.6.2. Random Forest (RF)

Random forest (RF) is an effective algorithm that fits the ensemble learning family. It outperforms individual decision trees. RF involves building multiple decision trees, and later, it combines their predictions. The significance of RF is a measure of feature importance based on the contribution of each feature involvement in reducing impurity (error) across all the developed trees. This flow helps in parameter tuning, interpreting results, and identifying the insights, thus leading to flexible handling of various kinds of data. RF supports the parallelization making the training faster as each tree developed is independent. In the case of prediction, the individual trees are averaged to attain the final outcome.

3. Results and Discussion

It is well known that surface roughness is one of the key parameters that decide the quality of the drilled holes and the cost of production. It largely depends on the kind of material to be drilled, the drill type to be used, and its cutting speed, feed, and the cutting force produced in drilling. Table 2 represents the regression equations for hybrid nano-composites and the neat CFRP composite for each drill type.

Table 2. Regression equations from RSM.

Composite Type	Drill Type	Regression Equation
Al_2O_3 hybrid nano-composite	Twist	0.9415 + 0.00.020018 × SS + 1.36 × F + 0.06897 × DD − 0.000000×S×SSS − 9. 4 × F × F − 0.002027 × DD × DD + 0.000088 × SS × F − 0.000001 × SS × DD − 0.1000 × F × DD
	Step	0.8863 + 0.000014 × SS + 1.36 × F + 0.06112 × DD − 0.000000 × SS × SS − 9.4 × F × F − 0.002027 × DD × DD + 0.000088 × SS × F − 0.000001 × SS × DD − 0.1000 × F × DD
	Core	1.1867 + 0.000016 × SS + 1.05 × F + 0.06012 × DD − 0.000000 × SS × SS − 9.4 × F × F − 0.002027 × DD × DD + 0.000088 × SS × F − 0.000001 × SS × DD − 0.1000 × F × DD
SiC hybrid nano-composite	Twist	1.1715 + 0.000011 × SS + 0.98 × F + 0.04702 × DD + 0.000000 × SS × SS − 1.5 × F × F − 0.000788 × DD × DD + 0.000010 × SS × F + 0.000000 × SS × DD + 0.0021 × F × DD
	Step	1.1656 + 0.000008 × SS + 0.66 × F + 0.03712 × DD + 0.000000 × SS × SS − 1.5 × F × F − 0.000788 × DD × DD + 0.000010 × SS × F + 0.000000 × SS × DD + 0.0021 × F × DD
	Core	1.4042 + 0.000010 × SS + 0.97 × F + 0.04317 × DD + 0.000000 × SS × SS − 1.5 × F × F − 0.000788 × DD × DD + 0.000010 × SS × F + 0.000000 × SS × DD + 0.0021 × F × DD
Neat CFRP composite	Twist	1.5652 + 0.000021 × SS + 1.23 × F + 0.07933 × DD − 0.000000 × SS × SS + 8.9 × F × F − 0.001777 × DD × DD + 0.000015 × SS × F − 0.000001 × SS × DD − 0.0479 × F × DD
	Step	1.3385 + 0.000020 × SS + 1.08 × F + 0.06688 × DD − 0.000000 × SS × SS + 8.9 × F × F − 0.001777 × DD × DD + 0.000015 × SS × F − 0.000001 × SS × DD − 0.0479 × F × DD
	Core	1.8969 + 0.000023 × SS + 1.10 × F + 0.07738 × DD − 0.000000 × SS × SS + 8.9 × F × F − 0.001777 × DD × DD + 0.000015 × SS × F − 0.000001 × SS × DD − 0.0479 × F × DD

From Table 3, it is observed that the surface roughness decreases with the addition of nanoparticles. The minimum surface roughness was noted for the Al_2O_3 hybrid nano-composite with an R_a value of 1.598 μm, followed by the SiC hybrid nano-composite with an R_a value being 1.783 μm, respectively. In contrast, the maximum R_a was observed for neat the CFRP composite (2.533 μm). This represents that the nanoparticle acts as a lubricant during the drilling of the composite by reducing the friction occurring at the tool–workpiece interaction. Furthermore, as declared in the previous work of the authors, it is proven that the addition of nanoparticles improves chemical bonding with polymer resin. The chemical bonding energy of oxygen atoms of Al_2O_3 with the hydrogen atoms of polymer chains is greater compared to the carbon atom of SiC combining with hydrogen atoms of polymer chains [11,12]. This good bonding between nanoparticles and resin molecules has shown improved mechanical properties of the hybrid nano-composites further indicating improved surface finish observed for hybrid nano-composites compared to the neat CFRP composite, as noted from Table 3. Also, from Table 3, the standard deviation values are less than 1, demonstrating that all the datasets are closer to mean values of the data set.

Table 3. Comparison of surface roughness (R_a) values obtained from experimental and RSM predictions.

Spindle Speed (rpm)	Feed (mm/rev)	Drill Dia (mm)	Drill Type	Experimental R_a (μm)			RSM-Predicted R_a (μm)			Error (%)		
				Al_2O_3	SiC	Neat	Al_2O_3	SiC	Neat	Al_2O_3	SiC	Neat
5500	0.01	8	Step	1.292	1.468	1.843	1.288	1.453	1.834	0.31	0.32	0.46
5500	0.01	8	Core	1.585	1.774	2.51	1.588	1.764	2.494	−0.21	0.54	0.65
5500	0.02	6	Core	1.543	1.723	2.416	1.544	1.710	2.412	−0.09	0.75	0.16
1500	0.01	4	Twist	1.216	1.391	1.893	1.216	1.373	1.891	0.01	1.27	0.10
3500	0.02	6	Step	1.224	1.411	1.751	1.226	1.402	1.745	−0.13	0.67	0.33
1500	0.01	4	Core	1.417	1.576	2.235	1.420	1.589	2.217	−0.19	−0.83	0.82
3500	0.01	6	Step	1.203	1.395	1.735	1.218	1.395	1.734	−1.23	0.01	0.05
1500	0.01	4	Step	1.134	1.317	1.612	1.123	1.320	1.612	0.94	−0.24	0.03
3500	0.02	6	Twist	1.343	1.481	2.067	1.342	1.484	2.053	0.08	−0.18	0.67
5500	0.02	6	Twist	1.371	1.512	2.097	1.369	1.506	2.084	0.11	0.39	0.63
5500	0.03	8	Twist	1.435	1.612	2.192	1.441	1.588	2.194	−0.45	1.49	−0.07
5500	0.03	4	Core	1.478	1.651	2.317	1.486	1.649	2.320	−0.56	0.13	−0.14
1500	0.01	8	Core	1.559	1.731	2.437	1.553	1.724	2.433	0.40	0.40	0.16
3500	0.01	6	Core	1.513	1.684	2.357	1.516	1.680	2.366	−0.20	0.24	−0.39
3500	0.03	6	Step	1.224	1.419	1.768	1.232	1.408	1.758	−0.62	0.79	0.56
1500	0.03	4	Core	1.432	1.61	2.249	1.428	1.608	2.242	0.30	0.14	0.29
5500	0.03	4	Twist	1.294	1.442	1.983	1.297	1.437	1.989	−0.22	0.32	−0.31
3500	0.02	6	Core	1.527	1.694	2.381	1.521	1.690	2.378	0.40	0.26	0.14
5500	0.01	4	Core	1.466	1.637	2.289	1.471	1.629	2.293	−0.35	0.46	−0.19
3500	0.02	6	Core	1.527	1.694	2.381	1.521	1.690	2.378	0.40	0.26	0.14
3500	0.02	8	Step	1.283	1.457	1.824	1.280	1.454	1.820	0.22	0.22	0.20
5500	0.03	8	Core	1.598	1.783	2.533	1.595	1.784	2.517	0.16	−0.06	0.64
3500	0.03	6	Twist	1.362	1.495	2.078	1.348	1.493	2.068	1.03	0.12	0.50
1500	0.03	8	Twist	1.395	1.547	2.149	1.391	1.543	2.140	0.30	0.28	0.43
3500	0.02	6	Core	1.527	1.694	2.381	1.521	1.690	2.378	0.40	0.26	0.14
3500	0.02	6	Twist	1.343	1.481	2.067	1.342	1.484	2.053	0.08	−0.13	0.67
1500	0.03	8	Step	1.275	1.442	1.815	1.267	1.444	1.807	0.64	−0.11	0.42
1500	0.01	8	Step	1.267	1.435	1.794	1.261	1.431	1.786	0.51	0.29	0.45
5500	0.01	8	Twist	1.423	1.585	2.175	1.428	1.568	2.168	−0.35	1.07	0.33
1500	0.02	6	Core	1.492	1.672	2.342	1.497	1.669	2.343	−0.36	0.17	−0.04
3500	0.02	6	Step	1.224	1.411	1.751	1.226	1.402	1.745	−0.13	0.67	0.33
1500	0.03	4	Step	1.146	1.326	1.633	1.138	1.333	1.637	0.73	−0.50	−0.24
3500	0.02	6	Core	1.527	1.694	2.381	1.521	1.690	2.378	0.40	0.25	0.14
1500	0.03	4	Twist	1.227	1.397	1.925	1.230	1.392	1.919	−0.26	0.34	0.29
1500	0.03	8	Core	1.535	1.742	2.454	1.553	1.743	2.455	−1.17	−0.05	−0.04
3500	0.02	8	Core	1.573	1.755	2.479	1.573	1.754	2.474	−0.02	0.06	0.22
3500	0.02	6	Core	1.527	1.694	2.381	1.521	1.690	2.378	0.40	0.26	0.14

Table 3. *Cont.*

Spindle Speed (rpm)	Feed (mm/rev)	Drill Dia (mm)	Drill Type	Experimental R_a (μm)			RSM-Predicted R_a (μm)			Error (%)		
				Al_2O_3	SiC	Neat	Al_2O_3	SiC	Neat	Al_2O_3	SiC	Neat
3500	0.02	6	Step	1.224	1.411	1.751	1.226	1.402	1.745	−0.13	0.67	0.33
3500	0.02	8	Twist	1.412	1.562	2.164	1.412	1.556	2.153	−0.01	0.40	0.50
3500	0.03	6	Core	1.535	1.712	2.397	1.524	1.699	2.391	0.73	0.76	0.26
5500	0.03	8	Step	1.302	1.484	1.862	1.301	1.477	1.857	0.05	0.49	0.26
3500	0.02	6	Twist	1.343	1.481	2.067	1.342	1.484	2.053	0.08	−0.18	0.67
3500	0.02	4	Core	1.451	1.625	2.261	1.452	1.619	2.267	−0.08	0.37	−0.28
5500	0.01	4	Step	1.169	1.362	1.682	1.167	1.353	1.676	0.18	0.69	0.35
3500	0.02	6	Step	1.224	1.411	1.751	1.226	1.402	1.745	−0.13	0.67	0.33
1500	0.02	6	Step	1.192	1.389	1.727	1.206	1.385	1.717	−1.19	0.28	0.60
3500	0.01	6	Twist	1.327	1.472	2.035	1.334	1.474	2.041	−0.53	−0.13	−0.27
3500	0.02	6	Core	1.527	1.694	2.381	1.521	1.690	2.378	0.40	0.26	0.14
5500	0.02	6	Step	1.258	1.429	1.781	1.245	1.418	1.774	1.02	0.78	0.40
3500	0.02	6	Twist	1.343	1.481	2.067	1.342	1.484	2.053	0.08	−0.18	0.67
3500	0.02	6	Twist	1.343	1.481	2.067	1.342	1.484	2.053	0.08	−0.18	0.67
5500	0.03	4	Step	1.187	1.371	1.708	1.188	1.366	1.703	−0.10	0.38	0.31
5500	0.01	4	Twist	1.289	1.423	1.962	1.275	1.418	1.960	1.05	0.37	0.12
3500	0.02	4	Twist	1.245	1.415	1.943	1.256	1.405	1.939	−0.85	0.68	0.21
1500	0.01	8	Twist	1.389	1.525	2.127	1.385	1.524	2.115	0.32	0.09	0.56
3500	0.02	6	Step	1.224	1.411	1.751	1.226	1.402	1.745	−0.13	0.67	0.33
3500	0.02	6	Step	1.224	1.411	1.751	1.226	1.402	1.745	−0.13	0.67	0.33
3500	0.02	6	Twist	1.343	1.481	2.067	1.342	1.484	2.053	0.08	−0.18	0.67
1500	0.02	6	Twist	1.312	1.457	1.994	1.314	1.461	2.023	−0.19	−0.30	−1.43
3500	0.02	4	Step	1.157	1.345	1.651	1.155	1.343	1.656	0.18	0.15	−0.30
	Standard deviation			0.133	0.131	0.270	0.133	0.132	0.270	0.494	0.412	0.367

3.1. Analysis of Main Effects Plot

The effect of drilling parameters is represented by the main effects plot for the hybrid nano-composites and the neat CFRP composite, as shown in Figure 4a–c. It can be observed that all the parameters significantly impact the surface roughness of the hole. For the Al_2O_3 hybrid nano-composite, the surface roughness is observed to be at a minimum for lower spindle speed, moderate feed, and lower drill diameter for step drill (Figure 4a). Similar results are noticed for the SiC hybrid nano-composite and neat CFRP composites with minimum surface roughness at lower spindle speed, feed, and drill diameter for step drill (Figure 4b). From the observation made, it is noted that the contribution of spindle speed, drill diameter, and drill type on surface roughness is more as compared to the feed of the drill tool. The surface roughness obtained is lower at a lower spindle speed and drill diameter (Figure 4c). The results are due to the lower heat generated at a lower spindle speed at the tool–workpiece interface.

3.2. ANOVA Analysis

ANOVA analysis (Table 4) shows the influence of process parameters on surface roughness. Table 4 shows that the contribution on surface roughness of drill type is 84.75, 84.70, and 92%, respectively, for the hybrid Al_2O_3 and SiC composites as well as the neat CFRP composite. The next effective parameter having an influence on surface roughness is drill diameter, with a 12.96, 13.10, and 7% contribution, respectively, followed by spindle speed, with a contribution of 1.57, 1.58, and 0.72%, respectively. The contribution of feed on surface roughness is less than 0.2% for all the composites.

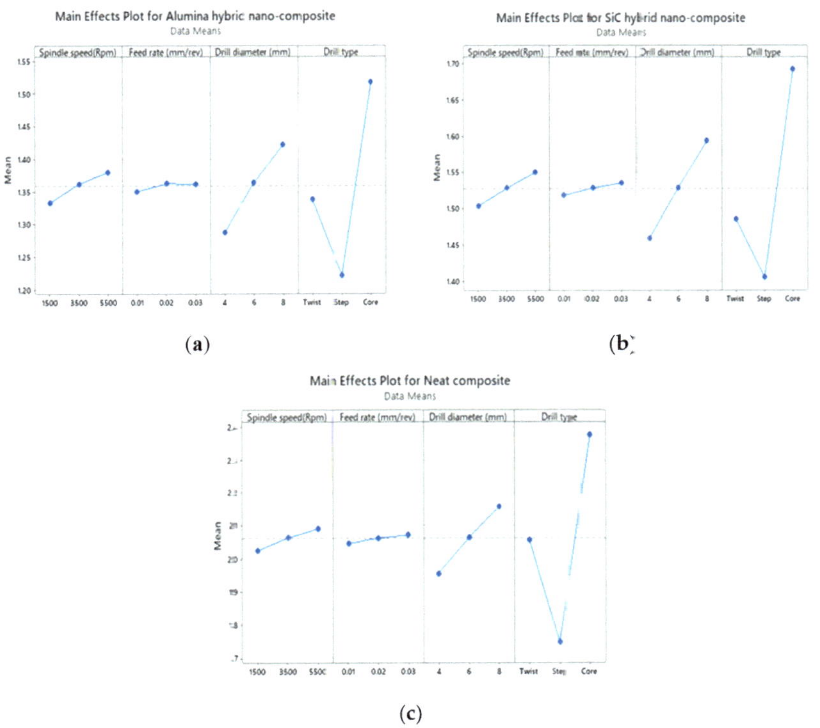

Figure 4. Main effects plot of surface roughness for (**a**) Al_2O_3 hybrid nano-composite, (**b**) SiC hybrid nano-composite, and (**c**) neat CFRP composite.

Table 4. ANOVA table for surface roughness at dry drilling condition.

Source	Al_2O_3 Hybrid Nano-Composite			SiC Hybrid Nano-Composite			Neat CFRP Composite		
	F-Value	*p*-Value	Contribution	F-Value	*p*-Value	Contribution	F-Value	*p*-Value	Contribution
SS (rpm)	271.56	0.000	1.57%	375.32	0.000	1.58%	439.08	0.000	0.72%
F (mm/rev)	17.07	0.000	0.10%	51.13	0.000	0.21%	67.15	0.000	0.11%
DD (mm)	2237.39	0.000	12.96%	3115.81	0.000	13.10%	4294.96	0.000	7.00%
D	7317.61	0.000	84.75%	10,072.15	0.000	84.70%	28,227.18	0.000	92.00%
Square	6.51	0.001	0.11%	0.96	0.423	0.01%	3.88	0.015	0.02%
SS (rpm) × S (rpm)	0.01	0.920	0.05%	0.09	0.768	0.00%	0.19	0.666	0.01%
F (mm/rev) × F (mm/rev)	0.12	0.730	0.01%	0.00	0.948	0.00%	0.09	0.761	0.00%
DD (mm) × DD (mm)	8.96	0.005	0.05%	1.89	0.177	0.01%	5.90	0.019	0.01%
2-Way Interaction	5.16	0.000	0.27%	5.65	0.000	0.21%	6.13	0.000	0.09%
SS (rpm) × F (mm/rev)	1.22	0.277	0.01%	0.02	0.878	0.00%	0.03	0.866	0.00%
SS (rpm) × DD (mm)	6.35	0.016	0.04%	0.22	0.644	0.00%	1.79	0.189	0.00%
SS (rpm) × D	2.58	0.088	0.03%	1.59	0.217	0.01%	1.00	0.378	0.00%
F (mm/rev) × DD (mm)	1.59	0.215	0.01%	0.00	0.975	0.00%	0.31	0.579	0.00%
F (mm/rev) × D	0.53	0.593	0.01%	0.76	0.473	0.01%	0.09	0.910	0.00%
DD (mm) × D	15.53	0.000	0.18%	22.96	0.000	0.19%	25.42	0.000	0.08%
R-square value		**99.76**			**99.82**			**99.93**	
R-square adjusted value		**99.66**			**99.75**			**99.32**	

SS—spindle speed; F—feed; DD—drill diameter; D—drill type.

3.3. Contour Plot Analysis

Contour plots were created to assess surface roughness for hybrid nano-composites and the neat CFRP composite. Figure 5a indicates that, in the case of the Al_2O_3 hybrid nano-composites, superior surface finish in drilled holes can be attained by employing a 4 mm drill diameter, spindle speeds below 1750 rpm, and a feed under 0.02 mm/rev. Similarly, for the SiC hybrid nano-composites (Figure 5b), optimal surface roughness results are achieved with a 4 mm drill diameter, spindle speeds lower than 2000 rpm, and a feed below 0.03 mm/rev. Likewise, for the neat CFRP composite (Figure 5c), a reduction in surface roughness is observed when maintaining spindle speeds below 2000 rpm, a feed under 0.025 mm/rev, and a 4 mm drill diameter.

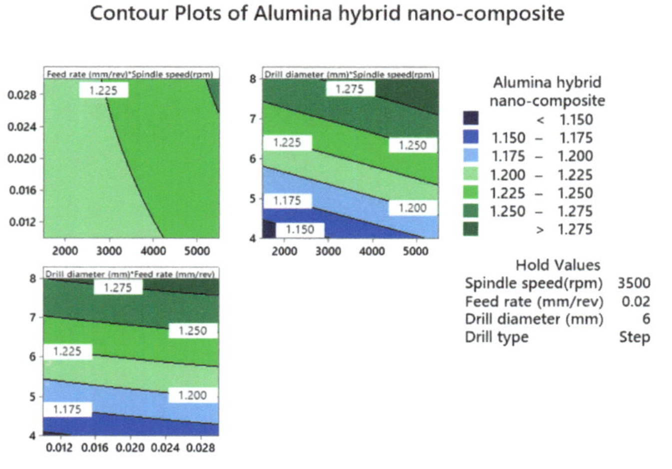

(**a**)

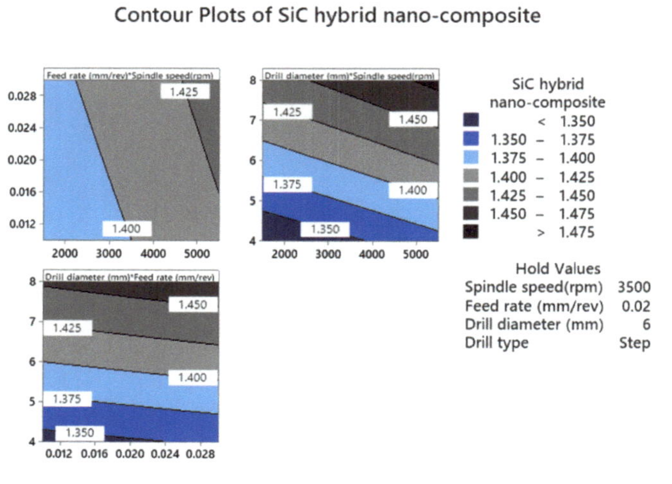

(**b**)

Figure 5. *Cont.*

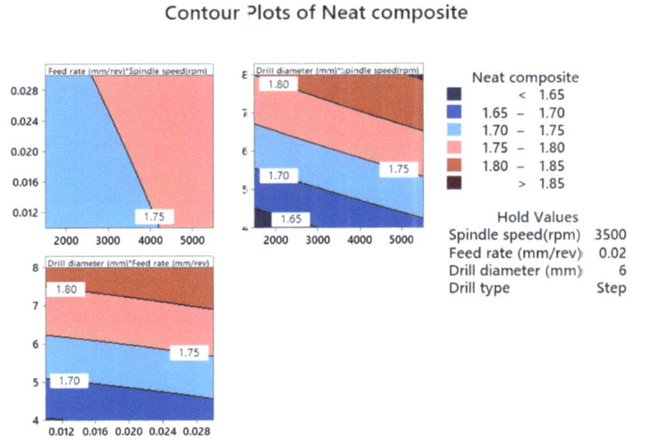

(c)

Figure 5. Contour plots of surface roughness for (**a**) Al_2O_3 hybrid nano-composite, (**b**) SiC hybrid nano-composite, and (**c**) neat CFRP composite.

3.4. Optimization of Process Parameters

Optimization techniques have greatly influenced the choice of diverse drilling process parameters. Therefore, optimizing these parameters is imperative in the composite drilling process. Through this optimization, researchers have observed enhancements in the quality of drilled holes and prolonged tool life. Optimization plots obtained for Al_2O_3 and SiC hybrid nano-composites as well as the neat CFRP composite are represented in Figure 6. The optimum cutting conditions for surface roughness are a spindle speed of 1500 rpm, feed of 0.01 mm/rev, drill diameter of 4 mm, and drill type step drill. From the optimization plots, it is observed that the overall desirability index (D) of the surface roughness is 0.9963.

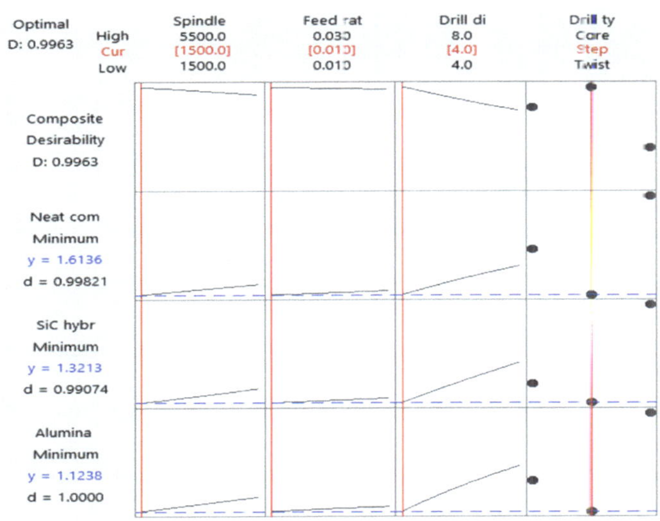

Figure 6. Optimization plots of surface roughness for hybrid nano-composites and neat CFRP composite.

3.5. Confirmation Test

Table 5 represents the confirmation test results, including the experimental and predicted values of the drilling process output variable (surface roughness) of the neat CFRP and hybrid nano-composites. The test was performed using the step, twist, and core drill at the above optimum input process parameters. In the comparison of experimental and predicted values from Table 5, it is observed that the deviation obtained is less than 3%. The values obtained from the confirmation test also showcase that the experimental values obtained are less than the predicted values.

Table 5. Comparison of experimental and RSM-predicted values of surface roughness from the confirmation test performed.

Optimum Input Process Parameters				Composite Type	Experimental Value	RSM-Predicted Value	Error (%)
SS (rpm)	F (mm/rev)	DD (mm)	D				
1500	0.01	4	Step	Al_2O_3 hybrid nano-composite	1.127	1.123	0.28
				SiC hybrid nano-composite	1.332	1.321	0.81
				Neat CFRP composite	1.642	1.613	1.76

SS—spindle speed; F—feed; DD—drill diameter; D—drill type.

3.6. Validation Test

To check the adequacy of the developed regression models from RSM, the validation test was performed based on the different sets of input process parameters, as shown in Table 6, that were not used in performing the experiment earlier but fall within the defined range of experiments. The results obtained from the validation experiment are displayed in Table 7. The experimental and RSM-predicted values appropriately agree with each other, and the percentage error obtained is also represented.

Table 6. Input process parameters selected for validation test.

Experiment No.	Spindle Speed (rpm)	Feed (mm/rev)	Drill Diameter (mm)	Drill Type
1	1500	0.02	4	Twist
2	5500	0.03	6	Step
3	1500	0.01	8	Core

Table 7. Validation test results of surface roughness performed for hybrid nano-composites and neat CFRP composite.

Composite Type	Optimum Input Process Parameters				Experimental Value	RSM-Predicted Value	Error (%)
	SS (rpm)	F (mm/rev)	DD (mm)	D			
Al_2O_3 hybrid nano-composite	1500	0.02	4	Twist	1.224	1.310	7.03
	5500	0.03	6	Step	1.253	1.285	2.55
	1500	0.01	8	Core	1.553	1.502	3.28
SiC hybrid nano-composite	1500	0.02	4	Twist	1.383	1.368	1.08
	5500	0.03	6	Step	1.424	1.345	5.55
	1500	0.01	8	Core	1.724	1.716	0.46
Neat CFRP composite	1500	0.02	4	Twist	1.904	1.853	2.68
	5500	0.03	6	Step	1.787	1.685	5.71
	1500	0.01	8	Core	2.433	2.498	2.67

SS—spindle speed; F—feed; DD—drill diameter; D—drill type.

3.7. Prediction Output

The results obtained from both models (ANN and RF) are presented in Table 8. These tables include predicted R_a values and the errors associated with each prediction for different machining parameters. By analyzing the results obtained from Table 8, the table shows predicted surface roughness (R_a) values from both the artificial neural network (ANN) and random forest (RF) models across diverse drilling scenarios, revealing commendable overall performance. Both models demonstrate their effectiveness in predicting surface roughness, underscoring their utility in understanding complex relationships between drilling parameters and material characteristics. While variations in prediction errors were observed across different combinations of spindle speed, feed, drill diameter, and drill type, the models generally exhibited robust predictive capabilities. This suggests the viability of ANN and RF models in capturing nuanced patterns and trends in surface roughness outcomes, offering valuable insights for optimizing drilling processes in composite materials. The prediction results that were attained showcase that RF outperforms ANN and RSM. Figures 7–9 represent the visual comparison of ANN and RF predictions for Al_2O_3, SiC, and neat CFRP composites. These figures clearly show how well each model performed in predicting surface roughness (R_a) across various conditions. Further, to assess the accuracy of the models, Figures 10–12 illustrate the ANN and RF error predictions concerning the RSM-predicted R_a values. This error analysis provides insights into areas where the models may require refinement and highlights potential areas for improvement. Table 8 also provides an insight into the relative error attained for R_a through the statistical RSM model and machine learning models (ANN and RF). The relative error attained through the implementation of RF is comparatively lower than with ANN and RSM. The reason for the best attainment of results is due to the nature of the support of parallelization (i.e., combining multiple decision trees leads to a reduction of overfitting of the model) and the nature of providing better insights into feature relationships. In this case, ANN is computationally expensive and very sensitive to hypermeters (hyperparameter tuning is challenging).

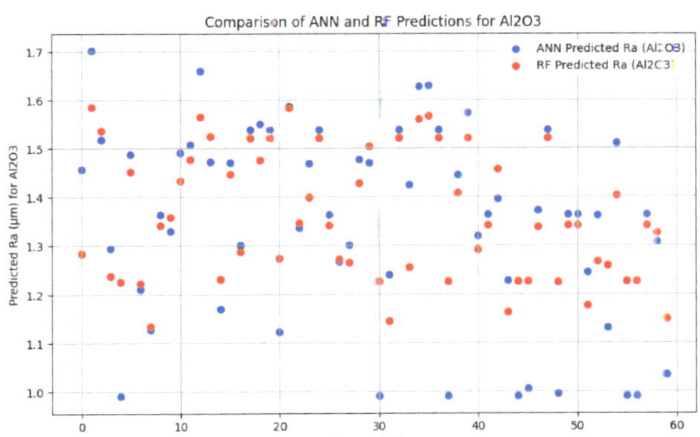

Figure 7. Comparison of ANN and RF predictions of R_a for Al_2O_3 hybrid nano-composite.

Table 8. Comparison of surface roughness values obtained from ANN and RF predictions and relative error scale with respect to RSM prediction.

Spindle Speed (rpm)	Feed (mm/rev)	Drill Diameter (mm)	Drill Type	ANN Predicted Ra (Al_2O_3)	ANN Predicted R_a (Sic)	ANN Predicted R_a (Neat)	RF Predicted R_a (Al_2O_3)	RF Predicted Ra (Sic)	RF Predicted Ra (Neat)	Error ANN (Al_2O_3)	Error ANN (Sic)	Error ANN (Neat)	Error RF (Al_2O_3)	Error RF (Sic)	Error RF (Neat)
5500	0.01	8	1	1.456352	1.571367	2.069188	1.28377	1.45562	1.82663	13.07	7.41	12.82	0.33	0.5	0.4
5500	0.01	8	0	1.702711	1.860966	2.782625	1.5858	1.76628	2.494	7.22	5.5	11.57	0.14	0.13	0
5500	0.02	6	0	1.517819	1.710853	2.380019	1.53677	1.70136	2.39969	1.7	0.05	1.33	0.47	0.51	0.51
1500	0.01	4	2	1.293901	1.492233	1.935694	1.23783	1.38834	1.91747	6.41	8.68	2.36	1.8	1.12	1.4
3500	0.02	6	1	0.990383	1.239939	1.328297	1.226	1.402	1.745	19.22	11.56	23.88	0	0	0
1500	0.01	4	0	1.486014	1.643242	2.497839	1.45184	1.62345	2.27398	4.65	3.41	12.67	2.24	2.17	2.57
3500	0.01	6	1	1.209693	1.279833	1.529487	1.22147	1.39773	1.73778	0.68	8.26	11.79	0.28	0.2	0.22
1500	0.01	4	1	1.127849	1.31431	1.777092	1.13339	1.32767	1.62793	0.43	0.43	10.24	0.93	0.58	0.99
3500	0.02	6	2	1.363549	1.498434	2.009733	1.34172	1.48377	2.0527	1.61	0.97	2.11	0.02	0.02	0.01
5500	0.02	6	2	1.328551	1.388244	2.140685	1.35766	1.49629	2.07034	2.95	7.82	2.72	0.83	0.64	0.66
5500	0.03	8	2	1.490139	1.638373	2.305222	1.43205	1.5773	2.18138	3.41	3.17	5.07	0.62	0.67	0.58
5500	0.03	4	0	1.506602	1.762022	2.394908	1.47622	1.64163	2.30721	1.39	6.85	3.23	0.66	0.45	0.55
1500	0.01	8	0	1.660435	1.980778	2.678755	1.56443	1.75276	2.47348	6.92	14.89	10.1	0.74	1.67	1.66
3500	0.01	6	0	1.470755	1.69469	2.362745	1.52385	1.69013	2.37947	2.98	0.87	0.14	0.52	0.6	0.57
3500	0.03	6	1	1.17039	1.26903	1.51443	1.2299	1.40707	1.75407	5	9.87	13.85	0.17	0.07	0.22
1500	0.03	4	0	1.470053	1.636276	2.388436	1.44558	1.62352	2.26901	2.94	1.76	6.53	1.23	0.97	1.2
5500	0.03	4	2	1.299501	1.497535	2.085393	1.28589	1.42999	1.97819	0.19	4.21	4.85	0.86	0.49	0.54
3500	0.02	6	0	1.53751	1.716213	2.35087	1.521	1.69	2.378	1.09	1.55	1.14	0	0	0
5500	0.01	4	0	1.550188	1.641287	2.362509	1.47563	1.63718	2.30393	5.38	0.75	3.03	0.31	0.5	0.48
3500	0.02	6	0	1.53751	1.716213	2.35087	1.521	1.69	2.378	1.09	1.55	1.14	0	0	0
3500	0.02	8	1	1.122149	1.378606	1.602812	1.27283	1.44604	1.80487	12.33	5.19	11.93	0.56	0.55	0.83
5500	0.03	8	0	1.586365	1.761309	2.602341	1.58391	1.77032	2.49542	0.54	1.27	3.39	0.7	0.77	0.86
3500	0.03	6	2	1.336592	1.438408	2.003405	1.34641	1.49066	2.06407	0.85	3.66	3.12	0.12	0.16	0.19
1500	0.03	8	2	1.468507	1.708831	2.232491	1.39783	1.55035	2.15086	5.57	10.75	4.32	0.49	0.48	0.51
3500	0.02	6	0	1.53751	1.716213	2.35087	1.521	1.69	2.378	1.09	1.55	1.14	0	0	0
3500	0.02	6	2	1.363549	1.498434	2.009733	1.34172	1.48377	2.0527	1.61	0.97	2.11	0.02	0.02	0.01
1500	0.03	8	1	1.26572	1.448932	1.942375	1.27066	1.44708	1.80806	0.1	0.34	7.49	0.29	0.21	0.06
1500	0.01	8	1	1.300653	1.442074	2.009445	1.26474	1.4374	1.79518	3.14	0.77	12.51	0.3	0.45	0.51
5500	0.01	8	2	1.475928	1.613553	2.284549	1.42778	1.56828	2.17075	3.36	2.91	5.38	0.02	0.02	0.13

Table 8. *Cont.*

Spindle Speed (rpm)	Feed (mm/rev)	Drill Diameter (mm)	Drill Type	ANN Predicted Ra (Al_2O_3)	ANN Predicted R_a (Sic)	ANN Predicted R_a (Neat)	RF Predicted R_a (Al_2O_3)	RF Predicted Ra (Sic)	RF Predicted Ra (Neat)	Error ANN (Al_2O_3)	Error ANN (Sic)	Error ANN (Neat)	Error RF (Al_2O_3)	Error RF (Sic)	Error RF (Neat)
1500	0.02	6	0	1.469299	1.639699	2.438071	1.50356	1.67446	2.3521	1.85	1.76	4.06	0.44	0.33	0.39
3500	0.02	6	1	0.990383	1.239939	1.328297	1.226	1.402	1.745	19.22	11.56	23.88	0	0	0
1500	0.03	4	1	1.238851	1.350103	1.749304	1.14411	1.33714	1.64055	8.86	1.28	6.86	0.54	0.31	0.22
3500	0.02	6	0	1.53751	1.716213	2.35087	1.521	1.69	2.378	1.09	1.55	1.14	0	0	0
1500	0.03	4	2	1.423592	1.554698	2.309336	1.25528	1.40284	1.94098	15.74	11.69	20.34	2.06	0.78	1.15
1500	0.03	8	0	1.627112	1.749065	2.587803	1.55896	1.75057	2.46465	4.77	0.35	5.41	0.38	0.43	0.39
3500	0.02	8	0	1.629802	1.823364	2.413243	1.56664	1.7523	2.47004	3.61	3.95	2.46	0.4	0.1	0.16
3500	0.02	6	0	1.53751	1.716213	2.35087	1.521	1.69	2.378	1.09	1.55	1.14	0	0	0
3500	0.02	6	1	0.990383	1.239939	1.328297	1.226	1.402	1.745	19.22	11.56	23.88	0	0	0
3500	0.02	8	2	1.444109	1.531701	2.131912	1.40664	1.5507	2.15089	2.27	1.56	0.98	0.38	0.34	0.1
3500	0.03	6	0	1.572901	1.665941	2.315238	1.52149	1.69308	2.38152	3.21	1.95	3.17	0.16	0.35	0.4
5500	0.03	8	1	1.319746	1.502715	2.03471	1.2895	1.46505	1.83691	1.44	1.74	9.57	0.88	0.81	1.08
3500	0.02	6	2	1.363549	1.498434	2.009733	1.34172	1.48377	2.0527	1.61	0.97	2.11	0.02	0.02	0.01
3500	0.02	4	0	1.395865	1.6597	2.264282	1.45666	1.62613	2.27745	3.87	2.51	0.12	0.32	0.44	0.46
5500	0.01	4	1	1.228172	1.294105	1.741528	1.16271	1.35276	1.67171	5.24	4.35	3.91	0.37	0.02	0.26
3500	0.02	6	1	0.990383	1.239939	1.328297	1.226	1.402	1.745	19.22	11.56	23.88	0	0	0
1500	0.02	6	1	1.005	1.062691	1.638218	1.22567	1.40241	1.74332	16.67	23.27	4.59	1.63	1.26	1.53
3500	0.01	6	2	1.371796	1.362519	1.885923	1.33754	1.47839	2.0446	2.83	7.56	7.6	0.27	0.3	0.18
3500	0.02	6	0	1.53751	1.716213	2.35087	1.521	1.69	2.378	1.09	1.55	1.14	0	0	0
5500	0.02	6	1	0.994668	1.141733	1.600672	1.22443	1.40565	1.74993	20.11	19.48	9.77	1.65	0.87	1.36
3500	0.02	6	2	1.363549	1.498434	2.009733	1.34172	1.48377	2.0527	1.61	0.97	2.11	0.02	0.02	0.01
3500	0.02	6	2	1.363549	1.498434	2.009733	1.34172	1.48377	2.0527	1.61	0.97	2.11	0.02	0.02	0.01
5500	0.03	4	1	1.243497	1.358665	1.831364	1.17602	1.36168	1.68777	4.67	0.54	7.51	1.01	0.32	0.89
5500	0.01	4	2	1.362128	1.524907	2.033442	1.26568	1.41161	1.95207	6.83	7.54	3.75	0.73	0.45	0.4
3500	0.02	4	2	1.130603	1.433997	2.015887	1.25715	1.40698	1.94287	9.98	2.06	3.97	0.09	0.14	0.2
1500	0.01	8	2	1.509835	1.539907	2.257786	1.40215	1.54727	2.14824	9.01	1.04	6.75	1.24	1.53	1.57
3500	0.02	6	1	0.990383	1.239939	1.328297	1.226	1.402	1.745	19.22	11.56	23.88	0	0	0
3500	0.02	6	1	0.990383	1.239939	1.328297	1.226	1.402	1.745	19.22	11.56	23.88	0	0	0
3500	0.02	6	2	1.363549	1.498434	2.009733	1.34172	1.48377	2.0527	1.61	0.97	2.11	0.02	0.02	0.01
1500	0.02	6	2	1.306254	1.408035	1.996562	1.32548	1.46827	2.0353	0.59	3.63	1.31	0.87	0.5	0.61
3500	0.02	4	1	1.034241	1.328674	1.478657	1.14936	1.34206	1.65323	10.46	1.07	10.71	0.49	0.07	0.17

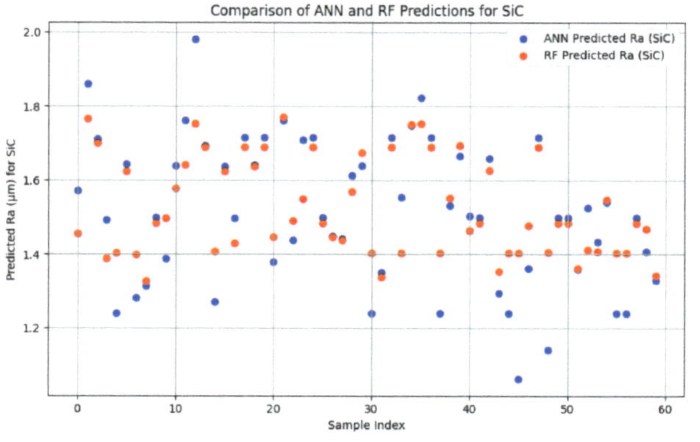

Figure 8. Comparison of ANN and RF predictions of R_a for SiC hybrid nano-composite.

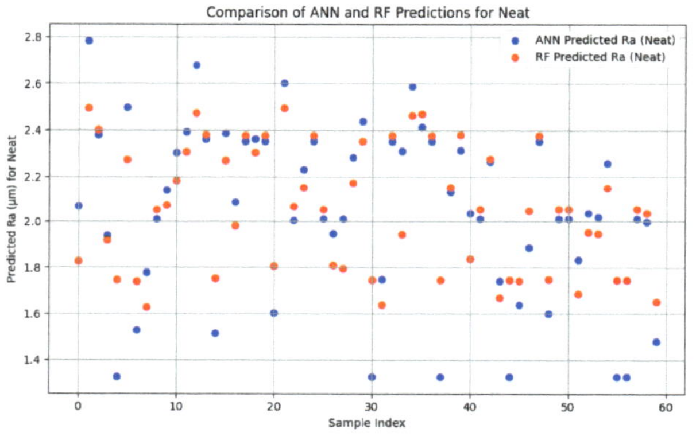

Figure 9. Comparison of ANN and RF predictions of R_a for neat CFRP composite.

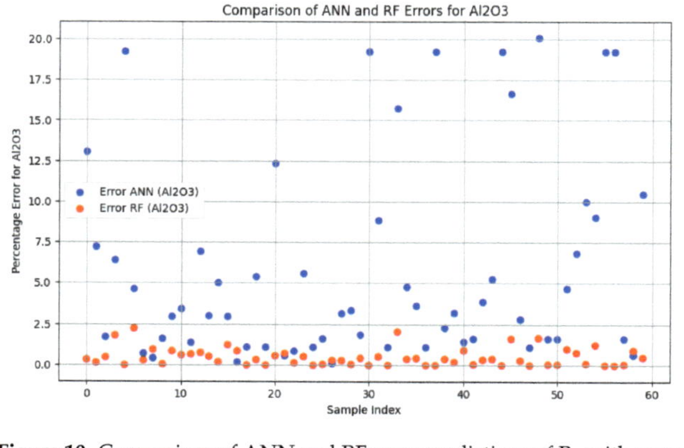

Figure 10. Comparison of ANN and RF error predictions of R_a with respect to RSM-predicted R_a for Al_2O_3 hybrid nano-composite.

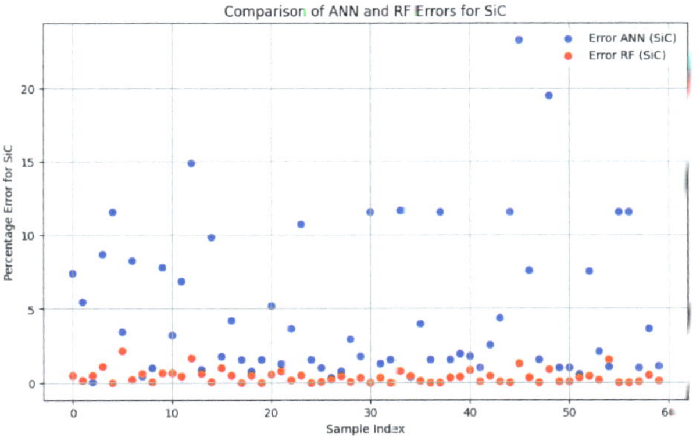

Figure 11. Comparison of ANN and RF error predictions of R_a with respect to RSM-predicted R_a for SiC hybrid nano-composite.

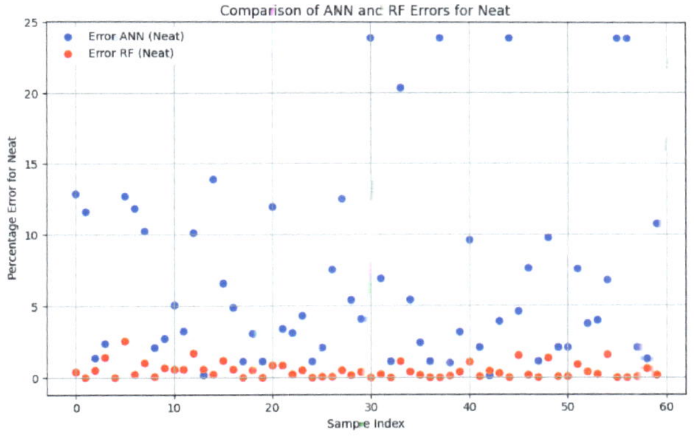

Figure 12. Comparison of ANN and RF error predictions of R_a with respect to RSM-predicted R_a for neat CFRP composite.

4. Conclusions

In this study, an investigation on drilling was conducted using three distinct drill types (step, twist, and core drill) on both neat CFRP and hybrid Al_2O_3 and SiC nano-composites. A total of 60 holes were drilled in each composite type. Based on the experimental findings, the following conclusions can be drawn:

- The input drilling parameters namely spindle speed, feed, drill diameter, and drill type significantly influenced the surface roughness of the investigated nano-composites. The maximum surface roughness value was observed for a higher drill diameter of 8 mm, followed by 6 and 4 mm drill diameters.
- The minimum surface roughness was observed for the Al_2O_3 hybrid nano-composite, followed by the SiC hybrid nano-composite, and maximum surface roughness was noted for the neat CFRP composite.
- Surface roughness increases with increasing spindle speed, feed, and drill diameter and the drill type step drill has shown better performance in reducing surface roughness.
- ANOVA results indicated that the drill type followed by drill diameter showed a higher percentage contribution to surface roughness.
- The optimization of surface roughness was evaluated using the desirability function approach. From the optimization plot, it was possible to determine the surface roughness by redefining the values of input process parameters within the experimental range.

- The experimental results obtained during the drilling of the hybrid nano-composites and the neat CFRP composite using a step drill, core drill, and twist drill adequately correspond with the RSM-predicted values.
- The comparative analysis of ANN and RF predictions for the hybrid nano-composites and the neat CFRP composite provides visual insights into the performance of each model across different materials.
- The relative error predictions of both ANN and RF concerning the RSM-predicted R_a values, while comparing the results, shows that RF outperforms ANN and RSM due to its interpretability nature.
- The optimized drilling parameters along with the machine learning approach can be employed to other composites, such as glass, Kevlar, and polyimide fiber, for determining their surface roughness qualities.

Author Contributions: Conceptualization, investigation, methodology, writing—original draft (T.N.); Conceptualization, investigation, methodology, writing—original draft (S.M.S.); Conceptualization, writing—review and editing (S.A.); Conceptualization, methodology, writing—review and editing, project administration (N.S.); Conceptualization, writing—review and editing (R.L.M.); Conceptualization, methodology (S.D.S.). All authors have read and agreed to the published version of the manuscript.

Funding: This research received no external funding.

Data Availability Statement: Data are contained within the article.

Conflicts of Interest: The authors declare no conflict of interest.

References

1. Ze, G.K.; Pramanik, A.; Basak, A.K.; Prakash, C.; Shankar, S.; Radhika, N. Challenges Associated with Drilling of Carbon Fiber Reinforced Polymer (CFRP) Composites—A Review. *Compos. Part C Open Access* **2023**, *11*, 100356. [CrossRef]
2. Xu, J.; Yin, Y.; Paulo Davim, J.; Li, L.; Ji, M.; Geier, N.; Chen, M. A Critical Review Addressing Drilling-Induced Damage of CFRP Composites. *Compos. Struct.* **2022**, *294*, 115594. [CrossRef]
3. Nuhiji, B.; Swait, T.; Bower, M.P.; Green, J.E.; Day, R.J.; Scaife, R.J. Tooling Materials Compatible with Carbon Fibre Composites in a Microwave Environment. *Compos. Part B Eng.* **2019**, *163*, 769–778. [CrossRef]
4. Kaybal, H.B.; Ulus, H.; Demir, O.; Şahin, Ö.S.; Avcı, A. Effects of Alumina Nanoparticles on Dynamic Impact Responses of Carbon Fiber Reinforced Epoxy Matrix Nanocomposites. *Eng. Sci. Technol. Int. J.* **2018**, *21*, 399–407. [CrossRef]
5. Li, N.; Li, Y.; Zhou, J.; He, Y.; Hao, X. Drilling Delamination and Thermal Damage of Carbon Nanotube/Carbon Fiber Reinforced Epoxy Composites Processed by Microwave Curing. *Int. J. Mach. Tools Manuf.* **2015**, *97*, 11–17. [CrossRef]
6. Sheikh-Ahmad, J.Y. *Machining of Polymer Composites*; Springer: Berlin/Heidelberg, Germany, 2009; Volume 387355391, ISBN 9780387355399.
7. Wetzel, B.; Haupert, F.; Qiu Zhang, M. Epoxy Nanocomposites with High Mechanical and Tribological Performance. *Compos. Sci. Technol.* **2003**, *63*, 2055–2067. [CrossRef]
8. De Cicco, D.; Asaee, Z.; Taheri, F. Use of Nanoparticles for Enhancing the Interlaminar Properties of Fiber-Reinforced Composites and Adhesively Bonded Joints—A Review. *Nanomaterials* **2017**, *7*, 360. [CrossRef]
9. Mohanty, A.; Srivastava, V.K.; Sastry, P.U. Investigation of Mechanical Properties of Alumina Nanoparticle-loaded Hybrid Glass/Carbon-fiber-reinforced Epoxy Composites. *J. Appl. Polym. Sci.* **2014**, *131*, 39749. [CrossRef]
10. Swain, P.T.R.; Biswas, S. Effect of Moisture Absorption on the Mechanical Properties of Ceramic Filled Jute/Epoxy Hybrid Composites. *IOP Conf. Ser. Mater. Sci. Eng.* **2017**, *178*, 12010. [CrossRef]
11. Shahabaz, S.M.; Shetty, P.K.; Shetty, N.; Sharma, S.; Divakara Shetty, S.; Naik, N. Effect of Alumina and Silicon Carbide Nanoparticle-Infused Polymer Matrix on Mechanical Properties of Unidirectional Carbon Fiber-Reinforced Polymer. *J. Compos. Sci.* **2022**, *6*, 381. [CrossRef]
12. Shahabaz, S.M.; Mehrotra, P.; Kalita, H.; Sharma, S.; Naik, N.; Noronha, D.J.; Shetty, N. Effect of Al_2O_3 and SiC Nano-Fillers on the Mechanical Properties of Carbon Fiber-Reinforced Epoxy Hybrid Composites. *J. Compos. Sci.* **2023**, *7*, 133. [CrossRef]
13. Davim, J.P.; Reis, P.; Antonio, C.C. Experimental Study of Drilling Glass Fiber Reinforced Plastics (GFRP) Manufactured by Hand Lay-Up. *Compos. Sci. Technol.* **2004**, *64*, 289–297. [CrossRef]
14. Tsao, C.C.; Hocheng, H.; Chen, Y.C. Delamination Reduction in Drilling Composite Materials by Active Backup Force. *CIRP Ann.* **2012**, *61*, 91–94. [CrossRef]
15. Rathod, D.; Rathod, M.; Patel, R.; Shahabaz, S.M.; Shetty, S.D.; Shetty, N. A Review on Strengthening, Delamination Formation and Suppression Techniques during Drilling of CFRP Composites. *Cogent Eng.* **2021**, *8*, 1941588. [CrossRef]

16. Palanikumar, K.; Davim, J.P. Mathematical Model to Predict Tool Wear on the Machining of Glass Fibre Reinforced Plastic Composites. *Mater. Des.* **2007**, *28*, 2008–2014. [CrossRef]
17. Ameur, M.F.; Habak, M.; Kenane, M.; Aouici, H.; Cheikh, M. Machinability Analysis of Dry Drilling of Carbon/Epoxy Composites: Cases of Exit Delamination and Cylindricity Error. *Int. J. Adv. Manuf. Technol.* **2017**, *88*, 2557–2571. [CrossRef]
18. Feng, Z.; Gani, H.; Damayanti, A.D.; Gani, H. An Explainable Ensemble Machine Learning Model to Elucidate the Influential Drilling Parameters Based on Rate of Penetration Prediction. *Geoenergy Sci. Eng.* **2023**, *231*, 212231. [CrossRef]
19. Solati, A.; Hamedi, M.; Safarabadi, M. Combined GA-ANN Approach for Prediction of HAZ and Bearing Strength in Laser Drilling of GFRP Composite. *Opt. Laser Technol.* **2019**, *113*, 104–115. [CrossRef]
20. Murthy, B.R.N.; Rodrigues, L.L.R.; Sharma, N.Y.; Anjaiah, D. Influence of Process Parameters on the Quality of Hole in Drilling of GFRP Composites-an Experimental Investigation Using DOE. In Proceedings of the 2010 International Conference on Mechanical and Electrical Technology, Singapore, 10–12 September 2010; IEEE: Piscataway, NJ, USA, 2010; pp. 87–90.
21. Montgomery, D.C. *Design and Analysis of Experiments*; John Wiley & Sons: Hoboken, NJ, USA, 2017; ISBN 1119113474.
22. Jayabal, S.; Natarajan, U. Optimization of Thrust Force, Torque, and Tool Wear in Drilling of Coir Fiber-Reinforced Composites Using Nelder–Mead and Genetic Algorithm Methods. *Int. J. Adv. Manuf. Technol.* **2010**, *51*, 371–381. [CrossRef]
23. Enemuoh, E.U.; El-Gizawy, A.S.; Okafor, A.C. An Approach for Development of Damage-Free Drilling of Carbon Fiber Reinforced Thermosets. *Int. J. Mach. Tools Manuf.* **2001**, *41*, 1795–1814. [CrossRef]
24. Karthikeyan, R.; Jaiganesh, S.; Pai, B.C. Optimization of Drilling Characteristics for Al/SiC p Composites Using Fuzzy/GA. *Met. Mater. Int.* **2002**, *8*, 163–168. [CrossRef]

Journal of
Manufacturing and
Materials Processing

MDPI

Article

Performance Analysis of Helical Milling and Drilling Operations While Machining Carbon Fiber-Reinforced Aluminum Laminates

Gururaj Bolar [1], Anoop Aroor Dinesh [2], Ashwin Polishetty [3], Raviraj Shetty [1,*], Anupama Hiremath [1,*] and V. L. Neelakantha [4]

[1] Department of Mechanical & Industrial Engineering, Manipal Institute of Technology, Manipal Academy of Higher Education, Manipal 576104, India; gururaj.bolar@manipal.edu
[2] Department of Mechanical Engineering, National Institute of Technology Goa, Ponda 403401, India
[3] School of Engineering, Computer and Mathematical Science, Auckland University of Technology, Auckland 1010, New Zealand; ashwin.polishetty@aut.ac.nz
[4] Department of Mechanical Engineering, Mangalore Institute of Technology and Engineering, Moodbidri, Mangalore 574225, India
* Correspondence: rr.shetty@manipal.edu (R.S.); anupama.hiremath@manipal.edu (A.H.)

Abstract: Being a difficult-to-cut material, Fiber Metal Laminates (FML) often pose challenges during conventional drilling and require judicious selection of machining parameters to ensure defect-free laminates that can serve reliably during their service lifetime. Helical milling is a promising technique for producing good-quality holes and is preferred over conventional drilling. The paper compares conventional drilling with the helical milling technique for producing holes in carbon fiber-reinforced aluminum laminates. The effect of machining parameters, such as cutting speed and axial feed, on the magnitude of cutting force and the machining temperature during conventional drilling as well as helical milling is studied. It was observed that the thrust force produced during machining reduces considerably during helical milling in comparison to conventional drilling at a constant axial feed rate. The highest machining temperature recorded for helical milling was much lower in comparison to the highest machining temperature measured during conventional drilling. The machining temperatures recorded during helical milling were well below the glass transition temperature of the epoxy used in carbon fiber prepreg, hence protecting the prepreg from thermal degradation during the hole-making process. The surface roughness of the holes produced by both techniques is measured, and the surface morphology of the drilled holes is analyzed using a scanning electron microscope. The surface roughness of the helical-milled holes was lower than that for holes produced by conventional drilling. Scanning electron microscope images provided insights into the interaction of the hole surface with the chips during the chip evacuation stage under different speeds and feed rates. The microhardness of the aluminum layers increased after processing holes using drilling and helical milling operations. The axial feed/axial pitch had minimal influence on the microhardness increase in comparison to the cutting speed.

Keywords: fiber metal laminate; drilling; helical milling; surface roughness; damage; temperature; force

1. Introduction

The need for lightweight, high-performance materials in the automotive and aerospace industries has led to an increased demand for new-age composite materials that amalgamate the merits of metals and composite materials. The emergence of Fiber Metal Laminates (FMLs) is a response to these intricate demands. They consist of Fiber Reinforced Polymers (FRPs) sandwiched between thin metal/alloy sheets, combining the strength of metals/alloys with the exceptional strength-to-weight ratio of FRPs. Presently, the skin panel of the upper fuselage of the aircraft is being built primarily by the use of FMLs [1]. Generally, FMLs are identified based on the FRPs used or on the type of metal used [2–4]. Based on the used metal/alloy sheets, the most widely used FMLs are aramid-reinforced aluminum laminate (ARALL), carbon fiber-reinforced aluminum laminates (CARALL), and glass-reinforced aluminum reinforced epoxy (GLARE) [5,6]. Identification of FMLs is also conducted based on the positioning of the metal layers within the multi-layer configuration, wherein the metal layers can be placed internally or externally during stacking. Another way in which the FMLs are identified is based on the direction of the fibers in the FRPs as unidirectional or cross-ply FMLs [7].

Even though FMLs are processed as near-net-shape structures, almost all practical applications require the fastening of FMLs during assembly, either through mechanical fasteners or adhesive bonding [8]. To date, mechanical fasteners that include temporary fasteners such as bolts or permanent fasteners like rivets [9] are the preferred joining method, as such methods do not require additional surface preparation and ease of disassembly during inspection [10]. In both cases, FMLs are subjected to secondary operations such as drilling to make holes within the FML structure. It is reported that the number of holes drilled for commercial aircraft assembly varies between 1.5 million and 3 million [11,12]. For the reliable functioning of the assembled part, it is imperative to produce superior surface quality holes, which will result in the mitigation of structural failure due to stress and fatigue. Almost all aircraft industries impose stringent norms, facilitating the requirement of a smooth surface finish of the drilled holes to reduce catastrophic failure when the aircraft is in operation. For commercial aircraft, the drilled holes in composites are mandated to be burr- and scratch-free and should have a surface roughness value that is less than 3.2 μm [13]. As a difficult-to-cut material, producing smooth, defect-free holes in FMLs through conventional drilling is often challenging. The interfaces generated between the layers of metal and FRP lead to increased machining process complexity. Higher thrust forces during drilling cause delamination of the FMLs. In fact, as per the published reports, drilling-induced delamination is the source of 60% of rejections in the aircraft industry. Deterioration of surface integrity during the drilling of FMLs is also linked to the abrasive nature of the reinforced anisotropic fibers. Studies have highlighted that the holes drilled in FMLs are attributed to varying surface roughness throughout the hole depth [14]. Also, the surface roughness of the metal layer is different from the FRP layer. While drilling metal/FRP stacks, the chip arising out of the hole drilled in the metal layer has the tendency to scrape the hole drilled in the FRP. Accordingly, the surface roughness of the hole in the FRP layer deteriorates compared to that in the metal layer. Additionally, FMLs based on aluminum alloys such as CARALL, ARALL, and GLARE tend to produce an excessive built-up edge (BUE) on the drill tool, leading to a higher rate of tool life deterioration [15]. This results in inadequate chip evacuation, and the chips left behind tend to erode the fibers and metal surface, consequently diminishing the surface finish of drilled holes [16]. The careful selection of the type of machining operation and the machining parameters becomes crucial for determining the final hole quality [17,18]. Higher feed and spindle speeds are reported to positively affect the surface quality of the drilled hole [19]. Boughdiri et al. [20] reported that during the drilling of GLARE laminates, the production of continuous chips may hinder the hole surface quality due to the prolonged interaction of the chips with the composite layer. Hence, it is advisable to have smaller broken chips while drilling FMLs. Moreover, at feeds lower than 0.08 mm/rev, chips adhere to or entangle with the flutes of the drill tool, in turn deteriorating the hole quality.

Delamination failure is one of the main types of failure occurring while machining FMLs. Delamination while machining FMLs occurs when the cutting force exceeds the interlaminar strength [21]. Thus, keeping the cutting forces to a minimum during the machining of FML is always desirable. Costa et al. [22] highlighted that during the drilling of CARALL laminates, cutting forces increase in magnitude when the tool encounters the aluminum metal layer, which reduces considerably while cutting the FRP layer. The phenomenon is attributed to the brittle nature of the FRPs. Kayihan et al. [23] performed drilling operations on Al-Ti-Carbon Fiber Reinforced Polymer (CFRP) laminates with different stacking orders and found that at higher speeds and lower feeds, the cutting forces are independent stacking sequences. Another important machining factor severely affecting the hole quality was the cutting temperature. It is found that higher cutting speeds in FMLs lead to accelerated tool wear [24] and resin burn-out [25], which in turn leads to fiber pull-out and also weakens the FRP and metal interfaces [26]. Thus, modern machining techniques make use of minimum quantity lubrication (MQL) and cryogenic liquid nitrogen to bring down the cutting zone temperature during the drilling of FMLs [27]. Studies have shown improvement in the accuracy of the drilled hole in FMLs when MQL and cryogenic cooling environments were used [28]. However, the use of such coolants leads to an increase in production costs. The operational difficulties involved in conventional drilling necessitate the requirement of an alternate machining process to produce accurate, good-quality holes in the FMLs.

Helical milling is one such machining technology that has been successfully employed to produce superior-quality holes in FMLs. This method of hole production has proven to mitigate the disadvantages involved in conventional drilling. Bolar et al. [29] reported that the holes produced in CARALL have minimum surface roughness and are characterized by smaller-sized burrs as opposed to the conventional drilling method. Ge et al. [30] highlighted that the fatigue life of Al 2024-T3/Ti-6Al-4V stacks with holes produced by helical milling improved two times in comparison to Al 2024-T3/Ti-6Al-4V stacks with holes produced by conventional drilling. Sun et al. [31] found in their study that helical milling of Ti/CFRP/Al stacks results in discontinuous chips, allowing for easy chip evacuation and improving the surface quality of the hole produced. Li et al. [32] reported an abrupt increase in axial force between layers as the tool transits from the CFRP layer to the aluminum metal layer, which results in burr formation at the interlayers. Accordingly, a tool microlifting technique was proposed to overcome the problem at the inter-layers of the CFRP/Al stacks. Jiaying et al. [33] deliberated on the influence of cooling on delamination during helical milling of CFRP/Ti-6Al-4V stacks. Helical milling under MQL and cryogenic cooling conditions results in smaller delamination with improved hole surface quality. Xu et al. [34] examined the surface quality of holes produced in CFRP/Ti6Al4V stacks and concluded that diamond-coated drills fare better than the tool coated with TiAlN. Boutrih et al. [35] emphasized that the drilling sequence has a bearing on hole quality while helical milling CFRP/Ti6Al4V stacks. Accordingly, hole quality was better when drilling Ti6Al4V/CFRP stacks in comparison to the CFRP/Ti6Al4V stack.

The literature review suggests that the helical milling process can be a preferred technology to produce superior-quality holes in hard-to-cut FML materials. A few studies have explored the machining performance of drilling and helical milling operations while machining CARALL and have analyzed the influence of process variables on cutting forces, machining temperature, surface roughness, and dimensional accuracies of the processed holes. In the machining of materials like FMLs, thrust force and machining temperature are very influential, as they affect the performance of stacked materials. Very high thrust loads and temperatures resulting from unoptimized process variable selection can lead to surface damage and, in some cases, result in the delamination of the stacked material. However, research highlighting the influence of the thrust load and machining temperature on the hole surface finish, surface damage, and microhardness subsurface damage is scant. Moreover, studies critically evaluating the surface integrity and surface damage in CARALL are limited. Therefore, the work will initially focus on evaluating the influence of process variables on the thrust load and machining temperatures while drilling and helical milling holes in CARALL. It will further explore the effect cutting force and temperatures have on surface integrity (surface roughness, surface damage, and microhardness) while drilling and helical milling holes in CARALL. Accordingly, the work aims to minimize critical damages like delamination and debonding generated during the drilling process and improve the surface quality of holes made in CARALL FMLs. The outcome of the work will provide academicians and industries with the framework to process defect-free holes in CARALL FMLs.

2. Materials and Methods

The machining experiments were conducted on CARALL laminates. The laminates were fabricated by sandwiching strips of Al 2024-T3 with 0.5 mm thickness (supplied by Mangaldeep Metals and Alloys) and unidirectional carbon fiber prepreg (CFP) with 0.2 mm thickness (supplied by Bhor Chemicals). The layup procedure followed is depicted in Figure 1a. The CFP was manufactured using epoxy resin (A-45) and has a resin content of 38 ± 3% and a fiber density of 200 GSM. The mechanical properties of carbon fibers used in the CFP system and Al 2024-T3 alloy are provided in Table 1. The aluminum layers were subjected to sulfuric acid anodizing (SAA) to improve the corrosion resistance and interfacial bonding strength at the metal and FRP interface. Figure 1a shows the surface of the aluminum sheet, which shows the formation of nanopores as a result of SAA.

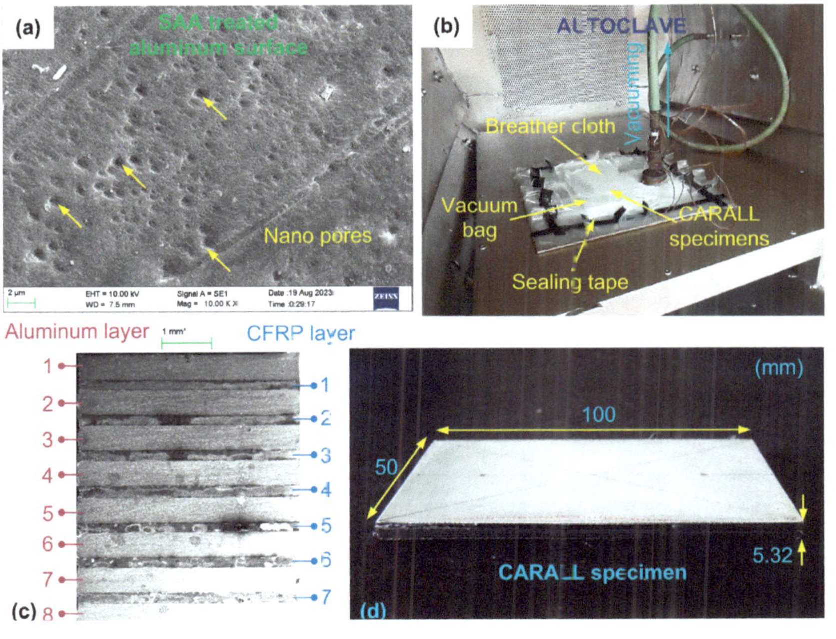

Figure 1. (**a**) Surface of aluminum sheet subjected to sulfuric acid anodizing; (**b**) specimen fabrication setup; (**c**) material stacking sequence; and (**d**) fabricated FML specimen.

Table 1. Mechanical properties of carbon fibers and 2024-T3 aluminum alloy.

	Carbon Fiber	Al2024-T3
Density (g/cm^3)	1.8	2.78
Filament diameter (μm)	7	-
Tensile strength (MPa)	4000	483
Tensile modulus (GPa)	240	73
Elongation (%)	1.7	18
Yield strength (MPa)	-	385
Shear strength (MPa)	-	283

The CARALL laminates were developed using the vacuum bagging method and autoclave cured using the setup shown in Figure 1b. The green FMLs were enclosed in a peel ply and kept on a mold plate. A breather cloth was placed above the peel ply to facilitate uniform vacuum pressure on the laminate. A sealant tape was used to prepare the vacuum bag, and the entire unit was then placed inside the autoclave for curing. A vacuum pump is used to evacuate the entrapped air from the vacuum bag. Subsequently, the temperature of the autoclave was raised to 80 °C, with a consistent heating rate of 2 °C/min maintained for 30 min. In the second cycle, the temperature was elevated to 130 °C and sustained for 90 min. Throughout both cycles, the curing pressure remained at 6 bars. Figure 1c illustrates the stacking arrangement, while Figure 1d showcases the cured CARALL laminates. The holes were then drilled into the CARALL work specimens affixed to a vertical machining center (AMS Spark), utilizing the setup depicted in Figure 2a. Drilling and helical milling operations were performed using the tools shown in Figure 2b. For comparison between the drilling and helical milling operations, the holes were machined considering the parity between the machining times. Accordingly, selected process variables and the corresponding machining times are listed in Table 2.

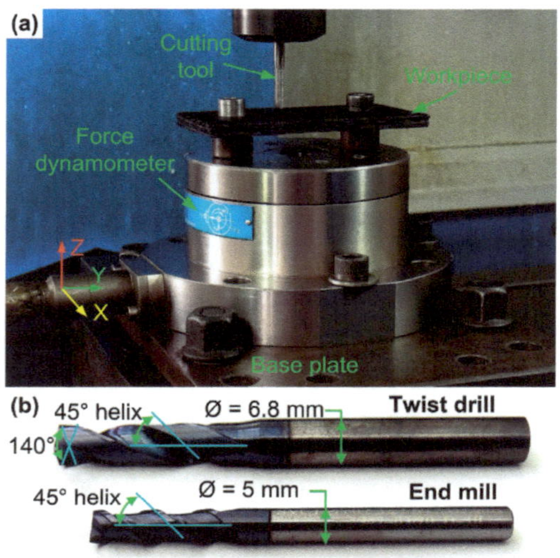

Figure 2. (**a**) Machining setup; (**b**) cutting tools for experimentation.

Table 2. Process variable combinations and corresponding machining time.

Case	Cutting Speed (V) (m/min)	Axial Feed (f_a) (mm/rev)		Tangential Feed (f_t) (mm/tooth)	Machining Time (t_m) (s)	
		Drilling	Helical Milling	Helical Milling	Drilling	Helical Milling
1	20	0.015	0.15	0.09	87	86
2	20	0.030	0.30	0.09	51	52
3	20	0.045	0.45	0.09	39	40
4	40	0.015	0.15	0.09	51	51
5	40	0.030	0.30	0.09	33	33
6	40	0.045	0.45	0.09	27	28
7	60	0.015	0.15	0.09	39	39
8	60	0.030	0.30	0.09	27	27
9	60	0.045	0.45	0.09	23	23

To measure the thrust force generated during machining, a dynamometer (Kistler 9272) connected to a charge amplifier (Kistler 5070A) and an analog-to-digital converter (type 5697A) are utilized, as depicted in Figure 3a. The force signal was acquired using Dynoware software (Type 2825D-02), and the sampling rate was set at 1000 Hz. Surface roughness was measured using a perthometer (Taylor Hobson—Form Talysurf 50) with a tip radius of 2.5 μm (see Figure 3b). The evaluation length of 3 mm was used, and the sampling length was chosen as 0.8 mm.

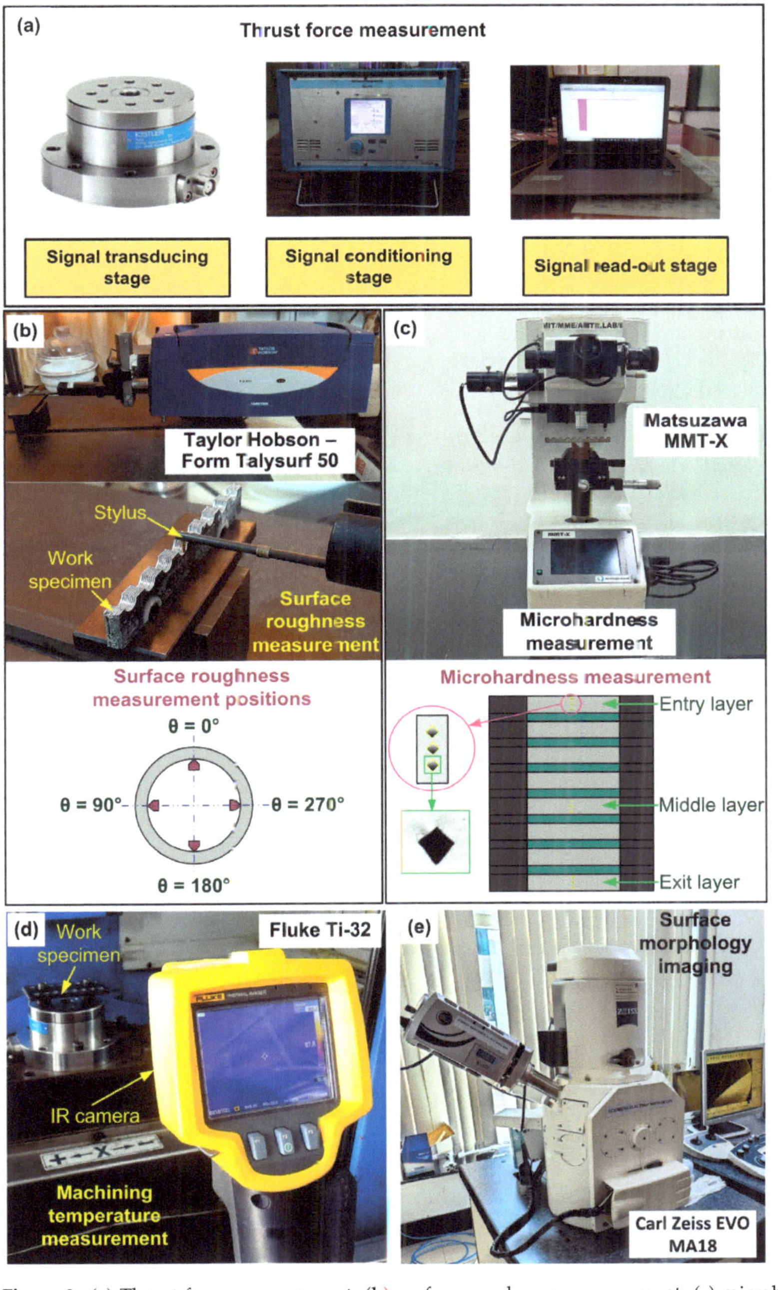

Figure 3. (**a**) Thrust force measurement; (**b**) surface roughness measurement; (**c**) microhardness measurement; (**d**) machining temperature measurement; (**e**) surface morphology imagining.

The rate of measurement was maintained at 0.5 mm/s, and the measurements were made at four locations. The microhardness measurements were taken at three different locations using the digital microhardness tester (Matsuzawa—MMT-X) with a diamond-shaped indenter by applying a 200 g load for 15 s, as shown in Figure 3c. The Fluke Ti32 infrared (IR) camera measured the machining temperature. The camera has a working range of 20 °C to 600 °C and an accuracy of ±2 °C. For accurate measurement, the surface of the laminates was painted black, which possesses a thermal emissivity of 0.95. The IR camera was positioned at a distance of 0.1 m from the work material (see Figure 3d). The surface morphology of the machined hole was analyzed with the help of a Carl Zeiss EVO MA18 scanning electron microscope (SEM), as shown in Figure 3e.

3. Results and Discussion

3.1. Thrust Force

Thrust force is a significant factor when machining materials like CARALL. A very high thrust load can influence the surface roughness, burr size, and geometric accuracy of the holes. It can also initiate damages like fiber breakage and microcracking and result in the delamination of stacks. Such damages can render the FML useless or lower its service life [20,21]. Therefore, the thrust force behavior of the two hole-making operations was investigated. Figure 4a displays the effect the cutting speed employed during drilling has on thrust force. Thrust force was reduced as higher cutting speeds were selected. For example, for a fixed axial feed of 0.015 mm/rev, a thrust force of magnitude 180.6 N was measured when a speed of 20 m/min was selected. With 40 m/min employment, thrust force reduced to 164.6 N, indicating an 8.9% decrease as cutting speed increased. A favorable lower thrust force of 145.2 N was noted when holes were drilled with a 60 m/min cutting speed, showing a further 11.8% fall in the force magnitude. The aluminum alloy is plastically deformed during the machining process, thereby releasing energy in the form of heat. With higher cutting speeds, the rate of shear deformation increases, thus increasing the energy dissipation and the heat. At higher temperatures, the material yield strength is lowered, thus lowering the resistance to shearing and reducing the thrust load. Moreover, the polymer resin is softened at higher temperatures, easing the cutting process and reducing the thrust force [20,29]. Figure 4a also depicts the thrust force vs. axial feed plot for the drilling process. Thrust force increased with the selection of a higher feed. For a fixed speed of 20 m/min, an average thrust load of 180.6 N was measured while drilling with a feed of 0.015 mm/rev. At a feed of 0.03 mm/rev, the force increased to 189.7 N. Finally, when the axial feed is at its highest (0.045 mm/rev), the thrust force of 212.5 N is recorded. In this particular case, an increase in feed value from 0.015 mm/rev to 0.045 mm/rev led to a 17.7% increase in thrust force. As axial feed increases, uncut chip thickness also increases. In such a situation, the resistance offered by the chip increases, and more energy is required to shear the material. Due to this, there is an increment in the thrust load as the axial feed increases. Additionally, the hardened FRPs offer higher resistance during cutting, leading to an increased thrust force [36].

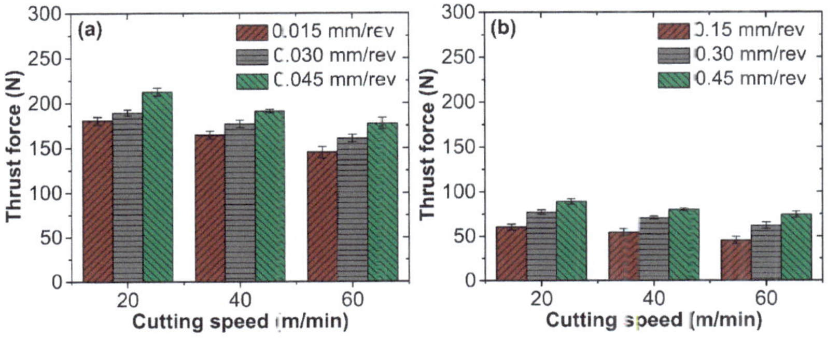

Figure 4. Thrust force variation with process variables in (**a**) drilling and (**b**) helical milling.

Thrust force fluctuation with the process variables during helical milling was similar to the trend observed during drilling. Thrust force is reduced with the increase in cutting speed, as seen in Figure 4b. At a feed of 0.15 mm/rev, the thrust force of 60.6 N was measured when the cutting speed was held at 20 m/min. At a speed of 40 m/min, the thrust force was reduced to 54.4 N. At the highest selected cutting speed of 60 m/min, a thrust force of 45.5 N was registered. A decrease of 24.9% in force magnitude was recorded as cutting speed escalated from 20 m/min to 60 m/min. Similar to the drilling operation, at higher cutting speeds, the rate of shear deformation increases, thus increasing the energy dissipation and the heat. At such high temperatures, the yield strength of the material reduces, thus lowering the resistance to shearing and reducing the thrust load [37]. The impact of axial feed on thrust force is visualized in Figure 4b. Thrust force increased as higher levels of axial feed were selected. With the use of a lower feed (0.15 mm/rev), the measured thrust force was 60.6 N. At an axial feed of 0.30 mm/rev, the thrust force rose to 77.4 N. At the highest selected axial speed of 0.45 mm/rev, a thrust force of 89.1 N was registered. Notably, there is a 35.1% increase in thrust force as axial feed increases from 0.15 mm/rev to 0.45 mm/rev. As axial feed increases, the thickness of uncut chips also increases. This necessitates applying higher energy for material shearing, thereby increasing the thrust force magnitude [38].

However, it is to be noted that the recorded thrust forces are significantly lower during helical milling. Figure 5 illustrates the force signals generated for drilling and helical milling processes at identical machining conditions (Case 3). The mean thrust force during drilling was 216.6 N, while an average thrust force of 92.5 N was recorded during helical milling. Reduced thrust load is attributed to the helical milling kinematics. Hole milling is carried out using an end milling cutter, where the cutting occurs at the frontal and peripheral cutting edges. Since the load is distributed between the two cutting edges, the thrust load is reduced compared to the drilling operation [29,39].

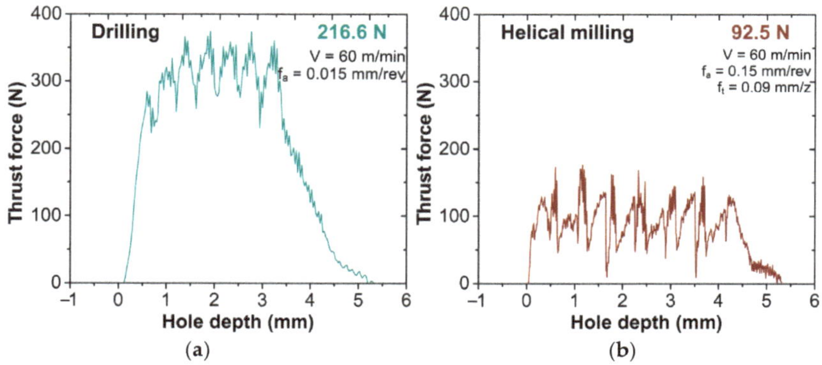

Figure 5. Raw force signal for (**a**) drilling and (**b**) helical milling.

3.2. Machining Temperature

The cutting temperatures attained while machining composites like FMLs are crucial as they influence the mechanical and failure behavior of the FMLs. Accordingly, the present work explores the development of machining temperatures under selected machining variables during the two hole-making operations. Figure 6a depicts how the machining temperature changes with the selected cutting speed and feed. At higher cutting speeds, the machining temperature increased. For instance, with a feed and speed held constant at 0.015 mm/rev and 20 m/min, respectively, a mean temperature of 100.5 °C was observed. However, when the speed increased to 40 m/min, an even higher temperature of 114.3 °C was registered. At the highest speed setting of 60 m/min, a mean temperature of 137.7 °C was observed. As noted, there was a 37.1% rise in cutting temperature when speed was raised from 20 m/min to 60 m/min. The rate at which the material deforms plastically increases at higher cutting speeds. In addition, the heat generated during the metal cutting will be concentrated at the cutting zone, thus increasing the magnitude of the measured temperature [29]. The machining temperature was also affected by the selected axial feed. At a constant cutting speed of 20 m/min, using an axial feed rate of 0.015 mm/rev, the recorded mean temperature was 100.5 °C. With the increment in feed to 0.03 mm/rev, the temperature reduced to 97.2 °C. At the highest level of axial feed (0.045 mm/rev), the measured mean temperature was 93.4 °C. With the axial feed increment, a reducing temperature trend was observed. The mean temperature reduction with the axial feed increment is ascribed to the cutting time. Employment of higher axial feed reduced the cutting time. This reduces the work–tool contact and the time available for the heat to dissipate to the workpiece. The chip takes up a major part of the heat, which lowers the temperature at the cutting zone. Additionally, chip breakability improves at higher feeds. This helps reduce contact friction and the associated heat generation and temperature [40].

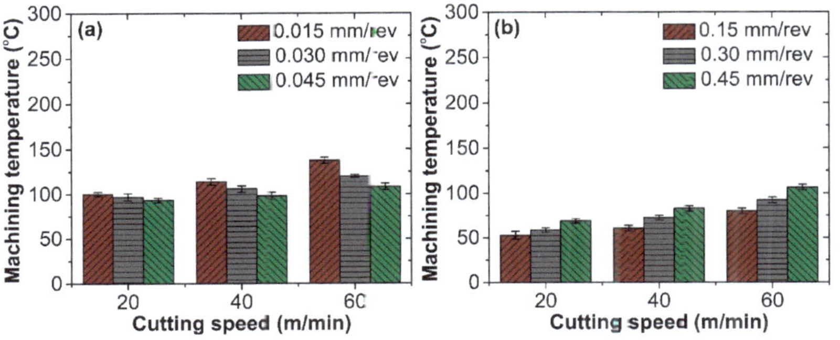

Figure 6. Machining temperature variation with process variables in (**a**) drilling and (**b**) helical milling.

Figure 6b shows the machining temperature variation with cutting speed during the helical milling process. When machining with a feed and speed of 0.15 mm/rev and 20 m/min, a mean cutting temperature of 52.9 °C was observed. At a speed of 40 m/min, an increased mean cutting temperature of 60.7 °C was recorded. Moreover, when the highest chosen speed of 60 m/min was applied, the temperature rose to 80.1 °C. Similar to the drilling process, the rate of material deformation increases as cutting speed increases. Consequently, heat generation and the machining temperature increase. Figure 6b also reveals the evolution of machining temperature with axial feed. In helical milling, with a machining speed of 20 m/min and a feed rate of 0.15 mm/revolution, a mean temperature of 52.9 °C was measured. When the feed rate was raised to 0.30 mm/revolution, the average machining temperature increased to 58.8 °C. Upon further increase to the highest axial feed level of 0.45 mm/revolution, the average temperature escalated to 68.9 °C. It is noteworthy that, in helical milling, the machining temperature rises in correlation with the axial feed increase. The trend is in contrast with that noticed during conventional drilling. Unlike drilling, helical milling is characterized by a higher work–tool contact. The end mill traverses a helical path and continuously interacts with the work material. Since the cutting occurs at the frontal and peripheral cutting edges, when the axial pitch is higher, the amount of material undergoing plastic deformation is significantly higher, thereby increasing the plastic deformation rate and the temperature [29].

When machining FMLs like CARALL, avoiding temperatures exceeding the glass transition temperature (T_g) and preventing thermal damage to the workpiece is preferable. In the present study, the selected process variables influence the heat generation rate and the magnitude of machining temperature in both hole-making operations. In the case of drilling, the maximum recorded temperature of 141.1 °C (Case 7) was higher than T_g = 130 °C for the epoxy resin used in the FMLs (see Figure 7). However, results show that helical milling for hole making was beneficial, as the recorded machining temperature of 82.7 °C was significantly lower than the T_g, indicating a lower probability of temperature-induced damages in CARALL.

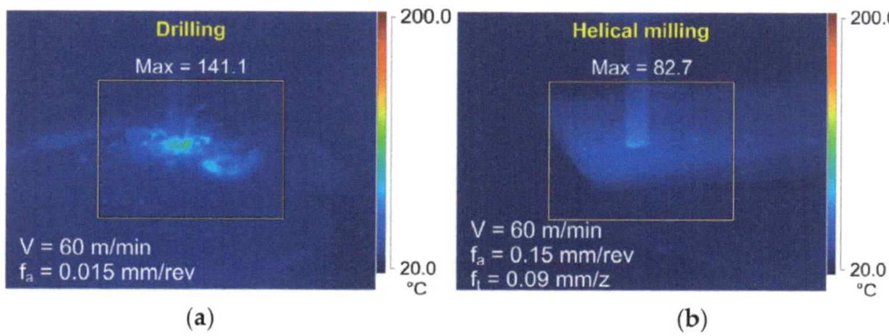

Figure 7. Comparison of machining temperature cartographs obtained for Case 7 during (**a**) drilling and (**b**) helical milling.

3.3. Surface Roughness

In the aerospace industry, the quality of the holes is essential, as it affects the functionality and service life of structural assemblies. Moreover, the choice of hole-making operations significantly affects the hole quality. Hence, an analysis of hole quality, specifically considering surface roughness, was conducted for the two hole-making operations utilized in this study. During drilling, variation in the surface roughness was observed for the selected speed and feed rates. Figure 8a presents the sway machining parameters that have surface roughness. As cutting speed increased, surface roughness decreased. At a constant feed of 0.015 mm/revolution, a mean roughness of 2.89 μm was observed when the speed was set at 20 m/min. At a 40 m/min cutting speed, a mean roughness of 2.65 μm was noted. At the highest employed speed of 60 m/min, the measured roughness was 2.25 μm. As observed, when the cutting speed increased to 40 m/min, the average surface roughness decreased by 8.3%. An increase in cutting speed to 60 m/min led to an additional reduction in measured surface roughness of 13.9%. The increase in surface roughness at reduced cutting speed can be attributed to chip size and built-up edge (BUE) formation [41]. Drilling at a lower cutting speed produces long, continuous chips. These work-hardened chips abrade the hole surface and erode the material from the CFRP layers as it evacuates through the tool flute [41]. Further, aluminum alloy tends to stick to the tool's cutting edge when the cutting speeds are low. However, as the cutting speed increased, surface roughness reduced. The noted enhancement in surface finish is linked to the decrease in chip size and the elevated temperatures developed at higher cutting speeds. At higher cutting speeds, small-sized chips are produced, thereby reducing the intensity of scratches on the drilled holes' surface. Higher cutting temperatures soften the resin material, smearing it across the machined surface and on top of the CFRP layers, generating a smoothening effect [29].

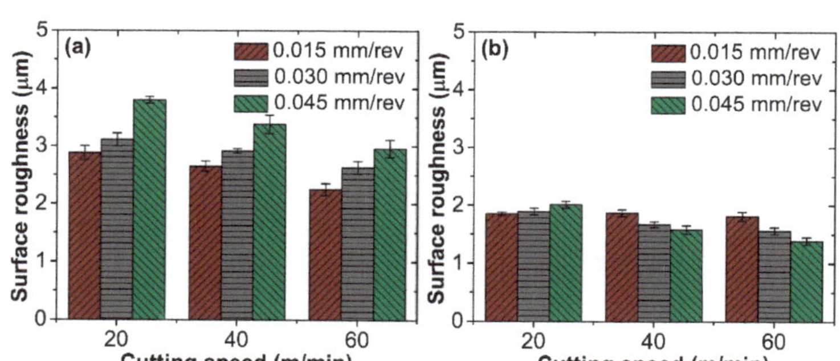

Figure 8. Surface roughness variation with process variables in (**a**) drilling and (**b**) helical milling.

Selected axial feed also affected the surface roughness in drilled holes. At a cutting speed of 20 m/min and a feed rate of 0.015 mm/rev, a mean roughness of 2.89 μm was recorded. Further, at a feed of 0.03 mm/revolution, the measured mean roughness was 3.11 μm. When utilizing the highest feed of 0.045 mm/revolution, surface roughness was increased to 3.8 μm. At lower feeds, the drilled holes indicated the presence of feed marks. However, critical damages like fiber pull-out and breakage were not observed at lower feeds, thereby producing holes with better surface quality. However, at higher feeds, fibers were exposed due to increased thrust loads, and fiber pull-out (Figure 9) was noted under extreme conditions, deteriorating the surface quality.

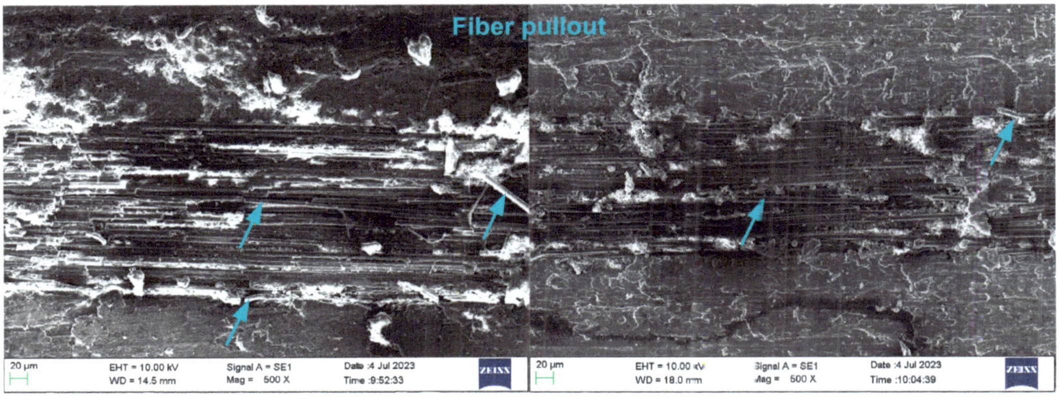

Figure 9. Fiber pull-out was observed at higher feed conditions during drilling.

Figure 8b shows the influence cutting variables have on surface roughness during helical milling. The surface roughness varied with cutting speed and axial feed. Milling with a lower feed, the axial feed, and the cutting speed interaction displayed a non-linear relationship. For a feed of 0.15 mm/rev, the mean surface roughness was 1.85 μm in holes cut at 20 m/min. When the cutting speed increased to 40 m/min, there was an increase in the mean roughness, which stood at 1.87 μm. A lower mean roughness of 1.81 μm was recorded with a 60 m/min speed. At a lower feed of 0.15 mm/rev, higher roughness at a smaller cutting speed is attributed to BUE formation and a higher force magnitude. BUE adhered to the bore surface, thereby deteriorating the surface quality. Also, material plowing was noted at the lower axial pitch, which generated feed marks. Additionally, due to the higher thrust load, carbon fibers in helical-milled holes were subjected to bending and eventual peeling. As a result, the CFRP layer exhibited poor surface quality, thus increasing the surface roughness of the FML as a whole [27]. However, at the higher selected cutting speed, the softened matrix material showed signs of smearing over the machined hole surface, thus reducing the surface unevenness and improving the surface finish.

However, with higher speeds of 40 m/min and 60 m/min, mean surface roughness decreased as the axial pitch increased. For example, for a 40 m/min cutting speed, a mean roughness of 1.87 μm was recorded when the axial pitch of 0.15 mm/rev was utilized. At a feed of 0.30 mm/rev, a mean roughness of 1.67 μm was recorded. Further, a 0.45 mm/rev feed helped produce surfaces with a mean roughness of 1.58 μm. A similar trend was obtained for a cutting speed of 60 m/min. Diminished surface finish at lower feed is ascribed to the formation of feed marks. Material removal resulted in material plowing and non-uniform plastic deformation at a lower pitch, producing feed marks on the machined surface. Additionally, the CFRP layers showed signs of fiber peeling and void formation, thereby increasing the roughness of the hole surface. However, the adaptation of a higher axial pitch produced defect-free surfaces, thus improving the machined surface quality. The resin material that underwent plasticization smeared and occupied the gaps available at the CFRP layers, thereby closing the gaps and improving the surface finish.

Closer observation reveals that the surface of the holes produced by milling operations showed an improved surface finish compared to the drilled ones (see Figure 10). Moreover, the magnitude of surface roughness of all the helical milled holes was considerably lower than the surface roughness requirements set by SANDVIK (less than 3.2 μm) [42], indicating the helical mill operation's capability to produce excellent-quality holes.

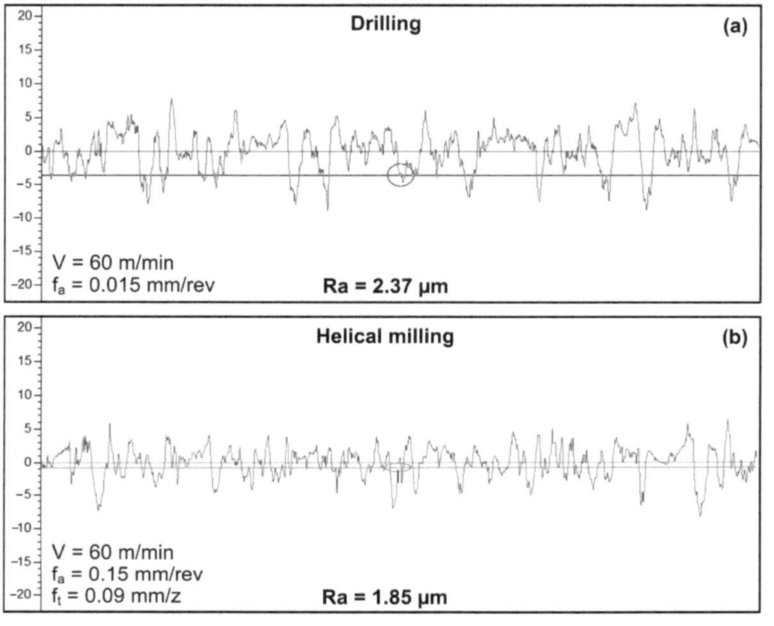

Figure 10. Comparison of surface roughness profiles obtained for Case 7 during (**a**) drilling and (**b**) helical milling.

3.4. Surface Morphology

The selected hole-making operations influence the hole quality. Figure 11 depicts the various forms of damage observed during the drilling process. The hole surface indicated the existence of deep groove marks created by strain-hardened continuous chips produced during the drilling operation at lower cutting speeds and feed conditions (see Figure 11a). The chips also interacted with the CFRP layer, resulting in material erosion. Moreover, fiber shearing, breakage, and fiber pull-out were noted in holes drilled at lower cutting speeds and higher feed conditions (see Figure 11b). Such damages result from the higher thrust force and temperatures [27]. Feed marks were observed on the hole surface drilled with lower axial feed (see Figure 11c). The hole surface showed signs of material smearing at higher cutting speeds during the drilling operation (see Figure 11d). As cutting speed increased during drilling, machining temperature also escalated. At higher temperatures, the resin material underwent thermal softening, smearing the material against the machined hole wall and improving the surface finish. Waste material ingression and interlayer burr formation are observed in the drill holes, as seen in Figure 11e. The selected feed influenced the surface finish of drilled holes. Moreover, adhered debris was commonly observed on the surface of the drilled holes (Figure 11f). As observed, the holes produced by drilling showed signs of surface anomalies in the form of feed marks, material smearing, fiber bundle exposure, and fiber pull-out, mostly attributed to the high thrust force and machining temperatures. Since helical milling aided in lowering thrust force and temperatures, the quality of milled holes was also studied to identify the presence of any surface irregularities [29].

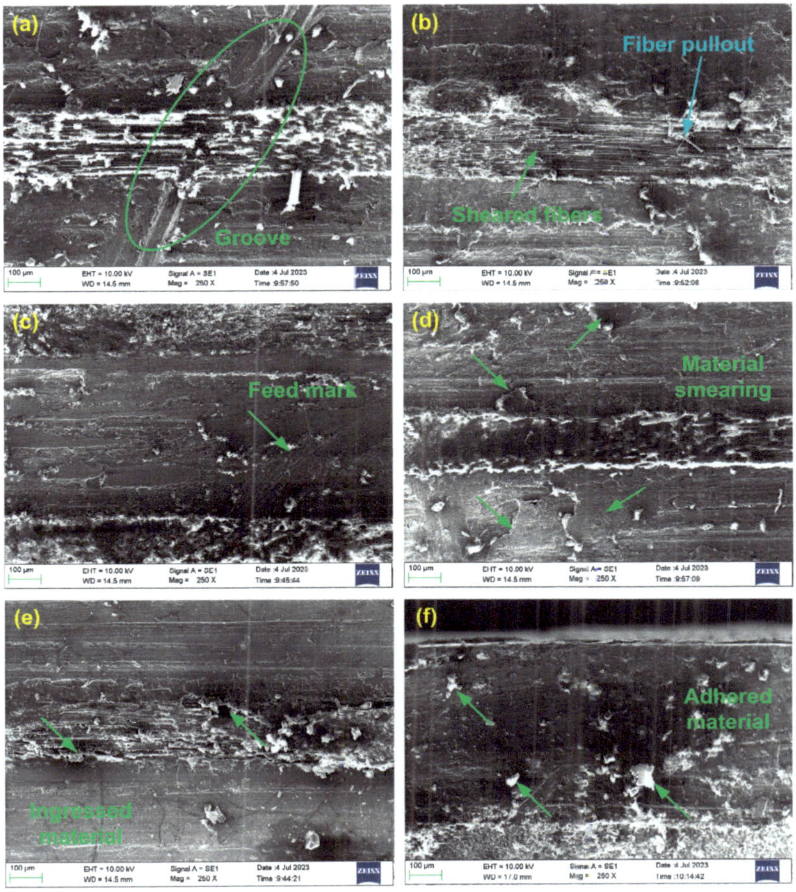

Figure 11. Surface damages in the form of (**a**) grooves, (**b**) fiber pullout, (**c**) feed marks, (**d**) material smearing, (**e**) material ingression, and (**f**) material adhesion were observed in the drilled holes.

The characterization of milled holes revealed the presence of chip debris adhering to the hole surface, as seen in Figure 12a. The kinematics of the helical milling operation produces broken, discontinuous chips. Due to their light weight and size, these chips find it challenging to extrude from the cutting area without cutting fluid, thus adhering to the wall of the milled hole. Material smearing was characterized by holes machined at higher cutting speed (see Figure 12b). The smearing is the result of high-temperature material softening occurring at higher cutting speeds [29]. Further inspection revealed the formation of feed marks on the metal surface (see Figure 12c). The feed marks were prominent in holes subjected to lower axial pitch cutting conditions. Feed mark formation is attributed to plowing action, which generates a non-uniform plastic flow of the material at the cutting edge. At lower axial pitch, the carbon fibers in helical milled holes were subjected to bending and eventual peeling, resulting from the advancement of the tool corner into the fiber bundle (see Figure 12d).

In general, hole drilling with low feed and higher speed generated a clean-cut surface (see Figure 13a). The phenomenon is credited to material smearing observed at such machining conditions, where the temperature is substantially higher than Tg. Under such conditions, the resin, which is in the plastic state, is recast on the hole surface, thereby improving the surface finish. However, in the case of helical milling, an adaptation of higher cutting speed and axial pitch produced a defect-free surface, as seen in Figure 13b. Moreover, in comparison to the drilling operation, far less damage was noted in the helical-milled holes. The quality of the hole surface was superior to that of the drilled ones, indicating the better utility of helical milling for processing holes for FMLs.

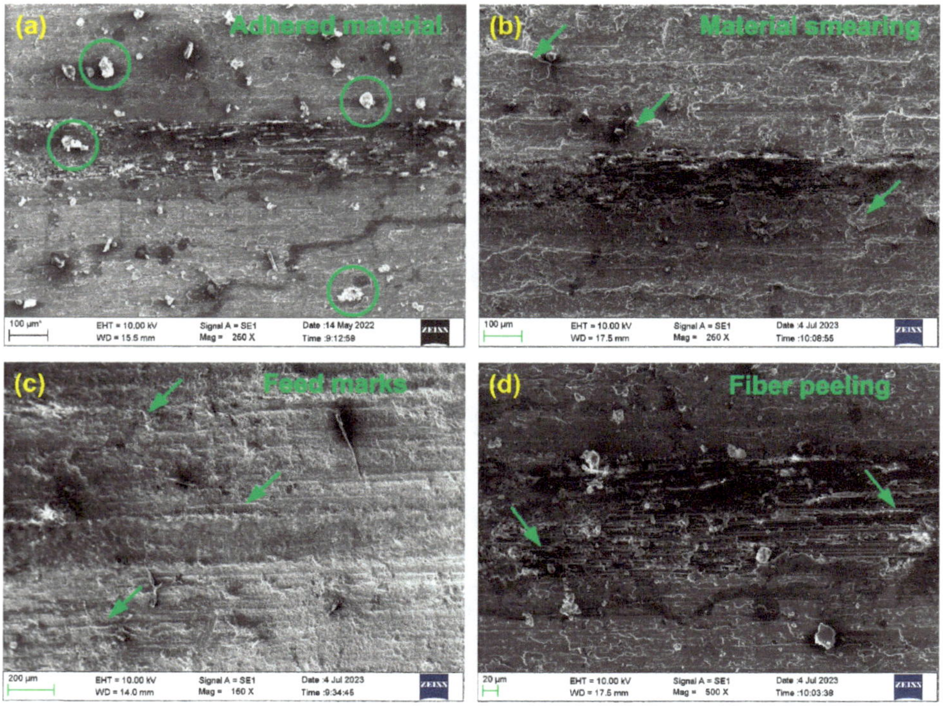

Figure 12. Surface damages in the form of (**a**) material adhesion, (**b**) material smearing, (**c**) feed marks, and (**d**) fiber peeling were observed in the helical milled holes.

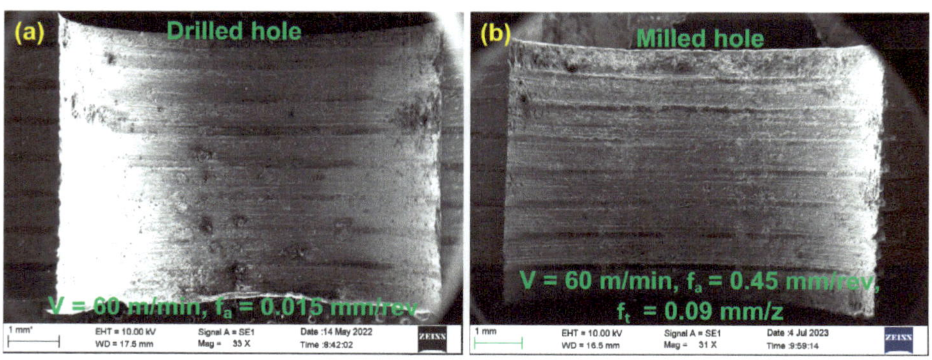

Figure 13. Holes with good surface finish were observed during (**a**) drilling and (**b**) helical milling.

3.5. Microhardness

The aluminum material is subjected to severe plastic deformation and temperatures during machining. Therefore, the microhardness of the holes processed using the drilling and helical milling processes was measured and evaluated. For the drilled holes, microhardness ranged between 140.8 and 158.2 HV, showing a 7.3% to 20.5% increase compared to the bulk material microhardness of 131.2 HV. In the case of milled holes, microhardness ranged between 152.7 and 168.3 HV, displaying a 16.4% to 28.2% increase. As noted, the increase in microhardness during drilling is less than during the helical milling process. As the material is plastically deformed in the drilling operation, an enormous quantity of heat is generated, resulting in very high machining temperatures. As a result, the aluminum material is subjected to thermal softening, thus lowering its microhardness. However, machining temperatures are considerably lower in helical milling than in drilling, and the material dominantly exhibits strain hardening. The alterations in microstructure due to strain hardening increase the microhardness.

Additionally, the effect of process variables on microhardness during drilling and helical milling was investigated. Figures 14a and 15a display the effect the two process variables have on the microhardness of drilled and milled holes. The feed rate increase showcased the minimal influence of the microhardness variation. With the rise in drilling feed from 0.015 mm/rev to 0.045 mm/rev, a minor increment in the microhardness (3.1% to 4.9%) was noted. Similarly, in the case of helical milling, as the pitch increased from 0.15 mm/rev to 0.45 mm/rev, a 0.9% to 2.7% increase was noted. The increase in microhardness with axial feed and axial pitch is attributed to the higher work-hardening tendency of the material at higher axial feed. The microhardness displayed a decreasing trend with the cutting speed for both processes. As the cutting speed increased from 20 m/min to 60 m/min, the microhardness decreased by 3.9% to 6.3%. In the case of helical milling, a 3.1% to 4.7% decrease was measured. The reduction in microhardness with higher cutting speeds is attributed to the temperature-induced softening of the work material.

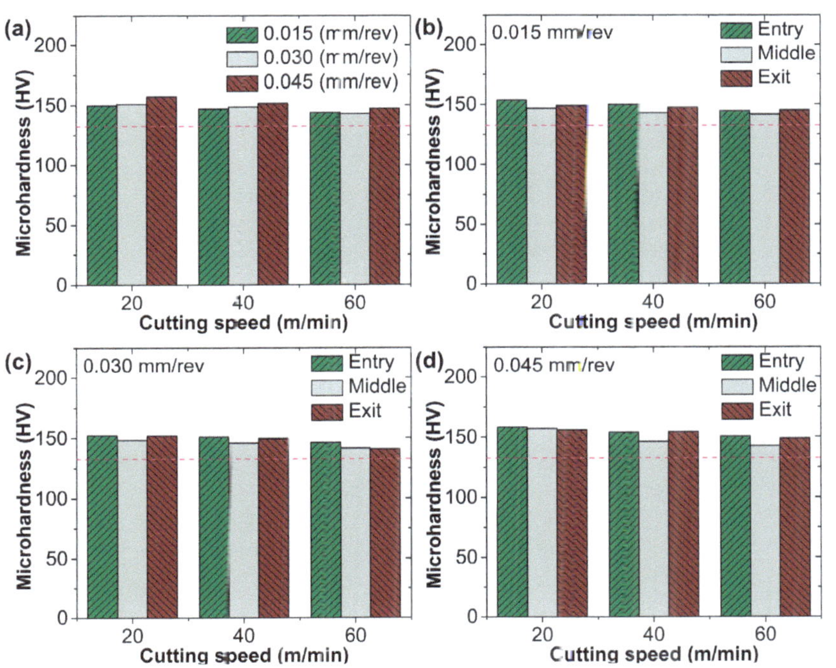

Figure 14. (**a**) Average microhardness vs. process variables for drilling operation; (**b**–**d**) microhardness at three different layers of the CARALL stack vs. process variables for drilling operation.

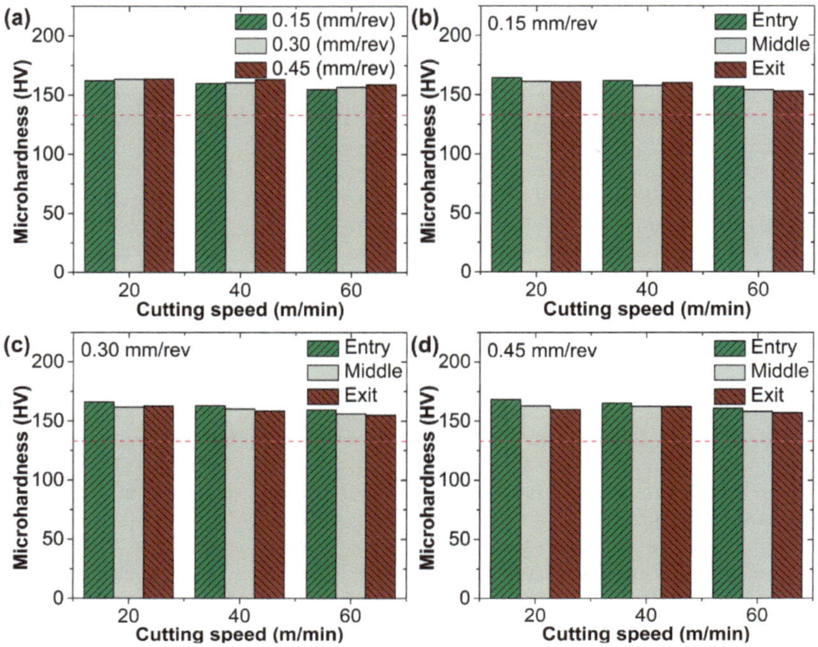

Figure 15. (**a**) Average microhardness vs. process variables for milling operation; (**b–d**) microhardness at three different layers of the CARALL stack vs. process variables for milling operation.

Furthermore, variation in the microhardness at three different layers was analyzed. In the case of drilling, the entry and exit layers exhibited similar microhardness values in the majority of the cases explored in the study. However, the microhardness at the middle layer of the CARALL stack was moderately lower compared to the extreme metal layers of the stack, as seen in Figure 14b,c. The variation in microhardness can be attributed to the machining temperatures the metal layers are subjected to during the drilling operation. The entry and exit layers are subjected to cutting temperatures. However, due to the higher thermal conductivity of the aluminum alloy, the heat is carried away, cooling the two outer layers quickly. However, the middle layer, which is subjected to cutting temperatures, cannot dissipate the heat quickly due to the surrounding epoxy, which has poor thermal conductivity. To some extent, the higher cutting temperatures during drilling soften the work material, thereby lowering its microhardness.

In most cases of helical milling operations (see Figure 15b,c), the microhardness was reduced at the exit layer compared to the entry and middle layers of the CARALL stack. In helical milling, at the entry layer, the material undergoes plastic deformation and work hardening, thus exhibiting higher microhardness. However, as the tool proceeds with the cutting and reaches the middle and exit layers, the frictional heat and the accumulated heat generated due to the plastic deformation can affect the microstructure, thus lowering the microhardness.

4. Conclusions

This work investigated the machining performance of helical milling and drilling operations while making holes in carbon fiber-reinforced aluminum laminates. The following important conclusions were noted:

- Thrust force increased with the axial feed, while a dropping trend was noted with increased cutting speed. The increase in thrust force with axial feed is attributed to the increase in uncut chip thickness, while the reduction in thrust force with cutting speed is linked to material thermal softening. Moreover, the thrust force recorded during helical milling is 97–124 N lower than the conventionally drilled force magnitude. The decrease is attributed to load distribution at the frontal and peripheral cutting edges of the end mill, which differs from the drill tool, where cutting solely takes place at the frontal cutting edges.

- Selected process variables affect the machining temperature. A maximum temperature of 136.7 °C while drilling was recorded with a higher cutting speed of 60 m/min and a lower axial feed of 0.015 mm/rev. The maximum temperature observed in the drilling was higher than the glass transition temperature of epoxy resin. In the case of helical milling, a temperature of 80.1 °C was recorded for similar productivity conditions, indicating a lower probability of temperature-induced damages in the FML.
- The selected hole-making operation influenced the roughness of the holes. Higher surface roughness was recorded in holes processed using the drilling process. A maximum average roughness of 3.8 µm was noted during the drilling process, while a roughness of 2.01 µm was observed in helical milled holes, indicating the helical mill operation's capability to produce excellent-quality holes.
- The morphology of the drilled holes indicated the presence of defects such as grooves, feed marks, material smearing, material ingression, interlayer burr formation, fiber bundle exposure, and fiber pull-out. Helical-milled hole surfaces were characterized by the presence of smeared material, feed marks, and buckled fibers. Principally, in comparison to drilled holes, helical-milled holes exhibited better surface quality even under dry cutting conditions.
- The selected levels of process variables were conductive and favorable for carrying out drilling and helical milling operations without inducing any critical defects like delamination and debonding in CARALL FMLs.
- The microhardness of the aluminum layers increased by 7.3% to 20.5% in the case of drilling and by 16.4% to 28.2% in the case of helical milling. The axial feed/axial pitch had minimal influence on the microhardness increase in comparison to the cutting speed. The position of the aluminum layer in the CARALL stacking sequence, to some extent, influenced the post-machining microhardness.

Author Contributions: Conceptualization and methodology, G.B. and A.A.D.; formal analysis and investigation, A.P., G.B. and A.A.D.; resources, G.B.; data curation, A.A.D. and V.L.N.; writing—original draft preparation, G.B. and A.A.D.; visualization, A.H. and V.L.N.; writing—review and editing, A.P., A.H. and R.S.; supervision, G.B. and R.S. All authors have read and agreed to the published version of the manuscript.

Funding: This research received no external funding.

Data Availability Statement: All the data included in the manuscript can be provided by the corresponding author upon reasonable request.

Acknowledgments: We acknowledge the support provided by the Manipal Academy of Higher Education for carrying out the research work.

Conflicts of Interest: The authors declare no conflict of interest.

References

1. Ng, L.F.; Yahya, M.Y.; Leong, H.Y.; Parameswaranpillai, J.; Muthukumar, C.; Syed Hamzah, S.M.S.A.; Dhar Malingam, S. State-Of-The-Art Review on Developing Lightweight Fiber-Metal Laminates Based on Synthetic/Natural Fibers. *Polym. Compos.* **2023**, *44*, 6275–6303. [CrossRef]
2. Xie, M.; Zhan, L.; Ma, B.; Hui, S. Classification of Fiber Metal Laminates (Fmls), Adhesion Theories and Methods for Improving Interfacial Adhesion: A Review. *Thin-Walled Struct.* **2024**, *11*, 17–44. [CrossRef]
3. Mahesh, V.; Mahesh, V.; Harursampath, D. Ballistic Characterization of Fiber Elastomer Metal Laminate Composites and Effect of Positioning of Fiber Reinforced Elastomer. *Proc. Inst. Mech. Eng. Part L J. Mater. Des. Appl.* **2022**, *236*, 663–673. [CrossRef]
4. Taherzadeh-Fard, A.; Liaghat, G.; Ahmadi, H.; Razmkhah, O.; Charandabi, S.C. Zarezadeh-Mehrizi, M.A.; Khodadadi, A. Experimental and Numerical Investigation of the Impact Response of Elastomer Layered Fiber Metal Laminates (Efmls). *Compos. Struct.* **2020**, *245*, 112–264. [CrossRef]

5. Seoane-Rivero, R.; Germán, L.; Santos, F.; Gondra, K. Development of New Hybrid Composites For High-Temperature Applications. *Polymers* **2023**, *15*, 4380. [CrossRef] [PubMed]
6. Costa, R.D.F.S.; Sales-Contini, R.C.M.; Silva, F.J.G.; Sebbe, N.; Jesus, A.M.P. A Critical Review on Fiber Metal Laminates (FML): From Manufacturing to Sustainable Processing. *Metals* **2023**, *13*, 638. [CrossRef]
7. Ghajar, R.; Ghadami, M. A Novel Experimental Method and Computational Micromechanical Model for In-Situ Damage Detection and Prediction of Stiffness Degradation In Cross-Ply FML. *Compos. Struct.* **2023**, *305*, 116–492. [CrossRef]
8. Ng, L.F.; Subramaniam, K. Composite Sandwich Panels with The Metallic Facesheets. In *Sandwich Composites*; CRC Press: Boca Raton, FL, USA, 2022; pp. 61–74.
9. Yang, Y.; Zhou, W.; Guo, Y.; Tong, Z.; Chen, L.; Ren, X.; Li, L. Effect of Laser Shock Peening without Protective Coating on Surface Integrity of Titanium-Based Carbon-Fibre/Epoxy Laminates. *Opt. Laser Technol.* **2023**, *167*, 109–685. [CrossRef]
10. Megahed, M.; Abd El-Baky, M.A.; Alsaeedy, A.M.; Alshorbagy, A.E. Synthesis Effect of Nano-Fillers on the Damage Resistance of GLARE. *Fibers Polym.* **2021**, *22*, 1366–1377. [CrossRef]
11. Giasin, K.; Hawxwell, J.; Sinke, J.; Dhakal, H.; Köklü, U.; Brousseau, E. The Effect of Cutting Tool Coating on the Form and Dimensional Errors of Machined Holes in GLARE®Fibre Metal Laminates. *Int. J. Adv. Manuf. Technol.* **2020**, *107*, 2817–2832. [CrossRef]
12. Giasin, K.; Dad, A.; Brousseau, E.; Pimenov, D.; Mia, M.; Morkavuk, S.; Koklu, U. The Effects of through Tool Cryogenic Machining on the Hole Quality in GLARE®Fibre Metal Laminates. *J. Manuf. Process.* **2021**, *64*, 996–1012. [CrossRef]
13. Zhang, Q.; Zheng, S.; Yu, C.; Wang, Q.; Ke, Y. Digital Thread-Based Modeling of Digital Twin Framework for the Aircraft Assembly System. *J. Manuf. Syst.* **2022**, *65*, 406–420. [CrossRef]
14. Sui, S.; Song, G.; Sun, C.; Zhu, Z.; Guo, K.; Sun, J. Experimental Investigation on the Performance of Novel Double Cone Integrated Tool in One-Shot Drilling of Metal Stacks. *Int. J. Adv. Manuf. Technol.* **2020**, *109*, 523–534. [CrossRef]
15. De Melo Silva, W.; Martins, P.S.; De Carvalho, V.E.; Da Cruz, N.C.; Claudino, E.; Carneiro, J.R.G. Improving Precision in Aluminum Alloy Machining Due to the Application of Diamond-Like Carbon Thin Film. *J. Tribol.* **2021**, *143*, 71403. [CrossRef]
16. Dougan, M.A.; Yazman, S.; Gemi, L.; Yildiz, M.; Yapici, A. A Review on Drilling of FML Stacks with Conventional and Unconventional Processing Methods under Different Conditions. *Compos. Struct.* **2022**, *297*, 115913. [CrossRef]
17. Logesh, K.; Moshi, A.A.M.; Bharathi, S.R.S.; Hariharasakthisudhan, P. A Multi-Objective Grey Relational Approach and Regression Analysis on Optimization of Drilling Process Parameters for GLARE Fiber Metal Laminates. *Surf. Rev. Lett.* **2022**, *29*, 2250066. [CrossRef]
18. Kumar, D.; Gururaja, S.; Jawahir, I.S. Machinability and Surface Integrity of Adhesively Bonded Ti/CFRP/Ti Hybrid Composite Laminates under Dry and Cryogenic Conditions. *J. Manuf. Process.* **2020**, *58*, 1075–1087. [CrossRef]
19. Sridhar, A.K.; Bolar, G.; Padmaraj, N.H. Comprehensive Experimental Investigation on Drilling Multi-Material Carbon Fiber Reinforced Aluminium Laminates. *J. King Saud Univ. Sci.* **2022**, *34*, 391–401. [CrossRef]
20. Boughdiri, I.; Giasin, K.; Mabrouki, T.; Zitoune, R. Effect of Cutting Parameters on Thrust Force, Torque, Hole Quality and Dust Generation during Drilling of GLARE 2B Laminates. *Compos. Struct.* **2021**, *261*, 113562. [CrossRef]
21. Seif, A.; Fathy, A.; Megahed, A.A. Effect of Drilling Process Parameters on Bearing Strength of Glass Fiber/Aluminum Mesh Reinforced Epoxy Composites. *Sci. Rep.* **2023**, *13*, 12143. [CrossRef]
22. Costa, R.; Duro, J.N.S.; Sousa, V.F.C.; Silva, T.E.F.; Figueiredo, D.A.; Jesus, A.M.P. Drilling of CFRP/Al Multi-Material Stacks Using WC-Co CVD Diamond Coated Tools. *Procedia Struct. Integr.* **2024**, *53*, 376–385. [CrossRef]
23. Kayihan, M.; Karaguzel, U.; Bakkal, M. Experimental Analysis on Drilling of Al/Ti/CFRP Hybrid Composites. *Mater. Manuf. Process.* **2021**, *36*, 215–222. [CrossRef]
24. Sorrentino, L.; Turchetta, S.; Parodo, G. Drilling of Glare Laminates: Effect of Cutting Parameters on Process Forces and Temperatures. *Int. J. Adv. Manuf. Technol.* **2022**, *120*, 645–657. [CrossRef]
25. Parodo, G.; Rubino, F.; Sorrentino, L.; Turchetta, S. Temperature Analysis In Fiber Metal Laminates Drilling: Experimental and Numerical Results. *Polym. Compos.* **2022**, *43*, 7600–7615. [CrossRef]
26. Chen, X.; Zhao, W.; Zhao, G.; Jamil, M.; He, N. Tool Wear and Surface Quality during Milling CFRP Laminates under Dry and LN2-Based Cryogenic Conditions. *Int. J. Adv. Manuf. Technol.* **2022**, *123*, 1785–1797. [CrossRef]
27. Giasin, K.; Ayvar-Soberanis, S. Microstructural Investigation of Drilling Induced Damage in Fibre Metal Laminates Constituents. *Compos. Part A Appl. Sci. Manuf.* **2017**, *97*, 166–178. [CrossRef]
28. Giasin, K. The Effect of Drilling Parameters, Cooling Technology, and Fiber Orientation on Hole Perpendicularity Error in Fiber Metal Laminates. *Int. J. Adv. Manuf. Technol.* **2018**, *97*, 4081–4099. [CrossRef]
29. Bolar, G.; Sridhar, A.K.; Ranjan, A. Drilling and Helical Milling for Hole Making in Multi-Material Carbon Reinforced Aluminum Laminates. *Int. J. Light. Mater. Manuf.* **2022**, *5*, 113–125. [CrossRef]
30. Ge, J.; Feist, T.; Elmore, A.; Reji, R.; Mclaughlin, B.; Jin, Y.; Sun, D. Open Hole Surface Integrity and Its Impact on Fatigue Performance of Al 2024-T3/Ti-6Al-4V Stacks. *Procedia CIRP* **2022**, *108*, 234–239. [CrossRef]
31. Sun, L.; Gao, H.; Wang, B.; Bao, Y.; Wang, M.; Ma, S. Mechanism of Reduction of Damage during Helical Milling of Titanium/CFRP/Aluminium Stacks. *Int. J. Adv. Manuf. Technol.* **2020**, *107*, 4741–4753. [CrossRef]
32. Li, X.; Jiao, A.; Liu, B.; Zhang, Y.; Liu, G.; Zhang, Z. Study on Helical Hole-Making Process of CFRP/Al Alloy Laminated Materials. *Int. J. Adv. Manuf. Technol.* **2022**, *121*, 6551–6568. [CrossRef]

33. Jiaying, G.E.; Guang, C.; Yongxiang, S.U.; Yunhe, Z.O.U.; Chengzu, R.E.N.; Xuda, Q.I.N.; Guofeng, W. Effect of Cooling Strategies on Performance and Mechanism of Helical Milling of CFRP/Ti-6Al-4 V Stacks. *Chinese J. Aeronaut.* **2022**, *35*, 388–403.
34. Xu, J.; Ji, M.; Chen, M.; El Mansori, M. Experimental Investigation on Drilling Machinability and Hole Quality of CFRP/Ti6Al4V Stacks under Different Cooling Conditions. *Int. J. Adv. Manuf. Technol.* **2020**, *109*, 1527–1539. [CrossRef]
35. Boutrih, L.; Makich, H.; Nouari, M.; Lanouar, B.E.N. Surface Quality In Dry Machining of CFRP Composite/Ti6Al4V Stack Laminate. *Procedia CIRP* **2022**, *108*, 758–763. [CrossRef]
36. Ramulu, M.; Branson, T.; Kim, D. A Study on the Drilling of Composite and Titanium Stacks. *Compos. Struct.* **2001**, *54*, 67–77. [CrossRef]
37. Sun, Y.; Sun, J.; Li, J.; Xiong, Q. An Experimental Investigation of the Influence of Cutting Parameters on Cutting Temperature in Milling Ti6Al4V by Applying Semi-Artificial Thermocouple. *Int. J. Adv. Manuf. Technol.* **2014**, *70*, 765–773. [CrossRef]
38. Giasin, K.; Barouni, A.; Dhakal, H.N.; Featherson, C.; Redouane, Z.; Morkavuk, S.; Koklu, U. Microstructural Investigation and Hole Quality Evaluation in S2/FM94 Glass-Fibre Composites under Dry and Cryogenic Conditions. *J. Reinf. Plast. Compos.* **2021**, *40*, 273–293. [CrossRef]
39. Amini, S.; Baraheni, M.; Hakimi, E. Enhancing Dimensional Accuracy and Surface Integrity by Helical Milling of Carbon Fiber Reinforced Polymers. *Int. J. Lightweight Mater. Manuf.* **2019**, *2*, 362–372. [CrossRef]
40. Zitoune, R.; Cadorin, N.; Collombet, F.; Šíma, M. Temperature and Wear Analysis in Function of the Cutting Tool Coating when Drilling of Composite Structure: In Situ Measurement by Optical Fiber. *Wear* **2017**, *376*, 1349–1858. [CrossRef]
41. Zitoune, R.; Krishnaraj, V.; Collombet, F. Study of Drilling of Composite Material and Aluminium Stack. *Compos. Struct.* **2010**, *92*, 1246–1255. [CrossRef]
42. Kumar, D.; Gururaja, S. Machining Damage and Surface Integrity Evaluation during Milling of UD-CFRP Laminates: Dry vs. Cryogenic. *Compos. Struct.* **2020**, *247*, 112–504. [CrossRef]

materials

MDPI

Article

Machinability and Surface Properties of Cryogenic Poly(methyl methacrylate) Machined via Single-Point Diamond Turning

Xiaoyu Wu [1,2,3], Qiang Kang [4], Xiaoxing Jiang [1,2] and Xudong Fang [1,2,*]

1 School of Mechanical Engineering, Xi'an Jiaotong University, Xi'an 710049, China; xymems@stu.xjtu.edu.cn (X.W.); xxjiang_2022@stu.xjtu.edu.cn (X.J.)
2 State Key Laboratory for Manufacturing Systems Engineering, International Joint Laboratory for Micro/Nano Manufacturing and Measurement Technology, Xi'an Jiaotong University, Xi'an 710056, China
3 School of Electrical Engineering, Xi'an Jiaotong University, Xi'an 710049, China
4 School of Mechano-Electronic Engineering, Xidian University, Xi'an 710071, China; kangqiang@xidian.edu.cn
* Correspondence: dongfangshuo30@xjtu.edu.cn

Abstract: Poly(methyl methacrylate) (PMMA), with a glass transition temperature (Tg) over 100 °C, shows good mechanical and optical properties and has broad applications after being machined with single-point diamond turning (SPDT) at room temperature. Because of the high Tg, current efforts mostly focus on optimizing machining parameters to improve workpiece precision without considering the modification of material properties. Cryogenic cooling has been proven to be an effective method in assisting ultra-precision machining for certain types of metals, alloys, and polymers, but has never been used for PMMA before. In this work, cryogenic cooling was attempted during the SPDT of PMMA workpieces to improve surface quality. The machinability and surface properties of cryogenically cooled PMMA were investigated based on the mechanical properties at corresponding temperatures. Nanoindentation tests show that, when temperature is changed from 25 °C to 0 °C, the hardness and Young's modulus are increased by 37% and 22%, respectively. At these two temperature points, optimal parameters including spindle speed, feed rate and cut depth were obtained using Taguchi methods to obtain workpieces with high surface quality. The surface quality was evaluated based on the total height of the profile (Pt) and the arithmetic mean deviation (Ra). The measurement results show that the values of Pt and Ra of the workpiece machined at 0 °C are 124 nm and 6 nm, respectively, while the corresponding values of that machined at 25 °C are 291 nm and 11 nm. The test data show that cryogenic machining is useful for improving the form accuracy and reducing the surface roughness of PMMA. Moreover, the relationship between temperature, material properties and machinability weas established with dynamic mechanical analysis (DMA) data and a theoretical model. This can explain the origin of the better surface quality of the cryogenic material. The basis of this is that temperature affects the viscoelasticity of the polymer and the corresponding mechanical properties due to relaxation. Then, the material property changes will affect surface profile formation during machining. The experimental results and theoretical analysis show that cryogenically cooled PMMA has good machinability and improved surface quality when using SPDT compared to that at ambient temperature.

Keywords: poly(methyl methacrylate) (PMMA); mechanical property; cryogenic; single point diamond of turning (SPDT); nanoindentation; surface roughness

1. Introduction

Single point diamond turning (SPDT) is one of the dominant processes used when producing ultra-precision parts. It can be used for various materials, including metals, ceramics, and polymers. Using this process, workpieces with a designed geometry and specific properties can be obtained for multiple applications like the aerospace industry, the automobile industry, etc. In recent decades, researchers have exerted great efforts in studying the effect of machining parameters, including the feed rate, spindle speed and cut depth, on component surface quality. Plenty of work has been completed on metals. For example, stainless steel, aluminum, and copper are processed using SPDT to reach micro-/nano-level accuracy with optimized machining conditions [1,2]. Nevertheless, investigations on the SPDT of polymers are quite limited. Only a few types of polymers can be machined using this process [3,4]. The hardness and Young's modulus of the polymers must reach a certain value to use SPDT; otherwise, the accuracy of the final part cannot be reached because of deformation induced by softness. A representative of the polymers which cannot be directly machined using SPDT is PDMS, which is too soft to maintain dimensional precision. Accordingly, the ultra-precision machining of polymers via SPDT is still a hot topic requiring further research.

PMMA, as a typical example of a polymer with excellent optical properties, has been used in many applications after being machined using SPDT. Its applications include biomedical instruments, consumer products, communication systems, and security instrumentation, depending on the material properties of PMMA, including its rigidity, toughness, light transmission, dimensional stability, and surface quality. Among the aforementioned properties, the surface quality of PMMA is a major challenge to address to extend its use to broader application fields. Researchers have attempted to use Taguchi orthogonal methods to optimize the machining parameters such as the depth of the cut, the spindle speed, the tool feed rate, the tool nose radius, and the coolant in order to obtain a high surface quality [5–8]. Bhaskar Goel et al. reported that the prominent factors affecting surface roughness were spindle speed, tool overhang and the feed rate of cutting, while the key factors affecting profile and waviness errors were spindle speed, cut depth and tool nose radius [9]. K. Jagtap et al. studied the influence of cutting parameters on the surface quality of PMMA obtained via SPDT [4]. They found that spindle speed is a major factor affecting surface flatness in the SPDT of PMMA. However, in the abovementioned published literature studying the effects of machining parameters, all the experiments were executed at room temperature, and the achievable minimum PV value was only about 0.83 μm, which needs to be lower in order to meet application requirements of PMMA optical components.

To further improve the surface accuracy of PMMA parts produced using SPDT, in addition to the machining parameters, the material properties need to be adjusted. The crystalline structure, glass transition temperature and viscoelasticity can cause significant form errors in the machined parts [10]. Yasuhiro Kakinuma and collaborators applied cryogenic cooling to soft PDMS in the cutting process to machine microgrooves [11]. They successfully changed the material from a rubbery state to a glassy state below the glass transition temperature in order to obtain a remarkable change in elasticity. With this material property change, an accurate shape can be obtained easily. Similar methods have been used for other materials to keep the temperature below their glass transition temperature in order to increase their hardness for machining [12–14]. The cryogenic temperature ensures that the polymeric material is ductile and elongational and enables it to withstand cutting forces without deformation.

Cryogenic cooling has been proven to be one effective way to change the material properties of polymers to make them suitable for ultra-precision machining. By quenching them below the glass transition temperature, their hardness can be considerably improved to avoid flexible deformation during machining. This method has been demonstrated by many researchers for soft polymers like PDMS and EVA. However, it has not been used for the SPDT of PMMA. One possible reason for this is that the glass transition temperature of this polymer is over 100 °C, far beyond ambient temperature. The material is in a glassy state when machining, and thus does not need further cooling. So far, no reports are available about the material property changes of PMMA for SPDT and the corresponding effects on the surface quality of the final components.

To obtain high-precision PMMA optical components with a superior surface quality, this work aims at changing the material properties of PMMA via SPDT with cryogenic cooling and investigating the effects on the machinability and surface quality of the final part. If it succeeds, many high-precision, lightweight parts with a smooth surface can be manufactured using PMMA, which can be widely used in aerospace, aviation, etc. PMMA was maintained at a temperature 100 °C lower than the Tg during machining. The Taguchi method was used to change the spindle speed, feed rate and depth of the cut to search for the optimal conditions for machining a flat surface. In addition, nanoindentation tests were performed to check the hardness and Young's modulus of the PMMA at different temperatures. The purpose was to obtain the relationship between the temperature, material properties, processing conditions and surface quality in order to fabricate PMMA components with high precision.

2. Experimental

2.1. Materials

PMMA materials were supplied by Mitsubishi Chemical Holdings (Beijing, China) Co., Ltd. Workpieces were obtained by means of turning using a lathe to a size of Ø10 mm × 10 mm. Cryogenic samples for nanoindentation tests were obtained by soaking the workpieces in liquid nitrogen for 5, 10, 15 or 60 min, while the workpieces for SPDT were cooled by flushing liquid nitrogen from a nozzle during machining.

2.2. Temperature Variation Test

In the tests regarding the hardness and Young's modulus of the materials, as the PMMA samples could not be maintained at a very low temperature during the tests due to the restrictions of the instrument, it was necessary to monitor the temperature on a flat surface by checking the time. The temperature as a function of time was obtained using the following procedure. The samples were soaked in liquid nitrogen for a certain time, then taken out and exposed to the air. The temperature shift with time on the flat surface was measured using an AT4516 multichannel temperature meter (Applent Instrument, Edison, NJ, USA).

2.3. Nanoindentation Test

First, the workpieces were polished using Mecatech 234 (PRESI, Paris, France) to obtain a relatively flat and smooth surface for test accuracy. Then, they were soaked in liquid nitrogen for 5, 10, 15 and 60 min before being taken out for nanoindentation. Hysitron TI950 (Bruker, Billerica, MA, USA) was used to test the mechanical properties of PMMA in a micro view for the hardness and Young's modulus. The indenter had a Berkovich tip with 142.3° incl. angle and 100 nm R with a standard holder. To save time and reduce test cost, an indentation depth of 1 μm was selected to ensure elastic deformation by referring to the literature [15]. As the hardness and Young's modulus of PMMA are sensitive to strain rates, following one recent published paper, a strain rate of 0.2/s was used to obtain a stable value of the Young's modulus [16]. If the rate was lower than this value, the Young's modulus among the tests would have a large variation. Tests were completed for the samples with a surface temperature of 0, 5, 10, 15, 20 and 25 °C. At each temperature, 5 points at different locations on each sample were tested to obtain the average value. With this method, the mechanical properties of PMMA as a function of temperature could be obtained.

2.4. SPDT of Cryogenically Cooled PMMA

Nanoform 700 ultra (Precitech, Keene, NH, USA) was used to complete SPDT of the workpieces with dimensions of Ø10 mm × 10 mm on the flat surface. They were divided into two groups: one group machined at room temperature as the control samples, and the other group machined at a low temperature by cooling with liquid nitrogen. The workpieces were fixed on the spindle using an air chuck and were rotated at the speed of the spindle. Machining parameters including the spindle speed, depth of the cut and the feed rate could be set with the movement of the x-axis and z-axis of the machine. The diamond tool used had a 0.491 mm radius, a 0° rake angle, and a 10° clearance angle. The setup for the SPDT of PMMA is shown in Figure 1.

Figure 1. Setup for the SPDT of PMMA with cryogenic cooling.

2.5. Measurement and Characterization Tools

A Taylor Hobson device (Ametak, Leicester, UK) was utilized to measure the form profile and surface roughness of the machined flat surface with the contacting method. A specific holder was designed to fix the PMMA workpiece during the test. The measurement setup is shown in Figure 2.

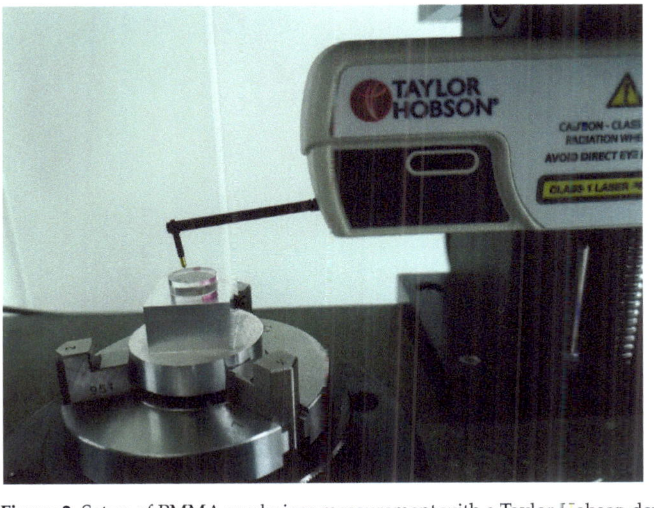

Figure 2. Setup of PMMA workpiece measurement with a Taylor Hobson device.

The viscoelasticity of PMMA was tested using DMA. A standard bar of PMMA with dimensions of 7.3 × 7.4 × 2.0 mm was tested following the frequency sweep and temperature sweep procedures using DMA 8000 (PerkinElmer, Waltham, MA, USA). With the frequency sweep as the first step, the frequency range for the linear viscoelastic region can be obtained. Then, at a fixed frequency of 1 Hz, a temperature sweep from 25 °C to −70 °C was completed. An oscillating force was applied and the resulting Young's modulus and tanδ could be calculated.

3. Results and Discussion

3.1. Temperature Change of Cryogenically Cooled PMMA in Air

No tool or setup is available to maintain the materials at a low temperature during the nanoindentation tests. Thus, the tests were generally completed at room temperature. The purpose of this experiment was to check the temperature shift of the material in air to determine the surface temperature at a specific time. Then, the material properties including the hardness and Young's modulus could be related to surface temperature by monitoring the time when the test was executed. Accordingly, the effect of material property change on the machinability of PMMA using SPDT could be investigated. This temperature shift curve is the basis for further studies on cryogenically cooled PMMA. The workpieces were soaked in liquid nitrogen for 5, 10, 15 or 60 min to test the temperature ramp rate in air. As observed in Figure 3, the workpieces could be cooled to −60 °C or even −70 °C. After they were exposed in air, the temperature on the surface increased to room temperature within around 12 min. Analysis of the data proves that the soaking time has a small effect on the temperature ramp rate. The samples with a soaking time of over 5 min show quite similar temperate shift curves, which can be expressed with the following fitted equation.

$$y = \mathrm{a} - \mathrm{bc}^x \tag{1}$$

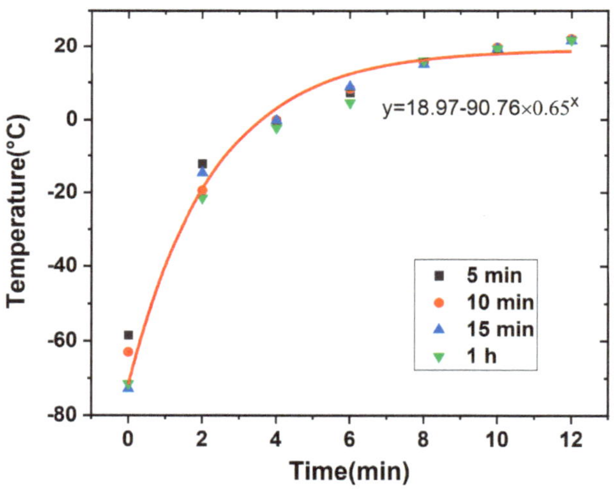

Figure 3. Temperature change of cryogenically cooled PMMA with time when exposed to air.

In Equation (1), y is the temperature on the workpiece surface and x is the time of exposure to air. a, b, and c are constants that are obtained by means of nonlinear fitting. In Figure 3, a fitting curve for 15 min is provided as an example. With this plot, the temperature in nanoindentation tests and other types of hardness tests for PMMA workpieces can be calculated by monitoring the time elapsed since leaving the liquid nitrogen bath.

3.2. Mechanical Property of Cryogenically Cooled PMMA

Examples of using the nanoindentation test to check the hardness and elastic modulus of PMMA and study the effects of loading rate, holding time, etc., can be found in the literature. However, all of the published experimental results were obtained at an ambient temperature. One possible reason for this phenomenon is that major challenges exist regarding testing at very low temperatures, including possible damage to the indentation tip and the whole instrument. Nevertheless, the mechanical properties at temperatures far below the Tg are very important for studying the structure, performance and processing of PMMA. Hence, efforts need to be made to investigate the hardness and elastic modulus of cryogenically cooled PMMA. The samples were soaked in liquid nitrogen for 15 min and then removed for testing. The tips were compressed into samples when the surface temperature was 0, 5, 10, 15, 20 and 25 °C by monitoring the time in the air. As the total process, including uploading, holding and unloading, was completed in about 15 s, the temperature was assumed to be constant during the process.

Typical load–depth curves of PMMA at 25 °C and 0 °C are shown in Figure 4. A maximum load of 6500 μN with a load rate of 1300 μN/s was applied to reach a depth of 1 μm to calculate the hardness and elastic modulus. At each temperature, indentation tests were repeated five times to obtain the average value of the target parameters. The indentation tests were performed five times on the exact same sample, including soaking in liquid nitrogen and measurement at a definite point. The soaking time and corresponding temperature were calculated based on the relationship obtained in Figure 3.

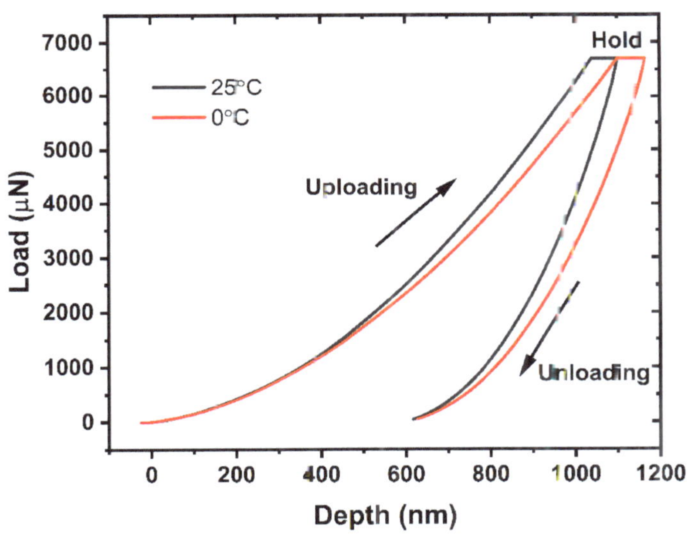

Figure 4. Representative load–depth curves of PMMA at 25 °C and 0 °C.

The average hardness and elastic modulus at the above temperatures are shown in Figures 5 and 6, respectively. As observed from the figures, the test data show excellent repeatability with small standard deviations. This is attributed to the homogenous microstructure and the maintenance of a stable state at temperatures far below the Tg. The nanoindentation hardness and elastic modulus of PMMA continuously increase when lowering the sample temperature by controlling the initial time of indentation. In particular, the hardness is significantly affected by temperature. It increases by 37% when the surface temperature is reduced from 25 °C to 0 °C. Meanwhile, the elastic modulus is increased by 22% when the temperature shifts from 25 °C to 0 °C.

As no tests for mechanical properties of cryogenically cooled PMMA are available in the literature, the hardness of the samples was retested with a micro hardness tester for authenticity. A micro hardness tester from Shanghai Shuming Optical Instrument (HXD-1000 TMC/LCD) (Shanghai, China) was used to check the HV values of the cryogenic samples. The HV values at the same temperature as tested in the nanoindentation tests are listed in Table 1. As obtained from Table 1, the HV value decreases with an increase in temperature. This is attributed to the material properties of PMMA, which is an amorphous polymer with viscoelasticity, as described in Section 3.4. Meanwhile, the hardness is positively related to the storage modulus. Accordingly, this is consistent with the result in Figure 12, showing that the ratio of the loss modulus to the storage modulus increases with temperature from 0 to 25 °C, which means that the hardness will show a decreasing trend following an increase in temperature. As observed in Figure 5, the unit of hardness obtained in the nanoindentation tests is GPa, while that obtained with the micro hardness tester is kgf/mm^2. The hardness decreases with increase in temperature, displaying the same trend as in the nanoindentation tests. In addition, the data show good repeatability with small errors, indicating uniform surface properties. After being converted to the same units, the values at the same temperature obtained with the two methods are different. One interesting phenomenon is that the difference is almost constant for the values at the five temperature points. The difference between the hardness value obtained with the micro hardness tester and that obtained from the nanoindentation tests was calculated with the same unit (GPa). The difference between the values at the five temperature points obtained with the two test methods was also calculated. For example, at 0 °C, the hardness obtained with the micro hardness tester was HV30.2 (0.29596 GPa), while that obtained with the nanoindentation tests is 0.38527 GPa, with the difference being 0.08931 GPa. The difference at the five temperature points was calculated and the average difference was 0.087965 GPa with a standard deviation error of 0.01030645 GPa. One possible reason for this is that a system error remains in the two test methods during measurement unit conversion. A representative indentation point of the PMMA produced by the micro hardness tester under a microscope is shown in Figure 7. The indented profile as observed in the figure is clear and large, which indicates that the hardness is low. We observed the trend that the harder the material is, the smaller the indented area. If the material is super rigid, the indented area will only cover one small area, like a point. The hardness value (HV) is also calculated based on the force applied and the cross-sectional area of the indented region. Hence, as observed from Figure 7, the PMMA material has a relatively low hardness. Comparing the data obtained using the two test methods, it can be confirmed that mechanical properties of PMMA are enhanced by lowering the temperature. Both the hardness and the elastic modulus experience a noticeable increase when the temperature is changed from 25 to 0 °C.

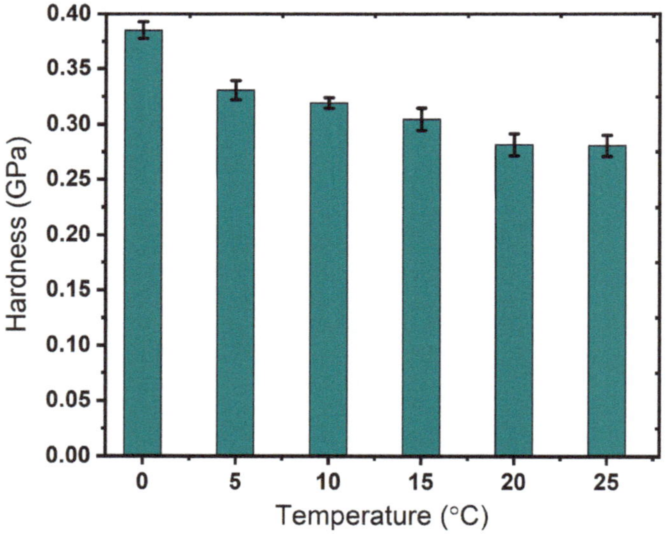

Figure 5. Nanoindentation hardness value of PMMA at different temperatures.

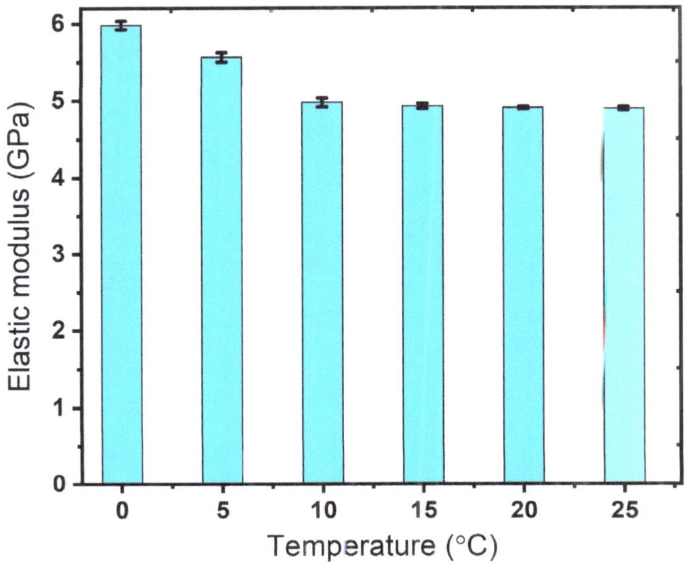

Figure 6. Nanoindentation elastic modulus value of PMMA at different temperatures.

Table 1. HV values of cryogenically cooled PMMA at different temperatures obtained using a micro hardness tester.

Temperature (°C)	Average HV (kgf/mm^2)	Standard Deviation
0	30.2	0.09
5	26.4	0.12
10	22.2	0.10
15	21.4	0.11
20	19.8	0.03
25	20.2	0.07

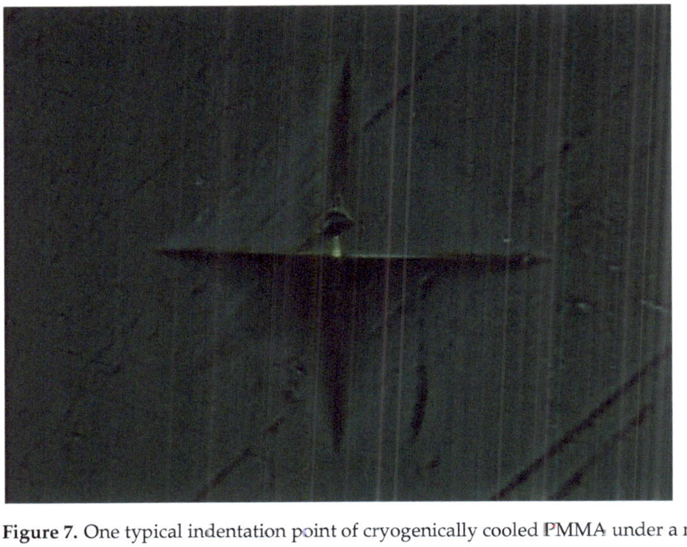

Figure 7. One typical indentation point of cryogenically cooled PMMA under a microscope.

3.3. Machinability and Surface Property of Cryogenically Cooled PMMA

PMMA is widely used in optical components in various fields requiring a high level of precision, and is commonly processed using SPDT. Thus, to meet the increasing demand for components with higher accuracy, it is important to study the processing methods of PMMA for higher precision and efficiency. In this work, liquid nitrogen cooling was applied to further investigate the machinability of cryogenically cooled PMMA using SPDT, which causes the temperature of the workpiece's surface to be 100 °C lower than the material's Tg. The setup for the SPDT of PMMA is displayed in Figure 1. A simple flat surface was machined in the SPDT experiments for the comparison of accuracy and efficiency, while a group of workpieces machined using SPDT at room temperature were used as references.

Machining parameters including the spindle speed, feed rate and depth of the cut were tested in groups following the Taguchi orthogonal method at the designed temperature to obtain a high surface quality. For the design of the experiment with the Taguchi method, according to the authors' pilot study in SPDT and the reported parameters in the literature, three input parameters including the spindle speed, feed rate, and cut depth were identified as factors and three levels of each factor were chosen for the surface quality study. The factors with the corresponding levels are listed in Table 2. The surface roughness (Ra, nm), and profile error (Pt, nm) were selected as the principal response parameters, which were measured via a Taylor Hobson device using the setup shown in Figure 2. To save time and reduce cost, only one temperature point, 0 °C, was used as a demo of cryogenically cooled PMMA for SPDT. This temperature was controlled by the adjusting flow rate of liquid nitrogen and was measured in real time using an infrared thermometer.

Table 2. Factors and levels in the SPDT of PMMA.

Factors	Levels		
	Level 1	**Level 2**	**Level 3**
Spindle speed (rpm)	1500	2000	2500
Feed rate (mm/min)	5	8	10
Cut depth (μm)	2	4	8

Optimized processing conditions, including spindle speed, feed rate and cut depth, were obtained for the specified temperature to obtain a high surface quality. The total height of the profile (Pt) and the arithmetic mean deviation (Ra) were used to evaluate form accuracy and surface roughness, respectively. For the workpieces machined at room temperature, a spindle speed of 1500 rpm, a feed rate of 5 mm/min, and a cut depth of 4 μm obtained the lowest values of Pt and Ra. The measured form accuracy and surface roughness are shown in Figures 8 and 9, respectively. As observed in the figures, a Pt value of 291 nm and an Ra of 11 nm were obtained. Following the same procedure, the optimal processing conditions for the SPDT of cryogenically cooled PMMA at 0 °C are a spindle speed of 2500 rpm, a feed rate of 10 mm/min and a cut depth of 2 μm. The form accuracy and surface roughness are shown in Figures 10 and 11, respectively. A Pt value of 124 nm and a surface roughness of 6 nm were obtained. Compared to the values at room temperature, Pt was decreased by 57.4%, and Ra was about 50% lower. It can be inferred that cryogenic machining is beneficial for improving the surface quality of PMMA. As the Tg of PMMA is around 105 °C, this work was completed at a temperature far below the Tg (>100 °C), which has not been achieved before. One reason for the initiation of this work was to investigate the effect of cryogenic cooling on machinability and surface quality for materials like PMMA with a high Tg. As observed from the results, the cryogenically cooled PMMA can be machined far below its Tg. With cryogenic cooling, a higher surface quality with low values of Pt and Ra can be obtained. Meanwhile, it is found that a higher spindle speed, a higher feed rate, and a smaller cut depth are needed to obtain a better surface quality at low temperatures.

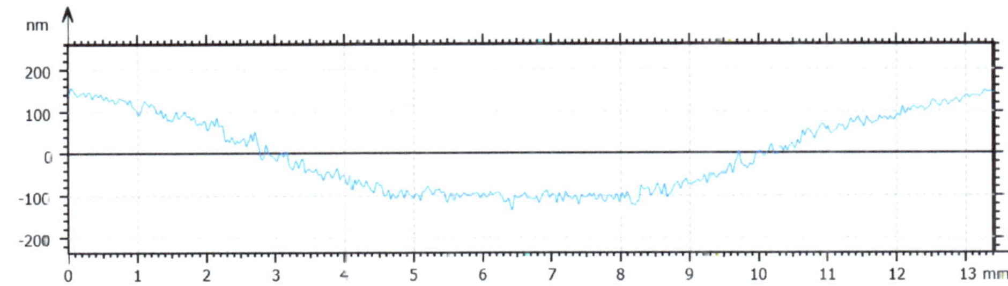

Figure 8. Surface profile of the PMMA workpiece machined via SPDT at room temperature under optimal processing conditions.

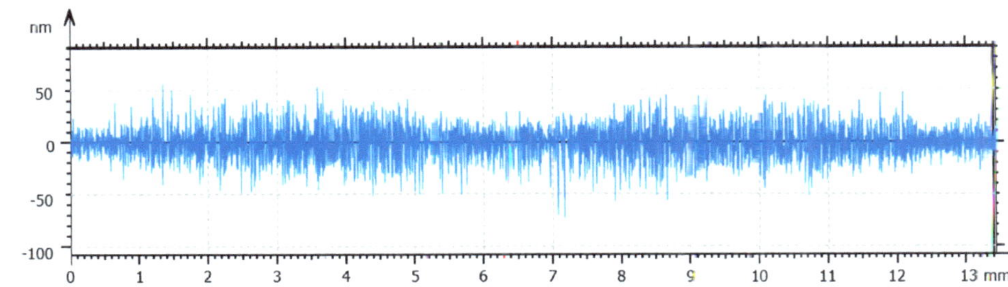

Figure 9. Surface roughness of the PMMA workpiece machined via SPDT at room temperature under optimal processing conditions.

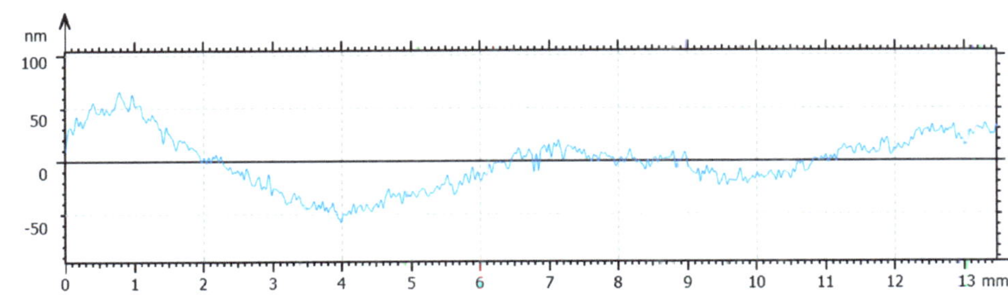

Figure 10. Surface profile of the PMMA workpiece machined via SPDT at 0 °C under optimal conditions.

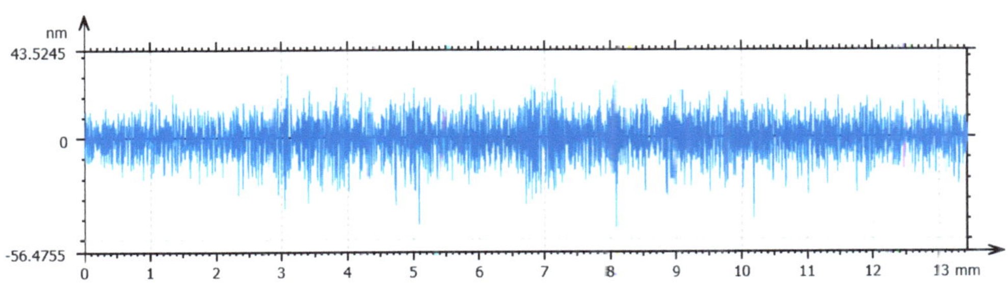

Figure 11. Surface roughness of the PMMA workpiece machined via SPDT at 0 °C under optimal conditions.

3.4. Relationship between Temperature, Material Property and Machinability

To understand the mechanism of the improved surface quality of cryogenically cooled PMMA using SPDT, the relationship between the material properties and the machined surface quality is investigated, with temperature as the connection. At low temperatures, the hardness and Young's modulus are significantly increased compared to those at room temperature. Accordingly, a better surface quality with lower values of Pt and Ra can be obtained. There is an inherent connection between them. PMMA, as an amorphous polymer, has typical viscoelastic properties, considerably affecting its mechanical properties and machinability. Thus, its viscoelasticity was studied using DMA. Tanδ is the ratio of the loss modulus to the storage modulus, and its change with temperature is shown in Figure 12. One peak is available around 20 °C, which has a high possibility of representing β relaxation attributed to the side group motion of (—COO—CH_3) [17]. With temperature decreasing from the peak point, γ relaxation mainly occurs with the rotation of the side group (—CH_3), possibly requiring some local main chain motion [18]. These are quite different from α relaxation, which describes large-scale cooperative rearrangements of the molecules in the region of glass transition. The glassy state relaxation behavior (α, β, and γ relaxation process) is closely related to the mechanical properties of polymers [19,20]. In this work, cooling from room temperature shows a typical suppression of the sub-Tg relaxation process. With a temperature decrease, the number of segments participating in the β relaxation (i.e., the number of segments dissipating mechanical energy) reduces and the transition into γ relaxation will be reflected in macroscopic properties such as the hardness and Young's modulus, as demonstrated in the nanoindentation tests.

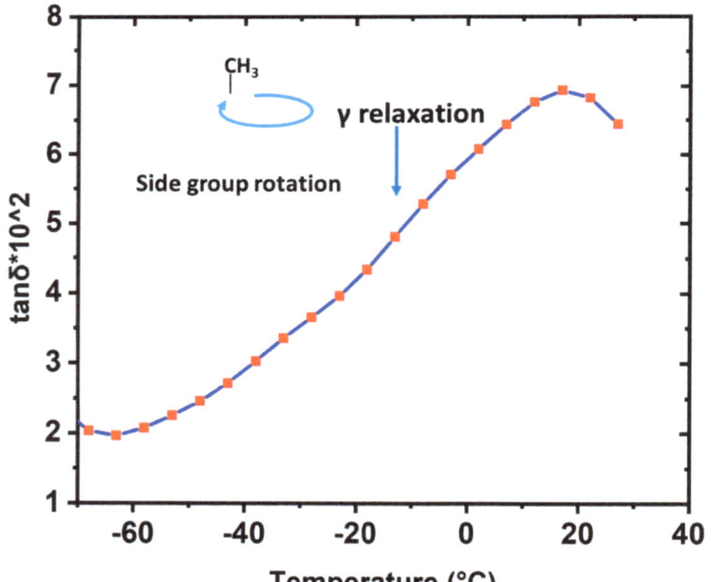

Figure 12. tanδ of PMMA measured at a fixed frequency of 1 Hz from 25 °C to −70 °C using DMA8000.

The temperature is 100 °C lower than the Tg, which is sufficient to obtain a non-equilibrium glassy state and maintain the orientation of the chain segments during machining. The α and β relaxation, which can be explained with free volume theory [21], describes the molecular change of the polymer, which can provide guidance for SPDT. At temperatures far below the Tg (Tg-100), no significant change in defect concentration occurs according to the quasi-point defect model. However, the molecular mobility of

PMMA is significantly decreased, as verified by the low-temperature part of the γ relaxation. The loss of chain mobility is mainly caused by local molecular rearrangements [22]. This reduction in chain mobility increases the stiffness of the workpiece, thus avoiding detrimental deformation.

Cryogenic cooling far below the Tg is beneficial for the improvement of mechanical properties, and its effect on the machinability of PMMA is also analyzed through surface profile modelling in SPDT. In diamond turning, the surface profile is commonly modelled per feed. As reported in previous studies, the profile is mainly affected by three components: the duplication effect of the diamond tool edge profile, material spring back and material plastic side flow [23,24]. Accordingly, the vertical distance between the highest peak and the lowest valley, i.e., F(x), can be expressed as

$$\mathrm{F}(x) = R_{tew}(x) + \frac{4(w_r - s_r)}{\mathrm{f}^2}\left(-\frac{\mathrm{f}}{2} \le x \le \frac{\mathrm{f}}{2}\right) \quad (2)$$

where $Rtew(x)$ is the active tool edge profile with consideration of tool edge waviness rate; w_r is the value of material plastic side flow in the center ($x = 0$), and s_r is the material spring back at the margin positions ($x = \pm f/2$) in one rate [25].

The surface profiles in this work can be analyzed using this empirical model. In Equation (1), $Rtew(x)$ is determined by the tool corner nose radius and edge waviness [26]. During the SPDT of PMMA in this work, the same tool was used. Therefore, this factor can be considered the same when comparing the profiles. The material plastic flow w_r is relatively complex and has insufficient calculation methods to obtain an analytical solution [25]. Nevertheless, for PMMA, it is mainly affected by the material's viscoelasticity. There exists a positive correlation between the plastic flow and the viscous behavior. For a qualitative comparison, according to the tanδ values at 0 °C and 25 °C in Figure 12, the loss modulus plays a more dominant role when temperature is higher, which represents viscous performance. This indicates that w_r will be larger at 25 °C. Meanwhile, material spring back s_r can be expressed as a proportional function of the tool edge radius and the ratio of a workpiece's material hardness to its Young's modulus (*H/E*) [27,28]. In this work, as proven by the available test data, the value of *H/E* at 0 °C is 1.12 times greater than that at 25 °C. Thus, considering the same diamond tool and value of *H/E*, the material spring back value s_r at 0 °C is larger. A comparison of the three factors at the two temperatures is summarized in Table 3. Substituting the comparison results into Equation (2), we can obtain a qualitative result indicating that the value of the surface profile at 0 °C is obviously smaller. This is consistent with the measurement results reflected by the Pt and Ra values.

Table 3. Comparison of three factors of the surface profile model at 25 °C and 0 °C.

Factors	0 °C	25 °C
Rtew(x)	=	
w_r		>
s_r	>	

A major effect of cryogenic cooling on PMMA is the control of its viscoelasticity. Lowering the temperature can suppress the viscous behavior of the material so that the relaxation time is longer, which makes the material more rigid with a higher hardness and Young's modulus. Meanwhile, the suppression of viscous behavior is useful for decreasing the material side plastic flow during SPDT, while the material spring back is enhanced as a proportional function of the ratio of hardness to the Young's modulus. Considering the above two factors, the machinability of PMMA using SPDT is improved by cryogenic cooling and a better surface quality can be achieved.

Cryogenic cooling has been widely attempted in machining materials including ceramics, polymers, metals, and alloys [29–32]. Nevertheless, no reports are available for the cryogenic cooling of PMMA during SPDT. In the previous reports, for example, the effect of cryogenic cooling on carbide–cobalt alloys mainly led to an increase in hardness compared to that at room temperature, while maintaining toughness, transverse rupture and impact strength, which can explain why the carbide tool materials are suitable for the cryogenic cooling method. Meanwhile, for some high-speed steels, cryogenic cooling does not affect hardness obviously, which means that cryogenic cooling will not affect their machining properties [32]. For PMMA, as tested in this work, cryogenic cooling is effective in increasing its hardness, even though its Tg is over 100 °C. The increase in hardness is effective in improving surface quality and workpiece precision during ultra-precision machining with optimized processing parameters. Accordingly, this work demonstrates that cryogenic cooling is effective for polymeric materials with a high Tg during ultra-precision machining, which opens a new avenue for manufacturing high-precision polymeric components. It contradicts the conventional wisdom that polymeric materials are soft and difficult to use as high-precision components compared to metal and alloys. The results in this work prove that it is feasible to fabricate high-precision polymeric components via cryogenic cooling.

4. Outlook

In this work, the effect of cryogenic cooling on the SPDT of PMMA is mainly studied from the view of material properties. During machining, the effects on tool wear, residual stress distribution, and chip formation can all contribute to the surface quality of the final workpiece. These effects will be studied in future work. Even though cryogenic cooling is helpful in assisting ultra-precision machining, some problems still need to be solved. One issue with cryogenic cooling is that the super-cold medium needs to be continuously supplied. The pipeline design and corresponding fixtures increase the cost of the final parts and consume time. In addition, the position of the output nozzle and volume flux have notable impacts on the machined parts. The above challenges took the authors plenty of time to address in this work. Not surprisingly, cryogenic cooling is generally used in research, but is not yet suitable for large-scale production. Further efforts need to be made before the cryogenic cooling process can be industrialized.

5. Conclusions

In this work, the machinability and surface properties of cryogenically cooled PMMA machined via SPDT were investigated. The cryogenic cooling was completed using liquid nitrogen. The optimal parameters for machining at room temperature and 0 °C were obtained using the Taguchi method. The form accuracy Pt and surface roughness Ra of the workpieces under optimal conditions were measured and compared. At room temperature, a Pt value of 291 nm and an Ra of 11 nm were obtained, while at 0 °C, a Pt value of 124 nm and an Ra of 6 nm were observed, displaying a significant improvement compared to the values at an ambient temperature. For the workpieces machined at room temperature, a spindle speed of 1500 rpm, a feed rate of 5 mm/min, and a cut depth of 4μm can be used to obtain lowest values of Pt and Ra. The optimal processing conditions for the SPDT of cryogenically cooled PMMA at 0 °C are a spindle speed of 2500 rpm, a feed rate of 10 mm/min and a cut depth of 2 μm. To reveal the mechanism of this process, the relationship between the temperature, material properties and machinability of PMMA was analyzed. The material property changes with temperature were studied via nanoindentation tests. As demonstrated by the experimental results, the hardness and Young's modulus at 0 °C increased by 37% and 22% compared to those at room temperature, respectively. The change in mechanical properties is mainly attributed to the viscoelasticity's variation with temperature. When temperature changed from 25 °C to 0 °C, the bulk material showed more elastic behavior with a decrease in tanδ, as verified by DMA tests. In addition, a model of the surface profile in SPDT was used to qualitatively explain the smoother surfaces at lower temperatures due to change in material properties. Decreasing the temperature can suppress sub-Tg relaxation, causing an enhancement of mechanical properties, which results in a better surface quality when using SPDT. It can be concluded that cryogenic cooling is effective in the ultra-precision machining of PMMA due to the material property changes caused by temperature. In future work, cryogenic machining can be attempted for materials like PMMA with a high Tg to achieve a high surface quality.

Author Contributions: Conceptualization, X.F., X.W. and Q.K.; Data curation, X.W. and X.J.; Formal analysis, X.W. and Q.K.; Funding acquisition, X.F.; Investigation, X.W. and Q.K.; Methodology, X.F.; Resources, X.F.; Software, Q.K. and X.J.; Supervision, X.F.; Validation, X.W. and Q.K.; Writing—original draft, X.W.; Writing—review and editing, X.F. All authors have read and agreed to the published version of the manuscript.

Funding: This research was funded by National Natural Science Foundation of China, grant number 52175517 & 51720105016; China Postdoctoral Science Foundation, grant number 2017M610634; Natural Science Foundation of Shaanxi Province, grant number 2018JQ5195; Shaanxi Province Postdoctoral Science Foundation, grant number 2017BSHEDZZ73; and Special scientific research plan project of Shaanxi Provincial Department of Education, grant number 13JS045.

Institutional Review Board Statement: Not applicable.

Informed Consent Statement: Not applicable.

Data Availability Statement: Data are contained within the article.

Conflicts of Interest: The authors declare no conflicts of interests regarding the publication of this article.

References

1. Gaidys, R.; Dambon, O.; Ostasevicius, V.; Dicke, C.; Narijauskaite, B. Ultrasonic tooling system design and development for single point diamond turning (SPDT) of ferrous metals. *Int. J. Adv. Manuf. Technol.* **2017**, *93*, 2841–2854. [CrossRef]
2. Chen, M.J.; Li, M.Q.; Cheng, J.; Xiao, Y.; Pang, Q.L. Study on the optical performance and characterization method of texture on KH_2PO_4 surface processed by single point diamond turning. *Appl. Surf. Sci.* **2013**, *279*, 233–244. [CrossRef]
3. Kobayashi, A. Ultra-Precision Machining of Plastics. In Proceedings of the SPIE 28th Annual Technical Symposium-Production Aspects of Single Point Machined Optics, San Diego, CA, USA, 21 August 1984. [CrossRef]
4. Jagtap, K.; Pawade, R. Experimental investigation on the influence of cutting parameters on surface quality obtained in SPDT of PMMA. *Int. J. Adv. Des. Manuf. Technol.* **2014**, *7*, 53–58.
5. Wang, H.; To, S.; Chan, C.Y. Investigation on the influence of tool-tip vibration on surface roughness and its representative measurement in ultra-precision diamond turning. *Int. J. Mach. Tools Manuf.* **2013**, *69*, 20–29. [CrossRef]
6. Liman, M.M.; Abou-El-Hossein, K.; Odedeyi, P.B. Modeling and Prediction of Surface Roughness in Ultra-High Precision Diamond Turning of Contact Lens Polymer Using RSM and ANN Methods. *Mater Sci. Forum.* **2018**, *928*, 139–143. [CrossRef]
7. Sakata, S.; Hayashi, A.; Terajima, T.; Nakao, Y. Influence of cutting condition on surface roughness in single point diamond turning of zr-based bulk metallic glass. In Proceedings of the Asme International Mechanical Engineering Congress and Exposition Phoenix, AZ, USA, 11–17 November 2016. [CrossRef]
8. Wang, H.; Zhang, Y.; Xie, Q. The Cutting Parameter Affecting to Surface Roughness in Single-point Diamond Turning. *Adv. Mater. Mater. Res.* **2014**, *887–888*, 1236–1239. [CrossRef]
9. Goel, B.; Singh, S.; Sarepaka, R.G.V. Precision Deterministic Machining of Polymethyl Methacrylate by Single-Point Diamond Turning. *Mater. Manuf. Process.* **2016**, *31*, 1917–1926. [CrossRef]
10. LEE, W.B.; Cheung, C.F. A dynamic surface topography model for the prediction of nano-surface generation in ultra-precision machining. *Int. J. Mech. Sci.* **2001**, *43*, 961–991. [CrossRef]
11. Kakinuma, Y.; Yasuda, N.; Aoyama, T. Micromachining of Soft Polymer Material applying Cryogenic Cooling. *J. Adv. Mech. Des. Syst. Manuf.* **2008**, *2*, 560–569. [CrossRef]
12. Dhokia, V.G.; Newman, S.T.; Crabtree, P.; Ansell, M.P. A methodology for the determination of foamed polymer contraction rates as a result of cryogenic CNC machining. *Robot. Comput. Integr. Manuf.* **2010**, *26*, 665–670. [CrossRef]
13. Kakinuma, Y.; Kidani, S.; Aoyama, T. Ultra-precision cryogenic machining of viscoelastic polymers. *CIRP Ann. Manuf. Technol.* **2012**, *61*, 79–82. [CrossRef]
14. Mishima, K.; Kakinuma, Y.; Aoyama, T. Pre-Deformation-Assisted Cryogenic Micromachining for Fabrication of Three-dimensional Unique Micro Channels. *J. Adv. Mech. Des. Syst. Manuf.* **2010**, *4*, 936–947. [CrossRef]
15. Lin, H.; Jin, T.; Lv, L.; Ai, Q.L. Indentation Size Effect in Pressure-Sensitive Polymer Based on A Criterion for Description of Yield Differential Effects and Shear Transformation-Mediated Plasticity. *Polymers* **2019**, *11*, 412. [CrossRef]
16. Jin, T.; Zhuo, Z.W.; Liu, Z.G.; Xiao, G.S.; Yuan, G.Z.; Shu, X.F. Sensitivity of PMMA nanoindentation measurements to strain rate. *J. Appl. Polym. Sci.* **2015**, *132*, 41896. [CrossRef]

17. Wittmann, J.C.; Kovacs, A.J. Influence de la Stereorégularite des Chaines sur les Transitions du Polyméthacrylate de Méthyle. *J. Polym. Sci. Part C Polym. Symp.* **1967**, *16*, 4443–4452. [CrossRef]
18. Gourari, A.; Bendaoud, M.; Lacabanne, C.; Boyer, R.F. Influence of tacticity on T_β, T_g, and T_{LL} in poly(methyl methacrylate)s by the method of thermally stimulated current (TSC). *J. Polym. Sci. Polym. Phys. Ed.* **1985**, *23*, 889–916. [CrossRef]
19. Mininni, R.M.; Moore, R.S.; Flick, J.R.; Petrie, S.E.B. The effect of excess volume on molecular mobility and on the mode of failure of glassy poly(ethylene terephthalate). *J. Macromol. Sci. Part B* **1973**, *8*, 343–359. [CrossRef]
20. Fukuhara, M.; Sampei, A. Low-temperature elastic moduli and internal dilational and shear friction of polymethyl methacrylate. *J. Polym. Sci. Part B Polym. Phys.* **1995**, *33*, 1847–1850. [CrossRef]
21. Yianakopoulos, G.; Vanderschueren, J.; Niezette, J.; Thielen, A. Influence of physical aging processes on electrical properties of amorphous polymers. *IEEE Trans. Electr. Insul.* **1990**, *25*, 693–701. [CrossRef]
22. Muzeau, E.; Vigier, G.; Vassoille, R.; Perez, J. Changes of thermodynamic and dynamic mechanical properties of poly(methyl methacrylate) due to structural relaxation: Low-temperature ageing and modelling. *Polymer* **1995**, *36*, 611–620. [CrossRef]
23. Gadelmawla, E.S.; Koura, M.M.; Maksoud, T.M.A.; Elewa, I.M.; Soliman, H.H. Roughness parameters. *J. Mater. Process. Technol.* **2002**, *123*, 133–145. [CrossRef]
24. Zong, W.J.; Huang, Y.H.; Zhang, Y.L.; Sun, T. Conservation law of surface roughness in single point diamond turning. *Int. J. Mach. Tools Manuf.* **2014**, *84*, 58–63. [CrossRef]
25. He, C.L.; Zong, W.J.; Xue, C.X.; Sun, T. An accurate 3D surface topography model for single-point diamond turning. *Int. J. Mach. Tools Manuf.* **2018**, *134*, 42–68. [CrossRef]
26. He, C.L.; Zong, W.J. Influencing factors and theoretical models for the surface topography in diamond turning process: A review. *Micromachines* **2019**, *10*, 288. [CrossRef]
27. Arcona, C. Tool Force, Chip Formation and Surface Finish in Diamond Turning. Ph.D. Thesis, North Carolina State University, Raleigh, NC, USA, 1996.
28. Arcona, C.; Dow, T.A. An Empirical Tool Force Model for Precision Machining. *J. Manuf. Sci. Eng.* **1998**, *120*, 700–707. [CrossRef]
29. Wang, Z.Y.; Rajurkar, K.P. Cryogenic machining of hard-to-cut materials. *Wear* **2000**, *239*, 168–175. [CrossRef]
30. Shokrani, A.; Dhokia, V.; Muñoz-Escalona, P.; Newman, S.T. State-of-the-art cryogenic machining and processing. *Int. J. Comput. Integr. Manuf.* **2013**, *26*, 616–648. [CrossRef]
31. Khanna, N.; Agrawal, C.; Pimenov, D.Y.; Singla, A.K.; Machado, A.R.; da Silva, L.R.R.; Gupta, M.K.; Sarikaya, M.; Krolczyk, G.M. Review on design and development of cryogenic machining setups for heat resistant alloys and composites. *J. Manuf. Process.* **2021**, *68*, 398–422. [CrossRef]
32. Yildiz, Y.; Nalbant, M. A review of cryogenic cooling in machining processes. *Int. J. Mach. Tools Manuf.* **2008**, *48*, 947–964. [CrossRef]

materials

MDPI

Article

Functional Technical Textile-Based Polymer Nanocomposites with Adsorbent Properties of Toxins and Dyes also Have Antibacterial Behavior

Marlene Andrade-Guel [1,*], Christian J. Cabello-Alvarado [1,2,*], Carlos Alberto Ávila Orta [1], Gregorio Cadenas-Pliego [1] and Brenda Cruz-Ortiz [3]

1 Centro de Investigación en Química Aplicada, Saltillo 25294, Coahuila, Mexico; carlos.avila@ciqa.edu.mx (C.A.Á.O.); gregorio.cadenas@ciqa.edu.mx (G.C.-P.)
2 México CONAHCYT-CIQA, Av. Insurgentes Sur 1562, Col. Credito Constructor, Alcaldía Benito Juárez, CDMX 03940, Mexico
3 Facultad de Ciencias Químicas, Universidad Autónoma de Coahuila, Saltillo 25280, Coahuila, Mexico; b.cruz@uadec.edu.mx
* Correspondence: marlene.andrade@ciqa.edu.mx (M.A.-G.); christian.cabello@ciqa.edu.mx (C.J.C.-A.)

Abstract: This is the first study of non-woven fabrics elaborated by melt-blowing from polymer nanocomposites made of Nylon 6 and nanoclay (Cloisite 20A) modified with an amine (1,4 diaminobutane dihydrochloride). Morphological and physical characteristics, adsorption capacity, and antibacterial properties are presented. From the X-ray diffraction (XRD) results, it was possible to observe a displacement of the signals to other 2θ angles, due to an α to ϒ phase shift. The scanning electron microscopy (SEM) images showed that the mean diameter of fiber decreased as the content of nanoclay increased. The mechanical tests showed that the tear strength force of neat nylon was 1.734 N, but this characteristic increased to 2.135 N for the sample with 0.5% modified nanoclay. The inulin adsorption efficiency of the Nylon 6/C20A 1.5% and Nylon 6/C20A 2% samples at 15 min was 75 and 74%, respectively. The adsorption capacity of Nylon 6/C20A 1.5% and Nylon 6/C20A 2% for methylene blue and methyl orange remained above 90% even after four adsorption cycles. In addition, non-woven fabrics present antibacterial activity against *E. coli*.

Keywords: Nylon 6; non-woven fabric; adsorption; toxins; dyes; *E. coli*

1. Introduction

Functional textiles are adapted to address the different necessities of human beings. Functional textiles showed a growth rate of 30% from 2015 to 2020 due to support and investment from different industries, which seek new technological developments in this area [1]. Textiles made from fibers with micrometric or nanometric sizes are attractive in different application areas. Three critical areas are energy generation, medicine, and environmental care [2].

On the other hand, adsorbent nanomaterials can have different characteristics and be incorporated into a polymer matrix to make composite materials and apply them in technical textiles.

Some adsorption studies with carbon-based materials from potato waste have had good results in the adsorption of dyes such as allura red, carmine, and tartrazine; these materials present different behaviors of monolayer and multilayer adsorption mechanisms [3]. It has also been seen that other materials designed and manufactured with nanometric dimensions (MXene nano sheets) effectively eliminate cationic and anionic azo dyes (methylene blue and Congo red) through adsorption processes [4,5]. These nanometric adsorbent materials can be incorporated into a polymeric matrix and form fibers for application as an adsorbent at an industrial level.

Moreover, functional textiles applied in medicine are promising products that help drug administration; specifically, they can improve the dermal penetration capacity of the active molecule and avoid the risk of toxicity. Another application is in housing medical devices for hemodialysis [6,7]. Koh et al. reported the fabrication of nanofibers for hemodialysis consisting of three layers of nanofibers: the top layer of poly(methyl methacrylate)-graft-poly(dimethylsiloxane) (PMMA-g-PDMS) nanofiber, the second layer of polyamide 6 (PA6), and the third one containing nanofibers of polyamide/zeolite for creatinine removal [8]. In the hemodialysis process, it is necessary to remove some nitrogenous molecules in the blood; one is urea because it causes kidney problems and affects the liver [9]. Abidin et al. prepared hollow fiber membranes to combine filtration and an adsorption process, removing 39.2% of urea with a double-layer fiber of polysulfone and polymethyl methacrylate (N-PMMA) [10].

Recent studies on non-woven fabrics incorporating nanoparticles like graphene nanoplatelets and carbon black have achieved 90% and 80% urea adsorption, respectively [11,12]. Some functional textiles have been used as filtration systems for separation and cleaning in the environmental care area; their production has increased due to the increasing environmental demands to remove contaminants from either air or water [13,14]. Various types of contaminants (organic and inorganic) are released into the water. Among these contaminants, dyes stand out because they are toxic. Dyes such as methylene blue and methyl orange are persistent pollutants in aquifers because conventional water treatment methods cannot remove them. Therefore, new methods for removing these contaminants are needed [15,16].

Some textile studies have reported the removal of dyes of different natures. Dao et al. studied non-woven polyester fabrics as a support substrate with rGO (reduced graphene oxide) and Cu2 particles that form a photocatalytic membrane for the degradation of methylene blue, under the influence of natural sunlight. The manufacturing process of this compound was carried out in four stages, achieving the elimination of 96% of methylene blue in 120 min [17]. Rana et al. studied the incorporation of glycidyl methacrylate into polyethylene non-woven fabric, and carried out a chemical modification of it and applied it in the adsorption of methylene blue, achieving the elimination of 52% in the first 5 h [18].

Andrade et al. reported on non-woven fabric based on Nylon 6 and ZnO. To remove methylene blue and reduce antimicrobial performance, one of the experiments involved accelerated washing, which consists of observing the effects on antimicrobial properties [19].

It has been seen that another advantage of using nanoparticles in textiles as antibacterial agents is that it reduces the side effects of contact with the skin when compared to agents that are currently used commercially.

On the other hand, in recent years, our research group has used the ultrasonic-assisted extrusion method for incorporating nanoparticles into a non-woven fabric. In this methodology, a nanocomposite containing a polymer matrix and the nanoparticle as a reinforcement is first prepared. Then, the non-woven fabric using the melt-blowing method is manufactured. This technique helps the dispersion of the nanoparticles and does not allow the nanoparticles to detach with washing. Our research group studied the non-woven fabric of Nylon 6 with different nanoparticles such as ZnO, carbon black, and graphene.

The novelty is that this is the first study of non-woven fabrics elaborated by melt-blowing from polymer nanocomposites made of Nylon 6 and nanoclay (Cloisite 20A) modified with an amine (1,4 diaminobutane dihydrochloride) to evaluate its effect on the physicochemical properties and organic compound adsorption. Finally, this study evaluates the amine's antibacterial properties.

2. Materials and Methods

2.1. Reagents

Nylon 6 from DuPont Zytel®7301 NC010, organoclay Cloisite C20A with a particle size < 10 μm and lamellar spacing of 2.7 nm. 1,4-diamino butane dihydrochloride purity of 99% was purchased from Sigma Aldrich products (St. Louis, MO, USA). Methylene blue (MB) (99%), methyl orange (MO) (99%), and inulin (98%) were purchased from Sigma Aldrich, urea (95%) was purchased from Faga Lab (Sinaloa, Mexico), and distilled water with a pH of 7 was used as a solvent to obtain the solutions.

2.2. Chemical Modification of Cloisite 20A by Ultrasonic Tip

Cloisite 20A modification with 1,4-diamino butane dihydrochloride was made following methodologies already reported by Andrade et al. [20]. The chemical modification treatment was carried out by dispersing 1 g of C20A nanoclay in 20 mL of distilled water, with 1,4-diaminobutane dihydrochloride, in a 1:1 ratio, using a homemade ultrasonic generator, with an output power of 750 W, at an amplitude of 50% and a variable frequency of 15 to 50 KHz, catenoidal ultrasonic tip (Branson Ultrasonics Corp., Brookfield, CT, USA; D, 51.27 cm). For safety reasons, the experiments were carried out in a soundproof cage. Ultrasound treatment time was 120 min and was used at room temperature. At the end of the experiments, the C20A clay was filtered and dried at 80 °C for 24 h.

2.3. Nanocomposite Preparation

The nanocomposites were prepared according to the method previously reported [21]. Polymer nanocomposite preparation was carried out using the ultrasound-assisted melt extrusion process (US) to homogenize the mixture of the particles within the polymer. For the extrusion process, a lab-size twin-screw extruder from Thermo Fisher Scientific (model, Prism TSE-24MC) with a screw diameter of 24 mm and L/D ratio of 40:1 was assisted by a catenoidal ultrasonic tip (Branson Ultrasonics Corp., CT; D, 51.27 cm). The extruder was connected to a homemade ultrasonic generator (15 to 50 kHz, 100% of 750 W). The temperature profile was a flat one, at 215 °C in all areas of the extruder, with a screw speed of 115 rpm. Nylon 6/C20A nanocomposites were prepared at different concentrations (0.25% wt, 0.5% wt, 0.75% wt, 1.5% wt, and 2% wt). These experimental conditions help increase the homogeneous size of the pellet. As a post-extrusion system, a cooling bath was used at the outlet of the die and a pelletizer (Thermo Fisher Scientific, Waltham, MA, USA).

2.4. Preparation of Non-Woven Fabric Materials

Nylon 6/C20A non-woven fabrics were elaborated by fiber extrusion technology (FET-UK) pilot machine equipment. The profile temperatures were 240 °C extrusion zone 1; 245 °C extrusion zone 2; 245 °C extrusion zone 3; 250 °C extrusion zone 4; 255 °C flange zone; 255 °C melt pump zone; 255 °C dual heat zone; 255 °C melt blow adapter zone; and 255 °C melt blow hot air zone. Figure 1 shows the identification of samples, for example, (Nylon 6/C20A 0.25%) polymer = Nylon 6; C20A = Cloisite 20A modified with 1,4-diamino butane dihydrochloride; 0.25% = concentration of the modified nanoclays.

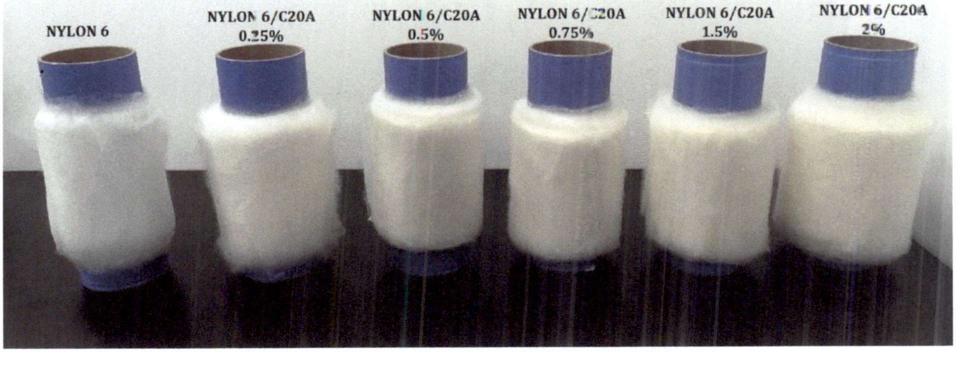

Figure 1. Identification of non-woven fabric samples.

2.5. Characterization

The morphology and elemental composition of the materials were characterized by scanning electron microscopy and energy-dispersive spectroscopy (SEM-EDS) through an electron microscope model JEOL model JSM-7401F (Jeol LTD., Akishima, Tokyo, Japan). The sample was coated with gold and palladium before analysis.

XRD (X-ray diffraction) was performed by a Rigaku model Ultima IV Smartlab (Rigaku Beijing, China) diffractometer operating at 40 kV and 40 mA with stability of 0.01%/8 h with a scanning interval in the 2θ scale from 10 to 80°.

The mechanical properties were measured following ASTM D2261 Standard norm (2013) standard test methods for tear strength properties of fabrics woven from stretch yarns. ASTM International standard tests were used to evaluate the tearing resistance of the non-woven fabric by the tongue (single rip) procedure. Tensile strength was carried out following ASTM D5034 Standard. These tests considered the machine direction (MD) and transversal direction (TD).

The equipment used for both the tear resistance and the resistance to breakage and elongation was a United Universal machine with an environmental chamber CEF-80, tensiometer model SFM-100KN from United Testing.

2.5.1. Adsorption of Uremic Toxins

The preparation of urea and inulin solutions and their measurement and calculations followed the method described in previous works [12]. First, Nylon 6 and Nylon 6/C20A, 30 cm × 30 cm non-woven fabric specimens were placed in a prototype (see Figure 6c; the prototype is made of glass and is the actual size of a hemodialysis dialyzer) with the solutions of uremic toxins. Then, the solution was prepared, which consisted of a mixture of urea and inulin at a concentration of 390 mg/dL. The simulation of the hemodialysis process lasted 4 h, and every 15 min, an aliquot was taken to analyze in the UV-Vis spectrometer.

All experiments were performed in triplicate. The removal percentage of each toxin was calculated with the following equation.

$$\%\,Removal = \frac{(Ci - Ce)}{Ci} \times 100 \tag{1}$$

where Ci is the initial concentration and Ce is the final concentration.

For the analysis of uremic toxin adsorption, the samples were analyzed in a UV-Vis spectrometer (Duetta Horiba Scientific, Beijing, China) at different wavelengths of 200 nm (urea) and 270 nm (inulin). The adsorption isotherms data were fitted and the correlation coefficient (R^2) was calculated using the trendline command in Microsoft Excel 2016 for each isotherm. The Langmuir isotherm was calculated using the following equation.

$$\frac{Ce}{q_e} = \frac{Ce}{q_m} + \frac{1}{K_{L\mathrm{qm}}} \tag{2}$$

where q_e ($mg \cdot g^{-1}$) and Ce ($mg \cdot L^{-1}$) are the concentrations of the solid and liquid phases of adsorbate in equilibrium, respectively, q_m is the maximum adsorption capacity, and K_L is the constant obtained from the graph of Ce/q_e against Ce.

The Freundlich isotherm was calculated using the following equation.

$$\ln qe = \ln K_F + \left(\frac{1}{n}\right) \ln Ce \tag{3}$$

where K_F ($mg \cdot g^{-1}$) ($L \cdot mg^{-1}$) and 1/n are the Freundlich constants related to the adsorption capacity and n is the heterogeneity calculated by linearly plotting Inqe against InCe [22].

2.5.2. Adsorption of Dyes (Methylene Blue and Methyl Orange)

Batch experiments followed the methodology reported previously [19]. The adsorption experiments were carried out by filtering 20 mL of the 200 mg/L solution (MO or MB) with non-woven fabric (circle 11.5 cm diameter) with a weight of 20 mg at room temperature for 60 min. The non-woven fabric was fixed in a glass funnel and the MB or MO solution was poured into the non-woven fabric slowly for 60 min. Every 10 min, an aliquot was taken and read in the UV-Vis spectrometer.

The residual concentration of the dye solution was determined using a calibration curve prepared at the corresponding maximum wavelength (465 nm MO) (590 nm MB) using a UV-Vis spectrometer (Duetta Horiba Scientific, Horiba, Beijing, China).

Adsorption capacity (mg/g) and removal efficiencies in the percentage of MB and MO dyes were calculated using Equations (1) and (2).

$$q_e = \frac{(Ci - Ce)V}{m} \tag{4}$$

$$\%adsorptionefficiency = \frac{(Ci - Ce)}{Ce} \times 100 \tag{5}$$

where Ci and Ce are the initial and equilibrium MO and MB dye (mg) concentrations. V and m are the volume of dye solution and amount of adsorbent (g). Langmuir and Freundlich isotherms were plotted using these data.

2.5.3. Reusability Study

Adsorbent materials must remain stable even after several adsorption cycles. The materials for the organic compounds adsorption were evaluated in four cycles, as described in Section 2.5.2, without prior material washing. The fabrics were dried at room temperature to conduct the following adsorption cycle to determine the material utilization.

2.5.4. Antibacterial Activity

The antibacterial activity was evaluated against Gram-negative bacteria (*Escherichia coli*). The Kirby–Bauer method was used. The bacteria were incubated at 37 °C for 18 h, with an *E. coli* concentration equal to 0.5 McFarland Standard. Grown cultures were prepared and smeared onto the surface of nutrient agar and soy trypticase in Petri plates. Specimens of approximately 3 mm × 3 mm in size were cut from the textile samples under aseptic conditions and placed on the agar surface and the plates were incubated at 37 °C for 18 h. The standard antibiotic was gentamicin, the negative control was DMSO, and the diameter of the formed zone of inhibition (in mm) was determined.

3. Results and Discussion

3.1. X-ray Diffraction (XRD)

Figure 2 shows Nylon 6 and Nylon 6 nanoclay C20A diffraction patterns with different clay contents (0.25, 0.5, 0.75, 1.0, and 2.0%). Neat Nylon 6 shows two strong signals at 20.33° and 24.19° corresponding to the α crystalline phase in the planes (2,0,0) and doublet in (2,0,2)/(0,0,2), respectively [23,24]. All Nylon 6 non-woven fabrics with C20A nanoclay at different concentrations show a pseudo-hexagonal Υ crystalline phase, consisting of parallel chains with hydrogen bonds causing torsion on the molecular chains in zigzag planes [25]. Therefore, it seems that the incorporation of modified nanoclay changes the crystalline structure of the polymer matrix of the non-woven fabrics, probably because nanoclay acts as a nucleating agent and increases the crystallization rate. This change is observed even at the lowest content of nanoclay (0.25%). Another intervening process is the fusion heat used for the manufacture of the non-woven fabric since some reports referred that a polymer transformation with an increase in temperature and then cooling such as that of film formation causes a change in the crystallinity of Nylon 6 [26]. The diffraction peak centered at 21.77° (Υ) in the non-woven fabric Nylon 6/C20A 1.5% shifts towards

21.23° as the nanoclay content increases. It is mainly due to the degree of intercalation of the nanoclay galleries in the polymer matrix, in addition to the dispersion that helps create a homogeneous distribution of the nanoclays in the nylon polymer matrix [20]. This peak for the non-woven fabric Nylon 6/C20A 2% appears at 21.5°. This observation agrees with Cabello et al., where the same shift in the crystalline phase was observed when the zeolite was incorporated [27].

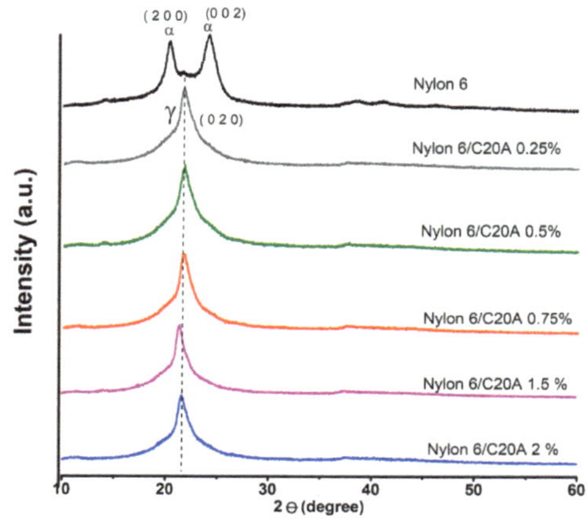

Figure 2. X-ray diffraction patterns of Nylon 6 non-woven fabric and Nylon 6/C20A non-woven fabric at different concentrations of nanoclay C20A (0.25, 0.5, 0.75, 1.5, and 2.0%).

3.2. Scanning Electron Microscopy (SEM)

The SEM-EDS analysis showed the morphological changes and elemental composition in the non-woven fabrics, for example, samples of neat Nylon 6 and Nylon 6 C20A (1.5 and 2.0%) (Figure 3). All samples are generally composed of uniform and smooth fibers randomly oriented, presenting a cylindrical shape (see Figure S1).

In addition, the fiber diameters distribution for the samples are as follows (Table 1). Figure 3 show frequency distribution for neat Nylon 6 fabric has a mean fiber diameter of 16 and 17 ± 1.5 μm. Nylon 6/C20A 1.5% has a mean fiber diameter of 12 ± 2.4 μm, and Nylon 6/C20A 2.0% has a main fiber diameter of 12.7 ± 1.76 μm. A decrease in the size of the fiber diameter is observed as the concentration of the nanoclay increases.

Table 1. Fiber diameter of non-woven fabric at different concentrations.

Sample	Fiber Diameter (μm)
Nylon 6	16 and 17 ± 1.5
Nylon 6/C20A 0.25%	16 ± 0.7
Nylon 6/C20A 0.5%	15 ±1.8
Nylon 6/C20A 0.75%	13 ± 2.1
Nylon 6/C20A 1.5%	12 ± 2.4
Nylon 6/C20A 2.0%	12.7 ± 1.76

A reduction in the fiber diameter can be observed as the modified nanoclay content increases because the nanoclays have amino groups (modifying groups) that can increase the polymer fluidity acting as a lubricant, reducing the fiber size when it is collected. These results agree with the results reported by Fukushima et al. These authors reported an increase in the fluidity of composites of PE-TiO_2 obtained by ultrasound-assisted extrusion attributed to the large exposed surface area of the particles which increased the mobility of the polymer chains [28].

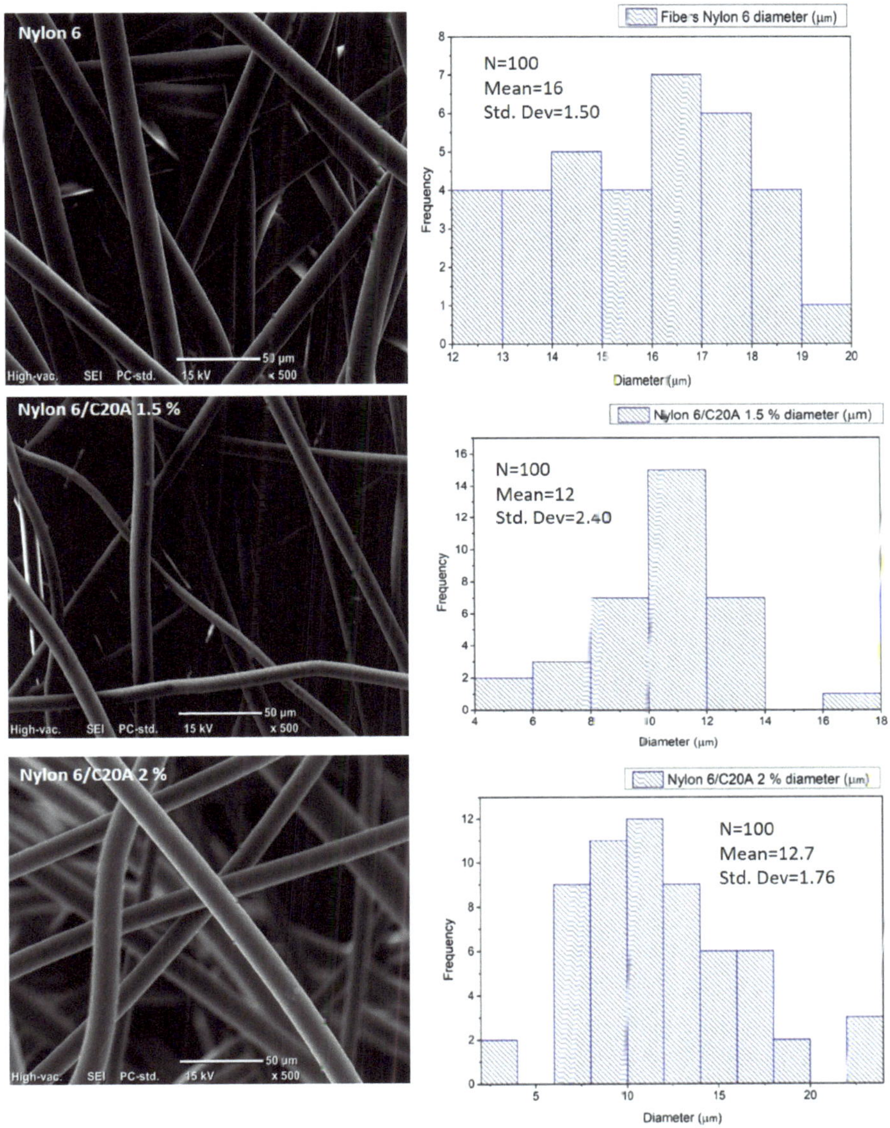

Figure 3. Frequency distribution and 500X SEM images of non-woven fabric of neat Nylon 6, Nylon 6/C20A 1.5%, and Nylon 6/C20A 2.0%.

EDS analysis was performed at 1000× magnification for neat Nylon 6 and Nylon 6/C20A 2.0% fabrics (Figure 4a,b). Neat Nylon only showed C and O elements. Nylon6/C20A 2.0% showed C at 0.27 KeV, N at 0.39 KeV, O at 0.52 KeV, Al at 1.48 KeV, Si at 1.73 KeV, Cl at 2.26 KeV, and Ca at 3.69 KeV. These elements are present in modified nanoclay and the polymer, as reported previously by Andrade-Guel et al. in PLA/C20A nanoclay nanocomposites [20].

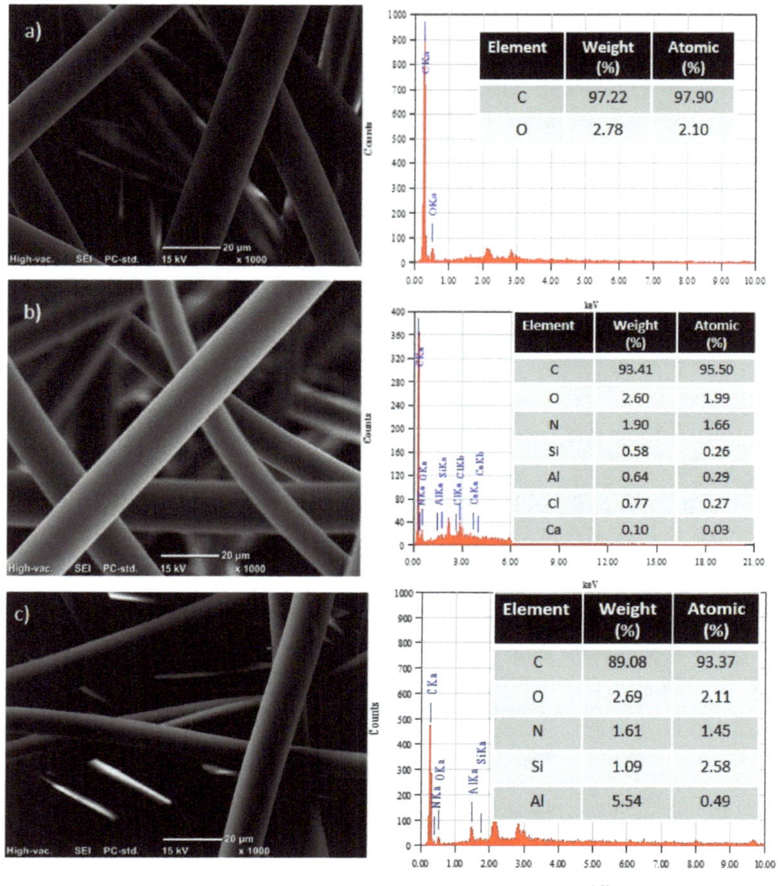

Figure 4. EDS and 1000X SEM images of non-woven fabric of (**a**) neat Nylon 6, (**b**) Nylon 6/C20A 1.5%, and (**c**) Nylon 6/C20A 2.00%.

3.3. Mechanical Tests

The mechanical properties are shown in Figure 5 and Table 2. Neat Nylon 6 non-woven fabric has a tear force of 1.734 N; when the clay is added, the tear force decreases.

Table 2. Mechanical properties of non-woven fabrics.

Sample	Tear Force (N)	Maximum Breaking Strength (N)
Nylon 6	1.734	13.24
Nylon 6/C20A 0.25%	1.334	1.19
Nylon 6/C20A 0.5%	2.334	4.73
Nylon 6/C20A 0.75%	1.4243	5.56
Nylon 6/C20A 1.5%	1.7348	3.66
Nylon 6/C20A 2.0%	1.7792	3.08

The Nylon 6/C20A 0.5% shows a higher tear force than neat Nylon. This behavior can be related to an effective dispersion at low concentrations. Nanoclay can inhibit the mobility of the polymer chains, causing a decrease in rigidity [12]. Also, previous studies have reported that the interactions between the layers of chemically modified silicates and the polymeric matrix can drive improvements in mechanical properties.

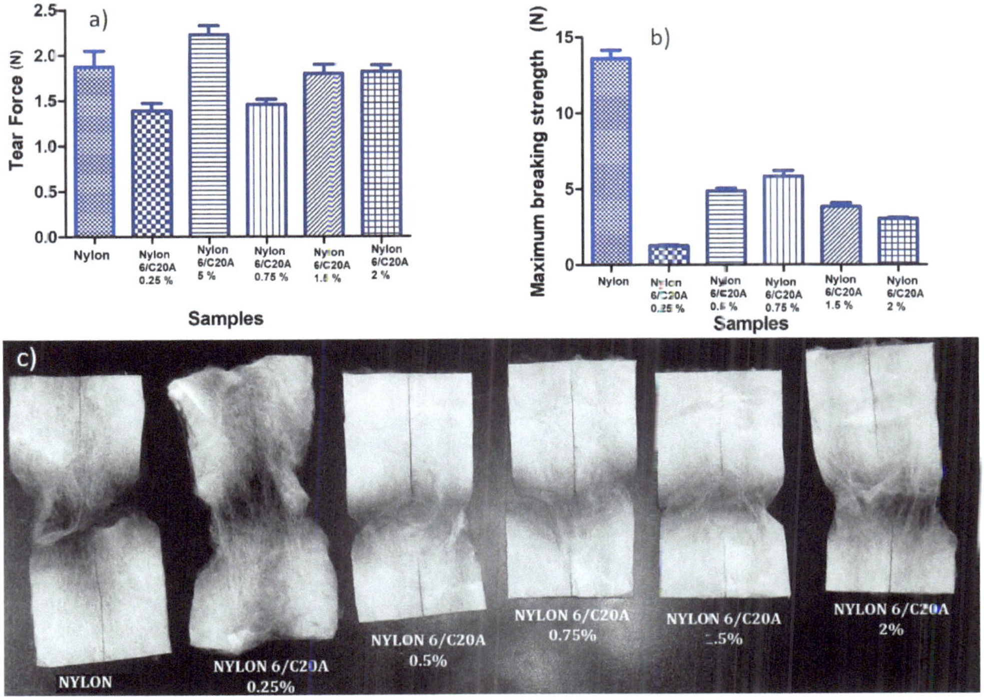

Figure 5. Mechanical properties of non-woven fabrics (**a**) maximum breaking strength, (**b**) tear force, and (**c**) images of fabrics after maximum breaking strength test.

Figure 5b shows the breaking strength of the non-woven fabrics, while Figure 5c shows the corresponding images after maximum breaking strength. In Table 2, it can be seen that the maximum breaking strength for Nylon 6 was 13.24 N, and the Nylon 6/C20A 0.25% sample reached its lowest point (1.19 N) when the modified nanoclay load increased to 0.75% wt, the maximum breaking strength increased to 5.56 N. This increase may be because at low concentrations, the nanoparticles act as breaking points and the polymer chains and cracks in the polymer flow faster, and by increasing the concentration of nanoclay, the polymer becomes more rigid and the cracks move with more difficulty at the crack sites. It has been shown that nanoparticles coated in a polymer matrix can prevent the formation of cracks or self-healing [29].

Regarding the morphological properties, they indicate that by increasing the concentration of modified nanoclay, the fiber diameter decreases; on the other hand, the mechanical properties benefited at concentrations of 0.5% and 0.75% of nanoclay, which indicates that there is a good performance when adding these additive contents.

3.4. *Toxin Adsorption of Non-Woven Nylon 6/C20A*

Figure 6 shows the adsorption of urea and inulin concerning contact time. The system equilibrium is achieved at the maximum adsorption efficiency. For urea (Figure 6a), an adsorption efficiency of 40% is achieved for the non-woven fabric of Nylon 6/C20A 1.5% and Nylon 6/C20A 2%. However, neat Nylon has an efficiency slightly above 47% because it has more significant active sites for the union with urea

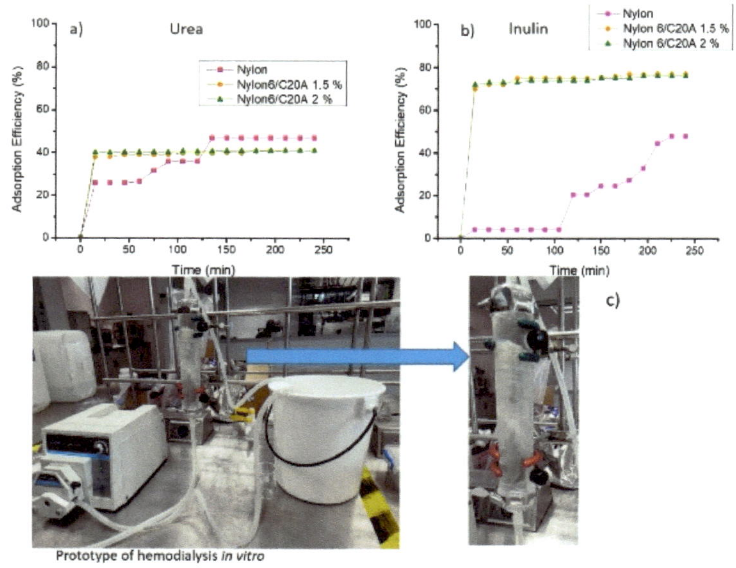

Figure 6. Adsorption profile over time of (**a**) urea and (**b**) inulin onto Nylon 6/C20A non-woven fabric (T = 37 °C, adsorbent dose= 380 mg/L), and (**c**) image of the hemodialysis in vitro prototype.

In a previous study, PLA-Cloisite 20A 5% presented an efficiency of 65%. This difference is due to the modified nanoclay amount. In the present study, we focus on evaluating non-woven fabric with a maximum nanoclay concentration of 2% [20]. Figure 6b shows the inulin adsorption: after 210 min, the maximum adsorption capacity of 48% is reached for neat Nylon 6. This result is in agreement with Andrade et al. [11]. Non-woven fabric Nylon 6/C20A 1.5% and Nylon 6/C20A 2% present similar behavior in the inulin adsorption process. After 15 min, 75 and 74% adsorption efficiencies are reached, respectively. These values indicate rapid adsorption because adsorption sites are generated by the modification of the nanoclay in the non-woven fabric surface. Figure 6c shows an image of the in vitro hemodialysis prototype. It is a continuous system where the non-woven fabric is placed into the device, and the adsorption efficiency is monitored for 4 h.

The experimental adsorption data of urea and inulin in Nylon 6/C20A nanocomposites were fitted with the Langmuir and Freundlich models; the results are presented in Tables 3 and 4, respectively. In Table 3, the correlation coefficient R^2 of the Freundlich model is more significant than that of the Langmuir model, indicating that the experimental data coincide with the Freundlich model in which the adsorption of urea onto Nylon and all samples of Nylon 6/C20A at different concentrations is carried out in a multilayer process [22]. All samples have heterogeneous behavior because the adsorption process takes place in different layers, the mass transport of adsorbate from the solution to the inner surface of the porous adsorbent where adsorption occurs. The structure of the amine-modified nanoclay allows electrostatic interactions with urea, carrying out a chemical adsorption, not only a physical one.

By contrast, the adsorption of inulin (Table 4) adjusts better to the Langmuir adsorption model, indicating a homogeneous surface, where the occupied sites form a monolayer [30]. All samples have homogeneous behavior because the adsorption process takes place in a single layer. The inulin structure presents hydroxyl groups that interact with the amino acid groups of the modified nanoclay, and a monolayer process occurs due to the fact that the adsorption sites are occupied quickly. The adsorption process in an aqueous solution can occur through the porosity of the interconnected fibers in the non-woven fabric or chemically between the functional groups of the inulin and the adsorbent.

Table 3. Isotherm constants and correlation coefficients for urea adsorption on Nylon 6/C20A non-woven fabric at different concentrations.

Sample	Langmuir			Freundlich		
	k	q_{max}	R^2	n	K_F	R^2
Nylon	0.02	3.67	0.9948	0.55	12.46	0.9994
Nylon 6/C20A 0.25%	0.04	8.11	0.9986	0.30	7.09	0.9989
Nylon 6/C20A 0.5%	0.4	8.10	0.9906	0.31	7.00	0.9939
Nylon 6/C20A 0.75%	0.11	1.44	0.9996	0.69	9.10	0.9997
Nylon 6/C20A 1.5%	0.01	1.60	0.9903	0.65	8.96	0.9995
Nylon 6/C20A 2%	0.01	1.47	0.9980	0.68	9.12	0.999

Table 4. Isotherm constants and correlation coefficients for inulin adsorption onto Nylon 6/C20A non-woven fabric at different concentrations.

Sample	Langmuir			Freundlich		
	k	q_{max}	R^2	n	K_F	R^2
Nylon	0.16	25.8	0.999	1.12	11.4	0.9996
Nylon 6/C20A 0.25%	0.06	0.94	0.9996	2.71	18.52	0.9994
Nylon 6/C20A 0.5%	0.006	0.7	0.9999	3.08	20.52	0.9992
Nylon 6/C20A 0.75%	0.04	1.13	0.9998	3.07	20.46	0.9994
Nylon 6/C20A 1.5%	0.059	66	0.9999	3.23	21.30	0.9994
Nylon 6/C20A 2%	0.042	1.24	0.9999	3.05	20.35	0.9995

3.5. Dye Adsorption

Figure 7a shows the adsorption of methylene blue dye with respect to time. In the case of neat Nylon, after 20 min, the adsorption efficiency reaches 13%. Nylon 6/C20A nanocomposites show rapid adsorption in the first 10 min. Then, the adsorption capacity increases, reaching the equilibrium after 60 min, removing 95 and 91% for Nylon 6/C20A 1.5% and Nylon 6/C20A 2%, respectively. In addition, the figure shows images from before and after methylene blue adsorption in the fabrics. Chen et al. studied the methylene blue adsorption on polypyrrole nanocomposites with metal oxide nanoparticles of SiO_2 and Al_2O_3. An efficiency of 80% of methylene blue at the equilibrium was reported at 60 min [31].

Figure 7b shows the adsorption of methyl orange (MO) in relation to time. Neat Nylon 6 fabric shows an adsorption efficiency of 22%. Nylon 6/C20A 1.5% shows a rapid adsorption in the first 10 min, reaching 60% of MO adsorption efficiency. In addition, at the equilibrium, 78% adsorption efficiency is reached. Nylon 6/C20A 2% nanocomposite reaches an adsorption efficiency of 92% at the equilibrium. This value is higher due to larger vacant sites being occupied by adsorbate molecules [32] because of the higher content of nanoclay. Additionally, Figure 7a,b show images from before and after dye adsorption of the fabric.

Tables 5 and 6 show the linear models of the Langmuir and Freundlich isotherms which determine the adsorption capacity and the type of adsorption. Both tables show that the R^2 of the Langmuir isotherm is higher than the Freundlich isotherm. Therefore, the adsorption process of Nylon 6/C20A non-woven fabric is better described by the monolayer isotherm. These results agree with Andrade et al., who studied the adsorption of "uremic toxins or MB and MO" on PLA/C20A nanocomposite. The data were better adjusted to the Langmuir model, indicating that the adsorption occurred in a single layer [20]. As shown in Table 5, the samples Nylon 6/C20A 0.25% and Nylon 6/C20A 0.5% show heterogeneous behavior; having a low concentration of the nanoclay, the dispersion allows multilayer adsorption.

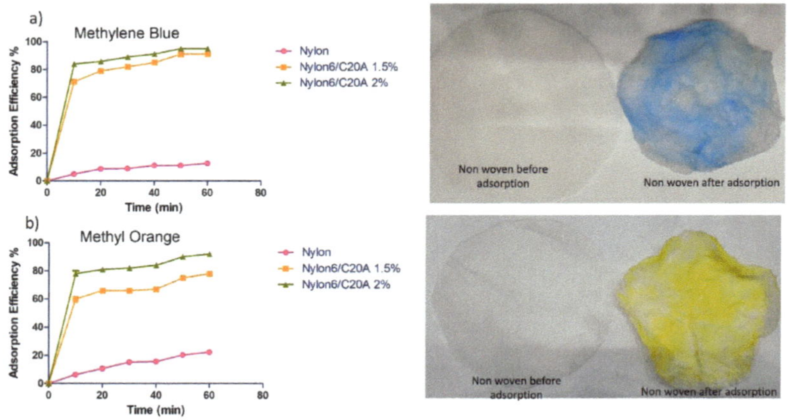

Figure 7. Time effect on the adsorption of (**a**) MB and (**b**) MO dye onto Nylon 6/C20A non-woven fabric (T = 25 °C, adsorbent dose = 200 mg/L); images of Nylon 6/C20A 2%.

Table 5. Adsorption isotherm parameters for MO dye adsorbed onto Nylon 6/C20A.

Sample	Langmuir			Freundlich		
	k	q_{max}	R^2	n	K_F	R^2
Nylon	9.3	312	0.9976	4.8	30.4	0.8799
Nylon 6/C20A 0.25%	8.3	113	0.9003	0.18	50.5	0.934
Nylon 6/C20A 0.5%	1.5	132	0.9317	4.9	52	0.9574
Nylon 6/C20A 0.75%	2.2	210	0.9886	4.0	51	0.9297
Nylon 6/C20A 1.5%	0.29	432	0.9994	2.6	8.11	0.9977
Nylon 6/C20A 2%	2	67.6	0.9998	6.5	14.77	0.9945

Table 6. Adsorption isotherm parameters for MB dye adsorbed onto Nylon 6/C20A.

Sample	Langmuir			Freundlich		
	k	q_{max}	R^2	n	K_F	R^2
Nylon	1.20	196.92	0.9387	9.61	52.17	0.9817
Nylon 6/C20A 0.25%	0.03	199	0.9990	0.52	2.77	0.9905
Nylon 6/C20A 0.5%	1.5	186	0.9953	1.25	9.47	0.9545
Nylon 6/C20A 0.75%	1.66	193	0.9973	1.4	6.09	0.9038
Nylon 6/C20A 1.5%	0.23	151.4	0.9933	4.6	11.32	0.9448
Nylon 6/C20A 2%	0.19	343	0.9985	9.7	20.11	0.9565

Table 6 shows the results of the adsorption isotherms. All the samples with different concentrations of nanoclay conform to the Langmuir isotherm, a monolayer is formed with the dye molecules on the surface of the non-woven fabric, as well as the MB. It is a cyclic structure that can form electrostatic interactions with the amino groups of the modified nanoclay. The Nylon 6 sample without nanoclay fits better with the Freundlich isotherm, forming multilayers of the MB on the surface of the polymer matrix.

3.6. Reusability of Non-Woven Fabric Nylon 6/C20A for the Adsorption of Dyes and Uremic Toxins

After the first process of adsorption, non-woven fabrics were recovered and dried at room temperature. This process does not require a regeneration with any solvent or acid solution, which is an advantage of the elaborated material, because most nanoparticle composites need a regeneration process. Figure 8a,b show the methylene blue and methyl orange adsorption capacity of Nylon 6/C20A non-woven fabric during four cycles. Even after four adsorption cycles, Nylon 6/C20A 1.5% and Nylon 6/C20A 2% maintain an adsorption efficiency above 90%. Figure 8c shows the adsorption efficiency after four cycles. In the case of neat Nylon, a decrease in adsorption is observed from cycle 3. However, the adsorption efficiency remains constant for Nylon 6/C20A 1.5% and Nylon 6/C20A 2% at 75 and 77%, respectively. Figure 8 shows the urea adsorption. The adsorption efficiency for neat Nylon is 47%, decreasing after the third cycle. By contrast, for Nylon 6/C20A 1.5% and Nylon 6/C20A 2%, the adsorption efficiencies do not decrease and remain constant after four cycles.

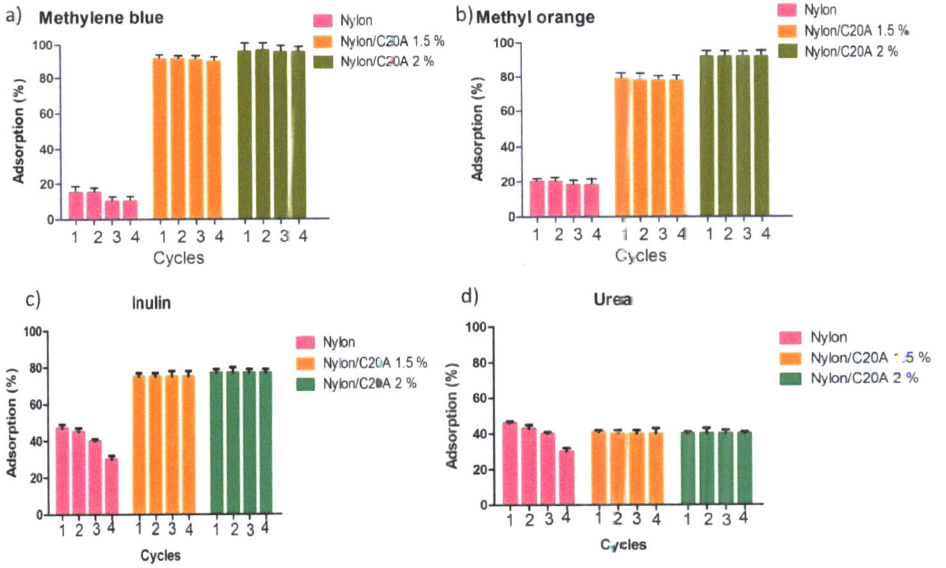

Figure 8. Reusability test for (**a**) MB, (**b**) MO, (**c**) inulin, and (**d**) urea adsorption.

The adsorption efficiency % of non-woven Nylon 6/C20A for uremic toxins and dyes was compared with other adsorbents based on nanocomposite reported in the literature (see Table 7). In the last study with carbon-based nanoparticles, the maximum adsorption efficiency was 80 to 95% for uremic toxins. This is because the surface area is greater and due to the chemical modification of the nanoparticles [11]. Andrade et al. reported an adsorption efficiency for urea of 65% and 97% for methylene blue with the same modified nanoclays used in this study. It is observed that compared to this study, the material is more efficient with the PLA polymer, which is due to the polymer structure that has carboxylic groups that allow efficient adsorption [20]. Le et al. studied the polyamide functionalized with Fe and observed an adsorption efficiency for urea of 85% [33]. Non-woven based on Nylon 6/ZnO and polyester-supported cuprous oxide/reduced graphene oxide both report above 90% for the adsorption of methylene blue [17,19]. Also, composites such as chitosan/polyvinyl alcohol/zeolite and polyethylene oxide/bentonite/polyaniline present values above 90% for MO adsorption, like this study [34,35].

Table 7. Comparison of the adsorption efficiency % for uremic toxins and dyes.

Material	Uremic Toxins (Adsorption %)		Dyes (Adsorption %)		References
	Urea	Inulin	Methylene Blue	Methyl Orange	
Nylon 6/CB	80–90	80–85	----	----	[11]
PLA/C20A nanoclay	65	----	97	----	[20]
Nylon 6/ZnO	----	----	93	----	[19]
Polyamide functionalized with Fe-based metal–organic	85	----	----	----	[33]
Non-woven polyester fabric-supported cuprous oxide/reduced graphene oxide	----	----	96	----	[17]
Chitosan/polyvinyl alcohol/zeolite electrospun composite	----	----	----	95	[34]
Polyethylene oxide/bentonite/polyaniline	----	----	96	94	[35]
Nylon 6/C20A 1.5%	40	75	90	78	This study
Nylon 6/C20A 2%	40	74	90	92	This study

3.7. Antibacterial Activity

Gram-negative bacteria such as *E. coli* are widely used to assess antibacterial activity and ecological safety in materials because their presence in environmental samples, food, or water indicates recent fecal contamination or poor sanitation practices. In addition to the materials obtained in this study, an antibiotic like gentamicin was used as a control.

Figure 9 shows the antibacterial activity of non-woven Nylon 6/C20A against *E. coli*. The antibiotic presents an inhibition zone of 13.5 mm; conversely, the non-woven Nylon 6/C20A fabric at different contents shows an inhibition zone of 4 $\pm$ 0.5, presenting slight antibacterial activity against *E. coli*. These results agree with those of Latwinska et al., who reported that PP and PLA non-woven fabrics have no antimicrobial activity against *E. coli* and *S. aureus*. However, PP/PLA/CuO.SiO_2 composites presented antibacterial activity [36].

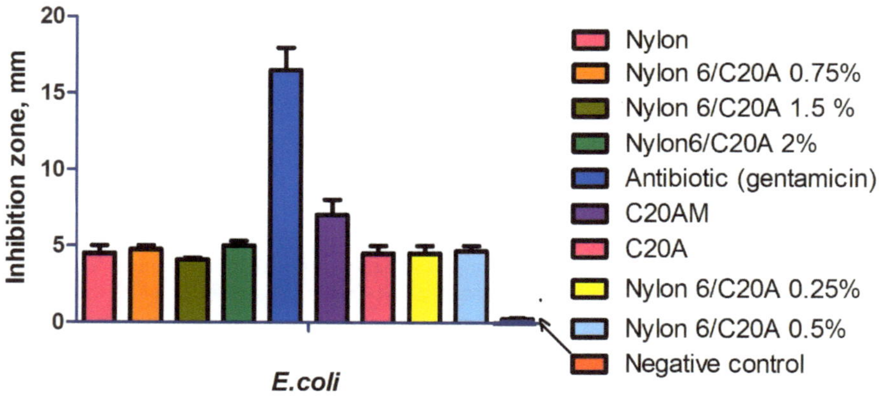

Figure 9. Effect of Nylon 6/C20A non-woven antibacterial activity against *E. coli*.

The antibacterial activity of neat Nylon 6 non-woven fabrics was presented in previous work [19]. The modified fabrics have a greater inhibition zone than the pristine nanoclay. This observation agrees with the study of Nigmatullin et al., where nanoclays Cloisite 20A, 15A, and 30A were modified with biocides and cationic surfactants. Although they attribute the antibacterial efficiency to modifying agents, the inhibition zones presented by these nanoclays were small [37].

The possible antibacterial mechanism of nanoclays may be due to damage to the cell wall and membrane and the bacteria proteins, as shown in Figure 10. The bacterial cell membrane has a negative surface charge and the nanoclays have a positive charge due to the amino groups with which it was chemically modified. Therefore, they bind to the membrane of the bacteria through electrostatic interactions, then cause damage to the membrane and interact with biomolecules of the bacteria such as proteins that cause the death of the bacteria [38].

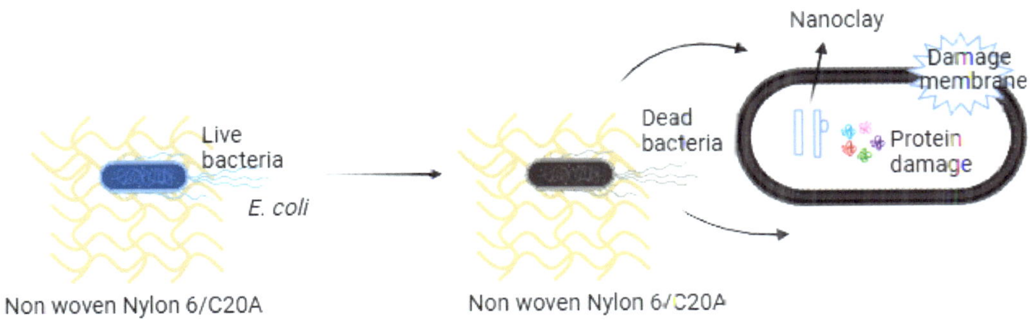

Figure 10. Diagram of the possible antibacterial mechanism of non-woven fabric.

4. Conclusions

Based on the results previously reported, this study produced non-woven fabrics based on polymer nanocomposites of Nylon 6 and modified nanoclay with amino groups. The XRD results showed a displacement of 2θ angles due to the phase shift of Nylon 6. The SEM images showed that the mean diameter of the fiber decreased as the nanoclay content increased. The mechanical tests showed that the sample with 0.5% of modified nanoclay augmented the fiber tear strength to 2.135 N. The inulin adsorption process occurred mainly after 15 min, presenting efficiencies of 75 and 74%. The adsorption capacity for methylene blue and methyl orange was 90%, even after four cycles, and the modified nanoclay non-woven fabrics showed antibacterial activity against *E. coli*.

As a future perspective, when developing this research on non-woven fabric with Nylon 6 and modified nanoclay, the objective is to obtain the necessary knowledge, to have a membrane that can have adsorption properties of uremic and antibacterial toxins, and then test it in environments that help us see their feasibility and selectivity for use in the medical field.

5. Patents

MX/a/2022/013972.

Supplementary Materials: The following supporting information can be downloaded at: https://www.mdpi.com/article/10.3390/ma17123007/s1, Figure S1: SEM of nonwoven fabric at 100x of nylon and nylon6/C20A a different concentrations.

Author Contributions: Conceptualization, C.J.C.-A. and M.A.-G.; methodology, B.C.-O.; software, G.C.-P.; validation, C.A.Á.O., G.C.-P. and C.J.C.-A.; formal analysis, B.C.-O.; investigation, M.A.-G.; resources, C.J.C.-A.; data curation, C.A.Á.O.; writing—original draft preparation, M.A.-G.; writing—review and editing, C.J.C.-A.; visualization, G.C.-P.; supervision, M.A.-G.; project administration, C.J.C.-A.; funding acquisition, C.J.C.-A. All authors have read and agreed to the published version of the manuscript.

Funding: This research was funded by CONAHCYT, key of project 320802, "Desarrollo de filtros selectivos nacionales a base de polímeros y nanopartículas modificadas, para mejorar el tratamiento de hemodiálisis y sustituir el acaparamiento del mercado de empresas extranjeras, Etapa 3".

Acknowledgments: The authors are also grateful to Jesús Ángel Cepeda Garza, María Guadalupe Méndez Padilla, Monica Aimee Ceniceros Reyes, Jesús Gilberto Rodríguez Velázquez, Myrna Salinas Hernández, Janett Valdez Garza, Antelemo R. Yasser Ruiz, and Julieta Sánchez Salazar for their technical support; and to the project Researchers from Mexico CONAHCyT no. 562 entitled Innovation and development of functional technical textiles based on advanced materials. The authors kindly acknowledge the postdoctoral scholarship (387368) provided by CONAHCyT-México. The authors thank Laboratorio Nacional de Materiales Grafenicos No. 321244.

Conflicts of Interest: The authors declare no conflicts of interest.

References

1. Fernandes, M.; Padrão, J.; Ribeiro, A.I.; Fernandes, R.D.; Melro, L.; Nicolau, T.; Mehravani, B.; Alves, C.; Rodrigues, R.; Zille, A. Polysaccharides and metal nanoparticles for functional textiles: A review. *Nanomaterials* **2022**, *12*, 1006. [CrossRef]
2. Cabello-Alvarado, C.J.; Andrade-Guel, M.L.; Medellín-Banda, D.I.; Melo-Lopez, L.; Ávila-Orta, C.A. Polymer Composites: Smart Synthetic Fibers Approach in Energy and Environmental Care. In *Handbook of Nanomaterials and Nanocomposites for Energy and Environmental Applications*; Springer International Publishing: Cham, Switzerland, 2020; pp. 1–26. [CrossRef]
3. Diao, Z.; Zhang, L.; Li, Q.; Gao, X.; Gao, X.; Seliem, M.K.; Dhaoudi, F.; Sellaoui, L.; Deng, S.; Bonilla-Petriciolet, A.; et al. Adsorption of food dyes from aqueous solution on a sweet potato residue-derived carbonaceous adsorbent: Analytical interpretation of adsorption mechanisms via adsorbent characterization and statistical physics modeling. *Chem. Eng. J.* **2024**, *482*, 148982. [CrossRef]
4. Wang, X.; Zhang, A.; Chen, M.; Seliem, M.K.; Mobarak, M.; Diao, Z.; Li, Z. Adsorption of azo dyes and Naproxen by few-layer MXene immobilized with dialdehyde starch nanoparticles: Adsorption properties and statistical physics modeling. *Chem. Eng. J.* **2023**, *473*, 145385. [CrossRef]
5. Wang, X.; Xu, Q.; Zhang, L.; Pei, L.; Xue, H.; Li, Z. Adsorption of methylene blue and Congo red from aqueous solution on 3D MXene/carbon foam hybrid aerogels: A study by experimental and statistical physics modeling. *J. Environ. Chem. Eng.* **2023**, *11*, 109206. [CrossRef]
6. Massella, D.; Argenziano, M.; Ferri, A.; Guan, J.; Giraud, S.; Cavalli, R.; Antonello, A.B.; Salaün, F. Bio-functional textiles: Combining pharmaceutical nanocarriers with fibrous materials for innovative dermatological therapies. *Pharmaceutics* **2019**, *11*, 403. [CrossRef]
7. Helmus, M.N.; Gibbons, D.F.; Cebon, D. Biocompatibility: Meeting a key functional requirement of next-generation medical devices. *Toxicol. Pathol.* **2008**, *36*, 70–80. [CrossRef]
8. Koh, E.; Lee, Y.T. Development of an embossed nanofiber hemodialysis membrane for improving capacity and efficiency via 3D printing and electrospinning technology. *Sep. Purif. Technol.* **2020**, *241*, 116657. [CrossRef]
9. Yavuz, A.; Tetta, C.; Ersoy, F.F.; D'intini, V.; Ratanarat, R.; De Cal, M.; Bordoni, V.; Salvatori, G.; Andrikos, E.; Yakupoglu, G.; et al. Uremic toxins: A new focus on an old subject. *Semin. Dial.* **2005**, *18*, 203–211. [CrossRef]
10. Abidin, M.N.Z.; Goh, P.S.; Said, N.; Ismail, A.F.; Othman, M.H.D.; Abdullah, M.S.; Ng, B.C.; Hasbullah, H.; Hamimah, S.; Abdul Kadir, S.; et al. Polysulfone/amino-silanized poly (methyl methacrylate) dual-layer hollow fiber membrane for uremic toxin separation. *Sep. Purif. Technol.* **2020**, *236*, 116216. [CrossRef]
11. Andrade-Guel, M.; Reyes-Rodríguez, P.Y.; Cabello-Alvarado, C.J.; Cadenas-Pliego, G.; Ávila-Orta, C.A. Influence of Modified Carbon Black on Nylon 6 Nonwoven Fabric and Performance as Adsorbent Material. *Nanomaterials* **2022**, *12*, 4247. [CrossRef]
12. Cabello-Alvarado, C.; Andrade-Guel, M.; Medellin-Banda, D.I.; Ávila-Orta, C.A.; Cadenas-Pliego, G.; Sáenz-Galindo, A.; Radillo-Radillo, R.; Lara-Sánchez, J.F.; Melo-Lopez, L. Non-woven fabrics based on Nylon 6/carbon black-graphene nanoplatelets obtained by melt-blowing for adsorption of urea, uric acid, and creatinine. *Mater. Lett.* **2022**, *320*, 132382. [CrossRef]
13. Bajaj, P.; Sengupta, A.K. Industrial applications of textiles: Textiles for filtration and coated fabrics. *Text. Prog.* **1985**, *14*, 1–30. [CrossRef]
14. Aldalbahi, A.; El-Naggar, M.E.; El-Newehy, M.H.; Rahaman, M.; Hatshan, M.R.; Khattab, T.A. Effects of technical textiles and synthetic nanofibers on environmental pollution. *Polymers* **2021**, *13*, 155. [CrossRef]
15. Nazir, M.A.; Khan, N.A.; Cheng, C.; Shah, S.S.A.; Najam, T.; Arshad, M.; Sharif, A.; Akhtar, S.; Rehman, A. Surface induced growth of ZIF-67 at Co-layered double hydroxide: Removal of methylene blue and methyl orange from water. *Appl. Clay Sci.* **2020**, *190*, 105564. [CrossRef]
16. Nguyen, C.H.; Fu, C.C.; Juang, R.S. Degradation of methylene blue and methyl orange by palladium-doped TiO_2 photocatalysis for water reuse: Efficiency and degradation pathways. *J. Clean. Prod.* **2018**, *202*, 413–427. [CrossRef]
17. Dao, M.U.; Nguyen, T.T.T.; Le, V.T.; Hoang, H.Y.; Le, T.T.N.; Pham, T.N.; Nguyen, T.T.; Akhmadullin, R.M.; Le, H.S.; Tran, H.V.; et al. Non-woven polyester fabric-supported cuprous oxide/reduced graphene oxide nanocomposite for photocatalytic degradation of methylene blue. *J. Mater. Sci.* **2021**, *56*, 10353–10366. [CrossRef]
18. Rana, M.S.; Rahman, N.; Chowdhury, T.A.; Dafader, N.C.; Sultana, S.; Sardar, M.N.; Kayser, M.N. Application of Sulfonated GMA-g-non Woven PE Fabric for the Efficient Removal of Methylene Blue Dye from Wastewater. *Am. J. Polym. Sci. Technol.* **2021**, *7*, 1–9. [CrossRef]

19. Andrade-Guel, M.; Ávila-Orta, C.A.; Cabello-Alvarado, C.; Cadenas-Pliego, G.; Esparza-González, S.C.; Pérez-Alvarez, M.; Quiñones-Jurado, Z.V. Non-woven fabrics Based on Nanocomposite Nylon 6/ZnO Obtained by Ultrasound-Assisted Extrusion for Improved Antimicrobial and Adsorption Methylene Blue Dye Properties. *Polymers* **2021**, *13*, 1888. [CrossRef]
20. Andrade-Guel, M.; Cabello-Alvarado, C.; Romero-Huitzil, R.L.; Rodríguez-Fernández, O.S.; Ávila-Orta, C.A.; Cadenas-Pliego, G.; Medellín-Banda, D.I.; Gallardo-Vega, C.; Cepeda-Garza, J. Nanocomposite PLA/C20A nanoclay by ultrasound-assisted melt extrusion for adsorption of uremic toxins and methylene blue dye. *Nanomaterials* **2021**, *11*, 2477. [CrossRef]
21. Cabello-Alvarado, C.; Andrade-Guel, M.; Ávila-Orta, C.A.; Gamero-Melo, P.; Reyes-Rodríguez, P.Y.; Quiñones-Jurado, Z.V.; Cadenas-Pliego, G.; Bartolo-Pérez, P.; Soriano-Corral, F.; Covarrubias-Gordillo, C. Composites based on nylon 6/clinoptilolite by ultrasound-assisted extrusion for enhanced flame retardant and mechanical properties. *Polym. Bull.* **2022**, *79*, 1803–1819. [CrossRef]
22. Reyes-Rodríguez, P.Y.; Ávila-Orta, C.A.; Andrade-Guel, M.; Cortés-Hernández, D.A.; Herrera-Guerrero, A.; Cabello-Alvarado, C.; Sánchez-Fuentes, J.; Ramos-Martínez, V.H.; Valdez-Garza, J.A.; Hurtado-López, G.F. Synthesis and characterization of magnetic nanoparticles $Zn_{1-x}Mg_xFe_2O_4$ with partial substitution of Mg^{2+} (x = 0.0, 0.25, 0.5, 0.75, and 1.0) for adsorption of uremic toxins *Ceram. Int.* **2020**, *46*, 27913–27921. [CrossRef]
23. Rusu, G.; Ueda, K.; Rusu, E.; Rusu, M. Polyamides from lactams by centrifugal molding via anionic ring-opening polymerization. *Polymer* **2001**, *42*, 5669–5678. [CrossRef]
24. Estrada-Flores, S.; Martínez-Luévanos, A.; Bartolo-Pérez, P.; García-Cerda, L.A.; Flores-Guia, T.E.; Aguilera-González, E.N. Facile synthesis of novel calcium silicate hydrated-nylon 6/66 nanocomposites by solution mixing method. *RSC Adv.* **2018**, *8*, 41818–41827. [CrossRef]
25. Araujo, E.M.; Leite, A.M.D.; da Paz, R.A.; Medeiros, V.D.N.; de Melo, T.J.A.; Lira, H.D.L. Polyamide 6 nanocomposites with inorganic particles modified with three quaternary ammonium salts. *Materials* **2011**, *4*, 1956–1966. [CrossRef]
26. Wu, T.M.; Liao, C.S. Polymorphism in nylon 6/clay nanocomposites. *Macromol. Chem. Phys.* **2000**, *201*, 2820–2825. [CrossRef]
27. Cabello-Alvarado, C.J.; Quiñones-Jurado, Z.V.; Cruz-Delgado, V.J.; Avila-Orta, C.A. Pigmentation and degradative activity of TiO_2 on polyethylene films using masterbatches fabricated using variable-frequency ultrasound-assisted melt-extrusion. *Materials* **2020**, *13*, 3855. [CrossRef]
28. Fukushima, K.; Tabuani, D.; Camino, G. Poly (lactic acid)/clay nanocomposites: Effect of nature and content of clay on morphology, thermal and thermo-mechanical properties. *Mater. Sci. Eng. C* **2012**, *32*, 1790–1795. [CrossRef]
29. Johnsen, B.B.; Kinloch, A.J.; Mohammed, R.D.; Taylor, A.C.; Sprenger, S. Toughening mechanisms of nanoparticle-modified epoxy polymers. *Polymer* **2007**, *48*, 530–541. [CrossRef]
30. Cabello-Alvarado, C.; Andrade-Guel, M.; Pérez-Alvarez, M.; Cadenas-Pliego, G.; Cortés-Hernández, D.A.; Bartolo-Pérez, P.; Ávila-Orta, C.A.; Cruz-Delgado, V.J.; Zepeda-Pedreguera, A. Graphene nanoplatelets modified with amino-groups by ultrasonic radiation of variable frequency for potential adsorption of uremic toxins. *Nanomaterials* **2019**, *9*, 1261. [CrossRef]
31. Chen, J.; Feng, J.; Yan, W. Influence of metal oxides on the adsorption characteristics of PPy/metal oxides for Methylene Blue. *J. Colloid Interface Sci.* **2016**, *475*, 26–35. [CrossRef]
32. Muslim, M.; Ali, A.; Neogi, I.; Dege, N.; Shahid, M.; Ahmad, M. Facile synthesis, topological study, and adsorption properties of a novel Co (II)-based coordination polymer for adsorptive removal of methylene blue and methyl orange dyes. *Polyhedron* **2021**, *210*, 115519. [CrossRef]
33. Le, T.; Esfahani, M.R. Thin-Film Nanocomposite Membranes Functionalized with Fe-Based Metal–Organic Frameworks for Enhanced Removal of Urea from Water. *ACS EST Water* **2024**, *4*, 2064–2075. [CrossRef]
34. Habiba, U.; Siddique, T.A.; Lee, J.J.L.; Joo, T.C.; Ang, B.C.; Afifi, A.M. Adsorption study of methyl orange by chitosan/polyvinyl alcohol/zeolite electrospun composite nanofibrous membrane. *Carbohydr. Polym.* **2018**, *191*, 79–85. [CrossRef]
35. Ali, H.; Mansor, E.S.; Taha, G.M. Microfiltration and adsorptive membranes for simultaneous removal of methyl orange and methylene blue using hybrid composites. *Polym. Bull.* **2022**, *79*, 7891–7908. [CrossRef]
36. Łatwińska, M.; Sójka-Ledakowicz, J.; Chruściel, J.; Piórkowski, M. PLA and PP composite nonwoven with antimicrobial activity for filtration applications. *Int. J. Polym. Sci.* **2016**, *2016*, 2510372. [CrossRef]
37. Nigmatullin, R.; Gao, F.; Konovalova, V. Polymer-layered silicate nanocomposites in the design of antimicrobial materials. *J. Mater. Sci.* **2008**, *43*, 5728–5733. [CrossRef]
38. Andra, S.; Balu, S.K.; Jeevanandam, J.; Muthalagu, M. Emerging nanomaterials for antibacterial textile fabrication. *Naunyn-Schmiedeberg's Arch. Pharmacol.* **2021**, *394*, 1355–1382. [CrossRef]

polymers

MDPI

Article

Anti-Microbial, Thermal, Mechanical, and Gas Barrier Properties of Linear Low-Density Polyethylene Extrusion Blow-Molded Bottles

Saleh Alkarri [1,*], Muhammed Naveed [1], Fatimah Alali [2], Jérôme Vachon [3], Aaron Walworth [1] and Abigail Vanderberg [4]

1 School of Packaging, Michigan State University, 448 Wilson Road, East Lansing, MI 48824-1223, USA
2 Almoosa College of Health Sciences, Ain Najm Rd, Al Mubarraz 36422, Saudi Arabia
3 SABIC, P.O. Box 319, 6160 AH Geleen, The Netherlands
4 Center for Advanced Microscopy, Michigan State University, 578 Wilson Road, CIPS Bldg, Rm B-6B, East Lansing, MI 48824-1223, USA
* Correspondence: alkarris@msu.edu

Abstract: Microbial contamination can occur on the surfaces of blow-molded bottles, necessitating the development and application of effective anti-microbial treatments to mitigate the hazards associated with microbial growth. In this study, new methods of incorporating anti-microbial particles into linear low-density polyethylene (LLDPE) extrusion blow-molded bottles were developed. The anti-microbial particles were thermally embossed on the external surface of the bottle through two particle deposition approaches (spray and powder) over the mold cavity. The produced bottles were studied for their thermal, mechanical, gas barrier, and anti-microbial properties. Both deposition approaches indicated a significant enhancement in anti-microbial activity, as well as barrier properties, while maintaining thermal and mechanical performance. Considering both the effect of anti-microbial agents and variations in tensile bar weight and thickness, the statistical analysis of the mechanical properties showed that applying the anti-microbial agents had no significant influence on the tensile properties of the blow-molded bottles. The external fixation of the particles over the surface of the bottles would result in optimum anti-microbial activity, making it a cost-effective solution compared to conventional compounding processing.

Keywords: anti-microbial activity; *E. coli* K-12 MG1655; anti-microbial agents; thermal embossing; extrusion blow molding

1. Introduction

Extrusion blow molding is a common manufacturing method for generating hollow plastic containers such as bottles and jars. These containers are utilized in many industries, including food and beverage, pharmaceutical, cosmetic, and household product production [1]. However, microbial contamination can occur on the surfaces of blow-molded bottles, necessitating the development and application of effective anti-microbial treatments to mitigate the hazards associated with microbial growth. For instance, the shelf life of a contained sensitive material stored in blow-molding plastic containers is highly dependent upon the level of sterility of the plastic. Sterilization techniques and aseptic filling are thus commonly employed [2]. Additionally, microorganisms can be transmitted onto the surfaces of blow-molded bottles through various means [3]. For instance, an individual may touch contaminated surfaces like door handles, countertops, or packaging materials and then inadvertently transfer microbes onto the bottles through direct contact. Microorganisms can also be disseminated through air movement within a production facility. In environments where microbial aerosols are prevalent, such as hospitals or crowded spaces, airborne microorganisms can also land on the surfaces of blow-molded bottles in their end use location. Airborne fungal spores in a room have been shown to settle on bottles, leading to contamination [4].

The surfaces of extrusion blow-molded bottles provide an ideal environment for bacterial growth due to their smooth texture and organic residues [1]. Colonization of surfaces by bacteria such as *Escherichia coli* (*E. coli*), *Staphylococcus aureus* (*S. aureus*), and *Pseudomonas aeruginosa* (*P. aeruginosa*) can result in rapid bacterial growth and the development of biofilms. Kim et al. (2020) state that biofilms may cause food degradation and perhaps spread diseases to customers. Biofilms can also serve as long-term bacterial contamination reservoirs [5]. To prevent bacterial colonization, creating anti-microbial surfaces for blow-molded bottles is crucial [6]. Growing demand for plastic containers in various industries has rapidly increased the use of blow molding in recent years [7,8]. Chadha et al. (2022) remark that population growth, urbanization, and the increased demand for packaged goods are all factors that affect the extrusion blow molding industry's market size. The extrusion blow molding market was estimated to be 100 billion USD in 2020, and according to Chadha et al. (2022), it is expected to increase at a CAGR of 5% between 2021 and 2028 [9]. Polyethylene (PE) and polyethylene terephthalate (PET) are two of the most widely used plastics in blow molding and account for a sizable share of global plastic consumption [10].

Advancements in the development of anti-microbial agents for plastics application have resulted in the emergence of two distinct categories, leachable and non-leachable agents, according to Gulati, Sharma, and Sharma (2021). Leachable agents can release anti-microbial compounds from the polymer matrix, offering sustained efficacy [11]. However, concerns about potentially harmful substance release arise. For example, silver nanoparticles embedded in the polymer gradually release silver ions when they come in contact with microorganisms, leading to the detection of silver in good it intends to protect [12]. In contrast, non-leachable agents remain fixed on the plastic surface, offering immediate and localized anti-microbial effects. Non-leachable agents include $Mg(OH)_2$ and chitosan, a natural biopolymer that has shown promise in anti-microbial packaging [12]. While non-leachable agents may be susceptible to wear and degradation, recent research has focused on formulating and evaluating these agents for enhanced performance and safety [11]. Since non-leachable agents require direct contact with microbes to have an effect, they are better suited as coatings rather than compounded articles, where the additive would be dispersed inside the matrix [13].

Various anti-microbial substances have been investigated in the field of extrusion blow molding. Incorporating copper (Cu) may be a viable method for doping in $Mg(OH)_2$, thereby enabling the modulation of the material's optical bandgap. CuO can modulate electron field emission characteristics owing to its low potential barrier [14]. However, more research is required to determine its effectiveness.

Anti-microbial agents have been incorporated into the process through different methods. One of these methods is melt-compounding with plastics like PE and PET [15]. Another method is the use of coating technologies applied to the exterior of the bottles [16].

In the case of melt-compounding, the anti-microbial agents are mixed with the plastic materials during the manufacturing process, creating a uniform matrix of the polymer and additive [17]. According to Huang et al., the coating process applies a layer of anti-microbial agent to the bottle surfaces, resulting in a thin layer, with thicknesses typically ranging from a few to tens of micrometers [16]. The adjustment of concentrations and thicknesses of anti-microbial agents is contingent upon the targeted degree of anti-bacterial efficacy and the particular demands of the application, as noted by Huang et al. [16]. Notably, the types of anti-microbial agents and the coating techniques employed may exhibit variations across different investigations, contingent upon factors such as the specific microorganisms being targeted and the intended duration of anti-microbial efficacy [16].

The study conducted by Hutasoit et al. has revealed that Cu-infused $Mg(OH)_2$ could exhibit robust anti-bacterial characteristics against a wide range of bacteria, including Gram-positive and Gram-negative strains, such as *Salmonella* spp., *E. coli*, and *S. aureus* because of the *Cu* and *Mg* contents [18]. Another study has shown that the alkyd resin nanocomposite derived from palm oil containing $Mg(OH)_2/MgO$ colloidal NPs has displayed catalytic performance and anti-microbial activity. Some bacteria, including methicillin-resistant *S. aureus* and *P. aeruginosa*, are killed by $Mg(OH)_2$ and $Cu(OH)_2$, respectively [19]. According to Birkett et al., the concentration and thickness of an anti-microbial coating greatly affect its efficiency [20]. Higher concentrations of anti-microbial compounds are typically associated with increased anti-microbial action. Darvish et al. pointed out that obtaining the optimum concentration is essential to avoid unintended implications, such as altering the polymer's physical characteristics or making leaching of the agent more likely [21]. Since the integrity of extrusion blow-molded bottles must be preserved during anti-microbial treatment, it is crucial to determine the concentration of the anti-microbial agent needed to achieve this goal [22]. Similarly, the thickness of the coating layer influences the anti-microbial performance. Thicker coatings can increase protection against microbial contamination [23]. However, excessively thick coatings may be prone to cracking or peeling, compromising their effectiveness [24]. Recent publications have highlighted the importance of optimizing the coating thickness to balance both anti-microbial activity and coating durability [25,26].

Among the various anti-bacterial agents mentioned, copper-infused $Mg(OH)_2$ has exhibited potential efficacy against Gram-positive and Gram-negative bacteria [27]. The broad-spectrum anti-bacterial activity of copper ions released from copper-infused $Mg(OH)_2$ targets DNA, proteins, and bacterial cell membranes [27]. This mechanism makes it effective against a wide range of bacteria, including those with varying cell wall structures [6,27].

The effectiveness of anti-microbial substances such as $Mg(OH)_2$, $Cu(OH)_2$, MgO, $CuCl_2$, and ZnO can differ, depending on the type of bacteria [28]. According to research findings, $CuCl_2$ exhibits noteworthy inhibitory properties against the proliferation of Gram-negative bacterial strains such as *E. coli* and *P. aeruginosa*. In contrast, it has been reported that MgO and ZnO exhibit greater efficacy against Gram-positive bacteria [29]. The observed variation in efficacy underscores the diverse antimicrobial properties of these compounds, as reported by Jakubovskis et al. [30].

The mechanisms by which anti-microbial particles induce cell death are multifaceted and contingent upon the particular agent utilized. Some examples of anti-microbial modes of action are the disruption of cell membranes, the production of reactive oxygen species (ROS), the suppression of enzymatic activities, or the induction of damage to DNA. According to Imani et al., the anti-microbial effectiveness of Cu-infused $Mg(OH)_2$, $Mg(OH)_2$, $Cu(OH)_2$, MgO, $CuCl_2$, and ZnO is often attributed to their multi-modal actions [31]. Imani et al. report that one particular mechanism entails the interference of bacterial cell membranes through the utilization of distinct anti-microbial nanoparticles, namely $Mg(OH)_2$, $Cu(OH)_2$, MgO, $CuCl_2$, and ZnO [31]. The NPs can interact with the bacterial cell membrane, thereby compromising its structural integrity and the consequent release of its cellular constituents. The disruption of the membrane structure and function results in the disturbance of crucial cellular processes and eventual cell death, as reported by Imani et al. [31].

An additional mechanism involves the production of ROS. According to Smaoui et al., specific anti-microbial particles, including $Mg(OH)_2$ and Cu-infused $Mg(OH)_2$, can produce ROS upon exposure to moisture or light. ROS, such as hydroxyl radicals and superoxide ions, elicit oxidative harm within bacterial cells, thus deactivating them [32].

The suppression of enzymatic activity represents another pivotal mechanism utilized by certain anti-microbial particles. Peters et al. have demonstrated the effectiveness of Cu-infused $Mg(OH)_2$ and $CuCl_2$ in impeding the function of crucial enzymes in bacterial cells [33]. This interference with enzymatic function leads to the impairment of crucial metabolic processes, ultimately culminating in the demise of the bacteria [33].

In addition, it has been observed that anti-microbial agents containing copper, such as Cu-infused $Mg(OH)_2$ and $Cu(OH)_2$, can potentially induce DNA damage in bacterial cells. According to Rojas et al., the agents interact with bacterial DNA, resulting in structural harm and disruption of its replication and transcription mechanisms [34]. The amalgamation of physical and chemical mechanisms in these particles effectively contributes to their anti-microbial properties, thereby enabling them to either inhibit bacterial growth or cause bacterial death [34].

According to Gumienna et al., the regulatory approval status of anti-microbial agents utilized in blow-molding applications may differ, based on the particular agent and its intended application, as determined by the Food and Drug Administration (FDA) [35]. The authors state that certain anti-microbial agents utilized in blow molding have not obtained approval from the FDA. Some anti-microbial agents, including copper and zinc compounds, have been generally recognized as safe (GRAS) by the FDA for diverse applications [36]. The substances that have been designated as GRAS have been deemed to meet the safety requirements set forth by the FDA and are therefore suitable for use in contact with pharmaceutical or food items, as per the findings of Mania et al. [36]. Evaluating anti-microbial agents' toxicity is critical due to its potential impact on human cells. The toxicity of various NPs, including Cu-infused $Mg(OH)_2$, $Mg(OH)_2$, $Cu(OH)_2$, MgO, $CuCl_2$, and ZnO, has been investigated in human cells through research conducted by Naz et al. [37] Their results indicate that nanoparticles typically demonstrate negligible cytotoxicity at the lower concentrations that fall within the anti-microbial range, and that they are well received by human cells [37]. However, high concentrations or prolonged exposure to specific anti-microbial agents may lead to adverse effects [38].

Furthermore, the durability and longevity of the anti-microbial effects are important aspects related to anti-microbial techniques in extrusion blow molding applications [22]. The influence of environmental conditions on the performance of anti-microbial coatings, as well as the development of sustainable and environmentally friendly anti-microbial agents, are also significant. These aspects are critical for the practical implementation and commercial viability of anti-microbial solutions in the extrusion blow molding industry [39].

Developing effective anti-microbial techniques in blow molding applications is crucial to ensuring product safety and effective protection from microbial contamination. The utilization of anti-microbial agents, including Cu-infused $Mg(OH)_2$, $Mg(OH)_2$, $Cu(OH)_2$, MgO, $CuCl_2$, and ZnO, has shown promising results in inhibiting bacterial growth on blow-molded bottle surfaces. These agents' concentrations, thicknesses, and mechanisms of action play essential roles in their anti-microbial efficacy. Comprehensive toxicity evaluations are necessary in the future to ensure the safety of these agents for human health.

2. Experimental

2.1. Materials

Linear low-density polyethylene (LLDPE) copolymer (DOWLEX 2045G grade) was obtained in pellet form from Dow Chemical Company (Midland, MI, USA). These pellets have the following characteristics: melting point = 120.0 °C, density = 920 kg/m^3, and melt flow index (MFI) = 1 g/10 min (190 °C/2.16 kg). $Mg(OH)_2$, Cu-infused $Mg(OH)_2$, MgO, and $Cu(OH)_2$ (purity: 99.99%) were gifted by Aqua Resources (Fort Walton Beach, FL, USA), both as dry powder and slurry (dispersed in water). ZnO (purity: 99.00%) was obtained from American Elements (Los Angeles, CA, USA) as a slurry (dispersed in water). Isopropyl alcohol (purity: 99.99%) was obtained from Macron Fine Chemicals (Center Valley, PA, USA).

2.2. Preparation of Anti-Microbial Suspensions for Internal Mold Cavity Spray

The $Mg(OH)_2$ NPs were obtained commercially in slurry form (7 wt.% $Mg(OH)_2$ and 93 wt.% water). The $Mg(OH)_2$ slurry (14.3 mL) was diluted with isopropyl alcohol (85.7 mL) to prepare an $Mg(OH)_2$ suspension at a concentration of 10 mg/mL. The Cu-infused $Mg(OH)_2$ NPs were obtained commercially as a slurry (7.47 wt.% Cu-infused $Mg(OH)_2$ and 92.53 wt.% water). The Cu-infused $Mg(OH)_2$ slurry (1.34 mL) was diluted with isopropyl alcohol (8.66 mL) to prepare a Cu-infused $Mg(OH)_2$ suspension at a concentration of 10 mg/mL. The MgO NPs (500 mg) were combined with isopropyl alcohol (50 mL) to prepare an MgO NPs suspension at a concentration of 10 mg/mL. The $Cu(OH)_2$

NPs were obtained commercially as a slurry (22.25 wt.% $Cu(OH)_2$ and 77.75 wt.% water). The $Cu(OH)_2$ slurry (1.75 mL) was diluted with isopropyl alcohol (48.25 mL) to prepare a $Cu(OH)_2$ NPs suspension at a concentration of 10 mg/mL. The ZnO NPs were obtained commercially as a slurry (20 wt.% ZnO and 80 wt.% water). The ZnO slurry (2 mL) was diluted with isopropyl alcohol (48 mL) to prepare a ZnO NPs suspension at a concentration of 10 mg/mL. The NP suspensions were vortexed at maximum speed for 30 s, and subsequently sonicated in an ultrasonic bath (Branson 2510 Ultrasonic Sonicator, Commack, NY, USA) at 23 °C for 10 min to ensure that the NPs were uniformly dispersed. After sonication, the suspension was vortexed once more at maximum speed for 30 s (Figure 1) [26].

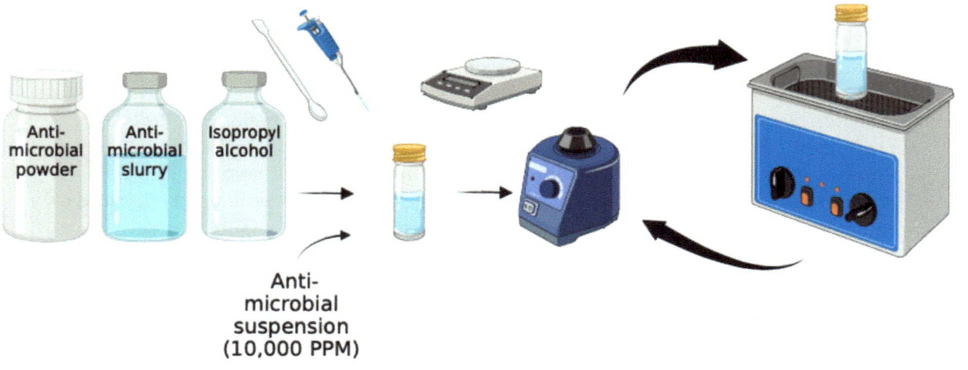

Figure 1. Preparation of the anti-microbial suspensions. Reprinted from ref. [26].

2.3. Blow Molding

The bottles were produced using a Bekum H111S extrusion blow molder (Serial 974948-5-056), outfitted with a chilled single cavity aluminum mold [40].The mold chiller was type BMB-II-B, manufactured by Fasti USA (Elgin, IL, USA). The mold was a 500 mL round bullet/cosmo-style bottle mold (please see Section S1 in the Supplementary Information (SI) Document). The blow molder was warmed up for at least one hour before each processing run. The internal cavity of the mold was cleaned with 100% isopropyl alcohol and non-woven polypropylene (PP) fabric, followed by compressed air, after each cycle when the anti-microbial suspension was applied. The first five containers retrieved from the machine were discarded as a method of purging the machine. Then five neat LLDPE bottles were produced, removed in order, and placed inverted (finish down) in a divided, numbered sample tray. For anti-microbial treatment, both sides of the mold cavity were sprayed evenly with the anti-microbial suspension five times (equivalent to approximately 1.8 mL for each cavity side) using 30 mL capacity fine mist spray bottles (Anyumocz brand, CN) and left to dry for 50 s before each cycle began (Figure 2).

The $Mg(OH)_2$ nanoplatelets were applied to the mold cavity as a powder using a cosmetic embossing powder tool (brand: BAOFALI) before the production cycle was initiated (please see Section S2 in the SI document).

The containers were stored inverted in divided trays to give them time to cool and to prevent the flash from becoming fused to the other containers. A total of 30 treated samples, plus 5 controls, were made per run. The bottles were laid out in the sample trays, as shown in Figure 3. Five minutes after production was complete, the flash was removed manually by twisting. The containers were then labeled by tray location and placed right-side up in a new sample tray. The sample tray was labeled with the date, run number, and manufacturing method.

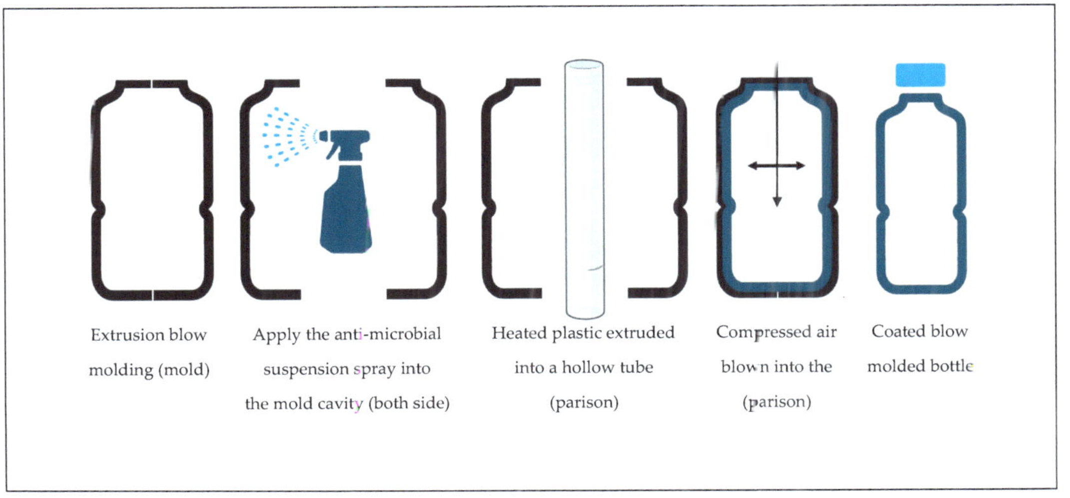

Figure 2. Illustration of extrusion blow molding.

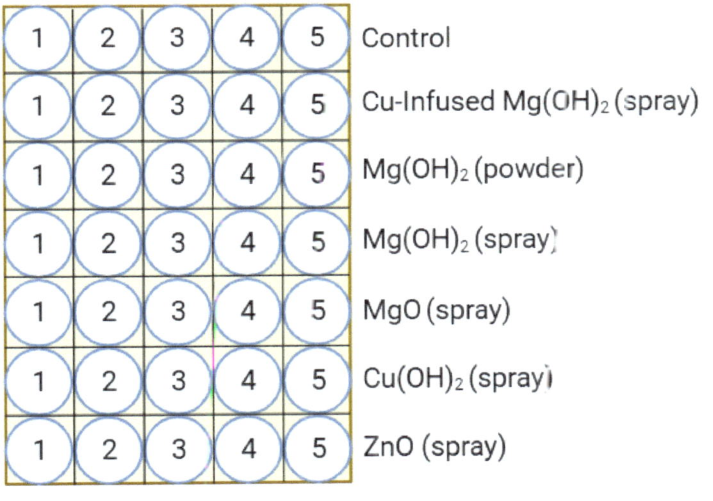

Figure 3. Sample tray layout.

2.4. Sample Preparation of Extrusion Blow Molding Bottles

2.4.1. Extrusion Blow-Molded Bottle Washing Techniques

Each set of extrusion blow-molded bottles (Figure 4) was dipped in a plastic container filled with a deionized water (diH_2O), and the outer surfaces were rubbed with a nitrile-gloved hand to assure the planned characterizations would show only the affixed particles that were adhered to the surfaces. The plastic container was washed and refilled with diH_2O after each set of treatments. After washing, the bottles were placed vertically on a piece of aluminum foil to dry for at least 48 h at room temperature.

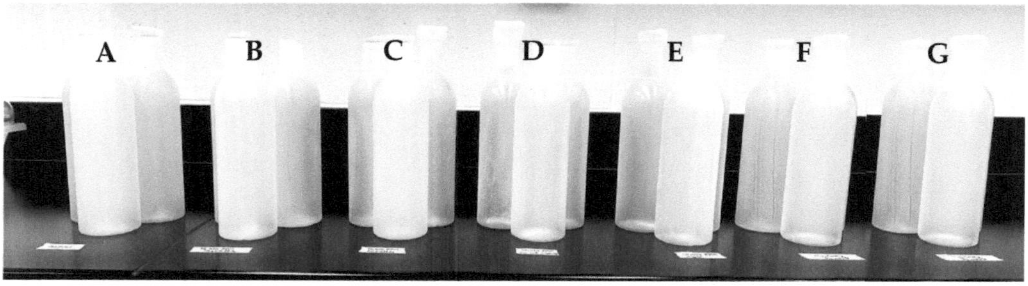

Figure 4. Various LLDPE extrusion blow-molded bottles; neat LLDPE (**A**) and thermally embossed LLDPE with Cu-infused $Mg(OH)_2$ spray (**B**), $Mg(OH)_2$ powder (**C**), $Mg(OH)_2$ spray (**D**), MgO spray (**E**), $Cu(OH)_2$ spray (**F**), and ZnO spray (**G**).

2.4.2. Sample Preparation of the Extruded Blow-Molded Bottles

The extrusion blow-molded bottles were cut down to different shapes and sizes as follows: (i) circular disk (dimension: 20 mm × 0.48 mm), used for anti-microbial, SEM, and EDX; (ii) square shape (dimension: 50 mm × 50 mm × 0.48 mm), used for barrier studies; and (iii) tensile bars (dimension: 127 mm × 25.4 mm × 0.48 mm), used to test the mechanical properties. A JDC PRECISION SAMPLE CUTTER (Thwing-Albert Instrument Company, Philadelphia, PA, USA, Model: JDC 1-10, Serial No 36757) was used to cut the strips of sheets into one inch widths.

2.5. Experimental Design and Statistical Analysis

The means, standard deviations, and percentage changes obtained from the investigation of various properties in the LLDPE extrusion blow-molded bottles thermally embossed with various anti-microbial nanoparticles were evaluated and compared using JMP software (JMP Pro 16.1.0 (539038), SAS Institute Inc., Cary, NC, USA). All experiments were independently replicated at least three times to properly evaluate the properties of the LLDPE extrusion blow-molded bottles.

2.6. Characterization

2.6.1. Scanning Electron Microscopy (SEM) and Energy Dispersive X-ray Spectroscopy (EDX)

SEM was used to characterize unprocessed anti-microbial nanoparticles and disks of blow-molded bottles thermally embossed with anti-microbial agents. The samples were imaged using a JEOL 7500F field emission SEM (JEOL Ltd., Tokyo, Japan). EDX was performed using an Oxford Instruments AZtec system (Oxford Instruments, High Wycomb, Bucks, UK) attached to the SEM. Prior to SEM and EDX analysis, dry powders were adhered to aluminum stubs using high vacuum carbon tabs (SPI Supplies, West Chester, PA, USA). Slurry suspensions of nanoparticles were dried by placing two drops of solution onto silicon wafers (Ted Pella, Inc., Redding, CA, USA), and then the wafers were attached to the stubs. The disks of blow-molded bottles were attached to the stubs using epoxy glue (System Three Resins, Inc., Aubur, WA, USA). The samples were coated with either iridium or osmium. Iridium coating was performed in a Quorum Technologies/Electron Microscopy Sciences Q150T sputter coater (Quorum Technologies, Laughton, East Sussex, UK). A Tennant20 CVD coater (Meiwafosis Co., Ltd., Osaka, Japan) was used for osmium coating.

2.6.2. Differential Scanning Calorimetry (DSC)

The TA Instrument, Model Q100 system, used DSC to determine the crystallization temperatures and melting points of the LLDPE samples. All samples were analyzed in triplicate. The temperature range for analysis was between −20 to 250 °C. The rate of temperature change was 10 °C min^{-1}. The process was then paused at $^{-1}$. The samples were then cooled at a rate of 10 °C min^{-1} to −20 °C and then re-heated at the same rate to

250 °C. All thermal responses were recorded to determine the crystallization temperature and melting point. The heat of fusion values were used to calculate the crystallinity degrees of the LLDPE samples. These values were determined from the second heating runs and evaluated according to Equation (1):

$$X_c(\%) = \left[\frac{\Delta H_c}{\Delta H_0 \cdot W}\right] \times 100 \quad (1)$$

where X_c is the crystallinity of the LLDPE samples, ΔH_0 is 100% crystalline LLDPE enthalpy of fusion [279 J/g] [41], ΔH_c is the heat of fusion, and W is the LLDPE fraction in the composite (weight).

2.6.3. Thermogravimetric Analysis (TGA)

The heat resistance and thermal decomposition of the LLDPE samples were evaluated using a Q-50 Thermogravimetric Analyzer (TGA) (TA Instruments, New Castle, DE, USA). Each sample, weighing between 6–10 mg, was subjected to heating at a rate of 10 °C/min up to a maximum of 600 °C under a nitrogen purge of 60 mL/min. A minimum of three replications was conducted for every sample, and the average result of the TGA data was used for analysis.

2.6.4. Tensile Properties

The test specimens (five replicates) were maintained at standard lab conditions (23 °C, 50% RH) for a minimum period of 40 h prior to testing for tensile measurements. Following the standards of ASTM D882-18 (Standard Test Method for Tensile Properties of Thin Plastic Sheeting), the tensile tests were conducted using an Instron tensile testing system (model 5565, Minneapolis, MA, USA). The specimens were held by pneumatic grips with an initial grip separation of 33 mm. The extension was controlled at a constant speed of 500 mm/min. The test concluded when the sample broke or when a decrease of 60% of peak force was reached.

2.6.5. Barrier Properties

Water Vapor Transition Rate (WVTR)

The determination of WVTR was carried out at 37.8 °C and 90% relative humidity using a Permatran-W system (Model 3/34, Mocon Inc., Minneapolis, MN, USA), adhering to the ASTM F1249 standards. The calculation of water vapor permeation was achieved by multiplying the thickness of the film specimen (two duplicates of each LLDPE sample) by the observed values for water vapor transmission. Specimens were masked using self-adhesive foil to provide an exposed surface area of 12.57 cm^2.

Oxygen Transition Rate (OTR)

The OTR of the LLDPE film samples was determined at 23 °C and 50% relative humidity using an Ox-Tran system (model 2/22, Mocon Inc., Minneapolis, MN, USA), in accordance with ASTM D3985. This process involved testing two duplicates of each LLDPE sample. Specimens were masked using self-adhesive foil to provide an exposed surface area of 12.57 cm^2. The test gas was 100% O_2.

2.6.6. Anti-Microbial Testing Method

The anti-microbial activities of the disks were evaluated sequentially for all experiments using the *E. coli* K-12 MG165. The stock culture was stored at −80 °C in a ThermoFisher TSX400 system. A streak containing the *E. coli* K-12 MG165 from the stock culture was then applied onto Tryptic Soy Agar (TSA) plates. A single colony was selected after a 24 h incubation at 37 °C and transferred to 5 mL of Tryptic Soy Broth (TSB). The broth was incubated for 18 h at 37 °C. After incubation, the cultures were centrifuged using a Fisher Scientific accuSpin micro 17 R centrifuge at 13 000× g. A total of 1 mL of culture was centrifuged for 5 min, and the supernatant was discarded. The cells were suspended in 1 mL of phosphate-buffered saline (PBS, Crystalgen, Innovation for Science, Commack, NY, USA), by vortexing. A total of 11.5 mL of PBS was added to the cell suspension after transferring it to a 15 mL tube.

Aliquots of this suspension were then exposed to the different types of disks. Every disk was placed separately in a pod using a contact lens cases manufactured by Bosch + Lomb. For each pod containing a single disk, 1 mL of bacterial suspension was added to submerge the disk into the culture broth; subsequently, the pods were closed. The pods were then attached to a mini rotator (Benchmark Scientific, Roto Mini Plus R 2024, Sayreville, NJ, USA) and rotated at 20 rpm (Figure 5) around the machine's horizontal axis to continuously agitate the broth and cause liquid renewal on the surface of the disks [26]. At 0, 4, and 24 h intervals, a 100 mL sample of the bacterial suspension was removed for the appropriate number of 1:10 dilutions, then incubated at 37 °C overnight after being plated on TSA. The bacterial cell density at each time point was determined by enumerating the colony-forming units (CFU) using a Scan300 (InterScience, Saint-Nom-la-Bretèche, France). Neat LLDPE disks were used as a "negative" control sample, while metallic copper disks were used as a "positive" control sample. In addition, the anti-microbial activity of different LLDPE composite disks was tested individually.

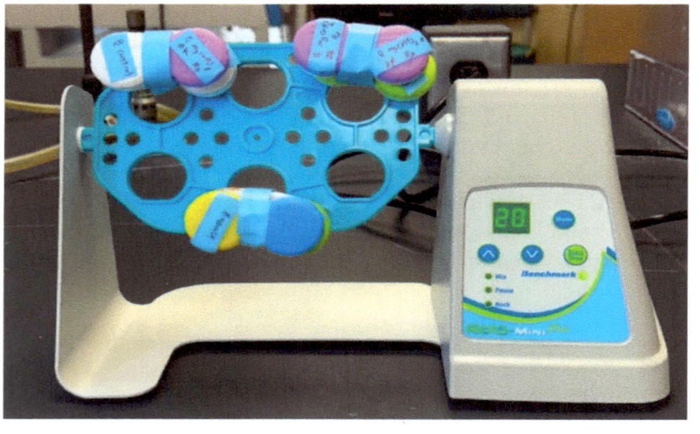

Figure 5. The pods were attached with tape to the mini rotator device for consistent agitation of the bacterial broth and surface renewal of the disks inside the pods. The blue tray holding the pods rotates around the horizontal axis. Reprinted from Ref. [26].

3. Results

3.1. Scanning Electron Microscopy (SEM) and Energy Dispersive X-ray Spectroscopy (EDX)

The SEM characterization for the various pure anti-microbial NPs is presented in Figure 6.

Figure 6. The SEM images of various pure anti-microbial NPs; Cu-infused $Mg(OH)_2$ spray (**A**), $Mg(OH)_2$ powder (**B**), $Mg(OH)_2$ spray (**C**), MgO spray (**D**), $Cu(OH)_2$ spray (**E**), and ZnO spray (**F**).

The SEM characterization of the studied extrusion blow-molded LLDPE bottles thermally embossed with variety of anti-microbial NPs is presented in Figure 7.

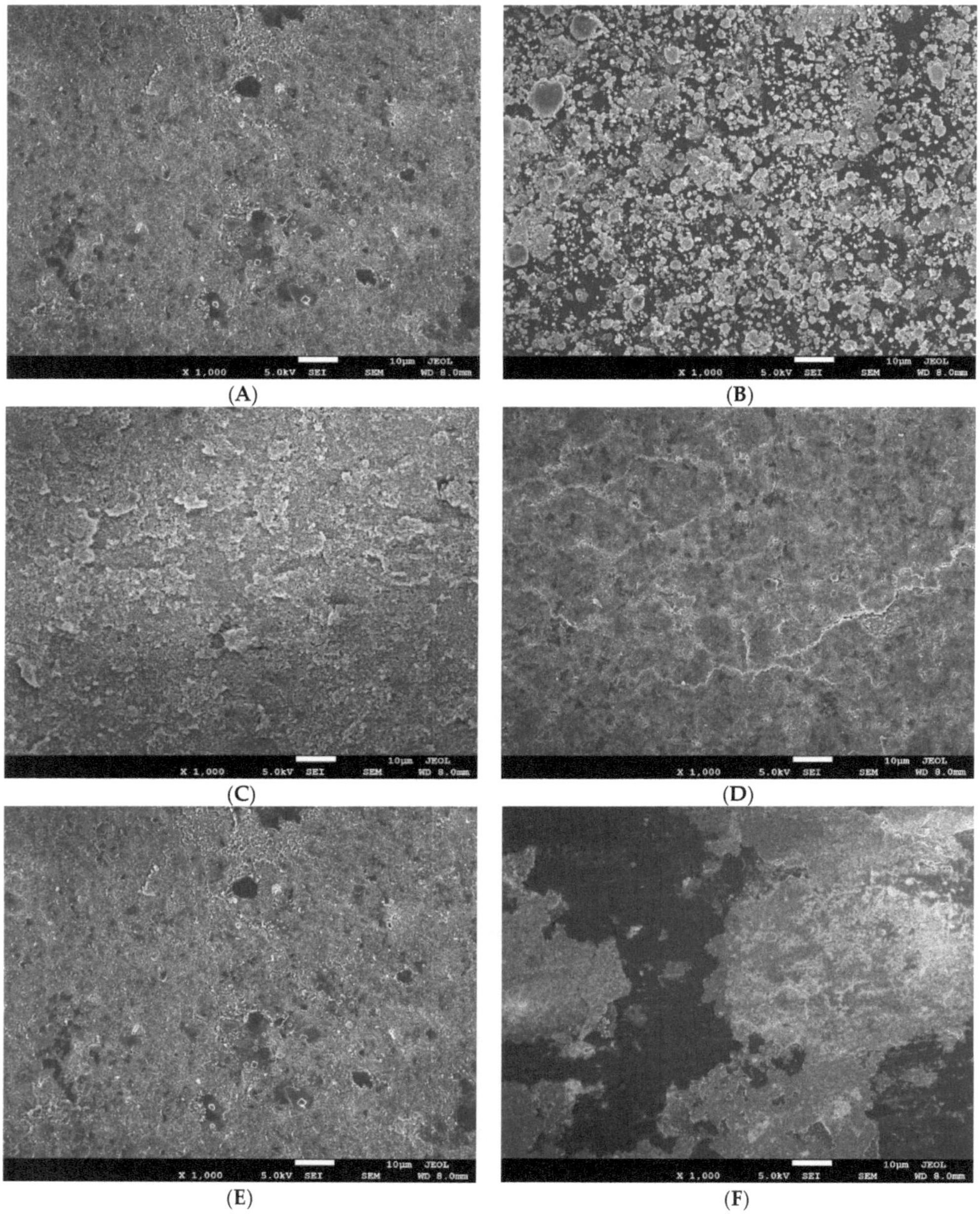

Figure 7. The SEM images of blow-molded bottles thermally embossed with Cu-infused $Mg(OH)_2$ spray (**A**), $Mg(OH)_2$ dry powder (**B**), $Mg(OH)_2$ spray (**C**), MgO spray (**D**), $Cu(OH)_2$ spray (**E**), and ZnO spray (**F**).

The EDX characterization for the studied extrusion blow-molded LLDPE bottles thermally embossed with variety of anti-microbial NPs is presented in Figure 8. To review a detailed report of the EDX mapping, please see Section S3 in the SI document.

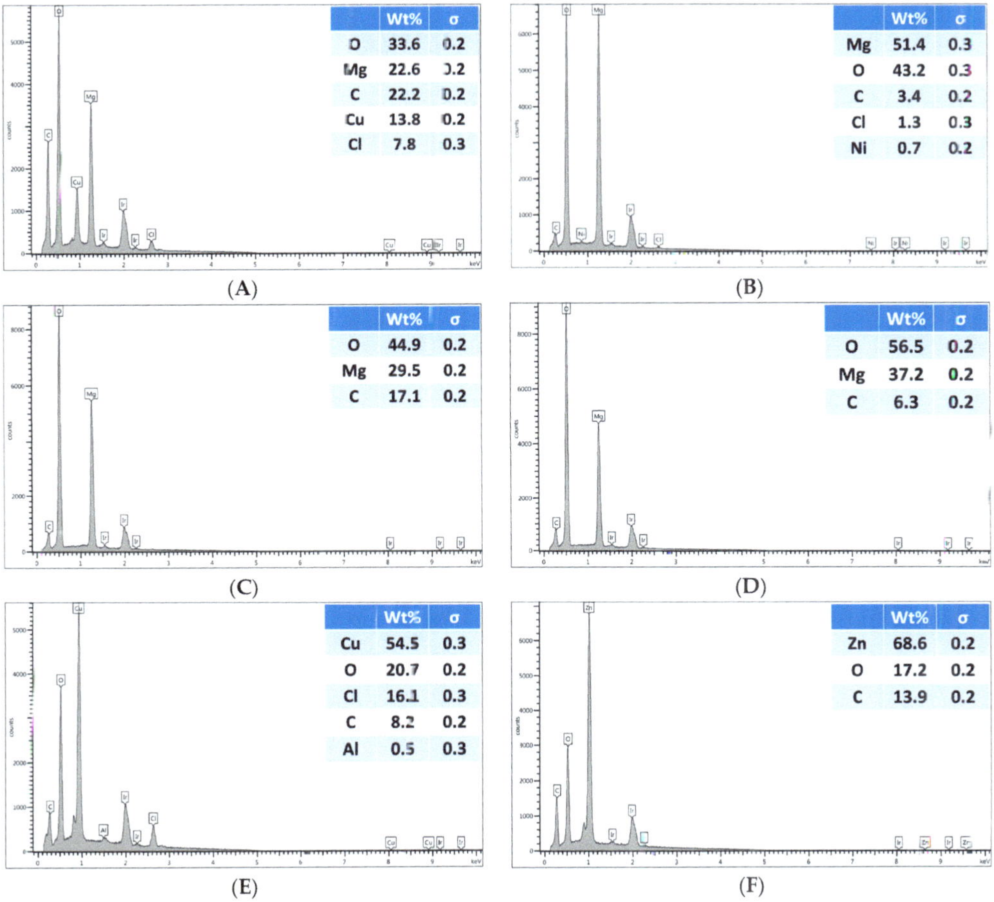

Figure 8. The EDX spectrums for blow-molded bottles thermally embossed with Cu-infused $Mg(OH)_2$ spray (**A**), $Mg(OH)_2$ dry powder (**B**), $Mg(OH)_2$ spray (**C**), MgO spray (**D**), $Cu(OH)_2$ spray (**E**), and ZnO spray (**F**).

3.2. Differential Scanning Calorimetry (DSC)

The thermal behaviors of the studied extrusion blow-molded LLDPE bottles thermally embossed with variety of anti-microbial NPs are presented in Figure 9, and the summarized values along with the crystallinity (%) are listed in Table 1.

Table 1. Thermal properties of various LLDPE samples obtained from DSC measurements.

Samples	ΔH_m [J/g]	T_m [°C]	T_c [°C]	*Crystallinity* [%]
Neat LLDPE	102.90 ± 1.43	124.87 ± 0.54	104.75 ± 1.32	37.01 ± 1.25
LLDPE Cu-infused $Mg(OH)_2$ (Spray)	96.81 ± 0.99	124.23 ± 0.10	105.87 ± 0.87	34.82 ± 1.77
LLDPE $Mg(OH)_2$ (Powder)	98.25 ± 1.87	123.97 ± 1.54	106.01 ± 1.44	35.34 ± 0.87
LLDPE $Mg(OH)_2$ (Spray)	106.80 ± 1.78	123.91 ± 0.76	106.12 ± 0.45	38.41 ± 0.93
LLDPE MgO (Spray)	99.09 ± 0.81	123.78 ± 0.37	105.75 ± 0.44	35.64 ± 1.34
LLDPE $Cu(OH)_2$ (Spray)	113.40 ± 1.12	124.58 ± 0.88	105.48 ± 1.61	40.79 ± 1.21
LLDPE ZnO (Spray)	103.00 ± 1.09	124.19 ± 0.66	105.84 ± 0.93	37.05 ± 0.82

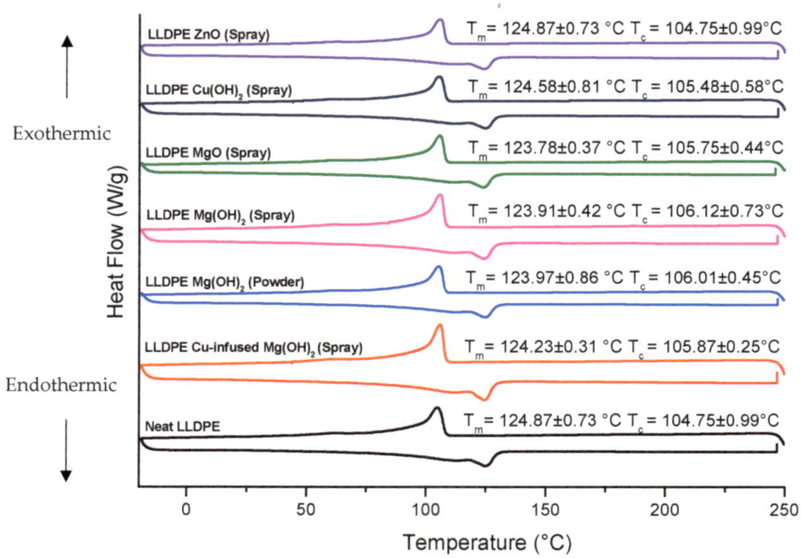

Figure 9. The thermal cycle of various extrusion blow-molded LLDPE bottles obtained from the DSC data.

3.3. Thermogravimetric Analysis (TGA)

The thermal stability of each of the studied extrusion blow-molded LLDPE bottles thermally embossed with variety of anti-microbial NPs is presented in Figure 10, and the summarized values of the temperatures at which 5% weight loss occurred and the estimated residual (wt.%) at 600 (°C) are listed in Table 2.

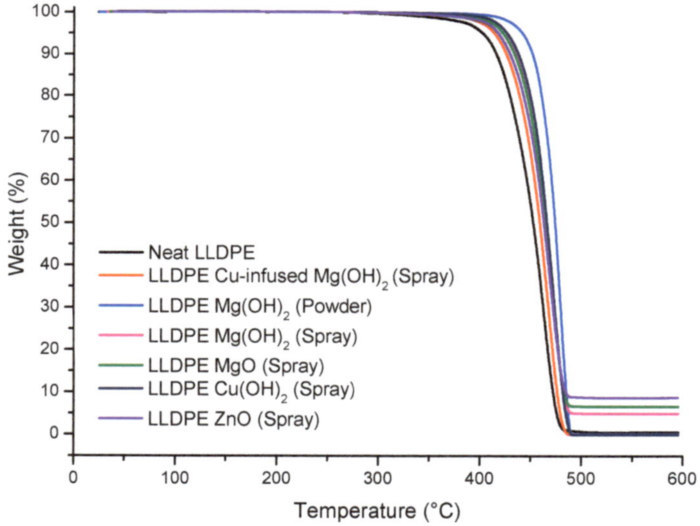

Figure 10. The thermal decomposition properties of various extrusion blow-molded LLDPE bottles obtained from the TGA data.

Table 2. The temperature at which various LLDPE extrusion blow-molded bottles underwent 5% weight loss, and the estimated residual at 600 (°C), as determined via TGA measurements.

Sample	The Temperature at Which 5% Weight Loss Occurred (°C)	Estimated Residual (wt.%) at 600 (°C)
Neat LLDPE	402.03 ± 1.62	0.61 ± 0.11
LLDPE Cu-infused $Mg(OH)_2$ (Spray)	413.59 ± 0.86	0.00 ± 0.43
LLDPE $Mg(OH)_2$ (Powder)	441.28 ± 1.98	0.12 ± 0.82
LLDPE $Mg(OH)_2$ (Spray)	426.01 ± 2.32	4.91 ± 0.08
LLDPE MgO (Spray)	422.12 ± 1.11	6.56 ± 0.45
LLDPE $Cu(OH)_2$ (Spray)	425.96 ± 0.95	0.03 ± 1.22
LLDPE ZnO (Spray)	417.58 ± 1.31	8.67 ± 0.65

3.4. Tensile Properties

The tensile properties of various extrusion blow-molded bottles thermally embossed with variety of anti-microbial NPs are presented in Figure 11.

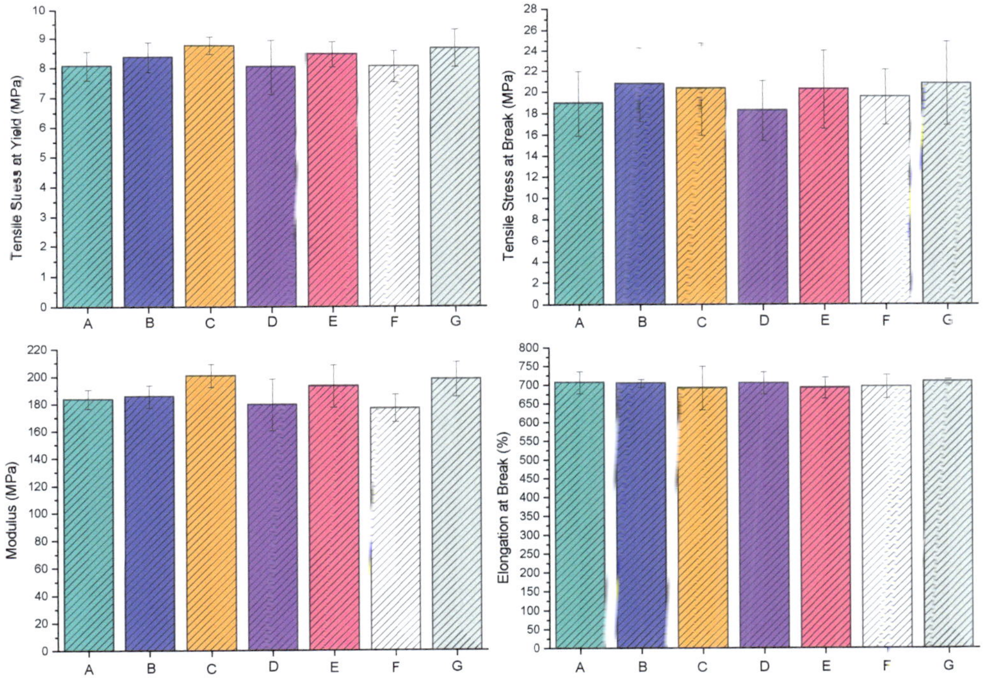

Figure 11. Tensile properties of various extrusion blow-molded bottles: Neat LLDPE (A), LLDPE Cu-infused $Mg(OH)_2$ spray (B), LLDPE $Mg(OH)_2$ powder (C), LLDPE $Mg(OH)_2$ spray (D), LLDPE MgO spray (E), LLDPE $Cu(OH)_2$ spray (F), and LLDPE ZnO spray (G).

3.5. Barrier Properties

The barrier properties of various extrusion blow-molded bottles thermally embossed with variety of anti-microbial NPs are presented in Figure 12.

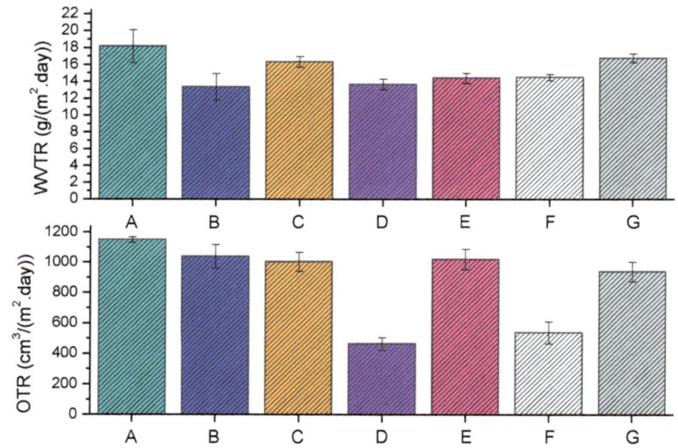

Figure 12. The WVTR and OTR of each graph represent the following samples on the x-axis; neat LLDPE (A), LLDPE Cu-infused $Mg(OH)_2$ spray (B), LLDPE $Mg(OH)_2$ powder (C), LLDPE $Mg(OH)_2$ spray (D), LLDPE MgO spray (E), LLDPE $Cu(OH)_2$ spray (F), and LLDPE ZnO spray (G).

3.6. Anti-Microbial Properties

The anti-microbial properties of various extrusion blow-molded bottles thermally embossed with variety of anti-microbial NPs are presented in Figure 13.

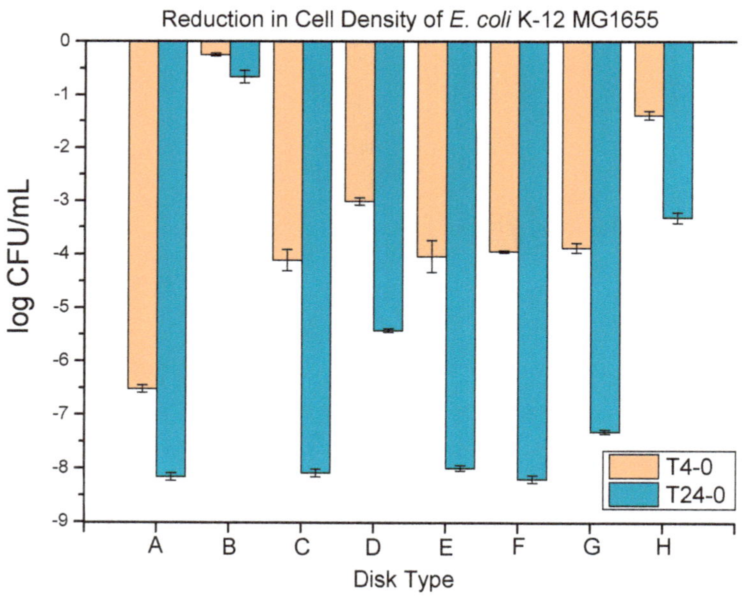

Figure 13. The anti-microbial data obtained for the extrusion blow-molded bottles thermally embossed with several types of anti-microbial agents at 4 and 24 h. Metallic copper—positive control (A), Neat LLDPE—negative control (B), LLDPE Cu-infused $Mg(OH)_2$ spray (C), LLDPE $Mg(OH)_2$ powder (D), LLDPE $Mg(OH)_2$ spray (E), LLDPE MgO spray (F), LLDPE $Cu(OH)_2$ spray (G), and LLDPE ZnO spray (H).

4. Discussion

SEM was used to characterize nanoparticle size and morphology, as shown in Figure 6. $Mg(OH)_2$ nanoparticles appeared as platelets ranging in size from 50–150 nm. Cu-infused $Mg(OH)_2$ contained similar platelets of $Mg(OH)_2$, with the addition of spherical, lobed Cu nanoparticles approximately 150 nm in diameter. MgO nanoparticles formed spherical agglomerates comprised of angular flakes. The flake size ranged from 50–200 nm; spherical agglomerates of MgO flakes measured 1–6 μm in diameter. $Cu(OH)_2$ appeared in two distinct sizes; large rectangular crystals greater than 500 nm in length were mixed with small rectangular nanoparticles of 20–50 nm in size. ZnO nanoparticles had a rounded shape measuring approximately 10–25 nm.

Blow-molded bottles thermally embossed with anti-microbial agents were analyzed using **SEM** and **EDX** to determine the uniformity of the coatings and to confirm the chemical composition of the nanoparticle layer. SEM images show uniform nanoparticle coverage in spray applied coatings of Cu-infused $Mg(OH)_2$, $Mg(OH)_2$, MgO, and $Cu(OH)_2$. However, ZnO spray provided uneven coating, and large areas of exposed bottle surface were visible. $Mg(OH)_2$ applied in powder form caused significant aggregation of nanoparticles, resulting in a spattered, irregular coating, with the bottle surface exposed throughout the disk, as shown in Figure 7. EDX analysis of uniformly coated areas of blow-molded bottles confirmed the presence and composition of the expected nanoparticles on the bottle surface, as shown in Figure 8. In a previous study, we showed that anti-microbial nanoparticles such as $Mg(OH)_2$ and $CuCl_2$ can be affixed to thermoplastic sheets through a thermal embossing process [42]. The $Mg(OH)_2$ and $CuCl_2$ nano crystals were coated on the sheet from a nano crystal suspension, dried, and then the coated sheet was heat pressed. The sheet successfully killed microbes, and the crystals that were thermally fixed on the surface were not affected by wiping or washing the surface [42]. A limitation of this approach is that such thermal embossing methods can only be applied to sheet substrates, and not to articles with complex shapes, as the heat pressing steps required to fix the crystals to the plastic's surface are difficult to achieve for non-flat shapes.

The DSC analysis of the LLDPE extrusion blow-molded bottles thermally embossed with Cu-infused $Mg(OH)_2$ (spray) showed the following **thermal properties**: a T_m of 124.23 °C ± 0.10, which was 0.51% lower compared to that of the neat LLDPE bottle (124.87 °C ± 0.54), as shown in Figure 9 and Table 1; a T_c of (105.87 °C ± 0.87), which was 1.07% higher compared to that of the neat LLDPE bottle (104.75 °C ± 1.32), as shown in Figure 9 and Table 1; a crystallinity (%) of 34.82% ± 1.77, which was 5.92% lower compared to that of the neat LLDPE bottle (37.01% ± 1.25), as shown in Table 1; the temperature at which 5% weight loss occurred was 413.59 °C ± 0.86, which was 2.88% higher than that of the neat LLDPE (402.03 °C ± 1.62), as shown in Figure 10 and Table 2, and the estimated inorganic residual (wt.%) at around 600 (°C) was 0.00 ± 0.43, which was 100% lower than that of the neat LLDPE (0.61 ± 0.11), as shown in Figure 10 and Table 2. Overall, the anti-microbial NPs did not significantly impact the thermal properties of the LLDPE extrusion blow-molded bottles. Cu-infused $Mg(OH)_2$ is a novel anti-microbial agent that had not been previously used for polymer application [43].

The LLDPE extrusion blow-molded bottles thermally embossed with LLDPE $Mg(OH)_2$ (powder) showed the following **thermal properties**: a T_m of 123 97 °C ± 1.54, which was 0.72% lower compared to that of the neat LLDPE bottle (124.87 °C ± 0.54), as shown in Figure 9 and Table 1; a T_c of 106.01 °C ± 1.44, which was 1.20% higher compared to that of the neat LLDPE bottle (104.75 °C ± 1.32), as shown in Figure 9 and Table 1. a crystallinity (%) of 35.34% ± 0.37, which was 4.51% lower compared to that of the neat LLDPE bottle (37.01% ± 1.25), as shown in Table 1; the temperature at which 5% weight loss occurred was 441.28 °C ± 1.98, which was 9.76% higher than that of the neat LLDPE (402.03 °C ± 1.62), as shown in Figure 10 and Table 2, and the estimated inorganic residual (wt.%) occurred at around 600 (°C) (0.12 ± 0.82), which was 80% lower than that of the neat LLDPE (0.61 ± 0.11), as shown in Figure 10 and Table 2 Overall, the anti-microbial NPs did not significantly impact the thermal properties of the LLDPE extrusion blow-molded bottles. The $Mg(OH)_2$ is a novel anti-microbial agent because of its unique nanoparticle sizes and shapes [44]. $Mg(OH)_2$ NPs are broad spectrum anti-microbial agents. Dong et al. demonstrated the anti-microbial activity of $Mg(OH)_2$ NPs against *E. coli* and the plant-associated bacterium *Burkholderia phytofirmans* [45]. Additional plant-associated pathogens, *Xanthomonas alfalfa* and *Pseudomonas syringae* (Huang et al.), were eliminated by $Mg(OH)_2$ NPs, as were the oral, caries-associated bacteria *Streptococcus mutans* [46] and *Streptococcus sobrinus* (Okamoto et al.) [47]. Additional work from our laboratories showed the $Mg(OH)_2$ and Copper Oxide NPs to be similar in their effectiveness against *E. coli* (Dong et al.).

The LLDPE extrusion blow-molded bottles thermally embossed with LLDPE $Mg(OH)_2$ (spray) showed the following **thermal properties**: a T_m of 123.91 °C ± 0.76, which was 0.77% lower compared to that of the neat LLDPE bottle (124.87 °C ± 0.54), as shown in Figure 9 and Table 1; a T_c of 106.12 °C ± 0.45, which was 1.31% higher compared to that of the neat LLDPE bottle (104.75 °C ± 1.32), as shown in Figure 9 and Table 1; a crystallinity (%) of 38.41% ± 0.93, which was 3.78% higher compared to that of the neat LLDPE bottle (37.01% ± 1.25), as shown in Table 1; the temperature at which 5% weight loss occurred was 426.01 °C ± 2.32, which was 5.96% higher than that of the neat LLDPE (402.03 °C ± 1.62), as shown in Figure 10 and Table 2; and the estimated inorganic residual (wt.%) at around 600 (°C) was 4.91 ± 0.08, which was 705% higher than that of the neat LLDPE (0.61 ± 0.11), as shown in Figure 10 and Table 2. Overall, the anti-microbial NPs did not significantly impact the thermal properties of the LLDPE extrusion blow-molded bottles.

The LLDPE extrusion blow-molded bottles thermally embossed with LLDPE MgO (spray) showed the following **thermal properties**: a T_m of 123.78 °C ± 0.37, which was 0.87% lower compared to that of the neat LLDPE bottle (124.87 °C ± 0.54), as shown in Figure 9 and Table 1; a T_c of 105.75 °C ± 0.44, which was 0.95% higher compared to that of the neat LLDPE bottle (104.75 °C ± 1.32), as shown in Figure 9 and Table 1; a crystallinity (%) of 35.64% ± 1.34, which was 3.70% higher compared to that of the neat LLDPE bottle (37.01% ± 1.25), as shown in Table 1; the temperature at which 5% weight loss occurred was 422.12 °C ± 1.11, which was 5.00% higher than that of the neat LLDPE (402.03 °C ± 1.62), as shown in Figure 10 and Table 2; and the estimated inorganic residual (wt.%) at around 600 (°C) was 6.56 ± 0.45, which was 975% higher than that of the neat LLDPE (0.61 ± 0.11), as shown in Figure 10 and Table 2. Overall, the anti-microbial NPs did not significantly impact the thermal properties of the LLDPE extrusion blow-molded bottles. Alwaan et al. showed that the crystallinity of the blends of mLLDPE compounded with MgO was continuously increased by the loading of MgO when compared with the neat material [48].

The LLDPE extrusion blow-molded bottles thermally embossed with LLDPE $Cu(OH)_2$ (spray) showed the following **thermal properties**: a T_m of 124.58 °C ± 0.37, which was 0.23% lower compared to that of the neat LLDPE bottle (124.87 °C ± 0.54), as shown in Figure 9 and Table 1; a T_c of 105.48 °C ± 1.61, which was 0.70% higher compared to that of the neat LLDPE bottle (104.75 °C ± 1.32), as shown in Figure 9 and Table 1; a crystallinity (%) of 40.79% ± 1.21, which was 10.21% higher compared to that of the neat LLDPE bottle (37.01% ± 1.25), as shown in Table 1; the temperature at which 5% weight loss occurred was 425.96 °C ± 0.95, which was 5.95% higher than that of the neat LLDPE (402.03 °C ± 1.62), as shown in Figure 10 and Table 2; and the estimated inorganic residual (wt.%) at around 600 (°C) was 0.03 ± 1.22, which was 95% lower than that of the neat LLDPE (0.61 ± 0.11), as shown in Figure 10 and Table 2. Overall, the anti-microbial NPs did not significantly impact the thermal properties of the LLDPE extrusion blow-molded bottles.

The LLDPE extrusion blow-molded bottles thermally embossed with LLDPE ZnO (spray) showed the following **thermal properties**: a T_m of 124.19 °C ± 0.66, which was 0.54% lower compared to that of the neat LLDPE bottle (124.87 °C ± 0.54), as shown in Figure 9 and Table 1; a T_c of 105.84 °C ± 0.93, which was 1.04% higher compared to that of the neat LLDPE bottle (104.75 °C ± 1.32), as shown in Figure 9 and Table 1; a crystallinity (%) of 37.05% ± 0.82, which was 0.11% higher compared to that of the neat LLDPE bottle (37.01% ± 1.25), as shown in Table 1; the temperature at which 5% weight loss occurred was

417.58 °C ± 1.31, which was 3.87% higher than that of the neat LLDPE (417.58 °C ± 1.31), as shown in Figure 10 and Table 2; and the estimated inorganic residual (wt.%) at around 600 (°C) was 8.67 ± 0.65, which was 1321% higher than that of the neat LLDPE (0.61 ± 0.11), as shown in Figure 10 and Table 2. Overall, the anti-microbial NPs did not significantly impact the thermal properties of the LLDPE extrusion blow-molded bottles.

The LLDPE extrusion blow-molded bottles thermally embossed with Cu-infused $Mg(OH)_2$ (spray) showed the following **tensile properties**: the tensile stress at yield (MPa) was 8.37 ± 0.51, which was 3.79% higher compared to that of the neat LLDPE (8.06 ± 0.48); the tensile stress at break (MPa) was 20.81 ± 3.54, which was 9.82% higher compared to that of the neat LLDPE (18.95 ± 3.01); the tensile stress modulus (MPa) was 185.51 ± 8.11, which was 1.09% higher compared to that of the neat LLDPE (183.59 ± 7.05); and the elongation at break (%) was 704.63 ± 10.40, which was 0.33% lower compared to that of the neat LLDPE (706.99 ± 29.79). Overall, the anti-microbial NPs slightly improved the tensile properties, as shown in Figure 11. In some cases, especially when dealing with epoxy coatings, the incorporation of inorganic NPs can improve the mechanical properties of the polymeric matrices [49]. Impact strength and stiffness, in particular, were enhanced by the filling of potential pinholes and voids in the matrix.

The LLDPE extrusion blow-molded bottles thermally embossed with $Mg(OH)_2$ (powder) showed the following **tensile properties**: the tensile stress at yield (MPa) was 8.77 ± 0.31, which was 8.80% higher compared to that of the neat LLDPE (8.06 ± 0.48); the tensile stress at break (MPa) was 20.40 ± 4.44, which was 7.65% higher compared to that of the neat LLDPE (18.95 ± 3.01); the tensile stress modulus (MPa) was 200.76 ± 8.31, which was 9.35% higher compared to that of the neat LLDPE (183.59 ± 7.05); and the elongation at break (%) was 692.03 ± 57.85, which was 2.12% lower compared to that of the neat LLDPE (706.99 ± 29.79). Overall, the anti-microbial NPs slightly improved the tensile properties, as presented in Figure 11.

The LLDPE extrusion blow-molded bottles thermally embossed with $Mg(OH)_2$ (spray) showed the following **tensile properties**: the tensile stress at yield (MPa) was 8.04 ± 0.92, which was 0.26% lower compared to that of the neat LLDPE (8.06 ± 0.48); the tensile stress at break (MPa) was 18.30 ± 2.82, which was 3.43% lower compared to that of the neat LLDPE (18.95 ± 3.01); the tensile stress modulus (MPa) was 179.38 ± 18.91, which was 2.29% lower compared to that of the neat LLDPE (183.59 ± 7.05); and the elongation at break (%) was 705.47 ± 30.15, which was 0.21% lower compared to that of the neat LLDPE (706.99 ± 29.79). Overall, the anti-microbial NPs slightly improved the tensile properties, as presented in Figure 11.

The LLDPE extrusion blow-molded bottles thermally embossed with MgO (spray) showed the following **tensile properties**: the tensile stress at yield (MPa) was 8.49 ± 0.44, which was 5.32% higher compared to that of the neat LLDPE (8.06 ± 0.48); the tensile stress at break (MPa) was 20.36 ± 3.76, which was 7.44% higher compared to that of the neat LLDPE (18.95 ± 3.01); the tensile stress modulus (MPa) was 193.21 ± 15.46, which was 5.24% higher compared to that of the neat LLDPE (183.59 ± 7.05); and the elongation at break (%) was 692.45 ± 28.40, which was 2.01% lower compared to that of the neat LLDPE (706.99 ± 29.79). Overall, the anti-microbial NPs slightly improved the tensile properties, as presented in Figure 11.

The LLDPE extrusion blow-molded bottles thermally embossed with $Cu(OH)_2$ (spray) showed the following **tensile properties**: the tensile stress at yield (MPa) was 8.07 ± 0.54, which was 0.11% higher compared to that of the neat LLDPE (8.06 ± 0.48); the tensile stress at break (MPa) was 19.62 ± 2.66, which was 3.54% higher compared to that of the neat LLDPE (18.95 ± 3.01); the tensile stress modulus (MPa) was 176.74 ± 9.96, which was 3.73% lower compared to that of the neat LLDPE (183.59 ± 7.05); and the elongation at break (%) was 695.96 ± 31.80, which was 1.56% lower compared to that of the neat LLDPE (706.99 ± 29.79). Overall, the anti-microbial NPs slightly improved the tensile properties, as presented in Figure 11.

The LLDPE extrusion blow-molded bottles thermally embossed with ZnO (spray) showed the following **tensile properties**: the tensile stress at yield (MPa) was 8.71 ± 0.66, which was 8.05% higher compared to that of the neat LLDPE (8.06 ± 0.48); the tensile stress at break (MPa) was 20.95 ± 4.01, which was 10.55% higher compared to that of the neat LLDPE (18.95 ± 3.01); the tensile stress modulus (MPa) was 198.10 ± 12.84, which was 7.90% higher compared to that of the neat LLDPE (183.59 ± 7.05); and the elongation at break (%) was 710.27 ± 7.11, which was 0.46% higher compared to that of the neat LLDPE (706.99 ± 29.79). Overall, the anti-microbial NPs slightly improved the tensile properties, as presented in Figure 11.

The effects of the anti-microbial agents, the variation in tensile bar weight, and the variation in the tensile bar thickness were statistically investigated to study their possible impacts on the tensile properties of the extrusion blow-molded bottles. The introduction of these anti-microbial agents at these loading levels (10,000 ppm and five sprays on each side of the mold cavity) can be achieved without any impact on the tensile properties, while providing a significant anti-microbial property to the bottles. The statistical analysis showed that after adjusting for the variation attributed to tensile bar thickness and bar weight, none of the six types of anti-microbial agents exhibited significantly different results to those of the control, as measured by tensile stress at yield, tensile stress at break, modulus, and elongation at break. A detailed statistical analysis is provided as part of the SI document (please see Section S4).

The LLDPE extrusion blow-molded bottles thermally embossed with Cu-infused $Mg(OH)_2$ (spray) showed the following **barrier properties**: the WVTR (g/(m^2·day)) and OTR (cm^3/(m^2·day)) were 13.34 ± 1.56 and 1037.56 ± 76.46, respectively, which were 26.54% and 9.63% lower, respectively, compared to that of the neat LLDPE (18.16 ± 1.93 and 1148.11 ± 18.84). Overall, the anti-microbial NPs improved the barrier properties, as presented in Figure 12.

The LLDPE extrusion blow-molded bottles thermally embossed with $Mg(OH)_2$ (powder) showed the following **barrier properties**: the WVTR (g/(m^2·day)) and OTR (cm^3/(m^2·day)) were 16.32 ± 0.63 and 1000.95 ± 62.25, respectively, which were 10.13% and 12.82% lower, respectively, compared to that of the neat LLDPE (18.16 ± 1.93 and 1148.11 ± 18.84). Overall, the anti-microbial NPs improved the barrier properties, as presented in Figure 12.

The LLDPE extrusion blow-molded bottles thermally embossed with Cu-infused $Mg(OH)_2$ (spray) showed the following **barrier properties**: the WVTR (g/(m^2·day)) and OTR (cm^3/(m^2·day)) were 13.67 ± 0.62 and 463.50 ± 41.29, respectively, which were 24.72% and 59.63% lower, respectively, compared to that of the neat LLDPE (18.16 ± 1.93 and 1148.11 ± 18.84). Overall, the anti-microbial NPs improved the barrier properties, as presented in Figure 12.

The LLDPE extrusion blow-molded bottles thermally embossed with MgO (spray) showed the following **barrier properties**: the WVTR (g/(m^2·day)) and OTR (cm^3/(m^2·day)) were 14.43 ± 0.59 and 1018.70 ± 67.63, respectively, which were 20.54% and 11.27% lower, respectively, compared to that of the neat LLDPE (18.16 ± 1.93 and 1148.11 ± 18.84). Overall, the anti-microbial NPs improved the barrier properties, as presented in Figure 12.

The LLDPE extrusion blow-molded bottles thermally embossed with $Cu(OH)_2$ (spray) showed the following **barrier properties**: the WVTR (g/(m^2·day)) and OTR (cm^3/(m^2·day)) were 14.52 ± 0.37 and 537.29 ± 70.92, respectively, which were 20.04% and 53.20% lower, respectively, compared to that of the neat LLDPE (18.16 ± 1.93 and 1148.11 ± 18.84). Overall, the anti-microbial NPs improved the barrier properties, as presented in Figure 12.

The LLDPE extrusion blow-molded bottles thermally embossed with ZnO (spray) showed the following **barrier properties**: the WVTR (g/(m^2·day)) and OTR (cm^3/(m^2·day)) were 16.81 ± 0.50 and 937.25 ± 64.30, respectively, which were 7.43% and 18.37% lower, respectively, compared to that of the neat LLDPE (18.16 ± 1.93 and 1148.11 ± 18.84). Overall, the anti-microbial NPs improved the barrier properties, as presented in Figure 12.

The improved barrier properties were likely due to the fixation of the inorganic crystals over the outer surface of the extrusion blow-molded bottles. Upon the incorporation of these particles, the porosities of the bottles were considerably narrowed. In all cases, the coating improved the gas barrier properties of the bottle. While the improvement of WVTR is only marginal in most cases, the improvement of OTR for sample (Figure 12D,F) is quite noticeable, with ~50% improvement. It is known that inorganic-based coatings can improve gas barrier performance [50]. In this case, the coating was not specifically tuned to improve gas barrier performance, hence the marginal improvement.

The **anti-microbial performance** of extrusion blow-molded bottles was tested against *E. coli* K-12 MG1655 (8.16 ± 0.10 log), as presented in Figure 13. The metallic copper disks (positive control) showed a 6.52 ± 0.07 and 8.16 ± 0.07 log reduction at 4 h and 24 h, respectively. The neat LLDPE disks (negative control) showed a 0.25 ± 0.03 and 0.66 ± 0.12 log reduction at 4 h and 24 h, respectively. The negative control's performance shows a bacterial reduction too low to be considered antibacterial and in combination with the positive control's result, this constitutes proof of the validity of the anti-bacterial test. The extrusion blow-molded bottles thermally embossed with Cu-infused $Mg(OH)_2$ particles (spray) showed a 4.11 ± 0.20 and 8.09 ± 0.07 log reduction at 4 h and 24 h, respectively, which presented a 99.999996 and 99.999996% reduction from the negative control, respectively. The extrusion blow-molded bottles thermally embossed with $Mg(OH)_2$ particles (powder) showed a 3.02 ± 0.07 and 5.43 ± 0.03 log reduction at 4 h and 24 h, respectively, which exhibited a 99.8333336 and 99.999996% reduction from the negative control, respectively. The extrusion blow-molded bottles thermally embossed with $Mg(OH)_2$ particles (spray) showed a 4.04 ± 0.30 and 8.00 ± 0.05 log reduction at 4 h and 24 h, respectively, which presented a 99.988886 and 99.999996% reduction from the negative control, respectively. The extrusion blow-molded bottles thermally embossed with MgO particles (spray) showed a 3.95 ± 0.02 and 8.20 ± 0.07 log reduction at 4 h and 24 h, respectively, which was a 99.988886 and 99.999996% reduction from the negative control, respectively. The extrusion blow-molded bottles thermally embossed with $Cu(OH)_2$ particles (spray) showed a 3.88 ± 0.09 and 7.32 ± 0.04 log reduction at 4 h and 24 h, respectively, which showed a 99.988886 and 99.999996% reduction from the negative control, respectively. The extrusion blow-molded bottles thermally embossed with ZnO particles (spray) showed a 1.40 ± 0.08 and 3.32 ± 0.10 log reduction at 4 h and 24 h, respectively, which was a 92.9222226 and 99.7888886% reduction from the negative control, respectively.

In all cases, the bacterial reduction after 24 h is highly increased compared to that at 4 h, showing that these coatings require several hours to a day to fully eradicate the initial incubated bacterial colonies. The ZnO reflects the lowest performance, with only a log 3 reduction after 24 h. ZnO NPs are known to be effective anti-bacterial agents, but superior additives, such as Ag NPs, are available [51]. For instance, $Mg(OH)_2$ powder provided better performance, reaching a log 5.5 bacterial reduction after 24 h. The performance was further enhanced when applying Cu- or Mg-based additives via the spray coating (Figure 13C,E–G); all have a bacterial reduction > to log 7 after 24 h. This level of sterility is far superior to that provided by disinfection and can only be achieved through sterilization techniques such as gamma radiation [52].

5. Conclusions

In this article, a novel method toward the development of anti-microbial extrusion blow-molded LLDPE bottles was reported, in which various types of anti-microbial agents (Cu-infused $Mg(OH)_2$, $Mg(OH)_2$, $Cu(OH)_2$, MgO, $CuCl_2$, and ZnO) were introduced onto the surface of the bottles. The produced samples were characterized via SEM and EDX and were evaluated for their thermal, mechanical, and anti-microbial properties via DSC, TGA, tensile, barrier, and anti-microbial testing. The results demonstrate significant improvement in anti-microbial activities, as well as barrier properties, while maintaining the thermal stability and mechanical performance of the neat polymer. This approach might be useful in industrial scale applications to help create environmentally friendly, cost-effective, time-efficient, and easily implemented anti-microbial systems. This study provides a promising alternative to the conventional melt-compounding process, in which the anti-microbial agents are mixed with plastic materials during the manufacturing process, requiring a higher load percentage of additives.

Supplementary Materials: The following supporting information can be downloaded at: https://www.mdpi.com/article/10.3390/polym16131914/s1, Figure S1: Illustration of the extrusion blow molding process; Figure S2: Extrusion blow molding machine and model specifications; Figure S3: Bottle mold design; Figure S4: Embossing powder tool used for $Mg(OH)_2$ NPs (dry powder) deposition into the mold cavity; Figure S5: The bottle mold after the application/deposition of the $Mg(OH)_2$ dry powder; Figure S6: EDX Mapping for LLDPE bottle thermally embossed with Cu-infused $Mg(OH)_2$ (Spray): SEM image of sample (A), O element mapping (B), Mg element mapping (C), Cl element mapping (D), and Cu element mapping (E); Figure S7: EDX Mapping for LLDPE bottle thermally embossed with $Mg(OH)_2$ (powder): SEM image of sample (A), C element mapping (B), O element mapping (C), Mg element mapping (D), and Al element mapping (E); Figure S8: EDX Mapping for LLDPE bottle thermally embossed with $Mg(OH)_2$ (spray): SEM image of sample (A), C element mapping (B), O element mapping (C), and Mg element mapping (D); Figure S9: EDX Mapping for LLDPE bottle thermally embossed with MgO (spray): SEM image of sample (A), C element mapping (B), O element mapping (C), Mg element mapping (D), and Ca element mapping; Figure S10: EDX Mapping for LLDPE bottle thermally embossed with Cu(OH)2 (spray): SEM image of sample (A), C element mapping (B), O element mapping (C), Al element mapping (D), Cl element mapping (E), and Cu element mapping (F); Figure S11: EDX Mapping for LLDPE bottle thermally embossed with ZnO (spray): SEM image of sample (A), C element mapping (B), O element mapping (C), and Zn element mapping (D); Figure S12: Comparison with control decision chart; Figure S13: Comparison with control decision chart; Figure S14: Comparison with control decision chart; Figure S15: Comparison with control decision chart; Table S1: Table S1: Results from stepwise regression on noise factors for each of the outcomes, tensile stress at yield, tensile stress at break, tensile modulus, and elongation at break. p-values less than 0.05 are in italics; Table S2: Parameter estimates; Table S3: Comparisons with control summary; Table S4: Parameter estimates; Table S5: Comparison with control summary; Table S6: Parameter estimates; Table S7: Comparisons with control summary; Table S8: Parameter estimates; Table S9: Comparisons with control summary.

Author Contributions: S.A. led the processing, performed the required characterization, and wrote the manuscript. S.A. and M.N. conducted the tensile testing. S.A., F.A. and J.V. performed the anti-microbial characterization. S.A. and A.W. conducted the extrusion blow molding trials. S.A. and A.V. completed the SEM and EDX characterization. All authors have read and agreed to the published version of the manuscript.

Funding: This research received no external funding.

Data Availability Statement: All data are available in the SI document.

Acknowledgments: Saleh Alkarri is thankful for SABIC, who generously funded his doctoral studies, as well as his research needs, at Michigan State University. We are all thankful for John Cairney and Lee Maddan from Aqua Resources Corp for (I) collaboratively developing and validating the anti-microbial testing protocol, (II) donating the anti-microbial agents, and (III) providing expertise regarding their killing mechanisms.

Conflicts of Interest: Author Jérôme Vachon was employed by SABIC. The remaining authors declare that the research was conducted in the absence of any commercial or financial relationships that could be construed as potential conflicts of interest.

References

1. Stanley, J.; John, A.; Pušnik Črešnar, K.; Fras Zemljič, L.; Lambropoulou, D.A.; Bikiaris, D.N. Active Agents Incorporated in Polymeric Substrates to Enhance Antibacterial and Antioxidant Properties in Food Packaging Applications. *Macromol* **2023**, *3*, 1–27. [CrossRef]
2. Ansari, I.; Datta, A.J. An overview of sterilization methods for packaging materials used in aseptic packaging systems. *Food Bioprod. Process.* **2003**, *81*, 57–65. [CrossRef]

3. Jabłońska-Trypuć, A.; Makuła, M.; Włodarczyk-Makuła, M.; Wołejko, E.; Wydro, U.; Serra-Majem, L.; Wiater, J. Inanimate surfaces as a source of hospital infections caused by fungi, bacteria and viruses with particular emphasis on SARS-CoV-2. *Int. J. Environ. Res. Public Health* **2022**, *19*, 8121. [CrossRef] [PubMed]
4. Mensah-Attipoe, J.; Toyinbo, O. Fungal growth and aerosolization from various conditions and materials. In *Fungal Infection*; IntechOpen: Rijeka, Croatia, 2019; pp. 1–10.
5. Kim, I.; Viswanathan, K.; Kasi, G.; Thanakkasaranee, S.; Sadeghi, K.; Seo, J. ZnO nanostructures in active antibacterial food packaging: Preparation methods, antimicrobial mechanisms, safety issues, future prospects, and challenges. *Food Rev. Int.* **2022**, *38*, 537–565. [CrossRef]
6. Alkarri, S.; Vachon, J. Relationship between Particle Size, Anti-Microbial Activity and Leachability of Copper Particles in Liquid Suspension and Compounded in Polypropylene. *Ann. Biomed. Sci. Eng.* **2024**, *8*, 021–031.
7. Walker, T.R.; McGuinty, E.; Charlebois, S.; Music, J. Single-use plastic packaging in the Canadian food industry: Consumer behavior and perceptions. *Humanit. Soc. Sci. Commun.* **2021**, *8*, 80. [CrossRef]
8. Alkarri, S.; Bin Saad, H.; Soliman, M. On Antimicrobial Polymers: Development, Mechanism of Action, International Testing Procedures, and Applications. *Polymers* **2024**, *16*, 771. [CrossRef]
9. Chadha, U.; Bhardwaj, P.; Selvaraj, S.K.; Arasu, K.; Praveena, S.; Pavan, A.; Khanna, M.; Singh, P.; Singh, S.; Chakravorty, A.; et al. Current Trends and Future Perspectives of Nanomaterials in Food Packaging Application. *J. Nanomater.* **2022**, *2022*, 2745416. [CrossRef]
10. Ncube, L.K.; Ude, A.U.; Ogunmuyiwa, E.N.; Zulkifli, R.; Beas, I.N. An Overview of Plastic Waste Generation and Management in Food Packaging Industries. *Recycling* **2021**, *6*, 12. [CrossRef]
11. Gulati, R.; Sharma, S.; Sharma, R.K. Antimicrobial textile: Recent developments and functional perspective. *Polym. Bull.* **2022**, *79*, 5747–5771. [CrossRef]
12. Metak, A.; Ajaal, T. Investigation on polymer based nano-silver as food packaging materials. *World Acad. Sci. Eng. Technol. Int. J. Chem. Mol. Eng.* **2013**, *7*, 1103–1109.
13. Jiménez-Gómez, C.P.; Cecilia, J.A. Chitosan: A natural biopolymer with a wide and varied range of applications. *Molecules* **2020**, *25*, 3981. [CrossRef] [PubMed]
14. Syed, M.R.; Naseem, S.S.; Adeel, T.; Yasmeen, B. Influence of Cu doping in Magnesium Hydroxide Nanoparticles for Bandgap Engineering. *J. Wuhan Univ. Technol. Sci. Ed.* **2023**, *38*, 485–489. [CrossRef]
15. Benyathiar, P.; Kumar, P.; Carpenter, G.; Brace, J.; Mishra, D.K. Polyethylene Terephthalate (PET) Bottle-to-Bottle Recycling for the Beverage Industry: A Review. *Polymers* **2022**, *14*, 2366. [CrossRef] [PubMed]
16. Huang, T.; Qian, Y.; Wei, J.; Zhou, C. Polymeric antimicrobial food packaging and its applications. *Polymers* **2019**, *11*, 560. [CrossRef] [PubMed]
17. Mizielińska, M.; Bartkowiak, A. Overview of Food Antimicrobial Packaging. In *Food Processing and Packaging Technologies—Recent Advances*; IntechOpen: Rijeka, Croatia, 2022.
18. Hutasoit, N.; Topa, S.H.; Javed, M.A.; Rahman Rashid, R.A.; Palombo, E.; Palanisamy, S.J.M. Antibacterial efficacy of cold-sprayed copper coatings against gram-positive staphylococcus aureus and gram-negative *Escherichia coli*. *Materials* **2021**, *14*, 6744. [CrossRef] [PubMed]
19. Fan, X.; Yahia, L.H.; Sacher, E.J.B. Antimicrobial properties of the Ag, Cu nanoparticle system. *Biology* **2021**, *10*, 137. [CrossRef] [PubMed]
20. Birkett, M.; Dover, L.; Cherian Lukose, C.; Wasy Zia, A.; Tambuwala, M.M.; Serrano-Aroca, Á. Recent advances in metal-based antimicrobial coatings for high-touch surfaces. *Int. J. Mol. Sci.* **2022**, *23*, 1162. [CrossRef] [PubMed]
21. Darvish, M.; Ajji, A. Effect of Polyethylene Film Thickness on the Antimicrobial Activity of Embedded Zinc Oxide Nanoparticles. *ACS Omega* **2021**, *6*, 26201–26209. [CrossRef]
22. Vasile, C.; Baican, M. Progresses in Food Packaging, Food Quality, and Safety-Controlled-Release Antioxidant and/or Antimicrobial Packaging. *Molecules* **2021**, *26*, 1263. [CrossRef]
23. Klapiszewski, Ł.; Bula, K.; Dobrowolska, A.; Czaczyk, K.; Jesionowski, T. A high-density polyethylene container based on ZnO/lignin dual fillers with potential antimicrobial activity. *Polym. Test.* **2018**, *73*, 51–59. [CrossRef]
24. Bharadishettar, N.; Bhat, K.U.; Bhat Panemangalore, D.J.M. Coating technologies for copper based antimicrobial active surfaces: A perspective review. *Metals* **2021**, *11*, 711. [CrossRef]
25. Alkarri, S. Investigate the Effect of Coating Concentration and Coating Thickness on the Anti-microbial Properties of Polycarbonate Sheet. *Ann. Biomed. Sci. Eng.* **2024**, *8*, 011–020.
26. Saleh, A.; Melinda, F.; John, C.; Lee, M.; Jin, H.K.; Jonathan, O.R. Investigating anti-bacterial and anti-COVID-19 virus properties and mode of action of pure $Mg(OH)_2$ and copper-infused $Mg(OH)_2$ nanoparticles and coated polypropylene surfaces. *Int. J. Clin. Virol.* **2024**, *8*, 008–023. [CrossRef]
27. Arendsen, L.P.; Thakar, R.; Sultan, A.H. The use of copper as an antimicrobial agent in health care, including obstetrics and gynecology. *Clin. Microbiol. Rev.* **2019**, *32*, e00125-18. [CrossRef]
28. Kharissova, O.V.; Torres-Martínez, L.M.; Kharisov, B.I. *Handbook of Nanomaterials and Nanocomposites for Energy and Environmental Applications*; Springer: Berlin/Heidelberg, Germany, 2021.
29. Domínguez, A.V.; Algaba, R.A.; Canturri, A.M.; Villodres, Á.R.; Smani, Y.J.A. Antibacterial activity of colloidal silver against gram-negative and gram-positive bacteria. *Antibiotics* **2020**, *9*, 36. [CrossRef] [PubMed]

30. Jakubovskis, R.; Ivaškė, A.; Malaiškienė, J.; Urbonavičius, J. Impact of Portland cement type on bacterial viability in biological concrete. *Cem. Concr. Compos.* **2022**, *127*, 104413. [CrossRef]
31. Imani, S.M.; Ladouceur, L.; Marshall, T.; Maclachlan, R.; Soleymani, L.; Didar, T.F.J.A. Antimicrobial nanomaterials and coatings: Current mechanisms and future perspectives to control the spread of viruses including SARS-CoV-2. *ACS Nano* **2020**, *14*, 12341–12369. [CrossRef] [PubMed]
32. Smaoui, S.; Chérif, I.; Hlima, H.B.; Khan, M.U.; Rebezov, M.; Thiruvengadam, M.; Sarkar, T.; Shariati, M.A.; Lorenzo, J.M.J.F.P.; Life, S. Zinc oxide nanoparticles in meat packaging: A systematic review of recent literature. *Food Packag. Shelf Life* **2023**, *36*, 101045. [CrossRef]
33. Peters, K.; Pazos, M.; Edoo, Z.; Hugonnet, J.-E.; Martorana, A.M.; Polissi, A.; VanNieuwenhze, M.S.; Arthur, M.; Vollmer, W. Copper inhibits peptidoglycan LD-transpeptidases suppressing β-lactam resistance due to bypass of penicillin-binding proteins. *Proc. Natl. Acad. Sci. USA* **2018**, *115*, 10786–10791. [CrossRef]
34. Rojas, B.; Soto, N.; Villalba, M.; Bello-Toledo, H.; Meléndrez-Castro, M.; Sánchez-Sanhueza, G.J.N. Antibacterial activity of copper nanoparticles (Cunps) against a resistant calcium hydroxide multispecies endodontic biofilm. *Nanomaterials* **2021**, *11*, 2254. [CrossRef] [PubMed]
35. Gumienna, M.; Górna, B.J.M. Antimicrobial food packaging with biodegradable polymers and bacteriocins. *Molecules* **2021**, *26*, 3735. [CrossRef] [PubMed]
36. Mania, S.; Cieślik, M.; Konzorski, M.; Święcikowski, P.; Nelson, A.; Banach, A.; Tylingo, R.J.P. The synergistic microbiological effects of industrial produced packaging polyethylene films incorporated with zinc nanoparticles. *Polymers* **2020**, *12*, 1198. [CrossRef] [PubMed]
37. Naz, S.; Gul, A.; Zia, M.; Mukhtiar, A. Toxicity of copper oxide nanoparticles: A review study. *IET Nanobiotechnol.* **2020**, *14*, 1–13. [CrossRef] [PubMed]
38. Alkarri, S.; Sharma, D.; Bergholz, T.M.; Rabnawaz, M. Fabrication methodologies for antimicrobial polypropylene surfaces with leachable and nonleachable antimicrobial agents. *J. Appl. Polym. Sci.* **2023**, *141*, e54757. [CrossRef]
39. Su, X.; Jia, C.; Xiang, H.; Zhu, M. Research progress in preparation, properties, and applications of medical protective fiber materials. *Appl. Mater. Today* **2023**, *32*, 101792. [CrossRef] [PubMed]
40. Valko, K.A. Effects of Extrusion Blow Molding Internal Cooling Technology on HDPE Container Performance. Master's Thesis, Michigan State University, East Lansing, MI, USA, 2004.
41. Oliveira, A.C.; Parra, D.F.; Ferreto, H.F.; Lugao, A.B. Characterization of injected linear low density polyethylene (LLDPE) irradiated by gamma-ray. In Proceedings of the INAC 2013: International Nuclear Atlantic Conference, Recife, Brazil, 24–29 November 2013.
42. Alkarri, S. Developing Methods for Incorporating Antimicrobial Biocidal Nanoparticles in Thermoplastics. Ph.D. Thesis, Michigan State University, East Lansing, MI, USA, 2023. Available online: https://d.lib.msu.edu/etd/51541 (accessed on 19 June 2024).
43. Lee, M. Hydroxides Monolayer Nanoplatelet and Methods of Preparing Same. WO2021011235A1, 7 July 2020.
44. Maddan, O.L. Nanoplatelet Magnesium Hydroxides and Methods of Preparing Same. US7736485B2, 9 April 2007.
45. Dong, C.; Cairney, J.; Sun, Q.; Maddan, O.L.; He, G.; Deng, Y. Investigation of $Mg(OH)_2$ nanoparticles as an antibacterial agent. *J. Nanoparticle Res.* **2010**, *12*, 2101–2109. [CrossRef]
46. Huang, Z.; Rajasekaran, P.; Ozcan, A.; Santra, S. Antimicrobial magnesium hydroxide nanoparticles as an alternative to Cu biocide for crop protection. *J. Agric. Food Chem.* **2018**, *66*, 8679–8686. [CrossRef]
47. Okamoto, K.; Kudo, D.; Phuong, D.N.D.; Iwamoto, Y.; Watanabe, K.; Yoshioka, Y.; Ariyoshi, W.; Yamasaki, R. Magnesium Hydroxide Nanoparticles Inhibit the Biofilm Formation of Cariogenic Microorganisms. *Nanomaterials* **2023**, *13*, 864. [CrossRef]
48. Alwaan, I.; Hassan, A.; Jawaid, M. Effect of natural rubber/epoxidized natural rubber (90/10) on mechanical and thermal properties of linear low density polyethylene. *J. Polym. Mater.* **2013**, *30*, 117–130.
49. Nguyen-Tri, P.; Nguyen, T.A.; Carriere, P.; Ngo Xuan, C. Nanocomposite coatings: Preparation, characterization, properties, and applications. *Int. J. Corros.* **2018**, *2018*, 1–19. [CrossRef]
50. Hedenqvist, M.S.; Johansson, K.J.S.; Technology, C. Barrier properties of SiOx-coated polymers: Multi-layer modelling and effects of mechanical folding. *Surf. Coat. Technol.* **2003**, *172*, 7–12. [CrossRef]
51. Vargas-Reus, M.A.; Memarzadeh, K.; Huang, J.; Ren, G.G.; Allaker, R.P. Antimicrobial activity of nanoparticulate metal oxides against peri-implantitis pathogens. *Int. J. Antimicrob. Agents* **2012**, *40*, 135–139. [CrossRef] [PubMed]
52. Singleton, E.V.; David, S.C.; Davies, J.B.; Hirst, T.R.; Paton, J.C.; Beard, M.R.; Hemmatzadeh, F.; Alsharifi, M. Sterility of gamma-irradiated pathogens: A new mathematical formula to calculate sterilizing doses. *J. Radiat. Res.* **2020**, *61*, 886–894. [CrossRef]

Journal of Composites Science

MDPI

Article

The Influence of Activated Carbon Particle Size on the Properties and Performance of Polysulfone Composite Membrane for Protein Separation

Gunawan Setia Prihandana [1,*], Aisyah Dewi Muthi'ah [1], Tutik Sriani [2] and Muslim Mahardika [3]

1 Department of Industrial Engineering, Faculty of Advanced Technology and Multidiscipline, Universitas Airlangga, Jl. Dr. Ir. H. Soekarno, Surabaya 60115, Indonesia; aisyahdm@ftmm.unair.ac.id
2 Department of Research and Development, PT. Global Meditek Utama—IITOYA, Sardonoharjo, Ngaglik, Sleman, Yogyakarta 55581, Indonesia; tsriani@gmail.com
3 Department of Mechanical and Industrial Engineering, Faculty of Engineering, Universitas Gadjah Mada, Jalan Grafika No. 2, Yogyakarta 55281, Indonesia; muslim_mahardika@ugm.ac.id
* Correspondence: gunawan.prihandana@ftmm.unair.ac.id; Tel.: +62-881-0360-00830

Abstract: The superiorities provided by polymeric composite membranes in comparison to the original membrane have generated increased attention, particularly in the field of protein separation applications. This work involved the fabrication of polysulfone composite membranes using variable loadings of activated carbon particle sizes, namely, 37 μm, 74 μm, 149 μm, and 297 μm. The membranes were fabricated via the phase-inversion method, employing water as the coagulant. In this study, the impact of the AC powder particle sizes on membrane morphology, water contact angle, porosity, average pore size, molecular weight cutoff, pure water flux, and protein rejection was examined. Different membrane morphologies and properties were achieved by incorporating a variety of AC particle sizes. A porous membrane with the maximum pure water flux was generated by the loading of finer AC particles. Concurrently, protein rejection is increasing as a result of the use of AC particles as an infill in the composite membrane. In comparison to all fabricated membranes, the AC filler with a particle size of 149 μm exhibited the highest rejection of the lysozyme protein, reaching up to 73.9%, with a relatively high water permeability of 33 LMH/Bar. In conclusion, this investigation provides recommendations for the selection of AC particle sizes for protein separation in conjunction with PSF ultrafiltration membranes.

Keywords: polysulfone; ultrafiltration; membranes; activated carbon; good health; protein separation; clean water

1. Introduction

Ultrafiltration has emerged as a highly efficient technique in the field of separation processes. The cheap investment cost and energy needs, combined with the high separation efficiency, make it highly advantageous for a range of applications, such as food and pharmaceutical processing, as well as the chemical products industry. High-purity proteins have become indispensable raw materials in the food and medical treatment industries [1]. As a result, there has been a huge increase in the need for strong and effective techniques for separating and purifying proteins. Various methods, such as membrane technology, centrifugation, precipitation, sedimentation, electrophoresis, and chromatography, have been created to obtain elevated degrees of protein purification [2–5]. Membrane technology is considered an efficient method for protein treatment because of its high separation efficiency. It preserves the structure, integrity, and efficacy of proteins while minimizing the need for chemical usage [6]. The operational principle of separation membranes is impeding molecules that exceed the size of the membrane pore. As a result, membranes of varying pore sizes can separate molecules according to their molecular weight. Membrane filtration can be used in a wide range of separation situations by choosing membranes with suitable pore diameters [7]. Ultrafiltration membranes are often effective for concentrating and separating proteins because their pore sizes are large enough to prevent protein molecules from passing through the membrane pores [8]. Membrane fouling, which involves the adsorption of solute on the membrane surface and the blocking of pores, greatly decreases the effectiveness of membrane separation [9]. Membranes are commonly produced using polymers such as polysulfone, polyethersulfone, polyamide, polyethylene, polypropylene, and polyvinyl fluoride. These polymers are chosen for their excellent mechanical, thermal, and chemical resistance properties [10].

Polysulfone (PSF) possesses exceptional thermal stability, elevated mechanical strength, and chemical resistance, rendering it an ideal substance for the production of filtration membranes [11,12]. Consequently, research endeavors on PSF membranes have experienced a substantial surge in the past decade. PSF membranes can be modified using techniques including physical mixing, chemical grafting, and coating. These processes are used to improve the performance of the membranes [13]. Although surface coating and chemical grafting onto premade membranes are convenient and flexible techniques, they encounter notable difficulties, such as pore structural deformation. This deformation results in a decrease in permeate flux, which ultimately restricts membrane performance [14]. In order to tackle these difficulties, the inclusion of inorganic elements in the solution is of great importance as it can improve the hydrophilicity, permeation, and antifouling properties [15,16]. The use of carbon-based materials in polymers to create ultrafiltration membranes has gained significant attention due to its potential to regulate pore shape, boost hydrophilicity, improve antifouling characteristics, and increase mechanical strength [17]. Nevertheless, the findings suggest that an excessive inclusion of inorganic particles has a negative impact on the selectivity, permeability, and shape of the membranes. The main goal is to employ appropriate inorganic materials with a comparatively lower proportion of inorganic fillers, improving the hydrophilicity of the membrane without compromising its permeability and selectivity [18].

Composite membranes composed of PSF and carbon nanotubes have been reported to demonstrate enhanced hydrophilicity, protein rejection, and water permeate flux [19]. Carbon-based nanoparticles are effective modifiers due to their hydrophilic surface functional groups, chemical stability, lack of toxicity, and excellent surface area [20]. Hosseini et al. [21] developed mixed matrix nanofiltration membranes based on polyethersulfone embedded with activated carbon nanoparticles (ACNPs) to remove sulfate and copper from water. The results indicate that the addition of super activated carbon nanoparticles to the casting solution significantly enhanced the rejection of sulfate and copper ions. Sharma et al. [22] fabricated polyethersulfone nanofiltration membranes by incorporating activated carbon nanoparticles at concentrations ranging from 0 to 1 wt.% to separate lignin from black liquor. The study's findings showed that AC-based membranes with a concentration of 0.1% had improved efficiency in separating lignin from black liquor, while also reducing membrane fouling. Nevertheless, while choosing additive materials, it is crucial to take into account not only their performance but also factors such as their local availability and cost-effectiveness [23]. Consequently, in light of the exorbitant expense associated with synthesizing nanoparticles, scientific investigation has prioritized the search for more economical alternatives in terms of powder size. Remy et al. [24] found that the addition of powdered activated carbon raised the critical flow by 10% and considerably prolonged the filtration time of the membrane.

Liu et al. [25] showed that the inclusion of powdered activated carbon resulted in longer membrane filtering time cycles, suggesting a decreased likelihood of membrane fouling. Torretta et al. [26] discovered that using a certain concentration of powdered activated carbon was the most efficient method for improving membrane flow, resulting in a notable 26% increase. To summarize, the use of powdered activated carbon as a filler has been found to improve the performance of composite membranes through multiple functions and mechanisms. Nevertheless, there is a dearth of research investigating the selectivity and permeability of composite membranes with different diameters of powdered activated carbon. The aim of this work is to manufacture a PSF composite membrane with varying loadings of AC particle sizes and investigate how these parameters impact the membrane characteristics and performance in protein separation.

The work involved incorporating activated carbon particles of different sizes (37 μm, 74 μm, 149 μm, and 297 μm) into a solution of PSF membrane. This was carried out to create composite ultrafiltration membranes using the wet phase inversion procedure. An assessment was conducted to determine the influence of activated carbon particle size on hydrophilicity, porosity, average pore size, pore structure, and molecular weight cutoff. The surface morphology of the membranes was examined using scanning electron microscopy (SEM). The evaluation of water permeability and protein separation efficacy of the membranes was performed using standardized experimental methodologies.

2. Materials and Methods

2.1. Materials

Polysulfone pellets (PSF, 80,000 g/mol, P.T. Solvay Chemicals Indonesia, Jakarta, Indonesia), N-methyl-2-pyrrolidone (NMP; PT. Samiraschem Indonesia, Jakarta, Indonesia), Activated Carbon (AC) powder, Filtrasorb 100 (Calgon Carbon Corporation, Pittsburgh, PA, USA), and an adjustable film applicator (Elcometer 3540/4 Four-Sided Film Applicators, Elcometer Limited, Manchester, UK) were purchased to prepare the PSF-AC membranes. Lysozyme (12 kDa), pepsin (35 kDa), and bovine serum al-bumin (BSA) (69 kDa) proteins from HiMedia Ltd. in Mumbai, India were purchased to prepare protein solutions. Pure water was used as a gelatinization medium for the membrane fabrication, and droplet for the water contact angle test and membrane permeability test. The Scanning Electron Microscopy (SEM) instrument, Phenom ProX (Thermo Fisher Scientific, Waltham, MA, USA) was employed to analyze the cross section and morphological properties of the membranes. AC powder was utilized as received, and its specifications are detailed in Table 1.

Table 1. Specification of AC powder.

Iodine Number, mg/g	850
Moisture by Weight	2%
Effective Size	0.8–1.0
Uniformity Coefficient	2.1
Abrasion Number	75
Screen Size by Weight, US Sieve Series	
On 8 mesh	15%
Through 30 mesh	4%
Apparent Density (tamped)	0.58 g/cc
Water Extractables	<1%
Non-Wettable	<1%

2.2. Membrane Fabrication

Activated carbon-polysulfone (AC-PSF) membranes for protein separation were prepared by dissolving PSF in NMP, incorporating AC carbon as an additive, casting a thin film from the resulting solution (containing PSF, NMP, and AC) at varying concentrations (as indicated in Table 2) onto a non-woven support, and immersing the non-woven substrate in a pure water bath using the wet phase inversion method. The casting solutions consisted of PSF with specific particle size distributions: pristine membrane, 297 μm, 149 μm, 74 μm, and 34 μm, all at a concentration of 1 wt.% as membrane filler.

Table 2. Membrane prepared in this study.

No.	Membrane Code	AC Powder Size (μm)
1	PSF-CA0	-
2	PSF-CA1	297
3	PSF-CA2	149
4	PSF-CA3	74
5	PSF-CA4	37

2.3. Water Contact Angle Test

A digital microscope (Dinolite Edge 3.0 AM73915MZTL, AnMo Electronics, New Taipei City, Taiwan) was employed to capture images of a 0.5 μL pure water droplet on the dried surface. Additionally, CAD software was utilized to measure the contact angle of the membranes. The contact angles for each membrane were measured three times and then averaged to ensure accuracy.

2.4. Ultrafiltration Experiments

2.4.1. Water Flux Test

The water flux of all membranes, including both pristine and PSF-AC membranes, was assessed using a dead-end cell setup, as depicted in Figure 1. This setup incorporates valves and pressure gauges to regulate the pressure within the feed chamber. Initially, the membrane sample was positioned at the bottom of the dead-end cell, with an active surface area of 14.6 cm^2. Above the tested membrane, there was a feed chamber with a capacity of 300 mL. A pressure regulator was employed to control the applied pressure from the nitrogen cylinder to the feed chamber. A working pressure of 2 bar was applied to the fabricated membranes. The permeate weight was automatically recorded by a computer every 2 min over a duration of 30 min in this filtration system. Subsequently, at least three replicate samples were assessed for each membrane, and the average value, along with its standard deviation, was reported.

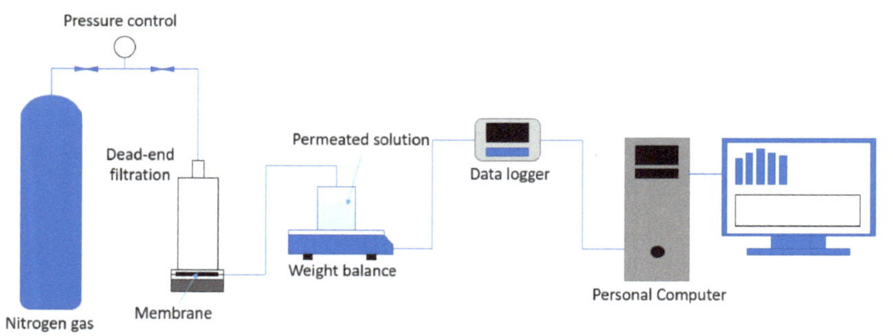

Figure 1. Experiment setup of the water flux test.

We employed the following equations [27] to calculate both the volumetric flux (J_v) and the permeability:

$$\text{Flux}(J_v) = \frac{Q}{A \times \Delta t} \quad (1)$$

$$\text{Permeability}(L_p) = \frac{J_v}{\Delta P} \quad (2)$$

In the context of membrane analysis, Q represents the quantity of permeate water (measured in liters) during the sampling time, Δt denotes the duration of the sampling period (in hours), A signifies the membrane area (measured in square meters), and ΔP represents the pressure difference (in bars).

2.4.2. Protein Separation

A solution containing 0.1 wt.% of BSA, pepsin, and lysozyme was carefully produced in a phosphate-buffered solution with a pH of 7.2. The ensuing protein separation experiment was carried out using a dead-end cell filtration test, with a fixed pressure of 2 bar. In order to measure the concentration of the protein solution that passed through, we used an UV–visible spectrophotometer set to a wavelength of 280 nm. The solute rejection (SR) was determined through meticulous analysis.

$$\%\text{SR} = \left[1 - \frac{C_p}{C_f}\right] \times 100 \quad (3)$$

In the context of protein concentration analysis, C_p represents the concentration of the permeated solution, while C_f denotes the concentration of the feed solution.

2.5. Porosity

The study focused on analyzing the membrane's porosity in order to assess the impact of including AC powder on the size of the membrane pores. This was carried ut by the gravimetric method, using the following equation [28]:

$$\varepsilon(\%) = \frac{W_w - W_d}{\rho H_2O \times A \times L} \times 100 \quad (4)$$

where W_w is the weight of the wet membrane, W_d is weight of the dry membrane, the effective area of the membrane is denoted by A, the thickness of the membrane is denoted by L, and the density of water (0.998 g/cm^3)is denoted by ρH_2O.

2.6. Average Pore Size

The average pore size of the surface membranes was estimated using the ultrafiltration experimental findings. In order to determine the average pore size of the membranes, we employed the molecular weight of the solute, which demonstrated a solute rejection (SR) greater than 80%, by utilizing the following equation:

$$\overline{R} = 100\left(\frac{\alpha}{\%SR}\right) \quad (5)$$

In this context, $\overline{R}$ indicates the mean pore size (radius), while α refers to the solute radius, which is calculated by the Stoke radius computed from the solute's molecular weight using Sarbolouki's approach [29].

2.7. Molecular Weight Cutoff (MWCO)

The molecular weight cutoff (MWCO) has a direct correlation with the pore size of the membrane. The objective of evaluating the MWCO of a membrane is to determine the minimum size of an unreactive solute that exhibits protein rejection within the range of 80% to 100% in an ultrafiltration experiment [30]. The BSA, pepsin, and lysozyme were chosen as the proteins to be evaluated in order to assess any decrease in pore size that may occur as a result of introducing activated carbon powder.

3. Results

3.1. Contact Angle Analysis

In order to confirm the hydrophobic nature of the membranes, we conducted additional measurements of the water contact angle for various composite membranes using pure water. A membrane surface that is more hydrophilic is indicated by a lower water contact angle value.

Figure 2 displays the static water contact angles of several PSF-CA composite membranes. The pristine PSF membrane (PSF-AC0) exhibits an initial contact angle of around 67.6° due to its intrinsic hydrophilicity. The introduction of a 1% AC particle at any range of powder size resulted in a lower contact angle of 59.5°, 59°, 58.2°, and 57.5° for PSF-AC1, PSF-AC2, PSF-AC3, and PSF-AC4, respectively, in comparison to pristine PSF membranes (67.6°). This indicates that the contact angle was further reduced by further reducing the size of AC particles. This decline indicates a significant rise in the hydrophilicity of the membranes.

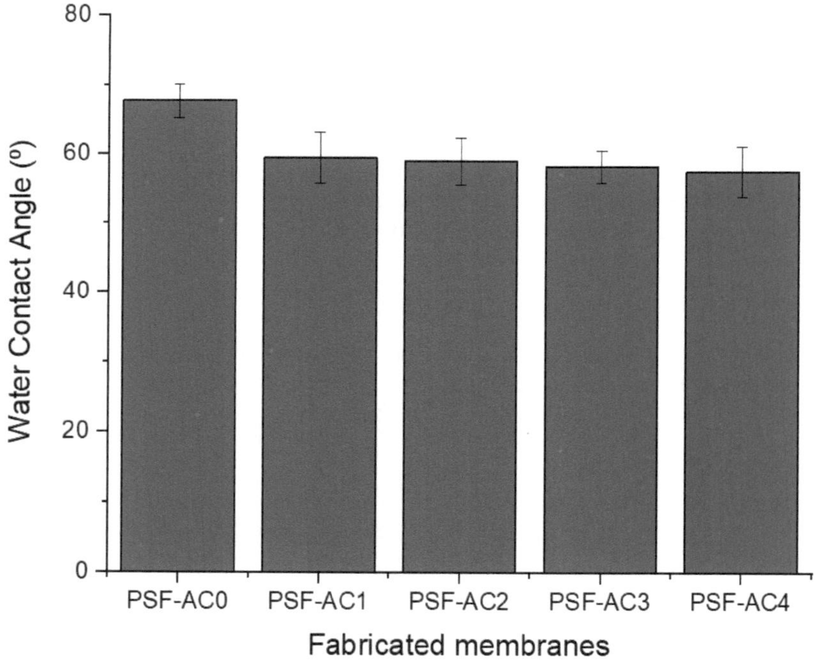

Figure 2. Water contact angle of the fabricated membranes.

The physical characteristics of powder, such as the size of particles, can affect how well it can be wetted. Powders with a greater particle size have superior wetting characteristics due to the enhanced ability of water to permeate the bigger interstitial gaps between the particles [31]. However, in this scenario, the interstitial gaps between larger particles are the surface of the membrane, which possesses a higher water contact angle. Consequently, this circumstance results in the PSF-AC membrane with larger particle sizes exhibiting more hydrophobicity.

3.2. Membrane Permeability Test

Figure 3 illustrates the water permeability of membranes with varying AC particle sizes incorporated. Upon comparing the water permeability of the pristine PSF membrane (14.75 LMH/Bar) with the fluxes of the composite membranes, it was observed that the fluxes increased as the AC particle size decreased, at the same AC concentration of 1 wt.%. The permeate flux with an activated carbon particle size of 297 microns is 32.5 LMH/bar, reaching a maximum value of 42.5 LMH/bar with activated carbon particles of 37 microns. According to Figure 2, it is evident that the improved water flux is a result of the heightened hydrophilicity, since the PSF-AC4 membranes exhibit the highest level of hydrophilicity. As stated in the contact angle analysis section, when the particle size is small, it aids in achieving a uniform distribution on the surface of the membrane, hence facilitating water to permeate through its porous multichannel structure, resulting in a higher water permeability.

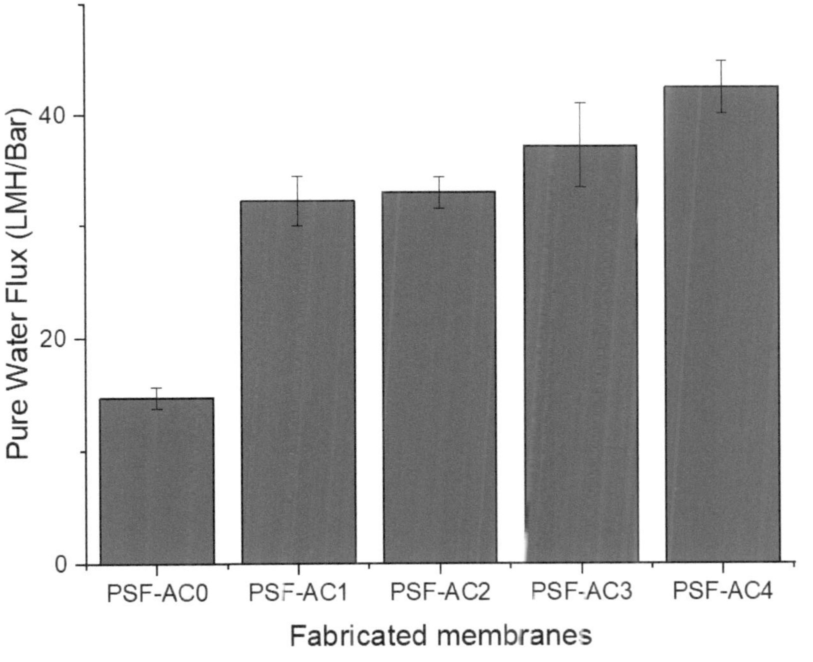

Figure 3. Water flux of the fabricated membranes.

Moreover, the addition of larger particle sizes increases the likelihood of aggregation, which, in turn, widens the space between particles and exposes the hydrophobic surface of the membrane. This leads to a decrease in effective porosity by blocking channels that allow vapor to pass through [32]. As a result, the water transfer facilitated by the porous multichannel structure of AC is compromised.

3.3. Effect of AC on Protein Rejection

The rejection and permeate capabilities of membranes depend on their morphological structure in conjunction with the parameters of the feed solution. In this study, three distinct proteins of varying sizes were employed as representative solutions for protein separation experiments. The solution containing BSA, pepsin, and lysozyme dissolved in water is used as a feed solution to study how the membranes reject these substances. Figure 4 displays the percentage of membrane protein rejection results, which were determined using Equation (3). The figure illustrates that the manufactured membranes exhibit a BSA rejection rate of 90%. All membranes have a pore size that is smaller than the molecular weight of BSA. The separation test for the membranes was extended using pepsin and lysozyme as solutes due to their smaller molecular sizes compared to BSA. Figure 4 demonstrates that the composite membrane exhibits a somewhat higher rejection of pepsin in comparison to the pristine membrane. A substantial disparity in protein rejection was seen when employing a lysozyme solution, with composite membranes exhibiting notable rejection of lysozyme. As shown in Figure 4, the PSF-AC2 membrane has the highest lysozyme rejection rate (73.9%), whereas the PSF-AC0 membrane has the lowest (49.3%). This can be elucidated by considering the size of the pores in the membrane. The membrane PSF-AC2 has the smallest pore size compared to all other produced membranes, resulting in the maximum rejection of lysozyme. On the other hand, the membrane PSF-AC0 has the biggest pore size and exhibits the lowest rejection of lysozyme. Hence, the membrane demonstrated superior size-exclusion separation properties, allowing it to reject protein more effectively than other membranes that were fabricated [33].

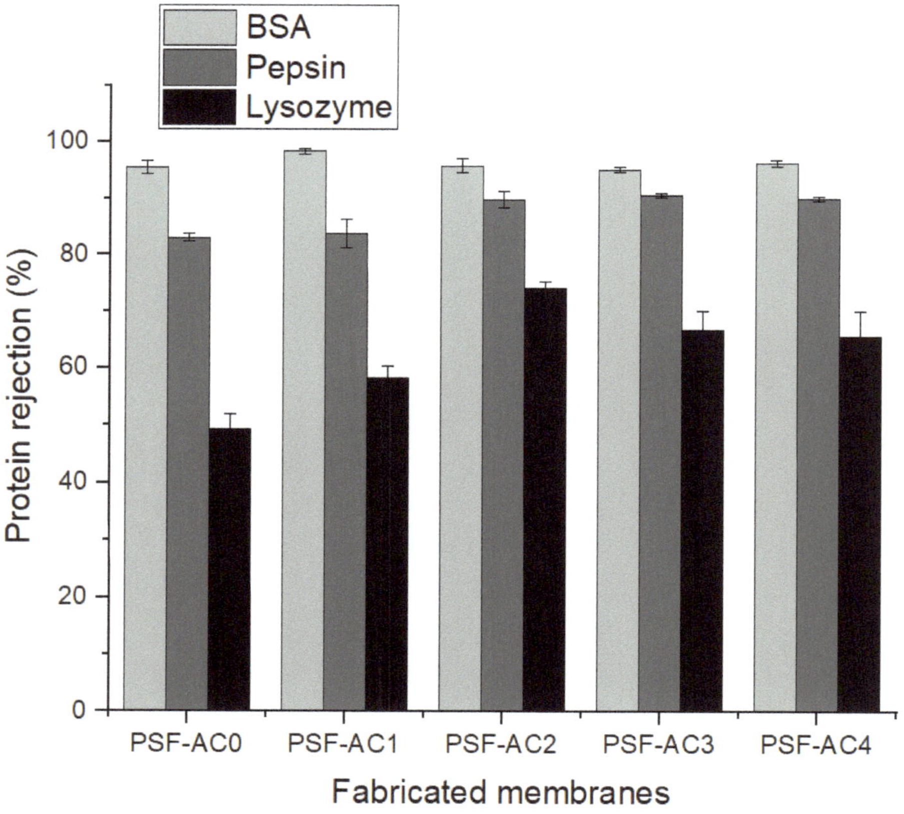

Figure 4. Protein rejection of the fabricated membranes.

3.4. Effect of AC on Membranes Porosity

An analysis of membrane porosity was used to evaluate the effect of nAC on the modified polymer membrane. The results of the apparent porosity of membranes that were incorporated with AC as an additive are presented in Table 3. The particulate size of the membranes was varied, with pristine membrane (PSF-AC0), 297 μm (PSF-AC1), 149 μm (PSF-AC2), 74 μm (PSF-AC3), and 37 μm (PSF-AC4). The pristine membrane had the lowest reported porosity values (63.41 8%) when AC was not present. The manufactured membrane reached its maximum porosity of 66.92% when the smallest AC powder of 37 μm was loaded. This phenomenon arises due to the fact that, when the concentration of AC is held constant at 1 wt.%, the decrease in particle size causes a larger quantity of particles, that, in turn, leads to a membrane with higher porosity and, as a result, an increased water flux [34].

Table 3. Porosity, pore radius, and MWCO of the fabricated membranes.

Membrane Code	Porosity (%)	Pore Radius, $\bar{R}(A^\circ)$	MWCO (kDa)
PSF-AC0	63.41	30.12	35
PSF-AC1	63.46	29.88	35
PSF-AC2	66.02	27.88	35
PSF-AC3	66.61	27.66	35
PSF-AC4	66.92	27.83	35

3.5. Measurement of Average Pore Size

The average pore size of the fabricated membranes was calculated using Equation (5) and is depicted in Table 3. The results suggest that the introduction of activated carbon particles into the membrane matrix resulted in a decrease in the size of the pores in composite membranes. Specifically, a smaller average pore radius was observed as the activated carbon powder size decreased from 297 μm to 37 μm in the case of composite membranes.

The pristine membrane, PSF-AC0, had a pore size of 30.12 A°, whereas the composite membranes had pore sizes ranging from 27.83 A° to 29.88 A°. Nevertheless, the membrane made with the smallest AC powder particles (37 μm) had a larger pore size (27.83 A°) compared to PSF-AC3, 74 μm (27.66 A°). The membranes exhibited the following order of pore size: PSF-AC0 (pristine membrane) > PSF-AC1 (297 μm) > PSF-AC2 (149 μm) > PSF-AC4 (74 μm) > PSF-AC3 (37 μm). The smaller powder will yield a reduced pore size since it tends to spread uniformly across the membrane structure, resulting in smaller pores. The introduction of the 37 μm AC particle led to a little increase in pore size, possibly attributable to the non-uniformity of the powder size, as the powder was separated using a sieve shaker, resulting in smaller particles under 37 μm remaining during the separation process.

3.6. Molecular Weight Cutoff Measurement (MWCO)

Protein solutions with molecular weights of 14 kDa for lysozyme, 35 kDa for pepsin, and 69 kDa for BSA were employed to ascertain the molecular weight cutoff measurement for both pristine and PSF-AC composite membranes, with varying powder sizes of AC. The result of MWCO is presented in Table 3. As shown in Table 3, the molecular weight threshold of all membranes was initially 35 kDa, which is higher than the MWCO of lysozyme (14 kDa). Nonetheless, whereas all membranes exhibit an identical molecular weight cutoff (MWCO) of 35 kDa, there are substantial variations in the rejection % of lysozyme across each membrane. For instance, as illustrated in Figure 4, AC with a powder size of 149 μm exhibits a gap of 73.9% relative to the lysozyme molecular weight cutoff standard of 80% [31], indicating that, despite the membranes possessing identical MWCO levels, they demonstrate varying percentages of solute rejection.

3.7. Membrane Morphology

In this study, we provide the surface morphology SEM images of the PSF pristine membranes and the membranes filled with AC powder as presented in Figure 5. Undoubtedly, the impact of AC powder on the surface morphology of the PSF membranes is substantial. Initially, the PSF composite membranes have a uniformly dispersed surface with an increased number of pores. This phenomenon can be elucidated by the capacity of the fine particles to infiltrate and establish a more compact structure with reduced pore size [35].

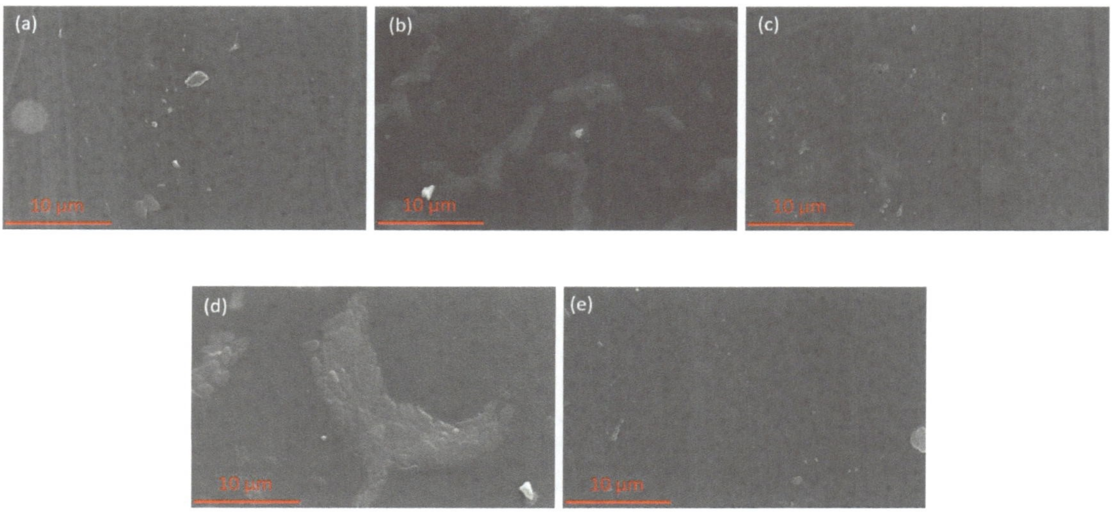

Figure 5. Surface morphology SEM images of the fabricated membranes: (**a**) PSF-AC0; (**b**) PSF-AC1; (**c**) PSF-AC2; (**d**) PSF-AC3; and (**e**) PSF-AC4.

The fabricated membranes are illustrated in cross-sectional SEM images in Figure 6. The PSF membranes are composed of a sponge-like layer and a finger-like layer. The selective layer is situated on the side of the membranes that contains the finger-like layer. The cross-sectional morphology of the modified membranes, which contain AC powder, is comparable to that of PSF pristine membranes. When coarse powder was utilized, the SEM images reveal that the membrane exhibited a greater tendency to form macro void structures rather than sponge-like structures [36]. The reduction of finger-like macro gaps by fine particles can be attributed to their capacity to efficiently fill the interstices, therefore enhancing the density and reducing the porosity of the structure [35].

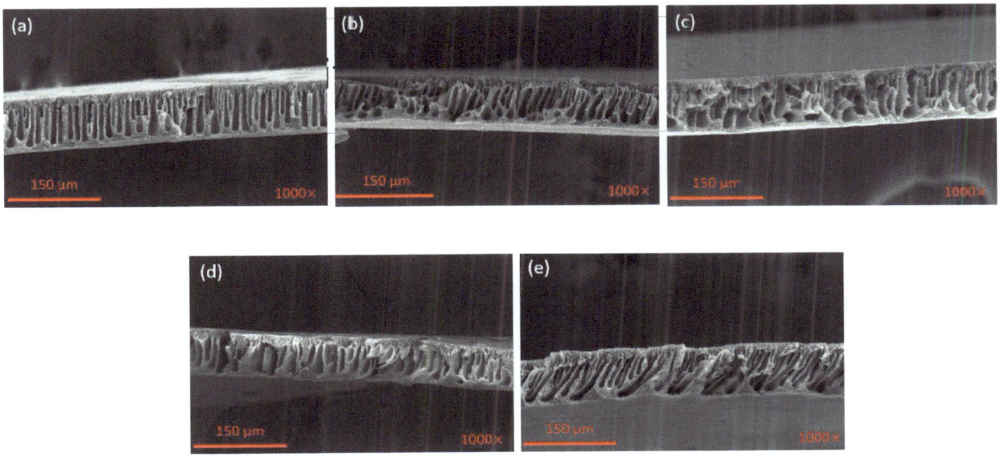

Figure 6. Cross section SEM images of the fabricated membranes: (**a**) PSF-AC0; (**b**) PSF-AC1; (**c**) PSF-AC2; (**d**) PSF-AC3; and (**e**) PSF-AC4.

Table 4 compares previous studies on PSF membrane performance that included AC and other additives as the principal modulator. In comparison to its modified counterpart, each column displays the value of the PSF membrane. AC, silver-nanoparticles, polyethylene glycol (PEG), polyvinylpyrrolidone (PVP), carbon nanotubes (CNT), CuCl2, FeCl3, and APS were among the additives that were incorporated. The majority of membrane modification investigations utilized PEG and PVP as pore formers. Additional additives at varying sizes and concentrations were employed to change the membrane. Each method improves the membranes' hydrophilicity, flux, porosity, average pore size, rejection capabilities, and morphology. Although the AC particles remained unmodified before application, the optimal membrane in this study exhibits a performance akin to comparable membranes from previous research regarding water permeability, protein rejection, and water contact angle. The membrane modification technique outlined in this work may serve as a pivotal factor for the expansion of membrane production in the protein separation sector.

Table 4. Overview of research on HA blending in polymeric membranes comparing the native membrane and modified membranes.

Polymer	Modifier	Permeability Flux, LMH/Bar	Pore Size, nm	Rejection (%)	ε (%)	WCA (°)	Ref.
PSF	AC	53.3/116.3	34.1/43.2	~93.2 (BSA)	-	-	[37]
PSF/PVP	Silver-nanoparticle	12/55	-	~48 (BSA)	-	81.2/60.9	[33]
PSF	PANI-CuCl$_2$, PANI-FeCl$_3$ and PSF/PANI/APS	30/450	92.12/122.94	96 (BSA)	-	41.8-80	[29]
PSF/PEG	AC	5/9.8	6.4/7.2	81–93 (total phenolic compounds)	83–86	48.2–66.7	[40]
PSF	CNT	1/8	2.82/10.18	98 (lignin)	74.4/75.5	57.6/74.4	[28]
PSF	AC	15/42.5	27.66/30.12	98 (BSA) 90 (Pepsin) 73.9 (Lysozyme)	63.41/66.92	57.5/67.6	This work

4. Conclusions

A series of polysulfone ultrafiltration composite membranes were fabricated by altering the size of AC particles via phase inversion processing. The membranes exhibited a unique shape and performance dependent on the various sizes of AC particles used as a filler. The inclusion of particles of varying sizes has an impact on the morphology of the membrane, namely, in relation to the finger-like and sponge structure of the membrane. All the composite membranes demonstrated a superior water flux and rejection of the lysozyme in comparison to the pristine membrane. The findings derived from this investigation are indicative of the significance of the AC particle size in influencing the structure and functionality of the composite PSF membrane. Hence, the appropriate choice of particle size is crucial in assessing the effectiveness of membrane performance for different applications, especially in the context of protein separation and purification.

Author Contributions: Conceptualization, G.S.P., M.M. and T.S.; methodology, G.S.P.; software, A.D.M. and T.S.; validation, G.S.P., M.M. and T.S.; formal analysis, T.S. and G.S.P.; investigation, G.S.P. and T.S.; resources, G.S.P., A.D.M. and M.M.; data curation, T.S.; writing—original draft preparation, G.S.P.; writing—review and editing, T.S. and M.M.; visualization, T.S.; supervision, M.M.; project administration, A.D.M.; funding acquisition, G.S.P., A.D.M. and M.M. All authors have read and agreed to the published version of the manuscript.

Funding: This research was funded by Hibah Direktorat Riset, Teknologi dan Pengabdian kepada Masyarakat 2023, Scheme: Penelitian Fundamental Reguler (Grant No. 0536/E5/PG.02.00/2023; 114/E5/PG.02.00.PL/2023; 1338/UN3.LPPM/PT.01.03/2023).

Data Availability Statement: The data presented in this study are available upon request from the corresponding author.

Acknowledgments: The authors would like to thank the staff of Lembaga Penelitian Dan Pengabdian Masyarakat, Universitas Airlangga, Indonesia, for their administrative support.

Conflicts of Interest: The authors declare no conflicts of interest.

References

1. Qiang, T.; Song, Y.; Zhu, R.; Yuan, W. Biomass material derived hierarchical porous TiO_2: Adjustable pore size for protein adsorption. *J. Alloys Compd.* **2020**, *829*, 154512. [CrossRef]
2. Kluszczynska, K.; Peczek, L.; Rozanski, A.; Czernek, L.; Duchler, M. U6/miR-211 expression ratio as a purity parameter for HEK293 cell-derived exosomes. *Acta Biochim. Pol.* **2022**, *69*, 409–415. [CrossRef] [PubMed]
3. Arbita, A.A.; Paul, N.A.; Cox, J.; Zhao, J. Amino acid sequence of two new milk-clotting proteases from the macroalga Gracilaria edulis. *Int. J. Biol. Macromol.* **2022**, *211*, 499–505. [CrossRef] [PubMed]
4. Ogata, M.; Sakamoto, M.; Yamauchi, N.; Nakazawa, M.; Koizumi, A.; Anazawa, R.; Kurumada, K.; Hidari, K.I.P.J.; Kono, H. Optimization of the conditions for the immobilization of glycopolypeptides on hydrophobic silica particulates and simple purification of lectin using glycopolypeptide-immobilized particulates. *Carbohydr. Res.* **2022**, *519*, 108624. [CrossRef]
5. Chai, M.; Ye, Y.; Chen, V. Separation and concentration of milk proteins with a submerged membrane vibrational system. *J. Membr. Sci.* **2017**, *524*, 305–314. [CrossRef]
6. Ding, L.; Zhang, W.; Ould-Dris, A.; Jaffrin, M.Y.; Tang, B. Concentration of milk proteins for producing cheese using a shear-enhanced ultrafiltration technique. *Ind. Eng. Chem. Res.* **2016**, *55*, 11130–11138. [CrossRef]
7. Xu, Z.; Ma, G.; Rana, D.; Matsuura, T.; Lan, C.Q. Engineering polyvinylidene fluoride membranes using charge tunable dendrimer polyamidoamine to enable microporous membranes for protein separation. *Sep. Purif. Technol.* **2025**, *354 Pt 4*, 128887. [CrossRef]
8. Kassa, S.T.; Hu, C.C.; Keshebo, D.L.; Ang, M.B.M.; Lai, J.Y.; Chu, J.P. Surface modification of high-rejection ultrafiltration membrane with antifouling capability using activated oxygen treatment and metallic glass deposition. *Appl. Surf. Sci.* **2020**, *529*, 147131. [CrossRef]
9. Jin, Z.; Shen, Y.; Chen, X.; Qiu, M.; Fan, Y. Construction of high-performance Ce-doped TiO_2 tight UF membranes for protein separation. *Appl. Surf. Sci.* **2023**, *610*, 155468. [CrossRef]
10. Simari, C.; Caprì, A.; Rehman, M.H.U.; Enotiadis, A.; Gatto, I.; Baglio, V.; Nicotera, I. Composite anion exchange membranes based on polysulfone and silica nanoscale ionic materials for water electrolyzers. *Electrochim. Acta* **2023**, *462*, 142788. [CrossRef]

11. Chen, S.; Li, L.; Feng, M.; Huang, T.; Zhang, N.; Wang, Y. UV-activated superwetting ability of electrospun polysulfone/titanium dioxide membranes toward highly efficient methylene blue removal and oil/water separation. *J. Membr. Sci.* **2024**, *695*, 122450. [CrossRef]
12. Mamah, S.C.; Goh, P.S.; Ismail, A.F.; Suzaimi, N.D.; Yogarathinam, L.T.; Raji, Y.O.; El-badawy, T.H. Recent development in modification of polysulfone membrane for water treatment application. *J. Water Process. Eng.* **2021**, *40*, 101835. [CrossRef]
13. Pagidi, A.; Saranya, R.; Arthanareeswaran, G.; Ismail, A.F.; Matsuura, T. Enhanced oil–water separation using polysulfone membranes modified with polymeric additives. *Desalination* **2014**, *344*, 280–288. [CrossRef]
14. Khan, A.; Sherazi, T.A.; Khan, Y.; Li, S.; Naqvi, S.A.R.; Cui, Z. Fabrication and characterization of polysulfone/modified nanocarbon black composite antifouling ultrafiltration membranes. *J. Membr. Sci.* **2018**, *554*, 71–82. [CrossRef]
15. Zhang, Y.; Cui, P.; Du, T.; Shan, L.; Wang, Y. Development of a sulfated Y-doped nonstoichiometric zirconia/polysulfone composite membrane for treatment of wastewater containing oil. *Sep. Purif. Technol.* **2009**, *70*, 153–159. [CrossRef]
16. Arthanareeswaran, G.; Thakur, R.S. Effect of inorganic particle on the performance of polyethersulfone-cellulose acetate ultrafiltration membranes. In *Sustainable Membrane Technology for Energy, Water, and Environment*; Ismail, A.F., Matsuura, T., Eds.; John Wiley & Sons: Hoboken, NJ, USA, 2012; p. 11.
17. Abid, Z.; Abbas, A.; Mahmood, A.; Rana, N.F.; Khan, S.J.; Duclaux, L.; Deen, K.M.; Ahmad, N.M. Water Treatment Using High Performance Antifouling Ultrafiltration Polyether Sulfone Membranes Incorporated with Activated Carbon. *Polymers* **2022**, *14*, 2264. [CrossRef]
18. Manawi, Y.; Kochkodan, V.; Hussein, M.A.; Khaleel, M.A.; Khraisheh, M.; Hilal, N. Can carbon-based nanomaterials revolutionize membrane fabrication for water treatment and desalination. *Desalination* **2016**, *391*, 69–88. [CrossRef]
19. Khalid, A.; Al-Juhani, A.A.; Al-Hamouz, O.C.; Laoui, T.; Khan, Z.; Atieh, M.A. Preparation and properties of nanocomposite polysulfone/multi-walled carbon nanotubes membranes for desalination. *Desalination* **2015**, *367*, 134–144. [CrossRef]
20. Bagheripour, E.; Moghadassi, A.R.; Hosseini, S.M.; Ray, M.B.; Parvizian, F.; Van der Bruggen, B. Highly hydrophilic and antifouling nanofiltration membrane incorporated with water-dispersible composite activated carbon/chitosan nanoparticles. *Chem. Eng. Res. Des.* **2018**, *132*, 812–821. [CrossRef]
21. Hosseini, S.M.; Amini, S.H.; Khodabakhshi, A.R.; Bagheripour, E.; Van der Bruggen, B. Activated carbon nanoparticles entrapped mixed matrix polyethersulfone based nanofiltration membrane for sulfate and copper removal from water. *J. Taiwan Inst. Chem. Eng.* **2018**, *82*, 169–178. [CrossRef]
22. Sharma, M.; Alves, P.; Gando-Ferreira, L.M. Effect of activated carbon nanoparticles on the performance of PES nanofiltration membranes to separate Kraft lignin from black liquor. *J. Water Process Eng.* **2023**, *52*, 103487. [CrossRef]
23. Noorani, N.; Barzegar, B.; Mehrdad, A.; Aghdasinia, H.; Peighambardoust, S.J.; Kazemian, H. CO_2 capture in activated pyrolytic coke/metal oxide nanoparticle composites. *Colloids Surf. A Physicochem. Eng. Asp.* **2023**, *679*, 132554. [CrossRef]
24. Remy, M.; Potier, V.; Temmink, H. Why low powdered activated carbon addition reduces membrane fouling in MBRs. *Water Res.* **2010**, *44*, 861–867. [CrossRef] [PubMed]
25. Wang, Y.; Xu, Y.; Zhang, S.; Li, Y.; Liu, W. Effects of powdered activated carbon dosage on the performance of membrane bioreactors treating biochemical tail water. *Sci. Total Environ.* **2023**, *874*, 162429. [CrossRef] [PubMed]
26. Torretta, V.; Urbini, G.; Raboni, M.; Copelli, S.; Viotti, P.; Luciano, A.; Mancini, G. Effect of powdered activated carbon to reduce fouling in membrane bioreactors: A sustainable solution.Case study. *Sustainability* **2013**, *5*, 1501–1509. [CrossRef]
27. Kanagaraj, P.; Nagendran, A.; Rana, D.; Matsuura, T.; Neelakandan, S.; Malaivizhi, K. Effects of Polyvinylpyrrolidone on the permeation and fouling-resistance properties of Polyetherimide ultrafiltration membranes. *Ind. Eng. Chem. Res.* **2015**, *54*, 4832–4838. [CrossRef]
28. Ferreira, I.; Alves, P.; Gil, M.H.; Gando-Ferreira, L.M. Lignin separation from black liquor by mixed matrix polysulfone nanofiltration membrane filled with multiwalled carbon nanotubes. *Sep. Purif. Technol.* **2021**, *260*, 118231.
29. Sarbolouki, M.N. A general diagram for estimating pore size of ultrafiltration and reverse osmosis membranes. *Sep. Sci. Technol.* **1982**, *17*, 381–386. [CrossRef]
30. Nagendran, A.; Mohan, D. Cellulose acetate and polyetherimide blend ultrafiltration membranes: II. Effect of additive. *Polym. Adv. Technol.* **2008**, *19*, 24–35. [CrossRef]
31. Fitzpatrick, J.J.; Salmon, J.; Ji, J.; Miao, S. Characterisation of the Wetting Behaviour of Poor Wetting Food Powders and the Influence of Temperature and Film Formation. *KONA Powder Part. J.* **2017**, *34*, 282–289. [CrossRef]
32. Wu, C.; Dai, X.; Sun, X.; Zhang, J. Preparation and characterization of fluoroalkyl activated carbons/PVDF composite membranes for water and resources recovery by membrane distillation. *Sep. Purif. Technol.* **2023**, *305*, 122519. [CrossRef]
33. Aziz, M.H.A.; Othman, M.H.D.; Hashim, N.A.; Rahman, M.A.; Jaafar, J.; Hubadillah, S.K.; Tai, Z.S. Pretreated aluminium dross waste as a source of inexpensive alumina-spinel composite ceramic hollow fibre membrane for pretreatment of oily saline produced water. *Ceram. Int.* **2019**, *45*, 2069–2078. [CrossRef]

34. Almanassra, I.W.; Jaber, L.; Chatla, A.; Abushawish, A.; Shanableh, A.; Ali Atieh, M. Unveiling the relationship between MOF porosity, particle size, and polyethersulfone membranes properties for efficient decontamination of dye and organic matter. *Chem. Eng. J.* **2023**, *471*, 144616–144634. [CrossRef]
35. Awang Chee, D.N.; Ismail, A.F.; Aziz, F.; Mohamed Amin, M.A.; Abdullah, N. The influence of alumina particle size on the properties and performance of alumina hollow fiber as support membrane for protein separation. *Sep. Purif. Technol.* **2020**, *250*, 117147. [CrossRef]
36. Tsung-Shou, Y.; Michael, D.S. Effect of particle size distribution on the sintering of alumina. *Commun. Am. Ceram. Soc.* **1988**, *487*, 484–487.
37. Ji, Y.; Ke, J.; Duan, F.; Chen, J. Preparation and characterization of modified activated carbon/polysulfone blended ultrafiltration membrane. *Desalination Water Treat.* **2017**, *98*, 78–84. [CrossRef]
38. Mollahosseini, A.; Rahimpour, A.; Jahamshahi, M.; Peyravi, M.; Khavarpour, M. The effect of silver nanoparticle size on performance and antibacteriality of polysulfone ultrafiltration membrane. *Desalination* **2012**, *306*, 41–50. [CrossRef]
39. Goel, V.; Tanwar, R.; Saikia, A.K.; Mandal, U.K. Separation characteristics of surface modified polysulfone ultrafiltration membrane using oxidative catalytic polymerization of aniline. *J. Polym. Mater.* **2022**, *39*, 283–305. [CrossRef]
40. Erragued, R.; Kujawski, W.; Kujawa, J.; Gando-Ferreira, L.M.; Bouaziz, M. Optimizing operating conditions for olive leaf valorization using activated carbon mixed matrix membrane. *J. Water Process Eng.* **2024**, *59*, 105036. [CrossRef]

polymers

MDPI

Article

Various Morphologies of Graphitic Carbon Nitride ($g\text{-}C_3N_4$) and Their Effect on the Thermomechanical Properties of Thermoset Epoxy Resin Composites

Dina Al Mais [1], Samir Mustapha [2,*], Yasmine N. Baghdadi [3], Kamal Bouhadir [4] and Ali R. Tehrani-Bagha [5,*]

1. B. & W. Bassatne Department of Chemical Engineering and Advanced Energy, American University of Beirut, Beirut P.O. Box 110236, Lebanon; dya08@mail.aub.edu
2. Department of Mechanical Engineering, American University of Beirut, Beirut P.C. Box 110236, Lebanon
3. Department of Chemical Engineering, Imperial College London, London SW7 2BX, UK; y.baghdadi20@imperial.ac.uk
4. Department of Chemistry, American University of Beirut, Beirut P.O. Box 110236, Lebanon; kb05@aub.edu.lb
5. School of Chemical Engineering, Aalto University, 02150 Espoo, Finland

* Correspondence: sm154@aub.edu.lb (S.M.); ali.tehrani@aalto.fi (A.R.T.-B.)

Abstract: This research aims to highlight the importance of diverse forms of graphitic carbon nitride ($g\text{-}C_3N_4$) as strengthening elements in epoxy composites. It explores the influence of three different forms of $g\text{-}C_3N_4$ and their concentrations on the mechanical properties of the epoxy composites. Various characterization techniques, such as scanning electron microscopy (SEM), dynamic light scattering (DLS), thermogravimetric analysis (TGA), and Fourier-transform infrared spectroscopy (FTIR), were utilized to comprehend the effects of $g\text{-}C_3N_4$ morphology and particle size on the physical and chemical characteristics of epoxy resin. Mechanical properties, such as tensile strength, strain, modulus, and fracture toughness, were determined for the composite samples. SEM analysis was performed to examine crack morphology in samples with different reinforcements. Findings indicate that optimal mechanical properties were achieved with a 0.5 wt% bulk $g\text{-}C_3N_4$ filler, enhancing tensile strength by 14%. SEM micrographs of fracture surfaces revealed a transition from brittle to rough morphology, suggesting increased toughness in the composites. While the TGA results showed no significant impact on degradation temperature, dynamic mechanical analysis demonstrated a 17% increase in glass transition temperature. Furthermore, the improvement in thermal breakdown up to 600 °C was attributed to reinforced covalent bonds between carbon and nitrogen, supported by FTIR results.

Keywords: epoxy resin; $g\text{-}C_3N_4$; mechanical properties; thermal properties; fracture toughness

1. Introduction and Literature Review

The exponential expansion of manufacturing industries worldwide has triggered a pressing need for materials that not only offer enhanced mechanical properties but also address concerns related to environmental sustainability and cost efficiency. Over the past few decades, polymers have emerged as key substitutes for traditional materials like metals across various applications. This shift is attributed to several advantages that polymers offer, including ease of processing, lightweight nature, heightened productivity, and cost reduction [1]. Polymers used in composite manufacturing are broadly classified into thermoplastics and thermosets based on their chemical bonding characteristics. Thermoplastic matrix materials consist of one- or two-dimensional molecules, enabling them to soften at elevated temperatures and regain stiffness upon cooling. In contrast, thermoset polymers are characterized by strong cross-linkages and undergo curing processes involving heat, pressure, and/or light irradiation. The inherent structure of thermoset polymers often results in superior performance in terms of strength and stiffness [2].

Polymer systems offer a versatile platform for enhancing characteristics like strength through the incorporation of both organic and inorganic fillers and fibers. Organic fillers, sourced from natural carbon-based compounds such as nanocrystalline cellulose, carbon nanotubes, graphene, and carbon fiber, find extensive applications across various disciplines, including bio-fabrication, aerospace, and manufacturing [3]. Conversely, inorganic fillers are commonly utilized in sectors like the food, pharmaceutical, and paper manufacturing industries to improve electrical conductivity [4]. The composition, type, quantity, dispersion, and interaction of fillers within the polymer matrix significantly influence reinforcement, enabling the production of a wide array of polymer composites with improved qualities for diverse industrial and technical applications [5]. Although significant progress has been made, issues such as brittleness, inadequate strength, poor stability at low temperatures, and low fracture toughness continue to present challenges [6].

Carbon nanomaterials are widely employed as reinforcing agents to improve the thermomechanical resistance, corrosion resilience, electrical conductivity, and flame-retardant attributes of polymer composites. This enhancement stems from their exceptional inherent properties and their ability to uniformly disperse within diverse polymeric matrices [7].

Rajsekhar et al. [8] conducted a study on the mechanical properties of epoxy resins with the addition of nano-alumina. They observed a notable enhancement in fracture toughness, which increased by 64% with 0.5 wt% nano-alumina samples.

Wang et al. [9] proposed the use of graphene oxide as a 2D reinforcing agent to improve the mechanical properties of epoxy and improve the dispersion of carbon nanotubes. Tang et al. [10] investigated the impact of graphene dispersion on the mechanical properties of graphene/epoxy composites. Highly dispersed graphene enhances fracture toughness by 52% at 0.2 wt.% loading, compared to a 24% improvement for poorly dispersed graphene. Tensile and flexural moduli showed no significant differences with varying dispersion levels. This study highlighted the importance of the good dispersion of fillers for optimizing the mechanical performance of epoxy composites. Furthermore, Zeng et al. investigated the impact of self-assembled carbon nanotube montmorillonite on the mechanical properties of the epoxy matrix, revealing significant improvements in both strength and toughness [11,12].

Kumar et al. [13] prepared epoxy composites with graphene-like nanocarbon sheets (GNCs) at weight fractions between 0.005 and 2 wt%. The maximum tensile strength and tensile modulus were obtained at 0.1 wt% GNCs. The specific tensile modulus increased by 14.7%, from 536 MPa to 615 MPa, and the ultimate tensile strength increased by 3.6%, from 55 MPa to 57 MPa. Maximum flexural strength and flexural modulus were achieved at 1 wt% filler. The flexural strength increased from 2.28 GPa to 2.9 GPa, and the flexural modulus increased from 43.85 GPa to 56.42 GPa at 1 wt% of GNCs. The reinforcement effect at relatively low GNC weight fractions were attributed to good dispersion and chemical interaction.

Khan et al. [14] reviewed the polymer composites filled with various nitride compounds, emphasizing their potential applications in electronics and thermal management. It covers the structure and properties of nitride compounds, such as boron nitride, silicon nitride, carbon nitride, aluminum nitride, and titanium nitride, and their effects on polymer composites. These nitride compounds enhance the mechanical, thermal, and electrical properties of polymers, making them suitable alternatives to expensive and heavy metals like aluminum and copper. Among carbon nitrides, graphitic carbon nitride (g-C_3N_4) stands out as the most stable allotrope at room temperature, representing one of the earliest polymers reported in the literature [15]. The synthesis of g-C_3N_4 through the polycondensation of nitrogen-rich organic precursors such as di-cyandiamide, melamine, and thiourea has attracted significant attention [16]. Various synthetic methodologies have been proposed and the resulting product in terms of shape, morphology, and reactivity were explored [16,17].

Song et al. [18] found that the interlaminar shear stress was enhanced by up to 37.2% after the incorporation of g-C_3N_4 onto the surface of carbon fiber. Another study showed that the addition of g-C_3N_4 to epoxy resin for strengthening carbon fiber led to an increase in tensile strength by approximately 19.5% [19]. These findings collectively underscore the significant potential of various nanomaterials in augmenting the mechanical properties of polymer composites for diverse engineering applications.

This study investigates the influence of utilizing three distinct morphologies of g-C_3N_4—nanosheets, bulk, and nanotubes—as reinforcing agents on the thermomechanical properties of epoxy resin. The research entails the synthesis of these diverse g-C_3N_4 morphologies derived from melamine. Furthermore, the study explores the impact of varying g-C_3N_4 morphologies and different weight percentages on the static mechanical properties of the epoxy as well as its thermal stability. By analyzing these parameters, the study aims to elucidate the most effective morphology and optimal weight percentage of g-C_3N_4 for enhancing the thermomechanical properties of epoxy resin.

2. Methodology

2.1. Materials

The epoxy resin utilized in this study, Araldite LY 564, along with its corresponding hardener, Aradur 2954, were procured from Huntsman Inc. (Texas, USA). Melamine, with a purity of 99%, was obtained from Sigma-Aldrich (Massachusetts, USA). Additionally, various other materials and chemicals were employed, including nitric acid solution, ethanol, ethylene glycol, and distilled water, to facilitate the synthesis and processing of the composite materials.

2.2. Preparation Process of the Particles

2.2.1. Bulk g-C_3N_4

A measure of 10 g of melamine was subjected to heating in a closed 30 mL ceramic crucible. The heating rate was maintained at 10 °C per minute, gradually increasing the temperature from 25 °C to 550 °C. This heating process was carried out in a well-ventilated furnace. Subsequently, once the temperature reached 550 °C, it was held constant for a duration of 4 h. This controlled heating and holding period facilitated the transformation of melamine into g-C_3N_4 powder (~7 g produced), as described in [20].

2.2.2. g-C_3N_4 Nanotubes

A solution comprising 1 g of melamine dissolved in 30 mL of ethylene glycol was combined with 60 mL of a 0.1 M aqueous nitric acid solution at a temperature of 25 °C for a duration of 20 min. Afterward, the resultant solution underwent centrifugation and was subjected to three washes with ethanol to remove any residual nitric acid and ethylene glycol. Following this, the obtained white precipitate was dried under reduced pressure at 60 °C for a period of 12 h. Subsequent to the drying process, the material was heated to 400 °C at a rate of 5 °C per minute and held at this temperature for 2 h. This process resulted in the formation of a yellow powder, indicative of the presence of g-C_3N_4 nanotubes, as reported in [21].

2.2.3. g-C_3N_4 Nanosheets

Nanosheets of g-C_3N_4 were synthesized through a hydrothermal treatment of melamine followed by calcination at elevated temperatures. Initially, 2 g of melamine was dispersed in 30 mL of water and stirred for 30 min. The resulting suspension was transferred to a 50 mL autoclave lined with Teflon and heated to 200 °C for a duration of 12 h. After cooling the white solid to room temperature, it was washed with distilled water and subsequently dried overnight under reduced pressure at 60 °C. The dried product was subsequently transferred into a 50 mL alumina crucible, which was fitted with a lid, and subjected to heating in a furnace at a rate of 2.3 °C per minute. It was maintained at a temperature of 550 °C for 4 h to form g-C_3N_4 nanosheets [22].

To insure uniformity among the three powders produced, the resulting yellow powder was crushed using a mortar and pestle before undergoing sieving to achieve a particle size of less than 38 mm, as detailed in [20].

2.3. Preparation of g-C_3N_4/Epoxy Composites

The required quantity of g-C_3N_4 for each experiment was determined based on a weight percentage relative to the combined mass of the polyamine hardener (Aradur 2594) and epoxy resin (Araldite LY 564).

Nine distinct sets of samples were prepared, each incorporating different weight percentages of g-C_3N_4 bulk, g-C_3N_4 nanosheets, and g-C_3N_4 nanotubes: specifically, 0.25%, 0.5%, and 1% by weight. The g-C_3N_4 powder was gradually introduced into a beaker containing 110 g of epoxy resin, followed by sonication for 30 min, and subsequent mixing at a speed of 10,000 rpm for 45 min. Due to the elevated mixing speed, the mixture experienced a rise in temperature, necessitating cooling to room temperature (RT) by placing the beaker on an ice pack. Once the mixture reached room temperature, 35.8 g of hardener was slowly added. Following degassing under vacuum at 25 °C for 2 h, the mixture was transferred into an aluminum mold and subjected to an additional 2 h of degassing at 25 °C. The samples were cured at 80 °C for 60 min, followed by post-curing at 140 °C for eight hours in accordance with the manufacturer's instructions, as outlined in [20].

2.4. Characterization

To evaluate the static mechanical properties of the produced composites, tensile testing and compact tension testing methods were employed. Mechanical tensile testing was conducted, utilizing a universal testing machine (UTM) equipped with a 100 kN load cell, operating at a loading speed of 1 mm/min using ASTM D638 standard test procedure. A laser extensometer was used to measure the strain. The aluminum mold used for preparing the tensile specimens had a gauge length of approximately 50 mm and a cross-sectional area of 13 × 7 mm^2 (see Figure 1a). From each batch, at least 5 samples were tested and analyzed [20]. The elastic modulus and the stress at failure were key parameters analyzed from the test results.

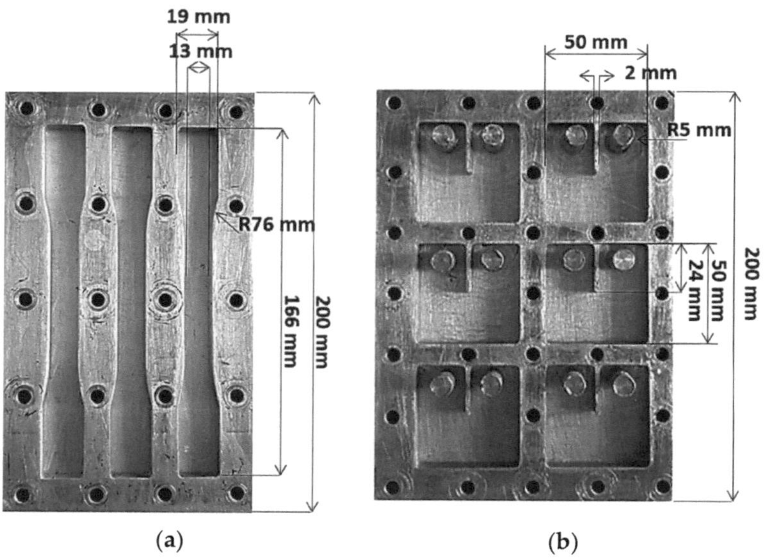

Figure 1. (**a**) Specifications of the molds used for fabricating the dog-bone samples intended for the mechanical tensile test and (**b**) dimensions of the mold utilized for the compact tension test [20].

The compact tension of the materials was completed using the same UTM, equipped with a 100 kN load cell, and operated at a cross-head speed of 0.5 mm/min. Square samples were prepared with dimensions of 40 mm in width, 10 mm in thickness, and featuring a 24 mm notch positioned at the center of the sample, using the ASTM standard D5045 (refer to Figure 1b). To conduct the test, a 1 mm crack was initiated at the tip of the notch using a blade. The results of the compact tension test were evaluated using the following equation derived from ASTM D5045 to determine the plane–strain fracture toughness, KI (in MPa.m$^{0.5}$):

$$K_I = \frac{P_Q}{B\sqrt{W}} \times \frac{(2+\alpha)}{(1-\alpha)^{\frac{3}{2}}} \times (0.866 + 4.64\,\alpha - 13.32\alpha^2 + 14.72\alpha^3 - 5.6\alpha^4)$$

The equation involves parameters such as the sample's width (W) and thickness (B), crack length to sample width ratio (α), and the change between the maximum and minimum load (ΔP) in MN.

Following tensile testing, the crack morphology of the samples was observed using a Tescan MIRA3 Field Emission Scanning Electron Microscope (FESEM). The samples were coated with a 15 nm thick layer of platinum prevent charging. Similarly, the as-prepared g-C_3N_4 samples were examined using the same SEM microscope by adding a small amount of the power to a stud with carbon tape. The samples were also coated with platinum to avoid charging. All samples were tested at a voltage of 10 kV using a secondary electron detector.

Dynamic light scattering (DLS) with the Q2000 instrument was employed to ascertain the particle size of the three variants of g-C_3N_4 (dispersed using different solvents). Specifically, a concentration of 1000 mg/L of bulk g-C_3N_4, nanotubes, and nanosheets were dispersed in two distinct solutions (ethanol and DMF) to identify the most suitable dispersing agent. Following the selection of the optimal dispersing solution, two concentrations of filler were sonicated for 3 min and subsequently analyzed using DLS at three different time intervals (t_1 = 0 min, t_2 = 15 min, and t_3 = 2 h), with four measurements conducted and averaged for each mixture.

The glass transition temperature of the epoxy composites was determined using a Q100 DSC (TA Instruments, New Castle, Delaware, United States) in a heat/cool/heat mode with heating and cooling rates set at 10 °C/min from −50 °C to 200 °C. Glass transition temperatures were calculated based on fluctuations in heat flow observed during the 3rd cycle (heating, cooling, and heating again).

To measure the thermal stability of the epoxy composites, TGA Q500 (TA Instruments) was used with nitrogen as the purge gas. Analysis was completed over a temperature range of 25 to 800 °C at a rate of 10 °C/min, using samples weighing between 7 mg and 15 mg. The decomposition temperature (T_d) was determined from the onset temperature observed in the obtained thermographs.

Furthermore, Fourier transform infrared (FTIR) spectroscopy was conducted on melamine (the starting material) and the synthesized g-C_3N_4 (bulk, nanotubes, and nanosheets) using an Agilent cary 630 FT-IR instrument (Santa Clara, CA, USA) in ATR mode.

3. Results and Discussion

3.1. Characterization of the Synthesized Particles

The synthesized particles underwent various analyses and characterizations to elucidate their physical–chemical properties and to discern differences in morphology and particle size among the different forms of g-C_3N_4. SEM, DLS, TGA, and FTIR were employed to perform the characterization.

SEM imaging of bulk g-C_3N_4, g-C_3N_4 nanosheets, and g-C_3N_4 nanotubes revealed substantial variations in morphology, as depicted in Figure 2. While pristine melamine particles ere known for their stone-like morphology, upon heating, melamine, a trimer of cyanamide, undergoes pyrolysis and polymerization to yield g-C_3N_4, resulting in the formation of flaky sheet-like structures stacked together (Figure 2a,b), as reported in the literature [23,24].

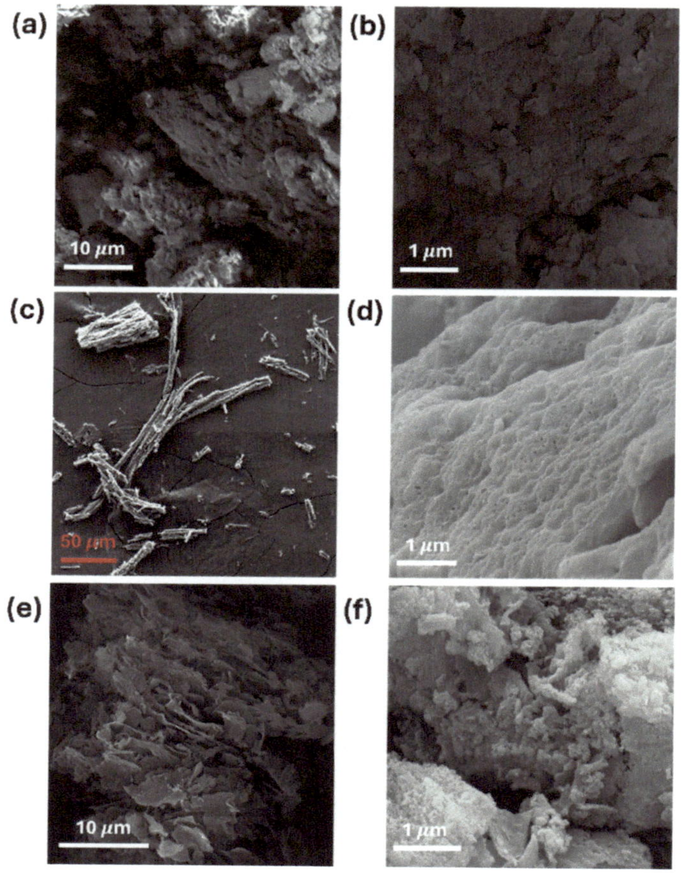

Figure 2. SEM micrographs of g-C_3N_4 (**a**,**b**) bulk, (**c**,**d**) nanotubes, and (**e**,**f**) nanosheets.

The morphology of g-C_3N_4 nanotubes exhibited structures characterized by open ends, as illustrated in Figure 2c,d [25]. Additionally, the synthesized g-C_3N_4 samples derived from melamine underwent transformation into 2D sheet like structures (Figure 2e,f), displaying a typical nanosheet structures with smooth surfaces, consistent with findings in the literature [22]. The TEM images of the g-C_3N_4 nanotubes and nanosheets are available in other papers [21,22], from which we have also adopted the synthesis methods.

Our preliminary investigation revealed that g-C_3N_4 demonstrates superior dispersion in dimethylformamide (DMF) compared to ethanol (refer to Figure S1). Table 1 presents the typical particle size distribution of the three g-C_3N_4 variants (nanosheets, bulk, and nanotubes) dispersed in DMF at various time intervals, as determined by DLS analysis. As indicated in Table 1, after 2 h, the size of g-C_3N_4 bulk and g-C_3N_4 nanosheets reaches a minimum value of 3.05 μm and 1.6 μm, respectively, while the size of g-C_3N_4 nanotubes remains largely unchanged. It is important to note that the observed average hydrodynamic diameter of the three forms of g-C_3N_4 from DLS analysis may be affected by multiple scattering effects, leading to potential errors [25].

Table 1. DLS measurements to determine the average particle size of g-C_3N_4 bulk, g-C_3N_4 nanotubes, and g-C_3N_4 nanosheets at various concentrations and time intervals.

Time (min)		**t_1 = 0**		**t_2 = 15**		**t_3 = 60**	
Concentration (g/L)		**0.25**	**0.125**	**0.25**	**0.125**	**0.25**	**0.125**
Size (µm)	**g-C_3N_4Bulk**	9.6 ± 0.1	9.3 ± 0.1	9.2 ± 0.2	9.2 ± 0.09	3.1 ± 0.09	2.8 ± 0.06
	g-C_3N_4 Nanotubes	3.7 ± 0.07	3.3 ± 0.07	4.9 ± 0.1	3.4 ± 0.06	3.1 ± 0.1	3.0 ± 0.1
	g-C_3N_4 Nanosheets	5.7 ± 0.08	4.2 ± 0.09	5.1± 0.06	3.9 ± 0.08	1.7 ± 0.06	1.9 ± 0.03

DLS quantifies the timescale of light fluctuations reaching the detector resulting from the random diffusion/movement of suspended particles within a sample cuvette. Smaller particles exhibit rapid diffusion, resulting in fast intensity fluctuations, while larger particles diffuse more slowly, leading to slower variations in scattered light intensity. In cases of multiple scattering, a photon of light is scattered by a diffusing particle, then re-scattered by one or more particles before reaching the detector, which can distort the observed particle size [26]. It is important to note that the size measured using DLS refers to the size of the agglomerates in the suspension rather than individual nanosheets or nanotubes.

The decrease in average particle size distribution over time can be attributed to the sedimentation of larger particles in the cuvette, leaving smaller particles dispersed for a longer duration. Consequently, DLS detects smaller particles more prominently over time, thus explaining the reduction in particle size observed for g-C_3N_4 bulk and g-C_3N_4 nanosheets. In the case of g-C_3N_4 nanotubes, their hollow structure enables them to float in the solution for longer periods of time. At higher concentrations (0.25 g/L), slight aggregation of the nanotubes (driven upwards) causes an increase in the detected size (4.9 ± 0.1 µm). For the following duration of 45 min, some of the nanotubes settled at the bottom of the vial causing the measured DLS size to drop back to 3.1 ± 0.1 µm. In spite of this decrease, the stability of the nanotubes in suspension proved superior to g-C_3N_4 fillers in bulk and as nanosheets.

Figure 3 illustrates the TGA curves thermograms of pure melamine, g-C_3N_4 bulk, g-C_3N_4 nanotubes, and g-C_3N_4 nanosheets. Pure melamine undergoes decomposition and sublimation at around 305 °C without any reduction in the weight prior to that temperature. This indicated that melamine did not undergo any polymerization reaction in that temperature range and reinforces the need to perform the synthesis of g-C_3N_4 in a closed-lid crucible to avoid this sublimation. On the other hand, the TGA curves thermograms of g-C_3N_4 bulk, g-C_3N_4 nanosheets, and g-C_3N_4 nanotubes indicated a slight weight loss of 2.5% within the range of 25–200 °C. While the weight loss below 100 °C is attributed to the desorption of water molecules on the surface, the change in weight above this temperature was an indication of additional polymerization taking place of the intermediates (melem and melan). This condensation reaction led to the loss of nitrogen and hydrogen from the samples in the form of ammonia gas. However, the direct thermal decomposition of g-C_3N_4 bulk leads to a significant weight loss of 97.5% between 600 °C and 760 °C, similar to that observed for nanotubes. At this temperature, covalent bonds between carbon and nitrogen begin to break leading to the full decomposition of the organic filler. This enhancement is typically attributed to the polymerization of melamine trimer, leading to the formation of extended 2D chains of g-C_3N_4. Notably, g-C_3N_4 nanosheets exhibited rapid disintegration without any residue within the range of 430–760 °C. The lower decomposition temperature of g-C_3N_4 nanosheets was attributed to the lower amount of NH_2 and NH groups, resulting in fewer carbon and nitrogen bonds [27].

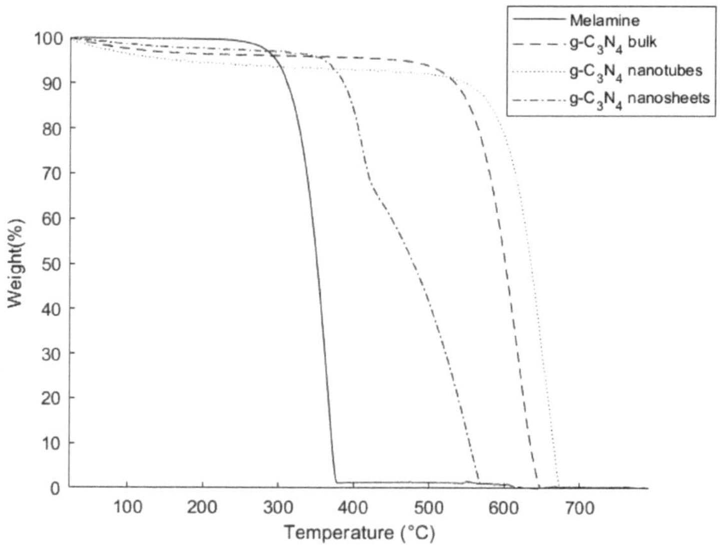

Figure 3. TGA thermograms of pure melamine, g-C_3N_4 bulk, g-C_3N_4 nanosheets, and g-C_3N_4 nanotubes.

The FTIR spectrum for melamine (Figure 4) reveals three main absorption regions in the chemical structures. Clear N-H stretching vibration bands were observed within 3300 and 3000 cm^{-1} and within 1650 and 1580 cm^{-1}, while peaks between 1020 and 1200 cm^{-1} correspond to the C-N bending vibration of amines, indicative of vibrations in melamine's aromatic rings, as evidenced by the peak around 810 cm^{-1}. A comparison with the one that is specific to the produced g-C_3N_4 bulk revealed that previously sharp and strong absorption peaks in the region of 3000–3600 cm^{-1} had merged into a broad peak centered at around 3100 cm^{-1}, corresponding to the NH or NH_2 stretching vibrations, suggesting the calcination of existing amino groups in melamine upon pyrolysis. Other signals were observed in the 1100–1600 cm^{-1} region, indicating the condensation of the triazine rings [27].

A shift in the major signals between g-C_3N_4 bulk and g-C_3N_4 nanotubes was also noted. The appearance of a signal at around 1600 cm^{-1} could be assigned to the C=N stretching vibration of the inherent structure of g-C_3N_4 nanotubes. Vibrations at 1641 cm^{-1}, 1710 cm^{-1}, and 3352 cm^{-1} could be assigned to the functionalized g-C_3N_4 nanotubes (carbonyl, carboxyl, and hydroxyl functional groups), as reported in the literature [28]. The chemical structure of g-C_3N_4 bulk and g-C_3N_4 nanosheets were presumed to be similar since their FTIR spectra were identical. Peaks related to the N-H stretch of amino groups, aromatic C-N stretching modes of the s-triazine ring were observed at 3000–3500 cm^{-1} and 1200–1700 cm^{-1}, respectively [23]. Peaks between 1200 and 1700 cm^{-1} for g-C_3N_4 nanosheets appeared sharper than those for g-C_3N_4 bulk powder, suggesting a more ordered packing of the s-triazine motifs in the nanosheet layers [24].

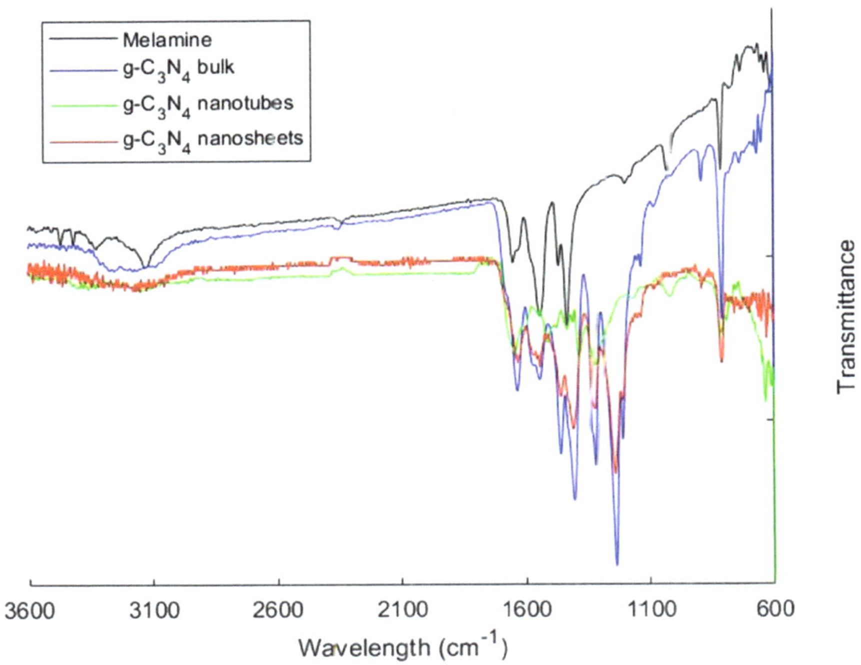

Figure 4. FTIR spectra of pure melamine, g-C_3N_4 bulk, g-C_3N_4 nanosheets, and g-C_3N_4 nanotubes.

3.2. Static Mechanical Properties and Fracture Morphology

The mechanical properties of epoxy composites incorporating various types of g-C_3N_4 at various concentrations were evaluated, encompassing tensile and compact tension tests. Table 2 summarizes the average tensile strength, strain, and modulus values, while Table S1 lists the KI values.

Table 2. Tensile strength, strain, and modulus of elasticity of g-$C_3N_{4\text{-reinforced epoxy}}$ composites.

Composite	Filler Concentration wt%	Tensile Strength (MPa)	% Change	Elastic Modulus (MPa)	% Change	Strain (mm/mm)	% Change
Pure Epoxy	0	44.61 ± 4.6	-	853.6 ± 83.6	-	0.07 ± 0.002	-
g-C_3N_4 bulk	0.2	48.04 ± 1.23	7	883.4 ± 11.5	3	0.07 ± 0.002	0
	0.5	50.82 ± 1.45	14	830.7 ± 76.3	−2	0.06 ± 0.009	−14
	1.0	48.73 ± 5.8	9	768.6 ± 95.4	−9	0.07 ± 0.002	0
g-C_3N_4nanosheets	0.25	46.19 ± 3.8	3	854.08 ± 22.5	0	0.09 ± 0.006	18
	0.5	35 ± 3.7	−21	983.5 ± 66.8	15	0.05 ± 0.006	−29
	1.0	45.82 ± 2.81	3	803.5 ± 38.2	−5	0.07 ± 0.002	0
g-C_3N_4nanotubes	0.25	44.12 ± 4.08	−1	921.7 ± 55	8	0.07 ± 0.002	0
	0.5	47.896 ± 5.55	8	956.8 ± 151	12	0.07 ± 0.002	0
	1.0	39.41 ± 5.7	−11	755.64 ± 13.2	−11	0.06 ± 0.004	−14

Figure 5 and Table 2 illustrate that the incorporation of g-C_3N_4 bulk and g-C_3N_4 nanotube particles led to an increase in the composites' tensile strength. Notably, at

0.5 wt.% of g-C_3N_4 bulk and g-C_3N_4 nanotubes, the tensile strength peaked at 14% and 8%, respectively. However, a gradual decline in mechanical properties was observed beyond this threshold, particularly evident at 1 wt.% g-C_3N_4 bulk and g-C_3N_4 nanotubes–epoxy composites, hinting at filler agglomeration, as reported in prior studies [27]. It is possible that the strong interactions between NH_2 and NH groups in g-C_3N_4 and the hydroxyl and epoxide groups of the resin contributed to these improvements.

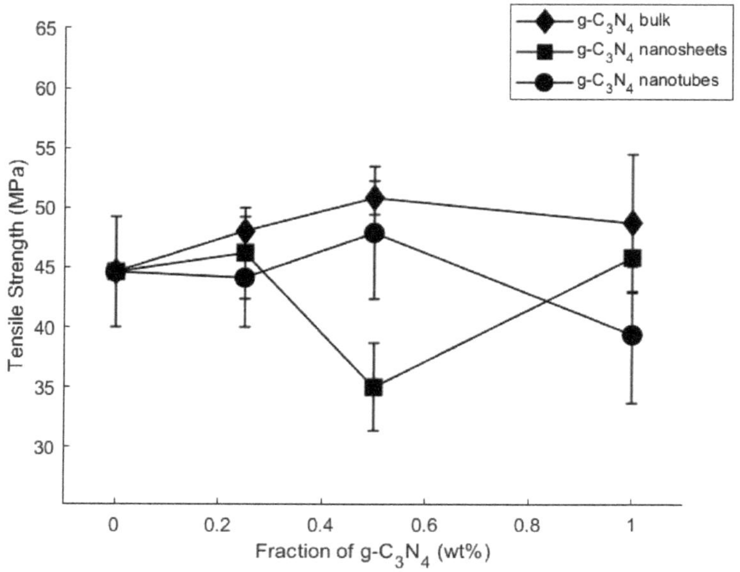

Figure 5. The average tensile strength of g-$C_3N_{4\text{-reinforced epoxy}}$ composites plotted against filler concentration (wt.%).

Introducing g-C_3N_4 nanosheet particles exhibited a distinct trend. While strength initially improved by 3% with 0.25 wt% and 1 wt% addition of g-C_3N_4 nanosheets, it sharply declined to −21% with a 0.5 wt% addition, suggesting hindered interactions between the filler and the matrix, possibly due to the formation of large g-C_3N_4 agglomerates.

Moreover, increasing the concentration of g-C_3N_4 bulk and g-C_3N_4 nanotubes led to a reduction in the modulus of elasticity, from 853 MPa of pure epoxy to 768 MPa for g-C_3N_4 bulk, 803 MPa for g-C_3N_4 nanosheets, and 755 MPa for g-C_3N_4 nanotubes–epoxy. This discrepancy may be attributed to the density difference, as g-C_3N_4 nanotubes and g-C_3N_4 nanosheets are smaller and less dense, potentially leading to increased overall agglomeration. It is important to note that these nanoparticles tend to aggregate when mixed with the hardener, and such aggregates may not uniformly disperse in the composite matrix, further influencing the mechanical response of the resulting composites.

Table S1 reveals that there was no enhancement in fracture toughness in any of the scenarios. Incorporating g-C_3N_4 bulk resulted in either a decrease (up to 16%) or no alteration in fracture toughness. Likewise, the fracture toughness of g-C_3N_4 nanosheets and g-C_3N_4 nanotube composites declined, with a maximum drop of 25% observed upon increasing the weight percentage. This decline in fracture toughness can be attributed to the type of crack formation. Previous studies have identified two distinct types of crack propagation: one involving continuous propagation with minimal fracture tip deformation (0.5–2 μm), and the other characterized by crack growth in a stick–slip manner. Consequently, the same material might exhibit both types of fracture behavior [29]. The mechanisms governing toughening are influenced by numerous factors such as the type, size, volume fraction, and particle–matrix bonding. The configurations of the inclusions, whether single or clustered, and their uniform dispersion within the matrix, also impact the material's fracture toughness [30].

SEM analysis was conducted on the cracked surface (of the samples) to gain deeper insights into the outcomes of mechanical testing, as illustrated in Figures 6–9. Pure epoxy exhibited distinctive brittle behavior, with the crack propagating smoothly and directly from the bottom left corner to the other corner with minimal deviation (Figure 6). Similar behavior was observed in samples containing 0.25 wt% of g-C_3N_4 bulk, g-C_3N_4 nanosheets, and g-C_3N_4 nanotubes (Figures 7, 8 and 9a), as well as in samples with 0.5 wt% of g-C_3N_4 nanosheets (Figure 8b). In all these cases, the fracture path was distinct and linear, indicates brittle behavior and weak interactions between g-C_3N_4 and epoxy.

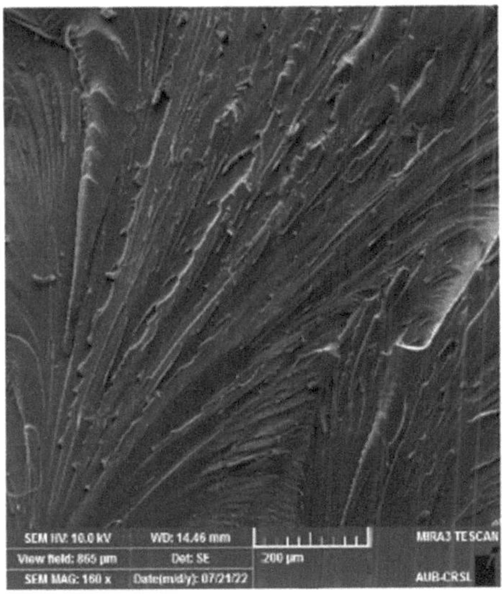

Figure 6. SEM micrograph of the fractured epoxy sample after tensile testing.

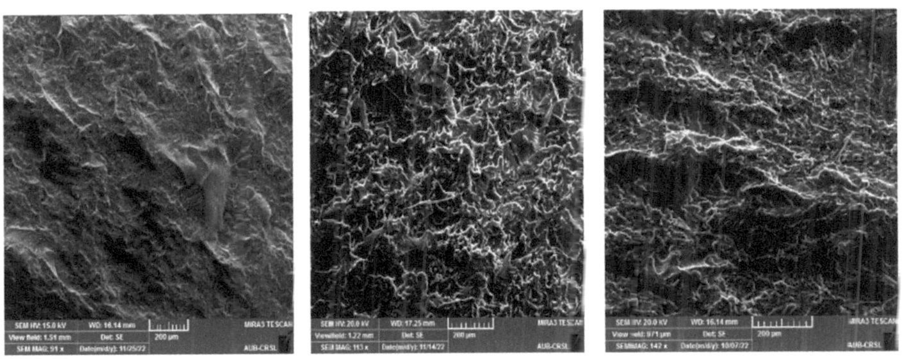

(a) (b) (c)

Figure 7. SEM micrograph of the fractured epoxy composite reinforced with g-C_3N_4 bulk: (**a**) 0.25 wt%, (**b**) 0.5 wt%, (**c**) 1 wt%.

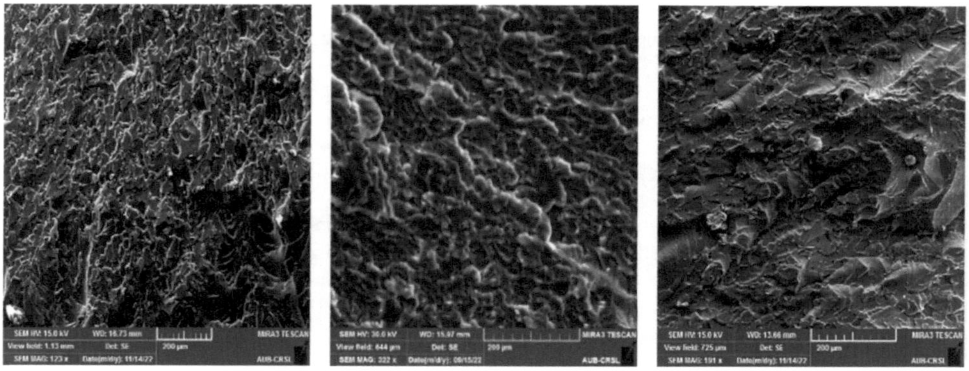

(**a**) (**b**) (**c**)

Figure 8. SEM micrograph of the fractured epoxy composite reinforced with g-C_3N_4 nanosheets: (**a**) 0.25 wt%, (**b**) 0.5 wt%, (**c**) 1 wt%.

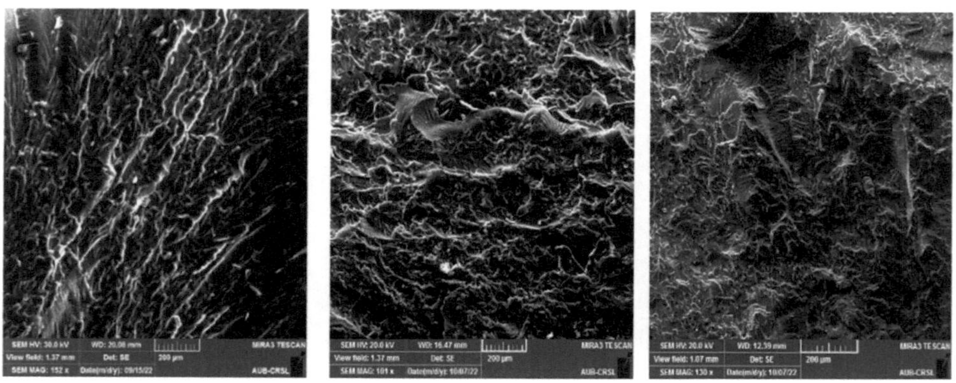

(**a**) (**b**) (**c**)

Figure 9. SEM micrograph of the fractured epoxy composite reinforced with g-C_3N_4 nanotubes: (**a**) 0.25 wt%, (**b**) 0.5 wt%, (**c**) 1 wt%.

However, the addition of 0.5 wt% of g-C_3N_4 bulk and g-C_3N_4 nanotubes resulted in a significantly rougher crack surface with an ambiguous crack trajectory. Due to the potential deflection of fractures by particles, multiple cracks were observed, extending in various directions. This observation suggests that the enhancement in mechanical properties with 0.5 wt% of g-C_3N_4 bulk and g-C_3N_4 nanotube particles is attributed to increased strain energy consumption and improved interfacial bonding strength between the filler and matrix. Achieving a homogeneous dispersion of the filler at 0.5 wt% led to improved mechanical properties, whereas filler loadings below 0.5 wt% did not yield significant enhancements in the matrix [20].

For samples with 1 wt%, a similar rough fracture morphology was observed; however, there is clear evidence of nanofiller agglomeration (as shown in Figures 7c, 8c and 9c). The increase in wt% of particles led to the formation of island-like structures, rich in filler within the resin, consistently with the literature [9,31]. These regions containing g-C_3N_4 agglomerates act as system flaws, promote the growth of additional cracks, and contribute to the premature failure of the test specimen by creating areas of higher stress concentration where cracks may initiate [32].

3.3. Thermal Properties

The thermal degradation behavior of epoxy composites reinforced with various types of g-C_3N_4 was investigated through TGA analysis. Figures S2–S4 show the TGA thermograms of the obtained composites and Table 3 provides a summary of the results. The average degradation temperature (Td) of pure epoxy resin was found to be 396.1 ± 0.4 °C, with a residue content of about 1.12% at 800 °C, in agreement with the literature [20]. Once reinforced, the decomposition profiles of the epoxy composites showed similar Td values ranging between 399.4 and 401.7 °C. This observed increase in decomposition temperature could be attributed to the so-called "pyrolysis barrier" of g-C_3N_4, which prevents the dispersion of gaseous molecules generated during the decomposition of the epoxy matrix. However, the increase in filler concentration did not significantly affect the Td values. The residual polymer concentration at 800 °C was unaffected by the addition of fillers. This was predicted from Figure 3, showing complete degradation of fillers before 800 °C. Interesting to note that in comparison to pure epoxy resin, all the reinforced composites demonstrated a lower rate of decomposition before the Td. As can be observed in the thermograms of Figures S2–S4, while the decomposition of all samples began at around 100 °C, at Td the observed weight of the samples was approximately 7% higher for all composites. Similarly, a decrease in the rate of decomposition was observed for the composites after 50% of the original weight was lost. These observations were an indication of an overall improvement in thermal stability due to stronger covalent bonding between the g-C_3N_4 filler and epoxy resin that prevented the rapid decomposition of the composites. It was concluded that a low concentration of the filler (up to 1 wt%) cannot enhance the thermal degradation behavior of the epoxy composites.

Table 3. The average degradation temperature (Td) and the residue % of g-C_3N_4-reinforced epoxy composites.

Composite	Filler Concentration wt%	Average Residue (%)	Average Td (°C)
Pure Epoxy	0	1.12	396.1 ± 0.4
g-C_3N_4 bulk	0.25	1.14	401.7 ± 0.1
	0.5	1.38	401.3 ± 0.5
	1.0	1.17	402.9 ± 0.8
g-C_3N_4 nanosheets	0.25	1.24	400.1 ± 0.4
	0.5	1.77	400.5 ± 0.9
	1.0	1.72	401.6 ± 0.3
g-C_3N_4 nanotubes	0.25	1.63	398.7 ± 0.3
	0.5	1.21	399.6 ± 0.6
	1.0	1.36	399.4 ± 0.4

DSC measurements were used to determine the glass transition temperatures (Tg), with corresponding thermograms provided in Figures S5–S7, and average Tg values summarized in Table 4. The introduction of fillers increased the Tg value, indicating improved thermal stability. No evidence of endothermic or exothermic peaks was observed in the tested temperature range tested, indicating the complete cure behavior of the composites. The Tg of pure epoxy (132 °C) increased with filler addition, reaching a maximum increase of about 17% for 1 wt% of g-C_3N_4 bulk. This could be attributed to the covalent bonds connections between the fillers and the epoxy matrix, which reduce matrix mobility and result in a larger shift in Tg. The -NH_2 groups on the surface of g-C_3N_4 particles can potentially react with epoxy monomers during polymerization [33].

Table 4. The average glass transition temperature (Tg) of g-C_3N_4-reinforced epoxy composites.

Composite	Filler Concentration wt%	Average Onset Tg (°C)
Pure Epoxy	0	132 ± 0.44
g-C_3N_4 bulk	0.25	152 ± 0.21
	0.5	153 ± 0.33
	1.0	155 ± 0.43
g-C_3N_4 nanosheets	0.25	153 ± 0.57
	0.5	154 ± 0.36
	1.0	151 ± 0.41
g-C_3N_4 nanotubes	0.25	145 ± 0.39
	0.5	149 ± 0.52
	1.0	152 ± 0.45

4. Conclusions

This study explored using different forms of graphitic carbon nitride (g-C_3N_4) to enhance epoxy resin composites. By varying the concentration and shape of the filler, we assessed their impact on the composite's strength. Techniques like microscopy revealed that adding 0.5 wt% bulk g-C_3N_4 increased composite strength by 14%. Fractured surfaces showed a transition from brittle to rough, indicating potential toughness enhancements due to an improvement in chemical boding between the epoxy matrix and g-C_3N_4 fillers. This was also reflected by the increase in glass transition temperature by up to 17%. Our work addresses the lack of literature on the effects of g-C_3N_4 morphology on epoxy composites. We are among the few to demonstrate how these morphologies influence mechanical properties, suggesting mechanisms where filler shape and dispersion affect stress distribution and crack propagation. Achieving uniform filler distribution remains a challenge, highlighting the need for ongoing research to optimize filler–matrix interactions. Additionally, consistent monitoring of epoxy composites is crucial due to their susceptibility to environmental factors. In summary, our study offers fundamental insights into the role of g-C_3N_4 morphology in epoxy composites, paving the way for future research to develop more effective composite materials.

Supplementary Materials: The following supporting information can be downloaded at: https://www.mdpi.com/article/10.3390/polym16131935/s1, Figure S1: Comparison between the 2 solutions. [1]: g-C_3N_4 bulk in DMF; [2]: g-C_3N_4 nanotubes in DMF; [3]: g-C_3N_4 nanosheets in DMF; [4]: g-C_3N_4 bulk in ethanol; [5]: g-C_3N_4 nanotubes in ethanol; [6]: g-C_3N_4 nanosheets in ethanol. Figure S2: TGA curves for pure epoxy and the manufactured composites with the g-C_3N_4 bulk. Figure S3: TGA curves for pure epoxy and the manufactured composites with the g-C_3N_4 nanosheets. Figure S4: TGA curves for pure epoxy and the manufactured composites with the g-C_3N_4 nanotubes. Figure S5: DSC graphs for pure epoxy and the manufactured composites with the g-C_3N_4 bulk. Figure S6: DSC graphs for pure epoxy and the manufactured composites with the g-C_3N_4 nanosheets. Figure S7: DSC graphs for pure epoxy and the manufactured composites with the g-C_3N_4 nanotubes. Table S1: The average and improvement values of the fracture toughness of g-C_3N_4 bulk, g-C_3N_4 nanosheets, and g-C_3N_4 nanotubes-Epoxy composites with different concentrations.

Author Contributions: Conceptualization, S.M. and K.B.; Methodology, D.A.M.; Formal analysis, D.A.M. and K.B.; Investigation, D.A.M., Y.N.B. and A.R.T.-B.; Resources, S.M.; Writing—original draft, D.A.M.; Writing—review & editing, S.M., Y.N.B., K.B. and A.R.T.-B.; Supervision, S.M., K.B. and A.R.T.-B.; Project administration, S.M.; Funding acquisition, S.M. All authors have read and agreed to the published version of the manuscript.

Funding: The financial support provided by the University Research Board at the American University of Beirut through their Award #104260 is gratefully acknowledged.

Data Availability Statement: Data are contained within the article and Supplementary Materials.

Conflicts of Interest: The authors declare no conflict of interest.

References

1. Gowda, T.Y.; Sanjay, M.; Bhat, K.S.; Madhu, P.; Senthamaraikannan, P.; Yogesha, B. Polymer matrix-natural fiber composites: An overview. *Cogent Eng.* **2018**, *5*, 1446667. [CrossRef]
2. Ticoalu, A.; Aravinthan, T.; Cardona, F. A review of current development in natural fiber composites for structural and infrastructure applications. In Proceedings of the Southern Region Engineering Conference (SREC 2010); 2010; pp. 113–117.
3. Khalil, H.P.S.A.; Chong, E.W.N.; Owolabi, F.A.T.; Asniza, M.; Tye, Y.Y.; Rizal, S.; Fazita, M.R.N.; Haafiz, M.K.M.; Nurmiati, Z.; Paridah, M.T. Enhancement of basic properties of polysaccharide-based composites with organic and inorganic fillers: A review. *J. Appl. Polym. Sci.* **2018**, *136*, 47251. [CrossRef]
4. Semaan, P. Micro-and Nano-Crystalline Cellulose for Enhancing the Mechanical Properties of Epoxy. Ph.D. Thesis, American University of Beirut, Beirut, Lebanon, 2021.
5. Kiran, M.; Govindaraju, H.; Jayaraju, T.; Kumar, N. Review-effect of fillers on mechanical properties of polymer matrix composites *Mater. Today Proc.* **2018**, *5*, 22421–22424. [CrossRef]
6. Wu, N.; Xu, D.; Wang, Z.; Wang, F.; Liu, J.; Liu, W.; Shao, Q.; Liu, H.; Gao, Q.; Guo, Z. Achieving superior electromagnetic wave absorbers through the novel metal-organic frameworks derived magnetic porous carbon nanorods. *Carbon* **2019**, *145*, 433–444. [CrossRef]
7. Vaithylingam, R.; Ansari, M.N.M.; Shanks, R.A. Recent advances in polyurethane-based nanocomposites: A review. *Polym. Technol. Eng.* **2017**, *56*, 1528–1541. [CrossRef]
8. Rajsekhar, V.; Gattu, M. Size-effect testing: Nano-alumina enhances fracture toughness of epoxy resins. *Theor. Appl. Fract. Mech.* **2023**, *125*, 103859. [CrossRef]
9. Wang, E.; Dong, Y.; Islam, M.Z.; Yu, L.; Liu, F.; Chen, S.; Qi, X.; Zhu, Y.; Fu, Y.; Xu, Z.; et al. Effect of graphene oxide-carbon nanotube hybrid filler on the mechanical property and thermal response speed of shape memory epoxy composites. *Compos. Sci. Technol.* **2019**, *169*, 209–216. [CrossRef]
10. Tang, L.-C.; Wan, Y.-J.; Yan, D.; Pei, Y.-B.; Zhao, L.; Li, Y.-B.; Wu, L.-B.; Jiang, J.-X.; Lai, G.-Q. The effect of graphene dispersion on the mechanical properties of graphene/epoxy composites. *Carbon* **2013**, *60*, 16–27. [CrossRef]
11. Zeng, S.; Shen, M.; Xue, Y.; Zheng, Y.; Zhang, K.; Han, Y.; Yang, L. Controllable mechanical properties of epoxy composites by incorporating self-assembled carbon nanotube–montmorillonite. *Compos. Part B Eng.* **2018**, *164*, 368–376. [CrossRef]
12. Zeng, S.; Shen, M.; Yang, L.; Xue, Y.; Lu, F.; Chen, S. Self-assembled montmorillonite–carbon nanotube for epoxy composites with superior mechanical and thermal properties. *Compos. Sci. Technol.* **2018**, *162*, 131–139. [CrossRef]
13. Kumar, A.; Chouhan, D.K.; Alegaonkar, P.S.; Patro, T.U. Graphene-like nanocarbon: An effective nanofiller for improving the mechanical and thermal properties of polymer at low weight fractions. *Compos. Sci. Technol.* **2016**, *127*, 79–87. [CrossRef]
14. Khan, A.; Puttegowda, M.; Jagadeesh, P.; Marwani, H.M.; Asiri, A.M.; Manikandan, A.; Khan, A.A.P.; Ashraf, G.M.; Rangappa, S.M.; Siengchin, S. Review on nitride compounds and its polymer composites: A multifunctional material. *J. Mater. Res. Technol.* **2022**, *18*, 2175–2193. [CrossRef]
15. Wang, X.; Blechert, S.; Antonietti, M. Polymeric Graphitic Carbon Nitride for Heterogeneous Photocatalysis. *ACS Catal.* **2012**, *2*, 1596–1606. [CrossRef]
16. Xiao, J.; Xie, Y.; Nawaz, F.; Wang, Y.; Du, P.; Cao, H. Preparation of short, robust and highly ordered titanium dioxide nanotube arrays and their applications as electrode. *Appl. Catal. B Environ.* **2016**, *183*, 417–425. [CrossRef]
17. Rono, N.; Kibet, J.K.; Martincigh, B.S.; Nyamori, V.O. A review of the current status of graphitic carbon nitride. *Crit. Rev. Solid State Mater. Sci.* **2020**, *46*, 189–217. [CrossRef]
18. Song, B.; Wang, T.; Sun, H.; Liu, H.; Mai, X.; Wang, X.; Wang, L.; Wang, N.; Huang, Y.; Guo, Z. Graphitic carbon nitride (g-C_3N_4) interfacially strengthened carbon fiber epoxy composites. *Compos. Sci. Technol.* **2018**, *167*, 515–521. [CrossRef]
19. Song, B.; Wang, T.; Wang, L.; Liu, H.; Mai, X.; Wang, X.; Wang, N.; Huang, Y.; Ma, Y.; Lu, Y.; et al. Interfacially reinforced carbon fiber/epoxy composite laminates via in-situ synthesized graphitic carbon nitride (g-C_3N_4). *Compos. Part B Eng.* **2018**, *158*, 259–268. [CrossRef]
20. Baghdadi, Y.N.; Sinno, J.; Bouhadir, K.; Harb, M.; Mustapha, S.; Patra, D.; Tehrani-Bagha, A.R. The mechanical and thermal properties of graphitic carbon nitride (g-C_3N_4)-based epoxy composites. *J. Appl. Polym. Sci.* **2021**, *138*, 51324. [CrossRef]
21. Tahir, M.; Mahmood, N.; Zhu, J.; Mahmood, A.; Butt, F.K.; Rizwan, S.; Aslam, I.; Tanveer, M.; Idrees, F.; Shakir, I.; et al. One Dimensional Graphitic Carbon Nitrides as Effective Metal-Free Oxygen Reduction Catalysts. *Sci. Rep.* **2015**, *5*, srep12389. [CrossRef]
22. Hong, Y.; Li, C.; Fang, Z.; Luo, B.; Shi, W. Rational synthesis of ultrathin graphitic carbon nitride nanosheets for efficient photocatalytic hydrogen evolution. *Carbon* **2017**, *121*, 463–471. [CrossRef]
23. Huang, L.; Li, J.; Zeng, H.; Zou, G.; Zhao, Y.; Huang, L.; Bi, J.; Gao, D.; Lin, Z. Surfactant-thermal synthesis of amino acid-templated zinc phosphates with 3-connected nets related to zeolite ABW. *Inorg. Chem.* **2019**, *58*, 4089–4092. [CrossRef] [PubMed]
24. Shi, Y.; Fu, L.; Chen, X.; Guo, J.; Yang, F.; Wang, J.; Zheng, Y.; Hu, Y. Hypophosphite/graphitic carbon nitride hybrids: Preparation and flame-retardant application in thermoplastic polyurethane. *Nanomaterials* **2017**, *7*, 259. [CrossRef]

25. Berne, B.J.; Pecora, R. *Dynamic Light Scattering: With Applications to Chemistry, Biology, and Physics*; Courier Corporation: North Chelmsford, MA, USA, 2000.
26. Zeng, C.; Lu, S.; Song, L.; Xiao, X.; Gao, J.; Pan, L.; He, Z.; Yu, J. Enhanced thermal properties in a hybrid graphene–alumina filler for epoxy composites. *RSC Adv.* **2015**, *5*, 35773–35782. [CrossRef]
27. Liao, H.; Zhang, B.; Huang, L.; Ma, D.; Jiao, Z.; Xie, Y.; Tan, S.; Cai, X. The utilization of carbon nitride to reinforce the mechanical and thermal properties of UV-curable waterborne polyurethane acrylate coatings. *Prog. Org. Coatings* **2015**, *89*, 35–41. [CrossRef]
28. Dai, K.; Zhang, X.; Fan, K.; Peng, T.; Wei, B. Hydrothermal synthesis of single-walled carbon nanotube–TiO2 hybrid and its photocatalytic activity. *Appl. Surf. Sci.* **2013**, *270*, 238–244. [CrossRef]
29. Kinloch, A.J.; Williams, J.G. Crack blunting mechanisms in polymers. *J. Mater. Sci.* **1980**, *15*, 987–996. [CrossRef]
30. Garg, A.C.; Mai, Y.-W. Failure mechanisms in toughened epoxy resins—A review. *Compos. Sci. Technol.* **1988**, *31*, 179–223. [CrossRef]
31. Chatterjee, S.; A Nüesch, F.; Chu, B.T.T. Comparing carbon nanotubes and graphene nanoplatelets as reinforcements in polyamide 12 composites. *Nanotechnology* **2011**, *22*, 275714. [CrossRef]
32. Tee, Z.Y.; Yeap, S.P.; Hassan, C.S.; Kiew, P.L. Nano and non-nano fillers in enhancing mechanical properties of epoxy resins: A brief review. *Polym. Technol. Mater.* **2021**, *61*, 709–725. [CrossRef]
33. Majdoub, M.; Anfar, Z.; Amedlous, A. Emerging chemical functionalization of g-C_3N_4: Covalent/noncovalent modifications and applications. *ACS Nano* **2020**, *14*, 12390–12469. [CrossRef]

polymers

MDPI

Article

Comparing End-of-Life Vehicle (ELV) and Packaging-Based Recyclates as Components in Polypropylene-Based Compounds for Automotive Applications

Markus Gall [1], Daniela Mileva [1], Wolfgang Stockreiter [1], Christophe Salles [2] and Markus Gahleitner [1,*]

[1] Borealis Polyolefine GmbH, Innovation Headquarters, St Peterstr. 25, 4021 Linz, Austria; markus.gall@borealisgroup.com (M.G.); daniela.mileva@borealisgroup.com (D.M.); wolfgang.stockreiter@borealisgroup.com (W.S.)

[2] Borealis Services S.A.S., 12 rue de Londres, 75015 Paris, France; christophe.salles@borealisgroup.com

* Correspondence: markus.gahleitner@borealisgroup.com; Tel.: +43-664-1227724

Abstract: Increasing recycled plastic content in cars to 25% by 2030 is one of the key measures for decarbonizing the automotive industry defined by the European Commission. This should include the recovery of plastics from end-of-life vehicles (ELVs), but such materials are hardly used in compounds today. To close the knowledge gap, two ELV recyclate grades largely based on bumper recycling were analyzed in comparison to a packaging-based post-consumer recyclate (PCR). The composition data were used to design polypropylene (PP) compounds for automotive applications with virgin base material and mineral reinforcement, which were characterized in relation to a commercial virgin-based compound. A compound with a 40 wt.-% ELV-based bumper recyclate can exceed one with just a 25 wt.-% packaging-based recyclate in terms of stiffness/impact balance. While the virgin reference can nearly be matched regarding mechanics, the flowability is not reached by any of the PCR compounds, making further development work necessary.

Keywords: polypropylene; recycling; automotive; morphology; mechanics

1. Introduction

Polypropylene (PP) is one of the thermoplastic polymers with the biggest global production volumes, reaching about 75 million tons (Mt) in 2020, with a clear trend towards circularity [1]. At the same time, even in Europe, only 35% of all polymer waste is being recycled, and according to Plastics Europe, only about 9% of all thermoplastic polymers used are based on recycled materials [2]. The proportion of mechanical recycling is at 45% higher for materials based on the collection of packaging materials, limiting the performance range of the thus-produced recyclates. While steel and aluminum recycling are pretty much standard elements of end-of-life vehicle (ELV) recovery, the contribution of these processes to plastics recycling is very limited.

The European Commission has defined a target for recycled plastic content in cars of 25% by 2030, of which 25% should come from closed-loop ELV treatment [3]. This is only one of a package of measures defined in the cited proposal aiming at an annual reduction of 12.3 million tons of carbon dioxide equivalents (CO2-eqs.) in 2035, which is a key contribution to decarbonizing the automotive industry. The measures can also be seen as one of the elements of the "ReShaping Plastics" program proposed by Systemiq [4] to achieve plastic circularity of 78% in Europe by 2050. Due to global trade connections, similar targets can be expected for other world regions as well in the future.

The role of PP-based materials is essential here due to their relatively large share of the overall plastic content of contemporary vehicles of about 30–40%, according to Matos et al. [5]. This is higher than in other application areas, and even though the usage period is longer in the automotive sector, the proportions of all polymers that have been recycled since their first market introduction in the 1960s are generally low. According to Geyer et al. [6], by 2015, only about 9% of all produced polymers had been recycled, while 12% were incinerated and 79%—more than 6 billion tons—ended up in landfills or the environment. More recent and detailed figures in this respect can be found in material flow analyses recently performed in Austria [7] and Europe [8]. The Austrian study focuses on application segments, highlighting the positive trend in plastic waste treatment between 1994 and 2010. In that period, the effective elimination of landfilling was achieved mostly by moving to incineration, while the mechanically recycled fraction still remained at 9%. In contrast, the European study covers the years 1950 to 2016 and presents a polymer-specific analysis. The fate of the 310 Mt of PP consumed in Europe in that period is given as follows: 48% landfilled, 16.5% incinerated (including the use as fuel substituent in the steel and cement industries), 13.7% "in-use stock" (including automotive usage), and 7.5% recycled or re-used. The balance is material exported outside Europe or that has an unclear fate. The recycling fraction for PP given in that study is clearly lower than that for poly(ethylene terephthalate) (PET) or any of the two polyethylene (PE) classes, high-density PE (HDPE) and low-density PE (LDPE). This analysis is certainly more comprehensive and far more detailed than the one used for the aforementioned Systemiq study [4].

Less data on polymer recycling in general are available for other world regions. A recent study for the US gives an overall recycling rate of just 6.2%, with 77% of all plastic waste going to landfill sites [9]; for China, the nation with the biggest polymer production worldwide, an average recycling rate of 29% for the period 1978–2019 is claimed [10], with 42% of polymers being landfilled and 17% discarded without any proper treatment. No specific data for ELV recycling are available in either case.

Another relevant angle for ELV recycling is the life cycle analyses (LCAs) of cars and the passenger transportation system in total, which will only be discussed regarding relevant aspects here. While the use phase of a vehicle with a combustion engine is generally responsible for the largest environmental impact of its life cycle, both construction and disposal must also be considered. In a Dutch study from 2003 [11], the increased use of polymer components for reducing the weight of passenger cars is highlighted, together with the fact that only a small fraction (<10%) of this ends up in disassembled parts, while the majority (~90%) becomes automotive shredder residue (ASR) as a side stream of steel recycling. A more recent Belgian study from 2016 [12] already considers an earlier version of the EC proposal, treating the polymer-related ASR fraction collectively as 'shredder light fraction' (SLF) and giving it little relevance in the overall LCA. The SLF is a collective light material fraction retrieved after milling the car body and, thus, comprises not only bumper elements but also interior and under-the-hood elements.

More relevant for the present study are papers that focus on LCAs of plastic components in various applications, mostly dealing with the recycling of short-lived products like packaging, but also considering components of electrical equipment and vehicles. The biggest potential for carbon footprint reduction is estimated for mechanical recycling [13], with the collective treatment of bigger material streams showing further advantages [14]. Few papers specifically deal with LCAs of plastics in the automotive sector: A study initiated by Renault Group from 2019 analyses the potential of closed-loop polypropylene recycling [15], concluding that increased PP recovery can generate a volume of 52 kt of recycled plastic for use in vehicle production. A more general screening of the automotive sector by Matos et al. [5] sees great potential for circularity, but also severe limitations resulting from design complexity and multi-polymer usage. An agreement with the more general LCA studies is that integrating different waste streams, like ELV and electronics recycling, may improve the situation.

As for the LCA studies, studies on practical aspects of mechanical recycling are mostly related to packaging or, more generally, to post-consumer plastic waste. Reviews in the field [16–18] consequently consider differentiated waste collection as the main source of post-consumer recyclates (PCRs), the composition of which reflects typical polymer selection for packaging applications. Polyolefins, i.e., the different types of PE and PP, dominate the composition next to PET, with a minority fraction of polystyrene (PS) and a significant sorting residue comprising other polymers and multilayer constructions [19–21]. While the latter are without doubt appropriate solutions for fulfilling complex requirements of mechanics and barrier, the respective polymer combinations like PET/PE with ethyl–vinyl alcohol copolymer (EVOH) as an oxygen barrier make mechanical recycling difficult to impossible due to the resulting multiphase systems with poor compatibility [22].

Cross-contamination effects between PP and PE [23], and especially between polyolefins and non-polyolefins [24], reduce the performance of PCR materials in cases of poor sorting quality. Phase structure, interfacial adhesion, and the relationship in mechanics between the phases are decisive factors, similar to multiphase polymer design in general. This calls for a close analysis of such recyclates prior to their application in designing compounds [25] and makes the use of specific compatibilizers and modifiers for improving phase adhesion necessary [26]. The latter is less of an issue in automotive compound design, as PP impact copolymers and PE-based plastomers are commonly used as components anyway.

Compared to the vast literature on packaging recycling, specific papers dealing with ELV recycling are rather rare. Probably, the earliest example of using ASR in part design is the study of Robson and Goodhead from 2003 [27], who applied a skin-core design with virgin material on the part surface and an ASR compound in the core. In up to five cycles, a massive drop in tensile strength and energy to break was found when compared to virgin PP. From the same time is the bumper recycling study of Luda et al. [28]. They put the focus on the degradation effects of PP-based bumper compositions in mechanical recycling, finding clearly positive effects of both paint stripping and re-stabilization prior to compounding.

Rabeau Epsztein et al. (2014) [29] compared the performances of three samples of recycled polyolefins from ELVs to be used in automobile applications, specifically as matrices for short glass fiber (SGF)-reinforced compounds. The PE content in the tested recyclates was rather high, limiting the stiffness and heat resistance of the final products. The work of Kozderka et al. [30] is rather a model study on the recycling of pure high-impact polypropylene (HIPP), the most common polymer in car body parts such as bumpers. HIPP mechanical properties were found to deteriorate from the first reprocessing step, albeit without significant effects before the sixth reprocessing step.

Yang et al. [31] used a rather special approach for incorporating ASR into compounds, applying solid-state shear milling (SSSM) and a mixing ratio of 70/30 to 10/90 (by mass) for the production of recycled PP/ASR compositions. They think that it is an advantage to avoid complete melting and found an acceptable performance of up to 50 wt.-% ASR content, increased further by adding a compatibilizer. Maleic anhydride-grafted styrene elastomer (SEBS-g-MAH) was found to give the best toughness. A very specific study related to the essential step of paint removal and its effects on PP bumper recyclate performance was carried out by Guo et al. [14]. Efficient paint stripping was especially found to improve the ductility of the recycled material significantly.

Recently, there have also been ideas about alternative fabrication techniques for automotive components like 3D printing [32]. While this technology is not likely to substitute high-speed conversion processes like injection molding for bigger parts of a car like bumpers, it has already been explored regarding its suitability for applying recycled polymers [33]. It is, however, difficult here as well to manage the balance between the flowability, solidification speed, and mechanical performance.

The present study had the following dual target:

- Analyzing semicommercially (i.e., as samples from pilot scale) available ELV recyclate grades based on bumper or 'shredder light fraction' (SLF) recycling in terms of composition and suitability for automotive compounds;
- Testing these as components in automotive compounds in direct comparison to a packaging-based post-consumer recyclate (PCR) already established as a compounding component

The performance targets for the final compounds were based on typical specifications of original equipment manufacturers (OEMs), i.e., car brand owners, for passenger vehicle bumpers. For this step, the performance was further compared to a virgin-based commercial compound.

2. Materials and Methods

2.1. Investigated Recyclates and Modifiers

Shredded flakes from bumper recycling in Germany (PCR-ELV-bumper) and the polymer shredder light fraction (SLF) in the Netherlands (PCR-ELV light fraction) were applied, both being produced in semicommercial scale and received under obligation of not disclosing the precise source. Prior to further characterization and compounding, the irregular-shaped flakes were pelletized on a twin-screw extruder ZK50 (Collin GmbH, Maitenbeth, Germany) with an L/D-ratio of 22, using a 200 μm screen-changer-type melt filter (same supplier), followed by strand pelletization. This was carried out not only to homogenize the material but also to remove impurities like paint and metal residues. In this compounding step, 0.30 wt% of a 1:1 mixture of pentaerythrityl-tetrakis(3-(3′,5′-di-tert. butyl-4-hydroxyphenyl)-propionate (Irganox 1010 from BASF SE, Ludwigshafen, Germany) and Tris (2,4-di-t-butylphenyl) phosphite (Irgafos 168 from BASF SE, Ludwigshafen, Germany) were added to re-stabilize the material. The third recyclate was a PCR-Packaging material based on Borealis' own pilot-scale operations. Table 1 summarizes the key analytical parameters and composition data for all PCRs (see Section 2 for details), the differences of which will be discussed in Section 3.

Table 1. The polymer characteristics of the different recyclates (PCRs).

		PCR-Packaging	PCR-ELV-Bumper	PCR-ELV-SLF
MFR [1]	g/10 min	25	12	8
Density	kg/m^3	910	1020	930
Ash [2]	wt.-%	1.5	18	3.0
DSC				
$T_{m,PE}$	°C	124	124	128
$H_{m,PE}$	J/g	2.5	0.4	56.6
$T_{m,PP}$	°C	163	165	163
$H_{m,PP}$	J/g	95.8	56.6	52.1
CRYSTEX				
SF	wt.-%	10	34	20
C2 total	wt.-%	5.8	23	29
C2(SF)	wt.-%	26	50	41
C2(CF)	wt.-%	3.5	10	27
IV total	dl/g	1.60	1.66	1.95
IV(SF)	dl/g	1.71	1.75	1.74
IV(CF)	dl/g	1.62	1.63	1.96
Mechanics				
Flexural modulus	MPa	1324	1587	637
NIS +23 °C	kJ/m^2	5.0	34	46
NIS −20 °C	kJ/m^2	2.2	6.6	8.8
IPT +23 °C E_{max}	J	3	18	18
IPT +23 °C E_{punct}	J	4	29	29
IPT −30 °C E_{max}	J	n.d.	16	13
IPT −30 °C E_{punct}	J	n.d.	17	14
CLTE +23/80 °C	μm/m°C	123	55	n.d.
CLTE −30/80 °C	μm/m°C	100	46	n.d.

[1] at 2.16 kg load, 230 °C; [2] TGA residue; n.d.—not determined.

The optical appearances of the flakes and the as-extruded pellets for the PCR-Packaging and PCR-ELV-Bumper materials are shown in Figure 1, highlighting the effect of homogenization.

Figure 1. Optical appearance of flakes before compounding and pellets after compounding for PCR-Packaging and PCR-ELV-Bumper materials.

One of the two PCR-ELV materials, the SLF type, was found to be too soft and too rich in PE, limiting its stiffness and heat resistance. Compounds were, therefore, only prepared based on two of the materials from Table 1. Depending on the PCR composition, different compositions of virgin base copolymers [34,35] with the addition of a PE-plastomer [36,37] were selected.

The melt flow rate (MFR, 230 °C/2.16 kg) and composition data for these two 'virgin mixes' are listed in Table 2. Due to the high mineral residue content in the PCR-ELV-Bumper, which was identified to be mostly talc via X-ray fluorescence spectroscopy (Mg, Al and Si as dominant elements), the reinforcing filler content was also adapted. As a reinforcing mineral filler, talc Steamic 1CA of Imerys, France, with a median diameter d_{50} of 1.8 µm and top cut diameter d95 of 6.2 µm, was used.

Table 2. The polymer characteristics of the virgin polyolefin mixes used for upgrading the recyclates (PCR materials).

		Virgin Polyolefin Mix 1	Virgin Polyolefin Mix 2
MFR [1]	g/10 min	7.7	9.6
CRYSTEX			
SF	wt.-%	39	30
C2 total	wt.-%	26	12
C2(SF)	wt.-%	55	37
C2(CF)	wt.-%	5.0	5.0
IV total	dL/g	1.8	1.9
IV(SF)	dL/g	2.9	3.53
IV(CF)	dL/g	1.5	1.5

[1] at 2.16 kg load, 230 °C.

A commercial virgin-based compound of Borealis AG (Vienna, Austria) was used as a reference material in terms of performance. It is characterized by an MFR of 22 g/10 min and a mineral filler content of 15 wt.-%.

2.2. Analytics and Mechanics

Standard characterization methods, as also used before for recyclates and their compositions [25], were used. MFR was measured according to ISO 1133 at 230 °C and a 2.16 kg load [38]. Differential scanning calorimetry (DSC) was run according to ISO 11357/part 3/method C2 [39] in a heat/cool/heat cycle with a scan rate of 10 °C/min in the temperature range of −30–26 to +225 °C with a TA Instrument Q200 performing differential scanning calorimetry (DSC) on 5 to 7 mg samples. The two melting temperatures corresponding to the PE ($T_{m,PE}$) and PP content ($T_{m,PP}$) and the two corresponding melting enthalpies ($H_{m,PE}$ and $H_{m,PP}$) were determined from the second heating step. Density was measured according to ISO 1183-1/2004 method A [40] on compression-molded specimens. Thermogravimetry (TGA) was used to determine the inorganic content of the PCR materials according to ISO 1172:1996 [41] with a Perkin Elmer (Waltham, MA, USA) TGA 8000. Approximately 10–20 mg of material was placed in a platinum pan, and temperature was equilibrated at 50 °C for 10 min and afterwards raised to 950 °C under nitrogen at a heating rate of 20 °C/min. The ash content was evaluated as the wt.-% at 850 °C.

For determining the crystalline (CF) and soluble (SF) fractions and their respective properties (intrinsic viscosity, IV, and ethylene content, C2), the CRYSTEX method [42] was applied. A CRYSTEX QC apparatus of PolymerChar Valencia, Spain, was used for this analysis, which has developed into an excellent alternative to the frequently used separation of PP copolymers and compositions into xylene cold soluble and insoluble fractions (XCS/XCI), followed by analysis of the fractions. The crystalline and amorphous fractions are separated through temperature cycles of dissolution at 160 °C, crystallization at 40 °C,0 and re-dissolution in 1,2,4-trichlorobenzene at 160 °C. The quantification of SF, CF, and ethylene content (C2) was achieved by means of an integrated infrared detector (IR4), and for the determination of the intrinsic viscosity (iV), an online 2-capillary viscometer was used.

All mechanical tests were carried out on injection-molded specimens prepared in line with ISO 19069-2 [43] on an Engel e-motion 310/55 machine (Enel Austria GmbH, Schwertberg, Austria) using a melt temperature of 230 °C for all materials, irrespective of the material melt flow rate. Flexural modulus was determined according to ISO 178/method A [44], and Charpy notched impact strength (NIS) was determined according to ISO 179 1eA [45], on test bars of 80 × 10 × 4 mm^3. Instrumented puncture tests (IPT) were conducted in accordance with ISO 6603-2:2023 [46] on plaques of 60 × 60 × 3 mm^3. The coefficient of linear thermal expansion (CLTE) was determined in accordance with ISO 11359-2 [47] on 10 mm long pieces cut from the centers of same injection-molded specimens used for the flexural modulus determination. The dimension of the CLTE specimen was 10 × 10 × 4 mm^3, testing only in machine direction (MD) for the present study. The measurement was performed in a temperature range from −30 resp. +23 °C to +80 °C at a heating rate of 1 °C/min.

The morphology of compounds was evaluated via scanning electron microscopy (SEM) on cryo-cut surfaces after staining with RuO_4 [48] on injection-molded specimens as for the flexural test. An Apreo S LoVac microscope of ThermoFisher Scientific (Waltham, MA, USA) was used.

2.3. Compounding

Compounds targeted at property profiles typical for automotive exterior components like bumpers or side trims were prepared based on two of the recyclates. A ThermoFisher (USA) TSE24 twin-screw extruder with an L/D ratio of 40, a high-intensity mixing screw configuration, and a temperature profile between 190 and 220 °C was used, followed by melt strand cooling in a water bath and strand pelletization. The following further additives were used in compounding: 0.20 wt.-% of the slip agent oleamide and 0.05 wt.-% of the antioxidant octadecyl 3-(3′,5′-di-tert. butyl-4-hydroxyphenyl)propionate. Both additives were dosed in a blend with powder of polypropylene homopolymer for better dispersion in the final compound, with the amount of said blend not exceeding 1 wt.-% of the total composition. Compound compositions and all analytical and mechanical characterization results are presented in Table 3.

Table 3. Conceptual PCR compounds and the resulting properties in comparison to the virgin compound.

		Packaging-PCR Compound 1	ELV-PCR Compound 2	Virgin Compound [1]
Virgin polyolefin mix 1	wt.-%	58.25		-
Virgin polyolefin mix 2	wt.-%		49.25	-
PCR-Packaging	wt.-%	25.0		-
PCR-ELV-Bumper	wt.-%		40.0	
Talc	wt-%	15.0	9.0	-
Additives	wt.-%	1.75	1.75	-
MFR [2]	g/10 min	10	9.0	22
Ash [3]	wt.-%	15.1	17.1	15.0
CRYSTEX				
SF	wt.-%	30	31	-
C2 total	wt.-%	18	16	-
C2 (SF)	wt.-%	50	42	-
C2 (CF)	wt.-%	5.0	7.0	-
IV total	dL/g	1.8	1.9	-
IV (SF)	dL/g	2.6	2.6	-
IV (CF)	dL/g	1.5	1.7	-
Mechanics				
Flexural modulus	MPa	1459	1508	1750
NIS +23 °C	kJ/m^2	54	43	40
NIS −20 °C	kJ/m^2	5.9	6	6
IPT +23 °C E_{max}	J	20	20	22
IPT +23 °C E_{punct}	J	37	34	36
IPT −30 °C E_{max}	J	35	17	21
IPT −30 °C E_{punct}	J	43	18	31
CLTE +23/80 °C	µm/m°C	90	76	71
CLTE −30/80 °C	µm/m°C	75	65	56

[1] composition not disclosed; [2] at 2.16 kg load, 230 °C; [3] TGA residue; n.d.—not determined.

3. Results

3.1. ELV and Packaging Recyclates

The compositions of the three PCR materials differ significantly, reflected by their analytical and mechanical properties in Table 1. Based on a combination of DSC (see Figure 2) and CRYSTEX analysis, one can calculate the approximate fractions of PP, PE and amorphous ethylene–propylene copolymer (EPC) in the respective types. For finding the crystalline PP and PE contents, the H_m-values related to the two polymers were normalized to the respective enthalpies for fully crystalline polymers, 293 J/g for PE [49] and 170 J/g for PP [50]. The result of this calculation, which assumes an equal degree of crystallinity of both polymers in the composition, was then related to the crystalline fraction (CF) from CRYSTEX, assuming SF to be identical to the EPC content in line with earlier work [39].

A rough idea of the relationship between crystalline PP and PE can already be obtained from the heating scans in Figure 2, including the fact that T_m also varies for both parts. Two of the three PCRs have a certain EP-random copolymer content, as indicated by the lower $T_{m,PP}$, and in none of the compositions is the PE really HDPE, which would have a $T_{m,PE}$ of >130 °C. The diagram in Figure 3 presents the full results of said calculation, already excluding the filler fraction, which is high only for the PCR-ELV-Bumper type (18 wt.-%). Both ELV-PCRs show higher EPC contents, but the 'light fraction' (SLF) material clearly sticks out in terms of PE content. While the other two PCRs bear at least some compositional resemblance to a virgin PP impact copolymer [51,52], this one is instead a PP/PE blend with the well-known limitations of such compositions [23,25].

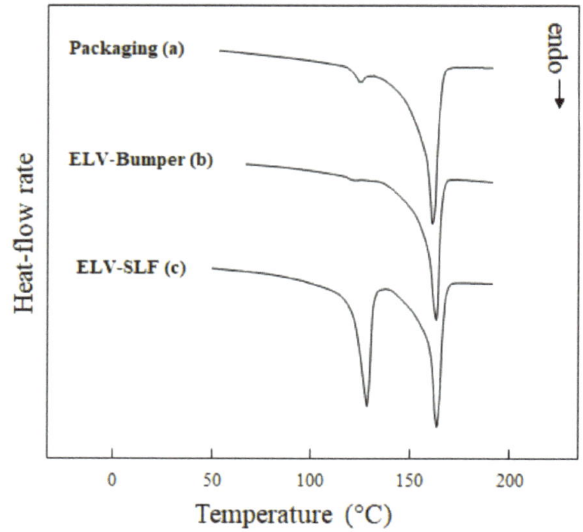

Figure 2. DSC heating scans (2nd heat) of the three PCR materials: (a) PCR-Packaging, (b) PCR-ELV-Bumper, and (c) PCR-ELV-SLF.

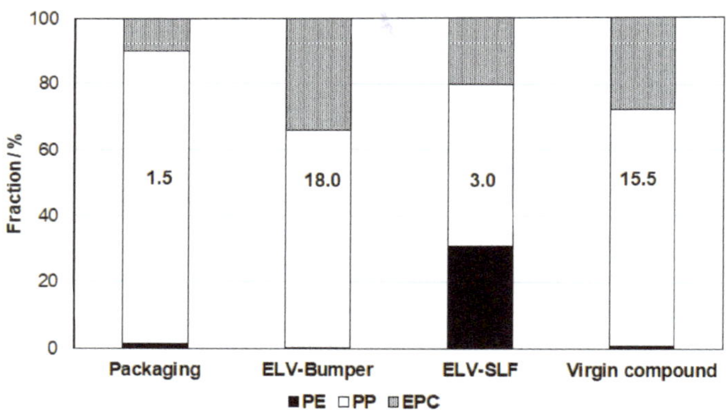

Figure 3. Relative composition of the polymer part of the PCR materials and the reference virgin compound based on DSC and CRYSTEX data (material designations as in Table 1; insert figures indicate ash content).

Three further important factors for the significantly different mechanical performances need to be considered:

- The molecular weight of elastomer-like EPC, as well as its composition and content, is decisive for the stiffness-impact balance of impact copolymers and similar compositions [35,51]. The PCR-Packaging type clearly has the lowest C2(SF), related to a rather high glass transition temperature in the EPC phase [52]. This, in combination with the lowest amount of SF, is limiting the impact strength. Both ELV types are similar in that respect, and the IV(SF) representing the molecular weight of the EPC phase is very similar for all PCRs.
- The molecular weight of the crystalline fractions is the dominant factor for the MFR of the compositions, but it is also decisive for room-temperature toughness. Only the

PCR-ELV-SLFtype shows a higher value for IV(CF), likely as a consequence of the high PE content. This is reflected by the highest Charpy NIS at 23 °C.

- Mineral filler content is decisive for the stiffness of the compositions, and the PCR-ELV-Bumper type also shows the highest flexural modulus. Secondary factors are EPC content and the PP/PE ratio, as the latter polymer is significantly less stiff.

Taken together, the differences essentially disqualified the PCR-ELV-SLF type for compound development, as no reasonable PCR content could be achieved with the low modulus of this composition. Balancing the EPC resp. overall elastomer and mineral filler content was viable for the other two PCR types, using the experience of normal virgin-based compound development [7,34,53]. The respective virgin polyolefin mixes described in Table 2 are both characterized by rather high SF contents and IV(SF) to improve the toughness and ductility of the resulting compounds, with their slightly different MFRs additionally balancing the overall processability. Variation in the mineral filler content can be seen as a standard measure not only to improve stiffness but also to reduce thermal expansion.

3.2. Automotive Compounds

A substantial number of automotive parts for exterior applications such as bumpers are typically composed of different kinds of polypropylenes in combination with external elastomers and fillers. The polypropylenes typically include homopolymers and heterophasic (impact) copolymers, in which the polypropylene matrix provides the stiffness and ethylene propylene elastomer (EPC) ensures the impact performance [34,35]. The external elastomers, mainly low-density ethylene-alpha-olefin copolymers, due to their low glass transition temperature and amorphous nature, contribute to the improvement of the impact behavior at both room and cold temperatures [36,37]. Using the molecular structure and amount of rubber phase, the dimensional stability and various aesthetic properties can be controlled [54]. The role of the inorganic filler is mainly to control the stiffness and thermal expansion, as mentioned above.

The various compound compositions resulted in respective unique phase morphologies controlling the final properties. Figure 4 represents a typical morphology analysis, as carried out by SEM to reveal the heterophasic nature of the blends together with the fine filler distribution. The bright round to slightly elongated objects in the SEM images represent the elastomer phase, composed from a combination of the reactor-based ethylene-propylene copolymer and the external plastomer. The elastomeric phase is well dispersed in the polymer matrix, which appears in black together with fine filler particles in light grey. The structures of all three compounds are similar, but in particular, Compound 2 comprising 40 wt.-% PCR-ELV-Bumper shows more heterogeneity in terms of both elastomer and filler particle size.

To compare the basic mechanical performance, Figure 5a shows the Charpy notched impact strength (NIS) (red bars) and the flexural modulus (blue bars) of the pure different recyclates and the resulting compounds measured at room temperature. Due to the nature of the ELV recyclate, it was possible to incorporate it in a higher amount in relation to the packaging-based material while preserving the stiffness-impact balance of the final material. Compound 2 contains 40 wt.-% of PCR-ELV-Bumper, while Compound 1 can accept just 25 wt.-% of PCR-Packaging in order to achieve a comparable stiffness/impact balance.

For the sake of comparison, the performance of a virgin-based compound, i.e., with 0 wt.-% PCR, is presented to demonstrate the mechanical requirements for the PCR-based compositions. It should be noted that the processability level of said reference in terms of MFR is not reached by any of the PCR compounds, meaning that further development work will be necessary. While less demanding applications are already reachable, this processability issue is still challenging, as increasing the MFR and/or reducing the elastomer-IV of a composition always results in reduced toughness [51,53]. The usual approach for MFR increase for PP is visbreaking, designating a peroxide-induced radical process to reduce the molecular weight, but this process is difficult in case of multiphase systems like impact copolymers and elastomer compounds [35]. As PE itself may branch and crosslink under the influence of radicals, visbreaking such compositions will change the viscosity ratio between the matrix and the elastomer phase, with negative consequences for mechanics and surface quality.

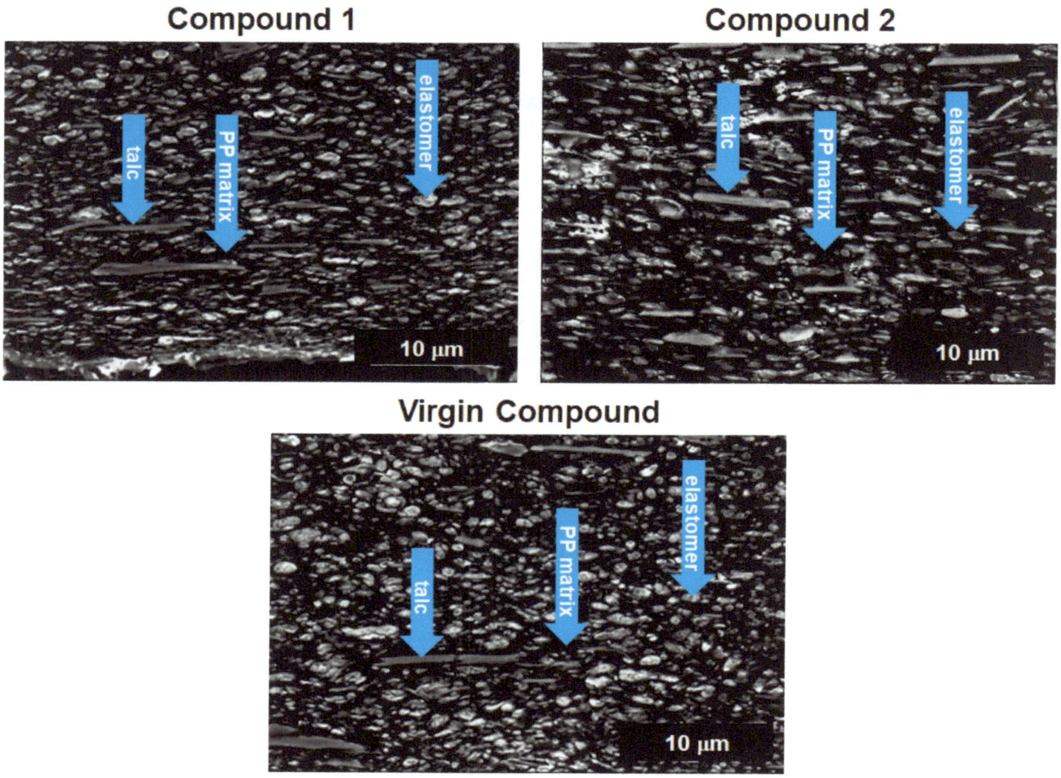

Figure 4. Morphology of tested compounds—SEM images collected from a cross-section of injection-molded specimens in flow direction; inset arrows indicate components.

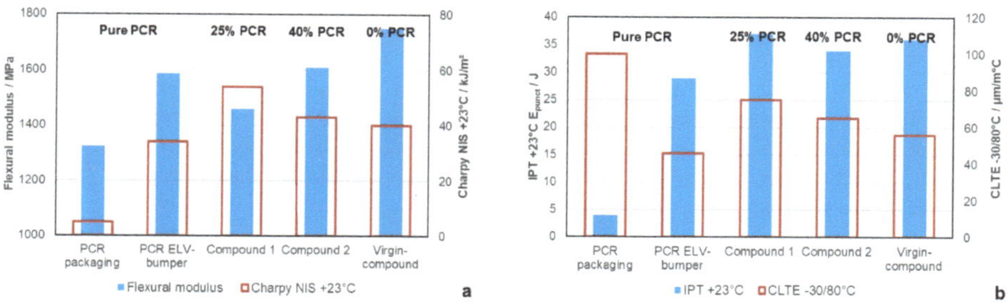

Figure 5. Application performance of the compounded PCRs and resulting compounds in direct comparison to a commercial virgin grade for automotive exterior applications: the (**a**) stiffness/impact balance and (**b**) puncture resistance and coefficient of thermal expansion.

Two further parameters of high relevance for automotive applications, puncture energy measured during the instrumented puncture test performed at 23 °C (blue bars) and the coefficient of linear thermal expansion (CLTE, red bars), are shown in Figure 5b. Compound 2 with the ELV-Bumper recyclate shows, in all cases, puncture energy and CLTE values closer to the virgin grade at higher contents of PCR than Compound 1. Table 3 further shows that the sub-zero impact properties of the recyclate-based compounds, at −20 °C (NIS) and −30 °C (IPT), are also comparable to the virgin reference.

4. Summary and Conclusions

The new ELV regulation for the European Union [3] aims at increasing circularity in the automotive industry by imposing mandatory PCR content targets. Mechanically recycled plastics from various waste streams can be used to achieve this target, but at least 25% of the PCR plastic has to originate from end-of-life vehicles. PP plays a major role in this as it is one of the most popular thermoplastics and the main plastic fraction in contemporary cars.

The results, collected in this study, demonstrate that packaging and end-of-life vehicle waste streams are valuable feedstocks to collect PCR PP for automotive applications (see Figure 6). It was found that a compound with a 40 wt.-% ELV-based bumper recyclate can even exceed one with just 25 wt.-% packaging-based recyclate in terms of stiffness/impact balance, which is certainly better than in any of the previous studies [14,28]. The virgin reference can nearly be matched regarding mechanics, but the flowability (MFR) is not reached by any of the PCR compounds, meaning that further development work will be necessary (as explained in detail above, simple visbreaking will not be sufficient).

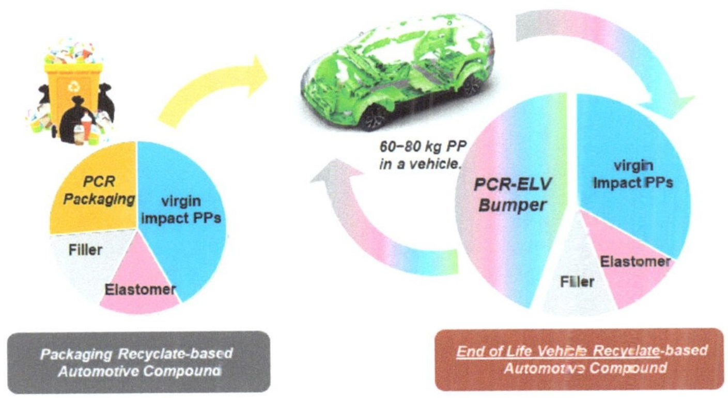

Figure 6. Scxhematic presentation of recyclate stream and contents in the PCR-Packaging based compound 1 and the PCR-ELV-Bumper based compound 2.

While the latest generation of mechanically recycled PCR PP from packaging waste is fit to be used in automotive applications already, ELV waste streams need further attention. In particular for the shredder light fraction (SLF), the level of purification needs to be improved before they can enter highly specified automotive applications. The PCR fraction originating from bumper applications seems suitable for closing the loop and being further re-used when it comes to achieve the basic mechanical performance.

Author Contributions: Conceptualization, D.M. and M.G. (Markus Gahleitner); methodology, M.G. (Markus Gall) and C.S.; investigation, M.G. (Markus Gall) and W.S.; resources, D.M.; writing—original draft preparation, M.G. (Markus Gahleitner); writing—review and editing, M.G. (Markus Gall) and M.G. (Markus Gahleitner); visualization, D.M.; project administration, D.M.; All authors have read and agreed to the published version of the manuscript.

Funding: This research received no external funding.

Institutional Review Board Statement: Not applicable.

Data Availability Statement: The original contributions presented in the study are included in the article, further inquiries can be directed to the corresponding author.

Conflicts of Interest: Markus Gall, Daniela Mileva, Wolfgang Stockreiter, and Markus Gahleitner were employed by the company Borealis Polyolefine GmbH, Linz, Austria. Christophe Salles was employed by the company Borealis Services S.A.S., Paris, France. All authors declare that the research was conducted in the absence of any commercial or financial relationships that could be construed as potential conflicts of interest.

References

1. Bernreitner, K.; Gahleitner, M.; Grestenberger, G.; Kahlen, S.; Niedersüß, P.; Tranninger, C. The Future of Polypropylene Lies in the Circular Economy. *Kunststoffe Intern* **2022**, *10*, 22–28.
2. Plastics Europe, Plastics—The Facts 2022. Available online: https://plasticseurope.org/knowledge-hub/plastics-the-facts-2022/ (accessed on 15 May 2024).
3. European Commission, 2023. Proposal for a Regulation of the European Parliament and of the Council on Circularity Requirements for Vehicle Design and on Management of End-of-Life Vehicles. Available online: https://environment.ec.europa.eu/publications/proposal-regulation-circularity-requirements-vehicle-design-and-management-end-life-vehicles_en (accessed on 15 May 2024).
4. Systemiq, 2022. ReShaping Plastics: Pathways to a Circular, Climate Neutral Plastics System in Europe. Available online: https://www.systemiq.earth/systems/circular-materials/reshaping-plastics/ (accessed on 15 May 2024).
5. Matos, J.; Santos, S.; Simões, C.L.; Martins, C.I.; Simoes, R. Practical application of circularity micro-indicators to automotive plastic parts in an industrial context. *Sustain. Prod. Consum.* **2023**, *43*, 155–167. [CrossRef]
6. Geyer, R.; Jambeck, J.R.; Law, K.L. Production, use, and fate of all plastics ever made. *Sci. Adv.* **2017**, *3*, e1700782. [CrossRef] [PubMed]
7. Van Eygen, E.; Feketitsch, J.; Laner, D.; Rechberger, H.; Fellner, J. Comprehensive analysis and quantification of national plastic flows: The case of Austria. *Resour. Conserv. Recycl.* **2017**, *117*, 183–194. [CrossRef]
8. Kawecki, D.; Wu, Q.; Gonçalves, J.S.V.; Nowack, B. Polymer-specific dynamic probabilistic material flow analysis of seven polymers in Europe from 1950 to 2016. *Resour. Conserv. Recycl.* **2021**, *173*, 105733. [CrossRef]
9. Di, J.; Reck, B.K.; Miatto, A.; Graede, T.E. United States plastics: Large flows, short lifetimes, and negligible recycling. *Resour. Conserv. Recycl.* **2021**, *167*, 105440. [CrossRef]
10. Jiang, X.; Wang, T.; Jiang, M.; Xu, M.; Yu, Y.; Guo, B.; Chen, D.; Hu, S.; Jiang, J.; Zhang, Y.; et al. Assessment of Plastic Stocks and Flows in China: 1978–2017. *Resour. Conserv. Recycl.* **2020**, *161*, 104969. [CrossRef]
11. Castro, M.B.G.; Remmerswaal, J.A.M.; Reuter, M.A. Life Cycle Impact Assessment of the Average Passenger Vehicle in the Netherlands. *Int. J. Life Cycle Assess.* **2003**, *8*, 297–304. [CrossRef]
12. Inghels, D.; Dullaert, W.; Raa, B.; Walther, G. Influence of composition, amount and life span of passenger cars on end-of-life vehicles waste in Belgium: A system dynamics approach. *Transp. Res. Part A Policy Pract.* **2016**, *91*, 80–104. [CrossRef]
13. Venkatachalam, V.; Pohler, M.; Spierling, S.; Nickel, L.; Barner, L.; Endres, H.-J. Design for Recycling Strategies Based on the Life Cycle Assessment and End of Life Options of Plastics in a Circular Economy. *Macromol. Chem. Phys.* **2022**, *223*, 2200046. [CrossRef]
14. Guo, W.; Li, K.; Fang, Z.; Feng, T.; Shi, T. A sustainable recycling process for end-of-life vehicle plastics: A case study on waste bumpers. *Waste Manag.* **2022**, *154*, 187–198. [CrossRef] [PubMed]
15. Gallone, T.; Zeni-Guido, A. Closed-Loop Polypropylene, an Opportunity for the Automotive Sector. *J. Field Actions Spec. Issue* **2019**, *19*, 48–53. Available online: https://journals.openedition.org/factsreports/5225 (accessed on 2 May 2024).
16. Yin, S.; Tuladhar, R.; Shi, F.; Shanks, R.A.; Combe, M.; Collister, T. Mechanical Reprocessing of Polyolefin Waste: A Review. *Polym. Eng. Sci.* **2015**, *55*, 2899–2909. [CrossRef]
17. Ragaert, K.; Delva, L.; Van Geem, K. Mechanical and chemical recycling of solid plastic waste. *Waste Manag.* **2017**, *68*, 24–58. [CrossRef] [PubMed]
18. Schyns, Z.O.G.; Shaver, M.P. Mechanical Recycling of Packaging Plastics: A Review. *Macromol. Rapid Commun.* **2020**, *42*, 2000415. [CrossRef] [PubMed]
19. Kaiser, K.; Schmid, M.; Schlummer, M. Recycling of Polymer-Based Multilayer Packaging: A Review. *Recycling* **2018**, *3*, 1. [CrossRef]
20. Roosen, M.; Mys, N.; Kleinhans, K.; Lase, I.S.; Huysveld, S.; Brouwer, M.; van Velzen, E.U.T.; Van Geem, K.M.; Dewulf, J.; Ragaert, K.; et al. Expanding the collection portfolio of plastic packaging: Impact on quantity and quality of sorted plastic waste fractions. *Resour. Conserv. Recycl.* **2022**, *178*, 106025. [CrossRef]
21. Seier, M.; Archodoulaki, V.-M.; Koch, T.; Duscher, B.; Gahleitner, M. Prospects for Recyclable Multilayer Packaging: A Case Study. *Polymers* **2023**, *15*, 2966. [CrossRef]
22. Seier, M.; Archodoulaki, V.-M.; Koch, T. The morphology and properties of recycled plastics made from multi-layered packages and the consequences for the circular economy. *Resour. Conserv. Recycl.* **2024**, *202*, 107388. [CrossRef]

23. Van Belle, A.; Demets, R.; Mys, N.; Van Kets, K.; Dewulf, J.; Van Geem, K.; De Meester, S.; Ragaert, K. Microstructural Contributions of Different Polyolefins to the Deformation Mechanisms of Their Binary Blends. *Polymers* **2020**, *12*, 1171. [CrossRef]
24. Demets, R.; Grodent, M.; Van Kets, K.; De Meester, S.; Ragaert, K. Macromolecular Insights into the Altered Mechanical Deformation Mechanisms of Non-Polyolefin Contaminated Polyolefins. *Polymers* **2022**, *14*, 239. [CrossRef] [PubMed]
25. Gall, M.; Freudenthaler, P.J.; Fischer, J.; Lang, R.W. Characterization of Composition and Structure–Property Relationships of Commercial Post-Consumer Polyethylene and Polypropylene Recyclates. *Polymers* **2021**, *13*, 1574. [CrossRef] [PubMed]
26. Mehrabi-Mazidia, M.; Sharif, H. Post-consumer recycled high density polyethylene/polypropylene blend with improved overall performance through modification by impact polypropylene copolymer: Morphology, properties and fracture resistance. *Polym. Int.* **2021**, *70*, 1701–1716. [CrossRef]
27. Robson, S.; Goodhead, T.C. A process for incorporating automotive shredder residue into thermoplastic mouldings. *J. Mater. Process. Technol.* **2003**, *139*, 327–331. [CrossRef]
28. Luda, M.P.; Ragosta, G.; Musto, P.; Acierno, D.; Di Maio, L.; Camino, G.; Nepote, V. Regenerative Recycling of Automotive Polymer Components: Poly(propylene) Based Car Bumpers. *Macromol. Mater. Eng.* **2003**, *288*, 613–620. [CrossRef]
29. Rabeau Epsztein, S.; Jevardat de Fombelle, M.-A.; Falher, T.; Jouannet, D.; Gallone, T.; Cauret, L. Substitution of Virgin material by recycled material from End-of-Life Vehicle (ELV). *Key Eng. Mater.* **2014**, *611–612*, 836–843. [CrossRef]
30. Kozderka, M.; Rose, B.; Bahlouli, N.; Kočí, V.; Caillaud, E. Recycled high impact polypropylene in the automotive industry—mechanical and environmental properties. *Int. J. Interact. Des. Manuf.* **2017**, *11*, 737–750. [CrossRef]
31. Yang, S.; Bai, S.; Duan, W.; Qi Wang, Q. Preparation of composites based on recycled polypropylene and automotive shredder residue. *Polym. Int.* **2018**, *67*, 936–945. [CrossRef]
32. Jahromi, F.T.; Nikzad, M.; Prasad, K.; Norén, J.; Isaksson, M.; Arian, A.; Sbarski, I. Additive manufacturing of polypropylene micro and nano composites through fused filament fabrication for automotive repair applications. *Polym. Adv. Technol.* **2023**, *34*, 1059–1074. [CrossRef]
33. Ghabezi, P.; Sam-Daliri, O.; Flanagan, T.; Walls, M.; Harrison, N.M. Circular economy innovation: A deep investigation on 3D printing of industrial waste polypropylene and carbon fibre composites. *Resour. Conserv. Recycl.* **2024**, *206*, 107667. [CrossRef]
34. Galli, P.; Milani, F.; Simonazzi, T. New Trends in the Field of Propylene Based Polymers. *Polym. J.* **1985**, *17*, 37–55. [CrossRef]
35. Gahleitner, M.; Doshev, P.; Tranninger, C. Heterophasic copolymers of polypropylene—Development, design principles and future challenges. *J. Appl. Polym. Sci.* **2013**, *130*, 3028–3037. [CrossRef]
36. Yu, T.C. Metallocene plastomer modification of polypropylenes. *Polym. Eng. Sci.* **2001**, *41*, 656–671. [CrossRef]
37. McNally, T.; McShane, P.; Nally, G.M.; Murphy, W.R.; Cook, M.; Miller, A. Rheology, phase morphology, mechanical and thermal properties of polypropylene/ metallocene catalysed ethylene 1-octene copolymer blends. *Polymer* **2020**, *43*, 3785–3793. [CrossRef]
38. *ISO 1133-1:2022*; Plastics—Determination of the Melt Mass-Flow Rate (MFR) and Melt Volume-Flow Rate (MVR) of Thermoplastics, Part 1: Standard Method. International Standard Organization: Geneva, Switzerland, 2019.
39. *ISO 11357-3:2018*; Plastics—Differential Scanning Calorimetry (DSC), Part 3: Determination of Temperature and Enthalpy of Melting and Crystallization. International Standard Organization: Geneva, Switzerland, 2018.
40. *ISO 1183-1:2019*; Plastics—Methods for Determining the Density of Non-Cellular Plastics, Part 1: Immersion Method, Liquid Pycnometer Method and Titration Method. International Standard Organization: Geneva, Switzerland, 2019.
41. *ISO 1172:1996*; Textile-Glass-Reinforced Plastics—Prepregs, Moulding Compounds and Laminates—Determination of the Textile-Glass and Mineral-Filler Content Using Calcination Methods. International Standard Organization: Geneva, Switzerland, 1996.
42. Jeremic, L.; Albrecht, A.; Sandholzer, M.; Gahleitner, M. Rapid characterization of high-impact ethylene–propylene copolymer composition by crystallization extraction separation: Comparability to standard separation methods. *Int. J. Polym. Anal. Charact.* **2020**, *25*, 581–596. [CrossRef]
43. *ISO 11359-2:2021*; Plastics—Thermomechanical analysis (TMA), Part 2: Determination of Coefficient of Linear Thermal Expansion and Glass Transition Temperature. International Standard Organization: Geneva, Switzerland, 2021.
44. *ISO 19069-2:2016*; Plastics—Polypropylene (PP) Moulding and Extrusion Materials, Part 2: Preparation of Test Specimens and Determination of Properties. International Standard Organization: Geneva, Switzerland, 2016.
45. *ISO 178:2019*; Plastics—Determination of Flexural Properties. International Standard Organization: Geneva, Switzerland, 2019.
46. *ISO 179-1:2010*; Plastics—Determination of Charpy Impact Properties, Part 1: Non-Instrumented Impact Test. International Standard Organization: Geneva, Switzerland, 2010.
47. *ISO 6603-2:2023*; Plastics—Determination of Puncture Impact Behaviour of Rigid Plastics, Part 2: Instrumented Impact Testing. International Standard Organization: Geneva, Switzerland, 2023.
48. Poelt, P.; Ingolic, E.; Gahleitner, M.; Bernreitner, K.; Geymayer, W. Characterization of modified polypropylene by scanning electron microscopy. *J. Appl. Polym. Sci.* **2000**, *78*, 1152–1162. [CrossRef]
49. Wunderlich, B.; Czornyj, G. A Study of Equilibrium Melting of Polyethylene. *Macromolecules* **1977**, *10*, 906–913. [CrossRef]
50. Lanyi, F.J.; Wenzke, N.; Kaschta, J.; Schubert, D.W. On the Determination of the Enthalpy of Fusion of α-Crystalline Isotactic Polypropylene Using Differential Scanning Calorimetry, X-Ray Diffraction, and Fourier-Transform Infrared Spectroscopy: An Old Story Revisited. *Adv. Eng. Mater.* **2020**, *22*, 1900796. [CrossRef]
51. Grein, C.; Bernreitner, K.; Hauer, A.; Gahleitner, M.; Neißl, W. Impact Modified Isotatic Polypropylene with Controlled Rubber Intrinsic Viscosities: Some New Aspects About Morphology and Fracture. *J. Appl. Polym. Sci.* **2003**, *87*, 1702–1712. [CrossRef]

52. Doshev, P.; Lohse, G.; Henning, S.; Krumova, M.; Heuvelsland, A.; Michler, G.; Radusch, H.-J. Phase Interactions and Structure Evolution of Heterophasic Ethylene–Propylene Copolymers as a Function of System Composition. *J. Appl. Polym. Sci.* **2006**, *101*, 2825–2837. [CrossRef]
53. Moritomi, S.; Watanabe, T.; Kanzaki, S. Polypropylene Compounds for Automotive Applications, Sumitomo Kagaku (English Edition), Report 1. 2010. Available online: https://www.sumitomo-chem.co.jp/english/rd/report/files/docs/01_2010-1e.pdf (accessed on 10 October 2023).
54. Grestenberger, G.; Potter, G.D.; Grein, C. Polypropylene/ethylene-propylene rubber (PP/EPR) blends for the automotive industry: Basic correlations between EPR-design and shrinkage. *Exp. Polym. Lett.* **2014**, *8*, 282–292. [CrossRef]

materials

MDPI

Article

Effect of Graphene on the Mechanical Properties of Recycled High-Density and High-Molecular-Weight Polyethylene Blends

Hniya Kharmoudi [1,2], Alae Lamtai [1,2], Said Elkoun [1,2,*], Mathieu Robert [1,2] and Carl Diez [3]

1 Center for Innovation in Technological Eco-Design (CITE), University of Sherbrooke, Sherbrooke, QC J1K 2R1, Canada; hniya.kharmoudi@usherbrooke.ca (H.K.); alae.lamtai@usherbrooke.ca (A.L.); mathieu.robert2@usherbrooke.ca (M.R.)
2 Research Center for High Performance Polymer and Composite Systems, CREPEC, Montréal, QC H3A 0C3, Canada
3 Soleno Inc. Maitrise de l'Eau Pluviale, Saint-Jean-sur-Richelieu, QC J2X 4B6, Canada; carl.diez@soleno.com
* Correspondence: said.elkoun@usherbrooke.ca

Abstract: This study uses an extrusion process to formulate blends based on recycled high-density and high-molecular-weight polyethylene (recHDPE, recHMWPE) for the manufacture of rainwater drainage pipes. The main objective of this project is to investigate the effects of incorporating graphene on the mechanical, thermal, and stress-cracking resistance properties of the recycled HDPE and HMWPE blends. Also, it aims to demonstrate that the addition of graphene may enable the use of different recycled polymers without compromising their properties. The effects of adding two amounts of graphene (0.5 and 1%) to recycled blends on the tensile and flexion properties, stress crack resistance (SCR) (using a notched crack ligament stress (NCLS) test), thermal behavior (using a differential scanning calorimeter (DSC) and a rheological plastometer) were investigated. The experimental results showed a significative enhancement when adding graphene in the SCR, some tensile properties (elongation at break and tensile strength), and flexural modulus. However, physical characterization showed that the samples containing 0.5% graphene exhibited lower crystallinity compared to the reference and, for the blend with 1% graphene, the fluidity also decreased for the blend filled with the graphene compared to the reference blend without any filler.

Keywords: recycled high-density polyethylene; recycled high molecular weight; graphene; mechanical; thermal properties; stress crack resistance

1. Introduction

Plastic products are omnipresent in our daily lives thanks to their advantages such as resilience, easy processing, low cost, durability, and impact resistance. The problem is that these multiple benefits can threaten the environment, since plastic waste is not reused. Worldwide, plastic production has grown over the last decade; it reached 400.3 million metric tons (Mt) in 2022 [1]. Unfortunately, most consumer plastics are intended to be single-use, with limited recyclability, leading to increased worldwide production and plastic waste consumption [2,3]. The treatment of plastic waste has become a real issue, especially after the COVID-19 crisis [4]. This is why the development of efficient recycling methods for plastic waste is necessary to promote the circular economy. Recycling is the result of the different stages of collection, sorting, washing, and processing of polymers [5].

Ensuring good properties for recycled products is not always guaranteed, since the properties of recycled plastics are generally not as good as those of virgin resins [6], which limits the reuse of these plastic wastes in various industrial applications [7,8]. Also, various problems are encountered during the recycling process due to polymer degradation and the incompatibility of different polymer types. Several additives are used to remedy these problems and improve the plastic's properties, such as antioxidants (AOs), which are chemical compounds that protect polymers against the thermal oxidation process [9].

They can interfere with the oxidative cycles to delay the oxidative degradation of polymer blends [8,9]. Fillers and modifiers are compounds added to the processing stage to improve the mechanical properties of the polymer blends. Fillers enhance the modulus and tensile strength but worsen the processability and elongation at break. In comparison, modifiers improve the elongation at break and impact strength [5]. Heat stabilizers, widely used in PVC blends, are added to prevent the thermal degradation of plastics [9]. Plasticizers, as organic substances of low volatility, are used to improve the flexibility and processability of plastic materials [9]. The use of compatibilizers is essential to exhibit interfacial activities in heterogeneous recycled polymer blends, and carbon black (CB) is considered as a reinforcing filler used to boost dimensional stability, as well as an antioxidant to extend service life [8,10]. Additives are crucial compounds of plastic materials, enabling the performance and modification of the polymer's properties and long-term use. The main objective of adding these components in the recycling stage is to prevent degradation phenomena during the recycling process and to improve the properties of the extruded materials generated in the recycling process [11].

To improve the mechanical properties of polymer blends based on recycled post-consumer high-density polyethylene (HDPE) and high-molecular-weight polyethylene (HMWPE), carbonaceous nanofillers such as carbon nanofibers, carbon nanotubes, and graphene nanoplatelets are widely used [5,12–15]. Since discovering graphene in 2004 [16], and thanks to its high mechanical performance, graphene has become a promising filler for the development of polymer composites with good mechanical properties [17].

Graphene is a two-dimensional (2D) allotrope of carbon consisting of a single layer of carbon atoms with a honeycomb lattice [18,19] bonded via weak Van der Waals forces [20] that has been found to have exceptional mechanical [19,21], electrical [19,21], and thermal properties [22,23]. Due to its exceptional two-dimensional structure, graphene has excellent mechanical properties, an ultimate tensile strength of 130 GPa, and a Young's modulus of 1TPa [24], with a high thermal conductivity of 5000 w/(m.k), displaying high mobility of charge carriers [17].

Incorporating graphene as a micro-filler in a polymer matrix improves the mechanical properties of polymer composites [25]. The addition of graphene to the polymer increased the Young's modulus and tensile strength of the polymer blend [26,27]. Moreover, Wang et al. compared the effect of adding graphite and carbon black to high-density polyethylene (HDPE) [28], and the results showed that HDPE/graphite improved tensile and impact strength better than HDPE/carbon black. Other papers have highlighted the effect of incorporating exfoliated graphite nanoplatelets into polypropylene (PP); this additive enhanced the dimensional stability and rheological behavior of the material.

Yassmin et al. studied the effect of adding graphene to an epoxy matrix; they concluded that the addition of this nanofiller increased the elastic modulus [29]. Another study, conducted by Gupta et al. [30], showed that the incorporation of graphene in vinyl ester nanocomposites significantly improve the storage modulus, loss modulus, and glass transition temperature of this material.

This study is conducted on an innovative industrial decontamination line. It investigates the mechanical and physical properties of a recycled HDPE and HMWPE blend reinforced with graphene. The composites were processed using an industrial twin-screw extruder. The effects of the processing parameters, such as filler concentration, were experimentally analyzed through thermal and rheological characterization, tensile and flexural tests to assess the mechanical behaviour, and notched crack ligament stress (NCLS) to evaluate the stress crack resistance of the PE blend (SCR).

2. Materials and Methods

2.1. Materials

2.1.1. Polyethylene

Recycled polyethylene blends were prepared via a twin-screw extruder using recycled HDPE (recHDPE) and recycled HMWPE (recHMWPE). The reference blend is based on 35% recHDPE and 65% recHMWPE.

The RecHDPE used in this study is a post-consumer recHDPE with a melt flow index of 0.5 g/10 min (190 °C/2.16 kg load) and a density of 0.951 g/cm^3. The RecHMWPE is also a post-consumer plastic with a melt flow of 0.1 g/10 min (190 °C/2.16 kg load) and a density of 0.948 (g/cm^3), Soleno inc., Saint-Jean sur Richelieu, QC, Canada, provided all these materials.

2.1.2. Graphene

The filler used in this study consists of graphene particles and is a multifunctional carbon additive formulated for across-the-board performance enhancements in thermoplastics and rubbers. The graphene was supplied in the form of a masterbatch based on medium-density PE (MDPE) filled with graphene Black 0X. Graphene pellets 0X are two-dimensional (2D) with 6 to 10 layers, a primary particle size of 0.5–1 µm, and a density of 0.2–0.3 g/cm^3.

Graphene was added to the reference blend in two different amounts (0.5 and 1%). Its composition and technical properties are presented in Tables 1–3 below:

Table 1. Graphene masterbatch composition.

Composition		Loading (wt%)
Carrier	MDPE	70
Filler	Graphene Black	30

Table 2. Chemical composition of graphene 0X.

Element	Value (wt%)
Carbon	>97
Oxygen	<1

Table 3. Technical properties of graphene masterbatch.

Property	Value (SI)	Test Method
Density	1.12 g/cm^3	ASTM D792 [31]
Fluidity	8 g/10 min	ASTM D1238 [32]

2.2. Methods

2.2.1. Preparation of Blends

The preparation of blends consists of extruding the pellets using a co-rotating twin-screw extruder at 1200 rpm, with a temperature profile between 190 and 210 °C. Initially, a premix of recHDPE with 5 wt% of the graphene masterbatch was prepared using an electric mixer. Subsequently, the prepared premix was added to the reference blend based on 35% recHDPE and 65% recHMWPE via a feeder. Two amounts of graphene masterbatch were tested (0.5 and 1 wt%) in the extrusion process. The main objective is to test the effect of adding (0.5 and 1 wt%) graphene on the thermomechanical and stress crack resistance properties of recHDPE and recHMWPE blends. The compositions of the extruded blends are illustrated in the Table 4 below:

Table 4. Blends composition.

Blend Number	recHDPE (wt%)	recHMWPE (wt%)	Graphene (wt%)
1	35	65	-
2	35	64.5	0.5
3	35	64	1

2.2.2. Samples Preparation

Samples for mechanical characterization (Tensile, Flexion, NCLS, IZOD) were prepared using a press machine at 180 °C for approximately 20 min. The molded plates were cooled to 80 °C; once the plates were ready, they were placed in a conditioning room (21 ± 2 °C) for 24 h before starting the characterization.

2.3. Physical Characterization

2.3.1. Mechanical Properties

The mechanical behavior was assessed using tensile and flexural tests. Tensile tests were performed based on the ASTM D638-14 [33] on five dog-bone-shaped specimens (specimen type IV: Figure 1) cut from the 3.2 mm thick molded plate. Tensile tests were performed on a Lab Integration machine with a crosshead speed of 50 mm/min at room temperature 23 °C; five specimens were tested for each test. Tensile strength, elongation at break, and Young's modulus were determined from stress–strain curves. Bending tests were carried out on the same machine as the tensile test in three-point bending mode with a crosshead speed of 10 mm/min according to ASTM D790 [34] on five rectangular specimens to determine the flexural modulus.

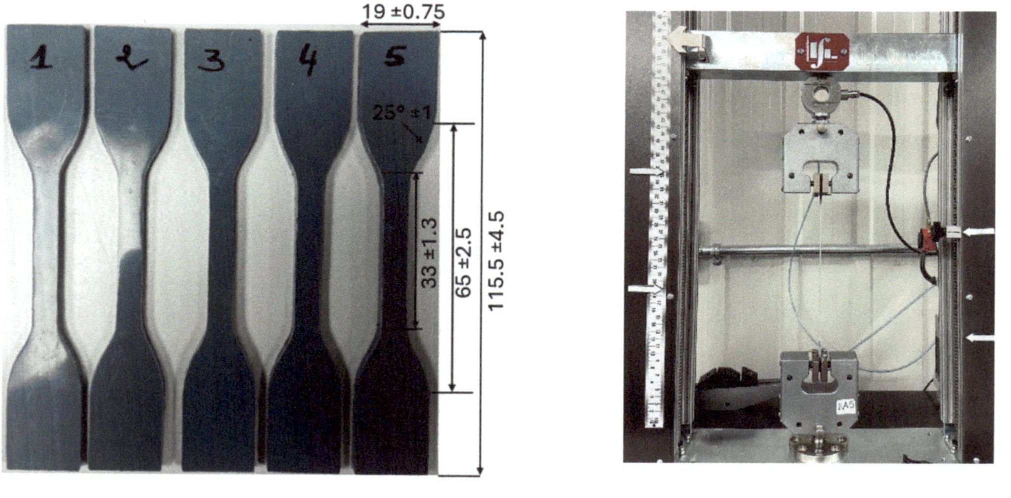

Figure 1. Dimensions (mm) of tensile specimens.

Resistance to crack propagation was also assessed using the notched constant ligament stress (NCLS) test according to the F2136-18 standard [35]. This test method is used to evaluate the crack resistance under constant ligament stress for high-density polyethylene resins or corrugated pipelines. The NCLS test subjects a dumbbell-shaped notched test specimen to constant ligament stress in the presence of a surface-active agent (lgepal) at an elevated temperature [35]. The purpose of this test is to measure the failure time associated with each dumbbell-shaped notched test specimen.

The NCLS test consists of notching five test specimens, then determining the load to be placed on each specimen and loading the weight tubes with a shot. Before attaching the shot tube to the lever arm, we must attach the specimens to the loading frame and place them into the bath conditioned at a temperature of 50 ± 1 °C for 30 min. Then, the weight is connected by a tube to the lever arm for each specimen and the specimen timer is begun immediately. The time to failure for each specimen was recorded.

The impact resistance of blends was evaluated using the IZOD test according to the ASTM D256-10 standard [36]; its main objective is to measure the energy absorbed by a material when a notched specimen is subjected to a sudden impact load. The IZOD test consists of notching and testing the 8 notched specimens with a V-shaped notch using a pendulum. Figure 2 illustrates (a) sample preparation, specimens (b) loading, (c) failure and (d) dimensions of the specimen. During the test, the energy absorbed by the specimen indicates the impact resistance and toughness of the material.

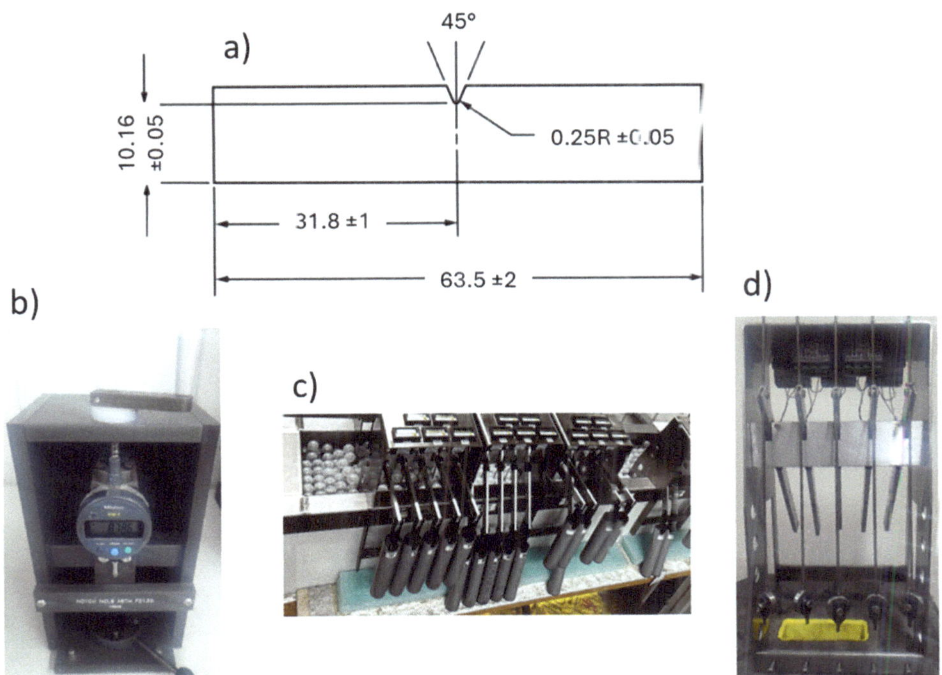

Figure 2. Steps of the NCLS test. Specimen (**a**) dimension [36], (**b**) notching, (**c**) loading in the test basin, and (**d**) failure.

2.3.2. Thermal Properties

Thermal analysis was realized using a differential scanning calorimeter (DSC) according to ASTM-D3895-14 [37] using a TA instruments DSC. Samples of 5–10 mg were encapsulated in an aluminium pan and heated from the ambient temperature to 80 °C in an inert gaseous environment (nitrogen) at a constant rate of 10 °C/min. Then, the samples were isothermally kept at 80 °C for 1 min before heating again up to 200 °C.

2.3.3. Rheological Properties

The melt flow index measures the fluidity of a polymer; this test method consists of determining the rate of extrusion of molten thermoplastic resins using an extrusion plastometer. The preheated resin is extruded through a die with a specific length and orifice diameter under a specific load, temperature, and piston position in the barrel. The fluidity was evaluated using a Lab Integration plastometer [32].

3. Results and Discussion

The composite blends were manufactured using 35% recHDPE and 65% recHMWPE as a matrix and graphene as an additive. Their mechanical, thermal, and stress cracking propagation properties were studied.

3.1. Melt Flow Index (MFI)

The fluidity of the extruded blends was realized to evaluate the effect of graphene on the MFI. After adding the graphene, the MFI decreases from 0.177 for the reference blend without any filler to 0.148 or 0.149 g/10 min for the blends with 0.5% graphene or 1% graphene, respectively. This decrease can be explained by the increase in the viscosity of the blends due to reduced chain mobility and enhanced molecular entanglement when adding the graphene. The graphene acts as a reinforcement filler in the polymer, providing a higher viscosity and therefore a lower MFI. It does not significantly affect the viscosity.

3.2. Tensile and Flexion Test

Additives play an important role in the processability and applications of plastic materials. Incorporating additives can improve the properties of the plastic materials and make them suitable for different applications [11].

It is important to evaluate the mechanical behavior by determining the strength and rigidity of the blends. To do this, tensile and flexion were realized. Figure 3 shows the tensile and flexural properties for the reference blend without graphene and the blends with 0.5% graphene and 1% graphene.

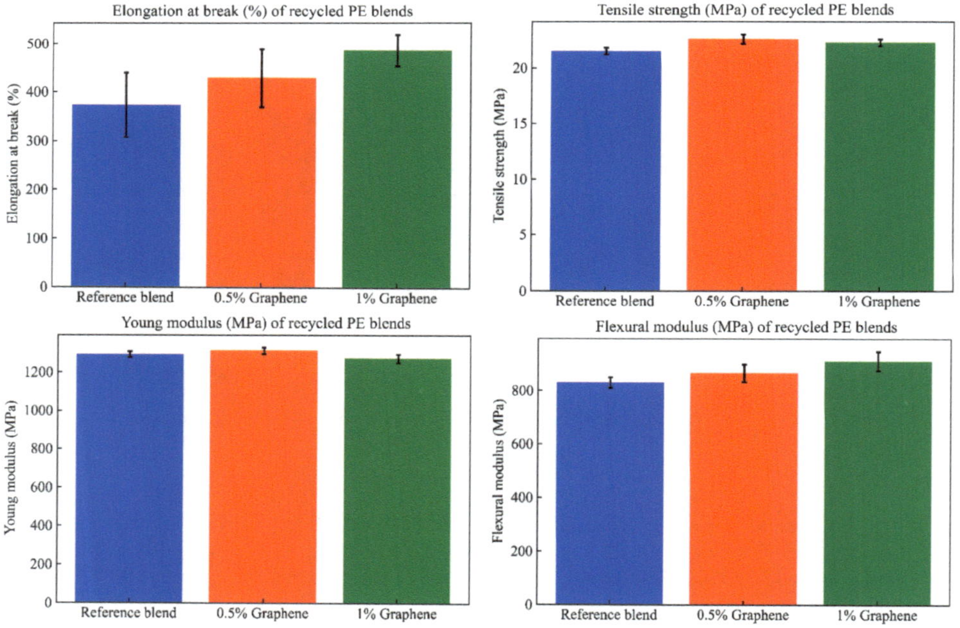

Figure 3. Tensile and flexural properties of PE blends with and without graphene.

It can be observed that the tensile properties of the reference blend without graphene are lower than those of the blends with graphene. Adding 0.5% or 1% graphene to the reference blend significantly improves the elongation at break by 15% or 30%. Similar results were obtained for the tensile strength and Young's modulus, which had increases of 5% and 4% for blends with 0.5 and 1% of graphene. The Young's modulus increases slightly by 2% for blends with 0.5% graphene and decreases by 2% for blends with 1% graphene.

These results show that the ductility of the recycled PE blends increases significantly when increasing the amount of graphene. Comparable effects were observed for the flexural properties, with increases of 4% and 10% in the flexural modulus for blends with 0.5% and 1% of graphene, respectively. The same results were reported by Diallo A.K. et al. in their study [6], where there was an improvement of 7% and 12% in the flexural modulus and 6% and 7% in the tensile strength, respectively, when incorporating 1% few-layer graphene (FLG) in a 1:4 ratio of prime/recHDPE and 1:1 ratio of prime/recHDPE [6]. The authors explained that this result is consistent with the SEM images, showing a more homogeneous and refined blend morphology in the presence of FLG. Also, smaller droplet domains help increase the mechanical performance [6].

Adding graphene to recycled HDPE/HMWPE blends increases the mechanical properties because microparticles transfer more tension and improve the connectivity of the composite components [38]. Thanks to the good mechanical properties of graphene and its very high surface area, even small amounts of graphene mixed with polymers can increase the mechanical properties of the blends [38]. The method of adding and choosing the required amount of graphene has a direct impact on the final properties of the blends, exceeding a specified quantity of graphene causes plates to form over each other and the reduction of adhesion between the components of the blends. In general, using low amounts of graphene is an advantage because it avoids agglomeration, which can decrease the mechanical properties. For example, adding up to 0.8 wt% graphene to a wood-fiber-recycled polypropylene composite increases its mechanical strength. On the other hand, the properties are reduced by the addition of 5 wt% graphene [39,40].

3.3. Stress Crack Resistance

Evaluating the stress crack resistance of the blends is critical, since toughness is a crucial mechanical property extremely relevant to pipeline applications. To meet the standards required by the BNQ (Bureau de la normalisation du Québec), the minimum break time is 24 h. The result from NCLS tests of the blends without and with graphene are illustrated in the graph below (Figure 4):

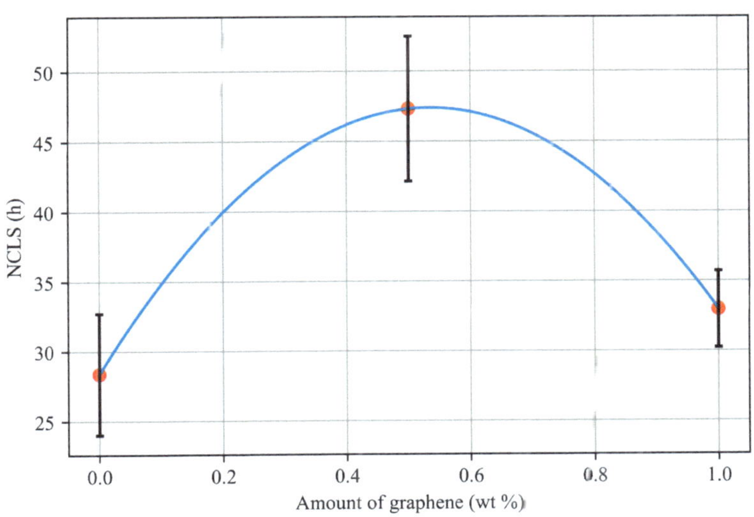

Figure 4. NCLS tests of PE blends as a function of graphene amount.

Crack resistance is an important factor affecting the long-term sustainability of polymers. The stress crack resistance is affected by the incorporation of graphene. It was observed that adding graphene increases the stress crack resistance by 73% and 20%, respectively, for 0.5 and 1% of graphene. The significant increase related to adding 0.5% of graphene can be explained by the reinforcing effect of graphene, which can potentially slow down crack propagation.

Using graphene as an additive enhances resistance to crack propagation because graphene hinders crack propagation, facilitating stress redistribution and delaying crack propagation. However, adding 1% graphene to the reference blend did not significantly improve the stress crack resistance. With the increase in graphene amount, the microcrack zones become closer to each other, resulting in the coalescence of microcracks, which facilitates the propagation of a major crack [17]. Several studies have proven that the addition of graphene simultaneously enhances stiffness, toughness, and ductility [41].

3.4. Impact Strength Resistance

The effect of adding graphene to the reference blend on the impact strength resistance is illustrated in the graphic bellow (Figure 5):

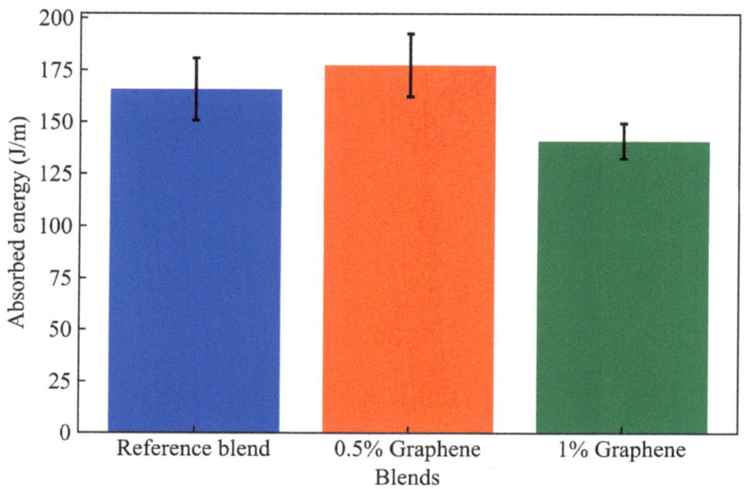

Figure 5. Impact strength resistance of recycled PE blends.

The impact strength resistance of the reference blend without graphene was 165.69 J/m. Adding 0.5% graphene to the reference blend slightly increased the impact strength resistance by 7%. In contrast, the addition of 1% graphene decreases the impact strength resistance by 15%. The observed decrease is due to the agglomeration of the filler. The addition of 1% graphene into reference blend increased the brittleness along with the proportion to rigidity, as shown in the flexural modulus results (Figure 3). Adding graphene to the reference blend may lead to more voids, which decreases the impact strength resistance [41]. The obtained results align with other studies [38,42].

3.5. Thermal Behavior

During the processing of polyolefin blends, thermo-oxidative degradation occurs within the polymeric chains, resulting in a finished product of inferior quality [6]. Some of the mechanisms activated during the thermo-oxidative degradation are cross-linking, chain scission, elimination of substituents, and formation of double bonds [43].

The thermal behavior of the blends and the effect of adding graphene were investigated using a DSC. The melt peak temperature and enthalpy were determined directly from the DSC curves, while the degree of crystallinity of blends was calculated from Equation (1). All the data are presented in Table 5.

Table 5. Thermal behavior of extruded blends.

Sample	Melt Peak Temperature (°C)	Meling Enthalpy (J/g)	Crystallinity (%)
Reference	129.4	135.3	46.2
0.5% Graphene	131.5	138.5	47.3
1% Graphene	130.6	144.7	49.4

ΔH_f^0 presents the enthalpy of fusion of a 100% crystalline polyethylene estimated in the literature at 293 J/g [44].

$$\chi_c = \frac{\Delta H_f}{\Delta H_f^0} \quad (1)$$

where ΔH_f is the melting enthalpy of the sample measured during the test (in J/g) and ΔH_f^0 represents the melting enthalpy of a 100% crystalline polyethylene estimated in the literature at 293 J/g [44].

The results show that the melting peaks of blends increased slighly (by 1 to 2 °C). As a result, the melting enthalpy increased by 3.2 to 9.4 °C, resulting in an enhancement of the degree of crystallinity by 2% and 7% for blends with 0.5% and 1% of graphene, respectively. Since the degree of crystallinity of these polymer blends is less than 50%, the amorphous regions are larger than the crystalline regions, which provides them more flexibility, allowing them to stretch further before breaking and resulting in increased elongation at break. Adding graphene reduces the degree of crystallinity and, contrarily, increases the elongation at break, as demonstrated by the tensile test. The degree of crystallinity affects the physical properties as well as the mechanical properties of semicrystalline polymers.

4. Conclusions

In this study, we investigated the effect of adding graphene to a reference blend based on recHDPE and recHMWPE. The purpose of this study was to assess the effect of adding graphene on the thermomechanical behavior and the stress crack resistance of the reference blend. The experimental tests showed that adding graphene improves tensile strength, elongation at break, and flexural modulus. Furthermore, stress crack and impact strength resistance are enhanced when using low amounts of graphene (0.5%). Adding graphene to recycled polyethylene blends could be an interesting alternative to ensure good mechanical properties and resistance to crack propagation that is comparable to virgin PE blends.

Moreover, rheological and microstructural experiments are being performed to evaluate the effect of adding graphene on the interfacial adhesion between the different phases of the blend in the presence of graphene.

Author Contributions: Literature review, H.K. and A.L.; methodology, H.K. and A.L.; validation, S.E.; extrusion tests, H.K.; laboratory tests, H.K. and A.L.; data analysis, H.K. and A.L.; original draft preparation, H.K. and A.L.; writing—review and editing, H.K., A.L. and S.E.; funding acquisition, C.D.; final reading, H.K., A.L., S.E., C.D. and M.R. All authors have read and agreed to the published version of the manuscript.

Funding: This research was funded by Soleno Inc. and Mitacs, grant number IT23689.

Institutional Review Board Statement: Not applicable.

Informed Consent Statement: Not applicable.

Data Availability Statement: The original contributions presented in the study are included in the article, further inquiries can be directed to the corresponding author.

Acknowledgments: The authors acknowledge Soleno Inc. for the equipment (extruder, laboratory instruments) and for financial support and the Mitacs program for their financial support.

Conflicts of Interest: The co-author Carl Diez is an employee of funding sponsor Soleno Inc. However, this sponsor has had no role in the design of the study; in the collection, analysis, or interpretation of data; in the writing of the manuscript; or in the decision to publish the results.

References

1. Organisation for Economic Co-Operation and Development. *Global Plastic Production from 1950 to 2022*; Organisation for Economic Co-Operation and Development: Paris, France, 2023.
2. Borrelle, S.B.; Ringma, J.; Law, K.L.; Monnahan, C.C.; Lebreton, L.; McGivern, A.; Murphy, E.; Jambeck, J.; Leonard, G.H.; Hilleary, M.A. Predicted growth in plastic waste exceeds efforts to mitigate plastic pollution. *Science* **2020**, *369*, 1515–1518. [CrossRef] [PubMed]
3. Walker, T.R.; Fequet, L. Current trends of unsustainable plastic production and micro(nano)plastic pollution. *TrAC Trends Anal. Chem.* **2023**, *160*, 116984. [CrossRef]
4. Vanapalli, K.R.; Sharma, H.B.; Ranjan, V.P.; Samal, B.; Bhattacharya, J.; Dubey, B.K.; Goel, S. Challenges and strategies for effective plastic waste management during and post COVID-19 pandemic. *Sci. Total Environ.* **2021**, *750*, 141514. [CrossRef]
5. Francis, R. *Recycling of Polymers: Methods, Characterization and Applications*; John Wiley & Sons: Hoboken, NJ, USA, 2016; ISBN 3-527-33848-9.
6. Diallo, A.K.; Helal, E.; Gutierrez, G.; Madinehei, M.; David, E.; Demarquette, N.; Moghimian, N. Graphene: A multifunctional additive for sustainability. *Sustain. Mater. Technol.* **2022**, *33*, e00487. [CrossRef]
7. Luzuriaga, S.; Kovářová, J.; Fortelný, I. Degradation of pre-aged polymers exposed to simulated recycling: Properties and thermal stability. *Polym. Degrad. Stab.* **2006**, *91*, 1226–1232. [CrossRef]
8. Patlolla, V.R.; Asmatulu, R. Recycling and Reusing Fiber-Reinforced Composites. *Environ. Res. J.* **2013**, *7*, 145.
9. Ambrogi, V.; Carfagna, C.; Cerruti, P.; Marturano, V. 4—Additives in Polymers. In *Modification of Polymer Properties*; Jasso-Gastinel, C.F., Kenny, J.M., Eds.; William Andrew Publishing: Norwich, NY, USA, 2017; pp. 87–108. ISBN 978-0-323-44353-1.
10. Blokhina, O.; Virolainen, E.; Fagerstedt, K.V. Antioxidants, oxidative damage and oxygen deprivation stress: A review. *Ann. Bot.* **2003**, *91*, 179–194. [CrossRef]
11. La Mantia, F.P. The role of additives in the recycling of polymers. In *Macromolecular Symposia*; Wiley Online Library: Hoboken, NJ, USA, 1998; Volume 135, pp. 157–165.
12. Huang, J.-C. Carbon black filled conducting polymers and polymer blends. *Adv. Polym. Technol. J. Polym. Process. Inst.* **2002**, *21*, 299–313. [CrossRef]
13. Korkees, F.; Aldrees, A.; Barsoum, I.; Alshammari, D. Functionalised graphene effect on the mechanical and thermal properties of recycled PA6/PA6,6 blends. *J. Compos. Mater.* **2021**, *55*, 2211–2224. [CrossRef]
14. Mittal, G.; Dhand, V.; Rhee, K.Y.; Park, S.-J.; Lee, W.R. A review on carbon nanotubes and graphene as fillers in reinforced polymer nanocomposites. *J. Ind. Eng. Chem.* **2015**, *21*, 11–25. [CrossRef]
15. Liu, M.; Kinloch, I.A.; Young, R.J.; Papageorgiou, D.G. Modelling mechanical percolation in graphene-reinforced elastomer nanocomposites. *Compos. Part B Eng.* **2019**, *178*, 107506. [CrossRef]
16. Novoselov, K.S.; Geim, A.K.; Morozov, S.V.; Jiang, D.; Zhang, Y.; Dubonos, S.V.; Grigorieva, I.V.; Firsov, A.A. Electric field effect in atomically thin carbon films. *Science* **2004**, *306*, 666–669. [CrossRef] [PubMed]
17. Wang, J.; Jin, X.; Li, C.; Wang, W.; Wu, H.; Guo, S. Graphene and graphene derivatives toughening polymers: Toward high toughness and strength. *Chem. Eng. J.* **2019**, *370*, 831–854. [CrossRef]
18. Verma, A.; Parashar, A.; Packirisamy, M. Effect of grain boundaries on the interfacial behaviour of graphene-polyethylene nanocomposite. *Appl. Surf. Sci.* **2019**, *470*, 1085–1092. [CrossRef]
19. Idowu, A.; Boesl, B.; Agarwal, A. 3D graphene foam-reinforced polymer composites—A review. *Carbon* **2018**, *135*, 52–71. [CrossRef]
20. Bianco, A.; Cheng, H.-M.; Enoki, T.; Gogotsi, Y.; Hurt, R.H.; Koratkar, N.; Kyotani, T.; Monthioux, M.; Park, C.R.; Tascon, J.M. *All in the Graphene Family–A Recommended Nomenclature for Two-Dimensional Carbon Materials*; Elsevier: Amsterdam, The Netherlands, 2013; Volume 65, pp. 1–6. ISBN 0008-6223.
21. Nieto, A.; Boesl, B.; Agarwal, A. Multi-scale intrinsic deformation mechanisms of 3D graphene foam. *Carbon* **2015**, *85*, 299–308. [CrossRef]
22. Wei, W.; Qu, X. Extraordinary physical properties of functionalized graphene. *Small* **2012**, *8*, 2138–2151. [CrossRef]
23. Liem, H.; Choy, H.S. Superior thermal conductivity of polymer nanocomposites by using graphene and boron nitride as fillers. *Solid State Commun.* **2013**, *163*, 41–45. [CrossRef]
24. Lee, C.; Wei, X.; Kysar, J.W.; Hone, J. Measurement of the Elastic Properties and Intrinsic Strength of Monolayer Graphene. *Science* **2008**, *321*, 385–388. [CrossRef]
25. Khan, Z.U.; Kausar, A.; Ullah, H.; Badshah, A.; Khan, W.U. A review of graphene oxide, graphene buckypaper, and polymer/graphene composites: Properties and fabrication techniques. *J. Plast. Film Sheeting* **2016**, *32*, 336–379. [CrossRef]
26. Sarker, F.; Potluri, P.; Afroj, S.; Koncherry, V.; Novoselov, K.S.; Karim, N. Ultrahigh performance of nanoengineered graphene-based natural jute fiber composites. *ACS Appl. Mater. Interfaces* **2019**, *11*, 21166–21176. [PubMed]
27. Sarker, F.; Karim, N.; Afroj, S.; Koncherry, V.; Novoselov, K.S.; Potluri, P. High-performance graphene-based natural fiber composites. *ACS Appl. Mater. Interfaces* **2018**, *10*, 34502–34512. [CrossRef]
28. Wang, L.; Hong, J.; Chen, G. Comparison study of graphite nanosheets and carbon black as fillers for high density polyethylene. *Polym. Eng. Sci.* **2010**, *50*, 2176–2181. [CrossRef]
29. Yasmin, A.; Luo, J.-J.; Daniel, I.M. Processing of expanded graphite reinforced polymer nanocomposites. *Compos. Sci. Technol.* **2006**, *66*, 1182–1189.

30. Gupta, S.; Raju Mantena, P.; Al-Ostaz, A. Dynamic mechanical and impact property correlation of nanoclay and graphite platelet reinforced vinyl ester nanocomposites. *J. Reinf. Plast. Compos.* **2010**, *29*, 2037–2047.
31. *ASTM D792-20*; Standard Test Methods for Density and Specific Gravity (Relative Density) of Plastics by Displacement. ASTM International: West Conshohocken, PA, USA, 2020.
32. *ASTM D1238-13*; Standard Test Method for Melt Flow Rates of Thermoplastics by Extrusion Plastometer. ASTM International: West Conshohocken, PA, USA, 2013.
33. *ASTM D638-14*; U.S. Department of Defense Standard Test Method for Tensile Properties of Plastics 2014. ASTM International: West Conshohocken, PA, USA, 2014.
34. *ASTM D790-17*; U.S. Department of Defense Standard Test Methods for Flexural Properties of Unreinforced and Reinforced Plastics and Electrical Insulating Materials 2017. ASTM International: West Conshohocken, PA, USA, 2017.
35. *ASTM F2136-18*; Standard Test Method for Notched, Constant Ligament-Stress (NCLS) Test to Determine Slow-Crack-Growth Resistance of HDPE Resins or HDPE Corrugated Pipe. ASTM International: West Conshohocken, PA, USA, 2024.
36. *ASTM D256-10*; Standard Test Methods for Determining the Izod Pendulum Impact Resistance of Plastics. ASTM International: West Conshohocken, PA, USA, 2018.
37. *ASTM D3895-14*; Standard Test Method for Oxidative-Induction Time of Polyolefins by Differential Scanning Calorimetry. ASTM International: West Conshohocken, PA, USA, 2014.
38. Beigloo, J.G.; Eslam, H.K.; Hemmasi, A.H.; Bazyar, B.; Ghasemi, I. Effect of nanographene on physical, mechanical, and thermal properties and morphology of nanocomposite made of recycled high density polyethylene and wood flour. *BioResources* **2017**, *12*, 1382–1394.
39. Farsi, M.; Younesi, K.H. Effect of single wall carbon nanotubes on physical and mechanical properties of wood fiber-LDPE composites. *J. For. Wood Prod.* **2012**, *65*, 169–186.
40. Sheshmani, S.; Amini, R. Preparation and characterization of some graphene based nanocomposite materials. *Carbohydr. Polym.* **2013**, *95*, 348–359. [CrossRef]
41. Papageorgiou, D.G.; Kinloch, I.A.; Young, R.J. Mechanical properties of graphene and graphene-based nanocomposites. *Prog. Mater. Sci.* **2017**, *90*, 75–127. [CrossRef]
42. Umesh, G.L.; Prasad, N.J.K.; Rudresh, B.M.; Devegowda, M. Influence of nano graphene on mechanical behavior of PA66/PA6 blend based hybrid nano composites: Effect of micro fillers. *Mater. Today Proc.* **2020**, *20*, 228–235. [CrossRef]
43. Slovokhotova, N.A.; Magrupov, M.A.; Kargin, V.A. Thermal degradation of polyethylene. *Polym. Sci. U.S.S.R.* **1964**, *6*, 2186–2192. [CrossRef]
44. Wunderlich, B. *Thermal Analysis*; Academic Press: New York, NY, USA, 1990; pp. 417–431.

MDPI AG
Grosspeteranlage 5
4052 Basel
Switzerland
Tel.: +41 61 683 77 34

MDPI Books Editorial Office
E-mail: books@mdpi.com
www.mdpi.com/books

www.ingramcontent.com/pod-product-compliance
Lightning Source LLC
Chambersburg PA
CBHW041430210926
51850CB00013B/3390
9783725842377